FUNDAMENTALS OF TRANSMISSION LINES AND ELECTROMAGNETIC FIELDS

S. R. SESHADRI
The University of Wisconsin

ADDISON-WESLEY PUBLISHING COMPANY
Reading, Massachusetts
Menlo Park, California · London · Amsterdam · Don Mills, Ontario · Sydney

This book is in the
ADDISON-WESLEY SERIES IN ELECTRICAL ENGINEERING

Consulting Editors:
David K. Cheng
Leonard A. Gould
Fred K. Manasse

ISBN 0-201-06722-6
EFGHIJKLMN-MA-89876543210

To the memory
of
my adopted parents

PREFACE

We are in the ordinary position of scientists of having to be content with piecemeal improvements: we can make several things clearer, but we cannot make anything clear

Frank Plumpton Ramsey

This is a textbook for an introductory course on the fundamentals of transmission lines and electromagnetic fields designed for undergraduate students. Part of the book was developed for second-semester junior students and part of it for first-semester senior students. There is enough material in the book for a full-year course. This book is intended for use by electrical engineering and applied physics students as well as by physics students.

The students who took the courses on which this book is based had an adequate knowledge of the engineering principles but they did not possess sufficient understanding of the necessary mathematics. They believed that there was need for extensive class notes, giving detailed explanations of the various derivations. With a view to fulfilling this need, detailed class notes were written to facilitate the learning process by the students. The enthusiastic reception accorded these notes by the students provided the necessary impetus for the expansion of an improved version into this book. Admittedly, there are many imperfections; nevertheless, I believe also that there are many features which will render the book useful to the academic community.

The distinguishing feature which I have strived to incorporate is a clarity of presentation such as to make the book suitable for self-study by the students. The necessary mathematical developments, instead of being avoided, are interwoven into the treatment without compromising unnecessarily on the rigor. Particular emphasis is placed on the physical interpretations.

The book contains ten chapters. As the table of contents shows, several new topics are included, in addition to those traditionally covered in undergraduate courses. A thorough treatment of progressive waves and of standing waves, of real power and reactive power, and of the concept of the reflection coefficient and its implications is given. In the entire treatment of the electromagnetic fields, both the kinematic and the dynamic aspects of the fields are discussed carefully. Electrical conduction is introduced as a diffusion phenomenon, and a clear account of the concept of skin depth and surface impedance of a good conductor for time-harmonic fields is included. In the treatment of transmission phenomena, the concepts of complex angle of transmission and of inhomogeneous plane waves are introduced. In the discussion of the theory of both closed and open waveguides, equal emphasis is placed on analysis and synthesis of the waveguide fields. The reflection symmetry of the potentials and the electromagnetic fields are stressed, and the image theory is developed in detail. An elementary discussion is given of the differences in the current distributions on a linear filament depending on whether it is used as a transmitter, a reflector, or a receiver. Clear and detailed presentations of the theory of the receiving antenna, and of the use of the combination of the antenna elements for the modification of the directional pattern and polarization of the radiation are included.

A noteworthy feature of the book is the set of problems, of which there are about 270. These problems are designed not only to clarify but also to extend the treatment of the various topics contained in the book.

It is usual to relegate several conceptually simple topics in electromagnetic fields to graduate-level courses because of the difficulty in obtaining physically interpretable numbers from the theory. The advent of digital computers has changed this situation. In the courses on which this book is based, I have encouraged students to use the digital computer for the solution of problems. To enable the students following this book to use digital computers for the solution of problems, I have made every effort to carry out the analyses to a stage which facilitates computer application. The design and the frequency response of a single stub impedance-matching device, the frequency response of the absorption coefficient of a lossy, planar dielectric slab, the mapping of the electric and the magnetic field lines in a parallel-plate waveguide, the mapping of the lines of surface current on the walls of a rectangular waveguide, the study of the dispersion relations of a dielectric slab waveguide, the directive gain of a linear antenna with an assumed current distribution in the form of a progressive wave having an associated propagation coefficient different from that of a uniform plane wave in the corresponding unbounded medium and the scanning of the beam of a simple uniform antenna array by changing the relative phase of the antenna currents are some of the topics which I have encouraged my students to investigate with the help of digital computers and whose study, with the help of a digital computer, I suggest for adoption.

The students are assumed to have a prior knowledge of electricity and magnetism as well as vector analysis. However, a brief review of vector analysis is included in an appendix for the sake of completeness.

A note on the method of numbering of the equations and the figures is in order. When reference is made to an equation as (9.8.7), the first two numbers indicate respectively the chapter and the section to which the equation pertains and the last gives the number of the equation. When reference is made to an equation in the same chapter, the first number is omitted. If reference is made to a figure as 6.5, the first number gives the chapter number and the second the figure number. A prefix P indicates that the figure appears in the problems given at the end of each chapter. Also, if reference is made to a problem as 9.1, it denotes Problem 1 in Chapter 9.

From my student days, I have received considerable help and encouragement from two of my teachers at Harvard University. I learned the various aspects of electromagnetic theory and their applications, as well as the importance in bridging the gap between theory and experiment, from Professor Ronold W. P. King, whose inspiring lectures and helpful advice kindled and sustained my interest in this branch of electrical engineering. Professor Tai Tsun Wu, through his illuminating and incisive criticisms, has helped me to obtain a deeper understanding of electromagnetic theory. To these two great teachers, I owe a deep debt of gratitude.

A number of my friends assisted me in the preparation of the manuscript of this book. Professor Gilbert H. Owyang of Worcester Polytechnic Institute helped in the drafting of Chapters 1, 2, 3, and 7. Professor Chin-Lin Chen of Purdue University and Professor Hang-Sheng Tuan of the State University of New York at Stony Brook read the entire manuscript and offered many suggestions for improvement. Doctor Cheng Chi Kao of Harvard University read the final manuscript and improved the technical accuracy of the book.

I am indebted to my senior colleague, Professor T. J. Higgins, who originally suggested the idea of writing the book, read the manuscript, and offered many valuable comments.

The numerical computations for the various figures were carried out in the Hybrid Computer Laboratory. The University administration extended sympathetic support and continued encouragement which made the writing of this book a pleasant task.

During the time of preparation of the manuscript of this book, I have read with advantage the publications of several previous authors whose works are listed at the end of this book.

The major portion of the manuscript was typed cheerfully and expertly by Mrs. Lynda Vaslow. Miss Shirley Quamme, Mrs. Sharyn Krumrei, and Mrs. Patricia Weis assisted in the typing of the manuscript in an able and good-humoured manner. My wife Susheela gave valuable assistance throughout the preparation of the book.

I dedicate this book respectfully to the memory of my adopted parents, S. N. Rangaswamy and Alangarammal.

Madison, Wisconsin S. R. S.
March 1971

CONTENTS

CHAPTER 1

PROPAGATION OF D.C. TRANSIENTS AND PULSES ON LOSSLESS TRANSMISSION LINES

1.1 INTRODUCTION

In electrical engineering, transmission lines are used both for conveying electrical power and, more or less imperfectly, for transmitting information. The transmission lines are used over frequencies ranging from approximately 60 Hz, in electrical power engineering, to 10^9 Hz and higher, in microwave engineering. Some of the common types of transmission lines are shown in Fig. 1.1. The two-wire transmission line, the coaxial transmission line, the parallel-plate waveguide, and the rectangular waveguide are all considered in this treatment. The two-wire transmission line consists of two closely spaced parallel wires of usually circular cross section. These transmission lines operate generally in free space and are supported mechanically at regular intervals by insulating dielectric spacers. The coaxial transmission line, as the name implies, consists of a coaxial dielectric region, usually free space, between the outer wall of a solid inner conductor and the inner wall of a hollow outer conductor. These conductors are of circular cross section. The parallel-plate waveguide is a parallel slab of dielectric or free space region sandwiched between two parallel plane conductors. A rectangular waveguide consists of a hollow rectangular metallic tube filled with a dielectric or just free space. In the case of the two-wire transmission line, the electromagnetic fields caused by the currents extend throughout space, but in all the other cases the electromagnetic fields are confined within metallic boundaries. The choice of the particular type of transmission line, among other factors, depends on the frequency of operation and the required power-handling capacity.

The theory of the two-wire transmission line can be developed either from the point of view of electromagnetic field theory or essentially from the point of view of electric circuit theory. In the latter case, which is the one adapted in the present treatment, the transmission line is considered as a uniform distributed parameter circuit consisting of certain values of series inductance and resistance and parallel capacitance and conductance. The values of these parameters depend on the geometry of the transmission line and have to be obtained necessarily from the application of electromagnetic field theory. There are certain approximations involved in treating the two-wire transmission line as a distributed parameter electric network and the more important of these approximations are stated without justification. Again, a systematic deduction of these approximations requires the application of electromagnetic field theory.

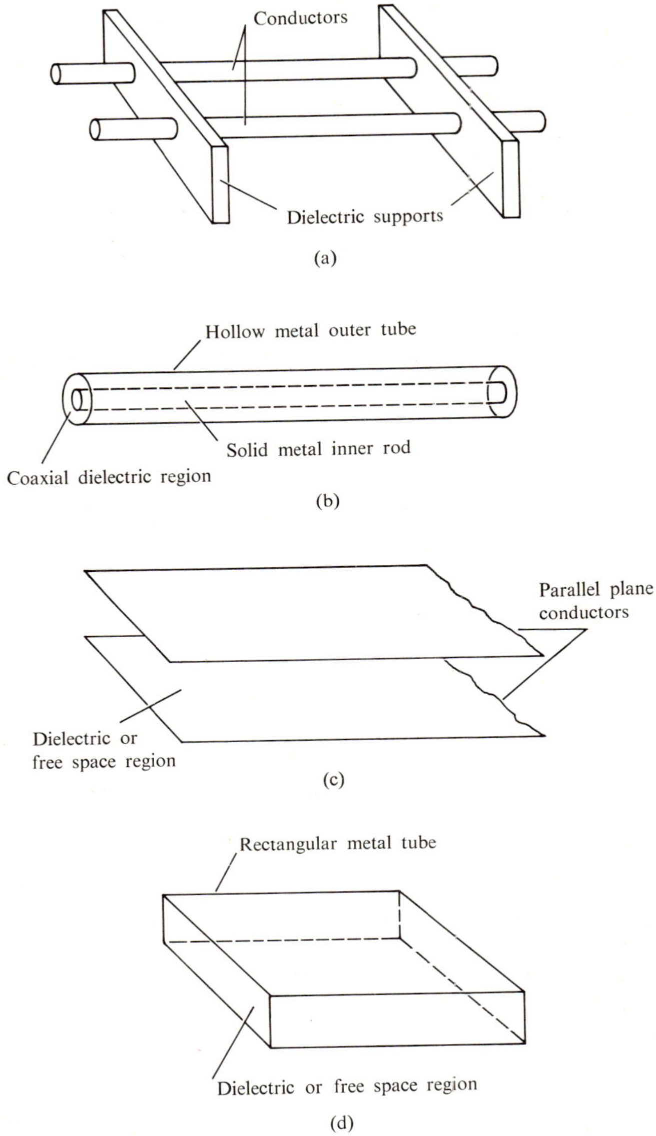

Fig. 1.1 Types of transmission lines: (a) two-wire transmission line; (b) coaxial transmission line; (c) parallel-plate waveguide; (d) rectangular waveguide.

The following theory of the transmission line, though developed with particular reference to the two-wire transmission line, is applicable also for certain modes of transmission in a coaxial transmission line and a parallel-plate waveguide.

1.2 DERIVATION OF THE TRANSMISSION-LINE EQUATIONS

Consider a transmission line consisting of two parallel wires of radius a and let the distance between the axes of the wires be denoted by b. The voltage and current on the transmission line are functions of both the distance along the transmission line and time. The usual form of the partial differential equations satisfied by the voltage and current are valid only under the following three main assumptions. (1) The separation distance b between the two wires and, therefore, also the radius a of the wires is small in comparison with the space scale length of variations of the voltage and current as well as of the associated electric and magnetic fields. The voltage and current with arbitrary time dependence can be considered as a suitable superposition of time-periodic quantities. Such time-periodic voltages and currents are also periodic in space and the space period is called the wavelength. It is established later that the wavelength $\lambda = v/f$, where f is the frequency or the inverse of the time period and v is the velocity of electromagnetic waves in the medium surrounding the transmission line. In this treatment of transmission-line theory, major emphasis is placed on the voltage and current which vary periodically in time and space, and for such a case the space scale length of variations is just the wavelength. Therefore, the first condition requires that

$$a < b \ll \lambda. \tag{2.1}$$

As an example, consider a transmission line in free space where $v = 3 \times 10^8$ m/sec. Let the separation distance be $b = 10$ cm. If the frequency of operation is $f = 3 \times 10^8$ Hz, then $\lambda = 1$ m and condition (2.1) is fulfilled. On the other hand, for a higher frequency of operation, namely $f = 3 \times 10^9$ Hz, $\lambda = 10$ cm, condition (2.1) is not fulfilled, and the usual form of the transmission-line equations is not valid. The reason for this stipulation is that the derivation of the transmission-line equations is based on circuit theory, which requires that the effect on one wire due to the current in the other take place instantaneously without any appreciable retardation in time. The time-retardation effect is negligible if the separation distance b is small in comparison with the space scale length for an arbitrary time-dependent phenomenon or the wavelength for a periodic phenomenon. (2) The second assumption is that the transmission line is uniform: every section of it is the same as every other section. This requirement of uniformity excludes transmission lines of finite length, since in that case a section near the end of the transmission line is not the same as the one in the middle. Thus, the transmission-line theory is strictly applicable only to transmission lines of infinite length. However, the present theory is applied to transmission lines of finite length with the necessary implication that the "end effects" are negligible. In most practical cases, the "end effects" are small enough to justify their omission. (3) The third assumption concerns the nature of the currents in the two wires. In any transverse cross section, let I_1 be the current in one wire and let I_2 be the current in the other. These currents are conveniently rewritten as follows:

$$I_1 = I_B + I_U, \qquad I_2 = -I_B + I_U, \tag{2.2}$$

where

$$I_B = \frac{I_1 - I_2}{2}, \qquad I_U = \frac{I_1 + I_2}{2}. \tag{2.3}$$

From (2.2) and (2.3), it follows that the transmission-line currents can be taken as consisting of two sets. For the first set, the currents in the two wires are equal in magnitude (I_B) but are opposite in direction. When this situation obtains, the transmission line is said to be balanced and is amenable to treatment from the point of view of circuit theory. For the second set, the currents in the two wires are equal in magnitude (I_U) and in direction. The transmission line is then said to be unbalanced and has to be treated from the point of view of electromagnetic theory. Since the present treatment of transmission lines is based on circuit theory, it follows that the transmission line is balanced or at any rate the unbalanced component I_U is negligible compared with the balanced component I_B. A transmission line becomes unbalanced (i) if it is situated unsymmetrically with respect to the environment such as other transmission lines or the earth, (ii) if it has an unsymmetrical load such as a transmission line connecting a linear antenna at some point other than at its center, and (iii) if it is excited in an unsymmetrical fashion. In most practical cases, the unbalanced part of the current on a transmission line is negligible and the present theory can be used to analyze the behavior of the transmission line.

Before proceeding to the derivation of the transmission-line equations, we emphasize a feature which distinguishes the transmission line from the conventional network. In an ordinary electric circuit, the geometrical dimensions of the circuit elements are so small in comparison with the wavelength that the elements may be reasonably considered to be lumped at a point. However, the length of a transmission line is usually several times the wavelength that its circuit elements require to be considered as distributed throughout its length. Thus, whereas the ordinary electric circuit consists of lumped elements, the transmission line has to be treated as a distributed parameter network.

Let L, C, R, and G be the inductance, capacitance, resistance, and conductance, all per unit length of the transmission line. We deduce the dependence of these parameters on the geometry of the transmission line, namely the radius a and the separation distance b, in Chapter 7. The inductance L and the resistance R are series elements; their units are respectively henrys per meter and ohms per meter. The capacitance C and the conductance G are shunt elements; their units are farads per meter and mhos per meter respectively. An equivalent circuit of the differential length Δx is shown in Fig. 1.2. Let $V(x, t)$ and $V(x + \Delta x, t)$ be the voltages at x and $x + \Delta x$ respectively. Similarly, let $I(x, t)$ and $I(x + \Delta x, t)$ be the currents at x and $x + \Delta x$ respectively. The sign conventions for the voltages and currents are included in Fig. 1.2. The values of the parameters for the length Δx of the transmission line

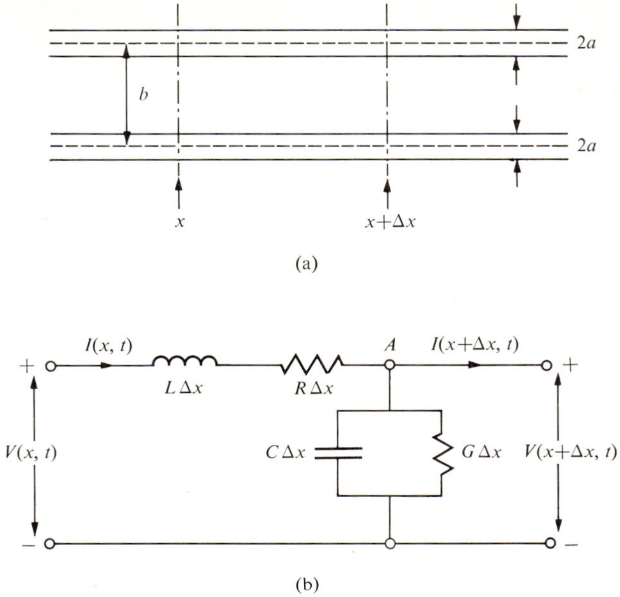

Fig. 1.2 Equivalent electric circuit of an infinitesimal length Δx of a two-wire transmission line.

shown in Fig. 1.2 are Δx multiplied by the parameter values per unit length. Kirchhoff's loop equation for the circuit shown in Fig. 1.2 is

$$V(x, t) - L\,\Delta x \frac{\partial}{\partial t} I(x, t) - R\,\Delta x I(x, t) - V(x + \Delta x, t) = 0. \tag{2.4}$$

On dividing (2.4) by Δx, taking the limit as Δx goes to zero, and noting that the limit $\{V(x + \Delta x, t) - V(x, t)\}/\Delta x$ as Δx tends to zero is $(\partial/\partial x)V(x, t)$, we find that

$$-\frac{\partial}{\partial x} V(x, t) = L \frac{\partial}{\partial t} I(x, t) + RI(x, t). \tag{2.5}$$

Kirchhoff's node equation for the node A in Fig. 1.2 is

$$I(x, t) - C\,\Delta x \frac{\partial}{\partial t} V(x + \Delta x, t) - G\,\Delta x V(x + \Delta x, t) - I(x + \Delta x, t) = 0. \tag{2.6}$$

A procedure similar to that used in (2.4), when applied to (2.6), yields

$$-\frac{\partial}{\partial x} I(x, t) = C \frac{\partial}{\partial t} V(x, t) + GV(x, t). \tag{2.7}$$

Since every section of the transmission line is the same as each of the other sections, it follows that (2.5) and (2.7) are the partial differential equations satisfied by the voltage and current at any point x on the transmission line and at any time t. These are first-order equations that contain both the voltage $V(x, t)$ and the current $I(x, t)$. We may obtain second-order partial differential equations containing either the voltage or the current in the following manner. We partially differentiate both sides of (2.5) with respect to x and replace the term $(\partial/\partial x)I(x, t)$, which then appears on the right-hand side of the equation, by its equivalent expression given in (2.7), obtaining

$$\frac{\partial^2}{\partial x^2} V(x, t) = LC \frac{\partial^2}{\partial t^2} V(x, t) + (LG + CR) \frac{\partial}{\partial t} V(x, t) + RGV(x, t). \tag{2.8}$$

Similarly, we partially differentiate (2.7) with respect to x and substitute for the term $(\partial/\partial x)V(x, t)$, which then appears on the right-hand side of the equation, its equivalent expression given in (2.5) and obtain

$$\frac{\partial^2}{\partial x^2} I(x, t) = LC \frac{\partial^2}{\partial t^2} I(x, t) + (LG + CR) \frac{\partial}{\partial t} I(x, t) + RGI(x, t). \tag{2.9}$$

We see that the second-order partial differential equations (2.8) and (2.9) satisfied by $V(x, t)$ and $I(x, t)$ respectively are the same.

It is difficult to solve the partial differential equations (2.8) and (2.9) in the general case. In the special case of lossless transmission lines corresponding to $R = G = 0$, we can solve them fairly simply, as shown in the next section. We shall explain the physical meaning of lossless transmission lines in Section 1.6.

Remember that the voltage at any point along the transmission line is the voltage difference between the two wires in the same transverse plane containing the point under consideration and that the voltage of the upper wire is measured with reference to that of the lower wire. Similarly, the current at any point along the transmission line is the current in the upper wire and the current in the lower wire is equal in magnitude and opposite in direction to that in the upper wire. The current in the upper wire in the direction left to right is taken as positive.

1.3 SOLUTION OF THE TRANSMISSION-LINE EQUATIONS FOR THE LOSSLESS CASE

For the lossless case corresponding to $R = G = 0$, the first-order partial differential equations (2.5) and (2.7) become

$$-\frac{\partial}{\partial x} V(x, t) = L \frac{\partial}{\partial t} I(x, t), \tag{3.1}$$

$$-\frac{\partial}{\partial x} I(x, t) = C \frac{\partial}{\partial t} V(x, t). \tag{3.2}$$

The second-order partial differential equations (2.8) for the voltage and (2.9) for the current become considerably simpler for the lossless case and we restate them here for convenience:

$$\frac{\partial^2}{\partial x^2} V(x, t) = \frac{1}{v^2} \frac{\partial^2}{\partial t^2} V(x, t), \tag{3.3}$$

$$\frac{\partial^2}{\partial x^2} I(x, t) = \frac{1}{v^2} \frac{\partial^2}{\partial t^2} I(x, t), \tag{3.4}$$

where

$$v = \frac{1}{\sqrt{LC}}. \tag{3.5}$$

We can verify that v has the dimensions of velocity and, as we shall establish presently, it is indeed the velocity of propagation of the voltage and current waveforms along a transmission line. The solutions of (3.3) or (3.4) are facilitated by the following change in variables:

$$\xi = x - vt; \qquad \eta = x + vt. \tag{3.6}$$

Suppressing for the present the arguments (x, t), or equivalently (ξ, η), and noting that the total infinitesimal change dV in V arises because of infinitesimal changes in both x and t (or equivalently ξ and η), we find that dV is related to the infinitesimal changes dx, dt, $d\xi$, and $d\eta$ as given by

$$dV = \frac{\partial V}{\partial x} dx + \frac{\partial V}{\partial t} dt = \frac{\partial V}{\partial \xi} d\xi + \frac{\partial V}{\partial \eta} d\eta. \tag{3.7}$$

From (3.6), we obtain

$$d\xi = dx - v\,dt; \qquad d\eta = dx + v\,dt. \tag{3.8}$$

We replace $d\xi$ and $d\eta$ in (3.7) by their equivalent expressions as given in (3.8) and equate the coefficients of dx and dt that occur on both sides of the equation to obtain

$$\frac{\partial V}{\partial x} = \frac{\partial V}{\partial \xi} + \frac{\partial V}{\partial \eta} \quad \text{or} \quad \frac{\partial}{\partial x} = \frac{\partial}{\partial \xi} + \frac{\partial}{\partial \eta}, \tag{3.9a}$$

$$\frac{1}{v}\frac{\partial V}{\partial t} = -\frac{\partial V}{\partial \xi} + \frac{\partial V}{\partial \eta} \quad \text{or} \quad \frac{1}{v}\frac{\partial}{\partial t} = -\frac{\partial}{\partial \xi} + \frac{\partial}{\partial \eta}. \tag{3.9b}$$

From (3.9a) and (3.9b), it is found that

$$\frac{\partial}{\partial x} + \frac{1}{v}\frac{\partial}{\partial t} = 2\frac{\partial}{\partial \eta}; \qquad \frac{\partial}{\partial x} - \frac{1}{v}\frac{\partial}{\partial t} = 2\frac{\partial}{\partial \xi}. \tag{3.10}$$

It is convenient to rearrange (3.3) as

$$\left(\frac{\partial}{\partial x} + \frac{1}{v}\frac{\partial}{\partial t}\right)\left(\frac{\partial}{\partial x} - \frac{1}{v}\frac{\partial}{\partial t}\right)V = 0. \tag{3.11}$$

Substitution of (3.10) in (3.11) yields

$$\frac{\partial^2}{\partial\xi\,\partial\eta}V = 0, \tag{3.12}$$

which is readily integrated twice, first with respect to ξ and then with respect to η, to give the following general solution of V in terms of the variables ξ and η or $x - vt$ and $x + vt$:

$$V(x, t) = f_1(x - vt) + f_2(x + vt) + f_0. \tag{3.13}$$

Note that f_1 and f_2 are any functions of their respective arguments and f_0 is a constant, independent of x and t. Similarly we can solve (3.4) to obtain:

$$I(x, t) = g_1(x - vt) + g_2(x + vt) + g_0, \tag{3.14}$$

where the g functions are not necessarily equal to the f functions given in (3.13). As before g_1 and g_2 are any functions of their respective arguments and g_0 is a constant, independent of x and t.

Note that we obtained $V(x, t)$ and $I(x, t)$ by solving separately the voltage and the current equations; this procedure has not placed in evidence the relation that exists between the f and g functions. Equations (3.1) and (3.2) indicate that $V(x, t)$ and $I(x, t)$ are related and therefore the f and g functions are related in some fashion, as established by the following procedure. Substitution of (3.13) in (3.2) yields

$$\frac{\partial}{\partial x}I(x, t) = Cvf_1'(x - vt) - Cvf_2'(x + vt), \tag{3.15}$$

where the primes indicate differentiation with respect to the argument. It is convenient to set

$$\frac{1}{Cv} = R_0 = \sqrt{\frac{L}{C}}, \tag{3.16}$$

which is verified as having the dimensions of a resistance. Noting that the differentiations with respect to the arguments in (3.15) are the same as the differentiations with respect to x, integrating both sides of (3.15) with respect to x, and using (3.16), we find

$$I(x, t) = \frac{1}{R_0}f_1(x - vt) - \frac{1}{R_0}f_2(x + vt) + f(t), \tag{3.17}$$

where $f(t)$ is some function of t. From (3.17), since $Lv/R_0 = 1$, we obtain

$$-L\frac{\partial}{\partial t}I(x, t) = f_1'(x - vt) + f_2'(x + vt) - Lf'(t). \tag{3.18}$$

Also, from (3.13),

$$\frac{\partial}{\partial x}V(x, t) = f_1'(x - vt) + f_2'(x + vt). \tag{3.19}$$

Since by (3.1), (3.18) and (3.19) are equal, it follows that $f'(t) = 0$ or $f(t) =$ constant, which without any loss of generality is denoted by g_0. Thus, from (3.17),

$$I(x, t) = \frac{1}{R_0} f_1(x - vt) - \frac{1}{R_0} f_2(x + vt) + g_0. \tag{3.20}$$

A comparison of (3.20) with (3.14) shows that

$$g_1(x - vt) = \frac{1}{R_0} f_1(x - vt), \tag{3.21}$$

$$g_2(x + vt) = -\frac{1}{R_0} f_2(x + vt). \tag{3.22}$$

We see from (3.21) and (3.22) that g_1 and f_1 functions and, similarly, g_2 and f_2 functions are related to each other. But there is no relationship between the constants f_0 and g_0 imposed by the differential equations satisfied by the voltage and current on a lossless transmission line.

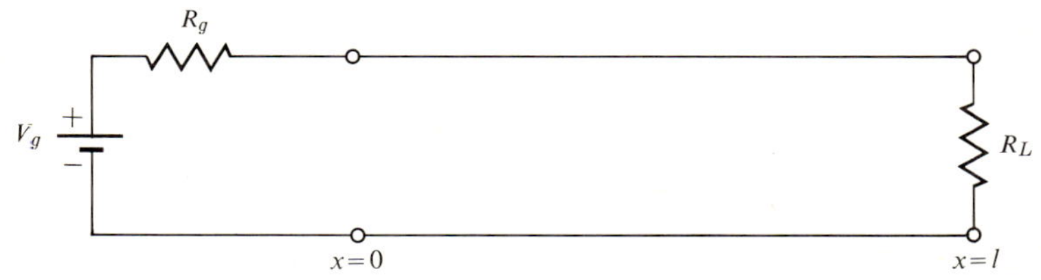

Fig. 1.3 Schematic diagram of a transmission line driven by a battery.

It is an easy matter to see that f_0 and g_0 are respectively the d.c. voltage and the d.c. current. For a lossless transmission line, since $R = G = 0$, the d.c. voltage and the d.c. current at any point on the transmission line and for all times are completely determined by the terminal or the boundary conditions. Consider the transmission line shown in Fig. 1.3 terminated in a load resistor R_L and excited by a d.c. generator of voltage V_g through a series resistor R_g. We find that the d.c. voltage and the d.c. current along the transmission line are given by

$$V = \frac{V_g}{(1 + R_g/R_L)} = f_0; \qquad I = \frac{V_g}{R_g + R_L} = g_0. \tag{3.23}$$

If R_g and R_L are varied so as to keep $(R_g + R_L)$ constant, it follows from (3.23) that the d.c. current is unchanged whereas the d.c. voltage can be made to assume any value less than V_g. Similarly, if R_g and R_L are varied so as to keep R_g/R_L constant, the d.c. voltage is unchanged, but the d.c. current can be varied to within certain limits. Thus, we can change the d.c. voltage and the d.c. current on a lossless transmission line independently, by suitably adjusting the terminal conditions, and no relationship between them is stipulated by the differential equations governing $V(x, t)$ and $I(x, t)$ on a lossless transmission line. On the other hand, (3.21) and (3.22) show that the time-varying voltage and current on a lossless transmission line are related through the factor R_0.

1.4 TRAVELING WAVES WITHOUT ATTENUATION AND DISTORTION

Each of the three terms on the right-hand side of (3.13) is a solution of the transmission-line equation and this fact enabled us to examine separately the dependence of the d.c. solutions on the terminal conditions in the last section. Here we discuss the properties of the other two terms in (3.13). Let

$$V_+(x, t) = f_1(x - vt). \tag{4.1}$$

Since the value of the function f_1 depends uniquely on its argument, it follows that the voltage at $x = x_1$ and $t = t_1$ is equal to the voltage at $x = x_2$ and a later time $t = t_2 > t_1$ only if

$$x_1 - vt_1 = x_2 - vt_2, \tag{4.2}$$

or

$$x_2 = x_1 + v(t_2 - t_1). \tag{4.3}$$

From (4.3), it follows that the voltage at $x = x_1$ and $t = t_1$ has traveled to a point $x = x_2$ (which is greater than x_1) in time $(t_2 - t_1)$; therefore, v gives the uniform velocity. In other words, $f_1(x - vt)$ represents a wave traveling in the positive x-direction with uniform velocity v. The subscript on V in (4.1) is chosen as + to indicate that the term on the right-hand side represents a wave traveling with uniform velocity v in the *positive* x-direction.

At time $t = t_1$, let the voltage wave $V_+(x, t)$ have, as shown in Fig. 1.4, nonvanishing values from $x = x_a$ to $x = x_b > x_a$. Consider a later time $t = t_2 > t_1$. In view of the foregoing discussion, we see that the voltages at $x = x_a$ and $x = x_b$ travel without changing their values, in the positive x-direction with uniform velocity v and, therefore, these voltages appear respectively at $x = x_a + v(t_2 - t_1)$ and $x = x_b + v(t_2 - t_1)$. At time $t = t_1$, consider an arbitrary point $x = x_1$ on the transmission line lying between $x = x_a$ and $x = x_b$. At time $t = t_2$, the voltage at $x = x_1$ also has moved the same distance $v(t_2 - t_1)$ in the positive x-direction without changing its value. As x_1 is varied from $x = x_a$ to $x = x_b$, it follows that every point on the voltage waveform at $t = t_1$ has moved at time $t = t_2$ the *same* distance $v(t_2 - t_1)$ in the positive x-direction without their values being changed; thus the waveform at $t = t_1$ *as a whole* has moved in the positive x-direction with uniform velocity v. Therefore, $V_+(x, t) = f_1(x - vt)$ represents a voltage waveform that travels as a whole in the positive x-direction with uniform velocity v and without any *distortion or attenuation*.

Similarly, we can show that

$$V_-(x, t) = f_2(x + vt) \tag{4.4}$$

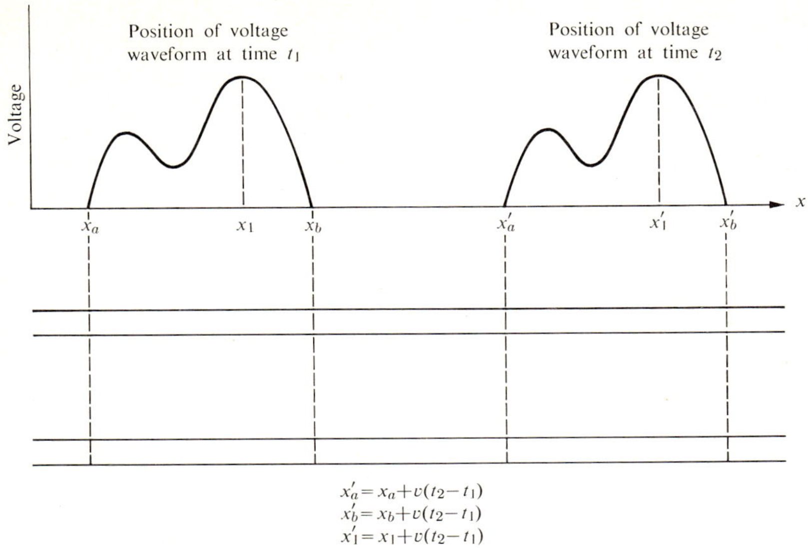

Fig. 1.4 Waveform moving uniformly as a whole in the positive x-direction with a velocity v.

represents a voltage waveform that travels as a whole in the negative x-direction with uniform velocity v and without any distortion or attenuation. The current waveforms given by the first two terms on the right-hand side of (3.20) are amenable to similar interpretations.

The absence of attenuation in a lossless transmission line is a generally valid result. In the case of a two-wire transmission line, the absence of losses also leads to distortionless propagation. But, it is possible to have a lossless linear system that gives rise to propagation with distortion. Also, the distortionless propagation does not imply the absence of losses, that is, it is possible to have propagation without distortion on a lossy transmission line. Such a transmission line is treated in Section 2.3.

1.5 SEMI-INFINITE TRANSMISSION LINE

The general solutions for the voltage and the current on a lossless transmission line are given by (3.13) and (3.20), respectively, where the f and the g functions are not determined except for the arguments. The terminal conditions determine the exact nature of these functions, as is shown here. For simplicity, we assume that the terminal conditions are such that there is no d.c. voltage or d.c. current on the transmission line. Consider a semi-infinite transmission line extending from $x = 0$ to $x = \infty$. Let a voltage generator $V_g(t)$ be connected through a series resistor R_g to the input $x = 0$ of the transmission line. The generator voltage $V_g(t)$ is assumed to have no d.c. part; therefore, it follows from (3.23) that the d.c. voltage and the d.c. current along the transmission line vanish. Since the generator is situated at $x = 0$, a voltage waveform starts at $x = 0$ and travels in the positive x-direction, and therefore $f_1(x - vt)$ is nonvanishing. Since the transmission line is uniform and is infinite in length, there is no mechanism to obtain a wave traveling in the negative x-direction; that is, $f_2(x + vt)$ is equal to zero. Therefore, the general expressions for the voltage and the current for the semi-infinite transmission line under consideration are given by

$$V(x, t) = f_1(x - vt), \tag{5.1}$$

$$I(x, t) = \frac{f_1(x - vt)}{R_0}. \tag{5.2}$$

From (5.1) and (5.2), we see that

$$\frac{V(x, t)}{I(x, t)} = R_0. \tag{5.3}$$

Thus, the ratio of the voltage to the current is the same at every point along the transmission line and is given by R_0, which is independent of the terminal conditions and is dependent only on the inductance L and the capacitance C, both per unit length. Since L and C characterize the transmission line, R_0 is usually called the characteristic resistance.

If the transmission line instead of extending from $x = 0$ to $x = \infty$ extends from $x = -\infty$ to $x = 0$ and is driven at $x = 0$ in exactly the same manner as in the previous case, it can be shown by the same arguments that the general expressions for the voltage and the current for the semi-infinite ($x = -\infty$ to $x = 0$) transmission line are given by

$$V(x, t) = f_2(x + vt), \tag{5.4}$$

$$I(x, t) = -\frac{f_2(x + vt)}{R_0}, \tag{5.5}$$

and, therefore,

$$\frac{V(x, t)}{I(x, t)} = -R_0. \tag{5.6}$$

It is found that the ratio of the voltage to the current is again the same at every point along the transmission line but is given by $-R_0$ which is the negative of that obtained in the previous case. A difference in sign of the ratio of the voltage to the current occurs even though the transmission line has the same characteristics in both cases. An examination near the input reveals that at the corresponding points in the two transmission lines the voltages have the same magnitude and sign. Consequently, the currents at the corresponding points in the two transmission lines are equal in magnitude but opposite in direction. These results are made explicit in (5.9) and (5.10), and the corresponding equations for the transmission line extending from $x = 0$ to $x = -\infty$. In the first case, the traveling wave is in the $+x$-direction and in the second case it is in the $-x$-direction. Thus, a change in the direction of the traveling wave results in a change in the sign of the ratio of the voltage to the current at every point along the transmission line.

Returning to the semi-infinite transmission line extending from $x = 0$ to $x = \infty$, we evaluate the as-yet-undetermined function $f_1(x - vt)$ as follows. In view of (5.3), the generator voltage operates in series with R_g and R_0 to produce the current $I(0, t)$ at the input $x = 0$ and (5.3) can then be used to obtain the input voltage $V(0, t)$. Hence,

$$V(0, t) = \frac{V_g(t)R_0}{R_g + R_0} = f_1(-vt), \tag{5.7}$$

$$I(0, t) = \frac{V_g(t)}{R_g + R_0} = \frac{f_1(-vt)}{R_0}. \tag{5.8}$$

Since $f_1(-vt)$ is known from (5.7), $f_1(x - vt)$ can be evaluated simply by replacing t by $t - (x/v)$, with the following result:

$$V(x, t) = f_1(x - vt) = \frac{R_0}{R_g + R_0} V_g\left(t - \frac{x}{v}\right), \tag{5.9}$$

$$I(x, t) = \frac{f_1(x - vt)}{R_0} = \frac{1}{R_g + R_0} V_g\left(t - \frac{x}{v}\right). \tag{5.10}$$

Since the generator voltage $V_g(t)$ is given, $V(x, t)$ and $I(x, t)$ are completely determined by (5.9) and (5.10).

1.6 ENERGY DISTRIBUTION ON A LOSSLESS TRANSMISSION LINE

For a lossless transmission line, all the energy supplied at the input $x = 0$ is merely redistributed along the transmission line with none of it lost irretrievably as heat. Basically, this phenomenon enables us to characterize the transmission line as lossless. Of course, some energy is lost in the resistor which is in series with the generator. Energy conservation requires that the difference between the energy supplied by the generator and that dissipated in the series resistor be merely redistributed along the transmission line. It is instructive to establish this energy conservation law. For this purpose, a simple type of generator is assumed, namely

$$V_g(t) = V_0 H(t), \tag{6.1}$$

where $H(t)$ is the Heaviside unit step function defined by

$$\begin{aligned} H(t) &= 0 \qquad t \le 0, \\ &= 1 \qquad t > 0. \end{aligned} \tag{6.2}$$

Hence, the generator is a battery of voltage V_0 applied at time $t = 0$. Thus, $V(x, t)$ and $I(x, t)$ along the transmission line, as obtained from (5.9) and (5.10), are given by

$$V(x, t) = \frac{R_0 V_0}{R_g + R_0} H\left(t - \frac{x}{v}\right), \tag{6.3}$$

$$I(x, t) = \frac{V_0}{R_g + R_0} H\left(t - \frac{x}{v}\right). \tag{6.4}$$

Let us consider the energy balance at time $t = T$. After $t = 0$, the current in the battery is obtained by setting $x = 0$ in (6.4) with the result

$$I(0, t) = \frac{V_0}{R_g + R_0}. \tag{6.5}$$

The total energy drawn from the battery of voltage V_0 in supplying the d.c. current (6.5) for the time T is

$$W_B = \frac{V_0^2 T}{R_g + R_0}. \tag{6.6}$$

The total energy dissipated in the resistor R_g by the d.c. current (6.5) for the time T is given by

$$W_D = \frac{R_g V_0^2 T}{(R_g + R_0)^2}. \tag{6.7}$$

Thus, the net energy supplied to the input of the transmission line is found to be

$$W_B - W_D = \frac{V_0^2 T}{R_g + R_0} - \frac{V_0^2 T R_g}{(R_g + R_0)^2} = \frac{V_0^2 T R_0}{(R_g + R_0)^2}. \tag{6.8}$$

At $t = T$, in view of (6.2), we see from (6.3) that $V(x, t) = 0$ for $T < x/v$ or $x > vT$ and $V(x, t) = R_0 V_0/(R_0 + R_g)$ for $x < vT$. In a similar manner, at $t = T$ we find from (6.4) that $I(x, t) = 0$ for $x > vT$ and $I(x, t) = V_0/(R_0 + R_g)$ for $x < vT$. It is noted that the voltage and the current are constants for $x < vT$ and zero for $x > vT$. The energy associated with the inductance of the transmission line from $x = 0$ to $x = vT$ is

$$W_L = \tfrac{1}{2}(LvT)\frac{V_0^2}{(R_0 + R_g)^2} = \tfrac{1}{2}T\frac{R_0 V_0^2}{(R_0 + R_g)^2}. \tag{6.9}$$

In a similar manner the energy associated with the capacitance of the transmission line from $x = 0$ to $x = vT$ is

$$W_C = \tfrac{1}{2}(CvT)\frac{R_0^2 V_0^2}{(R_0 + R_g)^2} = \tfrac{1}{2}T\frac{R_0 V_0^2}{(R_0 + R_g)^2}. \tag{6.10}$$

It is seen from (6.9) and (6.10) that the energy associated with the inductance and the capacitance are equal. Thus, the total energy appearing on the transmission line is

$$W_L + W_C = \frac{T R_0 V_0^2}{(R_0 + R_g)^2}, \tag{6.11}$$

which is found from (6.8) to be equal to the total energy supplied to the input of the transmission line. All that has happened is that the total energy supplied to the input, without being lost in any form, has redistributed itself along the transmission line.

At a later time $T_1 > T$, it is seen from (6.3) and (6.4) that the voltage and the current which previously had nonvanishing values only for $x < vT$, have nonvanishing values for an additional length $v(T_1 - T)$ of the transmission line. In the intervening time $(T_1 - T)$, the battery has supplied additional energy, some of which is lost in the resistor R_g. The remainder can be shown, as before, to be distributed along the transmission line from $x = vT$ to $x = vT_1$. Thus, as time progresses, energy is distributed over greater and greater lengths of the transmission line or energy is propagating along the transmission line with uniform velocity v.

1.7 FINITE TRANSMISSION LINE WITH A RESISTANCE TERMINATION: REFLECTION COEFFICIENT

Consider a finite transmission line extending from $x = 0$ to $x = l$ where a resistor R_L is connected across the transmission line. Let the transmission line be excited by some kind of a generator at $x = 0$. A voltage and a current wave travel in the positive x-direction and the ratio of the voltage to the current at every point before the wave reaches the termination is seen from (5.3) to be R_0. In particular, let V_+ be the voltage in front of the load resistor R_L due to the wave traveling in the positive x-direction and I_+ be the corresponding current. Let V_L and I_L be respectively the voltage across, and the current in, the resistor R_L. Since $V_+/I_+ = R_0$ is not equal to $V_L/I_L = R_L$, it follows that Kirchhoff's circuit equations near the load are not satisfied. In order that Kirchhoff's circuit equations be satisfied near the load, a voltage and a current wave should start at the load and travel in the negative x-direction. Just in front of the load, let V_- and I_- be the voltage and the associated current in this wave traveling in the negative x-direction. From (5.6), $V_-/I_- = -R_0$. When both incident and reflected waves are present, Kirchhoff's circuit equations yield

$$V_+ + V_- = V_L, \tag{7.1}$$

$$I_+ + I_- = I_L \quad \text{or} \quad \frac{V_+}{R_0} - \frac{V_-}{R_0} = \frac{V_L}{R_L}. \tag{7.2}$$

From (7.1) and (7.2), we obtain

$$\frac{V_-}{V_+} = \rho = -\frac{I_-}{I_+} = \frac{R_L - R_0}{R_L + R_0}, \tag{7.3}$$

$$\frac{V_L}{V_+} = \frac{2R_L}{R_L + R_0}; \quad \frac{I_L}{I_+} = \frac{2R_0}{R_L + R_0}. \tag{7.4}$$

The ratio $\rho = V_-/V_+$ is called the voltage reflection coefficient or simply the reflection coefficient. Similarly, the ratio I_-/I_+ is called the current reflection coefficient and is the negative of the voltage reflection coefficient. Unless otherwise mentioned ρ refers only to the voltage reflection coefficient. It follows that for a lossless transmission line, the magnitude of the reflection coefficient is always less than unity. Note that the reflection coefficient ρ is a real constant depending only on the characteristic resistance R_0 of the transmission line and the load resistor R_L and that the reflected voltage V_- equals ρ multiplied by the incident voltage V_+. This simplified result occurs only for a resistance termination and in the case of a more general termination, for example, by an inductance, the dependence of the reflected voltage V_- on the incident voltage V_+ is much more complicated, as shown later.

It is interesting to note from (7.3) that if $R_L = R_0, \rho = 0$ and the reflected voltage is zero. All the incident energy is absorbed by the load. The transmission line is then said to be matched. In the respect of having no reflected voltage, a finite transmission line terminated by its characteristic resistance is similar to a semi-infinite transmission line.

When the transmission line is short-circuited, that is $R_L = 0$, (7.3) yields

$$\rho = -1 \qquad \text{(short-circuit termination).} \tag{7.5}$$

In a similar manner, if the terminals are open-circuited, that is $R_L = \infty$, the following result is obtained:

$$\rho = 1 \qquad \text{(open-circuit termination).} \tag{7.6}$$

In these two special cases, magnitude-wise, the incident voltage and current are reflected back entirely by the termination.

1.8 PRINCIPLE OF SUPERPOSITION

Let $V_a(x, t)$ and $I_a(x, t)$ be one set of solutions for the voltage and the current along a transmission line. In a similar manner, let $V_b(x, t)$ and $I_b(x, t)$ be another set of solutions for the voltage and the current along the same transmission line. Then the fact that any linear combination of $V_a(x, t)$ and $V_b(x, t)$ is also a solution of the transmission-line equation follows from (2.8). From (2.5) and (2.7), it is verified that the same linear combination of $I_a(x, t)$ and $I_b(x, t)$ is the current associated with the assumed linear combination of $V_a(x, t)$ and $V_b(x, t)$. To clarify further this principle of superposition, consider the transmission line shown in Fig. 1.5, extending from $x = 0$ to $x = l$. Let a voltage generator $V_g(t)$ in series with a resistor R_g be connected to the terminals at $x = 0$ and let another voltage generator $V_L(t)$ in series with a resistor R_L be connected to the terminals at $x = l$. The principle of superposition enables us to obtain in the following manner the voltage and current distributions along the transmission line caused by the simultaneous presence of the two voltage

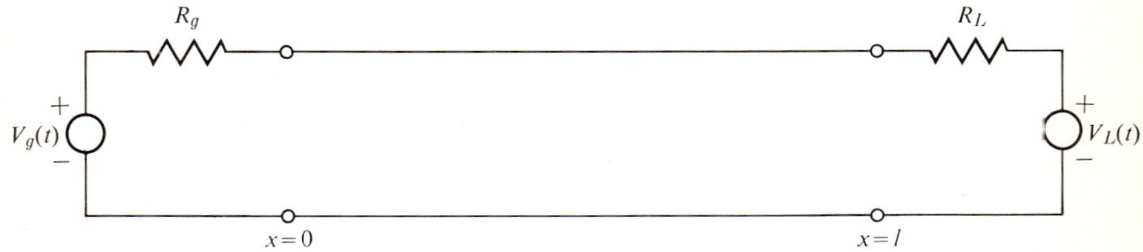

Fig. 1.5 Schematic diagram of a finite length transmission line driven by voltage generators at both the ends.

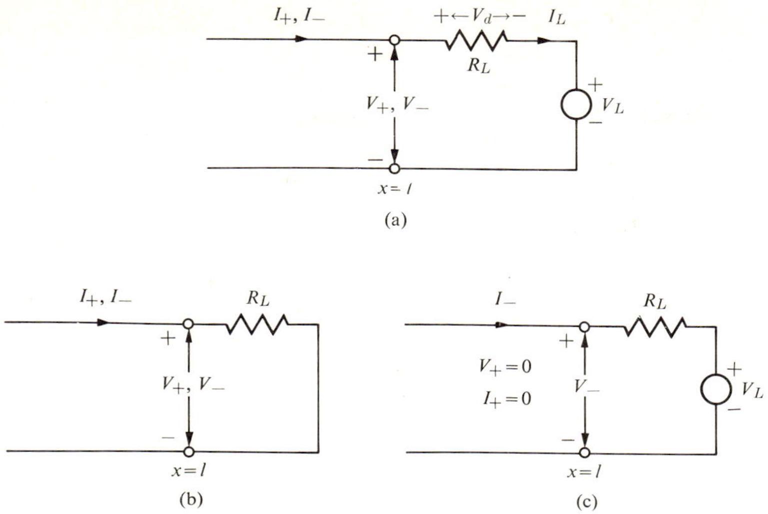

Fig. 1.6 Illustration of the superposition principle.

generators. First, remove the voltage generator $V_L(t)$, that is, replace it by a short-circuit, and evaluate the voltage distribution $V_a(x, t)$ and the current distribution $I_a(x, t)$. Then return the voltage generator $V_L(t)$ to its former position and replace the voltage generator $V_g(t)$ by a short-circuit. The new voltage and current distributions $V_b(x, t)$ and $I_b(x, t)$ are evaluated as before. By the principle of superposition, it follows that $V_a(x, t) + V_b(x, t)$ and $I_a(x, t) + I_b(x, t)$ are the voltage and current distributions resulting from the simultaneous presence of the two voltage generators. In the next section, we discuss a method of evaluating the voltage and current distributions along a lossless transmission line. For that purpose, a result capable of being derived most simply by the principle of superposition is established now.

Let V_+ and V_- be the voltages in front of the resistor R_L at $x = l$; these are associated with the waves traveling in the positive and the negative x-directions respectively. Let I_+ and I_- be the corresponding currents. Also, assume that V_d is the voltage across the resistor R_L due to the current I_L in it (Fig. 1.6a). Then Kirchhoff's circuit equations yield

$$V_+ + V_- = V_d + V_L(t), \tag{8.1}$$

$$I_+ + I_- = I_L \quad \text{or} \quad \frac{V_+}{R_0} - \frac{V_-}{R_0} = \frac{V_d}{R_L}. \tag{8.2}$$

The elimination of V_d from (8.1) and (8.2) yields

$$V_- = \rho V_+ + \frac{V_L(t) R_0}{R_0 + R_L}. \tag{8.3}$$

The substitution of $V_+ = I_+ R_0$ and $V_- = -I_- R_0$ in (8.3) gives

$$I_- = -\rho I_+ - \frac{V_L(t)}{R_0 + R_L}. \tag{8.4}$$

We can obtain the results (8.3) and (8.4) by using the superposition principle. First, let $V_L(t) = 0$ (Fig. 1.6b). Then, from (7.3), it is seen that

$$(V_-)_a = \rho V_+; \qquad (I_-)_a = -\rho I_+. \tag{8.5}$$

Note that V_+ and I_+ are caused by a generator situated at some point along the transmission line to the left of the load, say at $x = 0$. Now let this generator be short-circuited and the generator $V_L(t)$ be replaced in position. Then, from the equivalent circuit shown in Fig. 1.6(c) it is found that the voltage on the transmission line at $x = l$ associated with the wave traveling in the negative x-direction is given by

$$(V_-)_b = \frac{V_L(t)R_0}{R_0 + R_L}. \tag{8.6}$$

In a similar manner, the current at $x = l$ associated with the wave traveling in the negative x-direction is given by

$$(I_-)_b = -\frac{V_L(t)}{R_0 + R_L}. \tag{8.7}$$

Note that in obtaining (8.6) for the voltage and (8.7) for the current, the sign conventions are retained so as to be the same as used previously; these sign conventions are as in Fig. 1.2. When the incident wave and the generator are present simultaneously, the resulting voltage and current are the sum of the two separate results. Hence

$$V_- = (V_-)_a + (V_-)_b = \rho V_+ + \frac{V_L(t)R_0}{R_0 + R_L}, \tag{8.8}$$

and

$$I_- = (I_-)_a + (I_-)_b = -\rho I_+ - \frac{V_L(t)}{R_0 + R_L}. \tag{8.9}$$

A comparison of (8.8) with (8.3) and (8.9) with (8.4) shows that with the principle of superposition we can obtain, in a somewhat simpler manner, the same results as we deduced previously from detailed circuit calculations.

It is worthwhile to remember that the principle of superposition illustrated here is generally valid for any linear system.

1.9 REFLECTION DIAGRAM AND D.C. TRANSIENTS

Consider a lossless transmission line extending from $x = 0$ to $x = l$. Let R_0 be the characteristic resistance and v the velocity of the traveling waves. The transmission line is terminated by a resistor R_L. As shown in Fig. 1.7(a), a battery of voltage V_0 is connected through a series resistor R_g to the input terminals at time $t = 0$. Before

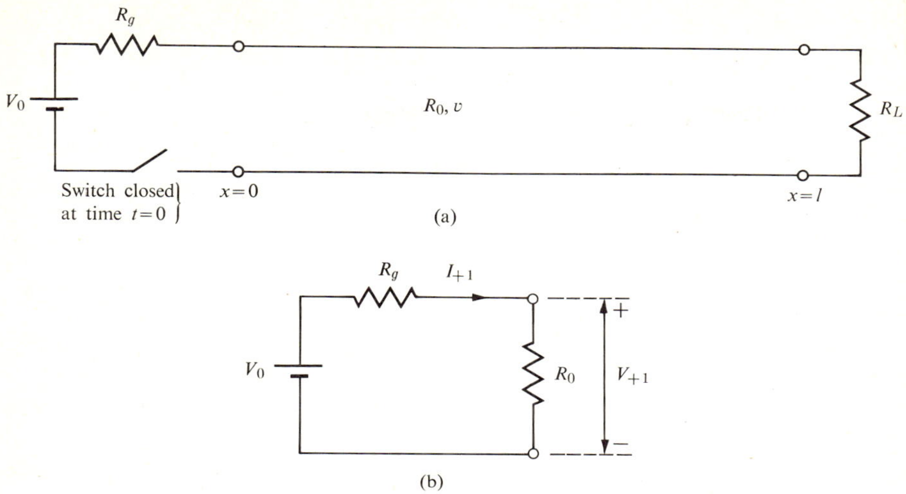

Fig. 1.7 D.C. transient on a finite length transmission line.

the battery is connected, the voltage and the current everywhere along the transmission line are zero. After the battery is connected and after the lapse of a very long time—in principle, infinite time—the d.c. conditions are attained. We see from Fig. 1.7(a) that the d.c. current and the d.c. voltage everywhere along the transmission line are given by

$$I_f = \frac{V_0}{R_g + R_L}; \qquad V_f = \frac{V_0 R_L}{R_g + R_L}. \tag{9.1}$$

We wish to examine how the final conditions as given in (9.1) are attained after the battery is connected to the terminals at $x = 0$ through a series resistor R_g.

As soon as the switch is closed, a wave starts to travel in the positive x-direction. This wave does not see that the transmission line is terminated. If V_{+1} and I_{+1} are the voltage and the current associated with this forward traveling wave, then $V_{+1} = R_0 I_{+1}$. Therefore, the equivalent circuit near the input terminals is as shown in Fig. 1.7(b), from which we deduce that

$$V_{+1} = \frac{V_0 R_0}{R_g + R_0}; \qquad I_{+1} = \frac{V_0}{R_g + R_0}. \tag{9.2}$$

Let us consider the voltage wave first. This wave travels down the transmission line with uniform velocity v and therefore reaches the termination in time $T = l/v$. The voltage reflection coefficient at the load is given by

$$\rho_L = \frac{R_L - R_0}{R_L + R_0}. \tag{9.3}$$

When the voltage V_{+1} is incident on the load resistor R_L, a voltage is reflected back and the value of this reflected voltage is $V_{-1} = V_{+1}\rho_L$, which now travels in the negative x-direction with the same uniform velocity v. At any time in the range $T < t < 2T$, a voltage discontinuity is traveling towards the generator. In front of the discontinuity, the voltage everywhere along the transmission line is V_{+1}, whereas everywhere behind the discontinuity the voltage is $V_{+1} + V_{-1}$. We shall clarify the meaning of the additional subscript 1 on V_+ and V_- subsequently. At time $t = 2T$, the voltage wave V_{-1} reaches the generator end and alters the circuit relations near the generator. With the additional voltage V_{-1} now present across the terminals, Kirchhoff's circuit equations can be satisfied only with the generation of an additional voltage wave V_{+2} which then travels in the positive x-direction. We can use the principle of superposition to obtain V_{+2}. The positively traveling voltages V_{+1} and V_{+2} at $x = 0$ are caused by the voltage generator V_0 and the negatively traveling voltage V_{-1}. First, the negatively traveling voltage V_{-1} is set equal to zero. We see that the voltage generator V_0 produces a positively traveling voltage and, since these conditions are the same as those obtained at $t = 0$, it follows that the generator V_0 produces the positively traveling voltage V_{+1}. Then, we assume that the negatively traveling voltage V_{-1} is present and the voltage generator V_0 is short-circuited. Note that the positively traveling voltage wave V_{+2} is produced entirely by V_{-1}. The voltage reflection coefficient for a transmission-line termination consisting of the resistor R_g is given by

$$\rho_g = \frac{R_g - R_0}{R_g + R_0}. \tag{9.4}$$

Therefore, the positively traveling wave generated by the incidence of the negatively traveling wave V_{-1} is given by $V_{+2} = \rho_g V_{-1}$. In other words, when the negatively traveling wave V_{-1} reaches the generator end, it produces an *additional* voltage wave V_{+2} traveling in the positive x-direction. The additional subscript on V_+ indicates the order of the positively traveling wave. For example, V_{+1} is the first and V_{+2} is the second positively traveling wave. Additional positively traveling waves are generated as time progresses and as additional negatively traveling waves reach the generator end.

In the time range $2T < t < 3T$, the voltage discontinuity is traveling in the positive x-direction. The voltage behind the discontinuity is $V_{+1} + V_{-1} + V_{+2}$, whereas ahead of the discontinuity the voltage is $V_{+1} + V_{-1}$. At time $t = 3T$, the discontinuity reaches the load and the presence of the additional positively traveling voltage V_{+2} produces the additional negatively traveling voltage $V_{-2} = \rho_L V_{+2}$, which travels toward the generator. Thus, at times which are multiples of the period T required for the wave to travel the entire length of the transmission line, additional reflected waves are generated. At each reflection, the additional reflected wave produced is equal to the appropriate reflection coefficient times the additional wave incident on the termination. These multiple reflections, in general, go on indefinitely. For the special case in which the terminating resistor is equal to the characteristic resistance R_0, the corresponding reflection coefficient vanishes. When a wave is incident on this termination, no reflected wave is produced.

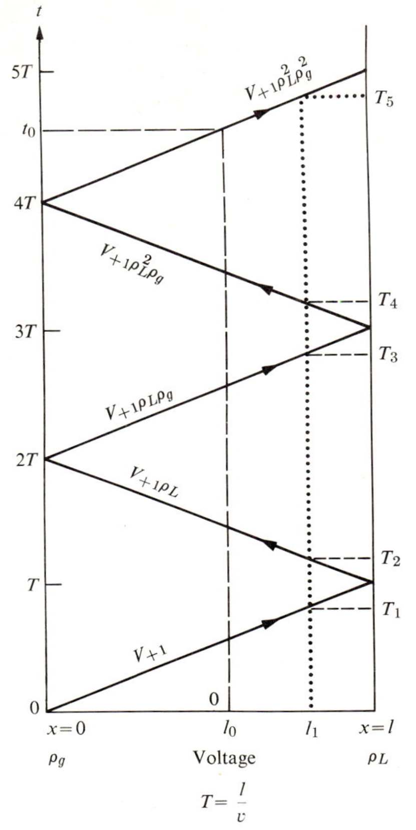

Fig. 1.8 Voltage reflection diagram.

We can best portray the development of the multiple reflections and the consequent building up of the voltage from its initial zero value to its final value by means of a reflection diagram as shown in Fig. 1.8. Since these reflection diagrams are applicable both for voltage and for current, it is usual to write "voltage" or "current", as the case may be, at the bottom of the reflection diagram. The abscissa denotes the length along the transmission line. In the present case the length extends from $x = 0$ to $x = l$. We have indicated the reflection coefficients that obtain at the terminals at the appropriate positions along the abscissa. The ordinate denotes the time. The locus of the position of the voltage discontinuity is indicated on this diagram. Since the discontinuity travels with uniform velocity v, the locus is a series of straight lines meeting at the abscissa points corresponding to $x = 0$ and $x = l$. In the time intervals $2(n - 1)T < t < (2n - 1)T$ where n is a nonvanishing positive integer, the wave travels in the positive x-direction and the slope of the straight lines, which are the locus of the position of the voltage discontinuities, is positive and is equal to $1/v$. But in the time intervals $(2n - 1)T < t < 2nT$, the wave travels in the negative x-direction and the slope of the straight lines is negative and is equal to $-1/v$. We can designate these sloping straight lines as locus straight lines for convenience in referring to them. The initial voltage along the transmission line, that is, the voltage at time $t = 0$, is marked at a point corresponding to the ordinate $t = 0$. The *additional* voltages produced at each of the terminations are marked on the locus straight lines. On the locus corresponding to the time range $0 < t < T$ there is marked V_{+1}, which is the voltage wave traveling in the positive x-direction and produced by the voltage generator V_0 at $x = 0$. In the present case, the value of V_{+1} is as given in (9.2). At $x = l$, $t = T$, an additional negatively traveling voltage V_{-1} is generated and this is marked on the locus straight line corresponding to the time range $T < t < 2T$. We obtain the value of V_{-1} by multiplying the additional incident wave that produces the reflection by the reflection coefficient pertaining to the appropriate reflection terminals; this is, $V_{-1} = \rho_L V_{+1}$. In a similar manner, we obtain the additional voltages corresponding to the various locus straight lines and these are shown in Fig. 1.8.

We can use the reflection diagram to obtain either the voltage distribution along the transmission line at a given time or the time dependence of the voltage at any point on the transmission line. Suppose that the voltage distribution along the transmission line is desired at time $t = t_0$ where, for the sake of definiteness, it is assumed that $4T < t_0 < 5T$. We draw a line, shown dashed in Fig. 1.8, corresponding to the ordinate $t = t_0$ and this line is seen to intersect the locus straight line at an abscissa point $x = l_0$. Therefore, at time $t = t_0$, the voltage discontinuity is situated at $x = l_0$ along the transmission line. For $0 \le x \le l_0$ along the transmission line the voltage has one value, and for $l_0 \le x \le l$ the voltage has another value. To find the total voltage at any point on the transmission line, we sum up all the incremental voltages which are indicated on the locus straight lines and which occur in time $t \le t_0$. Thus, for $0 \le x \le l_0$, in addition to the initial voltage which is zero in the present case, there are five incremental voltages to be summed up with the following result:

$$V(x, t_0) = V_{+1}(1 + \rho_L + \rho_L\rho_g + \rho_L^2\rho_g + \rho_L^2\rho_g^2) \qquad \text{for } 0 \le x \le l_0.$$

But for $l_0 \le x \le l$, in addition to the initial voltage there are only four incremental voltages to be summed up and hence the result is

$$V(x, t_0) = V_{+1}(1 + \rho_L + \rho_L\rho_g + \rho_L^2\rho_g) \qquad \text{for } l_0 \le x \le l.$$

Thus, for any time t_0 we can determine the voltage distribution along the transmission line with the help of the reflection diagram.

Suppose that we wish to obtain the time dependence of the voltage at a point $x = l_1$ on the transmission line. We draw a line, shown dotted in Fig. 1.8, corresponding to the abscissa point $x = l_1$, and this line is seen to intersect the successive locus straight lines at times T_1, T_2, T_3, etc. At each of the times, say $t = T_n$, a voltage discontinuity passes through the point $x = l_1$ on the transmission line and abruptly changes the voltage at the point by an amount representing the discontinuity in voltage that passes through the point. Note that this abrupt change in voltage can be either positive or negative depending on whether the traveling voltage discontinuity is positive or negative. For $0 \leq t \leq T_1$, the voltage at $x = l_1$ is zero, being the initial voltage on the transmission line. At $t = T_1$, a voltage discontinuity V_{+1} passes through $x = l_1$ and hence the voltage at $x = l_1$ changes to V_{+1}; this voltage continues to remain the same until $t = T_2$, when another voltage discontinuity equal to $V_{+1}\rho_L$ passes through $x = l_1$. The voltage at $x = l_1$ is therefore changed at $t = T_2$ to $V_{+1}(1 + \rho_L)$ which remains unchanged until $t = T_3$ when, due to the passage of another voltage discontinuity equal to $V_{+1}\rho_L\rho_g$, the voltage at $x = l_1$ is changed to $V_{+1}(1 + \rho_L + \rho_L\rho_g)$. Thus, we can obtain the voltage at $x = l_1$ as a function of time with the help of the reflection diagram.

We can see from the reflection diagram and the foregoing discussion that the voltage at any point on the transmission line is given by the following series expression

$$V(x, t) = V_{+1}[1 + \rho_L + \rho_L\rho_g + \rho_L^2\rho_g + \rho_L^2\rho_g^2 + \rho_L^3\rho_g^2 + \cdots]. \tag{9.5}$$

Each term in the series inside the square brackets in (9.5) comes into existence after a lapse of suitable time; the higher-order terms are due to the higher-order reflections at the terminals occurring at later times. It is convenient to regroup the alternate terms in the series as follows:

$$V(x, t) = V_{+1}[1 + \rho_L\rho_g + \rho_L^2\rho_g^2 + \cdots + \rho_L(1 + \rho_L\rho_g + \rho_L^2\rho_g^2 + \cdots)]. \tag{9.6}$$

After the lapse of infinite time, the series in (9.6) becomes an infinite series; in fact, it is a geometric series with a common ratio equal to $\rho_L\rho_g$. Since ρ_L and ρ_g both have magnitudes less than unity, it follows that $|\rho_L\rho_g| < 1$, a condition which permits the summation of the infinite geometric series to yield a finite value. Since

$$\sum_{n=0}^{\infty} x^n = \frac{1}{1 - x} \qquad \text{(for } |x| < 1\text{)}, \tag{9.7}$$

we can sum up the infinite geometric series that occurs in (9.6) after the lapse of infinite time to yield the following result:

$$V_f = \operatorname*{Limit}_{t\to\infty} V(x, t) = V_{+1}\frac{(1 + \rho_L)}{(1 - \rho_L\rho_g)}. \tag{9.8}$$

The substitution of V_{+1}, ρ_L, and ρ_g respectively from (9.2), (9.3), and (9.4) in (9.8) yields

$$V_f = \frac{V_0R_0}{(R_g + R_0)}\frac{1 + \dfrac{R_L - R_0}{R_L + R_0}}{\left[1 - \dfrac{(R_L - R_0)(R_g - R_0)}{(R_L + R_0)(R_g + R_0)}\right]}$$

$$= \frac{V_0R_02R_L}{[(R_g + R_0)(R_L + R_0) - (R_L - R_0)(R_g - R_0)]} = \frac{V_0R_L}{(R_g + R_L)}. \tag{9.9}$$

Note that the final d.c. voltage on the transmission line as obtained in (9.9) is in agreement with that given by (9.1), which was found by the solution of the d.c. circuit equations obtained after the lapse of infinite time.

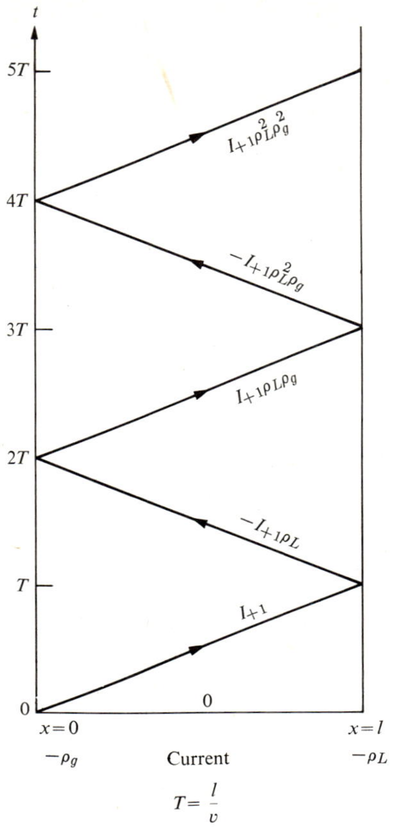

Fig. 1.9 Current reflection diagram.

We can also draw a reflection diagram for the current as shown in Fig. 1.9, the case to which the diagram pertains being identified at the bottom of the diagram. Since the reflection coefficient for the current is the negative of the reflection coefficient for the voltage, the reflection coefficients at $x = 0$ and $x = l$ on the current reflection diagram are noted as $-\rho_g$ and $-\rho_L$ respectively. The locus of the current discontinuity as a function of time is the same as that of the voltage discontinuity shown in Fig. 1.8. The initial current along the transmission line, that is, the current at $t = 0$, is marked at a point corresponding to the ordinate $t = 0$. The incremental currents produced at each of the terminations are marked appropriately on the locus straight lines. The voltage generator at $x = 0$ produces the current I_{+1} given in (9.2) and this is marked on the locus in the range $0 < t < T$. The additional reflected current wave produced at each termination is equal to the appropriate reflection coefficient times the current wave incident on it. Thus, the incremental current corresponding to the locus in the time range $T < t < 2T$ is $-I_{+1}\rho_L$. In a similar manner, we obtain the incremental currents corresponding to the successive time ranges of period T as shown in Fig. 1.9. As in the case of the voltage reflection diagram, the current reflection diagram is useful in obtaining the current distribution along the transmission line at a given time and the time dependence of the current at any point on the transmission line.

It is instructive to examine how the final current is reached on the transmission line. We can see from the current reflection diagram in Fig. 1.9 that the current at any point on the transmission line is given by

$$I(x, t) = I_{+1}[1 - \rho_L + \rho_L\rho_g - \rho_L^2\rho_g + \rho_L^2\rho_g^2 - \rho_L^3\rho_g^2, \ldots], \qquad (9.10)$$

where each additional higher-order term comes into existence after a lapse of suitable time. As before, we can regroup the alternate terms of the series in (9.10) as follows:

$$I(x, t) = I_{+1}[1 + \rho_L\rho_g + \rho_L^2\rho_g^2 + \cdots - \rho_L(1 + \rho_L\rho_g + \rho_L^2\rho_g^2 + \cdots)]. \quad (9.11)$$

After the lapse of infinite time, the geometric series in (9.11) becomes infinite and can be easily summed up using (9.7), since $|\rho_L\rho_g| < 1$, to yield the following result:

$$I_f = \operatorname*{Limit}_{t\to\infty} I(x, t) = I_{+1}\frac{(1 - \rho_L)}{(1 - \rho_L\rho_g)}. \qquad (9.12)$$

The substitution of I_{+1}, ρ_L, and ρ_g respectively from (9.2), (9.3), and (9.4) in (9.12) yields

$$I_f = \frac{V_0}{(R_g + R_0)}\frac{\left[1 - \dfrac{R_L - R_0}{R_L + R_0}\right]}{\left[1 - \dfrac{(R_L - R_0)(R_g - R_0)}{(R_L + R_0)(R_g + R_0)}\right]} = \frac{V_0}{R_L + R_g}, \qquad (9.13)$$

which is in agreement with the result given by (9.1). Note that (9.1) was obtained from the solution of the d.c. circuit equations that result after the lapse of infinite time.

1.10 NUMERICAL EXAMPLES OF TRANSIENTS ON LOSSLESS TRANSMISSION LINES

Consider a lossless transmission line extending from $x = 0$ to $x = 900$ m. Let the characteristic resistance R_0 be 50 ohms and the velocity v of the traveling waves be 3×10^8 m/sec. We connect a battery of 4 V at time $t = 0$ through a series resistor $R_g = 150$ ohms to the input terminals at $x = 0$ as shown in Fig. 1.10(a). The other terminals of the transmission line are open-circuited. We wish to find (i) the voltage distribution along the transmission line at time $t = 10 \times 10^{-6}$ sec and (ii) the time dependence of the voltage at $x = 600$ m up to time $t = 12 \times 10^{-6}$ sec.

The voltage reflection diagram pertaining to this problem is shown in Fig. 1.10(b). We find with the help of (7.3) and (7.6) that the voltage reflection coefficients ρ_g at the generator end and ρ_L at the load end are

$$\rho_g = \frac{150 - 50}{150 + 50} = \frac{1}{2}; \qquad \rho_L = 1.$$

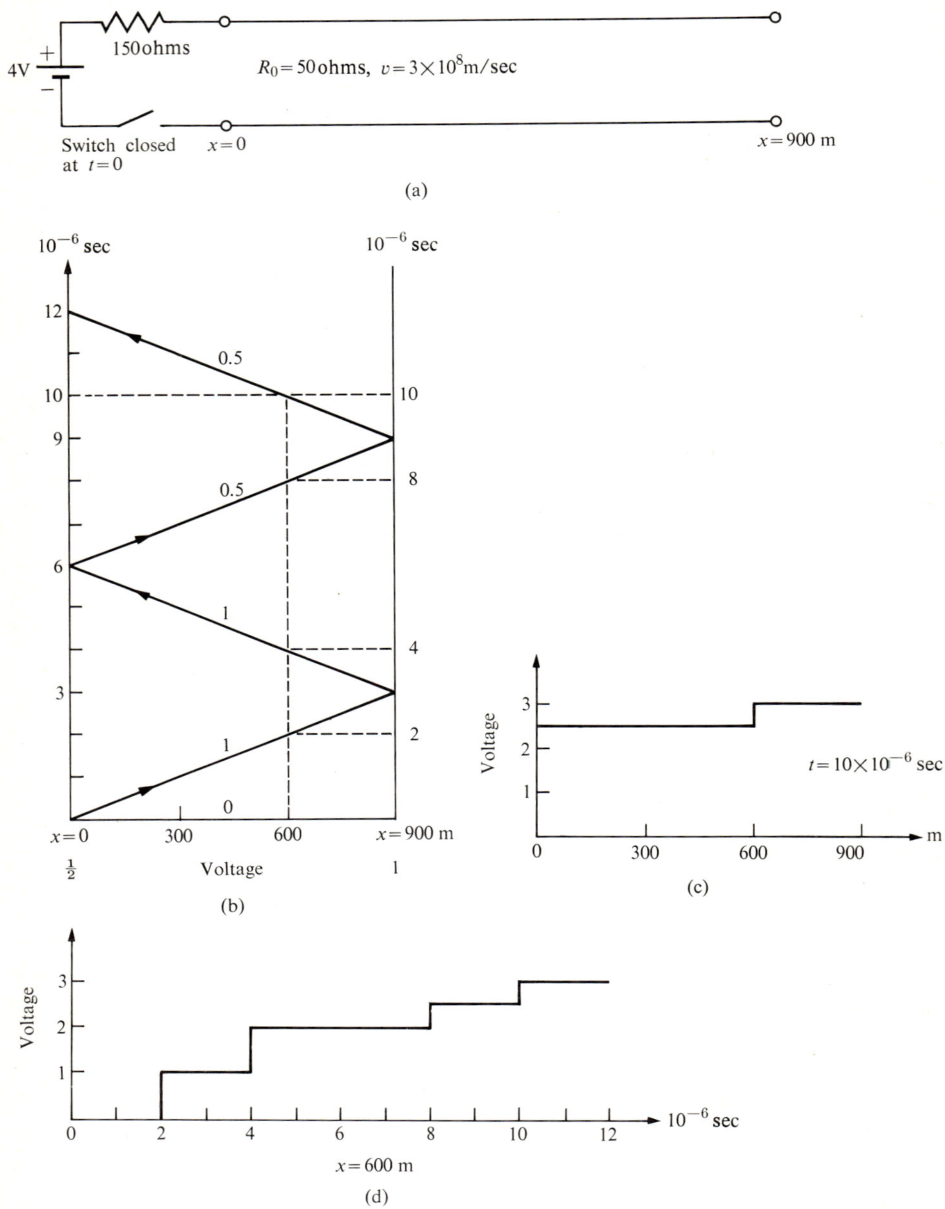

Fig. 1.10 D.C. transient on a finite length open-circuited transmission line.

The time T taken for the traveling wave to cover the entire distance of the transmission line once is

$$T = \frac{900}{3 \times 10^8} = 3 \times 10^{-6} \text{ sec.}$$

We draw the voltage reflection diagram up to a total time of 12×10^{-6} sec. We find from (9.2) that the voltage V_{+1} which starts at the generator end after the battery is connected and which travels in the positive x-direction is

$$V_{+1} = \frac{4 \times 50}{150 + 50} = 1 \text{ V.}$$

The incremental voltages associated with the various reflected traveling wave voltage discontinuities are indicated on the slanted straight lines which give the locus of the voltage discontinuity as a function of time. The initial voltage which is zero is marked at a point corresponding to $t = 0$. The voltage reflection coefficients are indicated at the appropriate points on the abscissa, which shows the distance along the transmission line.

To find the voltage distribution along the transmission line at $t = 10 \times 10^{-6}$ sec, we draw a dashed line corresponding to the ordinate $t = 10 \times 10^{-6}$ sec and this intersects the locus straight line at an abscissa point corresponding to $x = 600$ m. This indicates that the voltage discontinuity is at $x = 600$ m and is traveling toward the generator end. From the reflection diagram we find that the total voltage along the transmission line from $x = 0$ to $x = 600$ m is 2.5 V and that from $x = 600$ m to $x = 900$ m is 3 V. Thus, we determine the voltage distribution along the transmission line and it is as shown in Fig. 1.10(c).

To find the time dependence of the voltage at $x = 600$ m up to time $t = 12 \times 10^{-6}$ sec, we draw a dashed line corresponding to the abscissa point $x = 600$ m. This intersects the locus straight lines at 2×10^{-6}, 4×10^{-6}, 8×10^{-6}, 10×10^{-6}, ... sec. The voltage at $x = 600$ m is zero from $t = 0$ to $t = 2 \times 10^{-6}$ sec, when the voltage changes to 1 V. At each of the subsequent times indicated in the foregoing, the voltage at $x = 600$ m changes by an amount indicated on the appropriate locus straight line. In this manner, we obtain the voltage at $x = 600$ m as a function of time up to $t = 12 \times 10^{-6}$ sec. This is shown in Fig. 1.10(d).

Although we have considered only the propagation of d.c. transients so far, the same techniques can be utilized in the investigation of propagation of pulses on a finite transmission line. For example, consider the transmission line with the same parameters as before except that the voltage source, instead of being a battery connected at time $t = 0$, is a rectangular voltage pulse of peak voltage 4 V and time duration $T_0 = 4 \times 10^{-6}$ sec. Let the onset of the leading edge of the pulse be taken as the origin of time $t = 0$. As before, we wish to find the voltage distribution along the transmission line at time $t = 10 \times 10^{-6}$ sec.

First it is important to realize that a rectangular voltage pulse of amplitude V_0 and time duration T_0 is a suitable superposition of two step-voltages. The first step-voltage is of voltage V_0, switched on at time $t = 0$, and the second is of voltage $-V_0$,

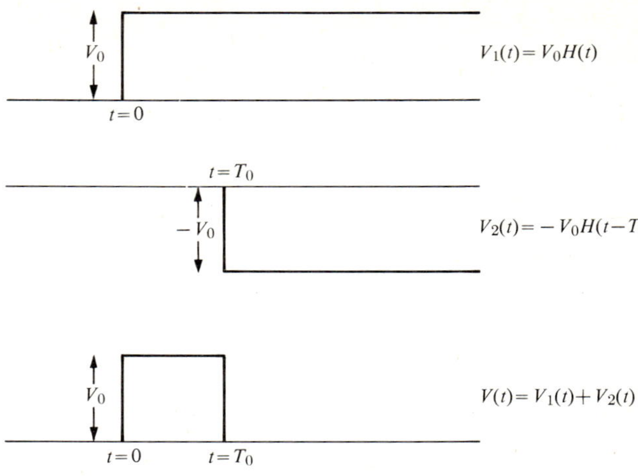

Fig. 1.11 Decomposition of a rectangular pulse into two step-voltages.

switched on at a later time $t = T_0$. From Fig. 1.11, we see that a superposition of these two step-voltages yields the rectangular pulse of amplitude V_0 and time duration T_0. Using the superposition principle we can obtain the response of the transmission line to pulse excitation by first successively determining the responses due to the two individual step-voltages and then superposing the results. The voltage reflection diagram corresponding to the step-voltage of 4 V switched on at $t = 0$ is the same as the one used previously (Fig. 1.10(b)) and we can obtain the voltage reflection diagram corresponding to the step-voltage of -4 V switched on at $t = 4 \times 10^{-6}$ sec in a similar manner. This is depicted in Fig. 1.12(a), in which the second reflection diagram is placed alongside the reflection diagram corresponding to the first case. The only differences between the second reflection diagram and the first are that the traveling wave voltage discontinuity corresponding to the trailing edge of the pulse occurs $T_0 = 4 \times 10^{-6}$ sec later than that corresponding to the leading edge of the pulse. Also, the voltage discontinuities for the various reflected waves corresponding to the trailing edge of the pulse are the negatives of the corresponding quantities pertaining to the leading edge of the pulse.

To find the voltage distribution along the transmission line at $t = 10 \times 10^{-6}$ sec, we draw in both the reflection diagrams a dotted line corresponding to the ordinate $t = 10 \times 10^{-6}$ sec. In the first reflection diagram, the dotted line intersects the locus straight line at an abscissa point corresponding to $x = 600$ m. This indicates that the voltage discontinuity is at $x = 600$ m and is traveling toward the generator. From the first reflection diagram we find that for $0 < x \leq 600$ m, the voltage is 2.5 V and for 600 m $\leq x \leq 900$ m the voltage is 3 V. The dotted line in the second reflection diagram intersects the locus straight line at an abscissa point corresponding to $x = 0$. Therefore, the voltage discontinuity due to the trailing edge of the pulse is at the generator end. From the second reflection diagram, we find that the voltage along the entire length of the transmission line is -2 V and this is due to the step-voltage of -4 V starting at $t = 4 \times 10^{-6}$ sec. To obtain the resultant voltage along the transmission line at $t = 10 \times 10^{-6}$ sec due to the rectangular pulse, we superpose the results obtained in the two cases to yield 0.5 V for $0 < x < 600$ m and 1.0 V for 600 m $< x <$ 900 m.

Since the voltages pertaining to the various reflected waves associated with the second step-voltage, which occurs $t = 4 \times 10^{-6}$ sec later than the first, are negative of the corresponding voltages associated with the first step-voltage, it is not even necessary to draw a second reflection diagram. The voltage discontinuity due to the second step-voltage always trails that due to the first by 4×10^{-6} sec and therefore

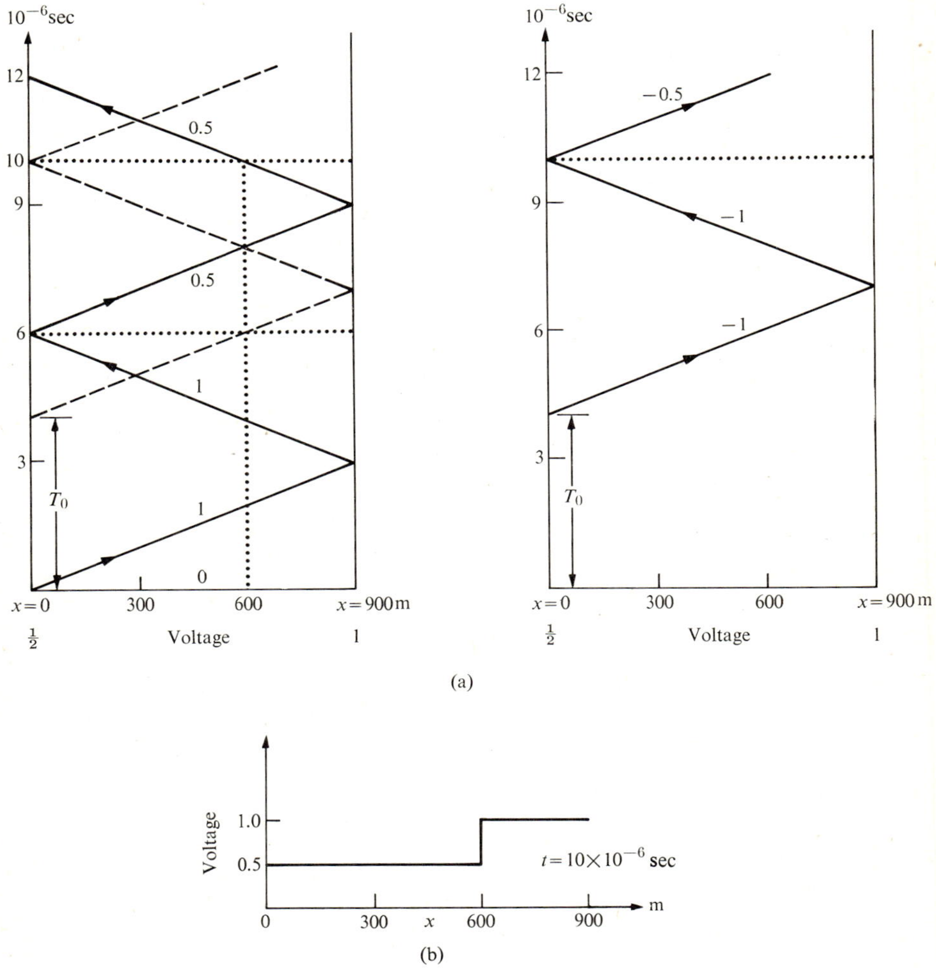

Fig. 1.12 Pulse excitation of a finite length open-circuited transmission line.

the position of the voltage discontinuity due to the second step-voltage can be obtained by drawing another dotted line on the first reflection diagram itself corresponding to an ordinate which is 4×10^{-6} sec less than that time for which we want the voltage distribution along the transmission line. In the present case, the second dotted line corresponds to an ordinate $t = 6 \times 10^{-6}$ sec. For $t < 6 \times 10^{-6}$ sec the various reflected voltages due to the second step-voltage exactly annul those caused by the first step-voltage, with the result that only the reflected voltages lying in time range 6×10^{-6} sec $< t < 10 \times 10^{-6}$ sec account for the resulting voltage distribution along the transmission line. From the reflection diagram we can verify that for $0 < x < 600$ m the voltage is 0.5 V, and for 600 m $< x <$ 900 m the voltage is 1.0 V. This result is the same as that obtained with the help of the two reflection diagrams and is depicted in Fig. 1.12(b).

1.11 REFLECTION DIAGRAM FOR A TRANSMISSION LINE WITH ADDITIONAL DISCONTINUITIES

The voltage and the current waves are reflected at the terminals as well as at any discontinuities that may be present along the transmission line. Near the discontinuities, the transmission line is not uniform and as mentioned earlier there are certain "end effects" which, in the present treatment, are assumed to be negligible. In the preceding two sections, we considered the reflection of waves occurring at the terminals with the help of the reflection diagrams. In this section, we shall show the usefulness of the reflection diagram in the investigation of reflection of waves at the discontinuities along a transmission line.

Consider a transmission line of characteristic resistance R_0 with a resistor R_p connected across the two wires at some point along the length of the transmission line. Let V_+ and I_+ be the voltage and the current incident on the resistor R_p from the left and let V_- and I_- be the corresponding reflected quantities. Let V_t and I_t be the transmitted voltage and current respectively. Also, as shown in Fig. 1.13(a), let I_p be the current in the resistor R_p. With the sign conventions for the voltages and

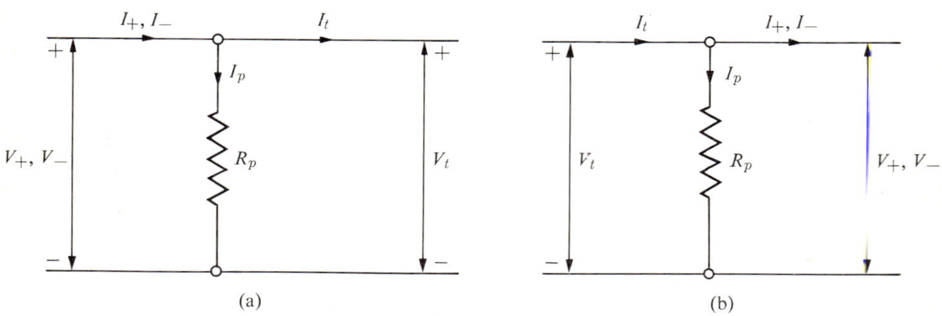

Fig. 1.13 Reflection of waves from a discontinuity along a transmission line.

the currents as used previously, Kirchhoff's circuit equations for the discontinuity are

$$V_+ + V_- = V_t, \tag{11.1}$$

$$I_+ + I_- = I_p + I_t, \tag{11.2}$$

$$V_+ = I_+R_0; \qquad V_- = -I_-R_0; \qquad V_t = I_tR_0 = I_pR_p. \tag{11.3}$$

With the help of (11.3), we first rewrite (11.2) in terms of V_+, V_-, and V_t. From the resulting equation and (11.1), we eliminate V_t and then V_-, to obtain the following results:

$$\rho_{12} = \frac{V_-}{V_+} = -\frac{I_-}{I_+} = \frac{-R_0}{R_0 + 2R_p}, \tag{11.4}$$

$$t_{12} = \frac{V_t}{V_+} = \frac{I_t}{I_+} = \frac{2R_p}{R_0 + 2R_p}. \tag{11.5}$$

The voltage reflection coefficient ρ_{12} and the voltage transmission coefficient t_{12}, as given in (11.4) and (11.5) respectively, correspond to a wave incident on the resistor from the left. We can evaluate the voltage reflection coefficient ρ_{21} and the voltage transmission coefficient t_{21}, corresponding to a wave incident on the resistor from the right, similarly with the help of Fig. 1.13(b). Thus,

$$\rho_{21} = \frac{V_+}{V_-} = -\frac{I_+}{I_-} = \frac{-R_0}{R_0 + 2R_p}, \tag{11.6}$$

$$t_{21} = \frac{V_t}{V_-} = \frac{I_t}{I_-} = \frac{2R_p}{R_0 + 2R_p}. \tag{11.7}$$

Note that for a wave incident on the resistor from the right, those with the subscripts $-$ and $+$ are respectively the incident and the reflected quantities. We notice another important property on comparing (11.4) and (11.5) with (11.6) and (11.7) respectively. This is the reciprocal property; the reflection and the transmission properties of a discontinuity along a transmission line are independent of the direction of the incident wave. However, if the transmission line had different characteristic resistances on either side of the resistor, then $\rho_{12} \neq \rho_{21}$ and $t_{12} \neq t_{21}$. In this case, there is reflection even without the parallel resistor R_p. An abrupt change in the characteristics of the transmission line is equally a discontinuity that is bound to cause reflection.

In order to understand the usefulness of the reflection diagram for systematically taking into account the additional reflected waves produced by the discontinuity caused by a resistor being placed across the two wires at some point along the transmission line, let us consider the modification of the transmission-line problem, by the insertion of a parallel resistor R_p at $x = x_p$ (where $0 < x_p < l$), that was discussed in Section 1.9. The voltage reflection diagram for this case of a transmission line with an additional discontinuity at $x = x_p$ is shown in Fig. 1.14, in which an

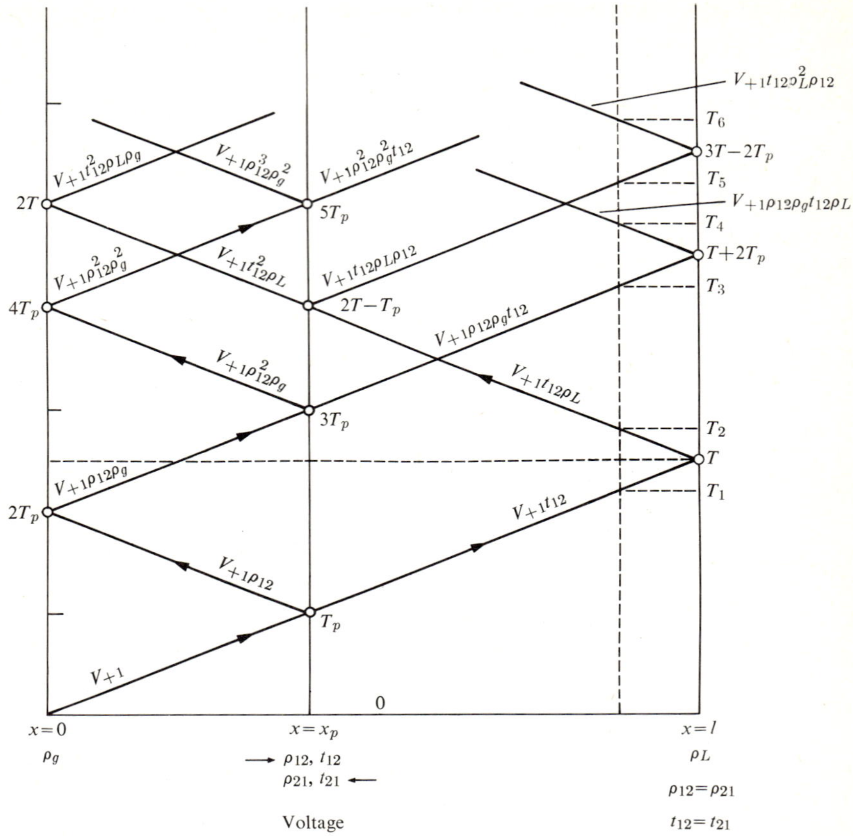

Fig. 1.14 Voltage reflection diagram for a transmission line with an additional discontinuity.

additional line is drawn corresponding to the abscissa $x = x_p$, from which additional reflected and transmitted waves can emanate. The reflection and the transmission coefficients for the voltage corresponding to the discontinuity, as well as the directions of incidence to which these coefficients pertain, are indicated below the abscissa point $x = x_p$ for both the directions of incidence. The locus of the position of the voltage discontinuities is a series of sloping straight lines with a slope equal to $1/v$ or $-1/v$, depending on whether the discontinuity is traveling in the positive or negative x-direction. Let $T_p = x_p/v$ where T_p is the time taken for the voltage discontinuity to travel from the generator end to the parallel resistor or vice versa. Similarly, $(T - T_p)$ is the time taken for the voltage discontinuity to travel from the parallel resistor to the load resistor or vice versa. As before, the initial voltage along the transmission line is zero and on the reflection diagram corresponds to the ordinate $t = 0$. We obtain the incremental voltage produced at each of the terminations or the discontinuity by multiplying the incident voltage by the appropriate reflection coefficient in the case of reflection or by the appropriate transmission coefficient in the case of transmission. These are marked on the locus straight lines. As in the problem discussed in Section 1.9 a voltage discontinuity of amplitude V_{+1} starts at the generator end as soon as the battery is switched on and travels in the positive x-direction with a velocity v. At time $t = T_p$, the voltage V_{+1} reaches the discontinuity, where it produces a reflected wave of voltage $V_{+1}\rho_{12}$ and a transmitted wave of voltage $V_{+1}t_{12}$. The reflected wave travels toward the generator and reaches it at time $t = 2T_p$, when it produces an additional voltage $V_{+1}\rho_{12}\rho_g$ which travels in the positive x-direction. This voltage discontinuity reaches the parallel resistor at $t = 3T_p$. Meanwhile, the transmitted wave of voltage $V_{+1}t_{12}$ travels in the positive x-direction and reaches the resistor R_L at time $t = T$, when a reflected wave of voltage $V_{+1}t_{12}\rho_L$ is produced. This reflected voltage discontinuity travels in the negative x-direction and reaches the parallel resistor at time $t = 2T - T_p$. The voltage discontinuities traveling from the generator and the load ends arrive at the parallel resistor at different times except in the special case when $x_p = l/2$, that is, when the parallel resistor is situated at the center of the transmission line. Each of the two waves arriving at the resistor produces a reflected wave and a transmitted wave. Thus, it is clear from the reflection diagram in Fig. 1.14 that two voltage discontinuities, one trailing the other in time by $2(T - 2T_p)$, travel toward the generator and two other voltage discontinuities, again one trailing the other by $2(T - 2T_p)$, travel toward the load resistor. These waves, in turn, produce additional reflected waves when they reach the generator and the load resistor R_L respectively. Except in the special case of $x_p = l/2$, the reflections at the generator end and the load resistor take place at different times. As mentioned earlier, we can obtain the voltages associated with the reflected and the transmitted waves with the knowledge of the incident voltage and the appropriate reflection and transmission coefficients. We can visualize the further evolution of the multiply reflected and transmitted waves with the help of the reflection diagram. This not only facilitates the systematic determination of the various reflected and transmitted waves produced with the passage of time, but also makes it possible to ascertain clearly the positions and directions of travel of the various voltage discontinuities at any time.

In order to determine the voltage distribution along the transmission line, we draw a dashed line corresponding to the ordinate of time for which we require the distribution. At every point on the transmission line, we can sum up the incremental voltages caused by the various reflected and transmitted waves lying below this dashed line to obtain the resultant voltage. Thus we find the voltage distribution along the transmission line for any given time. In order to determine the time dependence of the voltage at any point along the transmission line, we draw a dashed line corresponding to the given abscissa point and obtain the times corresponding to the intersections of this line with the various locus straight lines. The voltage at the given point on the transmission line changes at each of these times by an amount indicated on the corresponding locus straight line. Thus, the time dependence of the voltage at any point along the transmission line can be obtained even with an additional discontinuity present along the transmission line.

1.12 WAVES REFLECTED FROM A REACTIVE LOAD

At any point on a transmission line, such as for example near the load, the wave traveling in the positive x-direction or the incident wave is a function only of time. Similarly, the wave produced by the incident wave on reflection at the load and traveling in the negative x-direction or the reflected wave near the load is also a function only of time. We showed in Section 1.7 that, when the load is a resistor, the reflected voltage is the *same* function of time as the incident voltage or, in other words, the reflected voltage is proportional to the incident voltage for all times. We explained that the proportionality constant is designated the voltage reflection coefficient. In general, however, the reflected and the incident voltages have different functional dependences on time and the concept of voltage reflection coefficient no longer applies. This general state of affairs occurs when the voltage across the load at any instant depends not only on the current in it at that instant but also on the current at previous times as well, such as is the case when the load has a reactive component, that is, an inductance or a capacitance. To illustrate this phenomenon, we derive the reflected voltage produced by a constant voltage incident at time $t = 0$ on a load which consists of an inductance L_L in series with a resistor R_L.

Let V_+ and $V_-(t)$ be the values of the incident and the reflected voltages near the load. Similarly, let I_+ and $I_-(t)$ be the incident and the reflected currents. The voltage $V_L(t)$ across the load and the current $I_L(t)$ in the load are as shown in Fig. 1.15(a). Then we see that Kirchhoff's circuit equations for the load are given by

$$V_L(t) = V_+ + V_-(t), \tag{12.1}$$

$$I_L(t) = I_+ + I_-(t), \tag{12.2}$$

$$V_+ = R_0 I_+, \qquad V_-(t) = -R_0 I_-(t), \tag{12.3}$$

$$V_L(t) = L_L \frac{d}{dt} I_L(t) + R_L I_L(t). \tag{12.4}$$

If we substitute (12.1) to (12.3) in (12.4) and recognize that $(d/dt)V_+ = 0$ for $t > 0$, by rearranging the terms we obtain the following differential equation for determining $V_-(t)$:

$$L_L \frac{dV_-(t)}{dt} + (R_L + R_0)V_-(t) = V_+(R_L - R_0). \tag{12.5}$$

On arranging (12.5) as

$$\frac{L_L\, dV_-(t)}{-V_-(t)(R_L + R_0) + V_+(R_L - R_0)} = dt \tag{12.6}$$

and integrating both sides of (12.6) with respect to time, we obtain

$$-V_-(t)(R_L + R_0) + V_+(R_L - R_0) = A \exp\left[-\frac{(R_L + R_0)t}{L_L}\right], \tag{12.7}$$

where A is an as-yet-undetermined constant. For time slightly before $t = 0, I_L(t) = 0,$

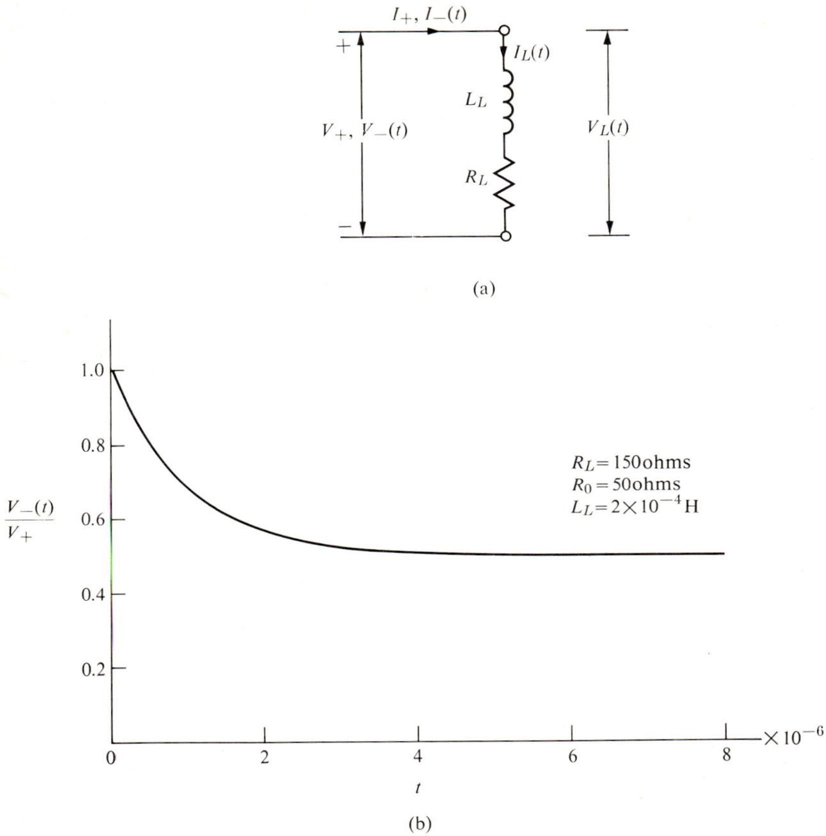

Fig. 1.15 Reflection from a reactive load.

and for time slightly after $t = 0$, $I_L(t)$ has a finite value. Thus, at $t = 0$, $(d/dt)I_L(t)$ is infinite and this implies that at $t = 0$ the load behaves like an open-circuit. Hence, $I_L(t) = 0$ for $t = 0$. From (12.2) and (12.3), it follows that

$$V_-(t) = V_+ \qquad \text{at } t = 0. \tag{12.8}$$

If we substitute the initial condition (12.8) in (12.7), we obtain

$$A = -2R_0V_+. \tag{12.9}$$

If we use (12.9) in (12.7), we find after some rearrangement that

$$V_-(t) = V_+ \frac{\left\{(R_L - R_0) + 2R_0 \exp\left[-\dfrac{(R_L + R_0)t}{L_L}\right]\right\}}{(R_L + R_0)} \qquad (\text{for } t > 0). \tag{12.10}$$

We can confirm the correctness of (12.10) by obtaining the values of $V_-(t)$ for certain special conditions for which simple physical arguments can be used to obtain the results. For $t = 0$, (12.10) yields $V_-(t) = V_+$, which is in accordance with the initial condition (12.8). For $L_L = 0$, (12.10) yields $V_-(t) = V_+(R_L - R_0)/(R_L + R_0)$, which reduces correctly to the result for a resistive load. After a considerable lapse of time, since the incident voltage is a constant, the reflection becomes equivalent to a d.c. voltage being reflected from the load. Therefore, the inductance has no effect and we obtain

$$V_-(t) = V_+(R_L - R_0)/(R_L + R_0).$$

This result is reproduced on setting $t = \infty$ in (12.10).

The important result to be noticed in (12.10) is that, whereas the incident voltage V_+ is constant as a function of time, the reflected voltage $V_-(t)$ has a certain functional dependence on time and therefore a reflection coefficient which is independent of time cannot be defined. In the more general case in which the incident voltage is a function of time, the dependence of $V_-(t)$ on $V_+(t)$ is more involved. Thus, in the case of voltages and currents which are arbitrary functions of time, the concept of a constant reflection coefficient is applicable only to the case of a resistive load.

The ratio of $V_-(t)/V_+$ is plotted in Fig. 1.15(b) as a function of time for certain values of the parameters R_0, R_L, and L_L. We see that the voltage $V_-(t)$ reduces monotonically from the value V_+ for $t = 0$ to the value $V_+\rho_L$ for $t = \infty$. More importantly, note that $V_-(t)$ has a discontinuity at $t = 0$ when $V_-(t)$ changes from zero to V_+. Hence, (12.5) and (12.10) are valid only for $t > 0$. In particular, (12.10) cannot be used to obtain $(d/dt)I_L(t)$ at $t = 0$. From (12.2), (12.3), and (12.10), it is easily verified that

$$\frac{d}{dt} I_L(t) = -\frac{1}{R_0}\frac{d}{dt} V_-(t) = \frac{2V_+}{L_L}$$

for t tending to zero from above. But the initial condition (12.8) is based on the fact that $(d/dt)I_L(t) = \infty$ at $t = 0$. Consequently, the validity of (12.10) is limited to the values of $t > 0$.

PROBLEMS

1.1 Consider the following voltage and current distributions:

$$V(x, t) = V_0 \cos \beta(x - vt); \qquad I(x, t) = \frac{V_0}{R_0} \cos \beta(x - vt),$$

where β is a constant, $v = 1/\sqrt{LC}$, and $R_0 = \sqrt{L/C}$. By actual substitution, verify that $V(x, t)$ and $I(x, t)$ satisfy the following transmission-line equations:

$$-\frac{\partial}{\partial x} V(x, t) = L\frac{\partial}{\partial t} I(x, t),$$

$$-\frac{\partial}{\partial x} I(x, t) = C\frac{\partial}{\partial t} V(x, t),$$

$$\frac{\partial^2}{\partial x^2} V(x, t) = \frac{1}{v^2} \frac{\partial^2}{\partial t^2} V(x, t),$$

$$\frac{\partial^2}{\partial x^2} I(x, t) = \frac{1}{v^2} \frac{\partial^2}{\partial t^2} I(x, t).$$

1.2 Show that $V_-(x, t) = f_2(x + vt)$ represents a voltage waveform that travels as a whole in the negative x-direction with uniform velocity v and without any distortion or attenuation.

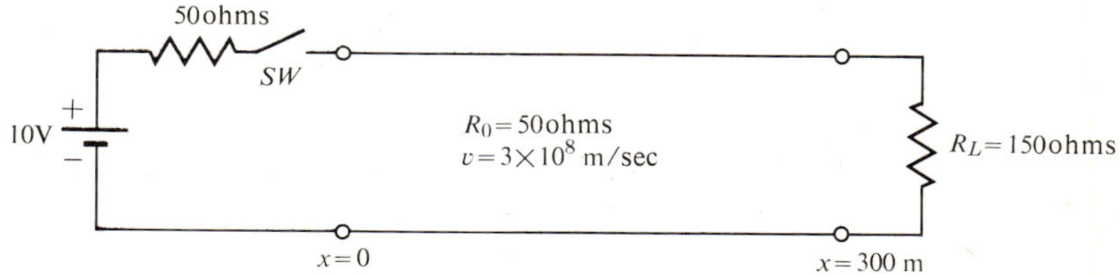

Fig. P1.1 Schematic diagram for the charging of the transmission line.

1.3 Consider the d.c. transient problem illustrated in Fig. P1.1. The switch SW is closed at time $t = 0$. Plot (a) the voltage distribution along the transmission line at time $t = 1.5 \times 10^{-6}$ sec and (b) the voltage across the load resistor R_L as a function of time.

1.4 In the d.c. transient problem illustrated in Fig. P1.1, after final conditions are reached, the battery is short-circuited at the new origin of time $t = 0$. Plot (a) the voltage distribution along the transmission line at time $t = 1.5 \times 10^{-6}$ sec and (b) the voltage across the load resistor R_L as a function of time.

1.5 A transmission line is initially charged to 10 V. The length l of the transmission line is 60 m and the velocity of propagation v is 3×10^8 m/sec. Verify that a rectangular pulse is formed on a load resistor suddenly connected to the transmission line provided that the load resistor is equal to the characteristic resistance of the transmission line. Find the amplitude and the time-duration of the pulse. Indicate a method for increasing the time-duration of the pulse.

1.6 There is a discontinuity along the length of the transmission line. Indicate a method for obtaining the location of the discontinuity.

1.7 Consider a transmission line charged initially by an arrangement as shown in Fig. P1.2. Find the voltage and current distributions along the transmission line. Suppose that the battery is short-circuited by closing the switch SW at time $t = 0$. Draw the voltage and current reflection diagrams that will facilitate the determination of the voltage and the current at any point along the transmission line and at any time after the switch is closed. What are the final voltage and current distributions along the transmission line? Determine how the final voltage and current distributions are reached with the help of the reflection diagrams.

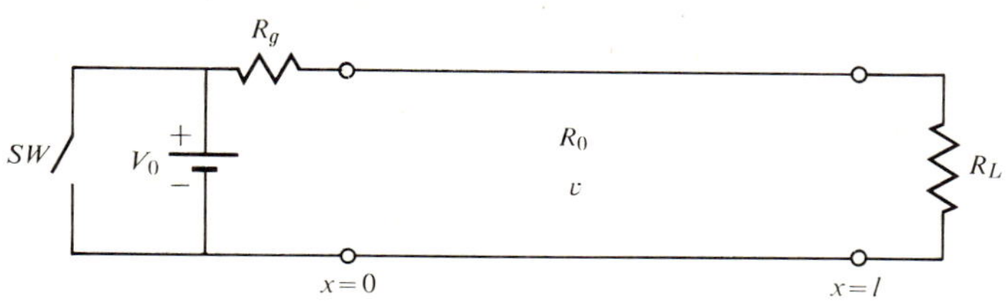

Fig. P1.2 Schematic diagram for the discharging of the transmission line.

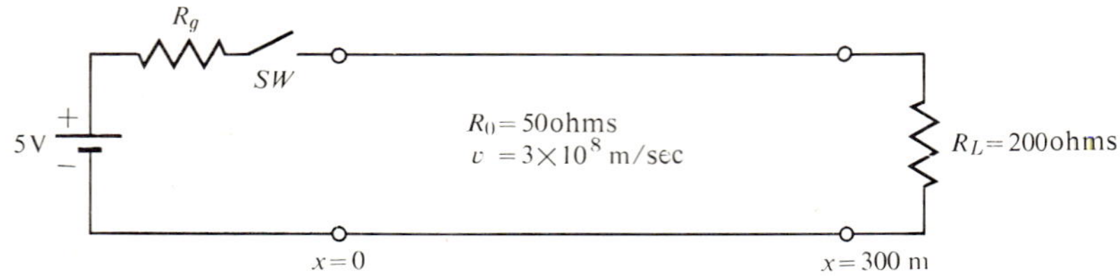

Fig. P1.3 Schematic diagram for the charging of the transmission line.

1.8 Consider the d.c. transient problem illustrated in Fig. P1.3. The switch SW is closed at time $t = 0$. For the following two cases, (i) $R_g = 25$ ohms and (ii) $R_g = 100$ ohms, plot (a) the voltage and current distributions along the transmission line at time $t = 2.5 \times 10^{-6}$ sec and (b) the voltage across and the current in the load resistor R_L as a function of time from $t = 0$ to $t = 2.5 \times 10^{-6}$ sec.

1.9 In the d.c. transient problem illustrated in Fig. P1.3, after the final conditions are reached suppose that the battery is short-circuited at the new origin of time $t = 0$. For the two cases, namely (i) $R_g = 25$ ohms and (ii) $R_g = 100$ ohms, plot (a) the voltage and current distributions along the transmission line at time $t = 2.5 \times 10^{-6}$ sec and (b) the voltage across and the current in the load resistor R_L as a function of time from $t = 0$ to $t = 2.5 \times 10^{-6}$ sec.

1.10 A transmission line shown in Fig. P1.4 is excited by a rectangular pulse of amplitude 10 V and time duration $T_0 = 1.5 \times 10^{-6}$ sec. Let the onset time of the leading edge of the pulse at the generator end be taken as the origin of time. Sketch the voltage and current distributions along the transmission line at time $t = 2 \times 10^{-6}$ sec. Also, find the voltage across and the current in the load resistor R_L as a function of time.

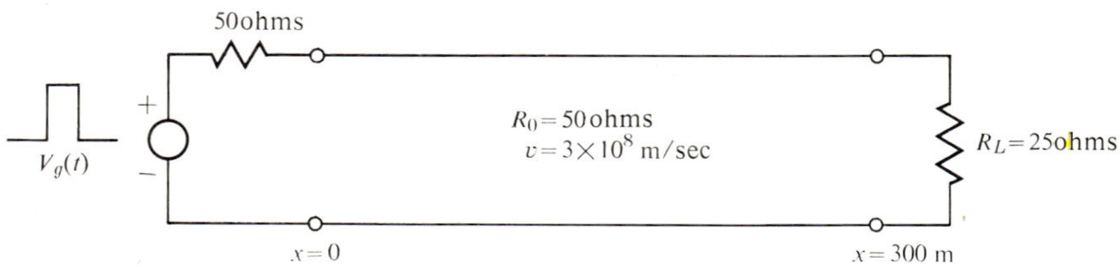

Fig. P1.4 A transmission line excited by a rectangular pulse of voltage.

1.11 Suppose that the transmission line shown in Fig. P1.4 is excited by a series of two rectangular pulses as illustrated in Fig. P1.5. Let the onset time of the leading edge of the pulse at the generator end be taken as the origin of time. Sketch the voltage and current distributions along the transmission line when time t is (i) 1.5×10^{-6} sec, (ii) 3.5×10^{-6} sec, and (iii) 5.5×10^{-6} sec. Also, find the voltage across and the current in the load resistor R_L as a function of time from $t = 0$ to $t = 6 \times 10^{-6}$ sec.

1.12 Consider the d.c. transient problem illustrated in Fig. P1.6. At the end $x = 0$, a step-voltage of amplitude 1 V is applied at $t = 0$ sec and, at the end $x = 600$ m, a step-voltage of amplitude

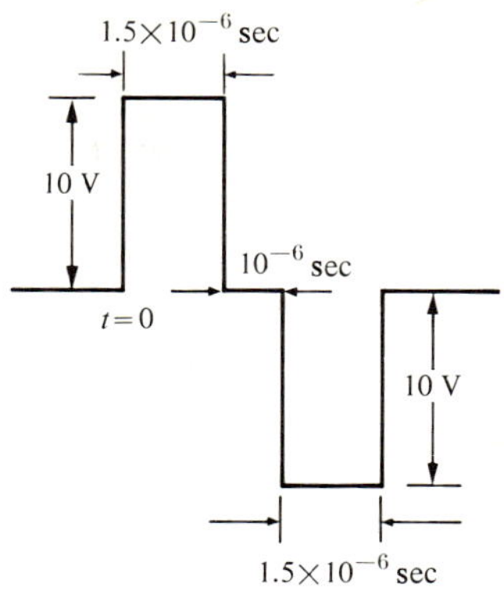

Fig. P1.5 A series of two rectangular pulses of voltage.

0.25 V is applied at $t = 2 \times 10^{-6}$ sec. Sketch the voltage and current distributions along the transmission line when time t is (i) 10^{-6} sec, (ii) 3×10^{-6} sec, and (iii) 5×10^{-6} sec. Also, find the voltage across and the current in the (i) 100 ohm and (ii) 50 ohm resistors as a function of time from $t = 0$ to $t = 5 \times 10^{-6}$ sec.

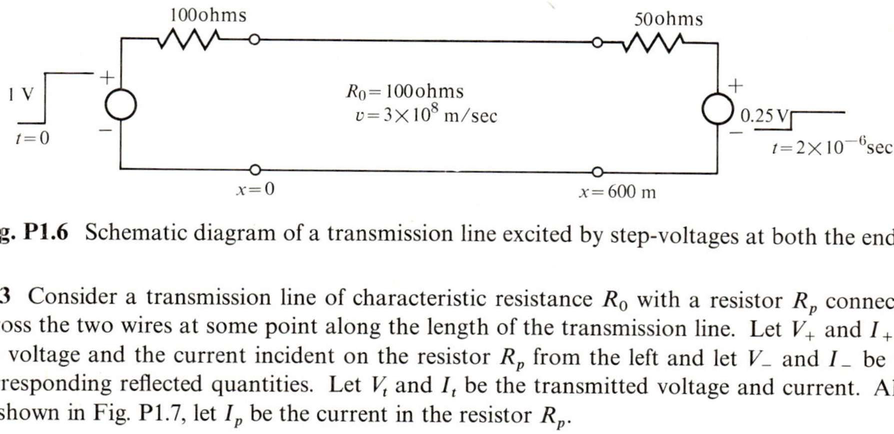

Fig. P1.6 Schematic diagram of a transmission line excited by step-voltages at both the ends.

1.13 Consider a transmission line of characteristic resistance R_0 with a resistor R_p connected across the two wires at some point along the length of the transmission line. Let V_+ and I_+ be the voltage and the current incident on the resistor R_p from the left and let V_- and I_- be the corresponding reflected quantities. Let V_t and I_t be the transmitted voltage and current. Also, as shown in Fig. P1.7, let I_p be the current in the resistor R_p.

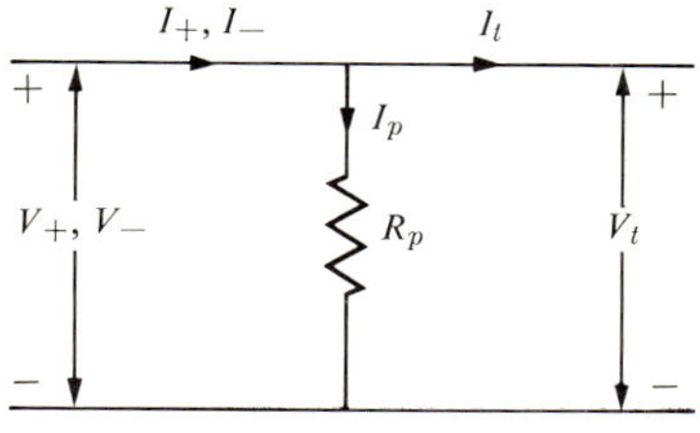

Fig. P1.7 Reflection and transmission of waves from a resistor connected across the two wires of a transmission line.

Calculate the powers associated with the incident (P_+), the reflected (P_-), and the transmitted (P_t) waves. Show that $P_+ - P_- - P_t$ is equal to the power dissipated in the parallel resistor R_p.

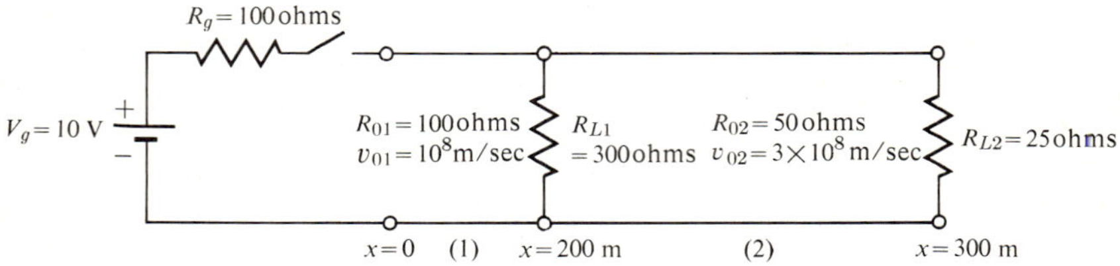

Fig. P1.8 Schematic diagram of two different transmission lines connected in tandem.

1.14 Consider a lossless transmission line 1 with characteristic resistance $R_{01} = 100$ ohms and associated propagation velocity $v_{01} = 10^8$ m/sec connected in tandem as shown in Fig. P1.8 to another lossless transmission line 2 with characteristic resistance $R_{02} = 50$ ohms and associated propagation velocity $v_{02} = 3 \times 10^8$ m/sec. Transmission lines 1 and 2 are terminated by resistances $R_{L1} = 300$ ohms and $R_{L2} = 25$ ohms respectively. Also, transmission lines 1 and 2 are of lengths $l_1 = 200$ m and $l_2 = 100$ m respectively. A battery of $V_g = 10$ V is connected at time $t = 0$ through a series resistance $R_g = 100$ ohms to the input end ($x = 0$) of the transmission line 1.

Find the final currents in the three resistances and determine how these final currents are reached with the help of the reflection diagram.

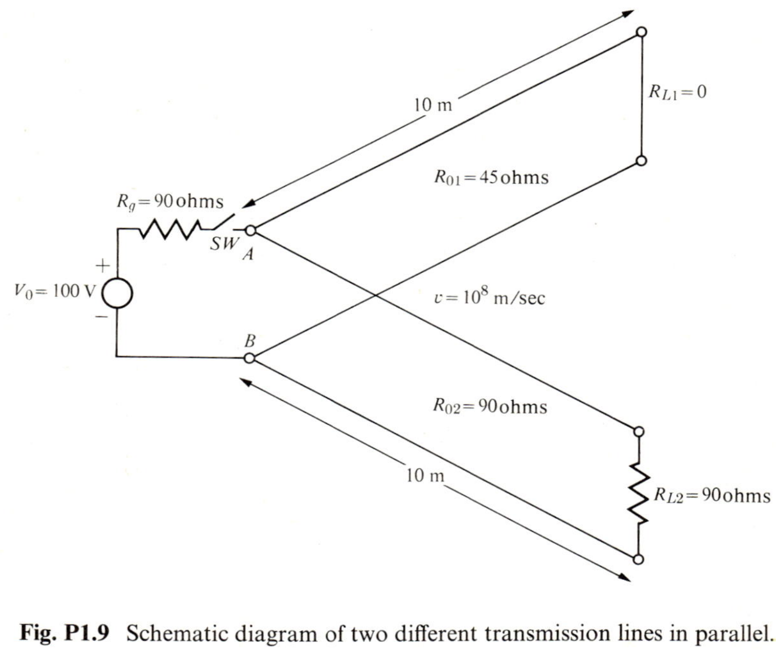

Fig. P1.9 Schematic diagram of two different transmission lines in parallel.

1.15 Consider a d.c. generator with an e.m.f. $V_0 = 100$ V connected at time $t = 0$ through a series resistance of $R_g = 90$ ohms to the terminals AB where two transmission lines join as shown in Fig. P1.9. Both the transmission lines are 10 m long and have the associated propagation velocity of 10^8 m/sec. The first transmission line has a characteristic resistance of $R_{01} = 45$ ohms and its terminals are short-circuited. The second transmission line has a characteristic resistance of $R_{02} = 90$ ohms and has a resistive load of 90 ohms.

Determine and plot the voltage $V_{R_{L2}}$ across the load resistor R_{L2} as a function of time.

Find the final currents in R_{L1} and R_{L2} and determine how these final currents are attained.

After the final state is reached, the switch SW is disconnected at the new origin of time $t = 0$. Find the voltage across the load resistor R_{L2} as a function of time. Also, determine how the final voltage across the load resistor R_{L2} is attained.

1.16 A transmission line with a characteristic resistance of R_0 ohms is terminated in a capacitance C_L farads in parallel with a resistance R_L ohms. A voltage wave of constant magnitude V_+ is incident on the termination at time $t = 0$.

Find the equation satisfied by the reflected voltage V_- across the terminals for $t > 0$.

Solve the equation and find the reflected voltage. Use the fact that the capacitance behaves like a short-circuit at $t = 0$.

Sketch V_- as a function of time using suitable values for V_+, R_0, C_L, and R_L.

How are all the previous results changed if the resistance R_L, instead of being parallel, is in series with the capacitor C_L?

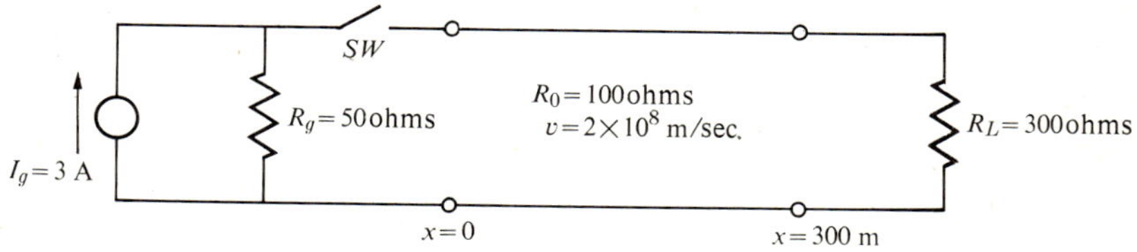

Fig. P1.10 Schematic diagram for the charging of the transmission line with a constant current generator.

1.17 Consider the d.c. transient problem illustrated in Fig. P1.10. A d.c. generator of current $I_g = 3$ A and an internal resistance of $R_g = 50$ ohms is connected to the transmission line by closing the switch SW at time $t = 0$. Plot (a) the current distribution along the transmission line at time $t = 7 \times 10^{-6}$ sec and (b) the input voltage as a function of time from $t = 0$ to $t = 7 \times 10^{-6}$ sec.

CHAPTER 2

DISTORTIONLESS TRANSMISSION, DIFFUSION, AND STEADY-STATE PROPAGATION ON A LOSSY TRANSMISSION LINE

2.1 INTRODUCTION

The imperfect conductivity of the wire gives rise to R, the resistance per unit length of the wire. In a similar manner, the imperfect insulation of the dielectric region between the wires gives rise to G, the conductance per unit length of the wire. In the case of a lossy transmission line, with all the parameters L, C, R, and G nonvanishing, the partial differential equations for the voltage or the current along the transmission line are difficult to solve when the time dependence is arbitrary. However, the solution can be obtained by the method of Fourier or Laplace transforms. If the voltage waveform impressed at the input terminals of the transmission line is broken down into its frequency components, we find that the attenuation coefficient and the phase-shift coefficient, or equivalently the phase velocity, are different for the various frequency components. In view of this, the voltage waveform as it propagates along the transmission line is both attenuated and distorted. In the special case of a lossless transmission line for which $R = G = 0$, it is found that the voltage waveform is neither attenuated nor distorted. We treated this special case previously by the method of transformation of variables without recourse to Fourier or Laplace transforms. There is another special case for which L, C, R, and G are all nonvanishing but their magnitudes are such as to satisfy the equality $LG = CR$. This special case, which can also be solved by a suitable transformation of variables, turns out to be very interesting in that it leads to a waveform that is attenuated but not distorted. The treatment of the "distortionless" transmission line is taken up in Section 2.3 following a brief consideration of d.c. propagation on a lossy transmission line in Section 2.2.

In the case of a lossless transmission line, the voltage waveform as a whole traveled along the transmission line with a finite velocity $v = 1/\sqrt{LC}$. In the case of a "distortionless" transmission line, the voltage waveform is shown to travel along the transmission line again with a finite velocity $v = 1/\sqrt{LC}$, the only difference being that every part of the waveform is attenuated exponentially with distance by the same factor or the waveform as a whole is attenuated. In the general case of a lossy transmission line in which L, C, R, and G are all nonvanishing, it is difficult to define a unique velocity since the waveform is not only attenuated but also distorted. Even in this complicated case, we see that the front of the waveform travels with the velocity $v = 1/\sqrt{LC}$. In all these cases, the voltage is said to be *propagated* along the transmission line. In 1855 Lord Kelvin, with a view to establishing the feasibility or otherwise of the Atlantic submarine cable, investigated the solution of the transmission-line equations for the case $L = G = 0$. In this leakage-free, noninductive submarine cable, we find that immediately after the application of a voltage at the input terminals, there is a voltage everywhere along the transmission line and in this case the voltage is said to be *diffused* along the transmission line. Since diffusion occurs in many physical phenomenon such as electrical conduction, we shall discuss the diffusion along the transmission line in Section 2.4.

Although it is difficult to solve the partial differential equations for the voltage and the current along a lossy transmission line for the case of general time dependence, we can solve these partial differential equations in a fairly simple manner for the special case in which the voltage and the current are periodic in time. The resulting solutions are known as time-harmonic solutions, which are important since in many practical applications the sources used to excite the transmission line are indeed time-harmonic. The time-harmonic solutions of the partial differential equations pertaining to a lossy transmission line as applied to both semi-infinite and finite transmission lines, as well as such concepts as propagation coefficient, characteristic impedance, input impedance, reflection coefficient, real and reactive power, resonance, and antiresonance in transmission-line circuits and group velocity, are treated in Sections 2.5 to 2.13.

2.2 D.C. PROPAGATION ON A LOSSY TRANSMISSION LINE

Since in the case of d.c. the voltage and the current do not vary in time, the first-order partial differential equations for the voltage and the current as well as the second-order partial differential equations for either the voltage or the current on a lossy transmission line are given by

$$-\frac{d}{dx} V(x, t) = RI(x, t), \tag{2.1}$$

$$-\frac{d}{dx} I(x, t) = GV(x, t), \tag{2.2}$$

$$\frac{d^2}{dx^2} V(x, t) = RGV(x, t), \tag{2.3}$$

$$\frac{d^2}{dx^2} I(x, t) = RGI(x, t). \tag{2.4}$$

We see that the two linearly independent solutions of (2.3) are given by

$$V_1(x, t) = V_1 e^{-\sqrt{RG}x}; \qquad V_2(x, t) = V_2 e^{\sqrt{RG}x}. \tag{2.5}$$

We obtain the two linearly independent solutions of (2.4) in a similar manner:

$$I_1(x, t) = I_1 e^{-\sqrt{RG}x}; \qquad I_2(x, t) = I_2 e^{\sqrt{RG}x}. \tag{2.6}$$

If the transmission line extends up to $x = \infty$, then $V_2(x, t)$ and $I_2(x, t)$ become infinite. Since this situation is untenable physically, it follows that $V_2 = I_2 = 0$. In a similar manner, if the transmission line extends up to $x = -\infty$, then $V_1 = I_1 = 0$. Thus, the first and the second solutions in (2.5) correspond respectively to those in (2.6). The first set of solutions in (2.5) and (2.6) gives the voltage and the current that decay in the positive x-direction and the second set gives the voltage and the current that decay in the negative x-direction. We find that substituting the first set of solutions in (2.1) or (2.2) yields

$$I_1 = \frac{V_1}{\sqrt{R/G}}. \tag{2.7}$$

In a similar manner, substituting the second set of solutions in (2.1) gives

$$I_2 = -\frac{V_2}{\sqrt{R/G}}. \tag{2.8}$$

As was mentioned previously, the difference in signs between the ratios I_1/V_1 and I_2/V_2 is merely due to the difference in the sign convention for the current in relation to the direction in which the current decays for the two cases. With the help of (2.5) to (2.8), we find that the general solutions for the d.c. voltage and the d.c. current on a lossy transmission line are given by

$$V(x, t) = V_1 e^{-\sqrt{RG}x} + V_2 e^{\sqrt{RG}x}, \tag{2.9}$$

$$I(x, t) = \frac{V_1}{\sqrt{R/G}} e^{-\sqrt{RG}x} - \frac{V_2}{\sqrt{R/G}} e^{\sqrt{RG}x}. \tag{2.10}$$

Let us consider a lossy transmission line extending from $x = 0$ to $x = \infty$. In order that the voltage and the current in (2.9) and (2.10) respectively be finite, it is necessary that $V_2 = 0$. Then we find the ratio $V(x, t)/I(x, t)$ to be a constant $\sqrt{R/G}$ which depends only on the characteristics of the transmission line and is therefore called the characteristic resistance. If this transmission line is driven by a generator of voltage V_g through a series resistor R_g, then we find with the help of Fig. 2.1, that

$$V_1 = \frac{V_g\sqrt{R/G}}{R_g + \sqrt{R/G}}. \tag{2.11}$$

Therefore, the voltage and the current on this semi-infinite transmission line are given by

$$V(x, t) = \frac{V_g\sqrt{R/G}}{R_g + \sqrt{R/G}} e^{-\sqrt{RG}x}, \tag{2.12}$$

$$I(x, t) = \frac{V_g}{R_g + \sqrt{R/G}} e^{-\sqrt{RG}x}. \tag{2.13}$$

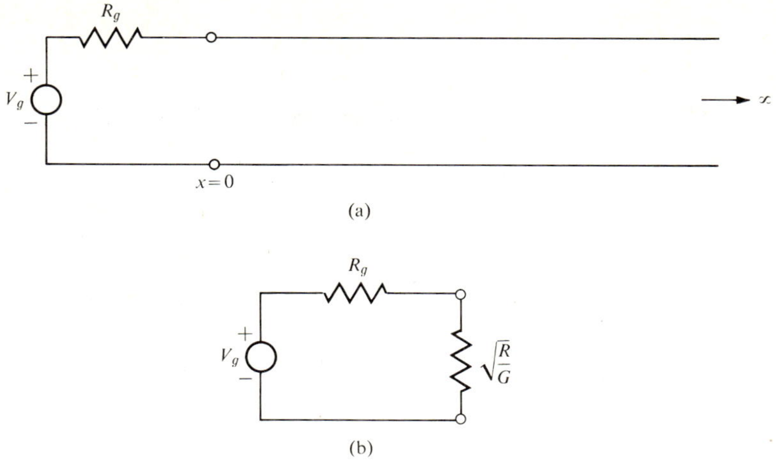

Fig. 2.1 (a) Semi-infinite lossy transmission line driven by a d.c. voltage generator V_g. (b) Equivalent electric circuit at the input.

For V and I, we retain both the arguments x and t for convenience even though for the d.c. case treated in this section, the voltage and the current are independent of time.

2.3 DISTORTIONLESS PROPAGATION ON A LOSSY TRANSMISSION LINE

It is convenient to recollect the first-order partial differential equations for the voltage and the current as well as the second-order partial differential equations for either the voltage or the current on a lossy transmission line, as follows:

$$-\frac{\partial}{\partial x} V(x, t) = RI(x, t) + L\frac{\partial}{\partial t} I(x, t), \tag{3.1}$$

$$-\frac{\partial}{\partial x} I(x, t) = GV(x, t) + C\frac{\partial}{\partial t} V(x, t), \tag{3.2}$$

$$\frac{\partial^2}{\partial x^2} V(x, t) = \frac{1}{v^2}\frac{\partial^2}{\partial t^2} V(x, t) + (LG + CR)\frac{\partial}{\partial t} V(x, t) + RGV(x, t), \tag{3.3}$$

$$\frac{\partial^2}{\partial x^2} I(x, t) = \frac{1}{v^2}\frac{\partial^2}{\partial t^2} I(x, t) + (LG + CR)\frac{\partial}{\partial t} I(x, t) + RGI(x, t), \tag{3.4}$$

where

$$v = 1/\sqrt{LC}. \tag{3.5}$$

Let the transmission line parameters L, C, R, and G be of such magnitudes as to satisfy the relation

$$R/L = G/C, \tag{3.6}$$

which is the requirement, as we shall show presently, for a lossy transmission line to be able to sustain distortionless propagation. We substitute the value of G from (3.6) in (3.3) with the following result:

$$\frac{\partial^2}{\partial x^2} V(x, t) = \frac{1}{v^2} \frac{\partial^2}{\partial t^2} V(x, t) + 2CR \frac{\partial}{\partial t} V(x, t) + \alpha^2 V(x, t), \tag{3.7}$$

where

$$\alpha = R\sqrt{C/L}. \tag{3.8}$$

The current $I(x, t)$ also satisfies an equation similar to (3.7). First let

$$y = -x. \tag{3.9}$$

Then, it follows that

$$\frac{\partial}{\partial y} V(x, t) = -\frac{\partial}{\partial x} V(x, t); \qquad \frac{\partial^2}{\partial y^2} V(x, t) = \frac{\partial^2}{\partial x^2} V(x, t). \tag{3.10}$$

With the help of (3.10), it is clear that the change of x to $-x$ does not change the differential equation (3.7). Therefore, if we find a solution of (3.7) and in that solution we change x to $-x$, then the resulting function is also a solution of (3.7).

Introduce the following change of variables:

$$\xi = x - vt; \qquad \eta = x. \tag{3.11}$$

Then,

$$dV = \frac{\partial V}{\partial x} dx + \frac{\partial V}{\partial t} dt = \frac{\partial V}{\partial \xi} d\xi + \frac{\partial V}{\partial \eta} d\eta, \tag{3.12}$$

and

$$d\xi = dx - v\, dt; \qquad d\eta = dx. \tag{3.13}$$

Note that in (3.12) we have omitted the arguments (x, t) or equivalently (ξ, η) of V for convenience. On substituting for $d\xi$ and $d\eta$ in (3.12) from (3.13) and equating the coefficients of dx and dt on either side of the resulting equation, it is found that

$$\frac{\partial}{\partial x} V = \left(\frac{\partial}{\partial \xi} + \frac{\partial}{\partial \eta}\right) V; \qquad \frac{1}{v} \frac{\partial}{\partial t} V = -\frac{\partial}{\partial \xi} V. \tag{3.14}$$

The substitution of (3.14) in (3.7) yields

$$\frac{\partial^2}{\partial \xi^2} V + 2 \frac{\partial^2}{\partial \xi\, \partial \eta} V + \frac{\partial^2}{\partial \eta^2} V = \frac{\partial^2}{\partial \xi^2} V - 2CRv \frac{\partial V}{\partial \xi} + \alpha^2 V. \tag{3.15}$$

Since $CRv = \alpha$, (3.15) may be recast into the form

$$2 \frac{\partial}{\partial \xi}\left\{\frac{\partial V}{\partial \eta} + \alpha V\right\} + \left\{\frac{\partial}{\partial \eta} - \alpha\right\}\left\{\frac{\partial V}{\partial \eta} + \alpha V\right\} = \left\{2 \frac{\partial}{\partial \xi} + \frac{\partial}{\partial \eta} - \alpha\right\}\left\{\frac{\partial V}{\partial \eta} + \alpha V\right\} = 0. \tag{3.16}$$

In view of the factor $\partial V/\partial \eta + \alpha V$, it follows that V can be expressed as a product of two functions: one is a function of η and the other of ξ. If the function of η is of the form $e^{-\alpha\eta}$, then (3.16) is satisfied for any function of ξ including a constant which is independent of both ξ and η. Thus a solution of (3.16) is of the form

$$\begin{aligned} V_1(x, t) &= e^{-\alpha\eta}\{f_1(\xi) + f_{01}\}, \\ &= e^{-\alpha x}\{f_1(x - vt) + f_{01}\}. \end{aligned} \tag{3.17}$$

We pointed out previously that if a solution of (3.7) is found and in that solution x is changed to $-x$, then the resulting function is also a solution of (3.7). Since $f_1(-x - vt) = f_1(-(x + vt))$, it can be set equal to another function of $x + vt$, say $f_2(x + vt)$; then another solution of (3.7) is

$$V_2(x, t) = e^{\alpha x}\{f_2(x + vt) + f_{02}\}. \tag{3.18}$$

In order that the solution (3.18) be as general as possible, we should note that the constant f_{02} in (3.18) need not necessarily be the same as in (3.17).

We find that the first and the second solutions, $I_1(x, t)$ and $I_2(x, t)$, corresponding to (3.17) and (3.18) respectively, which are obtained from the solutions of the second-order partial differential equation for the current similar to that given in (3.7) for the voltage, are given by

$$I_1(x, t) = e^{-\alpha x}\{g_1(x - vt) + g_{01}\}, \tag{3.19}$$

and

$$I_2(x, t) = e^{\alpha x}\{g_2(x + vt) + g_{02}\}. \tag{3.20}$$

In view of (3.1) and (3.2), it follows that there is a relationship between the voltage and the current distributions along the transmission line and therefore the f and the g functions in (3.17) to (3.20) are related. To obtain the relationship between the f and the g functions, it is instructive to consider first two special cases.

In (3.17) to (3.20), the solutions with the coefficients f_{01}, f_{02}, g_{01}, and g_{02} do not depend on time and therefore are the d.c. solutions which should be identical with those obtained in the previous section. If we compare these d.c. solutions in (3.17) to (3.20) with the corresponding solutions in the previous section we find

$$g_{01} = \frac{f_{01}}{\sqrt{R/G}}; \qquad g_{02} = -\frac{f_{02}}{\sqrt{R/G}}. \tag{3.21}$$

The remaining terms in (3.17) to (3.20) are the time-varying solutions, whose behavior we now consider in the limiting case of no losses. If $R = 0$, from (3.6) and (3.8), $G = 0$ and $\alpha = 0$. Then, the time-dependent solutions in (3.17) to (3.20) are the same as those we obtained previously for lossless transmission lines, for which case we showed that

$$g_1(x - vt) = \frac{f_1(x - vt)}{\sqrt{L/C}}; \qquad g_2(x + vt) = -\frac{f_2(x + vt)}{\sqrt{L/C}}. \tag{3.22}$$

Returning to the distortionless transmission line, we see from (3.6) that

$$\sqrt{R/G} = \sqrt{L/C} = R_0. \tag{3.23}$$

The results (3.21) and (3.22) for the two special cases together with (3.23) suggest the following relationships among the f and the g functions and therefore between the two sets of solutions for the voltage and the current:

$$g_1(x - vt) + g_{01} = [f_1(x - vt) + f_{01}]/R_0, \qquad (3.24a)$$

$$g_2(x + vt) + g_{02} = -[f_2(x + vt) + f_{02}]/R_0, \qquad (3.24b)$$

$$I_1(x, t) = V_1(x, t)/R_0; \qquad I_2(x, t) = -V_2(x, t)/R_0. \qquad (3.25)$$

The relations (3.24a,b) and (3.25) are consistent, as we have shown in the foregoing, with the solutions of the transmission-line equations for the special cases of d.c. propagation on a lossy transmission line and the time-dependent propagation on a lossless transmission line. That these relations are also consistent with (3.1) and (3.2) for a distortionless transmission line we can easily verify. From (3.17), we find that

$$-\frac{\partial}{\partial x} V_1(x, t) = \alpha e^{-\alpha x}\{f_1(x - vt) + f_{01}\} - e^{-\alpha x} f_1'(x - vt)$$

$$= Re^{-\alpha x} \frac{\{f_1(x - vt) + f_{01}\}}{R_0} - Lv \frac{e^{-\alpha x} f_1'(x - vt)}{R_0}, \qquad (3.26)$$

since

$$R/R_0 = R\sqrt{C/L} = \alpha; \qquad Lv/R_0 = L(1/\sqrt{LC})\sqrt{C/L} = 1.$$

From (3.19), we obtain

$$RI_1(x, t) + L\frac{\partial}{\partial t} I_1(x, t) = Re^{-\alpha x}\{g_1(x - vt) + g_{01}\} - Lve^{-\alpha x} g_1'(x - vt). \qquad (3.27)$$

In view of (3.24a), (3.26) and (3.27) are equal, showing that we have satisfied (3.1). In a similar manner, we can show that (3.17) and (3.19) satisfy (3.2) also. Also, $V_2(x, t)$ and $I_2(x, t)$ as given by (3.18) and (3.20) respectively satisfy both (3.1) and (3.2) in view of the relation (3.24b). On combining (3.17) with (3.18) and (3.19) with (3.20), and making use of the relations (3.24a) and (3.24b), we obtain the most general time-dependent solutions for the transmission-line equations subject to the constraint (3.6) as

$$V(x, t) = e^{-\alpha x} f_1(x - vt) + e^{\alpha x} f_2(x + vt), \qquad (3.28)$$

$$I(x, t) = \frac{e^{-\alpha x} f_1(x - vt)}{R_0} - \frac{e^{\alpha x} f_2(x + vt)}{R_0}, \qquad (3.29)$$

where R_0 is given in (3.23).

We showed previously that $f_1(x - vt)$ represents a waveform which travels as a whole in the positive x-direction with a uniform velocity v. As x becomes positively larger, $e^{-\alpha x}$ decreases exponentially with the result that $e^{-\alpha x} f_1(x - vt)$ represents a waveform which travels in the positive x-direction with a uniform velocity v and with every part of the waveform being diminished by the same factor in traveling through the same distance. Thus the relative shape of the waveform is preserved but its size is reduced exponentially as it travels with uniform velocity. It is usual to refer to this type of transmission as propagation without distortion but with attenuation. In a similar manner, $f_2(x + vt)$ represents a waveform which travels as a whole in the negative x-direction with uniform velocity v. Since $e^{\alpha x}$ decreases exponentially as x becomes negatively larger, it is clear that $e^{\alpha x}f_2(x + vt)$ represents a waveform propagating in the negative x-direction without distortion but with attenuation. Thus, in a lossy transmission line in which the magnitudes of the parameters are so as to satisfy the criterion (3.6) there is, in either direction, propagation with attenuation but without distortion.

If the transmission line extends from $x = -\infty$ to $x = 0$, the first set of solutions in (3.28) and (3.29) becomes infinite at $x = -\infty$ and, since this situation is physically untenable, the f_1 function should be set equal to zero. Similarly, if the transmission line extends from $x = 0$ to $x = \infty$, the f_2 functions should be set equal to zero to preclude any physically impossible solution. For the latter transmission line, (3.28) and (3.29) yield

$$V(x, t)/I(x, t) = R_0, \tag{3.30}$$

which, as before, we call the characteristic resistance. It is interesting to note that in both the lossless and the distortionless cases, the characteristic resistance has the same value provided L and C are the same. Note that the f_1 function is still undetermined and its form, as before, depends on the terminal conditions. For example, let us consider the following voltage generator,

$$V_g(t) = V_0H(t), \tag{3.31}$$

where $H(t)$ is the Heaviside unit step function, connected in series with a resistor R_g to the input terminals at $x = 0$ of the distortionless transmission line, as shown in

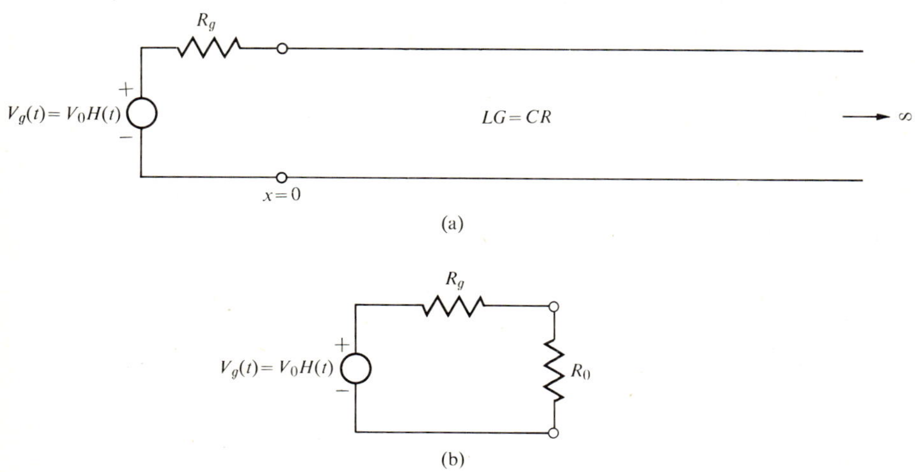

Fig. 2.2 (a) Semi-infinite distortionless ($LG = CR$) transmission line driven by a step-voltage $V_0H(t)$. (b) Equivalent electric circuit at the input.

Fig. 2.2. Then from (3.30) and Fig. 2.2, we see that

$$V(0, t) = \frac{V_0 R_0 H(t)}{R_0 + R_g} = f_1(-vt). \tag{3.32}$$

Since $f_1(-vt)$ is known, we can find $f_1(x - vt)$ by changing t to $t - (x/v)$ with the following result:

$$\frac{V_0 R_0 H\left(t - \dfrac{x}{v}\right)}{R_0 + R_g} = f_1(x - vt). \tag{3.33}$$

With the value of $f_1(x - vt)$ thus determined, we find the voltage and the current distribution on a semi-infinite transmission line $(0 < x < \infty)$ driven as shown in Fig. 2.2 to be given by

$$V(x, t) = \frac{V_0 R_0 e^{-\alpha x}}{R_0 + R_g} H\left(t - \frac{x}{v}\right), \tag{3.34}$$

$$I(x, t) = \frac{V_0 e^{-\alpha x}}{R_0 + R_g} H\left(t - \frac{x}{v}\right). \tag{3.35}$$

2.4 DIFFUSION ON A LOSSY TRANSMISSION LINE

Let us consider a noninductive $(L = 0)$, leakage-free $(G = 0)$ transmission line extending from $x = 0$ to $x = \infty$. We connect a voltage generator,

$$V_g(t) = V_0 H(t), \tag{4.1}$$

where $H(t)$ is the Heaviside unit step function, to the input terminals $x = 0$ of the transmission line as shown in Fig. 2.3. In other words, a d.c. voltage V_0 is switched on to the input terminals at time $t = 0$ with the result that the voltage across the input terminals is zero for $t \leq 0$ and is equal to V_0 for $t > 0$. The first-order partial differential equations satisfied by the voltage and the current distributions $V(x, t)$ and $I(x, t)$, as well as the second-order partial differential equations satisfied by either the voltage or the current, are given by

$$-\frac{\partial}{\partial x} V(x, t) = RI(x, t), \tag{4.2}$$

$$-\frac{\partial}{\partial x} I(x, t) = C \frac{\partial}{\partial t} V(x, t), \tag{4.3}$$

$$\frac{\partial^2}{\partial x^2} V(x, t) = CR \frac{\partial}{\partial t} V(x, t), \tag{4.4}$$

$$\frac{\partial^2}{\partial x^2} I(x, t) = CR \frac{\partial}{\partial t} I(x, t). \tag{4.5}$$

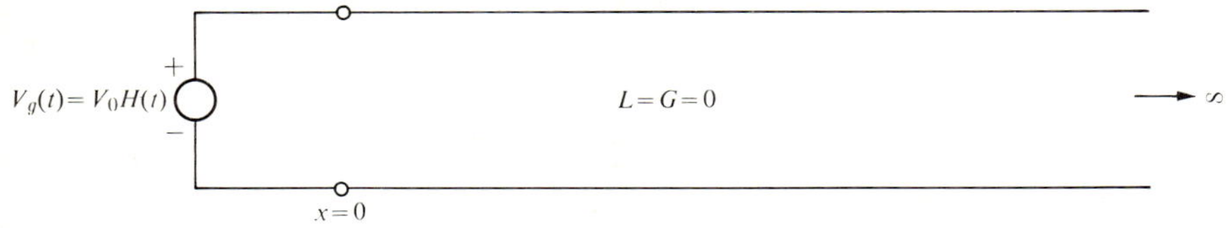

Fig. 2.3 Schematic diagram of a semi-infinite, noninductive ($L = 0$) and leakage-free ($G = 0$) transmission line driven by a step-voltage $V_0H(t)$.

We can deduce the solutions of (4.4) and (4.5) subject to the proper terminal conditions with the help of Laplace transforms. Here, we quote the solution of (4.5) obtained in that manner and verify that it satisfies (4.5). We can show that the solution of (4.5) is given by

$$I(x, t) = V_0H(t)\sqrt{C/\pi Rt}\, e^{-CRx^2/4t}. \tag{4.6}$$

Note that (4.6) reveals two important features. First, for $t < 0$, there is no current anywhere along the transmission line. We can expect this result since there can be no voltage or current along the transmission line before the source, which in the present case is a d.c. voltage V_0 switched on at time $t = 0$ and which is the cause of any voltage or current along the transmission line, comes into existence. As soon as t is slightly greater than zero that is, immediately after the source comes into existence, there is current and therefore voltage everywhere along the transmission line. The current and the voltage are said to be *diffused* along the transmission line. In contrast to the previous cases of lossless and distortionless transmission lines, the current and the voltage are not propagated with finite velocity.

For $t > 0$, since $H(t) = 1$, we find from (4.6) that

$$CR\frac{\partial}{\partial t} I(x, t) = V_0\sqrt{\frac{C}{\pi Rt}}\, e^{-CRx^2/4t}\left(\frac{-CR}{2t}\right) + V_0\sqrt{\frac{C}{\pi Rt}}\, e^{-CRx^2/4t}\,\frac{C^2R^2x^2}{4t^2}, \tag{4.7}$$

and

$$\frac{\partial}{\partial x} I(x, t) = V_0\sqrt{\frac{C}{\pi Rt}}\, e^{-CRx^2/4t}\left(-\frac{CRx}{2t}\right). \tag{4.8}$$

From (4.8), we find that

$$\frac{\partial^2}{\partial x^2} I(x, t) = V_0\sqrt{\frac{C}{\pi Rt}}\, e^{-CRx^2/4t}\left(-\frac{CR}{2t}\right) + V_0\sqrt{\frac{C}{\pi Rt}}\, e^{-CRx^2/4t}\,\frac{C^2R^2x^2}{4t^2}. \tag{4.9}$$

If we compare (4.7) with (4.9) we see that (4.6) satisfies (4.5).

With the help of (4.2), (4.3), and (4.6), the corresponding expression for $V(x, t)$ can be obtained. It still remains to show that $V(x, t)$ so obtained when specialized for the input terminals yields $V_g(t)$ as given by (4.1). From (4.2) and (4.6), we see that

$$\frac{\partial}{\partial x} V(x, t) = -V_0 H(t)\sqrt{CR/\pi t}\, e^{-CRx^2/4t}. \tag{4.10}$$

On integrating both sides of (4.10) with respect to x and changing the dummy variable of integration to z, we find that

$$V(x, t) = -\int^{x} V_0 H(t)\sqrt{CR/\pi t}\, e^{-CRz^2/4t}\, dz. \tag{4.11}$$

Since $I(x, t) = 0$ for $x = \infty$ and $t > 0$, as seen from (4.6), it follows that $V(x, t) = 0$ for $x = \infty$ and $t > 0$. In order to get $V(x, t) = 0$ for $x = \infty$ and $t > 0$ from (4.11), it is necessary to choose the lower limit of integration in (4.11) to be $+\infty$. We introduce the following change in variable of integration:

$$y = \sqrt{CR/4t}\, z; \qquad dy = \sqrt{CR/4t}\, dz. \tag{4.12}$$

Therefore, (4.11) becomes

$$\begin{aligned} V(x, t) &= -\int_{\infty}^{\sqrt{CRx^2/4t}} V_0 H(t)\sqrt{CR/\pi}\, e^{-y^2}\, dy \sqrt{4/CR} \\ &= V_0 H(t) \frac{2}{\sqrt{\pi}} \int_{\sqrt{CRx^2/4t}}^{\infty} e^{-y^2}\, dy. \end{aligned} \tag{4.13}$$

The function

$$\frac{2}{\sqrt{\pi}} \int_{y_0}^{\infty} e^{-y^2}\, dy = \operatorname{erfc}(y_0) \tag{4.14}$$

is well known and is called the complementary error function. It has the following properties

$$\operatorname{erfc}(0) = 1; \qquad \operatorname{erfc}(\infty) = 0. \tag{4.15}$$

We have sketched the complementary error function erfc (y) in Fig. 2.4; we can see that it is a monotonically decreasing function of its argument. We define the error function erf (y_0) as follows:

$$\frac{2}{\sqrt{\pi}} \int_{0}^{y_0} e^{-y^2}\, dy = \operatorname{erf}(y_0), \tag{4.16}$$

which has the properties

$$\operatorname{erf}(0) = 0; \qquad \operatorname{erf}(\infty) = 1. \tag{4.17}$$

Also, note that

$$\operatorname{erfc}(y_0) = 1 - \operatorname{erf}(y_0). \tag{4.18}$$

The error function erf (y) is also sketched in Fig. 2.4.

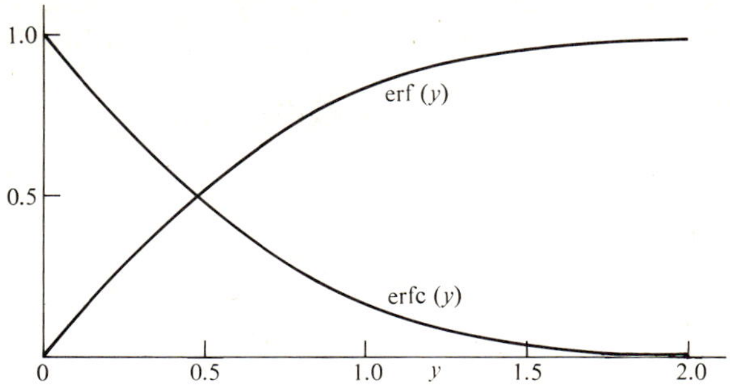

Fig. 2.4 Error function erf (y) and complementary error function erfc (y).

For $t > 0$ and $x = 0$, the lower limit of integration in (4.13) is zero and therefore, it follows from (4.14) and (4.15) that

$$V(0, t) = V_0 H(t). \tag{4.19}$$

If we compare (4.19) with (4.1) we see that $I(x, t)$ in (4.6) and $V(x, t)$ in (4.13) are consistent with the terminal condition; they have been already verified to satisfy the partial differential equations (4.2) to (4.5).

We can rewrite the voltage distribution $V(x, t)$ in (4.13) succinctly as follows:

$$V(x, t) = V_0 H(t) \operatorname{erfc}\left(\frac{1}{\sqrt{\tau}}\right); \qquad \tau = 4t/CRx^2. \tag{4.20}$$

In Fig. 2.5(a), we have depicted $V(x, t)$ as a function of τ and we can see that it increases monotonically from zero and attains the value V_0 for $\tau = \infty$. Note that $V(x, t)$ depends only upon τ for $t > 0$. At $x = x_0$, let the voltage attain a specified value less than V_0 in time $t = t_0$. Suppose that the distance is *doubled* to the value $x = 2x_0$. Then, the voltage at this new location attains the same specified value when τ reaches the same value $4t_0/CRx_0^2$ in the new location. Thus,

$$\frac{4t}{CR(2x_0)^2} = \frac{4t_0}{CRx_0^2} \quad \text{or} \quad t = 4t_0. \tag{4.21}$$

In other words, the time required for the voltage to reach the same value at the second location is *four* times the original time. We can generalize this result by the statement that the time required for the voltage at any location to reach a specified value is proportional to the square of the distance from the input terminals where the source is brought into existence at the origin of time.

At a location $x = x_0$ on a transmission line with the parameters C and R specified, we see from (4.20) that τ is proportional to time and hence for a given location $x = x_0$, we can interpret Fig. 2.5(a) as depicting the building up of voltage as a function of time. For a different location, the same τ value corresponds to a different

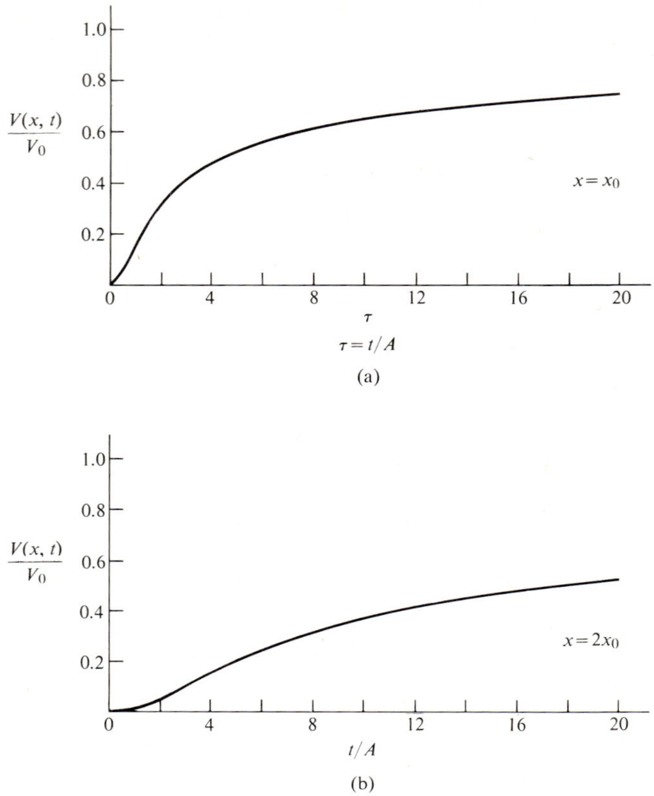

Fig. 2.5 Voltage distribution on a semi-infinite noninductive ($L = 0$) and leakage-free ($G = 0$) transmission line driven by a step-voltage V_0. $A = 0.25CRx_0^2$.

actual time t which, as indicated earlier, is proportional to the square of the distance from the input terminals. Thus, Fig. 2.5(a) can be interpreted as depicting the building up of voltage as a function of time for all locations on the transmission line provided the abscissa is understood to represent a time scale that is stretched as the distance along the transmission line is increased so as to be proportional to the square of the distance from the input terminals. To exhibit this phenomenon clearly, let Fig. 2.5(a) represent the time dependence of $V(x, t)$ at $x = x_0$. In Fig. 2.5(b), we show the time dependence of $V(x, t)$ at $x = 2x_0$. If we compare the two figures we see that Fig. 2.5(b) is the same as Fig. 2.5(a) with the time scale represented by the abscissa stretched by a factor of four.

Instead of a d.c. voltage V_0 switched on at time $t = 0$, let the input voltage be a rectangular pulse of amplitude V_0 and time duration t_1, switched on at time $t = 0$; that is, the leading edge of the pulse starts at time $t = 0$. We can consider this rectangular pulse as a superposition of a step-voltage V_0 switched on at $t = 0$ and another of voltage $-V_0$ switched on at time $t = t_1$. Thus, the input voltage may be expressed as

$$V_g(t) = V_0H(t) - V_0H(t - t_1). \tag{4.22}$$

Since the transmission line is a linear device and the superposition principle holds, we can obtain the voltage distribution $V(x, t)$ due to the source given by (4.22) by superposing the voltage distributions due to the two sources represented separately by the two terms in (4.22). Therefore, with the help of (4.1) and (4.20), we find the voltage distribution $V(x, t)$ due to the source (4.22) to be

$$V(x, t) = V_0H(t)\,\mathrm{erfc}\,(1/\sqrt{\tau}) - V_0H(t - t_1)\,\mathrm{erfc}\,(1/\sqrt{\tau - \tau_1}), \tag{4.23}$$

where

$$\tau_1 = 4t_1/CRx^2. \tag{4.24}$$

If we compare (4.20) and (4.23) we see that the result obtained by taking the mirror image of the curve in Fig. 2.5(a) about the abscissa, shifting it to the right by an amount $\tau = \tau_1$ and adding the ordinates of the resulting curve with those given in Fig. 2.5(a), is the voltage distribution $V(x, t)$ as given by (4.23) for the specified value of τ_1. In Fig. 2.6, the voltage distribution $V(x, t)$ as given by (4.23) and as obtained

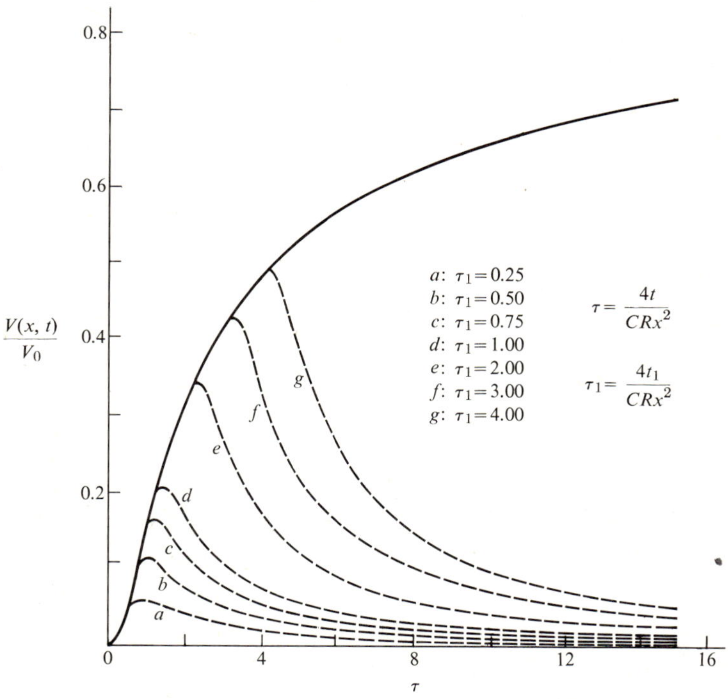

Fig. 2.6 Voltage distribution on a semi-infinite noninductive ($L = 0$) and leakage-free ($G = 0$) transmission line driven by a rectangular pulse of amplitude V_0 for various values of time duration t_1 sec. τ: scaled time; τ_1: scaled time duration of pulse.

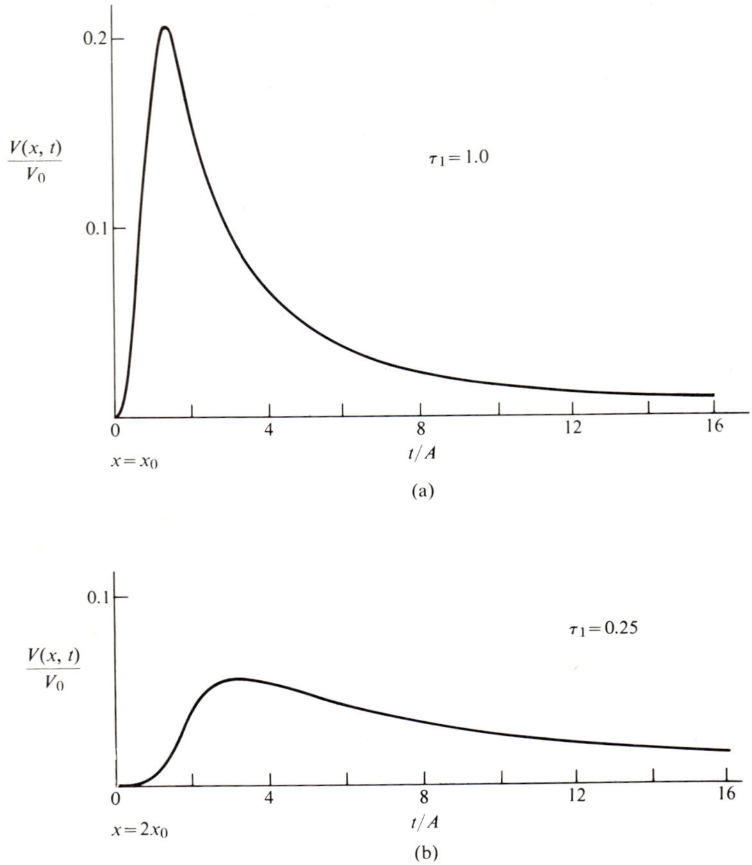

Fig. 2.7 Time dependence of voltage at two different locations on a semi-infinite, noninductive ($L = 0$) and leakage-free ($G = 0$) transmission line driven by a rectangular pulse of amplitude V_0 and time duration t_1 sec. $A = 0.25CRx_0^2$; $t_1 = 0.25CRx^2\tau_1$; τ_1: scaled time duration of pulse.

by the foregoing procedure is plotted for various values of τ_1 ranging from 0.25 to 4. The solid curve in Fig. 2.6 shows the voltage distribution as given in (4.20) and the dashed curves, together with the portion of the solid curve which joins the dashed curves to the origin, show the voltage distribution as given in (4.23) for various values of τ_1. The curves in Fig. 2.6 depict the time dependence of the voltage for all locations on the transmission line provided (1) the abscissa, as before, is understood to represent a time scale that is stretched proportionately to the square of the distance from the input terminals and (2) the scaled time duration τ_1 of the pulse is contracted by the same factor. Thus, if the distance is doubled, the time-scale is stretched by a factor of four and the scaled time duration τ_1 is reduced to one fourth of its former value. The curves in Fig. 2.6 also reveal the following two important features of the building up of voltage along a noninductive, leakage-free transmission line excited by a rectangular pulse. At any location on the transmission line, the maximum voltage reached is reduced as the scaled time duration τ_1 of the pulse is reduced. The building up of voltage caused by the first step-voltage representing the leading edge of the pulse is interrupted by the second step-voltage which represents the trailing edge of the pulse and which causes the voltage to decay since its amplitude is negative. As the time duration of the pulse is reduced, the time allowed for the building up of the voltage is reduced, and therefore the maximum voltage attained is reduced. As the distance along the transmission line is increased, since the time scale is stretched and since τ_1 is reduced, it follows that the maximum voltage reached is reduced and the time taken for this reduced maximum to be reached is increased. To illustrate the second feature of the voltage diffusion along the transmission line, we show in Fig. 2.7(a) $V(x, t)$ as a function of t for $\tau_1 = 1$ corresponding to $x = x_0$. In Fig. 2.7(b), $V(x, t)$ is shown as a function of t for $x = 2x_0$. In the second case $\tau_1 = \frac{1}{4}$ and the time scale is stretched by a factor of four. The voltage at $x = 2x_0$ we see in Fig. 2.7(b), compared with that we saw in Fig. 2.7(a), builds up at a slower rate and to a reduced maximum.

Consider now the case in which the transmission line is excited by a series of two pulses (Fig. 2.8d). The pulses are of the same duration of t_1 seconds and of amplitude V_0 volts. The time separation between the two pulses, that is the time difference between the trailing edge of the first pulse and the leading edge of the second, is $t = t_2$ seconds. Thus, $t = t_1 + t_2$ may be called the period of the pulse. The response of the transmission line to the second pulse is the same as that for the first pulse, except that the response starts $t_1 + t_2$ seconds later than that of the first pulse. The time dependence of the voltage for $\tau_1 = 1$ and $x = x_0$, as drawn in Fig. 2.7(a), is reproduced in Fig. 2.8(a), where the time dependence of the voltage due to the second pulse, also characterized by $\tau_1 = 1$, is included. Note that the voltage due to the second pulse starts at $\tau = \tau_1 + \tau_2$, where $\tau_2 = 4t_2/CRx^2$. We sum up the voltages due to the two pulses to yield the resultant voltage, which is also depicted in Fig. 2.8(a). The results of Fig. 2.8(a) corresponding to two other locations, which are further removed from the input terminals, are shown in Fig. 2.8(b) and (c). The curve of the resultant voltage of Fig. 2.8(a) shows two distinct humps indicative of the two separate adjacent pulses at the input terminals. The same is true to a lesser degree in Fig. 2.8(b) and (c), which correspond to more distant locations on the transmission line.

If the separation time t_2 between the two pulses is reduced, the two maxima in Fig. 2.8(a) get closer and as a result the minimum between them increases giving rise to poorer resolution of the two maxima. If the time duration $t = t_1$ of each pulse is reduced, the two maxima are reduced without their separation in time being altered substantially and this also leads to poorer resolution of the two maxima. Also, if the period $t_1 + t_2$ of the pulse is kept constant but the time duration $t = t_1$ is reduced, the resolution of the two maxima usually becomes poorer but not to the same degree as in the previous two cases. Thus, in order to be able to resolve the voltage response as due to two adjacent pulses, the scaled time duration τ_1 of each

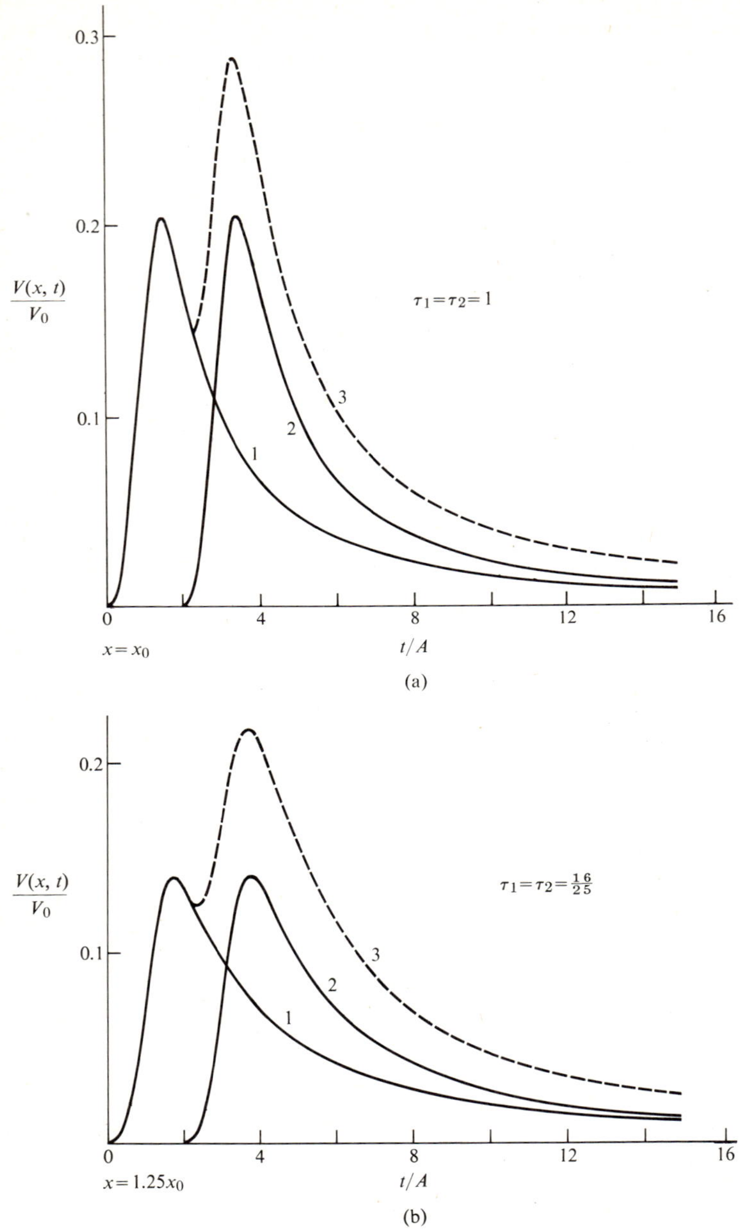

Fig. 2.8 Voltage response as a function of time at three different locations on a semi-infinite, noninductive ($L = 0$) and leakage-free ($G = 0$) transmission line driven by an input voltage consisting of a sequence of two pulses. $A = 0.25CRx_0^2$. Curve 1: response due to the first pulse alone; Curve 2: response due to the second pulse alone; Curve 3: response due to the sequence of two pulses.

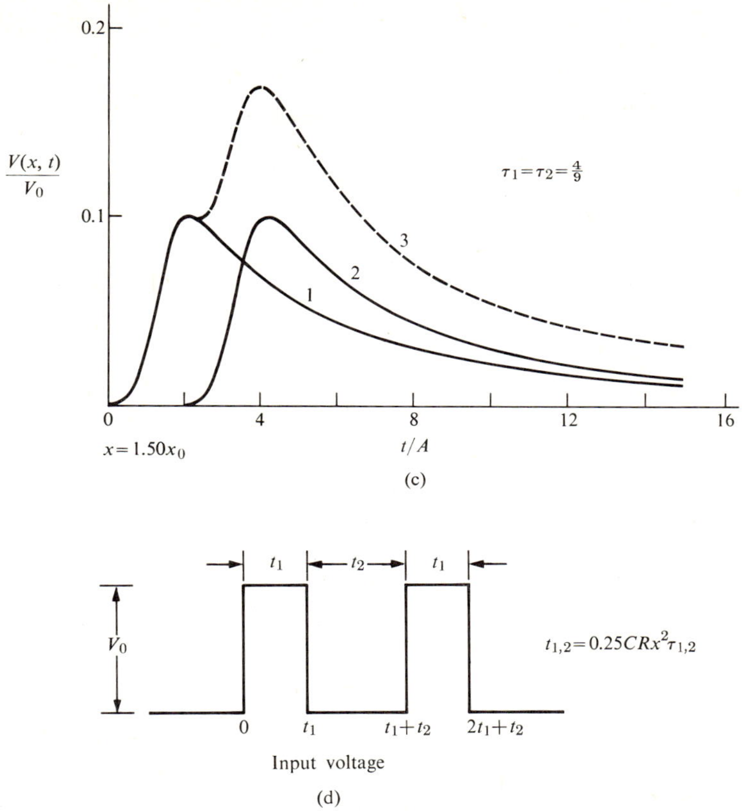

Fig. 2.8 (*continued*)

pulse as well as the scaled time separation τ_2 between the adjacent pulses should not fall below certain critical values which may be taken to be 1 for both the cases.

With the help of a train of pulses, information in the form of Morse code can be conveyed along the transmission line and this is known as "transmission-line keying". The speed of information transfer depends on the number of pulses per second and therefore, is inversely proportional to the pulse period $t_1 + t_2$. Since the scaled pulse period $\tau_1 + \tau_2$ has an optimum value for achieving adequate resolution between adjacent pulses, it follows from (4.24) that the speed of transmission-line keying is inversely proportional to CRx^2. Thus, with an increase in distance, the speed of transmission is automatically reduced. This law of speed of transmission-line keying is known as Kelvin's KR law. It was Lord Kelvin who in 1855, with a view to assessing the feasibility of Morse code communication through the trans-Atlantic submarine cable, approximated the cable by a noninductive leakage-free transmission line and arrived essentially at all the results discussed in this section.

2.5 FORM OF TIME-HARMONIC SOLUTIONS FOR A LOSSY TRANSMISSION LINE

We have not stipulated that the voltage and the current distributions along a lossy transmission line have any particular type of time dependence in the solutions that have been considered already. The particular case of time-periodic solutions for the voltage and the current is important, partly because in a vast majority of practical applications the voltage and the current are essentially time periodic and partly because the knowledge of the periodic solutions is helpful in building up solutions of arbitrary time dependence. For these reasons, we shall devote the remainder of the treatment of the transmission lines to the examination of the various aspects of the time-periodic solutions.

We can represent the voltage distribution $V(x, t)$ which is periodic in time as follows:

$$V(x, t) = |V(x)| \cos(\omega t + \theta_V), \tag{5.1}$$

$$= \text{Re}\left[V(x)e^{j\omega t}\right] \qquad \text{with} \qquad V(x) = |V(x)|\, e^{j\theta_V}, \tag{5.2}$$

where θ_V is also, in general, a function of x. A particular state or phase of the voltage distribution such as maxima or minima repeats itself every $2\pi/\omega$ seconds and thus $2\pi/\omega$ is the period. At a given location, the number of a particular phase of the voltage distribution that occurs every second is equal to $\omega/2\pi$ and is called the frequency f. It is usual to represent the time-periodic voltage in (5.1) as given in (5.2) without explicitly writing Re, which denotes "real part of". Since the voltage is a physical quantity and therefore real, whenever a representation similar to (5.2) is used, the "real part of" is to be understood. Also, the time factor $e^{j\omega t}$ is, by convention, never written down explicitly but is implied. The foregoing considerations also apply to the current distribution $I(x, t)$ which in the phasor notation is given by

$$I(x, t) = I(x) \qquad \text{where} \qquad I(x) = |I(x)|\, e^{j\theta_I}. \tag{5.3}$$

In (5.3), the time factor $e^{j\omega t}$ and the operation of taking the real part, though not written down explicitly, are implied.

It is important to bear in mind the following relevant remarks in connection with the so-called phasor representation of the voltage and the current. First, since the voltage is periodic in time, the existence of the voltage for all times from $t = -\infty$ to $t = \infty$ is automatically implied. In other words, the voltage is not localized in time. The voltage as given in (5.2) should not be mistaken to exist only for $t \geq 0$, for then $V(x, t)$ is clearly not periodic in time for all times and the analysis that we shall discuss presently would no longer be applicable. Secondly, before carrying out any operations with the phasor quantities, it is necessary for us to examine their validity. For example, since

$$V_1(x, t) + V_2(x, t) = \text{Re}\left[V_1(x)e^{j\omega t}\right] + \text{Re}\left[V_2(x)e^{j\omega t}\right]$$
$$= \text{Re}\left[\{V_1(x) + V_2(x)\}e^{j\omega t}\right], \tag{5.4}$$

it follows that the operations of taking the real part and addition commute with each other and, therefore, it is legitimate to add the phasor voltages $V_1(x)$ and $V_2(x)$, introduce the time factor and then take the real part instead of first introducing the time factor for each of the phasor voltages $V_1(x)$ and $V_2(x)$, taking their real parts individually and then adding the resultant quantities. On the other hand, since

$$\begin{aligned} V(x, t) \times I(x, t) &= \text{Re}\left[V(x)e^{j\omega t}\right] \times \text{Re}\left[I(x)e^{j\omega t}\right] \\ &\neq \text{Re}\left[V(x)e^{j\omega t} \times I(x)e^{j\omega t}\right], \end{aligned} \tag{5.5}$$

it is found that the operations of taking the real part and multiplication do not commute with each other. Consequently, when quadratic quantities are involved, it is necessary to first introduce the time factor for each of the phasor quantities, take their real parts individually and then multiply the resultant quantities. Thirdly, since

$$\begin{aligned} \frac{\partial}{\partial t} V(x, t) &= \frac{\partial}{\partial t} \text{Re}\left[V(x)e^{j\omega t}\right] \\ &= \text{Re}\left[j\omega V(x)e^{j\omega t}\right], \end{aligned} \tag{5.6}$$

it follows that the time derivative $\partial/\partial t$ is equivalent to multiplying the phasor quantity by $j\omega$.

In view of the foregoing, for the special case in which the voltage and the current are periodic quantities of period $2\pi/\omega$, the first-order partial-differential equations for the phasor voltage and current become

$$-\frac{dV(x)}{dx} = (R + j\omega L)I(x) = zI(x), \tag{5.7}$$

$$-\frac{dI(x)}{dx} = (G + j\omega C)V(x) = yV(x), \tag{5.8}$$

where

$$z = R + j\omega L \qquad y = G + j\omega C \tag{5.9}$$

are respectively the series impedance and the shunt admittance, both per unit length of the transmission line. In a similar manner, we can obtain the second-order partial-differential equations for either the voltage or the current:

$$\frac{d^2V(x)}{dx^2} - \gamma^2 V(x) = 0, \tag{5.10}$$

$$\frac{d^2I(x)}{dx^2} - \gamma^2 I(x) = 0, \tag{5.11}$$

where

$$\gamma = \sqrt{zy} = \sqrt{(R + j\omega L)(G + j\omega C)}. \tag{5.12}$$

We can easily obtain the general solutions of (5.10) and (5.11) with the following result:

$$V(x) = V_1 e^{-\gamma x} + V_2 e^{\gamma x}, \tag{5.13}$$

$$I(x) = I_1 e^{-\gamma x} + I_2 e^{\gamma x}, \tag{5.14}$$

where V_1, V_2, I_1, and I_2 are constants depending on the terminal conditions. The first and the second solutions in (5.13) correspond respectively to the first and the second solutions in (5.14). Also, the first and the second terms in (5.13) and (5.14) are independently the solutions of (5.10) and (5.11) respectively. The substitution of the first set in either (5.7) or (5.8) yields

$$\gamma V_1 = zI_1 \qquad \text{or} \qquad I_1 = V_1/Z_0, \tag{5.15}$$

where

$$Z_0 = \sqrt{z/y} = \sqrt{(R + j\omega L)/(G + j\omega C)}. \tag{5.16}$$

In a similar manner, the substitution of the second set in either (5.7) or (5.8) gives

$$I_2 = -V_2/Z_0. \tag{5.17}$$

Together with (5.15) and (5.17), the most general solution for the current (5.14) becomes

$$I(x) = \frac{V_1}{Z_0} e^{-\gamma x} - \frac{V_2}{Z_0} e^{\gamma x}. \tag{5.18}$$

We can see from (5.13) and (5.18) that the general expressions of the voltage and the current along the transmission line are given in terms of two arbitrary constants V_1 and V_2 which, as before, we can determine by the terminal conditions.

2.6 SEMI-INFINITE TRANSMISSION LINE

In general, γ is complex as seen from (5.12) and therefore, it is convenient to set

$$\gamma = \alpha + j\beta = \sqrt{(R + j\omega L)(G + j\omega C)} \qquad (\alpha \geq 0; \beta \geq 0), \tag{6.1}$$

where α and β are respectively the real and the imaginary parts of γ. If the transmission line extends from $x = -\infty$ to $x = 0$, $e^{-\gamma x}$ becomes infinite at $x = -\infty$ and therefore we should set V_1 equal to zero in (5.13) and (5.18) in order to get physically meaningful voltage and current on the transmission line. In a similar manner, if the transmission line extends from $x = 0$ to $x = \infty$, we should set V_2 equal to zero in (5.13) and (5.18) in order to get physically acceptable results. Thus, on a semi-infinite transmission line extending from $x = 0$ to $x = \infty$, the voltage and the current distributions are given by

$$V(x) = V_1 e^{-(\alpha + j\beta)x}; \qquad I(x) = \frac{V_1}{Z_0} e^{-(\alpha + j\beta)x}. \tag{6.2}$$

In order to get the real voltage distribution, we multiply $V(x)$ by $e^{j\omega t}$ and take the real part of the resulting quantity. For this purpose, let $V_1 = |V_1|\, e^{j\theta_{V_1}}$. Then, the real voltage distribution becomes

$$V(x, t) = |V_1|\, e^{-\alpha x} \cos(\omega t + \theta_{V_1} - \beta x). \tag{6.3}$$

The coefficient of the cosine term is known as the amplitude since it determines the maximum amplitude of variation of the voltage with time. As x increases, the amplitude decays exponentially and the rate of decay depends on α. The amplitude decays by a factor $1/e$ in a distance of $x = 1/\alpha$ and α is therefore called the *attenuation coefficient.* Its units are nepers per meter. The voltage varies from the maximum value $|V_1| e^{-\alpha x}$ to the minimum value $-|V_1| e^{-\alpha x}$ as time varies. The relative state or the phase of the voltage depends on the value of the cosine term and this value depends on the argument of the cosine function. Therefore, $\omega t + \theta_{V_1} - \beta x$ is called the phase factor. The phase factor changes as x increases and the rate of change of phase with x depends on β, which is therefore called the *phase-change coefficient.* Its units are radians per meter. Thus, α and β respectively control the amplitude and the phase of the voltage distribution and, therefore, together they determine the entire propagation characteristics of the voltage along the transmission line. Therefore, we call $\gamma = \alpha + j\beta$ *the propagation coefficient.*

As we stated earlier, the phase of the voltage distribution depends on the value of the cosine function. Therefore, the phase does not change if the phase factor changes by integer multiples of 2π. The phase factor is altered by changes in time as well as in distance along the transmission line. The phase of the voltage does not change if the time changes by integer multiples of $2\pi/\omega$, which we therefore call the time period or just the period. Then, we call $\omega/2\pi = f$ the frequency and ω the angular or the radian frequency. In a similar manner, the phase of the voltage does not change if the distance changes by integer multiples of

$$\lambda = 2\pi/\beta, \tag{6.4}$$

which we therefore call the space period or the wavelength.

Let us consider the change in phase due to the simultaneous change in time and distance along the transmission line. Let the phase of the voltage have some value at $t = t_1$ and $x = x_1$. At a later time $t = t_2 > t_1$, the phase is the same at $x = x_2$ provided that

$$\omega t_1 + \theta_{V_1} - \beta x_1 = \omega t_2 + \theta_{V_1} - \beta x_2,$$

or

$$v_{ph} = \frac{\omega}{\beta} = \frac{x_2 - x_1}{t_2 - t_1}. \tag{6.5}$$

In other words, a particular phase of the voltage has traveled a distance $(x_2 - x_1)$ in time $(t_2 - t_1)$. Therefore, from (6.5) we see that $v_{ph} = \omega/\beta$ gives the velocity of readjustment of phases along the transmission line or *the phase velocity.* Note that in (6.3) t varies from $-\infty$ to ∞ with the result that there is voltage distribution everywhere along the transmission line. Consequently, it is not possible for us to interpret v_{ph} as the velocity of propagation of any physical quantity of interest.

From (6.2), it is obtained that for any x

$$\frac{V(x)}{I(x)} = Z_0 = R_0 + jX_0 = \sqrt{\frac{R + j\omega L}{G + j\omega C}}, \tag{6.6}$$

which is called *the characteristic impedance* since it depends only on the parameters which are characteristic of the transmission line and the frequency of operation and not on any terminal conditions.

For a transmission line which extends from $x = -\infty$ to $x = 0$, the expressions analogous to (6.2), (6.3), (6.5), and (6.6) are

$$V(x) = V_2 e^{(\alpha + j\beta)x}; \qquad I(x) = -\frac{V_2}{Z_0} e^{(\alpha + j\beta)x}, \tag{6.7}$$

$$V(x, t) = |V_2|\, e^{\alpha x} \cos(\omega t + \theta_{V_2} + \beta x), \tag{6.8}$$

$$v_{ph} = \frac{\omega}{\beta} = -\frac{(x_2 - x_1)}{(t_2 - t_1)}, \tag{6.9}$$

$$\frac{V(x)}{I(x)} = -Z_0. \tag{6.10}$$

From (6.8), we see that the wave is attenuated by the same factor as in the previous case provided x becomes negatively large. Also, we see from (6.9) that the phase travels in the negative x-direction with the same speed. The results for the two semi-infinite transmission lines can be generalized by the statement that in the direction in which the phase progresses, the amplitude of the voltage and the current waves are attenuated. As we explained previously, the difference in sign in (6.6) and (6.10) is due to the sign convention for a positive current in relation to the direction in which the phase progresses or equivalently the direction in which the wave is attenuated. In (6.6) the directions of positive current and the phase progression are the same whereas these directions are opposite for (6.10). In other words, if the convention for the positive current is fixed, the sign of the ratio of the voltage to the current changes if the direction of progression of the phase fronts of the wave is reversed.

2.7 CHARACTERISTIC IMPEDANCE, PROPAGATION COEFFICIENT, PHASE VELOCITY, AND WAVELENGTH

It is instructive to examine the limiting values of the expressions for the characteristic impedance, propagation coefficient, phase velocity, and wavelength as given respectively by (5.16), (5.12), (6.5), and (6.4) under various special conditions.

1. Characteristic Impedance

a) Lossless case. For the lossless case, $R = G = 0$ and then from (5.16) we obtain

$$Z_0 = R_0 + jX_0 = \sqrt{L/C}; \qquad R_0 = \sqrt{L/C}; \qquad X_0 = 0. \tag{7.1}$$

The characteristic reactance X_0 vanishes for the lossless case and the characteristic resistance is as given by (7.1).

b) Distortionless case. For the distortionless case, $R/L = G/C$ and then (5.16) yields

$$Z_0 = R_0 + jX_0 = \sqrt{\frac{R + j\omega L}{(RC/L) + j\omega C}} = \sqrt{\frac{L}{C}}; \qquad R_0 = \sqrt{\frac{L}{C}}; \qquad X_0 = 0. \tag{7.2}$$

The characteristic reactance X_0 vanishes also for the distortionless case and the characteristic resistance becomes identical to that obtained for the lossless case.

c) Low-loss case. We define the low-loss case by the inequalities $R \ll \omega L$ and $G \ll \omega C$. For given values of L, C, R, and G, the low-loss case also corresponds to the limit of high frequency. We can rearrange the expression for the characteristic impedance as follows

$$Z_0 = R_0 + jX_0 = \sqrt{\frac{L}{C}}\left(1 + \frac{R}{j\omega L}\right)^{1/2}\left(1 + \frac{G}{j\omega C}\right)^{-1/2}. \tag{7.3}$$

The use of the binomial expansion formulae

$$(1 + y)^{1/2} = 1 + \tfrac{1}{2}y - \tfrac{1}{8}y^2 + \tfrac{1}{16}y^3, \tag{7.4}$$

$$(1 + y)^{-1/2} = 1 - \tfrac{1}{2}y + \tfrac{3}{8}y^2 - \tfrac{5}{16}y^3, \tag{7.5}$$

in (7.3) and the retention of terms only up to the order ω^{-2} yields

$$\begin{aligned} R_0 + jX_0 &= \sqrt{\frac{L}{C}}\left(1 + \frac{R}{2j\omega L} + \frac{R^2}{8\omega^2 L^2}\right)\left(1 - \frac{G}{2j\omega C} - \frac{3G^2}{8\omega^2 C^2}\right) \\ &= \sqrt{\frac{L}{C}}\left[1 + \frac{1}{2j\omega}\left(\frac{R}{L} - \frac{G}{C}\right) + \frac{R^2}{8\omega^2 L^2} + \frac{RG}{4\omega^2 LC} - \frac{3G^2}{8\omega^2 C^2}\right]. \end{aligned} \tag{7.6}$$

From (7.6), we can see that

$$R_0 = \sqrt{\frac{L}{C}}\left[1 + \frac{1}{8\omega^2}\left(\frac{R}{L} - \frac{G}{C}\right)\left(\frac{R}{L} + \frac{3G}{C}\right)\right], \tag{7.7}$$

$$X_0 = -\sqrt{\frac{L}{C}}\frac{1}{2\omega}\left(\frac{R}{L} - \frac{G}{C}\right). \tag{7.8}$$

The characteristic resistance as given by (7.7) is now a function of frequency and we can see that it is greater or less than that for the lossless or the distortionless case according to whether R/L is greater or less than G/C. Also, the characteristic reactance X_0 is nonvanishing. The condition $R/L > G/C$ is usually obtained in practice and then we see from (7.8) that X_0 is a capacitive reactance since it is negative and becomes less negative with the increase in frequency.

2. Propagation coefficient

a) Lossless case. For the lossless case, $R = G = 0$ and (5.12) becomes

$$\gamma = \alpha + j\beta = j\omega\sqrt{LC}; \qquad \alpha = 0; \qquad \beta = \omega\sqrt{LC}. \tag{7.9}$$

For the lossless case, there is no attenuation since the attenuation coefficient $\alpha = 0$. This result is to be expected. We can see that the phase-shift coefficient as given in (7.9) is a linear function of frequency.

b) Distortionless case. For the distortionless case, $R/L = G/C$ and then (5.12) reduces to

$$\gamma = \alpha + j\beta = \sqrt{(R + j\omega L)\left(\frac{RC}{L} + j\omega C\right)} = \sqrt{\frac{C}{L}}(R + j\omega L); \tag{7.10}$$

therefore,

$$\alpha = R\sqrt{C/L}; \qquad \beta = \omega\sqrt{LC}. \tag{7.11}$$

The attenuation coefficient given in (7.11) is nonvanishing and is independent of frequency. The phase-shift coefficient β as given by (7.11) is the same as for the lossless case and is a linear function of frequency.

c) Low-loss case. For the low-loss case, we can rearrange the expression for the propagation coefficient as follows:

$$\gamma = \alpha + j\beta = j\omega\sqrt{LC}\left(1 + \frac{R}{j\omega L}\right)^{1/2}\left(1 + \frac{G}{j\omega C}\right)^{1/2}. \tag{7.12}$$

The use of the binomial expansion formula (7.4) in (7.12) yields the following result if terms up to the order ω^{-3} are retained:

$$\alpha + j\beta = j\omega\sqrt{LC}\left[1 + \frac{R}{2j\omega L} + \frac{R^2}{8\omega^2L^2} - \frac{R^3}{16j\omega^3L^3}\right]$$

$$\times \left[1 + \frac{G}{2j\omega C} + \frac{G^2}{8\omega^2C^2} - \frac{G^3}{16j\omega^3C^3}\right]$$

$$= j\omega\sqrt{LC}\left[1 + \frac{R}{2j\omega L} + \frac{G}{2j\omega C} + \frac{R^2}{8\omega^2L^2} - \frac{RG}{4\omega^2CL} + \frac{G^2}{8\omega^2C^2}\right.$$

$$\left. - \frac{R^3}{16j\omega^3L^3} + \frac{R^2G}{16j\omega^3L^2C} + \frac{RG^2}{16j\omega^3LC^2} - \frac{G^3}{16j\omega^3C^3}\right]. \tag{7.13}$$

With the help of (7.13), we obtain

$$\alpha = \frac{\sqrt{LC}}{2}\left(\frac{R}{L} + \frac{G}{C}\right)\left[1 - \frac{1}{8\omega^2}\left(\frac{R}{L} - \frac{G}{C}\right)^2\right], \tag{7.14}$$

$$\beta = \omega\sqrt{LC}\left[1 + \frac{1}{8\omega^2}\left(\frac{R}{L} - \frac{G}{C}\right)^2\right]. \tag{7.15}$$

The attenuation coefficient α as given by (7.14) is nonvanishing and is a function of frequency. The phase-shift coefficient β as given by (7.15) is different from that obtained for the lossless and the distortionless cases. In particular, β as given by (7.15) is not a linear function of frequency.

With the help of (7.9), (7.11), and (7.15), we can easily obtain the following expressions for the phase velocity for the lossless, distortionless, and the low-loss cases:

$$v_{ph} = \frac{\omega}{\beta} = \frac{1}{\sqrt{LC}} \quad \text{(lossless case)}, \tag{7.16}$$

$$v_{ph} = \frac{\omega}{\beta} = \frac{1}{\sqrt{LC}} \quad \text{(distortionless case)}, \tag{7.17}$$

$$v_{ph} = \frac{\omega}{\beta} = \frac{1}{\sqrt{LC}}\left[1 - \frac{1}{8\omega^2}\left(\frac{R}{L} - \frac{G}{C}\right)^2\right] \quad \text{(low-loss case)}. \tag{7.18}$$

The fact that the phase velocity is independent of frequency for the lossless and the distortionless cases but is dependent on frequency for the low-loss case is worthy of notice. Also, the phase velocity for the low-loss case as given in (7.18) is less than that for the lossless or the distortionless case for the same parameter values for L and C.

With the help of (6.4) and (6.5), we see that

$$\lambda = v_{ph}/f. \tag{7.19}$$

The use of (7.16) to (7.18) in (7.19) gives us immediately the expressions for the wavelength for the lossless, distortionless and the low-loss cases.

2.8 TIME-AVERAGED, OR REAL, AND REACTIVE POWERS

The instantaneous power corresponding to a voltage and the associated current is equal to their product. As was discussed in Section 2.5, when the process of multiplication is involved the time factor $e^{j\omega t}$ has to be included first with the phasor voltage and current and their real parts taken separately before performing the multiplication of the voltage and the current. On setting $V_1/Z_0 = I_1 = |I_1|\, e^{j\theta_{I_1}}$, we obtain from (6.2) the real voltage and current distributions along a semi-infinite transmission line extending from $x = 0$ to $x = \infty$ as

$$V(x, t) = |V_1|\, e^{-\alpha x} \cos(\omega t + \theta_{V_1} - \beta x), \tag{8.1}$$

$$I(x, t) = |I_1|\, e^{-\alpha x} \cos(\omega t + \theta_{I_1} - \beta x). \tag{8.2}$$

In particular, we find the voltage and the current at the input $x = 0$ of the transmission line as

$$V(0, t) = |V_1| \cos(\omega t + \theta_{V_1}), \tag{8.3}$$

$$I(0, t) = |I_1| \cos(\omega t + \theta_{I_1}). \tag{8.4}$$

We obtain the instantaneous power delivered to the transmission line from (8.3) and (8.4) as

$$P(t) = V(0, t)I(0, t) = |V_1|\,|I_1| \cos(\omega t + \theta_{V_1}) \cos(\omega t + \theta_{I_1}), \tag{8.5}$$

which is found to vary with time periodically of period π/ω. Since $P(t)$ is periodic, its time average P_{real} can be determined by taking the time average over one period. Thus,

$$P_{\text{real}} = \frac{\omega}{\pi} \int_0^{\pi/\omega} P(t)\, dt. \tag{8.6}$$

With the help of the trigonometric formula

$$\cos x \cos y = \tfrac{1}{2} \cos (x + y) + \tfrac{1}{2} \cos (x - y), \tag{8.7}$$

we can write (8.5) as the sum of two terms and this expression for $P(t)$ is then inserted in (8.6) with the following result

$$P_{\text{real}} = \frac{\omega}{\pi} \int_0^{\pi/\omega} \frac{|V_1|\,|I_1|}{2} [\cos (2\omega t + \theta_{V_1} + \theta_{I_1}) + \cos (\theta_{V_1} - \theta_{I_1})]\, dt. \tag{8.8}$$

We see that the time average of the first term within the square brackets in (8.8) vanishes and on evaluating the time average of the second term within the square brackets in (8.8), we obtain the following result:

$$P_{\text{real}} = \frac{|V_1|\,|I_1|}{2} \cos (\theta_{V_1} - \theta_{I_1}). \tag{8.9}$$

We can rewrite (8.9) in terms of the phasor quantities V_1 and I_1 as follows:

$$P_{\text{real}} = \tfrac{1}{2} \operatorname{Re} (V_1 I_1^*), \tag{8.10}$$

where the asterisk denotes complex conjugate. Since $I_1 = V_1/Z_0$, we can express (8.10) alternatively as

$$P_{\text{real}} = \frac{1}{2} \frac{|V_1|^2}{|Z_0|^2} R_0, \tag{8.11}$$

where R_0, the real part of Z_0, is the characteristic resistance.

It is seen from (8.1) and (8.2) that the voltage and the current vanish at $x = \infty$. Therefore, no power is associated with the terminal voltage and current. In other words, there is no power output and all the time-averaged power that is delivered to the input of the transmission line is dissipated in the series resistance R and the shunt conductance G of the transmission line. It is instructive to compute the total time-averaged power dissipated along the transmission line and establish that it is equal to the time-averaged power input to the transmission line.

We can deduce the instantaneous power dissipated in the series resistor R and the shunt conductance G between x and $x + dx$ of the transmission line with the help of (8.1) and (8.2) respectively to be

$$[P_{\text{dis}}(x, t)]_R = R\, dx I_1^2(x, t) = R\, dx\, |I_1|^2\, e^{-2\alpha x} \cos^2 (\omega t + \theta_{I_1} - \beta x), \tag{8.12}$$

$$[P_{\text{dis}}(x, t)]_G = G\, dx V_1^2(x, t) = G\, dx\, |V_1|^2\, e^{-2\alpha x} \cos^2 (\omega t + \theta_{V_1} - \beta x). \tag{8.13}$$

We can obtain the time averages of (8.12) and (8.13) in accordance with (8.6). The procedure is similar to that used in deducing (8.9) and (8.10). The results of the time averages of (8.12) and (8.13) are respectively

$$[P_{\text{dis}}(x)]_R = \tfrac{1}{2}R\,|I_1|^2\,e^{-2\alpha x}\,dx, \tag{8.14}$$

$$[P_{\text{dis}}(x)]_G = \tfrac{1}{2}G\,|V_1|^2\,e^{-2\alpha x}\,dx. \tag{8.15}$$

We can evaluate the time-averaged powers dissipated in R and G throughout the length of the transmission line by integrating (8.14) and (8.15) from $x = 0$ to $x = \infty$. Hence,

$$P_{\text{dis},R} = \int_0^\infty \tfrac{1}{2}R\,|I_1|^2\,e^{-2\alpha x}\,dx = \frac{R\,|I_1|^2}{4\alpha}, \tag{8.16}$$

$$P_{\text{dis},G} = \int_0^\infty \tfrac{1}{2}G\,|V_1|^2\,e^{-2\alpha x}\,dx = \frac{G\,|V_1|^2}{4\alpha}. \tag{8.17}$$

Thus, we find the total time-averaged power dissipated throughout the length of the transmission line to be

$$P_{\text{dis}} = P_{\text{dis},R} + P_{\text{dis},G} = \frac{1}{4\alpha}[R\,|I_1|^2 + G\,|V_1|^2], \tag{8.18}$$

$$= \frac{|V_1|^2}{|Z_0|^2}\frac{1}{4\alpha}[R + G\,|Z_0|^2]. \tag{8.19}$$

In order that the total time-averaged power dissipated along the transmission line be equal to the total time-averaged power input, we see from (8.11) and (8.19) that the following relation needs to be established:

$$R_0 = [R + G\,|Z_0|^2]/2\alpha. \tag{8.20}$$

With the help of (6.1) and (6.6), we can prove the relation (8.20). However, (8.20) can be derived in a somewhat more elegant manner by starting from (5.7) to (5.9).

On multiplying both sides of (5.7) by $I^*(x)$, taking the complex conjugate of (5.8), multiplying both sides of the resulting equation by $V(x)$ and then adding the two equations, we obtain

$$-\left[I^*(x)\frac{d}{dx}V(x) + V(x)\frac{d}{dx}I^*(x)\right] = z\,|I(x)|^2 + y^*\,|V(x)|^2. \tag{8.21}$$

The left-hand side of (8.21) is simplified to $(d/dx)\{V(x)I^*(x)\}$. We integrate both sides of (8.21) with respect to x from $x = 0$ to $x = \infty$ and take their real parts. Then, on using (5.9), we obtain the following result:

$$\text{Re}\left\{-\int_0^\infty \frac{d}{dx}[V(x)I^*(x)]\,dx\right\} = -\text{Re}\,[V(x)I^*(x)]\Big|_{x=0}^{x=\infty},$$

$$= \int_0^\infty [R\,|I(x)|^2 + G\,|V(x)|^2]\,dx. \tag{8.22}$$

Since $V(x)$ and $I(x)$ vanish at $x = \infty$, we find from (6.2) that

$$-\text{Re}\,[V(x)I^*(x)]\Big|_{x=0}^{x=\infty} = \text{Re}\,[V(0)I^*(0)] = \frac{|V_1|^2}{|Z_0|^2}R_0. \tag{8.23}$$

With the help of (6.2), we can show that

$$\int_0^\infty R\,|I(x)|^2\,dx = R\int_0^\infty \frac{|V_1|^2}{|Z_0|^2}\,e^{-2\alpha x}\,dx = \frac{R\,|V_1|^2}{2\alpha\,|Z_0|^2}, \tag{8.24}$$

$$\int_0^\infty G\,|V(x)|^2\,dx = G\int_0^\infty |V_1|^2\,e^{-2\alpha x}\,dx = \frac{G\,|V_1|^2}{2\alpha}. \tag{8.25}$$

The substitution of (8.23) to (8.25) in (8.22) yields

$$R_0 = \frac{1}{2\alpha}\,[R + G\,|Z_0|^2], \tag{8.26}$$

which according to (8.20) is the required result. Thus, the time-averaged input power to the transmission line is entirely dissipated in the series resistor R and the shunt conductance G throughout the length of the transmission line.

From (6.2), we see that

$$\frac{V_1}{I_1} = Z_0 = R_0 + jX_0 = \frac{|V_1|}{|I_1|}\,e^{j(\theta_{V_1} - \theta_{I_1})}. \tag{8.27}$$

On equating separately the real and the imaginary parts of (8.27), we find that

$$R_0 = \frac{|V_1|}{|I_1|}\cos(\theta_{V_1} - \theta_{I_1}); \qquad X_0 = \frac{|V_1|}{|I_1|}\sin(\theta_{V_1} - \theta_{I_1}). \tag{8.28a,b}$$

In view of (6.6), the input current $I(0, t)$ as given in (8.4) is associated with an equivalent impedance Z_0 with a resistive part R_0 and a reactive part X_0. Consequently, the real or the irretrievable power is that dissipated in the resistor R_0 and is therefore given by

$$P_{irr}(t) = R_0 I^2(0, t) = R_0\,|I_1|^2\cos^2(\omega t + \theta_{I_1}). \tag{8.29}$$

With the help of (8.28a), we can replace $R_0\,|I_1|$ in (8.29) by $|V_1|\cos(\theta_{V_1} - \theta_{I_1})$ to obtain the following result:

$$P_{irr}(t) = |V_1|\,|I_1|\cos(\theta_{V_1} - \theta_{I_1})\cos^2(\omega t + \theta_{I_1}). \tag{8.30}$$

From (8.5) and (8.30), we see that

$$\begin{aligned} P(t) - P_{irr}(t) = |V_1|\,|I_1|\cos(\omega t + \theta_{I_1})[&\cos(\omega t + \theta_{V_1}) \\ &- \cos(\theta_{V_1} - \theta_{I_1})\cos(\omega t + \theta_{I_1})]. \end{aligned} \tag{8.31}$$

Since

$$\begin{aligned} \cos(\omega t + \theta_{V_1}) &= \cos(\omega t + \theta_{I_1} + \theta_{V_1} - \theta_{I_1}) \\ &= \cos(\omega t + \theta_{I_1})\cos(\theta_{V_1} - \theta_{I_1}) - \sin(\omega t + \theta_{I_1})\sin(\theta_{V_1} - \theta_{I_1}), \end{aligned} \tag{8.32}$$

(8.31) becomes

$$P(t) - P_{irr}(t) = -|V_1|\,|I_1|\sin(\theta_{V_1} - \theta_{I_1})\cos(\omega t + \theta_{I_1})\sin(\omega t + \theta_{I_1}). \tag{8.33}$$

Also, from (8.10)

$$P_{\text{real}} = \tfrac{1}{2}\,\text{Re}\,(V_1 I_1^*) = \frac{|V_1|\,|I_1|}{2}\cos(\theta_{V_1} - \theta_{I_1}). \tag{8.34}$$

In analogy with (8.34), we can set

$$P_{\text{reactive}} = \tfrac{1}{2}\,\text{Im}\,(V_1 I_1^*) = \frac{|V_1|\,|I_1|}{2}\sin(\theta_{V_1} - \theta_{I_1}), \tag{8.35}$$

where Im stands for the imaginary part. We shall clarify the physical meaning of P_{reactive} as given in (8.35) presently. With the help of (8.30), and (8.33) to (8.35), we can rearrange the instantaneous power $P(t)$ as

$$P(t) = P_{\text{real}}\,2\cos^2(\omega t + \theta_{I_1}) - P_{\text{reactive}}\,2\cos(\omega t + \theta_{I_1})\sin(\omega t + \theta_{I_1}), \tag{8.36}$$

which is amenable to the following physical interpretation. The first term on the right-hand side of (8.36) is the irretrievable power dissipated in the resistor and is always positive. We can obtain the time average of the irretrievable power by using (8.6) to be given by P_{real}, which is the time-averaged power input. Note also that P_{real} is equal to half the range of variation of the irretrievable power. The second term on the right-hand side of (8.36) can be either positive or negative; it represents the power delivered to or abstracted from the reactor and we therefore call it the reversible power. In analogy with P_{real}, one half the range of variation of the reversible power is equal to P_{reactive}, which is called the reactive power since it is a measure of the power oscillation. It is conventional to combine the time-averaged power P_{real} and the measure of the power oscillation, namely the reactive power P_{reactive}, into the following complex form:

$$\begin{aligned} P_{\text{complex}} = P_{\text{real}} + jP_{\text{reactive}} &= \tfrac{1}{2}\,\text{Re}\,(V_1 I_1^*) + j\tfrac{1}{2}\,\text{Im}\,(V_1 I_1^*), \\ &= \tfrac{1}{2} V_1 I_1^*, \end{aligned} \tag{8.37}$$

which is called the complex power.

2.9 TRANSMISSION LINE OF FINITE LENGTH

For convenience, we reproduce the general expression for the voltage as given in (5.13) and that for the current as given in (5.18) for a transmission line of finite length as follows:

$$V(x) = V_1 e^{-\gamma x} + V_2 e^{\gamma x}, \tag{9.1}$$

$$I(x) = \frac{V_1}{Z_0} e^{-\gamma x} - \frac{V_2}{Z_0} e^{\gamma x}, \tag{9.2}$$

where the propagation coefficient γ and the characteristic impedance Z_0 are given respectively in (6.1) and (6.6). Note that x denotes the distance along the transmission line. The constants V_1 and V_2 are arbitrary. We need two terminal conditions to determine them.

For example, if the input ($x = 0$) voltage V_0 and the input current I_0 are specified, we can determine V_1 and V_2. From (9.1) and (9.2), we obtain the input voltage

$V_0 = V\,(x = 0)$ and the input current $I_0 = I\,(x = 0)$ as:

$$V_1 + V_2 = V_0, \tag{9.3}$$

$$\frac{V_1}{Z_0} - \frac{V_2}{Z_0} = I_0. \tag{9.4}$$

The results of the solution of the simultaneous equations (9.3) and (9.4) are

$$V_1 = \tfrac{1}{2}(V_0 + Z_0 I_0); \qquad V_2 = \tfrac{1}{2}(V_0 - Z_0 I_0). \tag{9.5}$$

Substituting (9.5) in (9.1) and (9.2), we find

$$V(x) = \tfrac{1}{2}(V_0 + Z_0 I_0)e^{-\gamma x} + \tfrac{1}{2}(V_0 - Z_0 I_0)e^{\gamma x}, \tag{9.6}$$

$$I(x) = \frac{1}{2Z_0}(V_0 + Z_0 I_0)e^{-\gamma x} - \frac{1}{2Z_0}(V_0 - Z_0 I_0)e^{\gamma x}. \tag{9.7}$$

Alternatively, we can express (9.6) and (9.7) in terms of the hyperbolic functions as

$$V(x) = V_0 \cosh \gamma x - Z_0 I_0 \sinh \gamma x, \tag{9.8}$$

$$I(x) = -\frac{V_0}{Z_0} \sinh \gamma x + I_0 \cosh \gamma x, \tag{9.9}$$

where

$$\cosh y = \frac{e^y + e^{-y}}{2}; \qquad \sinh y = \frac{e^y - e^{-y}}{2}. \tag{9.10}$$

Suppose that the transmission line extends from $x = 0$ to $x = l$. Instead of the input voltage and current, let us specify the voltage and the current at the end $x = l$ of the transmission line. Assume $V_L = V\,(x = l)$ and $I_L = I\,(x = l)$. For convenience, we call V_L and I_L the load voltage and the load current respectively. From (9.1) and (9.2), we find the load voltage and the load current to be

$$V_1 e^{-\gamma l} + V_2 e^{\gamma l} = V_L, \tag{9.11}$$

$$\frac{V_1}{Z_0} e^{-\gamma l} - \frac{V_2}{Z_0} e^{\gamma l} = I_L. \tag{9.12}$$

The solution of the simultaneous equations (9.11) and (9.12) for V_1 and V_2 yields

$$V_1 = \tfrac{1}{2}(V_L + Z_0 I_L)e^{\gamma l}; \qquad V_2 = \tfrac{1}{2}(V_L - Z_0 I_L)e^{-\gamma l}. \tag{9.13}$$

Suppose that the impedance terminating the transmission line at $x = l$ is Z_L, which is called the load impedance. Then,

$$V_L = Z_L I_L. \tag{9.14}$$

In view of (9.14), instead of the load voltage V_L, we may specify the load impedance Z_L. Thus, if the two specified terminal conditions are the load current and the load impedance, with the help of (9.14), (9.13) becomes

$$V_1 = \frac{I_L}{2}(Z_L + Z_0)e^{\gamma l}; \qquad V_2 = \frac{I_L}{2}(Z_L - Z_0)e^{-\gamma l}. \tag{9.15}$$

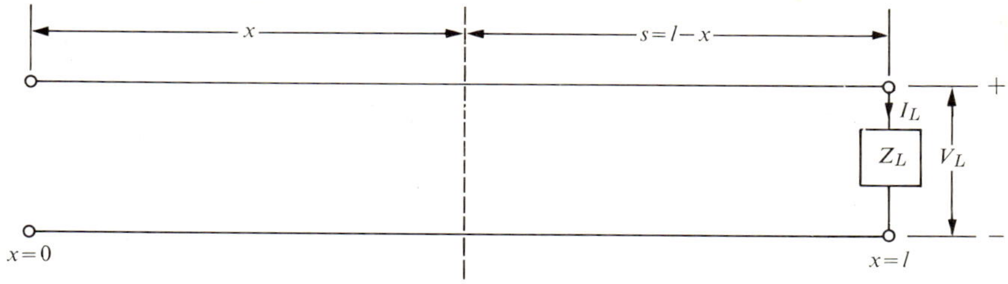

Fig. 2.9 Finite length transmission line with the load current I_L and the load impedance Z_L.

Substituting (9.15) in (9.1) and (9.2), we obtain the following results:

$$V(x) = \frac{I_L}{2}[(Z_L + Z_0)e^{\gamma(l-x)} + (Z_L - Z_0)e^{-\gamma(l-x)}], \tag{9.16}$$

$$I(x) = \frac{I_L}{2Z_0}[(Z_L + Z_0)e^{\gamma(l-x)} - (Z_L - Z_0)e^{-\gamma(l-x)}]. \tag{9.17}$$

We can simplify the writing of (9.16) and (9.17) by setting

$$s = l - x. \tag{9.18}$$

From Fig. 2.9, we can see that if x measures the distance to a particular point on the transmission line from the input ($x = 0$) terminals, then s measures the distance to the same point from the load ($x = l$) terminals. We can also write the voltage and the current distributions in (9.16) and (9.17) respectively in terms of the hyperbolic functions defined in (9.10) with the following results:

$$V(x) = I_L[Z_L \cosh \gamma s + Z_0 \sinh \gamma s], \tag{9.19}$$

$$I(x) = \frac{I_L}{Z_0}[Z_L \sinh \gamma s + Z_0 \cosh \gamma s]. \tag{9.20}$$

If the transmission line parameters γ, Z_0, and l are given and the load impedance Z_L and the load current I_L are specified, we can obtain the voltage and the current at any point along the transmission line with the help of (9.19) and (9.20).

2.10 INPUT IMPEDANCE

Let us consider a transmission line of finite length extending from $x = 0$ to $x = l$. Let the transmission line be terminated at $x = l$ by the load impedance Z_L. Let the current in the load impedance be I_L. Furthermore, assume that the transmission line is driven by a generator of voltage V_g and internal impedance Z_g in series with the input terminals, as shown in Fig. 2.10(a). If the impedance at the input terminals looking into the transmission line is Z_{in}, then the equivalent of the electric circuit near the input terminals is as shown in Fig. 2.10(b), with the help of which the input

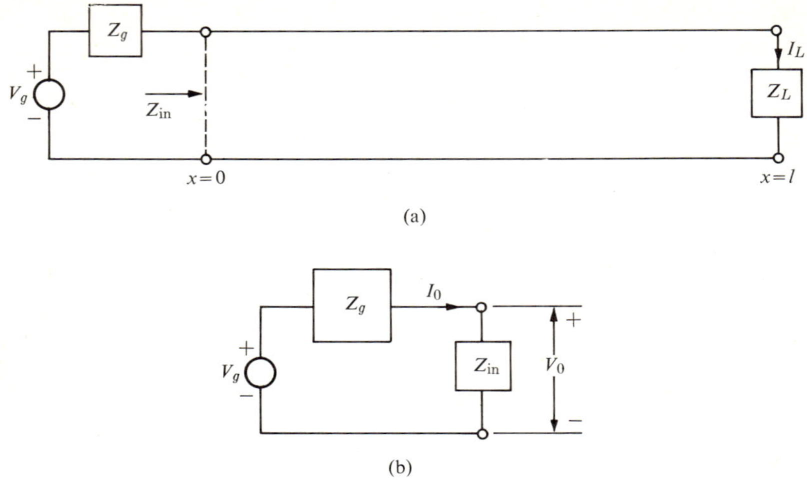

Fig. 2.10 (a) Finite length transmission line with a voltage generator V_g in series with the impedance Z_g connected to the input terminals at $x = 0$. The load impedance Z_L and the load current I_L are specified. (b) Equivalent electric circuit for finding the input voltage V_0 and the input current I_0.

current I_0 and the input voltage V_0 are obtained as follows:

$$I_0 = \frac{V_g}{Z_g + Z_{in}}; \qquad V_0 = \frac{V_g Z_{in}}{Z_g + Z_{in}}. \tag{10.1}$$

With the knowledge of I_0 and V_0, we can use (9.6) and (9.7) to completely determine the current and the voltage distributions along the transmission line. In view of the fact that the input impedance plays a major role in the determination of the input current and the input voltage as well as the voltage and the current distributions along the transmission line, the understanding of the characteristics of the input impedance assumes importance and warrants detailed consideration. From (9.16) to (9.18), we can obtain the impedance $Z(x) = V(x)/I(x)$ at any point on the transmission line looking towards the load:

$$Z(x) = Z_0 \frac{[(Z_L + Z_0)e^{\gamma s} + (Z_L - Z_0)e^{-\gamma s}]}{[(Z_L + Z_0)e^{\gamma s} - (Z_L - Z_0)e^{-\gamma s}]}. \tag{10.2}$$

Alternatively, we can use (9.19) and (9.20) to find an expression for $Z(x)$ in terms of the hyperbolic functions as follows:

$$Z(x) = Z_0 \frac{[Z_L \cosh \gamma s + Z_0 \sinh \gamma s]}{[Z_0 \cosh \gamma s + Z_L \sinh \gamma s]}. \tag{10.3}$$

We can also determine another equivalent expression for $Z(x)$ by dividing the numerator and the denominator of (10.3) by cosh γs with the following result:

$$Z(x) = Z_0 \frac{[Z_L + Z_0 \tanh \gamma s]}{[Z_0 + Z_L \tanh \gamma s]}, \tag{10.4}$$

where we obtain the hyperbolic tangent function from (9.10) as follows:

$$\tanh y = \frac{\sinh y}{\cosh y} = \frac{e^y - e^{-y}}{e^y + e^{-y}} = \frac{1 - e^{-2y}}{1 + e^{-2y}}. \tag{10.5}$$

The input impedance Z_{in} is the value of $Z(x)$ at $x = 0$. Since at $x = 0$, $s = l$, we can deduce the following equivalent expressions for the input impedance Z_{in} from (10.2) to (10.4):

$$Z_{\text{in}} = Z_0 \frac{[(Z_L + Z_0)e^{\gamma l} + (Z_L - Z_0)e^{-\gamma l}]}{[(Z_L + Z_0)e^{\gamma l} - (Z_L - Z_0)e^{-\gamma l}]}, \tag{10.6}$$

$$Z_{\text{in}} = Z_0 \frac{[Z_L \cosh \gamma l + Z_0 \sinh \gamma l]}{[Z_0 \cosh \gamma l + Z_L \sinh \gamma l]}, \tag{10.7}$$

$$Z_{\text{in}} = Z_0 \frac{[Z_L + Z_0 \tanh \gamma l]}{[Z_0 + Z_L \tanh \gamma l]}. \tag{10.8}$$

It is instructive to consider the characteristics of the input impedance for various special cases.

Case 1. $Z_L = Z_0$. When the load impedance is equal to the characteristic impedance, we see from (10.6) that

$$Z_{\text{in}} = Z_0. \tag{10.9}$$

It follows from (9.15) that $V_2 = 0$ for $Z_L = Z_0$ with the consequence that for this special case the voltage and the current distributions do not have the constituent whose amplitude decays, and whose phase progresses, in the negative x-direction. In this respect, the transmission line terminated at $x = l$ by the characteristic impedance is equivalent to a semi-infinite transmission line extending up to $x = \infty$. In the region from $x = 0$ to $x = l$, the voltage and the current distributions on the semi-infinite transmission line are the same as those on a finite transmission line terminated at $x = l$ by the characteristic impedance. Consequently, it can be verified from (6.6) and (10.9) that the input impedances of the two transmission lines are the same.

Case 2. $Z_L = \infty$. At $x = l$, if the terminals of the transmission line are left open-circuited, then from (10.8) we find that

$$Z_{\text{in}} = Z_0/\tanh \gamma l = Z_0 \coth \gamma l \qquad \text{(open-circuit termination)}, \tag{10.10}$$

where coth, the hyperbolic cotangent function, is obtained from (10.5) as

$$\coth y = 1/\tanh y = \frac{e^y + e^{-y}}{e^y - e^{-y}} = \frac{1 + e^{-2y}}{1 - e^{-2y}}. \tag{10.11}$$

Let us consider first the lossless case. It follows from (7.1) and (7.9) that

$$Z_0 = R_0 \text{ and } \gamma = j\beta; \qquad R_0 = \sqrt{L/C} \text{ and } \beta = \omega\sqrt{LC}. \tag{10.12}$$

Also, from (10.11)

$$\coth j\beta l = \frac{e^{j\beta l} + e^{-j\beta l}}{e^{j\beta l} - e^{-j\beta l}} = -j \cot \beta l. \tag{10.13}$$

Substituting (10.12) and (10.13) in (10.10), we find

$$Z_{in} = -jR_0 \cot \beta l. \tag{10.14}$$

Since $\cot \beta l$ is real, it follows from (10.14) that the input impedance of an open-circuited lossless transmission line is always reactive. When $\beta l = 2\pi l/\lambda \ll 1$, $\cot \beta l = 1/\beta l$ and (10.14) becomes

$$Z_{in} = -\frac{jR_0}{\beta l} = -j\sqrt{\frac{L}{C}}\frac{1}{\omega\sqrt{LC}\,l} = -\frac{j}{\omega Cl}. \tag{10.15}$$

Thus, a lossless transmission line of length l, very short in comparison to a wavelength and open-circuited at its ends, behaves like a capacitance of magnitude Cl. At high frequencies we can use such a transmission line as a circuit element to realize a small capacitive reactance.

In Fig. 2.11, we show $-\cot \beta l$ as a function of βl. If

$$(n - 1)\pi < \beta l < (2n - 1)\pi/2,$$

where n is an integer, $-\cot \beta l$ is negative and Z_{in} is a capacitive reactance. Since $\beta = 2\pi/\lambda$, we can also state the condition for realizing a capacitive reactance in terms of the wavelength as follows: $(n - 1)\lambda/2 < l < (2n - 1)\lambda/4$. In a similar manner, if $(2n - 1)\pi/2 < \beta l < n\pi$ or equivalently $(2n - 1)\lambda/4 < l < n\lambda/2$, $-\cot \beta l$ is positive and Z_{in} acts like an inductive reactance. Since $-\cot \beta l$ assumes all values from $-\infty$ to ∞, it follows that any reactance value can be realized by a suitable length of a lossless transmission line open-circuited at its terminals.

For $l = (2n - 1)\lambda/4$ the reactance becomes zero, which defines the condition for "resonance". Similarly, if $l = n\lambda/2$, the reactance becomes infinite which defines the condition for "antiresonance". For "antiresonant" lengths of the transmission line, since the reactance becomes infinite, we cannot ignore the losses. We shall show subsequently that the inclusion of losses forces the reactance to remain finite.

Let us consider a transmission line with very slight losses. Then, Z_0 has both a real part and an imaginary part. Since γ is complex, it follows from (10.11) that $\coth \gamma l$ is also complex. Consequently, Z_{in} is complex, that is, the input impedance of a lossy transmission line open-circuited at the terminals at $x = l$ has both a resistance R_{in} and a reactance X_{in}. Suppose $|\gamma|\, l \ll 1$, then $\coth \gamma l = 1/\gamma l$ and (10.10) becomes

$$Z_{in} = \frac{Z_0}{\gamma l} = \sqrt{\frac{R + j\omega L}{G + j\omega C}}\frac{1}{\sqrt{(R + j\omega L)(G + j\omega C)}\,l} = \frac{1}{(G + j\omega C)l}. \tag{10.16}$$

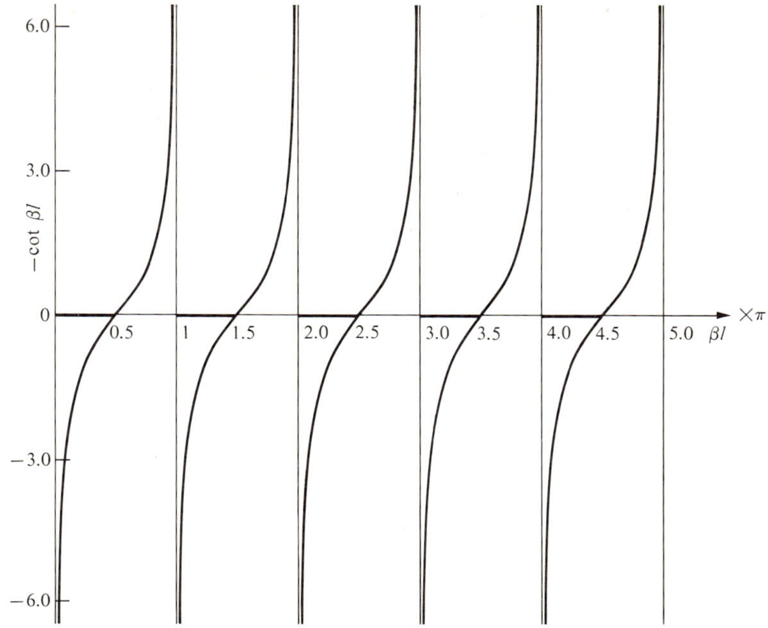

Fig. 2.11 −Cot βl as a function of βl.

Thus, a slightly lossy transmission line of length l, very short in comparison to a wavelength and open-circuited at its ends, behaves like a capacitance of magnitude Cl in parallel with a conductance Gl. In other words, the behavior is like a slightly leaky capacitance Cl.

When βl is not close to $(2n - 1)\pi/2$ or $n\pi$, that is, when the length of the transmission line is substantially different from that pertaining to "resonance" or "antiresonance" respectively, we can ignore the losses and the input impedance which is very close to that given by (10.14) is essentially reactive. When βl is very close to $(2n - 1)\pi/2$ or $n\pi$, that is, for resonant or antiresonant lengths respectively, we have to obtain the input impedance behavior directly from (10.10), which leads to many interesting features. The "resonance" and "antiresonance" features of an open-circuited, slightly lossy transmission line are discussed in the next section.

Case 3. $Z_L = 0$. At $x = l$, if the terminals of the transmission line are left short-circuited, (10.8) gives

$$Z_{\text{in}} = Z_0 \tanh \gamma l \qquad \text{(short-circuit termination)}, \tag{10.17}$$

where the hyperbolic tangent function is as defined in (10.5). For the lossless case, the relations given in (10.12) hold good. Also, from (10.5)

$$\tanh j\beta l = \frac{e^{j\beta l} - e^{-j\beta l}}{e^{j\beta l} + e^{-j\beta l}} = j \tan \beta l. \tag{10.18}$$

Hence, (10.17) reduces to

$$Z_{in} = jR_0 \tan \beta l. \tag{10.19}$$

Just as for the open-circuit case, we see from (10.19) that, for the short-circuit termination also, the input impedance of a lossless transmission line is always reactive. When $\beta l = 2\pi l/\lambda \ll 1$, $\tan \beta l = \beta l$ and (10.19) becomes

$$Z_{in} = jR_0\beta l = j\sqrt{(L/C)}\omega\sqrt{LC}\, l = j\omega Ll. \tag{10.20}$$

Therefore, a lossless transmission line of length l, very short in comparison to a wavelength and short-circuited at its ends, behaves like an inductance of magnitude Ll. At high frequencies, such a transmission line can be used as a circuit element to realize a small inductive reactance.

In Fig. 2.12, $\tan \beta l$ is shown as a function of βl. When

$$(n-1)\pi < \beta l < (2n-1)\frac{\pi}{2} \qquad \text{or} \qquad (n-1)\frac{\lambda}{2} < l < (2n-1)\frac{\lambda}{4},$$

$\tan \beta l$ is positive and Z_{in} is an inductive reactance. If $(2n-1)\pi/2 < \beta l < n\pi$ or $(2n-1)\lambda/4 < l < n\lambda/2$, $\tan \beta l$ is negative and Z_{in} acts like a capacitive reactance. Since $\tan \beta l$ assumes all values from $-\infty$ to ∞, it follows that any reactance value can be obtained. Therefore, for realizing any reactance value we can use a suitable

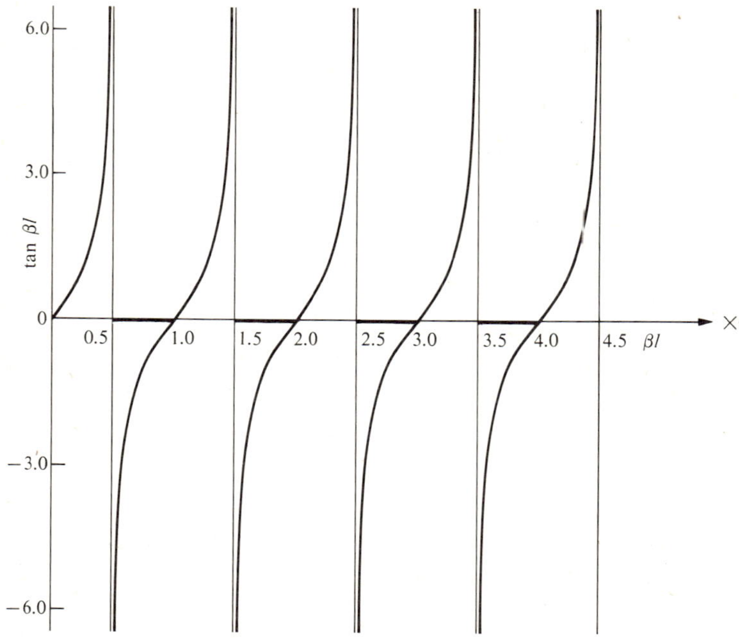

Fig. 2.12 Tan βl as a function of βl.

length of lossless transmission line either open-circuited (as discussed previously) or short-circuited at its ends.

For $\beta l = n\pi$ or $l = n\lambda/2$, there is "resonance" since the reactance becomes zero. Similarly, "antiresonance" is obtained, when $\beta l = (2n - 1)\pi/2$ or $l = (2n - 1)\lambda/4$. Let l_{OC} and l_{SC} be the lengths of the open-circuited and short-circuited lossless transmission line respectively. Their input impedances are obtained from (10.14) and (10.19) as

$$[Z_{in}]_{OC} = -jR_0 \cot \beta l_{OC}; \qquad [Z_{in}]_{SC} = jR_0 \tan \beta l_{SC}. \tag{10.21}$$

First, we can verify that the input impedances as given in (10.21) are not changed if the lengths are changed by $\lambda/2$. Let

$$l_{OC} = l_{SC} + \frac{\lambda}{4} \qquad \text{or} \qquad \beta l_{OC} = \beta l_{SC} + \frac{\pi}{2}. \tag{10.22}$$

Note that $\cot \beta l_{OC} = \cot [\beta l_{SC} + (\pi/2)] = -\tan \beta l_{SC}$. Therefore, $[Z_{in}]_{OC} = [Z_{in}]_{SC}$. Consequently, we see that the input impedances of the open-circuited and short-circuited transmission lines are the same provided their lengths differ by $\lambda/4$. The resonant and the antiresonant lengths in each case differ by $\lambda/4$. It is for this reason that the resonant and antiresonant lengths for the open-circuited termination correspond respectively to the antiresonant and resonant lengths for the short-circuited termination.

If the losses in the transmission line are included, as in the open-circuit termination, we can see from (10.17) that the input impedance of the short-circuit termination also has a resistive part R_{in} and a reactive part X_{in}. If $|\gamma|\, l \ll 1$, then $\tanh \gamma l = \gamma l$ and (10.17) becomes

$$Z_{in} = Z_0\gamma l = \sqrt{\frac{R + j\omega L}{G + j\omega C}} \sqrt{(R + j\omega L)(G + j\omega C)}\, l = (R + j\omega L)l. \tag{10.23}$$

Therefore, a slightly lossy transmission line of length l, very short in comparison to a wavelength and short-circuited at its ends, behaves like an inductance of magnitude Ll in series with a resistance Rl, that is, the behavior is like a lossy inductance Ll.

As in the case of the open-circuited termination, we can ignore the losses in the transmission line in the case of the short-circuited termination also, except in the neighborhood of "resonance" and "antiresonance". For the lossless case, the input impedance is given by (10.19) and it is reactive. For resonant and antiresonant lengths, the input impedance behavior has to be determined directly from (10.17). The "resonance" and "antiresonance" features of a short-circuited transmission line are similar to those obtained for the case of an open-circuited termination.

Two further special cases of the input impedance are important and these correspond to a lossless transmission line for which the relations in (10.12) hold good. Since

$$\cosh j\beta l = \frac{e^{j\beta l} + e^{-j\beta l}}{2} = \cos \beta l \tag{10.24a}$$

and

$$\sinh j\beta l = \frac{e^{j\beta l} - e^{-j\beta l}}{2} = j \sin \beta l, \tag{10.24b}$$

the expression for the input impedance of a lossless transmission line is obtained from (10.7) as

$$Z_{\text{in}} = R_0 \frac{[Z_L \cos \beta l + jR_0 \sin \beta l]}{[R_0 \cos \beta l + jZ_L \sin \beta l]}. \tag{10.25}$$

Case 4. $l = n\lambda/2$ *or* $\beta l = n\pi$. For a lossless transmission line whose length is integer multiples of $\lambda/2$, $\beta l = n\pi$ and $\sin \beta l = 0$ and $\cos \beta l = (-1)^n$. Therefore, from (10.25)

$$Z_{\text{in}} = Z_L \quad \text{for} \quad l = n\lambda/2 \quad \text{or} \quad \beta l = n\pi. \tag{10.26}$$

Thus, a lossless transmission line of length equal to integer multiples of $\lambda/2$ effectively transfers the load impedance as is to the input terminals.

Case 5. $l = (2n - 1)\lambda/4$ *or* $\beta l = (2n - 1)\pi/2$. Then, $\sin \beta l = (-1)^{n-1}$ and $\cos \beta l = 0$, and therefore, we find from (10.25) that

$$Z_{\text{in}} = R_0^2/Z_L \quad \text{for} \quad l = (2n - 1)\lambda/4 \quad \text{or} \quad \beta l = (2n - 1)\pi/2. \tag{10.27}$$

A lossless transmission line of length equal to odd integer multiples of $\lambda/4$, in effect, transfers the load impedance to the input terminals as its inverse multiplied by the square of the characteristic resistance and is usually referred to as a quarter-wave transformer.

The use of the transmission line as a circuit element at high frequencies and as an impedance transformer is borne out clearly from the foregoing treatment. We shall investigate further applications of the transmission line as a circuit element and as an impedance transformer later.

2.11 RESONANCE AND ANTIRESONANCE IN TRANSMISSION LINE CIRCUITS

Let us consider a resonant circuit consisting of an inductance in series with a capacitance. At resonance, the total reactance of the circuit vanishes. Unless we take the losses in the circuit into account, the impedance is zero at resonance. Consequently, we cannot ignore the losses at resonance, with the result that the impedance at resonance has a small residual resistance. We wish to find this resistance due to losses at the resonant condition of a slightly lossy transmission line open-circuited at its terminals. Using (10.10) and (10.11), we can express the input impedance of an open-circuited transmission line in the form:

$$Z_{\text{in}} = (R_0 + jX_0) \frac{[1 + e^{-2(\alpha + j\beta)l}]}{[1 - e^{-2(\alpha + j\beta)l}]}. \tag{11.1}$$

The result of multiplying the numerator and the denominator of (11.1) by the complex conjugate of the denominator, namely $1 - e^{-2(\alpha - j\beta)l}$, becomes on simplification

$$Z_{\text{in}} = (R_0 + jX_0)\frac{[1 - e^{-4\alpha l} - e^{-2\alpha l}2j \sin 2\beta l]}{[1 + e^{-4\alpha l} - e^{-2\alpha l}2 \cos 2\beta l]}. \tag{11.2}$$

The real and the imaginary parts of (11.2) are

$$\frac{R_{\text{in}}}{R_0} = \frac{(1 - e^{-4\alpha l}) + 2(X_0/R_0)e^{-2\alpha l} \sin 2\beta l}{[1 + e^{-4\alpha l} - e^{-2\alpha l}2 \cos 2\beta l]}, \tag{11.3}$$

$$\frac{X_{\text{in}}}{R_0} = \frac{(X_0/R_0)(1 - e^{-4\alpha l}) - 2e^{-2\alpha l} \sin 2\beta l}{[1 + e^{-4\alpha l} - e^{-2\alpha l}2 \cos 2\beta l]}. \tag{11.4}$$

Since the losses are small, it is legitimate to assume that $\alpha l \ll 1$ and (11.3) and (11.4) can then be considerably simplified. On expanding $e^{-2\alpha l}$ and $e^{-4\alpha l}$ up to terms of order $\alpha^2 l^2$, the denominator D in (11.3) and (11.4) becomes on rearrangement

$$\begin{aligned} D &= 1 + 1 - 4\alpha l + 8\alpha^2 l^2 - 2 \cos 2\beta l(1 - 2\alpha l + 2\alpha^2 l^2), \\ &= (1 - 2\alpha l + 2\alpha^2 l^2)2(1 - \cos 2\beta l) + 4\alpha^2 l^2. \end{aligned} \tag{11.5}$$

Since $1 - \cos 2\beta l = 2 \sin^2 \beta l$, (11.5) can be further simplified to give

$$D = 4[(1 - 2\alpha l + 2\alpha^2 l^2) \sin^2 \beta l + \alpha^2 l^2]. \tag{11.6}$$

It is convenient to rewrite (11.6) as

$$D = 4(D_0 \sin^2 \beta l + \alpha^2 l^2); \qquad D_0 = 1 - 2\alpha l + 2\alpha^2 l^2. \tag{11.7}$$

Note that for small losses X_0 is small in comparison with R_0, that is, $X_0/R_0 \ll 1$. We assume that αl and X_0/R_0 are small quantities of approximately the same order. The exponentials in the numerators of (11.3) and (11.4) are expanded into power series in αl with the following results:

$$(1 - e^{-4\alpha l}) + 2\frac{X_0}{R_0} e^{-2\alpha l} \sin 2\beta l = 2\left(2\alpha l + \frac{X_0}{R_0} \sin 2\beta l\right), \tag{11.8}$$

$$\frac{X_0}{R_0}(1 - e^{-4\alpha l}) - 2e^{-2\alpha l} \sin 2\beta l = 2\left(2\frac{X_0}{R_0}\alpha l - D_0 \sin 2\beta l\right). \tag{11.9}$$

In (11.8) we retain only the first nonvanishing terms which are linear in the small order quantities X_0/R_0 and αl. But, in (11.9), terms which are up to quadratic order in X_0/R_0 and αl are retained. With the help of (11.7) to (11.9), the approximate expressions for R_{in} and X_{in} in the limit of small losses become

$$\frac{R_{\text{in}}}{R_0} = \frac{2\alpha l + (X_0/R_0) \sin 2\beta l}{2(D_0 \sin^2 \beta l + \alpha^2 l^2)}, \tag{11.10}$$

$$\frac{X_{\text{in}}}{R_0} = \frac{2(X_0/R_0)\alpha l - D_0 \sin 2\beta l}{2(D_0 \sin^2 \beta l + \alpha^2 l^2)}. \tag{11.11}$$

First, we must verify that (11.10) and (11.11) reduce to the correct expressions in the limit of zero losses. When there are no losses $\alpha = X_0 = 0$ and (11.10) shows that $R_{\text{in}} = 0$. Since $\sin 2\beta l = 2 \sin \beta l \cos \beta l$, (11.11) shows that $X_{\text{in}} = -R_0 \cot \beta l$. These results are in conformity with (10.14). In the lossless case resonance corresponds to $\beta l = (2n - 1)\pi/2$, and with the inclusion of small losses, the resonance condition is changed slightly to $\beta l = (2n - 1)\pi/2 + \theta_r$ where θ_r is a small quantity. We can find the value of θ_r for which $X_{\text{in}} = 0$ in the following manner. Since

$$\begin{aligned}\sin 2\beta l &= \sin [(2n - 1)\pi + 2\theta_r] = \sin (2n - 1)\pi \cos 2\theta_r + \cos (2n - 1)\pi \sin 2\theta_r \\ &= -\sin 2\theta_r = -2\theta_r, \end{aligned} \tag{11.12}$$

we see from (11.11) that $X_{\text{in}} = 0$ occurs at

$$2(X_0/R_0)\alpha l - D_0 \sin 2\beta l = 0 = 2(X_0/R_0)\alpha l + 2\theta_r D_0.$$

Therefore

$$\theta_r = -\frac{X_0 \alpha l}{R_0 D_0} \simeq -\frac{X_0 \alpha l}{R_0}. \tag{11.13}$$

The resonance occurs at $\beta l = (2n - 1)\pi/2 - (X_0 \alpha l/R_0)$. For this value of βl, $\sin^2 \beta l = 1$ and therefore we obtain the input resistance at resonance from (11.10), (11.12), and (11.13) as

$$\frac{R_{\text{in}}}{R_0} = \frac{(2\alpha l + 2(X_0^2/R_0^2)\alpha l)}{2(D_0 + \alpha^2 l^2)}. \tag{11.14}$$

If (11.14) is expanded into a power series in X_0/R_0 and αl, the result of retaining only the leading term is

$$R_{\text{in}}/R_0 = \alpha l. \tag{11.15}$$

The inclusion of losses results in the slight shifting of the resonant condition and in a nonvanishing value (11.15) for the input resistance.

The antiresonance behavior of (11.10) and (11.11) obtained in the neighborhood of $\beta l = n\pi$ has many interesting features. At $\beta l = n\pi$, (11.10) shows that R_{in} has a large value equal to $R_0/\alpha l$. In fact, we can verify that $R_0/\alpha l$ is the maximum value of R_{in} and occurs at $\beta l = n\pi$. In view of the presence of the term $\alpha^2 l^2$ in the denominator of (11.11), X_{in} does not go to infinity at $\beta l = n\pi$. Indeed, we can show now that X_{in} goes to zero in the neighborhood of $\beta l = n\pi$. Let $\beta l = n\pi + \theta_a$, where θ_a is a small quantity, be the value of βl for $X_{\text{in}} = 0$. Since $\sin 2\beta l = \sin (2n\pi + 2\theta_a) = 2\theta_a$, it follows from (11.11) that $X_{\text{in}} = 0$ for $\theta_a = (X_0/R_0)\alpha l$, which is a second-order small quantity. Hence X_{in}, instead of becoming infinite at $\beta l = n\pi$ as is the case in the absence of losses, goes to zero in the close neighborhood of $\beta l = n\pi$, namely $\beta l = n\pi + (X_0/R_0)\alpha l$. In the absence of losses, X_{in} is positively very large for βl slightly less than $n\pi$ and is negatively very large for βl slightly greater than $n\pi$. We have shown that the inclusion of losses forces X_{in}, instead of going to infinity at $\beta l = n\pi$, to go to zero in its close neighborhood $\beta l = n\pi + (X_0 \alpha l/R_0)$. It follows therefore that X_{in} has a maximum for $\beta l < n\pi + (X_0 \alpha l/R_0)$ and a minimum for $\beta l > n\pi + (X_0 \alpha l/R_0)$. We can obtain the values of βl for which these extrema occur and the corresponding maximum and minimum values of X_{in} in the following manner.

Let $\beta l = n\pi + \theta_e$ correspond to the above-mentioned extrema in X_{in}. Note that θ_e is a small quantity. Therefore, $\sin \beta l = (-1)^n \theta_e$ and $\sin 2\beta l = 2\theta_e$ with the result that (11.11) reduces to

$$\frac{X_{in}}{R_0} = \frac{(X_0/R_0)\alpha l - \theta_e D_0}{(D_0\theta_e^2 + \alpha^2 l^2)}. \tag{11.16}$$

On differentiating (11.16) with respect to θ_e and setting the result equal to zero we find the equation for determining θ_e. We can easily obtain the equation satisfied by θ_e:

$$D_0\theta_e^2 - \theta_e 2(X_0/R_0)\alpha l - \alpha^2 l^2 = 0. \tag{11.17}$$

The two solutions of (11.17) are

$$[\theta_e]_{1,2} = \frac{\alpha l[(X_0/R_0) \mp \sqrt{(X_0^2/R_0^2) + D_0}]}{D_0}. \tag{11.18}$$

If we expand (11.18) into a power series in X_0/R_0 and αl, the result of retaining the terms which are up to quadratic order in X_0/R_0 and αl is

$$[\theta_e]_{1,2} = \alpha l[(X_0/R_0) \mp 1 \mp \alpha l]. \tag{11.19}$$

If we substitute (11.19) in (11.16) and the resulting expressions are expanded into power series in X_0/R_0 and αl, the results of retaining the first two orders of terms in X_0/R_0 and αl are

$$\frac{X_{in}}{R_0} = \frac{1}{2\alpha l}\left(1 - \alpha l + \frac{X_0}{R_0}\right) \qquad (\text{for } \theta_e = \theta_{e1}), \tag{11.20}$$

$$\frac{X_{in}}{R_0} = \frac{-1}{2\alpha l}\left(1 - \alpha l - \frac{X_0}{R_0}\right) \qquad (\text{for } \theta_e = \theta_{e2}). \tag{11.21}$$

From (11.20) and (11.21), we can find respectively the maximum and the minimum values of X_{in} in the neighborhood of antiresonance.

In practice, the leakage conductance G is negligible or at any rate $R/L > G/C$ with the result that X_0 is seen from (7.8) to be negative. Since the maximum and the minimum of X_{in}/R_0 near an antiresonance occur at slightly different values of βl, it is difficult to compare the magnitudes of these extrema, as given by (11.20) and (11.21).

The important feature of antiresonance is that the input resistance has a maximum and the input reactance has a rapid variation passing through zero. The resonance and the antiresonance behavior of a slightly lossy transmission line terminated in an open-circuit are shown in Fig. 2.13 for the case $\alpha/\beta = -X_0/R_0 = 0.01$. The curves are computed with the help of the relations (11.10) and (11.11). We can analyze the resonance and the antiresonance behavior of a slightly lossy transmission line with a short-circuit termination in a similar manner starting from (10.17). The results are shown in Fig. 2.14 for the same parameter values as used for the open-circuit

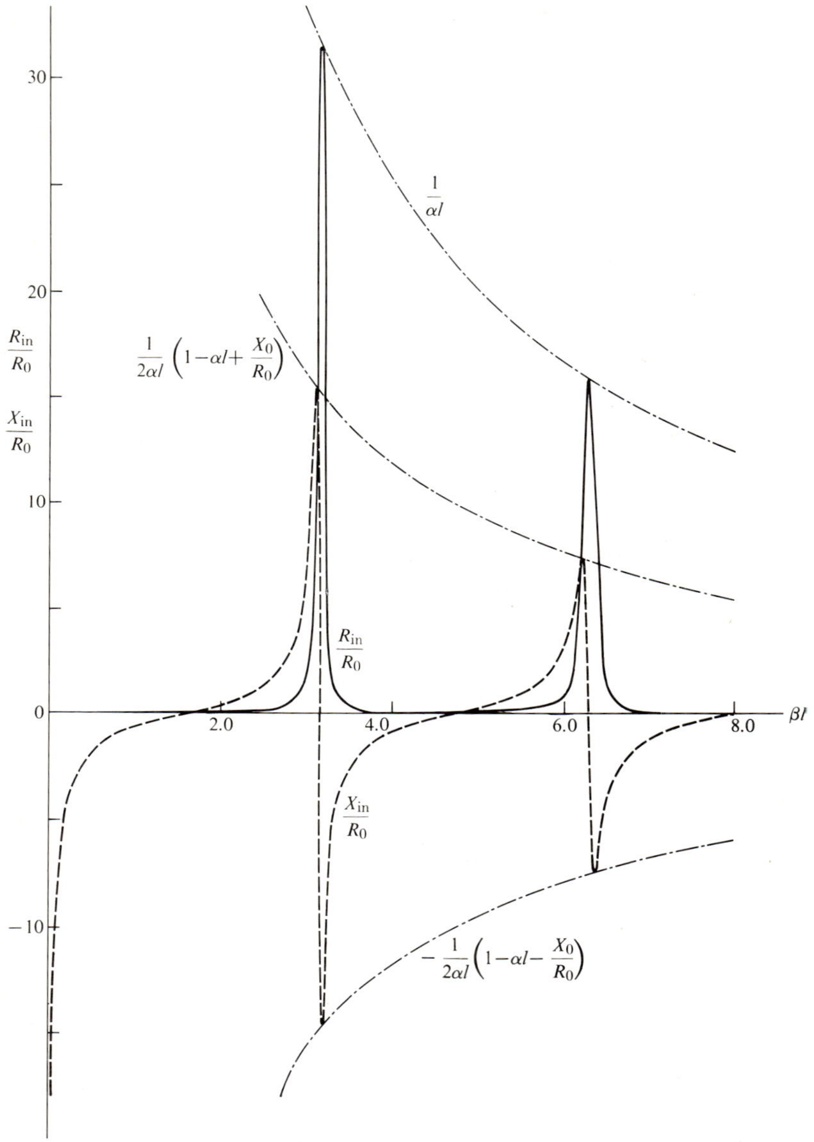

Fig. 2.13 Resonances and antiresonances of a low-loss, open-circuited transmission line $[\alpha/\beta = -X_0/R_0 = 0.01]$.

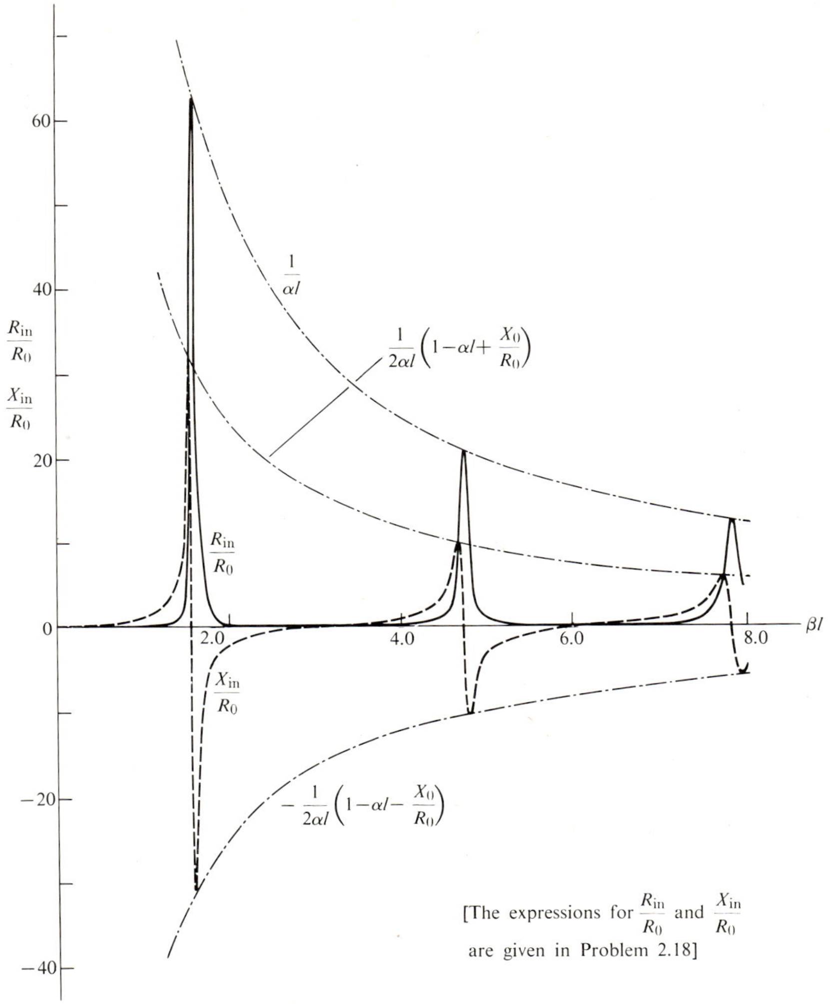

Fig. 2.14 Resonances and antiresonances of a low-loss, short-circuited transmission line $[\alpha/\beta = -X_0/R_0 = 0.01]$.

termination. We see that the curves for the short-circuit termination are similar to those for the open-circuit termination except that the curves appear translated to βl values differing by $\pi/2$.

Selectivity

So far we have discussed the behavior of R_{in} and X_{in} near antiresonance. But if we consider the magnitude of the input impedance $|Z_{in}|$ of a lossy transmission line with open-circuit termination, it has a maximum corresponding to $\beta l = n\pi$ and falls off in value on either side. If the transmission line is driven by a constant current generator applied to the input terminals, the magnitude of the resulting voltage and therefore the associated power have maxima near $\beta l = n\pi$ and fall off if βl is changed in either direction of the value that gives maximum voltage. Since $\beta l = 2\pi f l/v_{ph}$, we can change βl by keeping the length l fixed and changing the frequency. The frequency f_0 corresponding to the maximum voltage is given by $f_0 = nv_{ph}/2l$. The curve of power, which is proportional to $|Z_{\text{in}}|^2$, versus f has a peak at $f = f_0$. Let the "half-power" points, that is, the frequencies where $|Z_{\text{in}}|^2$ or equivalently the power is half the maximum value be given by $f = f_0 \pm (\Delta f/2)$. The ratio of the antiresonant frequency f_0 to the frequency range Δf between the "half-power" points is a measure of the sharpness of the peak and is called the selectivity Q. Hence

$$Q = f_0/\Delta f. \tag{11.22}$$

We wish to find the Q of the open-circuited slightly lossy transmission line.

We find from (10.10) that

$$\begin{aligned} Z_{\text{in}} &= Z_0 \frac{\cosh(\alpha + j\beta)l}{\sinh(\alpha + j\beta)l}, \\ &= Z_0 \frac{[\cosh \alpha l \cos \beta l + j \sinh \alpha l \sin \beta l]}{[\sinh \alpha l \cos \beta l + j \cosh \alpha l \sin \beta l]}. \end{aligned} \tag{11.23}$$

Since $\alpha l \ll 1$,

$$\cosh \alpha l = 1, \qquad \sinh \alpha l = \alpha l, \tag{11.24}$$

are good approximations. Let us consider a frequency $f = f_0 + \delta f$ in the neighborhood of the antiresonance frequency $f = f_0$. Then,

$$\beta l = \frac{2\pi(f_0 + \delta f)l}{v_{ph}} = n\pi + \frac{2\pi\, \delta f l}{v_{ph}} = n\pi + n\pi \frac{\delta f}{f_0}, \tag{11.25}$$

$$\cos \beta l = \cos n\pi \cos(n\pi\, \delta f/f_0) = (-1)^n, \tag{11.26}$$

$$\sin \beta l = \cos n\pi \sin(n\pi\, \delta f/f_0) = (-1)^n n\pi\, \delta f/f_0, \tag{11.27}$$

since $\delta f/f_0$ is a small quantity. The substitution of (11.24), (11.26), and (11.27) in (11.23) and the retention of only the leading terms in the expansions of the numerator and the denominator in powers of the small quantities αl and $n\pi\, \delta f/f_0$ yields

$$Z_{\text{in}} = \frac{Z_0}{\alpha l + jn\pi(\delta f/f_0)}. \tag{11.28}$$

Therefore,

$$|Z_{\text{in}}|^2 = \frac{|Z_0|^2}{\alpha^2 l^2 + (n\pi\, \delta f/f_0)^2}. \tag{11.29}$$

In (11.29) $|Z_{\text{in}}|^2$ has a maximum for $\delta f = 0$, that is for $f = f_0$, and the maximum value is obtained as $|Z_{\text{in}}|^2_{\text{max}} = |Z_0|^2/\alpha^2 l^2$. Hence we can conveniently rewrite

(11.29) as

$$\frac{|Z_{\text{in}}|^2}{|Z_{\text{in}}|^2_{\text{max}}} = \frac{1}{1 + (n\pi\ \delta f/f_0\alpha l)^2}. \tag{11.30}$$

The values of $\delta f = \pm \Delta f/2$ corresponding to the "half-power" points are deduced from (11.30) as

$$(\Delta f/2)n\pi/f_0\alpha l = 1 \quad \text{or} \quad \Delta f = 2f_0\alpha l/n\pi. \tag{11.31}$$

We obtain the expression for the selectivity Q from (11.22) and (11.31) as

$$Q = f_0/\Delta f = n\pi/2\alpha l = \beta/2\alpha \tag{11.32}$$

since at antiresonance $\beta l = n\pi$. For the low-loss case, with the help of only the leading terms in (7.14) and (7.15), we find that

$$Q = \frac{1}{[(R/\omega L) + (G/\omega C)]}. \tag{11.33}$$

We can also deduce the selectivity of the low-loss transmission line with short-circuit termination in a similar fashion and the same results (11.32) and (11.33) are obtained.

Previously, we showed that with suitable lengths, transmission lines with open-circuit or short-circuit terminations can be used to realize inductive or capacitive reactances. We have now established that with $\beta l = (2n - 1)\pi/2$ or $n\pi$ depending on whether the terminals are open- or short-circuited, the equivalents of series and parallel resonant circuits can be realized with the help of transmission lines, which thus have an important application as circuit elements at high frequencies.

2.12 REFLECTION COEFFICIENT AND REFLECTED POWER

Let us consider a transmission line extending from $x = 0$ to $x = l$. The general expressions for the voltage and the current along the transmission line are

$$V(x) = V_1 e^{-\gamma x} + V_2 e^{\gamma x}, \tag{12.1}$$

$$I(x) = I_1 e^{-\gamma x} + I_2 e^{\gamma x}, \tag{12.2}$$

$$I_1 = V_1/Z_0 \quad \text{and} \quad I_2 = -V_2/Z_0. \tag{12.3}$$

These expressions are the same as those given in (5.13), (5.14), (5.15), and (5.17). Z_0 and γ are the characteristic impedance and the propagation coefficient respectively. It is convenient to set

$$V_+ = V_1 e^{-\gamma l} \qquad V_- = V_2 e^{\gamma l}, \tag{12.4}$$

$$I_+ = I_1 e^{-\gamma l} \qquad I_- = I_2 e^{\gamma l}. \tag{12.5}$$

With the help of (12.3) to (12.5), we can show that

$$I_+ = V_+/Z_0; \qquad I_- = -V_-/Z_0. \tag{12.6}$$

On replacing V_1, V_2, I_1, and I_2 in (12.1) and (12.2) by their equivalent expressions as given in (12.4) and (12.5), we can deduce that

$$V(s) = V_+ e^{\gamma s} + V_- e^{-\gamma s}, \tag{12.7}$$

$$I(s) = I_+ e^{\gamma s} + I_- e^{-\gamma s}, \tag{12.8}$$

where

$$s = l - x, \tag{12.9}$$

measures the distance to any point on the transmission line from $x = l$, which is denoted as the load terminal. Assume that the transmission line is terminated at $x = l$ or $s = 0$ by the load impedance $Z_L = R_L + jX_L$. Also, let the voltage across, and the current in the load impedance Z_L be given by

$$V_L = V(s = 0); \qquad I_L = I(s = 0) = V_L/Z_L. \tag{12.10}$$

From (12.7), (12.8), and (12.10), Kirchhoff's equations for the circuit near the load are

$$V_+ + V_- = V_L; \qquad I_+ + I_- = I_L. \tag{12.11}$$

With the help of (12.6), (12.10), and (12.11), the following relations are deduced:

$$\frac{V_-}{V_+} = -\frac{I_-}{I_+} = \rho = |\rho|\, e^{j\theta_\rho} = \frac{Z_L - Z_0}{Z_L + Z_0}. \tag{12.12}$$

The ratio V_-/V_+ is the voltage reflection coefficient ρ which is complex; its magnitude is denoted by $|\rho|$ and its phase angle by θ_ρ. We see from (12.12) that the current reflection coefficient I_-/I_+ is the negative of the voltage reflection coefficient. As indicated previously, unless otherwise mentioned, we assume the reflection coefficient to refer only to the voltage.

In (12.12), we replace Z_L and Z_0 by $R_L + jX_L$ and $R_0 + jX_0$ respectively. The numerator and the denominator of (12.12) are each multiplied by the complex conjugate of $Z_L + Z_0$, that is by $(R_L + R_0) - j(X_L + X_0)$, to yield the following result:

$$\rho = [R_L^2 - R_0^2 + X_L^2 - X_0^2 + j2(X_L R_0 - R_L X_0)]/D_n, \tag{12.13}$$

where

$$D_n = (R_L + R_0)^2 + (X_L + X_0)^2 > 0. \tag{12.14}$$

From (12.13), it follows that

$$\operatorname{Im} \rho = 2(X_L R_0 - R_L X_0)/D_n. \tag{12.15}$$

On multiplying ρ by its complex conjugate, we find that

$$|\rho|^2 = \frac{(R_L - R_0)^2 + (X_L - X_0)^2}{(R_L + R_0)^2 + (X_L + X_0)^2} = 1 - \frac{4(R_L R_0 + X_L X_0)}{D_n}. \tag{12.16}$$

A passive load, in general, absorbs power. The sign of the time-averaged power $P_L = \frac{1}{2} \operatorname{Re} V_L I_L^*$ is determinate, that is, P_L is greater than or equal to zero. The limiting case of $P_L = 0$ corresponds to a reactive load. With the help of (12.6),

(12.11), and (12.12), it may be deduced that

$$P_L = \tfrac{1}{2}\,\mathrm{Re}\left[(V_+ + V_-)\frac{(V_+^* - V_-^*)}{Z_0^*}\right],$$

$$= \frac{1}{2}\left|\frac{V_+}{Z_0}\right|^2 R_0\left[1 - |\rho|^2 - 2\frac{X_0}{R_0}\,\mathrm{Im}\,\rho\right]. \tag{12.17}$$

Let us consider a semi-infinite transmission line extending from $s = 0$ to $s = \infty$. The input ($s = 0$) impedance of a semi-infinite transmission line is Z_0. As before, if V is the voltage at the input, the time-averaged power at the input is given by $\frac{1}{2}|V/Z_0|^2 R_0$, which must be positive for a passive transmission line since in that case the time-averaged power is delivered to and not abstracted from the transmission line. Only in the trivial case in which R and L vanish, or, C or G become infinite does R_0 become zero; otherwise it is nonvanishing and positive. Consequently, for a passive load, it follows from (12.17) that

$$|\rho|^2 \le |\rho|^2_{\max}, \tag{12.18}$$

where,

$$|\rho|^2_{\max} = 1 - 2(X_0/R_0)\,\mathrm{Im}\,\rho. \tag{12.19}$$

The use of (12.15) in (12.19) yields

$$|\rho|^2_{\max} = 1 - 4\frac{X_0}{R_0}\frac{(X_L R_0 - R_L X_0)}{D_n}. \tag{12.20}$$

With the help of (12.16) and (12.20), we deduce that

$$|\rho|^2 - |\rho|^2_{\max} = -\frac{4R_L}{D_n R_0}(R_0^2 + X_0^2). \tag{12.21}$$

The criterion (12.18) should be satisfied by all passive loads. Since a passive load necessarily implies that $R_L > 0$, it follows from (12.14) and (12.21) that $|\rho|^2$ given by (12.16) does indeed satisfy the passivity criterion (12.18).

We see that, for a passive load, $|\rho|$ as given in (12.16) is larger than unity provided that

$$R_L R_0 + X_L X_0 < 0. \tag{12.22}$$

Since R_0 and R_L are both positive, it follows that the necessary conditions for the fulfillment of the requirement (12.22) are that the reactive parts of the characteristic and the load impedances should both be nonvanishing and be of opposite signs. Thus, if $X_0 = 0$, $|\rho|$ can never be larger than unity for a passive load. If the imaginary parts of both sides of (6.6) are taken, we find that

$$X_0^2 = \frac{1}{2(G^2 + \omega^2 C^2)}\{-(\omega^2 LC + RG) + \sqrt{[(R^2 + \omega^2 L^2)(G^2 + \omega^2 C^2)]}\}, \tag{12.23}$$

from which X_0 can be shown to vanish only for the following nontrivial cases:

$$\begin{aligned} &\text{a) } \omega = 0 \quad \text{or} \quad L = C = 0 && \text{(D.C. Case)},\\ &\text{b) } R = G = 0 && \text{(Lossless Case)},\\ &\text{c) } R/L = G/C && \text{(Distortionless Case)}. \end{aligned} \tag{12.24}$$

Therefore, for a direct current, lossless and distortionless transmission lines $|\rho|$ corresponding to a passive load can never exceed unity. In the same manner, for a resistive load ($X_L = 0$), and for the case in which the reactive parts of the characteristic and the load impedances are of the same sign, $|\rho|$ is always less than unity. For $|\rho|$ to be greater than unity, the necessary and sufficient requirement is given by (12.22).

When $|\rho|^2 > 1$, we must not misinterpret it to mean that the "reflected" power is greater than the "incident" power. We can show now that the concept of incident and reflected powers has a limited range of validity. We can evaluate with the help of (6.1), (6.6), (12.6), (12.10), and (12.11) that the time-averaged power $P(s)$ at any point s on the transmission line, associated with the voltage and the current given by (12.7) and (12.8), is

$$P(s) = \tfrac{1}{2}\,\mathrm{Re}\, V(s)I^*(s) = P_i + P_r + P_c, \tag{12.25}$$

where,

$$P_i = \frac{1}{2}\left|\frac{V_+}{Z_0}\right|^2 R_0 e^{2\alpha s}, \tag{12.26}$$

$$P_r = -\frac{1}{2}\left|\frac{V_-}{Z_0}\right|^2 R_0 e^{-2\alpha s} = -P_i\,|\rho|^2\, e^{-4\alpha s}, \tag{12.27}$$

$$P_c = -X_0\,\mathrm{Im}\left[\frac{V_-}{Z_0}\left(\frac{V_+}{Z_0}\right)^* e^{-2j\beta s}\right] = -X_0\left|\frac{V_+}{Z_0}\right|^2 |\rho| \sin\left(\theta_\rho - 2\beta s\right). \tag{12.28}$$

As was discussed in Section 2.6, if the transmission line extends from $s = -\infty$ to $s = 0$, we should set V_- and therefore I_- in (12.7) and (12.8) equal to zero to obtain physically acceptable (finite) results. The remaining terms containing $e^{\gamma s}$ decay as s becomes negatively large. By replacing γ by $\alpha + j\beta$ and by including the time factor $e^{j\omega t}$, we may verify by using a procedure already discussed in Section 2.6, that in the remaining voltage and current waveforms, the phase moves with uniform velocity ω/β in the negative s-direction. It is therefore common to interpret $V_+e^{\gamma s}$ and $I_+e^{\gamma s}$ respectively as the voltage and the current waves progressing in the negative s-direction or the "incident" waves. The time-averaged power associated with this incident wave is P_i as given in (12.26) and is called the incident power. If, on the other hand, the transmission line extends from $s = 0$ to $s = \infty$, we should set V_+ and I_+ in (12.7) and (12.8) equal to zero to preclude the possibility of the voltage and the current becoming infinite for $s = \infty$. Then, the remaining terms $V_-e^{-\gamma s}$ and $I_-e^{-\gamma s}$ are respectively the voltage and the current waves, which decay as s becomes positively large and whose phase progresses with uniform velocity in the positive s-direction; these are identified as waves progressing in the positive s-direction or as the "reflected" waves. Then, P_r as given in (12.27) is shown to be the power associated with the reflected waves and is called the reflected power. When the transmission line is finite in extent, the voltage and the current waves as given in (12.7) and (12.8) consist of a linear superposition of waves progressing in either direction, that is, of the incident and the reflected waves. On a finite transmission line, the time-averaged power $P(s)$ associated with $V(s)$ and $I(s)$, which consist of a linear superposition of the incident and the reflected waves, is not in general equal to the sum of the "incident" and the "reflected" powers as may be seen from (12.25), which contains the additional term P_c. We see from (12.28) that for $X_0 = 0$, $P_c = 0$ for all s and then it is correct to separate the total power $P(s)$ associated with $V(s)$ and $I(s)$ as the sum of the incident and the reflected powers. For a direct current, lossless and distortionless transmission lines corresponding to $X_0 = 0$, $|\rho|$ is always less than unity and therefore the magnitude of the reflected power is seen to be always less than the incident power, which is conspicuously in accordance with what should obtain for a passive load. In other words, since a passive load absorbs power, the reflected power should be less than the incident power.

Also, $P_c = 0$ for

$$\theta_\rho - 2\beta s = 0, \pm n\pi \qquad (n \text{ is an integer}). \tag{12.29}$$

When (12.29) is satisfied, it is again meaningful to separate the total power as the sum of the incident and the reflected powers. In particular, at the position $s = 0$ of the load, (12.29) is satisfied if ρ is real, that is, if Z_L is equal to Z_0 multiplied by a positive real number. In this case, since the reactive parts of Z_0 and Z_L are of the same sign, $|\rho|$ is less than unity and then (12.27) shows that the magnitude of the reflected power is less than the incident power, as it should be for a passive load.

It is worthwhile to note that for $s = 0$, (12.29) is satisfied, in particular, for short- and open-circuit terminations. For these special cases, $|\rho| = 1$ and (12.27) shows that the magnitude of the reflected power is equal to the incident power, which result is to be expected for a passive, reactive load since such a load absorbs no power.

In summary, we emphasize that the magnitude of the voltage reflection coefficient being greater than unity is not inconsistent with the passivity of the load and that in situations where it is physically meaningful to separate the total power as the sum of the incident and the reflected powers it follows that the magnitude of the reflected power is less than the incident power, as it should be for a passive load.

2.13 GROUP VELOCITY

The phase velocity $v_{ph} = \omega/\beta$ is defined in (6.5) as the velocity of propagation of equiphase fronts in a single frequency wave. For the lossless and the distortionless cases, (7.9) and (7.11) show respectively that the phase-shift coefficient β is a linear function of ω with the result that the corresponding phase velocity is independent of frequency. For the lossy transmission line, we see from (7.18) that the phase velocity is a function of frequency. Instead of a single frequency, let us consider a group of frequencies around a central frequency. We can superpose linearly the voltages or the currents due to the constituent waves which are individually characterized by a single frequency to obtain the resultant wave shape. To determine the resultant wave shape further along the transmission line, we introduce appropriate phase shifts and attenuations in amplitude, if any, to evaluate the constituent waves in the new location and superpose the new constituent waves to obtain the resultant wave shape in the new location. For the lossless case, there is no attenuation and the time-delay equivalent x/v_{ph} of the phase shift where x is the change in distance and v_{ph} is the phase velocity, is the same for all the constituent waves, with the result that on superposition the original wave shape is recovered in the new location. Thus, as we have pointed out previously, v_{ph} represents the velocity of propagation of the wave shape as a whole. For the distortionless case, since α is independent of ω, all the constituent waves are attenuated by the same factor and since v_{ph} is also independent of the frequency, the time delay is the same for all the constituent waves with the result that on superposition, we recover a diminutive form of the original wave shape. The wave shape is not distorted but is attenuated. Therefore, even for the distortionless case, v_{ph} is the velocity of propagation of the whole wave. In the general case of lossy transmission lines, since α and v_{ph} are both functions of frequency, the wave shape in the new location is quite different from the original wave shape; it has both attenuation and distortion. In such a case, we cannot possibly define a significant velocity of propagation of the wave shape as a whole. When the phase velocity becomes a function of frequency, *dispersion* is said to occur. Even in the dispersive case, if the spread in the frequency is small and if the spread in the phase velocities is also small, we call the resulting wave a "wave packet" or "group" and in such a case a velocity of propagation of the amplitude maximum of the group known as the *group velocity* can be defined. An expression for the group velocity v_g can be obtained in the following manner.

Although a wave packet, in general, consists of a continuous variation of frequencies on either side of a central frequency, the expression for the group velocity v_g is derived here only on the basis of a very simple wave packet which consists of one frequency below the central frequency and one above. Consider a semi-infinite transmission line extending from $x = 0$ to $x = \infty$. Let the voltage across the input terminals at $x = 0$ be given by

$$V(0, t) = V_0[1 + m \cos \delta\omega t] \cos \omega t. \tag{13.1}$$

It is not difficult to realize that (13.1) represents an amplitude modulated wave of "carrier" angular frequency ω. The terms within the square brackets show that the amplitude of the wave, instead of remaining constant, varies very slowly at the angular frequency $\delta\omega$. With the help of the trigonometric relation

$$\cos A \cos B = \tfrac{1}{2} \cos (A + B) + \tfrac{1}{2} \cos (A - B), \tag{13.2}$$

we can recast (13.1) into the form

$$V(0, t) = V_0\left[\cos \omega t + \frac{m}{2} \cos (\omega - \delta\omega)t + \frac{m}{2} \cos (\omega + \delta\omega)t\right]. \tag{13.3}$$

Note that we call m, which is a constant usually less than 1, the modulation index. From (13.3), we find that there are only three frequencies, the central angular frequency ω and the so-called lower and upper sideband angular frequencies $(\omega - \delta\omega)$ and $(\omega + \delta\omega)$ respectively. Let α and β be the attenuation and the phase-shift coefficients corresponding to ω and let $\delta\alpha$ and $\delta\beta$ be the changes in the attenuation and the phase-shift coefficients corresponding to the change $\delta\omega$ in ω. Since the transmission line is a linear device, the voltage wave due to the excitation given by (13.1) can be found conveniently by first obtaining separately the voltage wave due to each one of the three constituent excitations as made explicit in (13.3) which is identical to (13.1) and then superposing the results. This procedure, together with (6.3), gives the voltage wave caused by the input voltage (13.1) as

$$V(x, t) = V_0\left[e^{-\alpha x}\cos(\omega t - \beta x) + \frac{m}{2}e^{-(\alpha-\delta\alpha)x}\cos\{(\omega - \delta\omega)t - (\beta - \delta\beta)x\}\right.$$
$$\left. + \frac{m}{2}e^{-(\alpha+\delta\alpha)x}\cos\{(\omega + \delta\omega)t - (\beta + \delta\beta)x\}\right]. \tag{13.4}$$

We assume that the transmission line is only very slightly lossy with the result that α is a small quantity. Therefore, the change $\delta\alpha$ in α due to the change in ω by $\delta\omega$ can be reasonably assumed to be a small quantity of higher order. Only propagation over moderate distances in which the wave shape is not drastically distorted is considered. Consequently $(\delta\alpha)x$ in (13.4) can be neglected in comparison with αx. In addition, the change $\delta\omega$ and the associated change $\delta\beta$ are assumed to be very small with the result that it is legitimate to assume that the change in β due to an increase $\delta\omega$ in ω is the same in magnitude as the change in β due to a decrease $\delta\omega$ in ω. With these approximations and (13.2), we may simplify (13.4) to yield

$$V(x, t) = V_0 e^{-\alpha x}\left[\cos(\omega t - \beta x) + \frac{m}{2}\cos\{(\omega - \delta\omega)t - (\beta - \delta\beta)x\}\right.$$
$$\left. + \frac{m}{2}\cos\{(\omega + \delta\omega)t - (\beta + \delta\beta)x\}\right],$$
$$= V_0 e^{-\alpha x}[1 + m\cos(\delta\omega t - \delta\beta x)]\cos(\omega t - \beta x). \tag{13.5}$$

As before, $\cos(\omega t - \beta x)$ is the phase term and $(\omega t - \beta x)$ is the phase factor, whose examination as indicated in Section 2.6 yields the phase velocity $v_{ph} = \omega/\beta$. The propagation characteristics of the amplitude of the wave is contained by the terms within the square brackets. A particular state, such as the maximum or the minimum, of the amplitude of the wave is defined by the term $(\delta\omega t - \delta\beta x)$. If we obtain the same state of the amplitude at time $t = t_1$ and $x = x_1$ and at later time $t = t_2 > t_1$ and $x = x_2$, then it follows that

$$\delta\omega t_1 - \delta\beta x_1 = \delta\omega t_2 - \delta\beta x_2; \qquad \frac{\delta\omega}{\delta\beta} = \frac{x_2 - x_1}{t_2 - t_1}. \tag{13.6}$$

Thus, a particular state of the amplitude is seen from (13.6) to travel with the velocity

$\delta\omega/\delta\beta$. This ratio in the limit of $\delta\omega$ approaching zero is called the group velocity:

$$v_g = \underset{\delta\omega \to 0}{\text{Limit}}\ \delta\omega/\delta\beta = d\omega/d\beta. \tag{13.7}$$

Note, therefore, that the concept of group velocity holds good only when the spread in frequencies in the wave packet is extremely small.

The following alternative expressions for the group velocity are useful. Since $v_{ph} = \omega/\beta$, we find that

$$v_g = \frac{1}{\dfrac{d\beta}{d\omega}} = \frac{1}{\dfrac{d}{d\omega}\left(\dfrac{\omega}{v_{ph}}\right)} = \frac{1}{\dfrac{1}{v_{ph}} - \dfrac{\omega}{v_{ph}^2}\dfrac{dv_{ph}}{d\omega}}$$

$$= \frac{v_{ph}}{\left[1 - \dfrac{\omega}{v_{ph}}\dfrac{dv_{ph}}{d\omega}\right]}. \tag{13.8}$$

Also, we may obtain:

$$v_g = \frac{d\omega}{d\beta} = \frac{d}{d\beta}(\beta v_{ph}) = v_{ph} + \beta\frac{d}{d\beta}v_{ph}. \tag{13.9}$$

Since $\beta = 2\pi/\lambda$, $\beta(d/d\beta) = -\lambda(d/d\lambda)$ and therefore (13.9) can be rewritten in terms of the wavelength λ to yield

$$v_g = v_{ph} - \lambda\frac{dv_{ph}}{d\lambda}. \tag{13.10}$$

From (13.8), we see that if the phase velocity is independent of frequency, that is, when there is no dispersion, the group velocity is equal to the phase velocity. On the other hand, if v_{ph} is a function of ω, there is dispersion and v_g is different from v_{ph}. From (7.16) and (7.17), since v_{ph} is independent of ω, it follows that

$$v_g = v_{ph} = 1/\sqrt{LC} \tag{13.11}$$

for lossless and distortionless cases. When v_{ph} is a function of ω, there are two possibilities. The phase velocity v_{ph} can decrease with frequency and therefore $dv_{ph}/d\omega < 0$, which together with (13.8) shows that the group velocity v_g is less than the phase velocity. This situation is referred to as *normal dispersion*. The phase velocity v_{ph} can increase with frequency, that is, $dv_{ph}/d\omega > 0$ for which case (13.8) shows that $v_g > v_{ph}$ and the corresponding state of affairs is called *anomalous dispersion*. For the low-loss transmission line, we deduce from (7.18) that

$$\frac{dv_{ph}}{d\omega} = \frac{1}{\sqrt{LC}}\frac{1}{4\omega^3}\left(\frac{R}{L} - \frac{G}{C}\right)^2 > 0, \tag{13.12}$$

and therefore there is anomalous dispersion. The substitution of (7.18) and (13.12) in (13.8) and repeated use of the approximation

$$\left[1 - \frac{1}{8\omega^2}\left(\frac{R}{L} - \frac{G}{C}\right)^2\right]^{-1} = 1 + \frac{1}{8\omega^2}\left(\frac{R}{L} - \frac{G}{C}\right)^2$$

yields the group velocity v_g for the low-loss case as

$$v_g = \frac{1}{\sqrt{LC}} \frac{1}{[1 - (1/8\omega^2)(R/L - G/C)^2]}. \tag{13.13}$$

From Section 7.3, we find that for the transmission line in free space, $1/\sqrt{LC} = c$, the velocity of electromagnetic waves in free space. From (7.18) and (13.13), we find that

$$v_g v_{ph} = c^2. \tag{13.14}$$

We see from (13.13) that the group velocity for the low-loss case is greater than the velocity of light. It is important, however, to remember that the results from the relativity theory are not violated by the fact that $v_g > c$. If, instead of a wave packet, we consider a disturbance really localized in time, then we can show that the initial disturbance, however small, or the wavefront arrives with the velocity of light. We established this result for the lossless and the distortionless cases but it can be shown to be true also for the general case of a lossy transmission line.

Subsequently, we introduce another velocity known as the energy transport velocity. For the lossless case, we show that the energy transport velocity is equal to the group velocity. But for a lossy case, especially in the case of anomalous dispersion such as is obtained in the low-loss transmission line, the group velocity which cannot be identified with the energy transport velocity does not have much physical significance.

PROBLEMS

2.1 The telegrapher's equations are given by

$$-\frac{\partial V}{\partial x} = RI + L\frac{\partial I}{\partial t}, \tag{a}$$

$$-\frac{\partial I}{\partial x} = GV + C\frac{\partial V}{\partial t}, \tag{b}$$

where x and t are the distance along the transmission line and time respectively. The other parameters have the usual meaning. Note that the arguments (x, t) of V and I have been omitted for convenience. Multiply (a) by I and (b) by V and add to obtain the following result:

$$-\frac{\partial}{\partial x}(VI) = RI^2 + GV^2 + \frac{\partial}{\partial t}[\tfrac{1}{2}LI^2 + \tfrac{1}{2}GV^2]. \tag{c}$$

Integrate (c) with respect to x from $x = x_1$ to $x = x_2 > x_1$ and verify the following result:

$$(VI)_{x=x_1} - \left[\int_{x_1}^{x_2} \{RI^2 + GV^2\}\,dx + \frac{\partial}{\partial t}\int_{x_1}^{x_2} \{\tfrac{1}{2}LI^2 + \tfrac{1}{2}CV^2\}\,dx\right] = (VI)_{x=x_2}. \tag{d}$$

Give a physical interpretation of the power-flow equation (d).

2.2 A semi-infinite high-frequency transmission line is used for d.c. transmission. The input resistance is found to be 5000 ohms. The voltage is found to decay to $1/e$ of its original value in one kilometer. Calculate the resistance and the conductance per unit length of the transmission line.

2.3 Consider a d.c. generator of voltage V_g connected at time $t = 0$, to the input end ($x = 0$) of a semi-infinite two-wire distortionless transmission line of the following parameters per unit length: inductance L, capacitance C, series resistance R and shunt conductance G. At any time t, ($t > 0$), find

a) The energy stored in the inductance.
b) The energy stored in the capacitance.

Show that these are equal.

Evaluate the difference between the rate at which energy is supplied by the generator and the rate of storage of energy in the inductance and the capacitance. Show that this difference is equal to the rate of dissipation of energy in the resistance and the conductance. Verify that the rates of energy dissipation in the resistance and the conductance are equal.

2.4 Show by actual substitution that the voltage distribution $V(x, t)$ given by

$$V(x, t) = V_0 \frac{2}{\sqrt{\pi}} \int_{z_0}^{\infty} e^{-z^2}\, dz, \qquad z_0 = \sqrt{\frac{CRx^2}{4t}},$$

satisfies the differential equation

$$\frac{\partial^2}{\partial x^2} V(x, t) = CR \frac{\partial}{\partial t} V(x, t).$$

Note that the differentiation of an integral with variable limits is carried out as follows:

$$\frac{d}{dy} \int_{a(y)}^{b(y)} f[x, y]\, dx = \int_{a(y)}^{b(y)} \frac{\partial f[x, y]}{\partial y}\, dx + f[b(y), y] \frac{\partial}{\partial y} b(y) - f[a(y), y] \frac{\partial}{\partial y} a(y).$$

2.5 Consider a leakage-free ($G = 0$), noninductive ($L = 0$) submarine cable extending from $x = 0$ to $x = \infty$. Assume that the voltage at the input is given by

$$V(0, t) = V_0 H(t),$$

where $H(t)$ is the Heaviside unit step function. The current in such a cable may be shown to be given by

$$I(x, t) = V_0 \sqrt{\frac{C}{\pi R t}} \exp\left(-\frac{CRx^2}{4t}\right) H(t).$$

For $t > 0$, show that the current at any point on the transmission line reaches a maximum value for $t = t_{\max} = 0.5\, CRx^2$. Verify that the maximum value of the current is given by

$$I_{\max}(x, t) = \frac{0.483\, V_0}{Rx}.$$

Show that the current as a function of time has inflexion points at

$$t_{\text{inf}} = 0.5(1 \pm \sqrt{\tfrac{2}{3}})CRx^2.$$

2.6 A two-wire transmission line has the following parameters:

$$R = 6.5 \times 10^{-2} \text{ ohm per meter,}$$
$$L = 2.4 \times 10^{-5} \text{ henry per meter,}$$
$$C = 7.8 \times 10^{-10} \text{ farad per meter,}$$
$$G = 9.2 \times 10^{-12} \text{ mho per meter.}$$

Find R_0, X_0, α and β for an operating frequency of 1 MHz using the exact expressions. State the units of these quantities. Repeat the calculations using the appropriate approximate expressions.

2.7 A coaxial transmission line has the following parameters:

$$R = 0.015 \text{ ohm per meter,}$$
$$L = 0.17 \times 10^{-6} \text{ henry per meter,}$$
$$C = 0.68 \times 10^{-10} \text{ farad per meter.}$$

Assuming that the equivalent shunt conductance G per unit length is negligible, determine the attenuation coefficient, the phase-shift coefficient, the characteristic impedance and the phase velocity for an operating frequency of 10^8 Hz using the exact expressions. Repeat the calculations using the appropriate approximate expressions.

2.8 The voltage reflection coefficient ρ_g at the generator end of a transmission line is given by

$$\rho_g = \frac{Z_g - Z_0}{Z_g + Z_0},$$

where Z_0 and Z_g are the characteristic and generator impedances respectively. If the transmission line is semi-infinite and is driven by a generator that has an open-circuit voltage V_g, show that the voltage at any point x on the transmission line is given by

$$V(x) = \frac{V_g}{2}(1 - \rho_g)e^{-\gamma x},$$

where γ is the propagation coefficient.

2.9 The characteristic impedance of a two-wire transmission line is given by

$$Z_0 = |Z_0| \exp(j\theta_{Z_0}) = \sqrt{\frac{R + j\omega L}{G + j\omega C}}.$$

In the limit of high frequencies ($\omega L \gg R, \omega C \gg G$) show that the following approximations obtain:

$$|Z_0| = \sqrt{\frac{L}{C}}\left\{1 + \frac{1}{4\omega^2}\left[\left(\frac{R}{L}\right)^2 - \left(\frac{G}{C}\right)^2\right]\right\},$$

$$\theta_{Z_0} = -\frac{1}{2\omega}\left(\frac{R}{L} - \frac{G}{C}\right).$$

2.10 The characteristic impedance Z_0 and the propagation coefficient γ are given by

$$Z_0 = R_0 + jX_0 = \sqrt{\frac{R + j\omega L}{G + j\omega C}},$$

$$\gamma = \alpha + j\beta = \sqrt{(R + j\omega L)(G + j\omega C)},$$

where all the parameters have the usual meaning. Obtain the approximate expressions for R_0, X_0, α, β, the phase velocity v_{ph} and the wavelength λ:

a) For the case of "unloaded cable" for which $R \gg \omega L$, $G \ll \omega C$.
b) In the limit of very low frequencies ($R \gg \omega L$, $G \gg \omega C$).

Show that the phase velocity in the limit of zero frequency cannot exceed $1/\sqrt{LC}$.

2.11 Obtain the following results:

$$R_0^2 = \frac{1}{2(G^2 + \omega^2 C^2)} \{\omega^2 LC + RG + \sqrt{[(R^2 + \omega^2 L^2)(G^2 + \omega^2 C^2)]}\}, \tag{a}$$

$$X_0^2 = \frac{1}{2(G^2 + \omega^2 C^2)} \{-(\omega^2 LC + RG) + \sqrt{[(R^2 + \omega^2 L^2)(G^2 + \omega^2 C^2)]}\}, \tag{b}$$

$$\alpha^2 = \tfrac{1}{2}\{-\omega^2 LC + RG + \sqrt{[(R^2 + \omega^2 L^2)(G^2 + \omega^2 C^2)]}\}, \tag{c}$$

$$\beta^2 = \tfrac{1}{2}\{\omega^2 LC - RG + \sqrt{[(R^2 + \omega^2 L^2)(G^2 + \omega^2 C^2)]}\}. \tag{d}$$

2.12 With the help of the expressions (a) to (d) in Problem 2.11 derive the following results:

$$R_0 = \frac{1}{2\alpha}[R + G\,|Z_0|^2] = \frac{\omega}{2\beta}[L + C\,|Z_0|^2],$$

$$X_0 = \frac{1}{2\beta}[R - G\,|Z_0|^2] = \frac{\omega}{2\alpha}[L - C\,|Z_0|^2].$$

2.13 A transmission line of length $x = l$ and characteristic impedance Z_0 is driven at $x = 0$ by a generator that has an open-circuit voltage V_g and an internal impedance Z_g. The transmission line is terminated at $x = l$ by an impedance Z_L. Let ρ_g and ρ_l be the reflection coefficients at $x = 0$ and $x = l$ respectively. Let $s = l - x$ and γ be the propagation coefficient. Show that the voltage distribution along the transmission line is given by

$$V(s) = \frac{V_g Z_0 e^{-\gamma l}}{(Z_g + Z_0)} \frac{\{e^{\gamma s} + \rho_l e^{-\gamma s}\}}{(1 - \rho_l \rho_g e^{-2\gamma l})}.$$

2.14 What is the required length of a (a) short-circuited and (b) open-circuited transmission line to provide (a) an inductive and (b) a capacitive reactance of 200 ohms? Assume that $\lambda = 75$ cm and $Z_0 = 80$ ohms.

2.15 Find the input impedance of a lossless transmission line of characteristic impedance 50 ohms and load impedance 150 ohms when the transmission line is (a) $\lambda/4$ long and (b) $\lambda/2$ long, where λ is the wavelength.

2.16 Find the input impedance of a transmission line when $Z_0 = 50$ ohms, $Z_L = 100 + j\,80$ ohms and $l = 1.12\lambda$, where λ is the wavelength.

2.17 The following impedance measurements are made at the input of a 100 km open-wire telephone line: (a) with the terminals short-circuited, the input impedance is $400 \angle 60°$ at 1000 Hz; (b) with the terminals open-circuited, the input impedance is $700 \angle -40°$ at 1000 Hz. Determine the characteristic impedance of the transmission line as well as the attenuation and the phase-shift coefficients of the transmission line. For the determination of the phase-shift coefficient, assume that the phase velocity v_{ph} is given approximately by $v_{ph} = 3 \times 10^8$ m/sec.

Also evaluate the parameters R, L, G, and C.

2.18 Show that for a low-loss transmission line with a short-circuit termination, the input resistance R_{in} and the input reactance X_{in} are given by the following expressions:

$$\frac{R_{in}}{R_0} = \frac{2\alpha l - (X_0/R_0)\sin 2\beta l}{2[D_0 \cos^2 \beta l + \alpha^2 l^2]},$$

and

$$\frac{X_{in}}{R_0} = \frac{2(X_0/R_0)\alpha l + D_0 \sin 2\beta l}{2[D_0 \cos^2 \beta l + \alpha^2 l^2]} \qquad (D_0 = 1 - 2\alpha l + 2\alpha^2 l^2),$$

where $\alpha + j\beta$ is the propagation coefficient, $R_0 + jX_0$ the characteristic impedance and l the length of the transmission line.

2.19 With the use of the results of Problem 2.18, show that the resonance of a short-circuited transmission line occurs for βl given by

$$\beta l = n\pi - \frac{X_0 \alpha l}{R_0}$$

where n is an integer. Prove that at resonance

$$R_{in}/R_0 = \alpha l.$$

2.20 With the help of the results of Problem 2.18 show that antiresonance of a short-circuited transmission line occurs for βl given by

$$\beta l = (2n - 1)\pi/2.$$

Deduce that at antiresonance

$$R_{in}/R_0 = 1/\alpha l.$$

Establish that near antiresonance the following results are obtained:

a) For $\beta l = (2n - 1)\dfrac{\pi}{2} + \alpha l \dfrac{X_0}{R_0}$; $\quad \dfrac{X_{in}}{R_0} = 0$;

b) For $\beta l = (2n - 1)\dfrac{\pi}{2} + \alpha l\left(\dfrac{X_0}{R_0} - 1 - \alpha l\right)$; $\quad \dfrac{X_{in}}{R_0} = \dfrac{1}{2\alpha l}\left(1 - \alpha l + \dfrac{X_0}{R_0}\right)$;

c) For $\beta l = (2n - 1)\dfrac{\pi}{2} + \alpha l\left(\dfrac{X_0}{R_0} + 1 + \alpha l\right)$; $\quad \dfrac{X_{in}}{R_0} = \dfrac{-1}{2\alpha l}\left(1 - \alpha l - \dfrac{X_0}{R_0}\right)$.

2.21 For a fixed characteristic impedance $Z_0 = R_0 + jX_0$ but a completely variable load impedance $Z_L = R_L + jX_L$ (both R_L and X_L being variable), show that the magnitude $|\rho|$ of the reflection coefficient is maximized when

$$R_L = 0 \quad \text{and} \quad X_L = -\frac{X_0}{|X_0|}|Z_0|.$$

Prove that for these values of R_L and X_L,

$$|\rho|^2 = 1 + \frac{4\,|X_0|\,|Z_0|}{R_0^2 + (|Z_0| - |X_0|)^2}.$$

2.22 For a given value of $|Z_0/Z_L|$, show that the magnitude $|\rho|$ of the voltage reflection coefficient has a maximum value given by

$$|\rho| = 1 + \sqrt{2}.$$

[*See* "Calculation of Power on a Transmission Line" by G. E. Stannard, *Proc. IEEE*, **55** (1), 132 (January, 1967).]

2.23 Find the expressions for the group velocity v_g of the waves on a two-wire transmission line for the following special cases:

a) No losses; $R = G = 0.$

b) No distortion; $\frac{R}{L} = \frac{G}{C}.$

c) High-frequencies; $\omega L \gg R$; $\omega C \gg G$.

d) "Unloaded cable"; $R \gg \omega L$; $\omega C \gg G$.

e) Low-frequencies; $R \gg \omega L$; $G \gg \omega C$.

CHAPTER 3

THEORY OF LOSSLESS TRANSMISSION LINES: VOLTAGE, CURRENT, IMPEDANCE, AND POWER

3.1 INTRODUCTION

At radio frequencies, the losses in the transmission line are usually very small with the result that the characteristics of transmission are approximately the same as those pertaining to the limiting case of lossless transmission lines. Therefore, the study of lossless transmission lines facilitates the understanding of the characteristics of the practical case of low-loss transmission lines. For this reason, we develop in some detail the theory of lossless transmission lines in this chapter. We treat the voltage and the current distributions in Section 3.2 and give a discussion of the differences between the progressive and the standing waves in Section 3.3. The study of the impedance variation along the transmission line as well as the method of determining the unknown impedance by the standing-wave measurements on the transmission line connected to the impedance is carried out in Sections 3.4 and 3.5. In Sections 3.6 and 3.7, we give the theory and the applications of the Smith chart for transmission-line calculations. The principles of operation of the quarter-wave transformer and the single stub impedance-matching device are analyzed in Section 3.8. Particular emphasis is placed in Section 3.9 on the important features of transport of power along lossless transmission lines. Finally, we present the definitions of the attenuation coefficient and the selectivity based on the energy considerations, with particular reference to the low-loss transmission lines, as well as discussing the concept of the so-called "conjugate-matching" in Section 3.10.

3.2 VOLTAGE AND CURRENT DISTRIBUTIONS ALONG A LOSSLESS TRANSMISSION LINE

At radio frequencies, the inequalities $\omega L \gg R$ and $\omega C \gg G$ are usually satisfied with the result that the transmission line operating at such frequencies belongs to the low-loss category. We can best understand the propagation characteristics of low-loss transmission lines by considering the limiting case of lossless transmission lines. For this reason, a detailed treatment of the voltage and the current distributions as well as impedance along a lossless transmission line is useful. We have shown that for the case of no losses,

$$\gamma = j\beta; \qquad Z_0 = R_0. \tag{2.1}$$

Then, it follows from (2.9.16 to 2.9.18) and (2.12.12) that the voltage $V(x)$ and the current $I(x)$ along a lossless transmission line may be written as

$$V(x) = \frac{I_L(Z_L + R_0)}{2} e^{j\beta s}[1 + |\rho| e^{j\phi}], \tag{2.2}$$

$$I(x) = \frac{I_L(Z_L + R_0)}{2R_0} e^{j\beta s}[1 - |\rho| e^{j\phi}], \tag{2.3}$$

where

$$s = l - x; \qquad \phi = \theta_\rho - 2\beta s. \tag{2.4}$$

The load impedance Z_L is connected across the terminals of the transmission line at $x = l$ or $s = 0$; I_L is the current in the load impedance. Even though the characteristic impedance of the transmission line is real, in view of the presence of a reactive part in the load impedance, the reflection coefficient ρ is, in general, complex and its phase angle θ_ρ is nonvanishing. We note from the discussion in Section 2.12 that for the lossless transmission line, the magnitude $|\rho|$ of the reflection coefficient is less than unity.

As the distance s of the observation point from the load terminals increases, the parameter ϕ which denotes the position along the transmission line decreases. When s is changed by $\lambda/2$, $2\beta s$ is changed by $2 \times (2\pi/\lambda) \times \lambda/2 = 2\pi$, with the result that from (2.2) to (2.4), we see that $|V(x)|$ and $|I(x)|$ are unchanged. In other words, the values of the voltage and the current along the transmission line repeat themselves every half-a-wavelength. If the distance s of the observation point from the load terminals is continuously changed, ϕ changes continuously and it follows, therefore, from (2.2) and (2.3) that the voltage $V(x)$ and the current $I(x)$ vary along the transmission line. We should remember that $V(x)$ and $I(x)$ are complex quantities. However, in practice, it is simpler to measure the magnitudes of the voltage $V(x)$ and the current $I(x)$ along the transmission line. Consequently, we analyze the variation of the magnitudes of the voltage $V(x)$ and the current $I(x)$ along the transmission line first.

From (2.2) and (2.3), we can deduce that

$$\frac{|V(x)|}{V_M} = [1 + |\rho|^2 + 2\,|\rho| \cos \phi]^{1/2}, \tag{2.5}$$

$$\frac{|I(x)|}{I_M} = [1 + |\rho|^2 - 2\,|\rho| \cos \phi]^{1/2}, \tag{2.6}$$

where

$$V_M = \left|\frac{I_L(Z_L + R_0)}{2}\right|; \qquad I_M = \left|\frac{I_L(Z_L + R_0)}{2R_0}\right| = \frac{V_M}{R_0}. \tag{2.7}$$

We should note that the real quantities V_M and I_M are independent of the position along the transmission line. The variations of $|V(x)|/V_M$ and $|I(x)|/I_M$ with distance along the transmission line are made perspicuous with the help of the so-called crank diagram shown in Fig. 3.1, where O is the center of a circle of radius equal to $|\rho|$. P is a point at a distance from O equal to unity. Since $|\rho| < 1$, P lies outside the

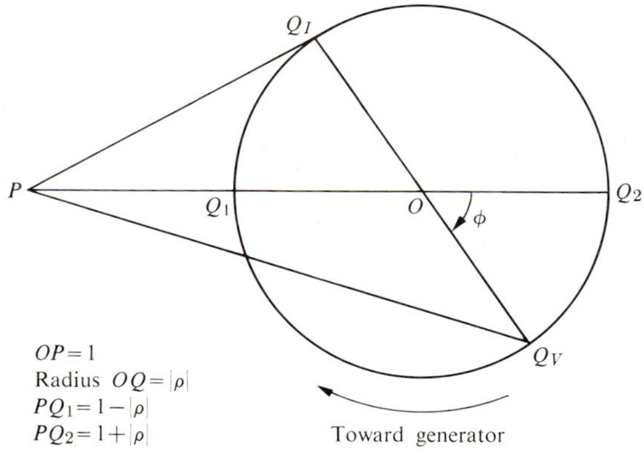

Fig. 3.1 The crank diagram for the magnitude of the voltage and the current.

circle. Let the straight line connecting P and O intersect the circle at Q_1 and Q_2. Let Q_IOQ_V be another diameter of the circle such that the angle $\measuredangle Q_2OQ_V = \phi$. From Fig. 3.1, we can deduce that

$$PQ_V = [1 + |\rho|^2 - 2|\rho| \cos(\pi - \phi)]^{1/2},$$

$$= [1 + |\rho|^2 + 2|\rho| \cos\phi]^{1/2} = \frac{|V(x)|}{V_M}, \tag{2.8}$$

and

$$PQ_I = [1 + |\rho|^2 - 2|\rho| \cos\phi]^{1/2} = \frac{|I(x)|}{I_M}. \tag{2.9}$$

As the distance of the observation point from the load terminals is increased, s is increased and ϕ is decreased. Therefore, the radius vectors OQ_V and OQ_I move in the clockwise direction. We see from (2.8) and (2.9) that PQ_V and PQ_I give, to within a multiplicative constant, the magnitudes of the voltage and the current respectively. As the terminal points Q_V and Q_I move around the circumference of the circle, the lengths PQ_V and PQ_I vary continuously between a maximum value and a minimum value. Therefore, the magnitudes of the voltage and the current vary periodically along the transmission line between the same maximum and minimum values.

When

$$\phi = -(2n - 1)\pi \qquad (n = \text{integer}) \text{ [voltage minimum; current maximum]}. \tag{2.10}$$

PQ_V attains the minimum value $PQ_1 = 1 - |\rho|$ and PQ_I attains the maximum value $PQ_2 = 1 + |\rho|$. Thus, the locations of voltage minima and current maxima coincide; these locations are given by (2.10). With the help of (2.8) to (2.10), we see that the minimum of the magnitude of the voltage and the maximum of the magnitude of the current are

$$|V(x)|_{\min} = V_M(1 - |\rho|), \tag{2.11}$$

$$|I(x)|_{\max} = \frac{V_M}{R_0}(1 + |\rho|). \tag{2.12}$$

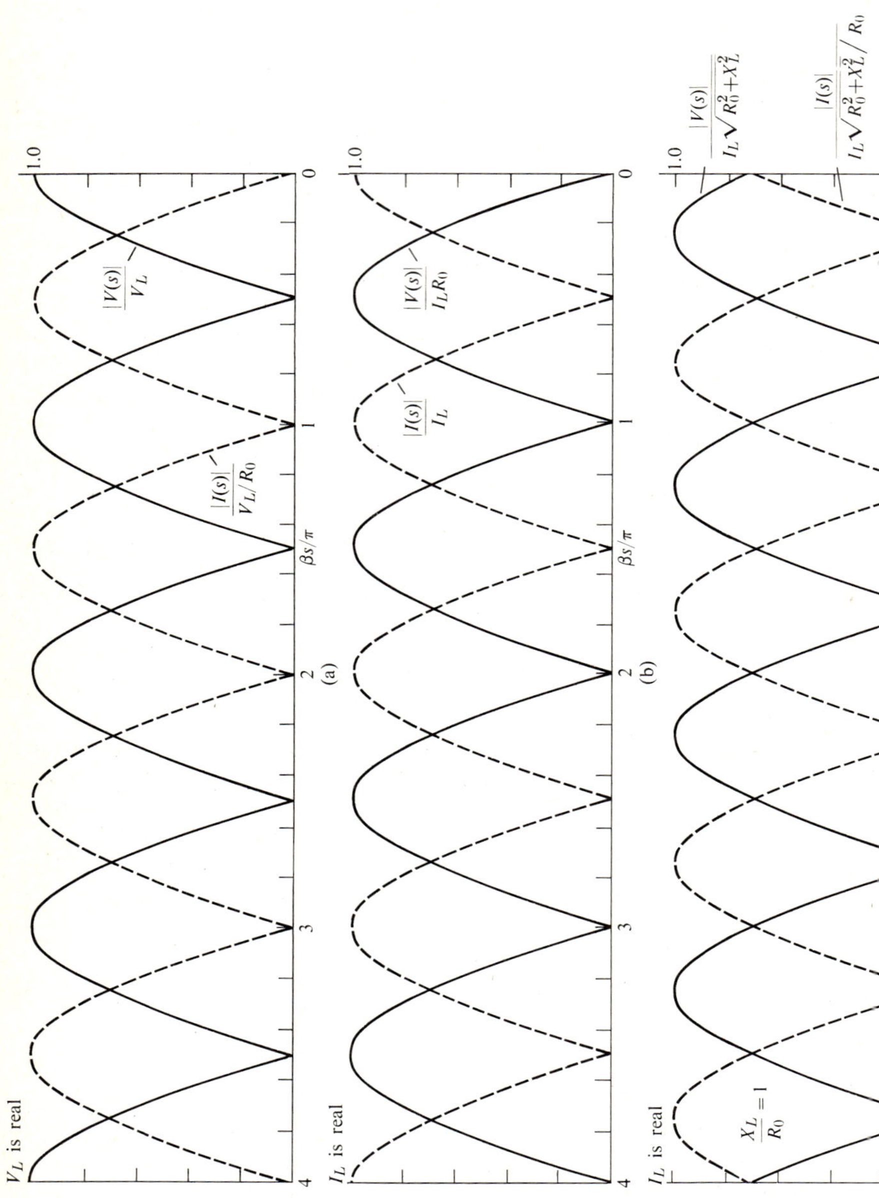

V_L is real
$\frac{|V(s)|}{V_L}$
$\frac{|I(s)|}{V_L/R_0}$
1.0
0
1
2
3
4
$\beta s/\pi$
(a)
I_L is real
$\frac{|V(s)|}{I_L R_0}$
$\frac{|I(s)|}{I_L}$
1.0
0
1
2
3
4
$\beta s/\pi$
(b)
I_L is real
$\frac{X_L}{R_0}=1$
$\frac{|V(s)|}{I_L\sqrt{R_0^2+X_L^2}}$
$\frac{|I(s)|}{I_L\sqrt{R_0^2+X_L^2}/R_0}$
1.0
0
1
2
3
4
$\beta s/\pi$
(c)

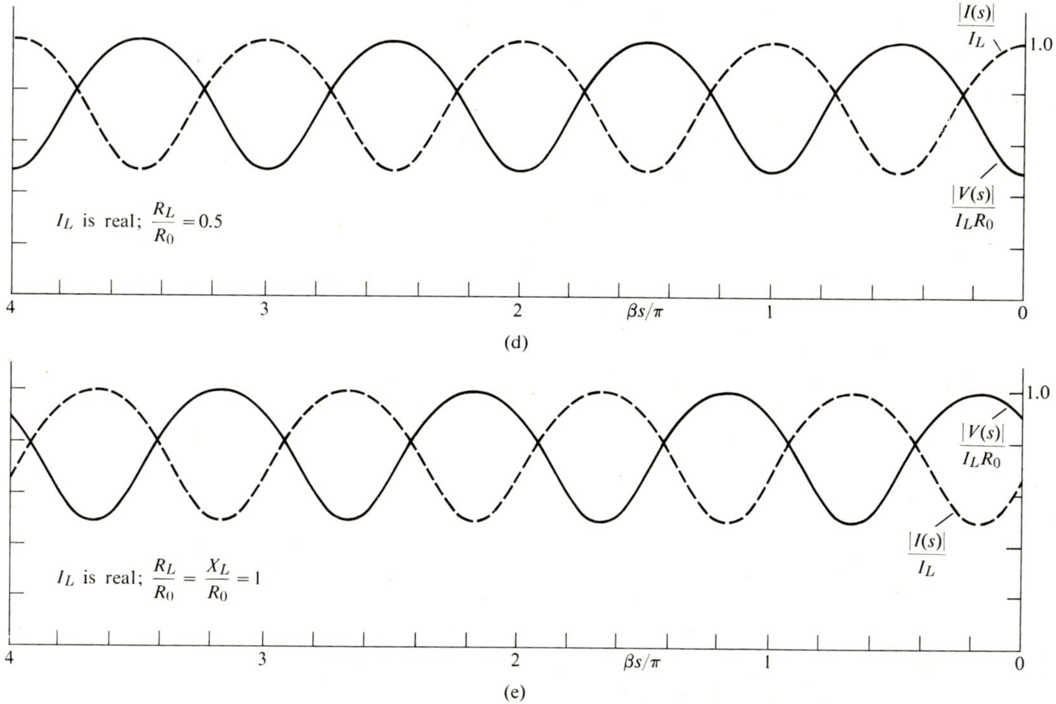

Fig. 3.2 The magnitudes of the voltage and the current on a lossless transmission line as a function of the electrical distance from the load in radians. (a) Open-circuit termination; (b) short-circuit termination; (c) reactive termination; (d) resistive termination; (e) impedance termination.

We have already pointed out that these maxima and minima repeat themselves every half-a-wavelength, which corresponds to a change of ϕ by 2π radians. When

$$\phi = -2n\pi \qquad (n = \text{integer}) \text{ [voltage maximum, current minimum]}, \tag{2.13}$$

PQ_V attains the maximum value $PQ_2 = 1 + |\rho|$ and PQ_I attains the minimum value $PQ_1 = 1 - |\rho|$. Therefore, the locations of the voltage maxima and the current minima also coincide; these locations are as given in (2.13). As before, with the help of (2.8), (2.9), and (2.13), we obtain the maximum of the magnitude of the voltage and the minimum of the magnitude of the current:

$$|V(x)|_{\text{max}} = V_M(1 + |\rho|), \tag{2.14}$$

$$|I(x)|_{\text{min}} = \frac{V_M}{R_0}(1 - |\rho|). \tag{2.15}$$

The distance between a maximum and an adjacent minimum is the same for the magnitudes of the voltage and the current and we obtain this distance from (2.10) and (2.13) as equivalent to a change of π radians in ϕ. From (2.4), we can deduce that a change of π in ϕ corresponds to a change of $\lambda/4$ in s or x. Thus, the distance between a maximum and a minimum of either the voltage magnitude or the current magnitude is equal to $\lambda/4$. Since the locations of the maxima and the minima of the magnitude of the voltage coincide respectively with those of the minima and the maxima of the magnitude of the current, it follows that the distance between a voltage maximum (or minimum) and the adjacent current maximum (or minimum) is also equal to $\lambda/4$. In addition to the foregoing results concerning the locations of the voltage and the current extrema, from (2.11), (2.12), (2.14), and (2.15) we can obtain the following results concerning the values of the voltage and the current extrema:

$$\frac{|V(x)|_{\text{min}}}{|I(x)|_{\text{min}}} = \frac{|V(x)|_{\text{max}}}{|I(x)|_{\text{max}}} = R_0. \tag{2.16}$$

That is, the ratio of the voltage minimum (or maximum) to the current minimum (or maximum) is equal to the characteristic resistance.

In Fig. 3.2, we have depicted $|V(x)|$ and $|I(x)|$ as a function of βs or the distance in radians of the observation point from the load terminals for various values of the load impedance. For the open-circuit and the short-circuit terminations as well as when the load impedance is reactive, we can establish with the help of the fact that $X_0 = 0$, (2.12.12), or (2.12.16) that $|\rho| = 1$ and hence the voltage and the current minima become equal to zero. When the load impedance is resistive, from (2.12.12) we see that ρ is real and therefore, $\theta_\rho = 0$ or $-\pi$. At the location $s = 0$ of the load, we find from (2.4) that $\phi = 0$ or $-\pi$. In view of (2.10) and (2.13), we see that the load terminals coincide with a voltage maximum or minimum.

In the analysis of the variation of the phase of the voltage and the current along the transmission line, it is convenient to start from expressions for $V(x)$ and $I(x)$ different from those given in (2.2) and (2.3). First, let us consider the case for which the transmission line has an open-circuit termination, that is, $Z_L = \infty$. Therefore, the load current $I_L = 0$ but the load voltage $V_L = Z_L I_L$ is finite. With the help of (2.1) and (2.10.24a,b), we may simplify (2.9.19) and (2.9.20) to yield the following expressions for $V(x)$ and $I(x)$:

$$V(x) = V_L \cos \beta s; \qquad I(x) = \frac{V_L}{R_0} e^{j(\pi/2)} \sin \beta s \qquad [\text{open-circuit termination}]. \quad (2.17)$$

It is convenient to refer the phase to that of V_L which is, therefore, assumed to be real. The phase angle θ_V of $V(x)$ is seen from (2.17) to remain constant as βs is varied, except when $\cos \beta s$ changes sign. At $\beta s = (2n - 1)\pi/2$, where n is an integer, $\cos \beta s$ changes sign and the phase angle θ_V increases abruptly by π radians. In a similar manner, we see from (2.17) that the phase angle θ_I of the current $I(x)$ starts with $\pi/2$, remains unchanged as βs is varied except for $\beta s = n\pi$ (n is an integer), when $\sin \beta s$ changes sign and θ_I increases abruptly by π radians. We depict the phase angles θ_V and θ_I of $V(x)$ and $I(x)$ respectively as a function of βs in Fig. 3.3(a), for the case of the open-circuit termination.

Suppose the transmission line has a short-circuit termination, that is, $Z_L = 0$. The load voltage $V_L = I_L Z_L = 0$, but the load current I_L is finite. Again, from (2.9.19) and (2.9.20), we obtain that

$$V(x) = I_L R_0 e^{j\pi/2} \sin \beta s; \qquad I(x) = I_L \cos \beta s \qquad [\text{short-circuit termination}]. \quad (2.18)$$

The phase is referred to that of I_L which is, therefore, taken to be real. The phase angle θ_V of $V(x)$ starts with $\pi/2$ for $\beta s = 0$, remains unchanged for variations in βs except for $\beta s = n\pi$, when $\sin \beta s$ changes sign and θ_V increases abruptly by π radians. In a similar manner, the phase angle θ_I of $I(x)$ starts with zero for $\beta s = 0$, remains unchanged for variations in βs except for $\beta s = (2n - 1)\pi/2$, when $\cos \beta s$ changes sign and θ_I increases abruptly by π radians. We depict the phase angles θ_V and θ_I corresponding to the short-circuit termination as a function of βs in Fig. 3.3(b).

Let us consider a transmission line which is terminated in a reactive load: $Z_L = jX_L$. Then, with the help of (2.1), and (2.10.24a,b), we may simplify (2.9.19) and (2.9.20) to yield

$$V(x) = I_L j[X_L \cos \beta s + R_0 \sin \beta s], \quad (2.19)$$

$$I(x) = \frac{I_L}{R_0}[-X_L \sin \beta s + R_0 \cos \beta s]. \quad (2.20)$$

On setting,

$$R_0 = \sqrt{R_0^2 + X_L^2} \cos \theta_0, \qquad X_L = \sqrt{R_0^2 + X_L^2} \sin \theta_0, \quad (2.21)$$

we may still further simplify (2.19) and (2.20) to give

$$V(x) = I_L \sqrt{R_0^2 + X_L^2}\, e^{j\pi/2} \sin (\beta s + \theta_0), \quad (2.22)$$

$$I(x) = \frac{I_L}{R_0} \sqrt{R_0^2 + X_L^2} \cos (\beta s + \theta_0), \quad (2.23)$$

where θ_0 is obtained from (2.21) as

$$\theta_0 = \tan^{-1} (X_L/R_0). \quad (2.24)$$

In view of (2.22) and (2.23), the phase is conveniently referred to that of the load current I_L, which is, therefore, assumed to be real. For the sake of definiteness, we assume the reactance X_L to be positive. The phase angle θ_V of $V(x)$ is seen from (2.22) to start with $\pi/2$ for $s = 0$, remains unchanged as βs is varied except for $\beta s = n\pi - \theta_0$ (n is an integer), when $\sin \beta s$ changes sign and θ_V increases abruptly by π radians. In a similar manner, the phase angle θ_I of $I(x)$ starts with zero for $\beta s = 0$, remains unchanged as βs is varied except for $\beta s = (2n - 1)(\pi/2) - \theta_0$, when $\cos \beta s$ changes sign and θ_I increases abruptly by π radians. We depict the phase angles θ_V and θ_I corresponding to a reactive load as a function of βs in Fig. 3.3(c).

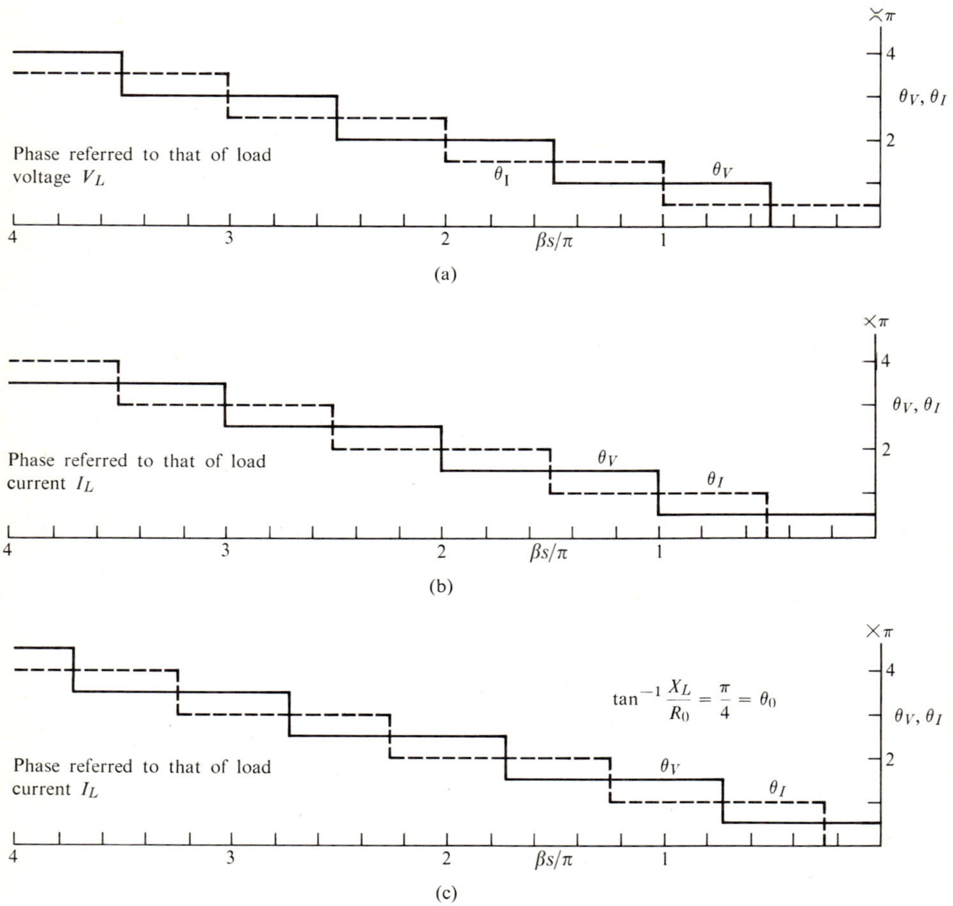

Fig. 3.3 The phase angles of the voltage and the current on a lossless transmission line as a function of the electrical distance from the load in radians. (a) Open-circuit termination; (b) short-circuit termination; (c) reactive termination; (d) resistive termination; (e) impedance termination.

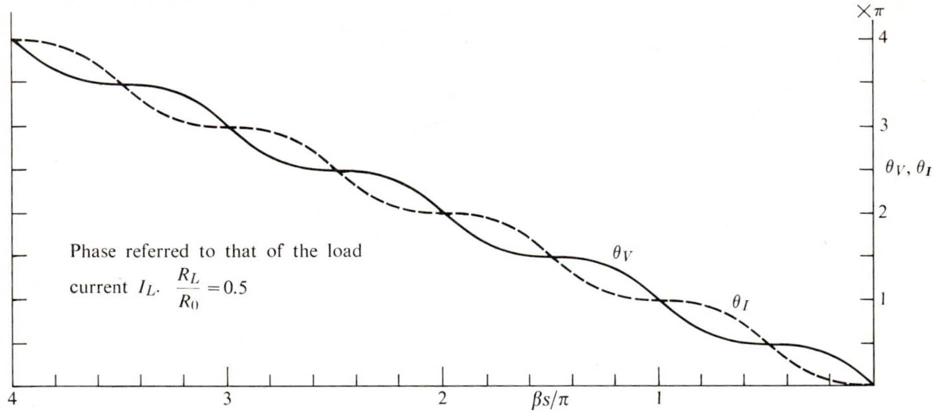

(d)

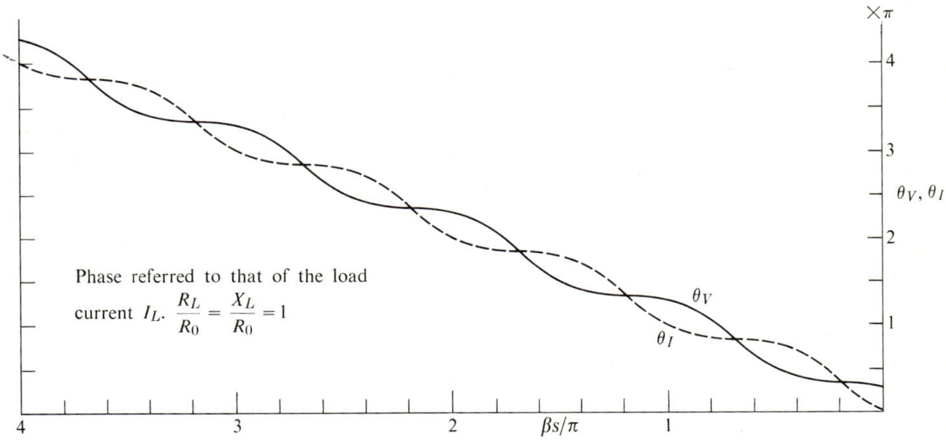

(e)

Fig. 3.3 *(continued)*

Let us consider a transmission line which is terminated in the load impedance $Z_L = R_L + jX_L$. Then, from (2.9.19) and (2.9.20), we obtain

$$V(x) = I_L[R_L \cos \beta s + j\{X_L \cos \beta s + R_0 \sin \beta s\}], \tag{2.25}$$

$$I(x) = \frac{I_L}{R_0}[\{-X_L \sin \beta s + R_0 \cos \beta s\} + jR_L \sin \beta s]. \tag{2.26}$$

An examination of (2.25) and (2.26) reveals that the phases are conveniently referred again to that of the load current I_L which is taken to be real. We determine the phase angles θ_V of $V(x)$ and θ_I of $I(x)$ from (2.25) and (2.26); the results are stated first for the special case of a resistive load corresponding to $X_L = 0$ and then for the general case with both R_L and X_L present.

Resistive Load

$$\theta_V = \tan^{-1}\left\{\frac{R_0}{R_L}\tan\beta s\right\}. \tag{2.27}$$

$$\theta_I = \tan^{-1}\left\{\frac{R_L}{R_0}\tan\beta s\right\}. \tag{2.28}$$

Load with Both R_L and X_L Present

$$\theta_V = \tan^{-1}\left\{\frac{X_L}{R_L} + \frac{R_0}{R_L}\tan\beta s\right\}. \tag{2.29}$$

$$\theta_I = \tan^{-1}\left[\frac{1}{\dfrac{-X_L}{R_L} + \dfrac{R_0}{R_L}\cot\beta s}\right]. \tag{2.30}$$

We evaluate the phase angles θ_V and θ_I as a function of βs from (2.27) to (2.30) both for the case of a resistive load and for the general case with both R_L and X_L present. These results are shown in Fig. 3.3(d) for the resistive load and in Fig. 3.3(e) for the load containing both a resistive and a reactive part. In Fig. 3.3(d) and 3.3(e), we see that the phase angles vary continuously as a function of βs. In contrast, for the case of open-circuit and short-circuit terminations as well as for the case of a reactive load, the phase angles vary discontinuously as a function of βs. We note that for the cases in which the phase angles vary discontinuously as a function of βs, the magnitude $|\rho|$ of the reflection coefficient is unity.

We find from Fig. 3.3(d) that the curves of the phase angles θ_V and θ_I corresponding to a resistive termination possess certain symmetries. It is interesting to examine these symmetries. Let $\beta s_0 < \pi/2$. Let us consider two locations on the transmission line such that one of them is located at a distance βs_0 away from the load and the other at the same distance away from the location $\beta s = \pi$ in the direction towards the load. Thus, the two locations on the transmission line are specified by

$$\begin{aligned} &1)\ \beta s_1 = \beta s_0,\\ &2)\ \beta s_2 = \pi - \beta s_0. \end{aligned} \tag{2.31}$$

Let

$$\theta = \tan^{-1}\left\{\frac{R_0}{R_L}\tan\beta s\right\} \tag{2.32}$$

be an acute angle. Also, let all the phase angles be referred to the phase angle βs. Therefore, for the first location, we obtain

$$(\theta_V - \beta s_1) = \theta - \beta s_0. \tag{2.33}$$

We can show that

$$\tan^{-1}\left\{\frac{R_0}{R_L}\tan\beta s_2\right\} = \pi - \theta. \tag{2.34}$$

Therefore, for the second location, we find that

$$(\theta_V - \beta s_2) = \pi - \theta - \beta s_2 = \beta s_0 - \theta, \tag{2.35}$$

which is the negative of that obtained for the first location. We can argue that the foregoing results imply that if the phase angle curve θ_V in the range $0 < \beta s < \pi/2$ is reflected about the straight line $\theta_V = \beta s$ and the resulting curve reflected again about the straight line $\theta_V = \pi - \beta s$, which passes through the points $\beta s = \pi$, $\theta_V = 0$, and $\beta s = \pi/2$, $\theta_V = \pi/2$, we obtain the phase angle curve θ_V corresponding to the range $\pi/2 < \beta s < \pi$. Since the curve $\theta_V - \beta s$ is periodic of period π, it follows that if the phase angle curve in the range $(n - 1)\pi < \beta s < (2n - 1)\pi/2$ is reflected about the straight line $\theta_V = \beta s$ and the resulting curve reflected again about the straight line $\theta_V = (2n - 1)\pi - \beta s$, which passes through the points $\beta s = (2n - 1)\pi$, $\theta_V = 0$, and $\beta s = (2n - 1)\pi/2$, $\theta_V = (2n - 1)\pi/2$, we obtain the phase angle curve θ_V corresponding to the range $(2n - 1)\pi/2 < \beta s < n\pi$. These results hold good also for the phase angle curve θ_I corresponding to the resistive termination.

From (2.17), the magnitudes of the voltage and the current distribution corresponding to the open-circuit termination are obtained as follows.

Open-circuit Termination

$$\frac{|V(x)|}{V_L} = |\cos \beta s|; \qquad \frac{|I(x)|}{V_L/R_0} = |\sin \beta s|. \tag{2.36}$$

We take the terminal voltage to be real and use it for normalization. In all the other cases, we assume the terminal current I_L to be real and use that for normalization. In a similar manner, we deduce from (2.18), (2.22) to (2.24), (2.25), and (2.26), the magnitudes of the voltage and the current distribution corresponding to the other types of terminations with the following results.

Short-circuit Termination

$$\frac{|V(x)|}{I_L R_0} = |\sin \beta s|; \qquad \frac{|I(x)|}{I_L} = |\cos \beta s|. \tag{2.37}$$

Reactive Termination

$$\frac{|V(x)|}{I_L\sqrt{R_0^2 + X_L^2}} = |\sin (\beta s + \theta_0)|, \qquad \frac{|I(x)|}{I_L\sqrt{R_0^2 + X_L^2}/R_0} = |\cos (\beta s + \theta_0)|,$$

where

$$\tan \theta_0 = X_L/R_0. \tag{2.38}$$

Resistive Termination

$$\frac{|V(x)|}{I_L R_0} = \left[\frac{R_L^2}{R_0^2}\cos^2 \beta s + \sin^2 \beta s\right]^{1/2}, \tag{2.39a}$$

$$\frac{|I(x)|}{I_L} = \left[\cos^2 \beta s + \frac{R_L^2}{R_0^2}\sin^2 \beta s\right]^{1/2} \tag{2.39b}$$

Impedance Termination

$$\frac{|V(x)|}{I_L R_0} = \left[\frac{R_L^2}{R_0^2}\cos^2 \beta s + \left\{\frac{X_L}{R_0}\cos \beta s + \sin \beta s\right\}^2\right]^{1/2}, \tag{2.40a}$$

$$\frac{|I(x)|}{I_L} = \left[\left\{-\frac{X_L}{R_0}\sin \beta s + \cos \beta s\right\}^2 + \frac{R_L^2}{R_0^2}\sin^2 \beta s\right]^{1/2} \tag{2.40b}$$

We evaluate the normalized magnitudes of the voltage and the current distributions with the help of (2.36) to (2.40) for the various types of termination and use them in the construction of Fig. 3.2(a) to 3.2(e).

3.3 PROGRESSIVE AND STANDING WAVES

Let us consider a lossless, semi-infinite ($0 < x < \infty$) transmission line or a lossless, finite ($0 < x < l$) transmission line terminated by its characteristic resistance. The lossless transmission line is always considered as the limiting case of a low-loss transmission line. Then, it follows from the requirement of finiteness or (2.9.15) that V_2 in (2.9.1) and (2.9.2) is equal to zero. We obtain, therefore, the real voltage and the real current distributions from (2.9.1) and (2.9.2) as in (2.6.3) with the following result:

$$V(x, t) = |V_1| \cos(\omega t + \theta_{V_1} - \beta x), \tag{3.1}$$

$$I(x, t) = \frac{|V_1|}{R_0} \cos(\omega t + \theta_{V_1} - \beta x). \tag{3.2}$$

Equations (3.1) and (3.2) represent the so-called *progressive waves* which are characterized by the following two features:

a) The *amplitude* of the voltage and the current distributions are *independent* of the location on the transmission line.
b) The *phase* of the voltage and the current distributions vary *linearly* with distance along the transmission line, the proportionality constant being β.

Moreover, at a given time, the phenomena repeat themselves along the transmission line every $2\pi/\beta$ m, as a consequence of which the space period is $2\pi/\beta$.

When a lossless, finite-length ($0 < x < l$) transmission line is terminated in an impedance which differs from the characteristic resistance, the phasor voltage and the phasor current are as given in (2.5.2) and (2.5.3) and may be rewritten here as

$$V(x) = |V(x)|\, e^{j\theta_V(x)}, \tag{3.3}$$

$$I(x) = |I(x)|\, e^{j\theta_I(x)}. \tag{3.4}$$

We include the argument x for the phase angles θ_V and θ_I to indicate their dependence on the location along the transmission line. From (3.3) and (3.4), it is an easy matter to obtain the expressions for the real voltage and the real current distributions along the transmission line. The results are

$$V(x, t) = |V(x)| \cos [\omega t + \theta_V(x)], \tag{3.5}$$

$$I(x, t) = |I(x)| \cos [\omega t + \theta_I(x)]. \tag{3.6}$$

We depict in Fig. 3.3 for various types of load impedances the phase angles $\theta_V(x)$ and $\theta_I(x)$, as given in (2.29) and (2.30) respectively. We note that the phase does *not vary linearly* with distance along the transmission line. The deviation from linearity of the phase change with distance is quite marked for the open-circuit and the short-circuit terminations as well as when the load impedance is reactive. In these special cases the phase remains constant with distance along the transmission line except at certain discrete locations where the phase changes abruptly by π radians. In these cases, the reflection is complete in the sense that the magnitude $|\rho|$ of the reflection coefficient has the maximum possible value of unity. We illustrate in Fig. 3.2 for various types of termination the amplitudes $|V(x)|$ and $|I(x)|$ as given in (2.5) and (2.6) respectively. We note that the amplitudes $|V(x)|$ and $|I(x)|$ are not *constants* along the transmission line, but vary periodically between the same maximum and minimum values. In such a case, *standing waves* are said to be formed on the transmission line. In contrast to the progressive waves, the standing waves are characterized by the following two features:

a) The *amplitude* of the voltage and the current distributions *vary* with the location on the transmission line, oscillating between the same maximum and minimum values.
b) The *phase* of the voltage and the current distributions do *not vary linearly* with distance along the transmission line. In the limiting case of fully developed standing waves corresponding to $|\rho| = 1$, the phase remains constant with distance except at certain discrete locations where it changes abruptly by π radians.

An examination of the *amplitude* of the standing waves as shown in Fig. 3.2 appears to indicate that at a given time, in the standing waves, the phenomenon repeats itself every π/β m. However, examination of the *phase* as shown in Fig. 3.3 reveals that in the standing waves, at a given time, the phenomenon repeats itself every $2\pi/\beta$ m. Since every wave phenomenon is described completely only by its amplitude and phase, it follows that the space period of the standing waves is the same as that of the progressive waves and is equal to $2\pi/\beta$ m. We must emphasize the fact that the space period is the same both for the progressive and the standing waves.

The parameter which is indicative of the existence as well as the extent of the development of the standing waves is called the *standing-wave ratio* (SWR), which is defined as the ratio of the maximum to the minimum amplitude of either the voltage or the current. From (2.11), (2.12), (2.14), and (2.15), we obtain the SWR as

$$S = \frac{|V(x)|_{max}}{|V(x)|_{min}} = \frac{|I(x)|_{max}}{|I(x)|_{min}} = \frac{1 + |\rho|}{1 - |\rho|}. \tag{3.7}$$

Since $|\rho|$ varies between 0 and 1, it follows from (3.7) that S varies between 1 to ∞ and is always positive. For the progressive waves, since the amplitude does not vary along the transmission line, the standing-wave ratio $S = 1$. Just as the lower limit $S = 1$ of the standing-wave ratio indicates the absence of the standing waves, so the upper limit $S = \infty$ denotes the fully developed standing waves. Note that $S = \infty$ corresponds to $|\rho| = 1$. Consequently, open-circuit and short-circuit terminations as well as a reactive load impedance, all of which yield $|\rho| = 1$, lead to fully developed standing waves.

3.4 IMPEDANCE

From (2.2) and (2.3), we obtain the impedance $Z(x) = V(x)/I(x)$ along the transmission line as

$$Z(x) = R_0 \frac{[1 + |\rho|\, e^{j\phi}]}{[1 - |\rho|\, e^{j\phi}]}. \tag{4.1}$$

At the locations of voltage minima or current maxima, we can use (2.10) and (3.7) in (4.1) to give

$$Z(x) = R_0 \frac{(1 - |\rho|)}{(1 + |\rho|)} = \frac{R_0}{S}, \tag{4.2}$$

since $\phi = -(2n - 1)\pi$ and $e^{j\phi} = -1$. In the same manner, at the locations of voltage maxima or current minima, we can use (2.13) and (3.7) in (4.1) to obtain

$$Z(x) = R_0 \frac{(1 + |\rho|)}{(1 - |\rho|)} = R_0 S, \tag{4.3}$$

since $\phi = -2n\pi$ and $e^{j\phi} = 1$. Thus, at the locations of voltage minima the impedance is a minimum and is resistive as given in (4.2). At the locations of voltage maxima, the impedance is a maximum and is again resistive, as given in (4.3).

It is convenient to separate the general expression (4.1) for the impedance looking towards the load at any location on the transmission line into its real and imaginary parts. For that purpose, we set

$$|\rho|\, e^{j\phi} = u + jv; \qquad \text{i.e. } u = |\rho| \cos\phi; \qquad v = |\rho| \sin\phi. \tag{4.4}$$

Together with (4.4), (4.1) becomes

$$\frac{Z(x)}{R_0} = \frac{1 + u + jv}{1 - u - jv}. \tag{4.5}$$

On multiplying the numerator and the denominator of (4.5) by the complex conjugate of the denominator, we obtain

$$\frac{Z(x)}{R_0} = \frac{1 - u^2 - v^2 + j2v}{(1 - u)^2 + v^2}. \tag{4.6}$$

When the real and the imaginary parts of (4.6) are written down separately, we find that

$$r = \frac{R(x)}{R_0} = \frac{1 - u^2 - v^2}{(1 - u)^2 + v^2}; \qquad x = \frac{X(x)}{R_0} = \frac{2v}{(1 - u)^2 + v^2}. \tag{4.7a,b}$$

The ratio $R(x)/R_0$ of the resistance to the characteristic resistance is called *per unit resistance* r and, in a similar manner, the ratio $X(x)/R_0$ of the reactance to the characteristic resistance is called *per unit reactance* x. The substitution of u and v in terms of $|\rho|$ and ϕ from (4.4) into (4.7) gives the following expressions for the per unit resistance and reactance:

$$\frac{R(x)}{R_0} = \frac{1 - |\rho|^2}{(1 + |\rho|^2 - 2\,|\rho| \cos \phi)},$$

$$\frac{X(x)}{R_0} = \frac{2\,|\rho| \sin \phi}{(1 + |\rho|^2 - 2\,|\rho| \cos \phi)}. \tag{4.8a,b}$$

We note that the denominator in (4.8), since it is equal to $(1 - u)^2 + v^2$, is always positive.

We see from (4.8) that the reactance is proportional to $\sin \phi$. At the locations of voltage minima $\phi = -(2n - 1)\pi$ and at the locations of voltage maxima $\phi = -2n\pi$. At these locations of voltage extrema, $\sin \phi = 0$ and therefore the reactance vanishes; the impedance becomes resistive. We can recover (4.2) and (4.3) corresponding to the locations of voltage minima and maxima respectively from the expression (4.8a) for the resistance.

The nature of the reactance, that is whether it is inductive or capacitive, depends on the sign of $\sin \phi$. Let us consider the following two cases:

Case (a).

$$-2(n - 1)\pi > \phi > -(2n - 1)\pi,$$

Case (b).

$$-(2n - 1)\pi > \phi > -2n\pi. \tag{4.9}$$

For the Case (a) $\sin \phi$ is negative and the reactance is capacitive and for the Case (b) $\sin \phi$ is positive and the reactance is inductive. There is an interesting connection between the nature of the reactance at a given location on the transmission line and the slope of the voltage standing-wave pattern at the same location. The voltage standing-wave pattern is the curve of the magnitude of the voltage as a function of βs. From (2.4) and (2.5), we easily obtain

$$\frac{d}{d(\beta s)}\left\{\frac{|V(x)|}{V_M}\right\} = \frac{2\,|\rho| \sin \phi}{[1 + |\rho|^2 + 2\,|\rho| \cos \phi]^{1/2}}. \tag{4.10}$$

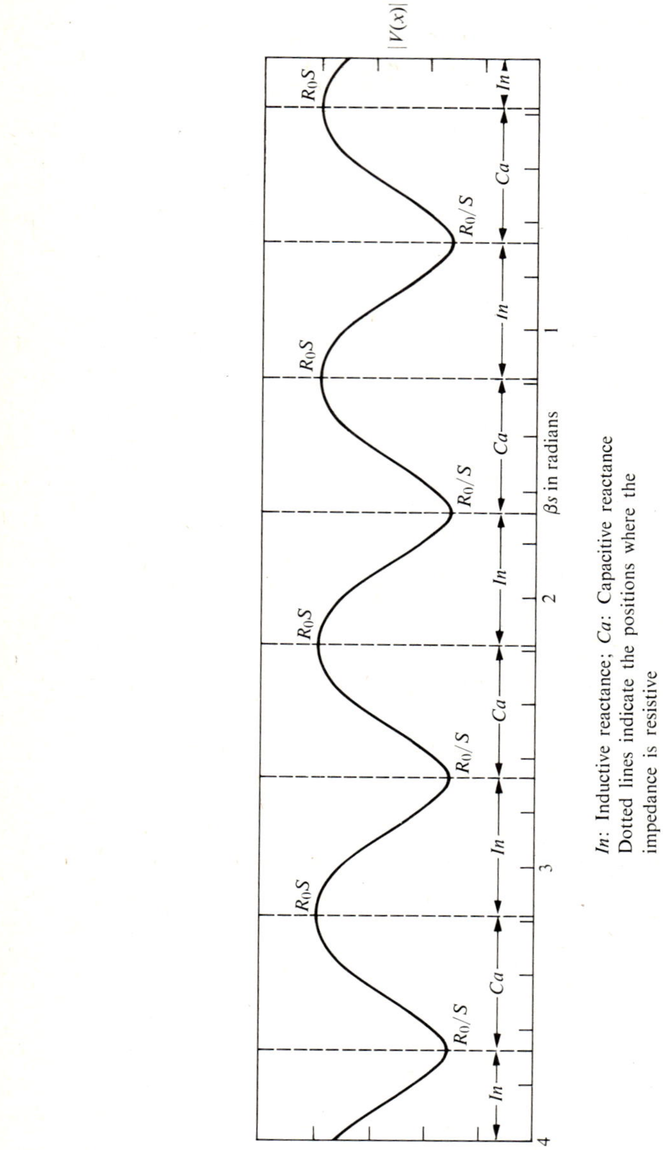

Fig. 3.4 Voltage standing-wave pattern showing the locations where the impedance is a resistance and the nature of the reactance at the other locations.

When (4.10) is positive, the voltage standing-wave pattern slopes *down* towards the load. This is because βs increases as the observation point moves away from the load terminals. When (4.10) is negative, the voltage standing-wave pattern slopes *up* towards the load. We should note that the sign of (4.10) is controlled by that of $\sin \phi$, whose sign is also that of the reactance as given in (4.8). Therefore, it follows that if the voltage standing-wave pattern slopes *down towards the load* the corresponding reactance is *inductive*, and if it slopes *up towards the load* the corresponding reactance is *capacitive*.

In Fig. 3.4, we have indicated on a voltage standing-wave pattern the locations where the impedance is a resistance and also the nature of the reactance at the other locations.

3.5 DETERMINATION OF UNKNOWN IMPEDANCE

We wish to determine the unknown impedance Z_L connected to the terminals ($s = 0$) of a given transmission line. Since the value of the impedance depends on the angular frequency ω, the angular frequency ω for which we want the impedance value is assumed to be given. We also assume that the characteristics of the transmission line, taken to be lossless, are known; that is, the characteristic resistance R_0 as well as the velocity v associated with the transmission line are specified. Since ω and v are known, it follows that we also know $\beta = \omega/v$.

The unknown impedance Z_L is the value of the impedance in (4.1) evaluated at $s = 0$. Hence, from (4.1) and (2.4), we find that

$$Z_L = R_0 \frac{[1 + |\rho| \, e^{j\theta_\rho}]}{[1 - |\rho| \, e^{j\theta_\rho}]} . \tag{5.1}$$

In order to be able to determine Z_L, we see from (5.1) that the amplitude $|\rho|$ and the phase θ_ρ of the reflection coefficient have to be obtained by measurement. The following measurements are usually made to determine $|\rho|$ and θ_ρ.

a) We obtain the standing-wave ratio S by the measurement of the maximum and the minimum of the magnitude of the voltage distribution along the transmission line. From (3.7), it follows that we can then deduce $|\rho|$ through the relation

$$|\rho| = \frac{S - 1}{S + 1} . \tag{5.2}$$

b) We measure the distance $s_{1,\min}$ of the first minimum of the magnitude of the voltage from the load terminals. It follows from (2.4) and (2.10) that for the first ($n = 1$) minimum

$$\phi = -\pi = \theta_\rho - 2\beta s_{1,\min} \quad \text{or} \quad \theta_\rho = -\pi + 2\beta s_{1,\min} . \tag{5.3}$$

Since β is known, we can evaluate θ_ρ with the help of (5.3).

Once we have obtained $|\rho|$ and θ_ρ from the foregoing two measurements, since R_0 is known, we can evaluate Z_L with the help of (5.1). It is usual to make another measurement and determine β also.

c) We measure the distance $s_{2,\min}$ from the load terminals, of the minimum of the magnitude of the voltage adjacent to the first one. We obtain from (2.4) and (2.10) that

$$-3\pi = \theta_\rho - 2\beta s_{2,\min}. \tag{5.4}$$

From (5.3) and (5.4), we find that

$$\beta = \pi/(s_{2,\min} - s_{1,\min}). \tag{5.5}$$

Thus, β can be obtained.

We should note that in obtaining β, it is not necessary for us to measure the distances of the first and the second voltage minima from the load terminals. All that is necessary is for us to measure the separation distance between two successive voltage minima. Since they are usually sharper, the voltage minima can be located more precisely than the voltage maxima. Hence, we use the voltage minima for obtaining β.

Example. A lossless transmission line of characteristic resistance $R_0 = 50$ ohms is terminated in an impedance Z_L whose value we wish to determine. We find the standing-wave ratio S for the voltage distribution to be 3. The locations of the first and the second voltage minima from the load terminals are given by $s_{1,\min} = 2$ m and $s_{2,\min} = 7$ m.

From (5.2), we deduce $|\rho|$ as

$$|\rho| = \frac{S-1}{S+1} = \frac{3-1}{3+1} = 0.5. \tag{5.6}$$

We obtain the value of β from (5.5) as

$$\beta = \frac{\pi}{7-2} = \frac{\pi}{5}. \tag{5.7}$$

In view of (5.7),

$$\frac{\lambda}{4} = \frac{1}{4} \times \frac{2\pi}{\beta} = \frac{1}{4} \times \frac{2\pi}{\pi/5} = 2.5 \text{ m}.$$

Since the first voltage minimum occurs at a distance less than $\lambda/4$ from the load, it follows that there is no voltage maximum between the first voltage minimum and the load. Consequently, the voltage standing-wave pattern slopes up towards the load, at the load terminals. From the discussion of the previous section, we conclude that the load impedance has a capacitive reactance.

From (5.3), we obtain

$$\theta_\rho = -\pi + 2 \times \frac{\pi}{5} \times 2 = -0.2\pi. \tag{5.8}$$

The substitution of (5.6) and (5.8) in (5.1) yields the value of the load impedance as

$$Z_L = 50\frac{[1 + 0.5\, e^{-j0.2\pi}]}{[1 - 0.5\, e^{-j0.2\pi}]} = 85.1 - j66.7 \text{ ohms}. \tag{5.9}$$

The reactance as evaluated in (5.9) is capacitive, confirming the earlier deduction on the nature of the load reactance.

Often times, it is difficult for us to make meaningful measurements on the transmission line very near the load. As a consequence, we have to ascertain the distance $s_{1,\text{min}}$ from the load to the location of the first minimum of the magnitude of the voltage in a different manner. However, the measurements of the voltage standing-wave ratio S and the distance $(s_{2,\text{min}} - s_{1,\text{min}})$ between two successive voltage minima need not necessarily be carried out near the load and can be conveniently performed further away from the load terminals. We can determine the distance $s_{1,\text{min}}$ from the load to the first voltage minimum in the following manner. We replace the load impedance by a short-circuit. We note that one of the voltage minima coincides with the location of the load. Along the transmission line, at a convenient distance from the load, we mark the location of one of the voltage minima with a short-circuit termination. The short-circuit termination is then replaced by the impedance to be measured. The voltage minimum, in general, does not occur at the location of the load but moves away through a distance $s_{1,\text{min}}$. Since the distance between successive voltage minima depends only on β and not on the terminal impedance, we see that the distance between successive voltage minima is not altered when the short-circuit termination is replaced by the actual impedance. Therefore, at a distance $s_{1,\text{min}}$ in the direction *away from the load* from every location of the voltage minima with a short-circuit termination, there are voltage minima when the short-circuit is replaced by the actual unknown impedance, as we may see with the help of Fig. 3.5. We measure the distance from the location, which we have already marked, of one of the voltage minima with a short-circuit termination in the direction *away from the load* to the location of the voltage minimum corresponding to the transmission line terminated by the actual unknown impedance to give $s_{1,\text{min}}$. We evaluate the unknown impedance Z_L using the knowledge of S and $s_{1,\text{min}}$, as we have already explained.

Suppose, instead of measuring the distance in the direction *away from the load*, we measure it in the direction *towards the load* and denote it by $\bar{s}_{1,\text{min}}$. Since the successive voltage minima occur at a distance of $\lambda/2$, we find that

$$s_{1,\text{min}} = (\lambda/2) - \bar{s}_{1,\text{min}}, \tag{5.10}$$

as we can see from Fig. 3.5.

One of the common measurements using the procedure which we have discussed in the foregoing is that of the input impedance at a given frequency of an antenna driven by a transmission line. Knowledge of the variation of the input impedance of an antenna as a function of frequency is important in evaluating its performance. We can obtain the input impedance of an antenna at any frequency by performing the standing-wave measurements on the transmission line feeding it.

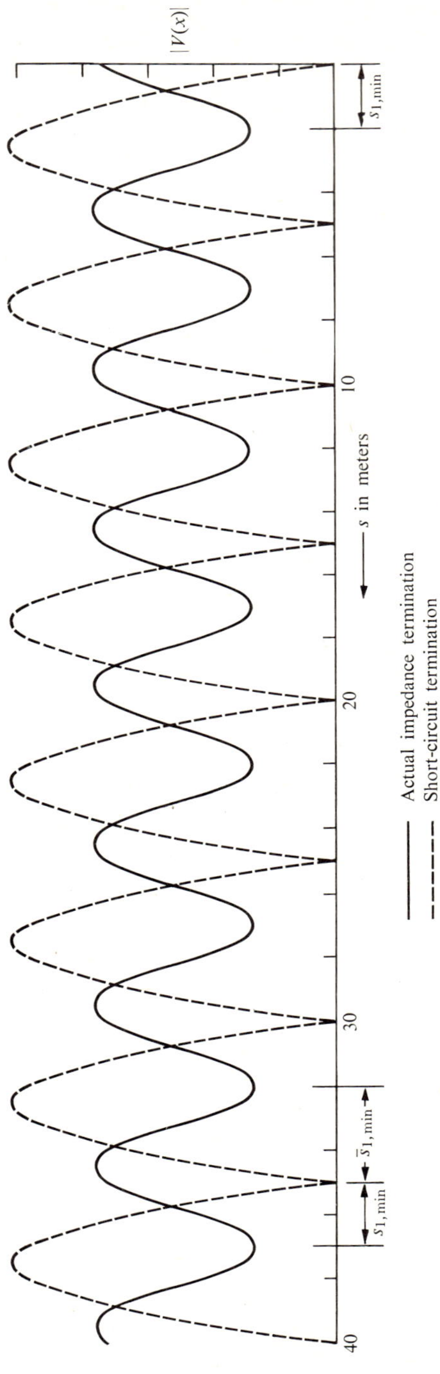

Fig. 3.5 Voltage standing-wave pattern for the determination of the unknown load impedance.

3.6 THEORY OF SMITH CHART FOR IMPEDANCE AND ADMITTANCE

We define the voltage reflection coefficient ρ for the load impedance Z_L from (2.12.12) as

$$\rho = |\rho|\, e^{j\theta_\rho} = \frac{Z_L - R_0}{Z_L + R_0}. \tag{6.1}$$

For convenience, we designate (6.1) as the *terminal* reflection coefficient. In the same manner as for the load impedance Z_L, we can obtain a *general* reflection coefficient for the impedance $Z(x)$ looking towards the load at any point along the transmission line with the help of (4.1):

$$\frac{Z(x) - R_0}{Z(x) + R_0} = |\rho|\, e^{j\phi}; \qquad \phi = \theta_\rho - 2\beta s. \tag{6.2}$$

Note that (6.1) is a special case of (6.2) corresponding to $s = 0$. If we compare (6.1) and (6.2) we see that the magnitude $|\rho|$ of the *general* reflection coefficient is the same as that for the terminal reflection coefficient and hence can be determined entirely by the load impedance. We obtain the phase ϕ of the general reflection coefficient in a very simple manner from that of the terminal reflection coefficient by subtracting twice the electrical distance of the observation point from the load terminals.

The following important observations are worthy of emphasis.

a) As the observation point moves away from the load, the *magnitude* of the general reflection coefficient remains *unchanged*. (6.3)

b) As the observation point moves away from the load, the *phase* of the general reflection coefficient *decreases* by *twice* the electrical distance covered in the direction away from the load. (6.4)

From (6.2), we see that once the general reflection coefficient $[|\rho|, \phi]$ is given, we can obtain the impedance $[R(x), X(x)]$ at any point along the transmission line looking towards the load or vice versa. We see from (6.1), as a special case, that once the terminal reflection coefficient $[|\rho|, \theta_\rho]$ is given, the load impedance $[R_L, X_L]$ can be obtained or vice versa. In 1939, P. H. Smith* devised a widely used graphical chart, now known as the Smith chart, for effecting this calculation. Essentially, the Smith chart portrays contours of constant per unit resistance and per unit reactance or per unit conductance and per unit susceptance on the general reflection coefficient plane, thus establishing a one-to-one correspondence between the general reflection coefficient and the impedance or the admittance. Although omitted hereafter for the sake of brevity, we should understand that the impedance or the admittance values correspond to those looking towards the load.

It is instructive for us to consider first the general reflection coefficient plane and examine the procedure for obtaining the general reflection coefficient given the terminal reflection coefficient and the distance s of the observation point from the load terminals. We can describe a point on the general reflection coefficient plane

*P. H. Smith, "Transmission-line Calculator", *Electronics*, **12,** 29–31 (January, 1939).

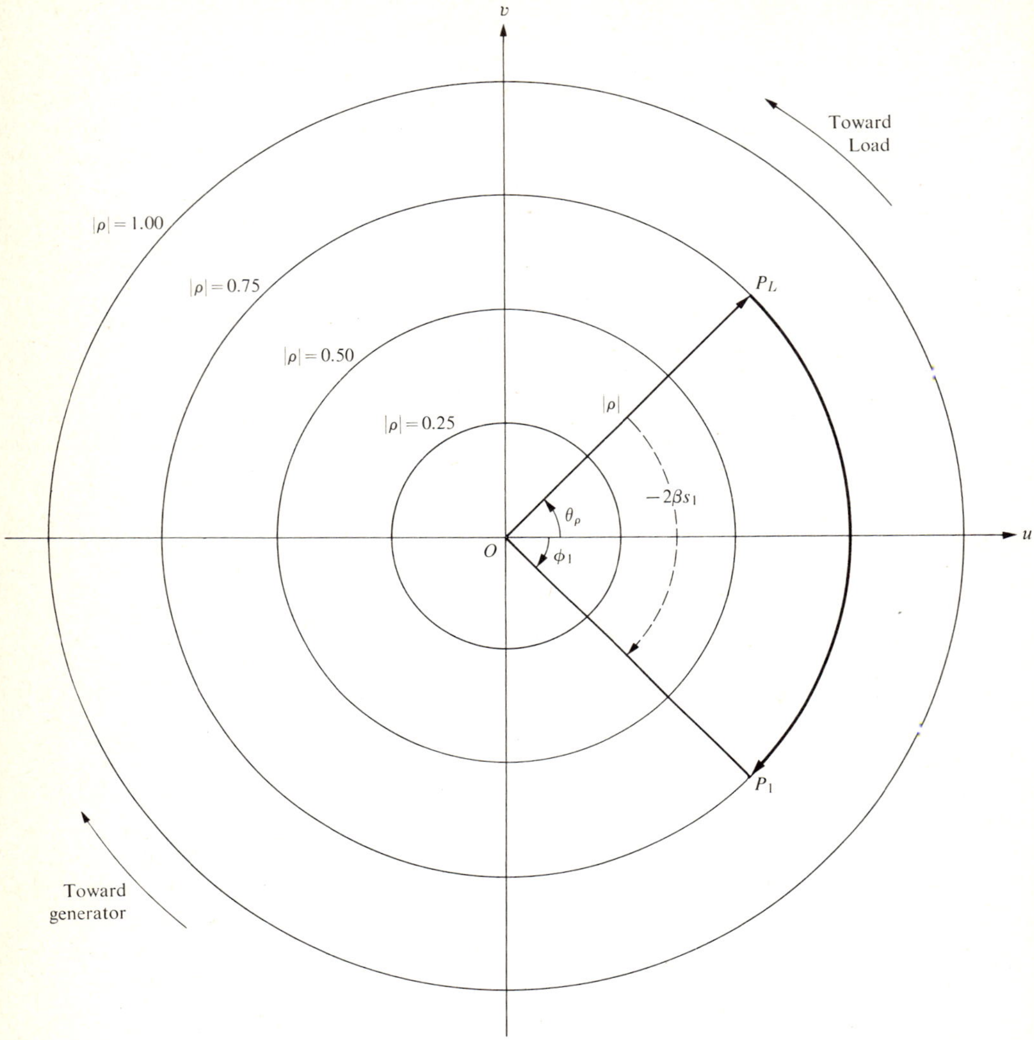

Fig. 3.6 Determination of the representative point on the general reflection coefficient plane.

either in terms of the polar coordinates $|\rho|$ and ϕ or in terms of the equivalent rectangular coordinates u and v. The relationships of u and v to $|\rho|$ and ϕ are as expressed in (4.4). Contours of constant $|\rho|$ are circles in the reflection coefficient plane. In Fig. 3.6, we show circles corresponding to $|\rho| = 0.25, 0.50, 0.75,$ and 1.00. For a lossless transmission line $|\rho|$ never exceeds unity, as a consequence of which the representative point on the reflection coefficient plane always lies within the circle corresponding to $|\rho| = 1$. We represent a terminal reflection coefficient by the point P_L in Fig. 3.6. Evidently, $OP_L = |\rho|$ and for the case shown in Fig. 3.6, $OP_L = 0.75$. Also,

$\sphericalangle P_L O u = \theta_\rho$, the phase of the terminal reflection coefficient. For the case shown in Fig. 3.6, $\sphericalangle P_L O u = \pi/4$ radians. Suppose that we wish to find the general reflection coefficient at an observation point at a distance $s = s_1 = \lambda/8$ from the load terminals. From (6.3), we see that as the observation point moves from the load terminals, the representative point on the reflection coefficient plane moves around the circumference of a circle of radius equal to $|\rho|$ pertaining to the load impedance. For the case shown in the figure, the point P_L moves along the circumference of the circle of radius $|\rho| = 0.75$. From (6.4), we see that as the observation point moves away from the load (or equivalently towards the generator), P_L moves in the *clockwise* direction through an angle $2\beta s_1 = 2 \times (2\pi/\lambda) \times \lambda/8 = \pi/2$ radians. We should note that the motion of the representative point on the reflection coefficient plane in the *clockwise* (*counter-clockwise*) direction corresponds to the motion of the observation point along the transmission line in the direction *towards the generator* (*towards the load*) and we have included these results in Fig. 3.6 for convenience. The required general reflection coefficient is therefore represented by the point P_1. The $\sphericalangle P_L O P_1 = 2\beta s_1$ radians and the $\sphericalangle u O P_1 = \phi_1 = \theta_\rho - 2\beta s_1$ radians. Clearly, $OP_1 = |\rho|$. Suppose now that we wish also to find the general reflection coefficient at an observation point which is at a distance s_2 *towards the load* from the first observation point. The representative point P_1 should be moved along the circumference of the same circle but now in the *counter-clockwise* direction through an angle $2\beta s_2$ radians to obtain the new representative point P_2 (not shown in Fig. 3.6) giving the general reflection coefficient for the second observation point. The magnitude $|\rho|$ of the general reflection coefficient at the second observation point is equal to OP_2 and the phase ϕ_2 is equal to $\sphericalangle u O P_2 = \theta_\rho - 2\beta s_1 + 2\beta s_2$. In this manner, we obtain the general reflection coefficient at any location along the transmission line from the knowledge of its value at a given location as well as the position of the new location relative to the given location.

We pointed out earlier that there is a one-to-one correspondence between the general reflection coefficient and the per unit impedance. The values of the per unit resistance r and the per unit reactance x corresponding to every point on the reflection coefficient or the u–v plane can be obtained in the following manner. On multiplying both sides of (4.7a) by the denominator on the right-hand side of the equation we find after rearrangement

$$u^2(r + 1) - 2ur + v^2(r + 1) = 1 - r. \tag{6.5}$$

If we divide (6.5) throughout by $(r + 1)$ and complete the square in u, we obtain

$$\left(u - \frac{r}{r + 1}\right)^2 + v^2 = \frac{1}{(1 + r)^2}. \tag{6.6}$$

In a similar manner, if we multiply both sides of (4.7b) by the denominator on the right-hand side of the equation we obtain after some rearrangement

$$u^2 x - 2ux + v^2 x - 2v = -x. \tag{6.7}$$

If we then divide (6.7) throughout by x and complete the squares in u and v, we deduce that

$$(u - 1)^2 + \left(v - \frac{1}{x}\right)^2 = \frac{1}{x^2}. \tag{6.8}$$

From (6.6), we see that a given r corresponds to a circle centered at $u = r/(r + 1)$, $v = 0$, and radius equal to $1/(r + 1)$. As an example, $r = 0$ corresponds to a circle centered at $u = 0$, $v = 0$, and radius equal to 1. We should note that this circle is identical with that corresponding to $|\rho| = 1$. The circles corresponding to all the other values of r greater than zero lie within the unit circle. The circle corresponding to $r = 1$ is centered at $u = \frac{1}{2}$, $v = 0$, and radius equal to $\frac{1}{2}$ and it therefore passes through the origin. Since $u = 1, v = 0$ satisfies (6.6) for any value of r, it follows that all the resistance circles pass through the point $u = 1$, $v = 0$. Since the centers of all the resistance circles lie on the same straight line $v = 0$ and since their radii decrease with increasing r, it follows that a circle corresponding to a larger r lies completely within that for a smaller r. Since $r \geq 0$, it follows that all the representative points lie within the unit circle corresponding to $r = 0$. This result is in agreement with that derived via the fact that $|\rho|$ never exceeds unity for a passive load. In Fig. 3.7, we show the resistance circles corresponding to $r = 0, 0.5, 1, 2$, and 4.

In a similar manner, we see from (6.8) that a given x also corresponds to a circle centered at $u = 1$, $v = 1/x$, and radius equal to $1/|x|$. Since x can be positive or negative, the centers of all the reactance circles lie on the line $u = 1$ on either side of the u-axis. For example, the circle corresponding to $x = 0$ is centered at $u = 1$, $v = \infty$, and radius equal to ∞. It is not difficult for us to see that this circle degenerates into a straight line coinciding with the u-axis. Since $u = 1, v = 0$, satisfies (6.8) for any x value, it follows that all the reactance circles also pass through the point $u = 1, v = 0$. Since the centers of all the reactance circles lie on the line $u = 1$, at a distance $1/x$ from $u = 1, v = 0$, and since their radii are equal to $1/|x|$, it follows that the reactance circles lie either above or below the u-axis depending on whether x is positive or negative. Only the portions of the reactance circles lying within the unit circle corresponding to $r = 0$ are required, since for a passive, lossless transmission line with a passive load, the representative point never lies outside the unit circle. The reactance circles corresponding to $x = 0$, $\pm\frac{1}{2}$, ± 1, ± 2, and ± 4 are shown in Fig. 3.7.

A resistance and a reactance circle pass through every point P on the reflection coefficient plane. The r and the x values corresponding to these circles give the per unit resistance and the per unit reactance associated with this point. If the characteristic resistance R_0 of the transmission line is known, we can obtain the actual resistance and reactance values. The general reflection coefficient associated with any point P can be determined in the manner we discussed earlier. Thus, with the help of Fig. 3.7, we can evaluate the impedance value corresponding to any general reflection coefficient or vice versa. In Fig. 3.7, the circles corresponding to $|\rho| = 0.25$, 0.50, 0.75, and 1.0, as well as the directions corresponding to the movement of the observation point on the transmission line towards the generator and towards the

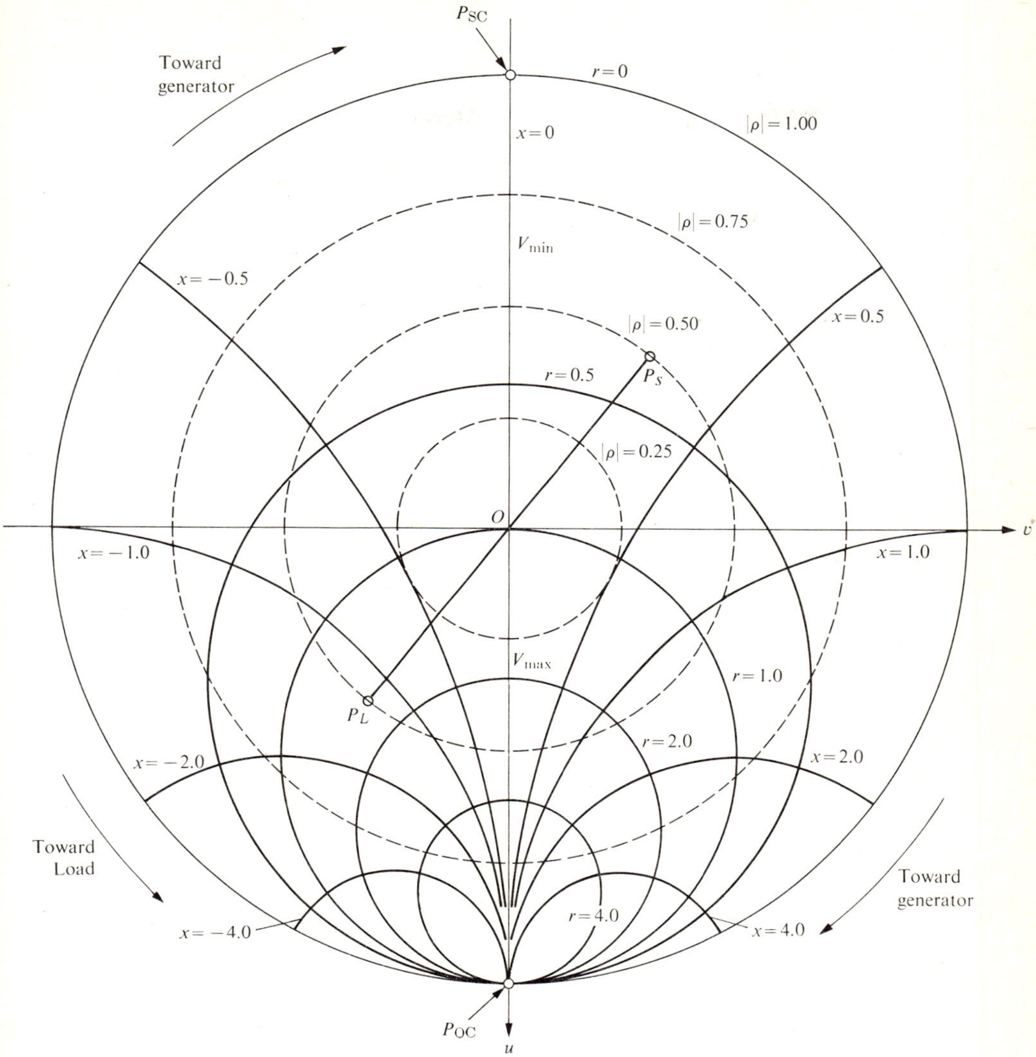

Fig. 3.7 The Smith impedance chart.

load are included. Figure 3.7, with many additional r and x circles, without the constant $|\rho|$ circles which can be easily inserted for the required $|\rho|$ value and with the additional information on the distance along the transmission line both in fractions of a wavelength and in radians marked outside the unit circle, constitutes what has come to be known as the *Smith chart.*

Certain special features of the Smith chart are important enough to be enumerated in detail. We pointed out that the impedances corresponding to the locations of extrema of magnitude of voltage are resistances, with the result that these locations are represented in Fig. 3.7 by the line $P_{SC}OP_{OC}$ since the reactance associated with any point on this line is zero. Also, the per unit resistance at the locations of voltage minima is equal to $1/S$ whereas that at the locations of voltage maxima is equal to S. Since the per unit resistance at the locations of voltage minima is *smaller* than at the locations of voltage maxima and since the per unit resistance associated with the points on the line OP_{SC} is *smaller* than that associated with the points on the line OP_{OC}, it follows that the lines OP_{SC} and OP_{OC} represent respectively the locations of voltage minima and maxima. The contour traced by the representative point on the reflection coefficient plane, as the observation point moves along the transmission line, is a circle of constant $|\rho|$ centered at $u = 0$ and $v = 0$ and this circle intersects each of the lines OP_{SC} and OP_{OC} once. The distance between these intersection points corresponds to $2\beta s = \pi$ or $s = \lambda/4$. This verifies the fact that the distance between a voltage minimum and a successive voltage maximum is equal to $\lambda/4$.

The per unit resistance at the location of a voltage maximum is equal to the standing-wave ratio S. This fact provides a simple method for determining the standing-wave ratio S with the help of the Smith chart. We read off the per unit resistance associated with the point of intersection of the circle of appropriate $|\rho|$ with the voltage maximum line OP_{OC} to obtain the standing-wave ratio S.

A short-circuit corresponds to the location of a voltage minimum and zero impedance. Thus, the point P_{SC} ($u = -1, v = 0$) corresponds to a short-circuit. In a similar manner, an open-circuit corresponds to the location of a voltage maximum and infinite impedance. Therefore, the point P_{OC} ($u = 1, v = 0$) corresponds to an open-circuit.

As the observation point moves through a distance of $\lambda/2$ along the transmission line, the representative point on the Smith chart moves around the constant $|\rho|$ circle through an angle $2\beta s = 2 \times (2\pi/\lambda) \times \lambda/2 = 2\pi$ and hence returns to the original location. This is to be expected since the locations on the transmission line separated by distances which are integer multiples of $\lambda/2$ have the same impedances associated with them and therefore are represented by the same point on the Smith chart.

It follows from (2.10.27) that if the per unit impedance at any location on the transmission line is $z_L = Z_L/R_0$, then the per unit impedance $z_s = Z_s/R_0$ at another location separated from the first by an odd integer multiple of $\lambda/4$ is given by

$$z_s = 1/z_L = y_L, \tag{6.9}$$

where y_L is the per unit admittance at the first location and is equal to $Y_L/G_0 = Y_L R_0$. This observation enables us to use the Smith chart for finding the admittance associated with any given impedance. We first convert the given impedance into a per unit value by dividing it by an appropriate characteristic resistance R_0 which is chosen in such a manner as to be able to represent the value of the per unit impedance accurately on the Smith chart. Suppose P_L in Fig. 3.7 represents the given per unit impedance. Then, the circle with the center at the origin and radius equal to OP_L gives the locus of the representative point as the observation point moves along the transmission line. The movement of the observation point through a distance $\lambda/4$ corresponds to the movement of the representative point on the Smith chart through an angle $2\beta s = 2 \times (2\pi/\lambda) \times \lambda/4 = \pi$ radians. Therefore, the point P_s diametrically opposite to P_L on the same $|\rho|$ circle corresponds to the per unit impedance given in (6.9) and hence corresponds to the per unit admittance value associated with the given per unit impedance. We read off the per unit resistance and the per unit reactance associated with the point P_s from the Smith chart and multiply them by the characteristic conductance $G_0 = 1/R_0$, to yield the admittance value Y_L corresponding to the given impedance Z_L.

So far, we have used the Smith chart as an impedance chart. We can equally use it as an admittance chart. Let $Y_L = 1/Z_L$, $G_0 = 1/R_0$, and $Y(x) = 1/Z(x)$. With these relations we can manipulate (6.1) and (6.2) to yield the following results:

$$|\rho|\, e^{j\theta'_\rho} = \frac{Y_L - G_0}{Y_L + G_0}, \qquad \theta'_\rho = \theta_\rho + \pi, \tag{6.10}$$

and

$$|\rho|\, e^{j\phi'} = \frac{Y(x) - G_0}{Y(x) + G_0}, \qquad \phi' = \theta'_\rho - 2\beta s. \tag{6.11}$$

Note that (6.10) and (6.11) are identical with (6.1) and (6.2) respectively, except for the replacement of θ_ρ and ϕ by θ'_ρ and ϕ' respectively. On setting

$$u = |\rho| \cos \phi', \qquad v = |\rho| \sin \phi', \tag{6.12}$$

we can derive results identical to (6.6) and (6.8) from (6.11) except for the replacement of r and x by g and b respectively, where g is the per unit conductance and b is the per unit susceptance. We have illustrated the Smith admittance chart in Fig. 3.8.

It is necessary for us to note certain differences between the impedance and the admittance charts. Note that the points P_{OC} and P_{SC} are interchanged. This is to be expected since the admittance value corresponding to a short-circuit is infinite and that corresponding to an open-circuit is zero. Also, the admittances corresponding to the locations of extrema of magnitude of voltage are pure conductances. Since the per unit conductance at the locations of voltage minima is *greater* than that at the locations of voltage maxima and since the per unit conductance associated with the points on the line OP_{SC} in Fig. 3.8 is *greater* than that associated with the points on the line OP_{OC}, it follows that the lines OP_{SC} and OP_{OC} in Fig. 3.8 represent respectively the locations of voltage minima and maxima. We should note that the per unit conductance at the locations of voltage minima is equal to the standing-wave ratio S. Consequently, the standing-wave ratio S can be obtained by reading off the per unit conductance value associated with the point of intersection of the circle of appropriate $|\rho|$ with the voltage minimum line OP_{SC} in Fig. 3.8. In the matter of finding the standing-wave ratio, we can combine the results of the impedance and the admittance charts by the statement that the per unit resistance or the per unit conductance associated with the point of intersection of the circle of appropriate $|\rho|$ with the positive u-axis gives the standing-wave ratio S.

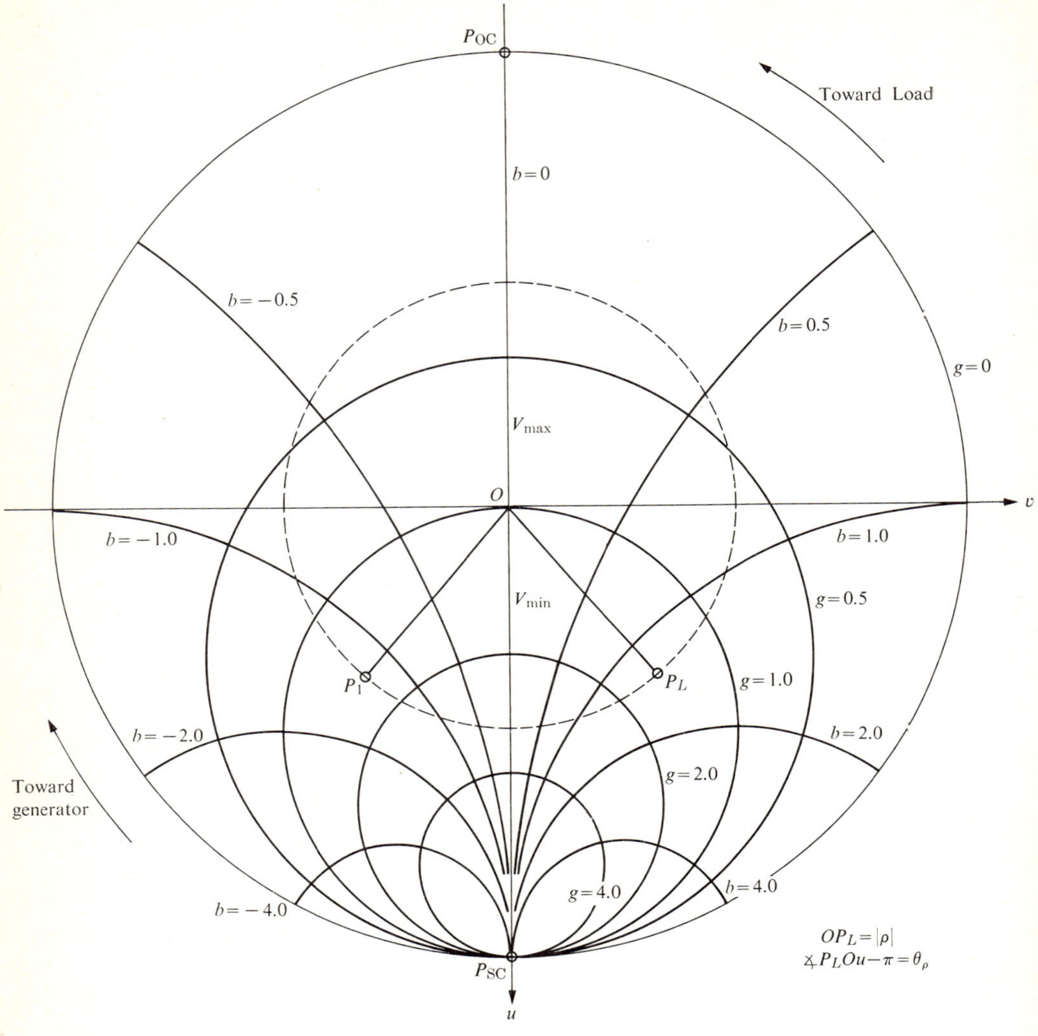

Fig. 3.8 The Smith admittance chart.

Suppose that the point P_L in Fig. 3.8 represents the per unit load admittance. Then, the magnitude $|\rho|$ of the terminal reflection coefficient is equal to OP_L but the phase θ_ρ of the terminal reflection coefficient is not given by $\measuredangle P_LOu$. From (6.10), we see that $\measuredangle P_LOu = \theta_\rho + \pi$. Therefore, we have to subtract π from the angle $\measuredangle P_LOu$ in radians to obtain the phase θ_ρ of the terminal reflection coefficient. In a similar manner, if the point P_1 in Fig. 3.8 represents the per unit admittance at some arbitrary location along the transmission line, then the magnitude $|\rho|$ of the associated general reflection coefficient is given by OP_1 and in view of (6.10) and (6.11), we have to subtract π from the angle $\sphericalangle P_1Ou$ in radians, to obtain the phase ϕ of the general reflection coefficient. We have to note the foregoing feature of the phase of the reflection coefficient when using the Smith chart on an admittance basis.

3.7 APPLICATIONS OF THE SMITH CHART

We shall give a few examples of the use of the Smith chart in simplifying lossless transmission-line calculations in this section. Additional applications of the Smith chart for somewhat more involved lossless transmission-line calculations are given in the next section.

Example 1. A lossless transmission line of characteristic resistance $R_0 = 50$ ohms and of length $s = 0.2\,\lambda$ is terminated in a short-circuit. Find by using the Smith chart (a) $|\rho|$ and θ_ρ of the terminal reflection coefficient and (b) the input impedance.

The short-circuit is represented on the Smith chart in Fig. 3.9 by the point P_L which coincides with $P_{SC}\,(u = -1, v = 0)$. The magnitude $|\rho|$ of the terminal reflection coefficient is seen from Fig. 3.9 to be equal to $OP_L = 1$. The phase θ_ρ of the terminal reflection coefficient is given by the angle $\sphericalangle uOP_L = \pi$ radians. The locus of the representative point on the Smith chart is the circle centered at the origin and radius equal to unity. In other words, it is the circle corresponding to $r = 0$. As the observation point moves from the load terminals in the direction of the generator a distance of $0.2\,\lambda$ to reach the input terminals, the representative point on the Smith chart moves in the clockwise direction along the circumference of the unit circle through an angle $2\beta s = 2 \times (2\pi/\lambda) \times 0.2\,\lambda = 144$ degrees to reach the point P_1, which therefore represents the location of the input terminals. The r value corresponding to P_1 is equal to zero and the x value corresponding to P_1 is read off from the chart to be equal to approximately 3.1. Therefore, we obtain the input impedance as $Z_{in} = (r + jx)R_0 = (0 + j3.1)50 = j155$ ohms. Thus, the input impedance is a reactance. Whenever the point representing the load lies on the $r = 0$ circle, the locus of the representative point is also on the $r = 0$ circle, with the result that the impedance associated with any point along the transmission line has zero resistance and therefore is a reactance. We can see from Fig. 3.9 that this situation holds for the short-circuit and the open-circuit terminations as well as for a reactive load. We should note that we have used the Smith chart in Fig. 3.9 on an impedance basis.

Example 2. A lossless transmission line of characteristic resistance $R_0 = 50$ ohms is terminated in a load impedance $Z_L = 50 + j50$ ohms. With the help of the Smith chart, find (a) $|\rho|$ and θ_ρ of the terminal reflection coefficient, (b) the distance in fractions of a wavelength of the location of the first voltage maximum from the load terminals, and (c) the standing-wave ratio S on the transmission line.

The per unit load impedance z_L is given by $z_L = Z_L/R_0 = (50 + j50)/50 = 1 + j1$; therefore $r = 1$ and $x = 1$. The point P_L corresponding to $r = 1$ and $x = 1$ is entered on the Smith chart in Fig. 3.10. We see from Fig. 3.10 that the magnitude $|\rho|$ of the terminal reflection coefficient is equal to approximately $OP_L = 0.45$. The

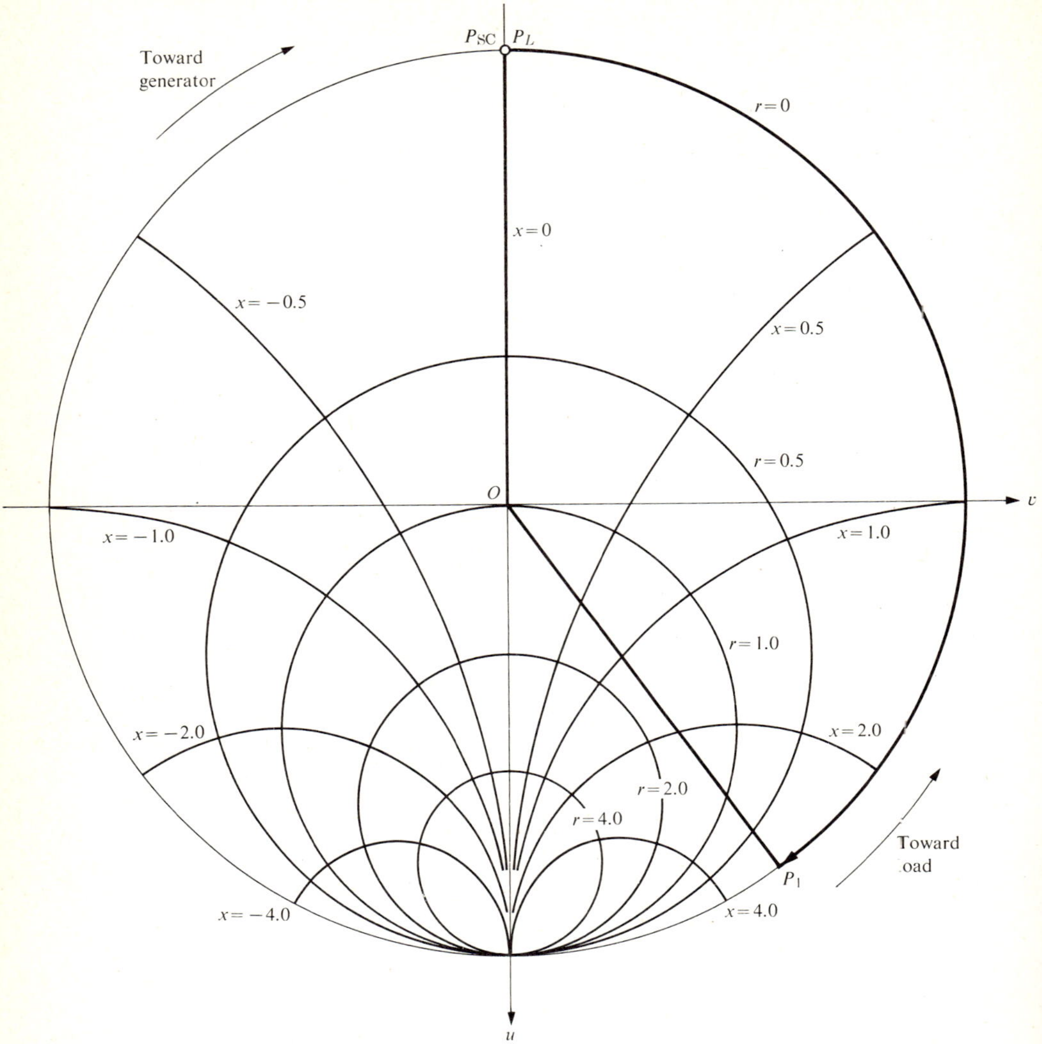

Fig. 3.9 The Smith impedance chart for the determination of the input impedance.

phase θ_ρ of the terminal reflection coefficient is given by the angle $\sphericalangle uOP_L = 64$ degrees or 1.12 radians. The locus of the representative point on the Smith chart is the circle centered at the origin and radius equal to OP_L. Let this circle intersect the voltage maximum line OP_{OC} at P_1. If s is the distance from the load terminals to the location of the first voltage maximum, then from Fig. 3.10, we can see that $2\beta s$ is equal to the angle $\sphericalangle uOP_L$ in radians. Therefore, $2\beta s = 1.12$ or $s = 0.089\ \lambda$. The value of the standing-wave ratio S is equal to the per unit resistance corresponding to the voltage maximum point P_1. From Fig. 3.10, we read off S as approximately 2.6.

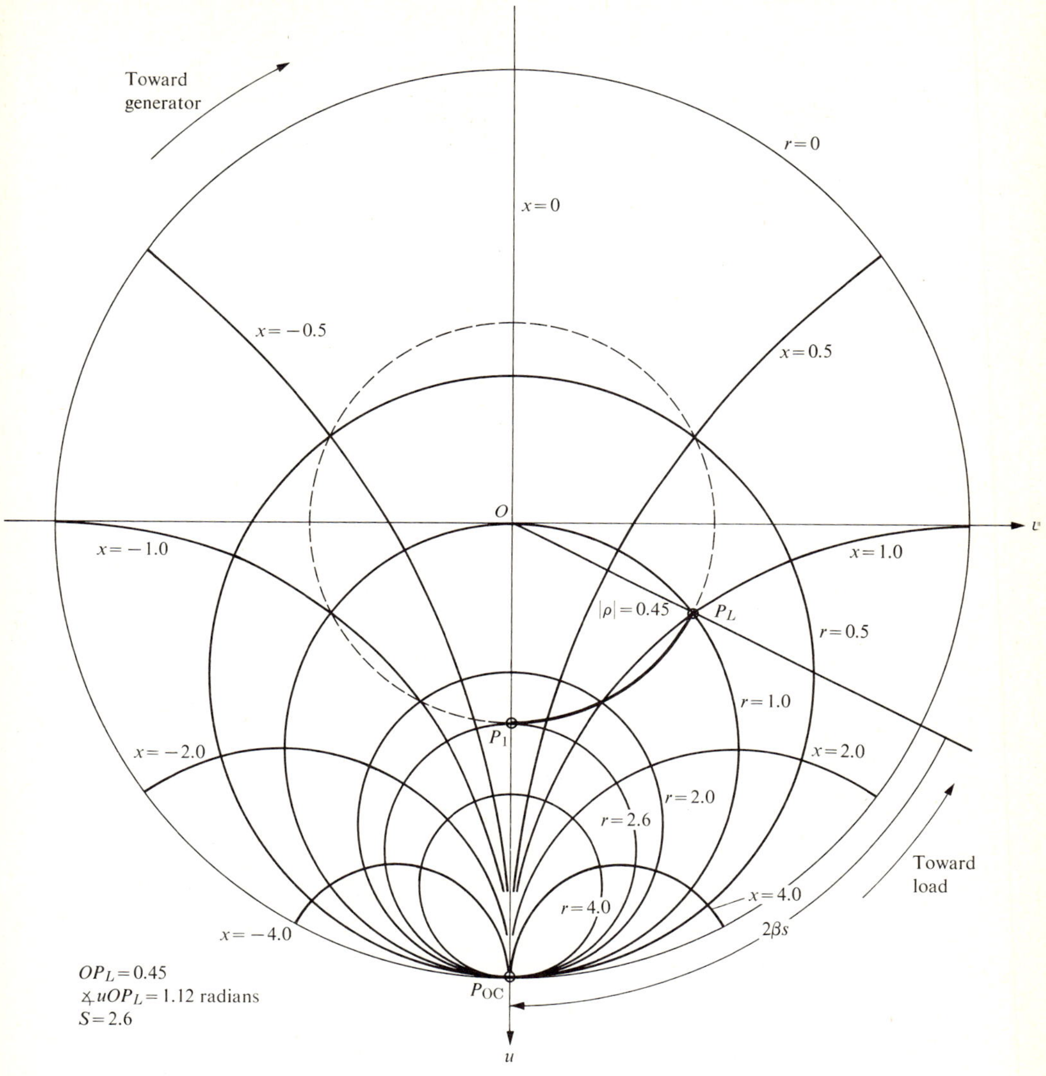

Fig. 3.10 The Smith impedance chart for the determinations of the location of the voltage maximum and the standing-wave ratio S.

The impedance at the voltage maximum point P_1 is therefore $2.6 \times 50 = 130$ ohms. In a similar manner, the impedance at the voltage minimum point which is at a distance of $0.25\,\lambda$ from the voltage maximum point is equal to approximately $50/2.6 = 19.2$ ohms.

Example 3. It is instructive to rework, with the help of the Smith chart, the problem in Section 3.5 concerning the determination of the unknown impedance using the data obtained from the standing-wave measurements on the transmission line connected to the load impedance.

Since the standing-wave ratio is equal to the per unit resistance at the location of the voltage maximum, it follows that the point of intersection P_1 of the positive u-axis (the voltage maximum line) with the resistance circle corresponding to $r = 3$ is a point on the locus of the representative point on the Smith chart. (See Fig. 3.11.)

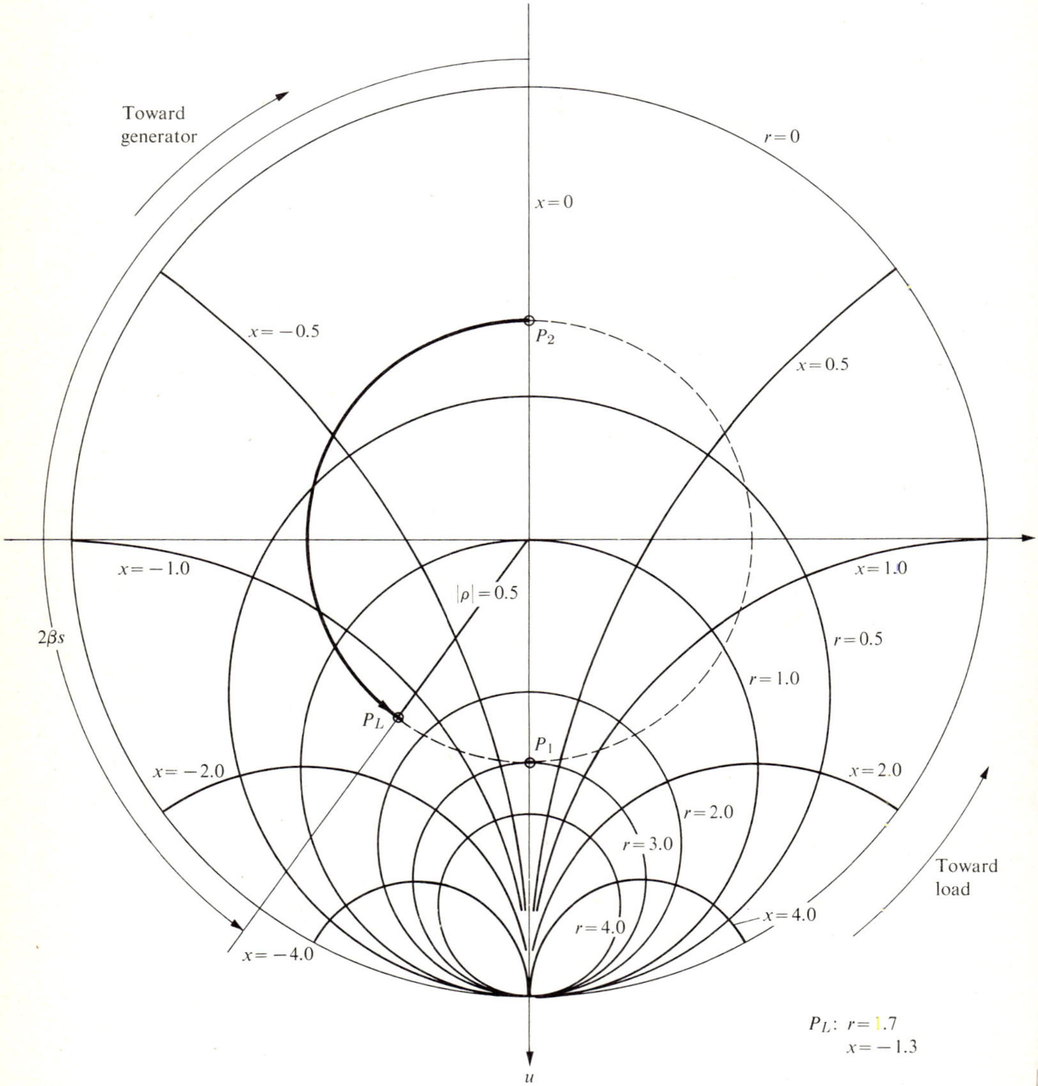

Fig. 3.11 The Smith impedance chart for the determination of the unknown load impedance.

Then, OP_1 is equal to $|\rho|$ which remains constant as the observation point moves along the transmission line. Consequently, the circle centered at the origin and radius equal to OP_1 is the locus we referred to previously. It is worthwhile for us to note in passing from Fig. 3.11 that $|\rho| = 0.5$. Incidentally, this is the method by which we can obtain with the help of the Smith chart $|\rho|$ given S or vice versa. Let the locus circle intersect at P_2 the negative u-axis (the voltage minimum line). As discussed in Section 3.5, $\beta = \pi/5$. The distance of the load terminals from the position of the voltage minimum is $s = 2$ m. Therefore, the representative point P_2 has to be moved through an angle $2\beta s = 2 \times (\pi/5) \times 2 = 144$ degrees in the counter-clockwise direction along the locus circle to obtain the point P_L representing the load impedance. We must note that the movement of the representative point in the counter-clockwise direction corresponds to the movement of the observation point towards the load. The per unit resistance and the per unit reactance corresponding to the point P_L are read off from the Smith chart as $r = 1.7$ and $x = -1.3$ approximately. We can then evaluate the load impedance as $Z_L = R_0(r + jx) = 50(1.7 - j1.3) = 85 - j65$ ohms.

If we compare the results obtained for this problem using the Smith chart with those evaluated using the relevant formulas we find that whereas the Smith chart clearly enables the answer to be obtained in a simpler manner, it entails a certain loss in accuracy. However, the accuracy provided by the Smith chart is sufficient for most practical purposes.

3.8 TRANSMISSION-LINE IMPEDANCE MATCHING

One of the main functions of a transmission line is to transfer power from the generator to the load. It is therefore natural to require that as much of the power from the generator as possible be delivered to the load with a minimum of losses along the transmission line. At radio frequencies, the transmission line has some small losses. Consequently, due to the resistance of the wires of the transmission line as well as due to the leakage in the dielectric region between them, there is power loss in transmission. For long transmission lines, that is long in comparison with a wavelength, it becomes necessary to contend with the power losses along the transmission line. We can show that the transmission-line losses are reduced if the standing-wave ratio is decreased and that in the limiting case for which there are no standing waves the transmission-line losses are reduced to a minimum value. Hence in a long transmission line (hereinafter distinguished as the main line) feeding a given load, a short section consisting of a series, a parallel, or a series-parallel combination of transmission lines is added between the load and the main line. This short section is usually designated as the matching section of the transmission line. The purpose of the matching section is to transform the load impedance and present to the main line a terminating impedance equal to its characteristic impedance. As a consequence there are no standing waves on the main line and the transmission-line losses are thus minimized. This technique of inserting a matching section to transform the load impedance to the characteristic impedance of the main transmission line is known as transmission-line impedance matching.

For a low-loss transmission line, the characteristic impedance is predominantly resistive with a very small reactive part. The matching section is then required to transform the load impedance into another impedance equal to the characteristic impedance of the main line. But, since the reactive part of the characteristic impedance of the main line is generally very small in comparison with its resistive part, the parameters of the actual matching section are approximately the same as those pertaining to a matching section, which transforms the load impedance into a pure resistance equal to the characteristic resistance of the main line. Consequently, in the following theoretical analysis, we focus attention only on the case for which the matching section is required to transform the load impedance into the characteristic resistance of the main line. Minor adjustments of the calculated parameters of the matching section are necessary in practice to bring the main line into a perfect impedance match resulting in no standing waves.

In the following theoretical analysis, we consider only two very simple impedance-matching devices. The first device consists only of a series section of a transmission line known as a quarter-wave transformer. The second device, known as a single stub matching device, consists of a parallel section of a transmission line connected to the main line at an appropriate distance from the load which is directly connected to the main line. As we mentioned earlier, in so far as the design of the matching section is concerned, we assume the transmission line to be lossless and hence its characteristic impedance has only a resistive part.

Quarter-wave Transformer

This device is useful in matching a resistive load R_L to a transmission line having a characteristic resistance R_0 which is different from the load resistor R_L. Let A represent the terminals of a long transmission line whose characteristic resistance is R_0 and let AB represent a section of a series transmission line, one pair of whose terminals is connected to the load resistor R_L and the other pair to the terminals of the main line (see Fig. 3.12). Let all the quantities referring to the matching section AB of the transmission line be denoted by the additional subscript m. Thus, R_{0m} is

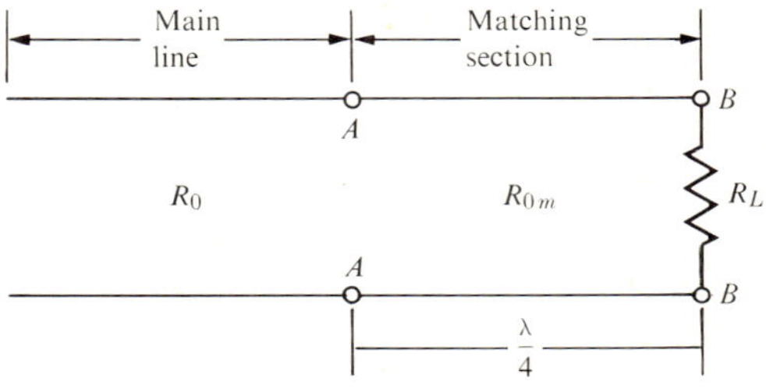

Fig. 3.12 Quarter-wave transformer.

the characteristic resistance and s_m is the length of the matching section of the transmission line. As the name of this impedance-matching device implies, we choose s_m to be equal to $\lambda/4$ at the frequency for which we want the impedance match. Then, it follows from (2.10.27) that the input impedance Z_t at the junction A between the main line and the matching section looking towards the load is given by

$$Z_t = R_{0m}^2/R_L. \tag{8.1}$$

In order for the equivalent load resistance terminating the main line to be equal to R_0, we see from (8.1) that the characteristic resistance R_{0m} of the matching section should be

$$R_{0m} = \sqrt{R_L R_0}. \tag{8.2}$$

Thus, if we insert between the main line and the load a quarter-wavelength of a transmission line having a characteristic resistance equal to the geometric mean of the load resistance and the characteristic resistance of the main line, the terminating resistance of the main line is equal to its characteristic resistance with the result that there are no standing waves on the main line.

We should bear in mind that the impedance match is achieved at only a single (designed) frequency for which the length of the matching section is a quarter-wavelength. At other frequencies, the length of the matching section is different from a quarter-wavelength. As a consequence, Z_t is different from R_0 and the standing waves are formed on the main line. Since the load impedance R_L does not change with the frequency, the terminal reflection coefficient and therefore the standing-wave ratio pertaining to the matching section does not change as the frequency is varied. But, since the impedance Z_t terminating the main line changes with the frequency, the standing-wave ratio S pertaining to the main line changes as the frequency is varied. We can calculate the standing-wave ratio S in the following manner. From (2.10.25), Z_t is obtained as

$$Z_t = R_{0m}\frac{[R_L \cos \beta s_m + jR_{0m} \sin \beta s_m]}{[R_{0m} \cos \beta s_m + jR_L \sin \beta s_m]}, \tag{8.3}$$

where

$$\beta = \omega/v. \tag{8.4}$$

The voltage reflection coefficient pertaining to the main line is

$$\rho = \frac{Z_t - R_0}{Z_t + R_0}, \tag{8.5}$$

and hence S is given by

$$S = \frac{1 + |\rho|}{1 - |\rho|}. \tag{8.6}$$

Let ω_0 be the angular frequency for which the matching section is designed. Therefore, $\beta_0 s_m = \pi/2$, where $\beta_0 = \omega_0/v$.

Let us consider the following example where $R_0 = 50$ ohms and $R_L = 200$ ohms. We obtain the characteristic resistance of the matching section of the transmission

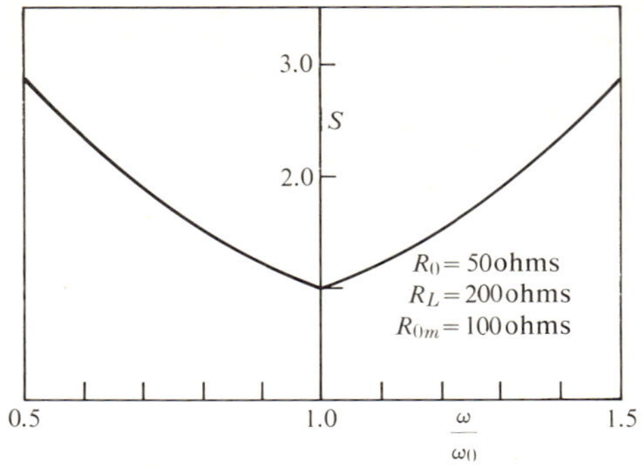

Fig. 3.13 The frequency sensitivity of a quarter-wave transformer; design frequency is given by $\omega/\omega_0 = 1$.

line from (8.2) as $R_{0m} = \sqrt{50 \times 200} = 100$ ohms. At the designed angular frequency $\omega = \omega_0$, $s_m = \lambda/4$. The voltage reflection coefficient ρ_m pertaining to the matching section is $\rho_m = (R_L - R_{0m})/(R_L + R_{0m}) = \frac{1}{3}$. Hence the standing-wave ratio S_m pertaining to the matching section is $S_m = (1 + |\rho_m|)/(1 - |\rho_m|) = 2$ and this does not change as the operating frequency is changed from the designed frequency. As we discussed previously, the standing-wave ratio S pertaining to the main line changes as the operating frequency is changed from the designed frequency. On writing βs_m in (8.3) as $\beta s_m = \beta_0 s_m(\beta/\beta_0) = (\pi/2)(\omega/\omega_0)$, we can obtain the standing-wave ratio S as a function of the ratio ω/ω_0 with the help of (8.3), (8.5), and (8.6). In Fig. 3.13, S is plotted as a function of ω/ω_0 for the range $0.5 \leq \omega/\omega_0 < 1.5$. From Fig. 3.13, we see S to be equal to 1 for $\omega/\omega_0 = 1$, indicating perfect impedance match. But we find that S increases as ω/ω_0 is changed from unity in either direction, indicating impedance mismatch. We thus see the impedance match obtained with the use of a quarter-wave transformer to be frequency sensitive.

Single-stub Impedance Matching

In the single-stub impedance-matching device, a section of a transmission line is connected in parallel to the main line and since admittances add at the junction of parallel transmission lines, it is convenient for us to use the admittance basis for the analysis of the single-stub impedance-matching device. Let us consider a long transmission line of characteristic conductance G_0, terminated in a load admittance Y_L which is different from G_0 (Fig. 3.14). There are standing waves on the transmission line. At the locations of voltage minimum, the input admittance is a conductance which we obtain from (4.2) as G_0S, where S is the standing-wave ratio. In a similar manner, at the locations of voltage maximum, the input admittance again is a conductance which we obtain from (4.3) as G_0/S. The conductances at the locations of voltage minima and maxima are respectively greater or less than the characteristic conductance G_0. Since the conductance varies smoothly along the transmission

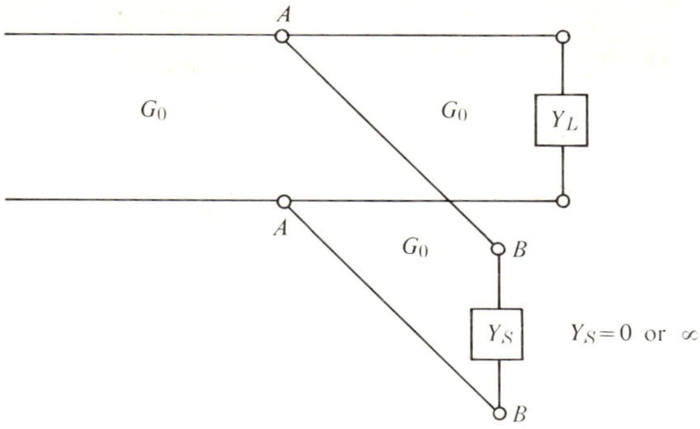

Fig. 3.14 Schematic diagram of a single-stub impedance-matching device.

line, it follows that there are locations such as A in Fig. 3.14 situated between the locations of the voltage extrema, where the conductance is equal to the characteristic conductance. The admittances are real only at the locations of voltage extrema. Therefore, the admittance corresponding to the location A in Fig. 3.14 consists of a conductance equal to the characteristic conductance and a nonvanishing susceptance. We have pointed out that the input admittance of a lossless transmission line with open- or short-circuit terminations is a susceptance. We can obtain any value of susceptance by suitably adjusting the length of the open- or short-circuited transmission line. A section of this transmission line usually called the stub is connected in parallel to the main line at A and the length of the stub is adjusted so as to yield at the junction an input susceptance equal in magnitude but opposite in sign to the susceptance of the main line at the junction. The combined admittance at the junction due to the main line and the stub is just equal to the characteristic conductance G_0 with the result that there are no standing waves on the main line between the generator end and the above-mentioned junction A where the stub is connected to the main line. Usually, the stub also has the same characteristic conductance as the main line.

We can use the Smith chart with advantage in the evaluation of the parameters of a single-stub impedance-matching device. It is instructive to consider the following numerical example for illustrating the usefulness of the Smith chart for this purpose.

Example. A single-stub impedance-matching device is to be designed to match a transmission line of characteristic resistance $R_0 = 400$ ohms to a load impedance $Z_L = 800 + j300$ ohms. The wavelength is 1.50 m. Determine the distance from the load to the location of the stub and the proper length for the stub.

The per unit impedance $z_L = Z_L/R_0$ of the load is $z_L = (800 + j300)/400 = 2 + j0.75$. The point P_L on the Smith chart in Fig. 3.15 represents z_L since the r and the x values corresponding to P_L are $r = 2$ and $x = 0.75$. We draw a circle centered at the origin ($u = 0, v = 0$) and radius equal to OP_L. This circle represents the locus of the representative point on the Smith chart as the observation point

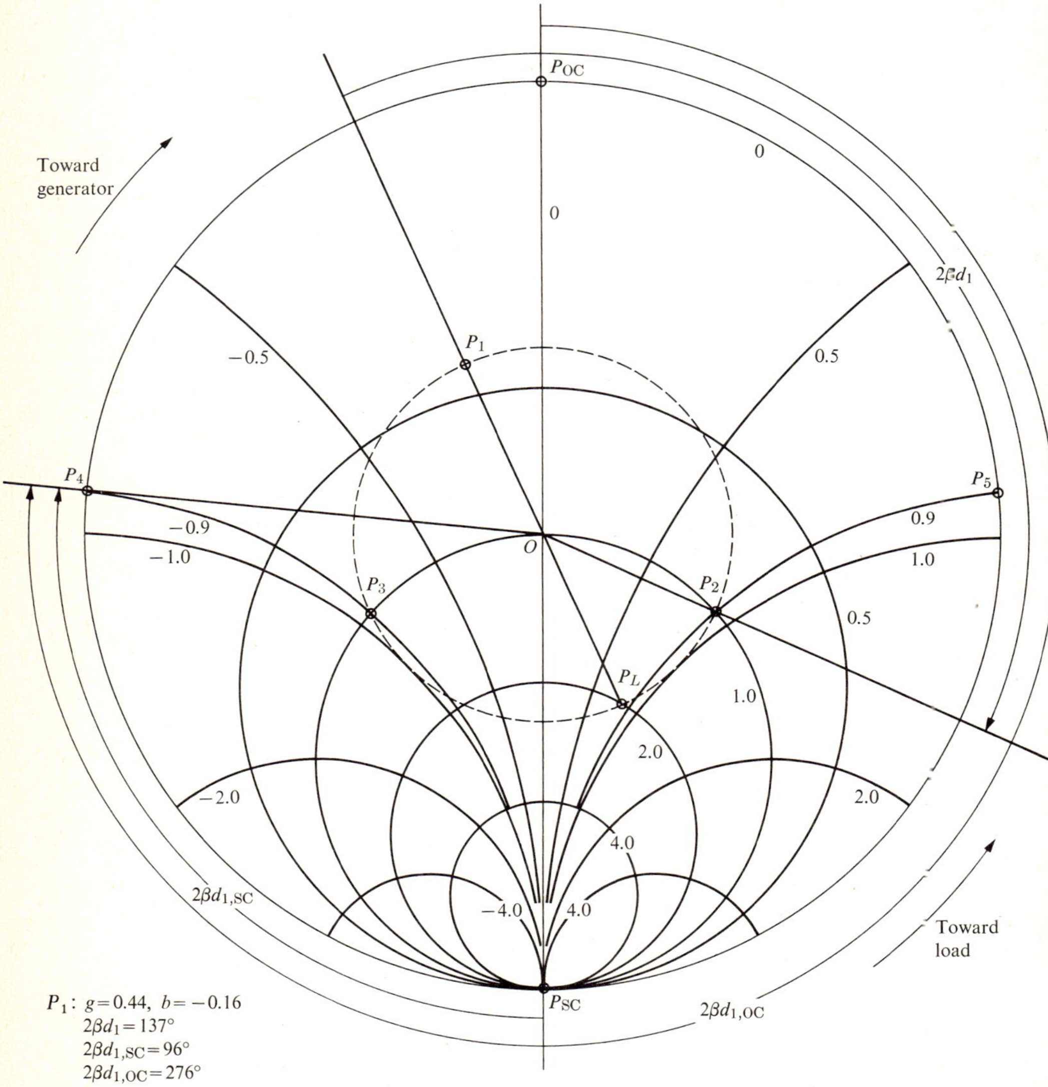

Fig. 3.15 The Smith chart for the determination of the parameters of the single-stub impedance-matching device.

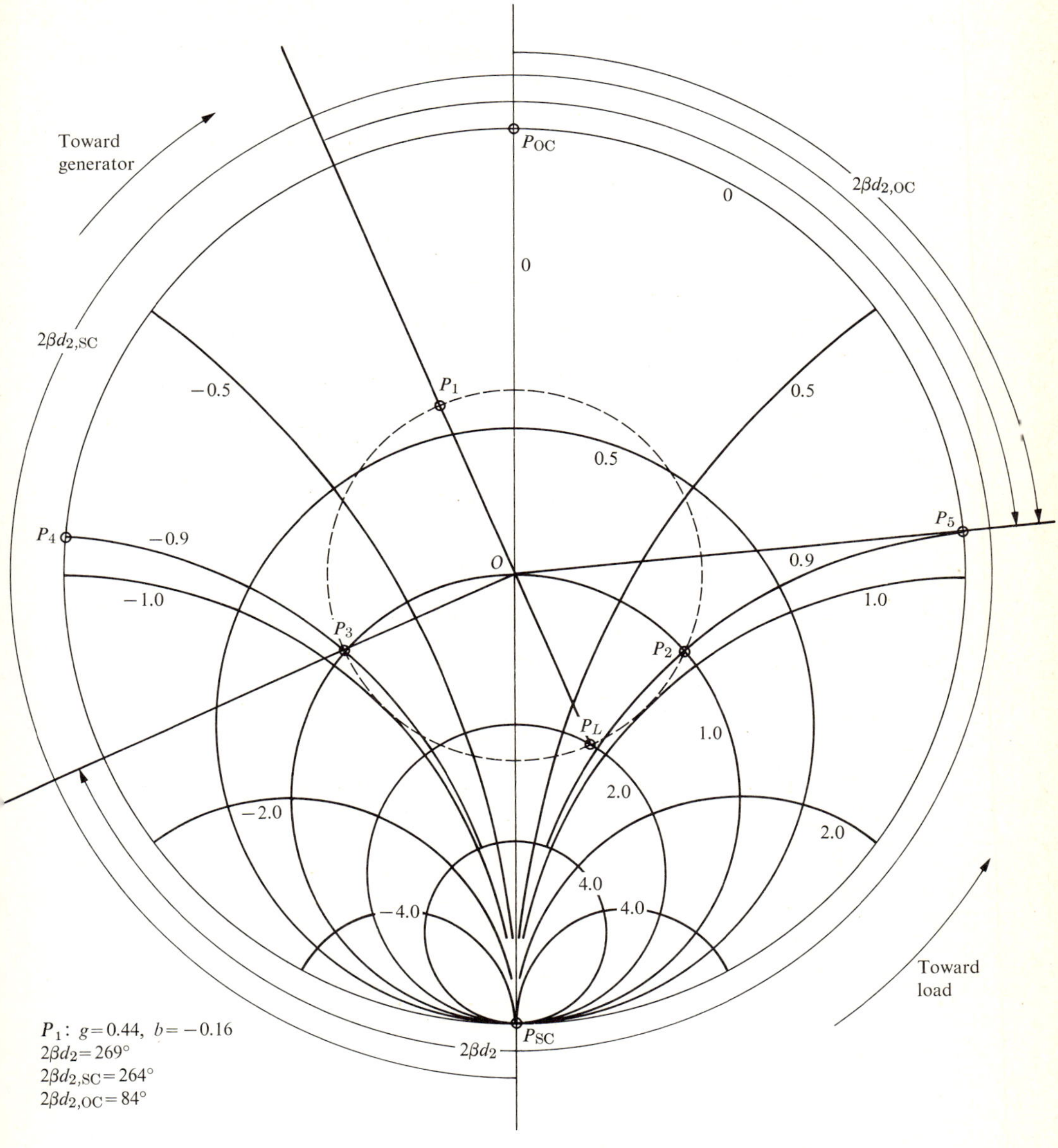

Fig. 3.15 (*continued*)

moves along the main transmission line. We find the point P_1 which is on the locus circle diametrically opposite to P_L. As we explained in Section 3.6, the r and the x values corresponding to P_1 give the per unit admittance $y_L = g_L + jb_L = Y_L/G_0$ corresponding to the load impedance. From Fig. 3.15, we find that $g_L = 0.44$ and $b_L = -0.16$. For the remaining part of this example, the Smith chart in Fig. 3.15 is used on an admittance basis. Since the per unit conductance on the main line at the location of the stub equals unity, it follows that the location A for the stub should be on the conductance circle corresponding to $g = 1$. We have already shown that all points along the main transmission line are on the so-called locus circle, that is, the circle centered at the origin and radius equal to OP_L. Therefore, the location A for the stub should correspond to the point or points of intersection of the locus circle with the $g = 1$ circle. From Fig. 3.15, we see that there are two points of intersection denoted by P_2 and P_3 resulting in two possible locations for the stub. Let the distances from the load to the two possible locations for the stub be denoted by d_1 and d_2 where d_1 is the smaller of the two distances. Since the representative point on the Smith chart moves in the clockwise direction as the observation point on the main line moves from the load in the direction towards the generator, it follows that P_2 corresponds to d_1, the smaller of the two distances, and P_3 corresponds to d_2, the larger of the two distances.

The two possible locations for the stub are considered in turn. From Fig. 3.15 and the discussions in Sections 3.6 and 3.7, we see that $2\beta d_1 = (4\pi/\lambda)d_1 = \measuredangle P_1OP_2 = 137$ degrees. Therefore, $d_1 = 28.5$ cm since $\lambda = 1.5$ m. The susceptance on the main line at the location A of the stub is that pertaining to the point P_2 and hence is equal to $j0.9$. Therefore, the input susceptance of the stub should be equal to $-j0.9$ in order to be able to annul exactly the susceptance on the main line at the location of the stub. There are two possible terminations for the stub, namely, short-circuit and open-circuit terminations and both of these terminations are considered. The short-circuit is represented by the point P_{SC} $(u = 1, v = 0)$ on the Smith chart. The locus of the representative point on the Smith chart corresponding to the movement of the observation point along the stub transmission line is the circle centered at the origin $(u = 0, v = 0)$ and radius equal to $OP_{SC} = 1$, that is, the $g = 0$ circle. The point P_4 on this circle corresponds to a susceptance of $-j0.9$. Let $d_{1,SC}$ be the length of the short-circuited stub that has a per unit input susceptance equal to $-j0.9$. Since the movement of the observation point on the stub transmission line in the direction away from the load corresponds to the movement of the representative point on the Smith chart in the clockwise direction, it follows that $2\beta d_{1,SC} = (4\pi/\lambda)d_{1,SC} = \measuredangle P_{SC}OP_4 = 96$ degrees and therefore $d_{1,SC} = 20$ cm. The per unit input admittance of the short-circuited stub of length $d_{1,SC} = 20$ cm is equal to $-j0.9$ for $\lambda = 1.5$ m. The per unit admittance on the main line at the location of the stub is equal to $1 + j0.9$. The resultant per unit admittance at the location of the stub due to the main line and the stub is equal to unity and therefore perfect impedance match is achieved.

Let us consider now the case for which the stub has an open-circuit termination. The open-circuit is represented by the point P_{OC} $(u = -1, v = 0)$ on the Smith chart. Let $d_{1,OC}$ be the length of the open-circuited stub that has the same per unit input

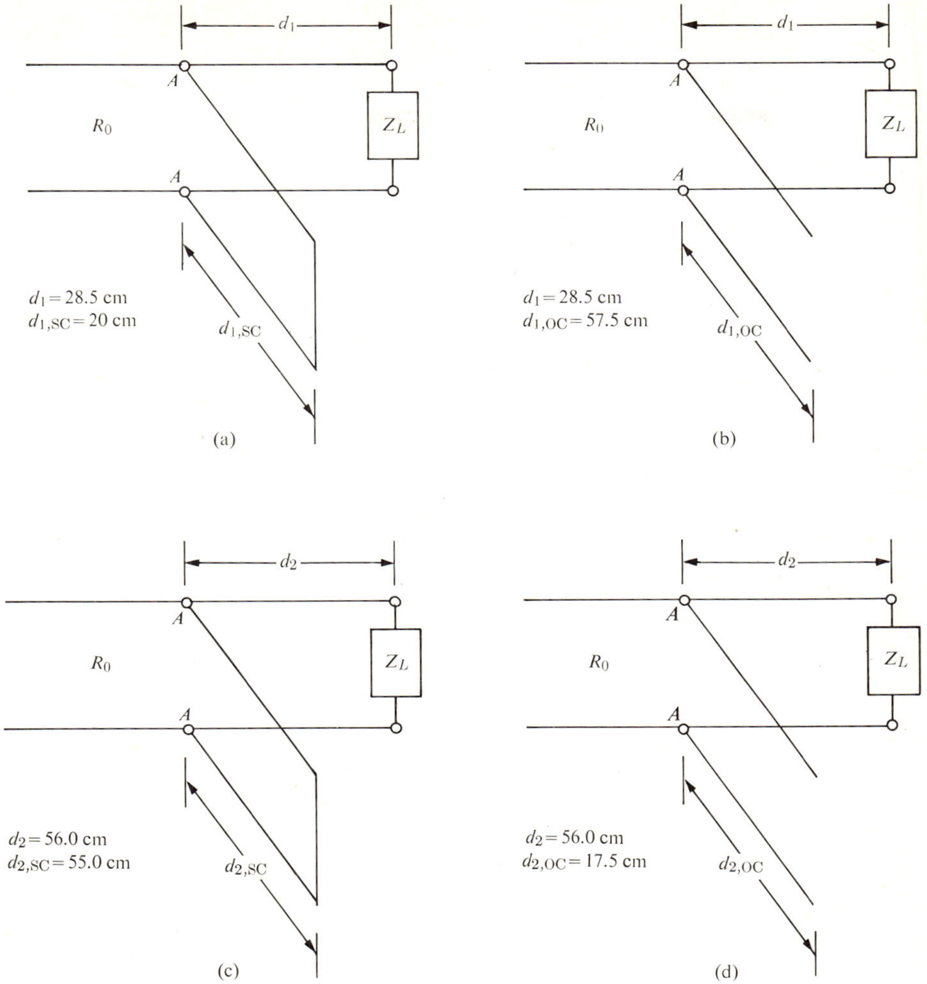

Fig. 3.16 Schematic diagram of the four possible sets of parameters of a single-stub impedance-matching device.

susceptance ($-j0.9$) as the short-circuited stub discussed previously. By the same arguments as those used in obtaining the length of the short-circuited stub, we can deduce that $2\beta d_{1,\text{OC}} = (4\pi/\lambda)d_{1,\text{OC}} = \sphericalangle P_{\text{OC}}OP_4 = 276$ degrees and therefore $d_{1,\text{OC}} =$ 57.5 cm. Note that $d_{1,\text{OC}} - d_{1,\text{SC}} = 37.5$ cm $= \lambda/4$. In Section 2.10, we proved that the lengths of lossless transmission line having the same input susceptance but with either short-circuit or open-circuit terminations differ by $\lambda/4$. The values of $d_{1,\text{SC}}$ and $d_{1,\text{OC}}$ obtained here are in accordance with the above general result. In Figs. 3.16(a) and 3.16(b), we summarize the parameters for the single-stub impedance-matching device for the first location of the stub both for the short-circuited and for the open-circuited stubs.

Returning to the second possible location for the stub, we find from Fig. 3.15 that $2\beta d_2 = (4\pi/\lambda)d_2 = \measuredangle P_1OP_3 = 269$ degrees. Therefore $d_2 = 56.0$ cm. The per unit susceptance on the main line at the location of the stub is now equal to $-j0.9$ and therefore the input susceptance of the stub should be equal to $j0.9$, as represented by the point P_5 on the Smith chart. Let $d_{2,\text{SC}}$ and $d_{2,\text{OC}}$ be respectively the lengths of the short-circuited and the open-circuited stubs whose per unit input susceptance is equal to $j0.9$. Then, as before, $2\beta d_{2,\text{SC}} = (4\pi/\lambda)d_{2,\text{SC}} = \measuredangle P_{\text{SC}}OP_5 = 264$ degrees and hence $d_{2,\text{SC}} = 55.0$ cm. Similarly, $2\beta d_{2,\text{OC}} = (4\pi/\lambda)d_{2,\text{OC}} = \measuredangle P_{\text{OC}}OP_5 = 84$ degrees and therefore $d_{2,\text{OC}} = 17.5$ cm. Note that $d_{2,\text{SC}}$ and $d_{2,\text{OC}}$ again differ by $\lambda = 37.5$ cm. In Fig. 3.16(c) and 3.16(d), we summarize the parameters of the single-stub impedance-matching device for the second location of the stub, both for the short-circuited and for the open-circuited stubs.

The evaluation of the parameters of the single-stub impedance-matching device has been carried out with the help of the Smith chart. From the time of its introduction until the recent advent of the electronic computers, the Smith chart has served as a valuable calculational aid. However, at the present time, the electronic computers are being used increasingly and effectively in performing every kind of transmission-line calculation. For that purpose, analytical expressions for the parameters of the single-stub impedance-matching device are useful and we therefore derive them here.

Let us consider a lossless transmission line of characteristic conductance G_0 terminated in a load admittance Y_L. We can obtain the admittance $Y(s)$ at a distance s from the load from (2.10.25) as

$$Y(s) = G_0 \frac{[Y_L + jG_0 \tan \beta s]}{[G_0 + jY_L \tan \beta s]}. \tag{8.7}$$

In terms of the per unit values, we can simplify (8.7) to

$$y(s) = \frac{[y_L + j \tan \beta s]}{[1 + jy_L \tan \beta s]}. \tag{8.8}$$

Let us consider the short-circuited stub first. Referring to Fig. 3.14, we obtain the per unit admittance y at A on the main line looking towards the load from (8.8) as

$$y = \frac{[y_L + j \tan \beta d]}{[1 + jy_L \tan \beta d]}, \tag{8.9}$$

where d is the distance of the location A of the stub from the load. The per unit admittance y_m at A on the stub transmission line looking towards the short-circuit $[y(0) = \infty]$ is obtained from (8.8) to be

$$y_m = -j \cot \beta d_{\text{SC}}, \tag{8.10}$$

where d_{SC} is the length of the short-circuited stub. The total per unit admittance at A due to the main line and the stub should be equal to unity for achieving impedance match. Therefore, we obtain the requirement for impedance match from (8.9) and (8.10) as

$$\frac{[y_L + j\tan\beta d]}{[1 + jy_L\tan\beta d]} - j\cot\beta d_{SC} = 1. \tag{8.11}$$

Let $y_L = g_L + jb_L$. On multiplying (8.11) by the denominator of the first term and equating the real and the imaginary parts of the resulting equation separately, we find

$$g_L + g_L\cot\beta d_{SC}\tan\beta d = 1 - b_L\tan\beta d, \tag{8.12}$$

$$b_L + \tan\beta d\{1 - g_L + b_L\cot\beta d_{SC}\} = \cot\beta d_{SC}. \tag{8.13}$$

From (8.12), we obtain that

$$\tan\beta d = \frac{1 - g_L}{b_L + g_L\cot\beta d_{SC}}. \tag{8.14}$$

If we substitute the value of $\tan\beta d$ from (8.14) in (8.13) we get, after some simplification,

$$\cot^2\beta d_{SC} = [(1 - g_L)^2 + b_L^2]/g_L. \tag{8.15}$$

It follows from (8.15) that

$$\cot\beta d_{SC} = \pm\sqrt{\frac{(1 - g_L)^2 + b_L^2}{g_L}} \quad \text{or} \quad \tan\beta d_{SC} = \pm\sqrt{\frac{g_L}{(1 - g_L)^2 + b_L^2}}. \tag{8.16}$$

We should note from (8.16) that there are two possible values for βd_{SC}. Let them be $\beta d_{1,SC}$ and $\beta d_{2,SC}$ and let them correspond respectively to the upper and lower signs in (8.16). Since $\cot\beta d_{SC}$ has two possible values, it is found from (8.14) that βd has also two possible values, βd_1 and βd_2 corresponding respectively to $\beta d_{1,SC}$ and $\beta d_{2,SC}$. We deduce easily from (8.14) and (8.16) that

$$\tan\beta d_1 = \frac{1 - g_L}{b_L + \sqrt{g_L[(1 - g_L)^2 + b_L^2]}}, \tag{8.17}$$

$$\tan\beta d_2 = \frac{1 - g_L}{b_L - \sqrt{g_L[(1 - g_L)^2 + b_L^2]}}. \tag{8.18}$$

In a similar manner, we find from (8.16) that

$$\beta d_{1,SC} = \tan^{-1}\sqrt{\frac{g_L}{(1 - g_L)^2 + b_L^2}}, \tag{8.19}$$

and

$$\beta d_{2,SC} = \pi - \tan^{-1}\sqrt{\frac{g_L}{(1 - g_L)^2 + b_L^2}}. \tag{8.20}$$

Let us suppose now that the stub has an open-circuit termination. We find from (8.8) that the per unit admittance y_m at A on the stub transmission line looking towards the open-circuit $[y(0) = 0]$ is

$$y_m = j\tan\beta d_{OC}, \tag{8.21}$$

where d_{OC} is the length of the open-circuited stub. For the open-circuited stub, the equations corresponding to (8.11) to (8.13) become

$$\frac{[y_L + j \tan \beta d]}{[1 + jy_L \tan \beta d]} + j \tan \beta d_{OC} = 1, \tag{8.22}$$

$$g_L - g_L \tan \beta d \tan \beta d_{OC} = 1 - b_L \tan \beta d, \tag{8.23}$$

$$b_L + \tan \beta d\{1 - g_L - b_L \tan \beta d_{OC}\} = -\tan \beta d_{OC}. \tag{8.24}$$

From (8.23), we can show that

$$\tan \beta d = \frac{1 - g_L}{b_L - g_L \tan \beta d_{OC}}. \tag{8.25}$$

The substitution of the value of tan βd from (8.25) in (8.24) gives after some simplification

$$\tan \beta d_{OC} = \mp \sqrt{\frac{(1 - g_L)^2 + b_L^2}{g_L}}. \tag{8.26}$$

There are two possible values for βd_{OC}. Let them be $\beta d_{1,OC}$ and $\beta d_{2,OC}$, corresponding respectively to the upper and the lower signs in (8.26). Since tan βd_{OC} has two possible values, we see from (8.25) that βd has two possible values also, βd_1 and βd_2 corresponding respectively to $\beta d_{1,OC}$ and $\beta d_{2,OC}$. The substitution of (8.26) in (8.25) yields for tan βd_1 and tan βd_2 the same expressions as those given in (8.17) and (8.18) respectively. This is to be expected since the discussion based on the Smith chart showed that there are only two possible locations (within a distance of $\lambda/2$ from the load) for the stub. The distances of the location of the stub from the load are d_1 and d_2. The expressions for finding d_1 and d_2 are given by (8.17) and (8.18) respectively. Also, we find from (8.26) that

$$\beta d_{1,OC} = \pi - \tan^{-1} \sqrt{\frac{(1 - g_L)^2 + b_L^2}{g_L}}, \tag{8.27}$$

and

$$\beta d_{2,OC} = \tan^{-1} \sqrt{\frac{(1 - g_L)^2 + b_L^2}{g_L}}. \tag{8.28}$$

The results contained in (8.17) to (8.20), (8.27), and (8.28) are conveniently regrouped as follows:

1. First Location of the Stub (Distance d_1 from the Load)

$$\tan \beta d_1 = (1 - g_L)\{b_L + \sqrt{g_L[(1 - g_L)^2 + b_L^2]}\}^{-1}. \tag{8.17}$$

Short-circuited Stub (length $d_{1,SC}$)

$$\beta d_{1,SC} = \tan^{-1} \sqrt{\frac{g_L}{(1 - g_L)^2 + b_L^2}}. \tag{8.19}$$

Open-circuited Stub (length $d_{1,\mathrm{OC}}$)

$$\beta d_{1,\mathrm{OC}} = \pi - \tan^{-1}\sqrt{\frac{(1 - g_L)^2 + b_L^2}{g_L}}. \tag{8.27}$$

2. *Second Location of the Stub (Distance d_2 from the Load)*

$$\tan \beta d_2 = (1 - g_L)\{b_L - \sqrt{g_L[(1 - g_L)^2 + b_L^2]}\}^{-1}. \tag{8.18}$$

Short-circuited Stub (length $d_{2,\mathrm{SC}}$)

$$\beta d_{2,\mathrm{SC}} = \pi - \tan^{-1}\sqrt{\frac{g_L}{(1 - g_L)^2 + b_L^2}}. \tag{8.20}$$

Open-circuited Stub (length $d_{2,\mathrm{OC}}$)

$$\beta d_{2,\mathrm{OC}} = \tan^{-1}\sqrt{\frac{(1 - g_L)^2 + b_L^2}{g_L}}. \tag{8.28}$$

With the help of the foregoing six expressions, we can obtain all the four possible sets of parameters for the single-stub impedance-matching device provided the characteristic conductance and the load admittance are given. We can ascertain from (8.19) and (8.27), and (8.20) and (8.28), that the lengths of the short-circuited and the open-circuited stubs for a given stub location differ by $\lambda/4$.

We should remember that there are standing waves on the stub as well as on the main line between the location of the stub and the load. At the frequency for which the single-stub impedance-matching device is designed, there are no standing waves on the main line between the generator and the location of the stub. However, if the frequency of operation is different from the designed frequency, impedance mismatch occurs resulting in standing waves on the main line even between the generator and the load. The impedance mismatch occurs due to the following three reasons: (1) the characteristic impedance of the transmission line is changed as the frequency is varied, (2) the value of the load impedance changes with the frequency, and most importantly (3) the impedance transforming properties of the matching section undergo changes with the changes in frequency. We examine the effect of the last factor on the frequency sensitiveness of the single-stub impedance-matching device.

For this purpose, we take the load admittance to be a conductance G_L and assume the transmission lines to be lossless, with the result that they have only a characteristic conductance G_0. We assume that the values of both these conductances are independent of frequency. The standing-wave ratio pertaining to the stub is infinite and that pertaining to the main line between the location of the stub and the load can be shown to be equal to G_0/G_L or G_L/G_0, whichever yields a value greater than unity. These two standing-wave ratios do not change with the frequency. But the standing-wave ratio on the main line between the generator and the location of the stub changes with the frequency and is therefore a measure of the frequency sensitiveness of the impedance-matching device. We can evaluate this standing-wave ratio in the following manner. We obtain the per unit admittance on the main line at the location of the stub from (8.9) as

$$y = \frac{[g_L + j\tan\beta d_n]}{[1 + jg_L\tan\beta d_n]}, \tag{8.29}$$

where $n = 1$ or 2 corresponding to the two possible locations for the stub. We find

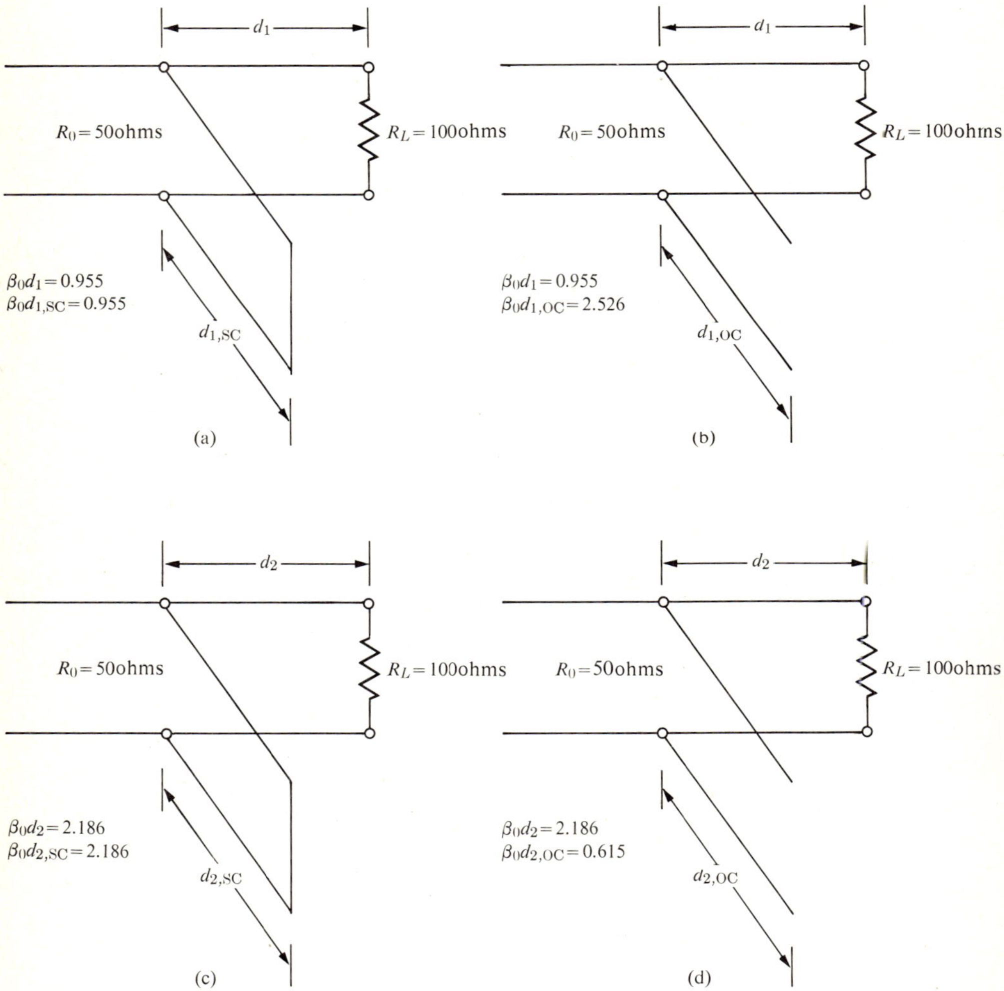

Fig. 3.17 The frequency sensitivity of a single-stub impedance-matching device; design frequency is given by $\omega/\omega_0 = 1$. $S_m = 2.0$; $S_{\text{stub}} = \infty$; and S = standing-wave ratio on the main transmission line between stub location and generator.

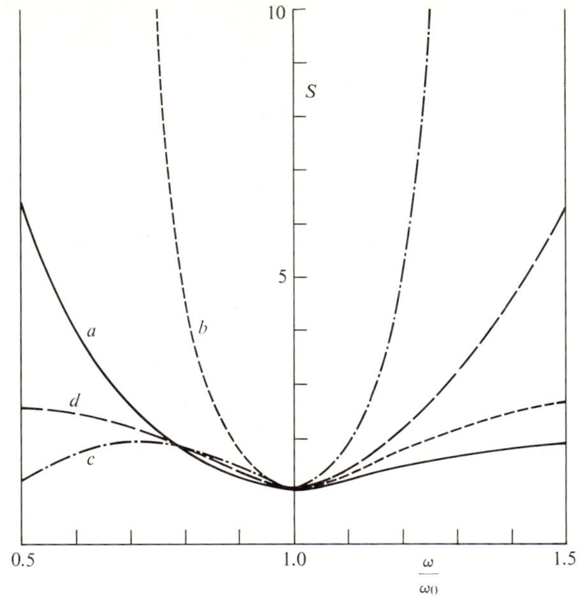

Fig. 3.17 *(continued)*

from (8.10) and (8.21) that the per unit admittance y_m at A on the stub transmission line is

$$y_m = -j \cot \beta d_{n,\mathrm{SC}} \qquad \text{(short-circuited stub)}, \tag{8.30a}$$

or

$$= j \tan \beta d_{n,\mathrm{OC}} \qquad \text{(open-circuited stub)}. \tag{8.30b}$$

Hence, the total per unit admittance y_t terminating the main line is

$$y_t = y + y_m. \tag{8.31}$$

We can deduce the voltage reflection coefficient pertaining to the main line from (8.5) to be

$$\rho = \frac{1 - y_t}{1 + y_t}, \tag{8.32}$$

and we can then obtain S from (8.6). Let $\omega = \omega_0$ be the angular frequency for which the stub is designed. It is convenient for us to let $\beta d = \beta_0 d(\beta/\beta_0) = \beta_0 d(\omega/\omega_0)$, where β_0 is the phase-shift coefficient at the designed frequency and d stands for d_n, $d_{n,\mathrm{SC}}$, and $d_{n,\mathrm{OC}}$. With the help of (8.29) to (8.32) and (8.6), we can obtain the standing-wave ratio S as a function of the ratio ω/ω_0 and we show it in Fig. 3.17 for the four different sets of parameters of the single-stub impedance-matching device. As we mentioned previously, Fig. 3.17 pertains to lossless transmission lines and frequency-independent conductance for the load. From Fig. 3.17, we see S to be equal to 1 for $\omega/\omega_0 = 1$, indicating perfect impedance match. But we find S to increase as ω/ω_0 is changed from unity in either direction, indicating impedance mismatch. The frequency sensitivity of the single-stub impedance-matching device is evident from Fig. 3.17.

3.9 POWER TRANSPORT ON LOSSLESS TRANSMISSION LINES

The energy transport on lossless transmission lines has many interesting features. We discuss some of the more important features in this section.

1. Energy Transport Velocity

In the case of lossless transmission lines, we can introduce the concept of velocity of energy propagation. The energy transport velocity v_E is defined as the ratio of the time-averaged power flux P to the time-averaged energy W associated with the inductance and the capacitance per unit length of the transmission line:

$$v_E = P/W. \tag{9.1}$$

Since P has the dimensions of energy divided by time and W has the dimensions of energy divided by length, it follows that (9.1) has the proper dimensions of velocity. We can derive the value of v_E in terms of the transmission-line parameters in a straightforward manner. We should note that (9.1) refers to a progressive wave only. For a progressive wave, which exists, for example, on a semi-infinite ($0 < x < \infty$) lossless transmission line, the voltage and the current are given by

$$V(x) = V_1 e^{-j\beta x}, \qquad I(x) = I_1 e^{-j\beta x} = \frac{V_1}{R_0} e^{-j\beta x}, \tag{9.2}$$

where R_0 is the characteristic resistance. With the help of (2.8.10), and (9.2), we obtain the time-averaged power flux P as

$$P = \tfrac{1}{2} \operatorname{Re} V(x)I^*(x) = \tfrac{1}{2} \operatorname{Re} V_1 I_1^* = \frac{1}{2}\frac{|V_1|^2}{R_0}. \tag{9.3}$$

We find the real voltage and the real current from (9.2) to be

$$V(x, t) = |V_1| \cos(\omega t + \theta_{V_1} - \beta x); \qquad I(x, t) = \frac{|V_1|}{R_0} \cos(\omega t + \theta_{V_1} - \beta x). \tag{9.4}$$

The time-averaged energy W_L associated with the inductance L per unit length of the transmission line is

$$W_L = \frac{\omega}{2\pi} \int_0^{2\pi/\omega} \tfrac{1}{2} L I^2(x, t)\, dt. \tag{9.5}$$

On substituting $I(x, t)$ from (9.4) into (9.5) and carrying out the resulting integration by the method employed in obtaining (2.8.10), we see that

$$W_L = \tfrac{1}{4} \operatorname{Re} L I(x) I^*(x) = \tfrac{1}{4} L |I_1|^2 = \tfrac{1}{4} L \frac{|V_1|^2}{R_0^2} = \tfrac{1}{4} C |V_1|^2, \tag{9.6}$$

where the relation $R_0^2 = L/C$ has been used. In a similar manner, the time-averaged energy W_C associated with the capacitance C per unit length of the transmission line is

$$W_C = \tfrac{1}{4} \operatorname{Re} CV(x)V^*(x) = \tfrac{1}{4}C\,|V_1|^2. \tag{9.7}$$

Note that $W_L = W_C$. Hence, the total time-averaged energy W associated with unit length of the transmission line is

$$W = W_L + W_C = \tfrac{1}{2}C\,|V_1|^2. \tag{9.8}$$

Therefore, from (9.1), (9.3), and (9.8), we derive v_E as

$$v_E = \frac{P}{W} = \frac{1}{2}\frac{|V_1|^2}{R_0}\frac{1}{\tfrac{1}{2}C\,|V_1|^2} = \frac{1}{\sqrt{LC}} = v_{ph}. \tag{9.9}$$

We thus establish the interesting result that on a lossless transmission line the energy transport velocity is equal to the phase velocity.

In general, for a lossless medium, the energy transport velocity is equal to the group velocity. For a lossless transmission line, we can establish this result in the following manner. Substituting (9.2) into (2.5.7) and (2.5.8), and taking into account the fact that $R = G = 0$ for a lossless transmission line, we get

$$\beta V_1 = \omega L I_1, \tag{9.10}$$

$$\beta I_1 = \omega C V_1. \tag{9.11}$$

We should note that in general β is a function of ω and therefore vice versa. In the same manner, V_1 and I_1 are also functions of β. On differentiating (9.10) with respect to β and multiplying the resultant equation by I_1^*, we obtain

$$V_1 I_1^* + \beta I_1^* \frac{\partial V_1}{\partial \beta} = L\,|I_1|^2 \frac{\partial \omega}{\partial \beta} + (\omega L I_1^*) \frac{\partial I_1}{\partial \beta}. \tag{9.12}$$

Since $\omega L I_1^* = \beta V_1^*$ from (9.10), we can simplify (9.12) to yield

$$V_1 I_1^* + \beta I_1^* \frac{\partial V_1}{\partial \beta} = L\,|I_1|^2 \frac{\partial \omega}{\partial \beta} + \beta V_1^* \frac{\partial I_1}{\partial \beta}. \tag{9.13}$$

Similarly, on differentiating (9.11) with respect to β and multiplying the resultant equation by V_1^*, we find that

$$V_1^* I_1 + \beta V_1^* \frac{\partial I_1}{\partial \beta} = C\,|V_1|^2 \frac{\partial \omega}{\partial \beta} + (\omega C V_1^*) \frac{\partial V_1}{\partial \beta}. \tag{9.14}$$

Since $\omega C V_1^* = \beta I_1^*$ from (9.11), (9.14) reduces to

$$V_1^* I_1 + \beta V_1^* \frac{\partial I_1}{\partial \beta} = C\,|V_1|^2 \frac{\partial \omega}{\partial \beta} + \beta I_1^* \frac{\partial V_1}{\partial \beta}, \tag{9.15}$$

which when added to (9.13) gives the following result:

$$V_1 I_1^* + V_1^* I_1 = 2 \operatorname{Re}(V_1 I_1^*) = \{L\,|I_1|^2 + C\,|V_1|^2\} \frac{\partial \omega}{\partial \beta}. \tag{9.16}$$

The substitution of (9.3), (9.6), (9.7), and (9.8) in (9.16) yields

$$P = W\,\partial\omega/\partial\beta \qquad \text{or} \qquad v_E = P/W = \partial\omega/\partial\beta = v_g. \tag{9.17}$$

We have proved the energy transport velocity v_E to be equal to the group velocity. This is a general result valid for any lossless medium. But in a transmission line, when there are no losses, the phase velocity becomes independent of the frequency and therefore is equal to the group velocity. It is in view of this special circumstance that in (9.9) the energy transport velocity has been found to be equal to the phase velocity.

2. Power Delivered by a Matched Generator to the Input of a Lossless Transmission Line

Let us consider a lossless transmission line extending from $x = 0$ to $x = l$ (Fig. 3.18a). Let a load impedance $Z_L = R_L + jX_L$ be connected to the terminals at $x = l$. Also, let us assume that a voltage generator V_g in series with a resistor R_g is connected to the input terminals at $x = 0$. The voltage generator is said to be matched to the transmission line if R_g is equal to the characteristic resistance R_0 of the transmission line. We wish to find the time-averaged power $P(x = 0)$ delivered to the input terminals of the transmission line. If $Z_{\text{in}} = R_0 z_{\text{in}}$ is the input

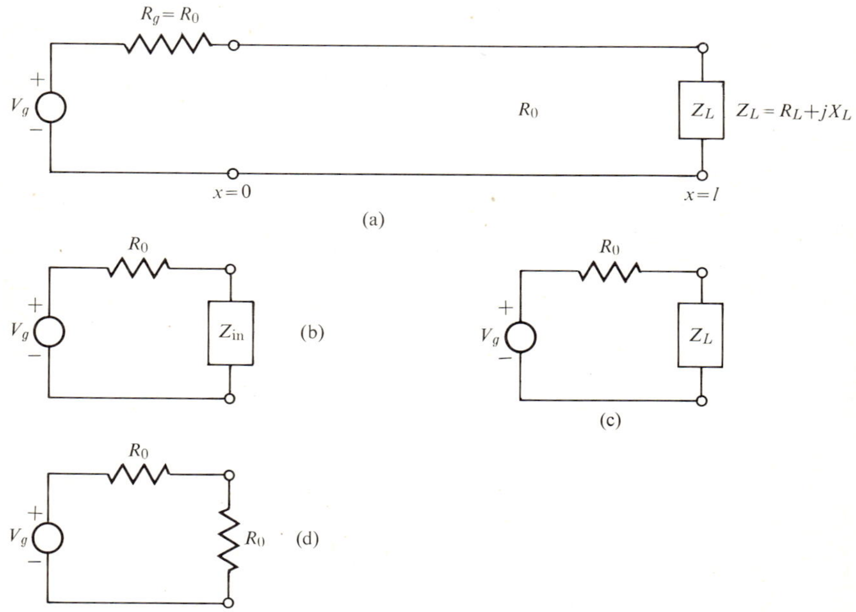

Fig. 3.18 (a) Schematic diagram of a lossless transmission line driven by a matched generator and terminated by an arbitrary load impedance. (b) Equivalent electric circuit at the input. (c) Equivalent electric circuit at the input with $l = 0$. (d) Equivalent electric circuit at the input for the incident progressive wave only.

impedance, we find from the equivalent electric circuit at the input as shown in Fig. 3.18(b), that the input current I_1 and the input voltage V_1 are

$$I_1 = \frac{V_g}{R_0 + R_0 z_{\text{in}}}, \qquad V_1 = \frac{V_g R_0 z_{\text{in}}}{R_0 + R_0 z_{\text{in}}}. \tag{9.18}$$

Therefore, the time-averaged power delivered to the input terminals of the transmission line is

$$P\,(x = 0) = \tfrac{1}{2}\,\text{Re}\, V_1 I_1^* = \frac{|V_g|^2\, r_{\text{in}}}{2R_0(1 + z_{\text{in}})(1 + z_{\text{in}}^*)}, \tag{9.19}$$

where r_{in} is the per unit input resistance whose expression is given in (4.8a). From (4.1), the per unit input impedance z_{in} is

$$z_{\text{in}} = \frac{1 + |\rho|\, e^{j\phi}}{1 - |\rho|\, e^{j\phi}}. \tag{9.20}$$

Therefore,

$$\begin{aligned}(1 + z_{\text{in}})(1 + z_{\text{in}}^*) &= \frac{2}{(1 - |\rho|\, e^{j\phi})} \times \frac{2}{(1 - |\rho|\, e^{-j\phi})}, \\ &= 4[1 + |\rho|^2 - 2\,|\rho| \cos \phi]^{-1}. \end{aligned} \tag{9.21}$$

The substitution of (9.21) and the value of r_{in} from (4.8a) in (9.19) yields

$$\begin{aligned} P\,(x = 0) &= \frac{|V_g|^2\,(1 - |\rho|^2)}{2R_0[1 + |\rho|^2 - 2\,|\rho| \cos \phi]4[1 + |\rho|^2 - 2\,|\rho| \cos \phi]^{-1}}, \\ &= \frac{|V_g|^2\,[1 - |\rho|^2]}{8R_0}. \end{aligned} \tag{9.22}$$

The phase ϕ of the general reflection coefficient contains the length l of the transmission line but it has disappeared in (9.22) showing the interesting result that the time-averaged power delivered by a matched generator to a lossless transmission line with an arbitrary load impedance is independent of the length of the transmission line.

It is instructive for us to confirm (9.22) by deriving the result after setting $l = 0$ and thus demonstrate that $P\,(x = 0)$ is independent of the length of the transmission line. The equivalent electric circuit at the input for the present case is as shown in Fig. 3.18(c). The input current I_1 and the input voltage V_1 are then given by

$$I_1 = \frac{V_g}{R_0 + Z_L}; \qquad V_1 = \frac{V_g Z_L}{R_0 + Z_L}. \tag{9.23}$$

Therefore

$$P\,(x = 0) = \tfrac{1}{2}\,\text{Re}\, V_1 I_1^* = \frac{|V_g|^2\, R_L}{2(R_0 + Z_L)(R_0 + Z_L^*)}. \tag{9.24}$$

For the case $X_0 = 0$, we see from (2.12.14) and (2.12.16) that

$$1 - |\rho|^2 = 4R_L R_0/[(R_L + R_0)^2 + X_L^2]. \tag{9.25}$$

Also,

$$(R_0 + Z_L)(R_0 + Z_L^*) = (R_0 + R_L)^2 + X_L^2. \tag{9.26}$$

Hence, from (9.25) and (9.26), it follows that

$$\frac{R_L}{(R_0 + Z_L)(R_0 + Z_L^*)} = \frac{(1 - |\rho|^2)}{4R_0}, \tag{9.27}$$

which when substituted in (9.24) yields

$$P\,(x = 0) = \frac{|V_g|^2}{8R_0}(1 - |\rho|^2). \tag{9.28}$$

The result obtained in (9.28) is identical to that given in (9.22).

Since the transmission line is lossless, it is to be expected that the power delivered at the input is transferred as is, without getting diminished, along the transmission line. Though this result is intuitively obvious, it is worthwhile for us to establish it rigorously. For $R = G = 0$, (2.8.21) becomes

$$-\frac{d}{dx}[V(x)I^*(x)] = j\omega L\,|I(x)|^2 - j\omega C\,|V(x)|^2. \tag{9.29}$$

Since $P(x) = \frac{1}{2}\text{Re}\ V(x)I^*(x)$, the real part of (9.29) yields

$$\frac{d}{dx}P(x) = 0 \quad \text{or} \quad P(x) = \text{constant}, \tag{9.30}$$

showing that the time-averaged power $P(x)$ associated with the voltage and the current is independent of the location on the transmission line. Therefore, the power delivered at the input is propagated as is along the transmission line.

It is now possible for us to give a physical interpretation of (9.22). In Section 2.12, we showed that the time-averaged power associated with the voltage and the current along a lossless transmission line whose characteristic impedance has only a real part can be legitimately separated into the incident power $P_{\text{inc}}(x)$ and the reflected power $P_{\text{ref}}(x)$, which are the powers associated separately with the two progressive waves. The incident power $P_{\text{inc}}(x)$ is associated with the progressive wave whose phase fronts travel away from the generator and the reflected power $P_{\text{ref}}(x)$ with that whose phase fronts travel towards the generator. For the incident progressive wave, the equivalent circuit at the input is as shown in Fig. 3.18(d). Therefore, we obtain the incident current at the input $I_{\text{inc}}\,(x = 0)$ and the incident voltage at the input $V_{\text{inc}}\,(x = 0)$ as

$$I_{\text{inc}}\,(x = 0) = V_g/2R_0; \qquad V_{\text{inc}}\,(x = 0) = V_g/2. \tag{9.31}$$

Hence, the time-averaged incident power is

$$P_{\text{inc}}\,(x = 0) = \tfrac{1}{2}\,\text{Re}\,V_{\text{inc}}\,(x = 0)\,I^*_{\text{inc}}\,(x = 0) = |V_g|^2/8R_0. \tag{9.32}$$

In view of (9.30),

$$P_{\text{inc}} = P_{\text{inc}}\,(x = l) = P_{\text{inc}}\,(x = 0) = |V_g|^2/8R_0. \tag{9.33}$$

Let $I_{\text{inc}}\,(x = l)$ and $V_{\text{inc}}\,(x = l)$ be the values of the incident current and the incident voltage at the location of the load. Then, the reflected current $I_{ref}\,(x = l)$ and the reflected voltage $V_{ref}\,(x = l)$ at the location of the load are given by

$$I_{ref}\,(x = l) = -\rho I_{inc}\,(x = l), \qquad V_{ref}\,(x = l) = \rho V_{inc}\,(x = l), \tag{9.34}$$

where ρ is the voltage reflection coefficient corresponding to the load impedance Z_L. Consequently, we find the reflected power $P_{ref}\,(x = l)$ at the location of the load with the help of (9.32) and (9.33) to be

$$P_{ref}\,(x = l) = \tfrac{1}{2}\,\mathrm{Re}\; V_{ref}\,(x = l)\, I^*_{ref}\,(x = l) = -\,|\rho|^2\, P_{inc}. \tag{9.35}$$

The minus sign in (9.35) indicates that the power is now flowing in the opposite direction, that is, towards the generator. Again, in view of (9.30),

$$P_{ref} = P_{ref}\,(x = 0) = P_{ref}\,(x = l) = -\,|\rho|^2\, P_{inc}. \tag{9.36}$$

Since the generator is matched to the transmission line, the entire power incident on it is absorbed. Thus, the generator delivers the time-averaged power P_{inc} in the incident wave and the time-averaged power P_{ref} is returned to the generator by the reflected wave. There are no further reflections at the generator end since it is matched to the transmission line. Consequently the net time-averaged power delivered to the input terminals of the transmission line is

$$P\,(x = 0) = P_{inc} + P_{ref} = P_{inc}(1 - |\rho|^2) = \frac{|V_g|^2}{8R_0}(1 - |\rho|^2), \tag{9.37}$$

which reproduces the result in (9.22).

The foregoing analysis is used to work out the following numerical example.

Example. Consider a long lossless transmission line of characteristic resistance $R_0 = 50$ ohms, as shown in Fig. 3.19. The load impedance is given by $Z_L = 50 + j50$ ohms. A voltage generator $V_g = 20$ V in series with a matching resistor $R_g = R_0 = 50$ ohms is connected to the input terminals of the transmission line. Find the time-averaged power delivered to the transmission line. Also, determine the magnitudes of the current and the voltage at the locations of (a) a voltage minimum, (b) a voltage maximum and (c) the load impedance.

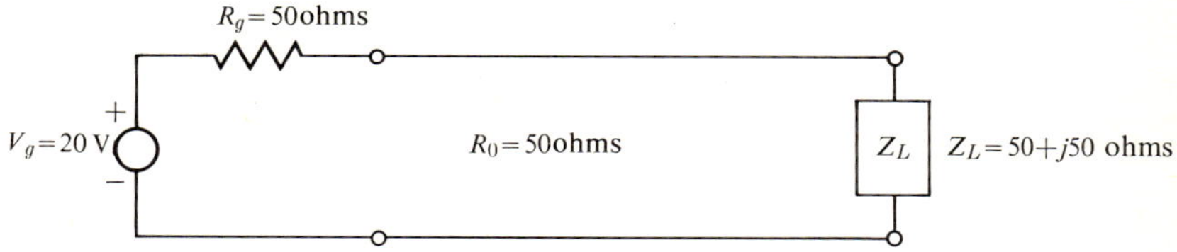

Fig. 3.19 Power delivered by a matched generator to the input terminals of a lossless transmission line.

The voltage reflection coefficient corresponding to the given load impedance is

$$\rho = \frac{Z_L - R_0}{Z_L + R_0} = \frac{j50}{100 + j50},$$

and therefore,

$$|\rho| = 1/\sqrt{5} \qquad \text{and} \qquad |\rho|^2 = \tfrac{1}{5}.$$

The time-averaged power P_{inc} in the incident wave is

$$P_{\text{inc}} = \frac{V_g^2}{8R_0} = \frac{20 \times 20}{8 \times 50} = 1 \text{ W}.$$

The time-averaged power P delivered to the input terminals of the transmission line is, therefore,

$$P = P_{\text{inc}}(1 - |\rho|^2) = \tfrac{4}{5} = 0.8 \text{ W}.$$

The standing-wave ratio S pertaining to the given load impedance is

$$S = \frac{1 + |\rho|}{1 - |\rho|} = \frac{\sqrt{5} + 1}{\sqrt{5} - 1} = 2.62.$$

At the location of the voltage minimum, the impedance is resistive and has a minimum value equal to $R_0/S = 50/2.62$ ohms. Similarly, at the location of the voltage maximum, the impedance is resistive and has a maximum value equal to $R_0S = 50 \times 2.62$ ohms. At the locations of the voltage minimum, the voltage maximum, and the load impedance, the associated time-averaged power P is the same, namely 0.8 W.

Let V_{min} be the voltage at the location of the voltage minimum which, as we have discussed previously, corresponds to the location of the current maximum I_{max}. Consequently,

$$\tfrac{1}{2}\,|I_{\text{max}}|^2 \frac{R_0}{S} = 0.8 \quad \text{or} \quad |I_{\text{max}}| = 0.29 \text{ A},$$

and

$$|V_{\text{min}}| = \frac{R_0}{S}\,|I_{\text{max}}| = \frac{50 \times 0.29}{2.62} = 5.53 \text{ V}.$$

Let V_{max} be the voltage at the location of the voltage maximum which corresponds also to the location of the minimum current I_{min}. From (2.16), we see that

$$|V_{\text{max}}| = R_0\,|I_{\text{max}}| = 14.5 \text{ V},$$

and

$$|I_{\text{min}}| = |V_{\text{min}}|/R_0 = 0.11 \text{ A}.$$

The time-averaged power associated with V_{max} and I_{min} is

$$P = \tfrac{1}{2} \times 14.5 \times 0.11 = 0.8 \text{ W},$$

which is equal to the time-averaged power delivered to the transmission line, as it should be.

Let V_L and I_L be the voltage and the current at the load terminals. Then,

$$\tfrac{1}{2}R_L\,|I_L|^2 = \tfrac{1}{2} \times 50 \times |I_L|^2 = 0.8, \qquad |I_L| = 0.18 \text{ A},$$

and

$$\tfrac{1}{2}\,|V_L|^2 \frac{R_L}{|Z_L|^2} = \tfrac{1}{2}\,|V_L|^2 \times \frac{50}{2 \times 50 \times 50} = 0.8, \qquad |V_L| = 12.6 \text{ V}.$$

Thus, in the case of a *lossless* transmission line driven by a *matched* generator, we can determine the magnitudes of the voltage and the current at various locations along the transmission line provided the impedance values at those locations are known.

3.10 POWER TRANSPORT ON LOW-LOSS TRANSMISSION LINES

1. Attenuation Coefficient

For a low-loss transmission line which is either semi-infinite or finite in length with a characteristic impedance termination, we can obtain the phasor voltage and the phasor current distributions from (2.6.2) as

$$V(x) = V_1 e^{-(\alpha + j\beta)x}; \qquad I(x) = \left(\frac{V_1}{Z_0}\right) e^{-(\alpha + j\beta)x}. \tag{10.1}$$

We find the time-averaged power $P(x)$ at any location along the transmission line associated with the voltage and the current given in (10.1) to be

$$P(x) = \tfrac{1}{2} \operatorname{Re} V(x)I^*(x) = \frac{|V_1|^2}{2\,|Z_0|^2} R_0 e^{-2\alpha x}. \tag{10.2}$$

We can see from (10.2) that the power decays exponentially with the distance. This is due to the dissipation of energy that takes place along the transmission line. We obtain the power loss P_L per unit length from (10.2) as

$$P_L(x) = -\partial P(x)/\partial x = 2\alpha P(x). \tag{10.3}$$

We can deduce the attenuation coefficient α from (10.3) to be

$$\alpha = P_L(x)/2P(x). \tag{10.4}$$

The formula (10.4) for the attenuation coefficient is quite general in the sense that it is applicable to any system where the power transmission obeys the exponential decay law.

The attenuation coefficient α for the transmission line with loss can be evaluated with the help of (10.4) in the following manner. We obtain the time-averaged power loss P_L per unit length of the transmission line from (2.8.14) and (2.8.15) as

$$P_L(x) = \frac{1}{2}\frac{|V_1|^2}{|Z_0|^2} e^{-2\alpha x}(R + G\,|Z_0|^2). \tag{10.5}$$

The substitution of (10.2) and (10.5) in (10.4) gives

$$\alpha = \frac{1}{2R_0}\,[R + G\,|Z_0|^2], \tag{10.6}$$

which is in agreement with the expression for α obtained in (2.8.26). For a low-loss transmission line, we find from (2.7.7) and (2.7.8) that the leading term of the expansion of the expression for Z_0 in powers of the small quantities $R/\omega L$ and $G/\omega C$ is $Z_0 = R_0 \approx \sqrt{L/C}$. Therefore, for a low-loss transmission line (10.6) reduces to

$$\alpha = \tfrac{1}{2}\left[R\sqrt{C/L} + G\sqrt{L/C}\right]. \tag{10.7}$$

It is interesting to note that (10.7) is exactly equal to the leading term of the expansion of α in powers of the small quantities $R/\omega L$ and $G/\omega C$, as given in (2.7.14).

2. Selectivity

Previously, the selectivity Q for the antiresonant transmission-line circuit was defined as the ratio of the antiresonant frequency to the frequency range Δf between the "half-power" points. In order to evaluate Q by this definition, we require a knowledge of the variation of the input impedance with frequency. It is desirable to have a more fundamental definition of Q based on energy considerations, since such a definition has a more general applicability. With particular reference to the low-loss transmission lines under consideration, we can define the selectivity Q based on energy considerations as

$$Q = \omega_0(W/P_L), \tag{10.8}$$

where W is the time-averaged energy associated with the series inductance and the shunt capacitance of the transmission line and P_L is the time-averaged power dissipated in the series resistance and the shunt conductance of the transmission line. Both W and P_L have to be evaluated in the neighborhood of the antiresonant frequency $f_0 = \omega_0/2\pi$. We have to show that Q as defined in (10.8) gives the same value as we obtained previously with the help of the definition contained in (2.11.22).

For this purpose, we consider a finite length $(0 < x < l)$ transmission line with open-circuit termination. We assume that the losses in the transmission line are so small that for the evaluation of W and P_L, the voltage and the current distributions along the transmission line may be reasonably taken to be the same as those on the corresponding lossless transmission line. For a transmission line with open-circuit termination, antiresonance occurs for $\beta l = n\pi$, where n is an integer. The voltage and the current distributions on a lossless transmission line with open-circuit termination are, as given in (2.17),

$$V(x) = V_L \cos \beta(l - x), \qquad I(x) = j(V_L/R_0) \sin \beta(l - x), \tag{10.9}$$

where V_L is the voltage at $x = l$. We obtain the time-averaged energy W_L associated with the inductance throughout the entire length of the transmission line as

$$\begin{aligned} W_L &= \int_0^l \tfrac{1}{4} L I(x) I^*(x)\, dx = \int_0^l \tfrac{1}{4} L \frac{|V_L|^2}{R_0^2} \sin^2 \beta(l - x)\, dx, \\ &= \int_0^l \frac{1}{8} \frac{L}{R_0^2} |V_L|^2 \left[1 - \cos 2\beta(l - x)\right] dx, \\ &= \tfrac{1}{8} C\, |V_L|^2\, l. \end{aligned} \tag{10.10}$$

In the derivation of (10.10), we have used the relation $R_0^2 = L/C$. In a similar manner, we find the time-averaged energy W_C associated with the capacitance of the entire transmission line to be

$$W_C = \int_0^l \tfrac{1}{4}CV(x)V^*(x)\,dx = \int_0^l \tfrac{1}{4}C\,|V_L|^2 \cos^2 \beta(l-x)\,dx,$$
$$= \tfrac{1}{8}C\,|V_L|^2\,l. \tag{10.11}$$

Hence, from (10.10) and (10.11), we obtain the total time-averaged energy associated with the reactive elements of the transmission line as

$$W = W_L + W_C = \tfrac{1}{4}C\,|V_L|^2\,l. \tag{10.12}$$

We deduce the time-averaged power dissipated in the resistance of the entire transmission line as

$$P_{L,R} = \int_0^l \tfrac{1}{2}R\,|I(x)|^2\,dx = \int_0^l \tfrac{1}{2}R\,\frac{|V_L|^2}{R_0^2}\sin^2 \beta(l-x)\,dx,$$
$$= \tfrac{1}{4}R\,\frac{|V_L|^2}{R_0^2}\,l. \tag{10.13}$$

Similarly, we evaluate the time-averaged power dissipated in the conductance of the entire transmission line as

$$P_{L,G} = \int_0^l \tfrac{1}{2}G\,|V(x)|^2\,dx = \int_0^l \tfrac{1}{2}G\,|V_L|^2 \cos^2 \beta(l-x)\,dx,$$
$$= \tfrac{1}{4}G\,|V_L|^2\,l. \tag{10.14}$$

Hence, we see from (10.13) and (10.14) that the total time-averaged power loss in the entire transmission line is

$$P_L = P_{L,R} + P_{L,G} = \frac{1}{4}\frac{|V_L|^2}{R_0^2}\,l(R + GR_0^2). \tag{10.15}$$

In the limit of very small losses, the attenuation coefficient α as given in (10.6) becomes

$$\alpha = \frac{1}{2R_0}[R + GR_0^2]. \tag{10.16}$$

Together with (10.16), we can simplify (10.15) to

$$P_L = \frac{1}{2}\frac{|V_L|^2\,l\alpha}{R_0}. \tag{10.17}$$

We use (10.12) and (10.17) in (10.8) to obtain the following expression for the selectivity:

$$Q = \frac{\omega_0 \frac{1}{4}C\,|V_L|^2\,l}{\frac{1}{2}(|V_L|^2\,l\alpha/R_0)},$$
$$= \frac{\omega_0 C R_0}{2\alpha} = \frac{\beta}{2\alpha}, \tag{10.18}$$

since $\omega_0 C R_0 = \omega_0 C\sqrt{L/C} = \omega_0\sqrt{LC} = \beta$. The expression (10.18) for Q is the same as in (2.11.32) which we obtained using a different definition for Q, namely that in (2.11.22).

3. Energy Transport Velocity

In the case of a lossy transmission line, since power is lost irretrievably along the transmission line, it is difficult for us to define unambiguously a velocity of energy transport and at any rate there is no clear connection between the velocity of energy transport and the group velocity, as in the case of a lossless transmission line.

4. Power Delivered by a Matched Generator to the Input of a Lossy Transmission Line

For a lossy transmission line, if we attempt to carry out the calculations similar to those contained in (9.18) to (9.22), which pertain to a lossless transmission line, we find that the power delivered by a matched generator to the input terminals of a lossy transmission line with an arbitrary load impedance does depend on the length of the transmission line. The dependence of the input power on the length of the transmission line is quite complicated except in the special case of a distortionless transmission line for which we obtain the same result as in (9.22) with $|\rho|^2$ replaced by $|\rho|^2 e^{-4\alpha l}$.

5. Impedance Matching for Maximum Power Transfer

Let a generator of voltage V_g with a series impedance $Z_g = R_g + jX_g$ be connected to the input terminals of a lossy transmission line. Let $Z_{\text{in}} = R_{\text{in}} + jX_{\text{in}}$ be the input impedance of the transmission line. From the equivalent electric circuit corresponding to the input as shown in Fig. 3.20, we obtain the input current I_1 and the input voltage V_1 as

$$I_1 = \frac{V_g}{Z_g + Z_{\text{in}}}; \qquad V_1 = \frac{V_g Z_{\text{in}}}{Z_g + Z_{\text{in}}}. \tag{10.19}$$

Therefore, the time-averaged power delivered to the input terminals of the transmission line is

$$P = \tfrac{1}{2}\,\text{Re}\, V_1 I_1^* = \frac{|V_g|^2 R_{\text{in}}}{2[(R_g + R_{\text{in}})^2 + (X_g + X_{\text{in}})^2]}. \tag{10.20}$$

We wish to find the proper values of R_{in} and X_{in} which give a maximum input power for a given R_g and X_g. We obtain the conditions for maximum value of P by simultaneously satisfying the relations

$$\frac{\partial P}{\partial R_{\text{in}}} = 0; \qquad \frac{\partial P}{\partial X_{\text{in}}} = 0. \tag{10.21}$$

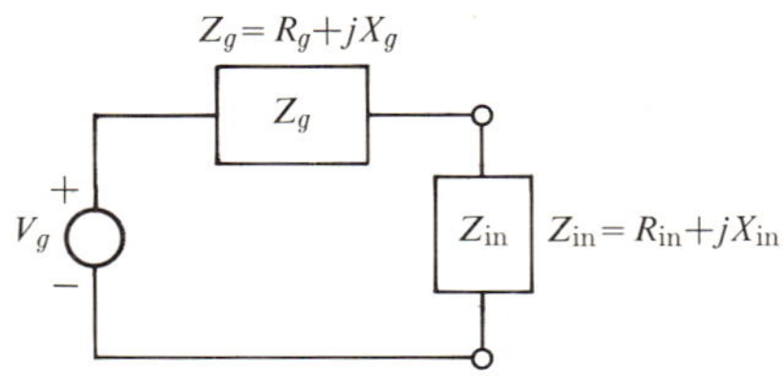

Fig. 3.20 Equivalent electric circuit for determining the value of Z_{in} which yields the maximum power input.

The application of (10.21) to (10.20) yields the following results:

$$R_g^2 - R_{\text{in}}^2 + (X_g + X_{\text{in}})^2 = 0, \tag{10.22}$$

and

$$X_g = -X_{\text{in}}. \tag{10.23}$$

From (10.23), we find that for maximum power transfer, the input reactance should be equal and opposite to that of the generator. The substitution of (10.23) in (10.22) and the fact that all resistances are positive, shows that

$$R_g = R_{\text{in}}. \tag{10.24}$$

Therefore, the maximum input power is obtained when the input impedance is the complex conjugate of the generator impedance. This is known as "conjugate matching". On the other hand, when the input impedance is equal to the generator impedance, no reflections occur at the generator end and the transmission line is said to be "impedance matched" to the generator. In general, the "conjugate matching" and the "impedance matching" are different. But in the special case of vanishing generator reactance, both types of matching become identical.

PROBLEMS

3.1 A lossless transmission line of length 25 m has a characteristic impedance of $Z_0 = 100$ ohms. The terminating impedance is equal to the characteristic impedance. A generator whose open-circuit voltage is $V_g = 50$ V and whose internal impedance is $Z_g = 50$ ohms is connected to the input terminals of the transmission line. The phase velocity for propagation along the transmission line is $v_{ph} = 3 \times 10^8$ m/sec. The frequency of operation is $f = 10^8$ Hz. Find the magnitudes and the phase angles of the current in and the voltage across the load impedance.

3.2 Suppose that in Problem 3.1, the terminating impedance, instead of being equal to the characteristic impedance, is given by $Z_L = 50 + j25$ ohms. All the other parameters remain unchanged. Find the magnitudes and the phase angles of the current in and the voltage across the load impedance.

3.3 A lossless transmission line of length 20 m has a characteristic impedance of $Z_0 = 50$ ohms. The load impedance is given by $Z_L = 25 - j75$ ohms. A generator whose open-circuit voltage is $V_g = 10$ V and whose internal impedance is $Z_g = 100$ ohms is connected to the input terminals of the transmission line. The phase velocity for propagation along the transmission line is $v_{ph} = 2 \times 10^8$ m/sec. The frequency of operation is $f = 6.7 \times 10^7$ Hz. Find the magnitudes and the phase angles of the current in and the voltage across the load impedance.

Also, determine the phasor voltage and current at (a) the input terminals of the transmission line and (b) midway along the length of the transmission line.

3.4 A lossless transmission line of characteristic resistance $R_0 = 50$ ohms is terminated in a load impedance $Z_L = 50 - j100$ ohms. Find (a) the magnitude $|\rho|$ and (b) the phase angle θ_ρ of the terminal reflection coefficient ρ. What is the standing-wave ratio S on the transmission line?

Sketch the crank diagram for examining the variation of the magnitude of the voltage along the transmission line. With the help of the crank diagram find the distance in fractions of a wavelength from the load terminals to the location of (a) the first voltage maximum and (b) the first voltage minimum.

Also, find the nearest distance in fractions of a wavelength from the load terminals to the location where the magnitude of the voltage is the average of its maximum and minimum values.

3.5 Consider a lossless transmission line of length 20 m with short-circuit terminations at both the ends. Find (a) the largest wavelength and (b) the second largest wavelength of the standing waves that can be formed on the transmission line. If the phase velocity of the waves on the transmission line is $v_{ph} = 3 \times 10^8$ m/sec, what are the corresponding frequencies of the two standing waves?

3.6 A transmission line of characteristic impedance $Z_0 = 50$ ohms is terminated in a load impedance $Z_L = 150 + j75$ ohms. Find the magnitude $|\rho|$ and the phase angle θ_ρ of the terminal reflection coefficient ρ. What is the standing-wave ratio S on the transmission line?

3.7 A lossless transmission line is terminated in a load admittance $Y_L = G_L + jB_L$ in which B_L is a variable. The characteristic admittance of the transmission line is G_0 mhos. Find the value of B_L which minimizes the standing-wave ratio S. If the minimum value of $S = S_{min}$, what are the two possible values of G_L in terms of S_{min}?

3.8 A dissipationless transmission line has a characteristic resistance of R_0 ohms. The length of the transmission line is $l = \lambda/8$ where λ is the wavelength. The load impedance $Z_L = R_L + jX_L$ connected to the terminals of the transmission line has a magnitude equal to R_0. Show that the input impedance of the transmission line is a resistance.

3.9 a) A lossless transmission line is terminated in an unknown impedance. At two locations along the transmission line, the impedances are measured with the following results: Location 1, $Z(x) = 250$ ohms; Location 2, $Z(x) = 10$ ohms. Identify the two locations. What is the separation distance between these two locations? Find the characteristic resistance R_0 and the voltage reflection coefficient ρ as far as can be determined. Indicate the additional measurements that have to be performed in order to be able to determine ρ completely.

b) It is desired to impedance match the transmission line in part (a) by connecting a resistor parallel to the transmission line at one of the two above-mentioned locations. Determine the location and the value of the resistor.

Is it possible to obtain impedance matching by connecting parallel to the transmission line at a point in between the above-mentioned two locations (i) a resistor, (ii) a reactance? Give reasons.

3.10 Consider a lossless transmission line of characteristic resistance $R_0 = 50$ ohms. The standing-wave ratio is $S = 3$. Can the load impedance be completely determined? If not, find the two values of the resistive load R_L which yield the given standing-wave ratio.

3.11 A lossless transmission line of characteristic resistance $R_0 = 50$ ohms is terminated in a load impedance which is to be determined. The standing-wave ratio S on the transmission line is measured as 5. A maximum of the voltage standing wave is found to occur at the termination of the transmission line. Determine the load impedance Z_L. If instead of a maximum, a minimum of the voltage standing wave occurred at the termination, what is the load impedance?

3.12 a) The standing-wave ratio S on a lossless transmission line of characteristic impedance $R_0 = 50$ ohms is 3.3. A voltage minimum is observed 0.19 wavelength in front of the load impedance. Find the load impedance.

b) Examine the slope of the magnitude of the voltage on the transmission line near the load impedance and hence ascertain the nature of the load reactance.

c) Repeat the solution of part (a) using the Smith chart.

3.13 A lossless transmission line of characteristic impedance $R_0 = 50$ ohms is terminated in a load impedance Z_L. The observed standing-wave ratio on the transmission line is $S = 9$. The distance between successive minima in the voltage standing wave is 0.2 m and the distance from the load impedance to the first minimum in the voltage standing wave is 0.08 m. Find the load impedance.

3.14 A lossless transmission line of characteristic impedance $R_0 = 50$ ohms is terminated in an impedance $Z_L = 120 - j75$ ohms. The wavelength on the transmission line is 2 m. Find the standing-wave ratio S and the maximum and the minimum impedances on the transmission line. Find the distance between the locations of the load impedance and the first minimum in the voltage standing wave.

If the transmission line is to transmit a power of 100 W, find the maximum and the minimum of the magnitudes of the voltage and the current as well as the magnitude of the voltage across the load impedance.

3.15 A lossless transmission line of characteristic impedance $R_0 = 75$ ohms is terminated by a load impedance $Z_L = 110 + j70$ ohms. Find the position of the minimum of the voltage standing wave nearest to the load impedance and the standing-wave ratio S. If the voltage at the minimum point is 10 V, what are the values of (a) the minimum current and (b) the maximum current and where do these occur? State all the distances in fractions of a wavelength.

3.16 A high-frequency, lossless transmission line of characteristic impedance $R_0 = 100$ ohms is terminated in a load impedance $Z_L = 70 + j36$ ohms. The transmission line is 0.692 wavelength long. With the help of the Smith chart on an admittance basis, determine the reactance which, if connected across the input terminals, makes the input impedance of the combination a resistance. Under this condition, what is the input impedance of the combination?

Suggest a method for realizing the required reactance.

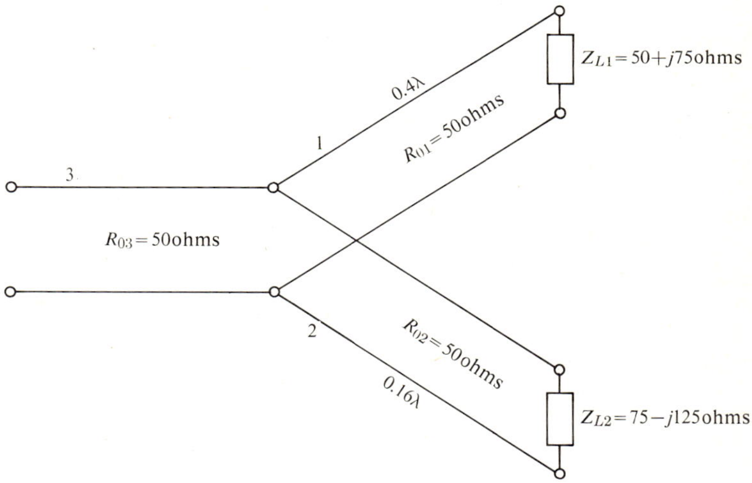

Fig. P3.1 Schematic diagram of parallel connected transmission lines.

3.17 Two transmission lines are connected in parallel to a third transmission line feeding the first two as shown in Fig. P3.1. All the transmission lines have the same characteristic resistance of 50 ohms. Transmission line 1 is 0.4 λ long and is terminated in a load impedance $Z_{L1} = 50 + j75$ ohms. Transmission line 2 is 0.16 λ long and is terminated in a load impedance $Z_{L2} = 75 - j125$ ohms.

a) Use the Smith chart to the maximum extent possible. State whether or not it is desirable to use the Smith chart on an impedance basis. Justify the choice.

b) What are the standing-wave ratios on transmission lines 1 and 2?

c) What is the length of transmission line 3 in fractions of a wavelength in order that the input corresponds to a minimum of the voltage standing wave? What is the input impedance? Find the standing-wave ratio on transmission line 3 from the knowledge of its input impedance.

d) If 50 W of power is supplied to transmission line 3, what is the maximum value of the current on transmission line 3 and where does it occur?

e) Find the values of the power absorbed in the loads Z_{L1} and Z_{L2}.

3.18 Design a quarter-wave transformer to impedance match a 400 ohms lossless transmission line to a 100 ohms resistive load at a frequency of 100 MHz. What are the voltage standing-wave ratios on (i) the main transmission line and (ii) the matching section. Assume the phase velocity to be 3×10^8 m/sec.

The transmission line is operated at only $\frac{3}{4}$ of the designed frequency.

a) What is the operating wavelength?

b) State, including the reasons, whether or not the standing-wave ratio S is changed on (i) the main transmission line and (ii) the matching section?

c) With the help of the Smith chart, find the actual values of the standing-wave ratios.

d) State, including the reasons, whether or not the Smith chart is used on an impedance or an admittance basis.

3.19 A single-stub tuner is to be designed to match a 100 ohms transmission line to a load impedance of $100 - j200$ ohms. The wavelength is 4 m. Determine the shortest distance from the load impedance to the tuning stub and the proper length of the stub. Use the Smith chart.

3.20 A single-stub tuner is to be designed to match a 400 ohms transmission line to a load impedance of $700 + j400$ ohms. The wavelength is 1.50 m. Determine the distance from the load impedance to the tuning stub and the proper length of the stub. Examine both the possible locations for the stub. Use the Smith chart.

3.21 A dissipationless transmission line of characteristic impedance $R_0 = 400$ ohms is feeding a load impedance $Z_L = 200$ ohms. The voltage across the load impedance is 100 V. Find the voltage and the current on the transmission line at (a) one-quarter wavelength and (b) one-half wavelength from the load impedance.

3.22 Consider a finite length ($0 < x < l$) transmission line with short-circuit termination. The losses in the transmission line are assumed to be small. With the help of the formula (3.10.8), show that the selectivity of this transmission line near antiresonance is given by $Q = \beta/2\alpha$. Note that $\alpha + j\beta$ is the propagation coefficient.

3.23 Consider a distortionless transmission line of characteristic resistance R_0 extending from $x = 0$ to $x = l$. A load impedance $Z_L = R_L + jX_L$ is connected to the terminals at $x = l$. A voltage generator V_g in series with a resistor R_0 is connected to the input terminals at $x = 0$. Find the time-averaged power P ($x = 0$) delivered to the input terminals of the transmission line. Give a physical interpretation of the result in terms of the incident and the reflected powers associated respectively with the two progressive waves.

3.24 A dissipationless transmission line of characteristic impedance $Z_0 = 200$ ohms is connected to a load impedance $Z_L = 100 + j300$ ohms. The length of the transmission line is $l = 65$ cm. A generator of voltage $V_g = 1$ V and of frequency $f = 3 \times 10^8$ Hz is connected in series with an adjustable impedance Z_g to the input terminals of the transmission line. The phase velocity of propagation is $v_{ph} = 3 \times 10^8$ m/sec. Find the value of Z_g for maximum power transfer to the transmission line and determine the value of this maximum power.

What are the magnitudes of the voltage across and the current in the load impedance Z_L?

3.25 A lossless transmission line of length l has a characteristic impedance of $Z_0 = 50$ ohms. The terminating impedance has a resistive part of 25 ohms and an inductive reactive part whose value is unknown. The phase velocity for propagation along the transmission line is $v_{ph} = 3 \times 10^8$ m/sec. The frequency of operation is $f = 10^8$ Hz. The standing-wave ratio is found to be 3. With the help of the Smith chart determine the magnitude $|\rho|$ and the phase angle θ_ρ of the terminal reflection coefficient ρ. Find the reactive part of the terminating impedance. If the first maximum of the voltage standing-wave pattern occurs at the input terminals, find the length l of the transmission line. What is the value of the input admittance?

If the magnitude of the input voltage is 1 V, determine the magnitude of the voltage across the terminating impedance.

3.26 A lossless transmission line of characteristic impedance $Z_0 = 50$ ohms is terminated in an unknown load impedance Z_L. The frequency of operation is 5×10^8 Hz, the phase velocity is $v_{ph} = 3 \times 10^8$ m/sec and the standing-wave ratio is $S = 2.5$. If the first maximum of the voltage standing-wave pattern occurs at a distance of 10 cm from the load, find without using the Smith chart the terminal reflection coefficient and the load impedance.

What is the value of the input impedance if the length of the transmission line is 35 cm?

Determine the value of the load impedance in order for a minimum of the voltage standing-wave pattern to occur at the load terminals.

3.27 A lossless transmission line of length $l = 16.5$ m has a characteristic impedance of $Z_0 = 50$ ohms. The terminal reflection coefficient is $\rho = j0.5$ and the wavelength is 2 m. The incident voltage at the load terminals is given by $V_i\,(s = 0) = 1$ V. Find the magnitude and the phase angle of (a) the reflected voltage $V_r\,(s = 0)$ at the load terminals, (b) the incident voltage $V_i\,(s = l)$ at the input terminals, (c) the reflected voltage $V_r\,(s = l)$ at the input terminals, (d) the reflected current $I_r\,(s = 0)$ at the load terminals and (e) the voltage midway between the load and the input terminals.

3.28 A lossless transmission line of length $l = 30$ m has a characteristic impedance of $Z_0 = 50$ ohms. The load impedance is $Z_L = 25$ ohms. The frequency of operation is $f = 10^7$ Hz and the phase velocity is $v_{ph} = 3 \times 10^8$ m/sec. Find the terminal reflection coefficient and the standing-wave ratio. Determine the input impedance.

If a generator of voltage $V_g = 12$ V and an internal resistance of $R_g = 75$ ohms is connected to the input terminals, find the magnitude of the voltage at a distance of 7.5 m from the load terminals. For the same location on the transmission line, sketch the voltage as a function of time from $t = 0$ to $t = 2 \times 10^{-7}$ sec assuming that the phase angle of the generator voltage is zero.

3.29 A lossless transmission line of length $l = 7.5$ m has a characteristic impedance of $Z_0 = 50$ ohms. The load impedance is given by $Z_L = 25$ ohms. The phase velocity is $v_{ph} = 3 \times 10^8$ m/sec. A generator of voltage $V_g(t) = 20 \cos(6\pi f t) \cos(2\pi f t)$ where $f = 10^7$ Hz is connected in series with an impedance of $Z_g = 100$ ohms to the input terminals of the transmission line. Find the expressions for the input and the load currents.

3.30 A lossless transmission line of characteristic impedance $Z_0 = 50$ ohms is terminated by an unknown load impedance. The standing-wave ratio is found to be $S = 7$. The magnitude of the voltage across the load terminals is found to be five times that of the minimum value of the magnitude of the voltage along the transmission line. The first extremum from the load terminals of the voltage standing-wave is a minimum. Without using the Smith chart, determine the magnitude and the phase angle of the terminal reflection coefficient as well as the value of the load impedance.

CHAPTER 4

FUNDAMENTALS OF ELECTROMAGNETIC THEORY

4.1 INTRODUCTION

Electromagnetic theory is a study of the nature of the electromagnetic fields as governed by Maxwell's equations. Since it is beyond the scope of this treatment, we shall not systematically deduce Maxwell's equations but only present them here as basic postulates. A variety of macroscopic electromagnetic phenomenon has been successfully explained on the basis of Maxwell's equations, lending support to their introduction as postulates.

We shall first briefly review electrostatics with emphasis on Coulomb's law of force and Gauss's law and develop the concepts of polarization and permittivity. An introductory treatment of magnetostatics is given with particular emphasis on Ampère's law of force, Biot-Savart's law and Ampère's circuital law. We shall then discuss the magnetization and the concept of permeability for the macroscopic description of magnetic materials.

We shall introduce the experimental law of induction due to Faraday and generalize Ampère's circuital law for the time-varying case to include the displacement current. The researches of Coulomb, Gauss, Ampère, Biot-Savart and Faraday are interwoven into a plausible development of Maxwell's equations. We shall then give both the integral and the differential forms of Maxwell's equations. Since the validity of the integral forms of Maxwell's equations is not impaired by the presence of boundaries separating two different media, we shall use them to obtain the boundary conditions that the fields must satisfy on the interfaces between different media.

We shall introduce the concepts of electric and magnetic energy densities as well as the power flux density. We shall mention the important case in which the electromagnetic fields have a periodic time dependence and present the simplified forms of Maxwell's equations and the boundary conditions for the special case of the time-harmonic fields. We shall then deduce the time-averaged values of the energy and the power flux densities. Finally, we shall include a brief discussion of the frequency response of dielectric materials.

4.2 ELECTROSTATICS

1. Electrostatic Field

A field is a region in which every point is associated with a physical function. If the physical function is a force, the corresponding field is a vector field of force. Electrostatic field is a vector field of force. The law of force experimentally obtained by Coulomb is the basis for the concept of the electrostatic field. Coulomb's law of force states that the force acting on a charge q_2 due to a charge q_1 is given by

$$\boldsymbol{F}_{12} = \frac{q_1 q_2}{4\pi\varepsilon r_{12}^2}\,\hat{\boldsymbol{r}}_{12} \quad \text{N}, \tag{2.1}$$

where r_{12} is the distance between the two charges and $\hat{\boldsymbol{r}}_{12}$ is a unit vector along the line joining the two charges as shown in Fig. 4.1. The force is different in different media whose electrical property is characterized by the constant ε known as the permittivity. We shall discuss the concept of permittivity of material media in the following section. The permittivity of vacuum is denoted by ε_0 and the permittivity of material media is expressed in terms of the vacuum permittivity as $\varepsilon = \varepsilon_0\varepsilon_r$ where ε_r is known as the relative permittivity. In the rationalized MKSQ units which we shall use throughout this treatment $\varepsilon_0 = 8.854 \times 10^{-12}$ F/m or approximately, $\varepsilon_0 \approx (1/36\pi) \times 10^{-9}$ F/m. Note that q_1 and q_2 are signed charges with the result that the force between the charges of the same sign is repulsive whereas that between the charges of the opposite sign is attractive. In the rationalized MKSQ units, the factor 4π appears in the force equation and is therefore absent from the differential equations satisfied by the fields.

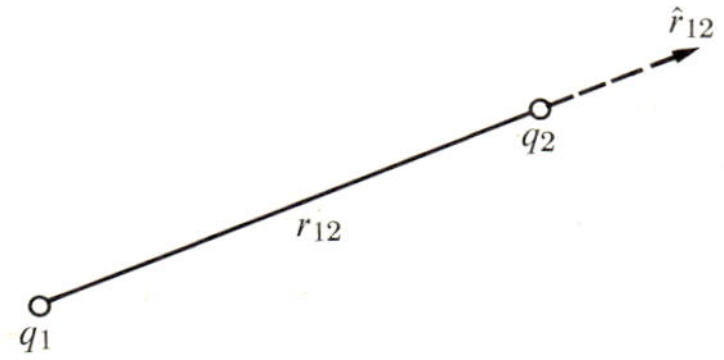

Fig. 4.1 Coulomb's law of force.

Let the position P' of the charge q be denoted by the position vector $\boldsymbol{r}'$ and let an observation point P be specified by the position vector $\boldsymbol{r}$. The electric field $\boldsymbol{E}(\boldsymbol{r})$ at the point P in the region containing the charge is the force acting on a unit charge placed at the point in question. Therefore,

$$\boldsymbol{E}(\boldsymbol{r}) = q\hat{\boldsymbol{r}}_{P'P}/4\pi\varepsilon r_{P'P}^2, \tag{2.2}$$

where $r_{P'P}$ is the distance between the points P' and P and $\hat{\boldsymbol{r}}_{P'P}$ is the unit vector lying along the straight line $P'P$ and directed away from the charge. In the Cartesian coordinate system, if x', y', and z' are the coordinates of the *source point* specified by the position vector $\boldsymbol{r}'$ where the charge is located and x, y, and z are the coordinates of the *observation point* specified by the position vector $\boldsymbol{r}$ where the electric field is observed, then we can ascertain from Fig. 4.2 that

$$\boldsymbol{r}_{P'P} = \boldsymbol{r} - \boldsymbol{r}' = \hat{\boldsymbol{r}}_{P'P} r_{P'P} = \hat{x}(x - x') + \hat{y}(y - y') + \hat{z}(z - z'). \tag{2.3}$$

From (2.3), we see that

$$r_{P'P} = [(x - x')^2 + (y - y')^2 + (z - z')^2]^{1/2}, \tag{2.4}$$

and

$$\hat{\boldsymbol{r}}_{P'P} = [(x - x')^2 + (y - y')^2 + (z - z')^2]^{-1/2}[\hat{x}(x - x') + \hat{y}(y - y') + \hat{z}(z - z')]. \tag{2.5}$$

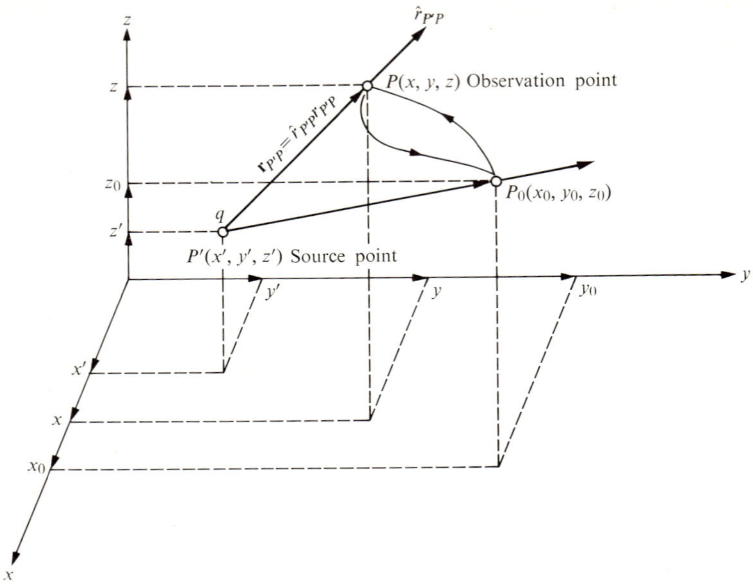

Fig. 4.2 Geometry for the displacement of a unit charge in an electric field due to the point charge q.

We can verify from (2.4) and (2.5) that

$$\nabla\left(\frac{1}{r_{P'P}}\right) = \nabla\left(\frac{1}{|\boldsymbol{r} - \boldsymbol{r}'|}\right) = \hat{x}\,\frac{\partial}{\partial x}\left(\frac{1}{|\boldsymbol{r} - \boldsymbol{r}'|}\right) + \hat{y}\,\frac{\partial}{\partial y}\left(\frac{1}{|\boldsymbol{r} - \boldsymbol{r}'|}\right)$$
$$+ \hat{z}\,\frac{\partial}{\partial z}\left(\frac{1}{|\boldsymbol{r} - \boldsymbol{r}'|}\right) = -\frac{\hat{r}_{P'P}}{|\boldsymbol{r} - \boldsymbol{r}'|^2}. \tag{2.6}$$

Although established with reference to the Cartesian coordinates, the result $\nabla(1/|\boldsymbol{r} - \boldsymbol{r}'|) = -\hat{r}_{P'P}/|\boldsymbol{r} - \boldsymbol{r}'|^2$ is true generally for all coordinate systems. Together with (2.6), we may rewrite (2.2) as

$$\boldsymbol{E}(\boldsymbol{r}) = -\frac{q}{4\pi\varepsilon}\,\nabla\left(\frac{1}{|\boldsymbol{r} - \boldsymbol{r}'|}\right), \tag{2.7}$$

or equivalently

$$\boldsymbol{E}(\boldsymbol{r}) = -\nabla\phi(\boldsymbol{r}), \tag{2.8}$$

and

$$\phi(\boldsymbol{r}) = \frac{q}{4\pi\varepsilon\,|\boldsymbol{r} - \boldsymbol{r}'|}. \tag{2.9}$$

2. Scalar Potential

In order to understand the physical meaning of $\phi(\boldsymbol{r})$, let us consider a unit charge located at P_0 specified by the position vector $\boldsymbol{r}_0$. The force $\boldsymbol{F}$ acting *on* the unit charge is equal to the electric field at P_0 and we therefore obtain it from (2.7) as

$$\boldsymbol{F} = -\frac{q}{4\pi\varepsilon}\nabla_0\left(\frac{1}{|\boldsymbol{r}_0 - \boldsymbol{r}'|}\right) = -\frac{q}{4\pi\varepsilon}\nabla_0\left(\frac{1}{r_{P'P_0}}\right), \tag{2.10}$$

where ∇_0 is an operation with respect to the position vector $\boldsymbol{r}_0$. Suppose that the unit charge has an infinitesimal displacement

$$d\boldsymbol{r}_0 = \hat{x}\,dx_0 + \hat{y}\,dy_0 + \hat{z}\,dz_0, \tag{2.11}$$

which gives both the magnitude and the direction. Then, the work done by the electric field on the charge is equal to $\boldsymbol{F} \cdot d\boldsymbol{r}_0$, that is the magnitude of the displacement multiplied by the component of the force in the direction of the displacement. Therefore, the work dW done *by* the charge *on* the electric field is equal to $-\boldsymbol{F} \cdot d\boldsymbol{r}_0$ and hence, from (2.10),

$$dW = \frac{q}{4\pi\varepsilon}\nabla_0\left(\frac{1}{|\boldsymbol{r}_0 - \boldsymbol{r}'|}\right) \cdot d\boldsymbol{r}_0. \tag{2.12}$$

The total infinitesimal change $d(1/r_{P'P_0})$ in $1/r_{P'P_0}$ arises due to the changes dx_0, dy_0, and dz_0 in the coordinates x_0, y_0, and z_0 respectively. Hence, with the help of (2.6) and (2.11), we find that

$$\begin{aligned} d\left(\frac{1}{r_{P'P_0}}\right) &= \frac{\partial}{\partial x_0}\left(\frac{1}{r_{P'P_0}}\right)dx_0 + \frac{\partial}{\partial y_0}\left(\frac{1}{r_{P'P_0}}\right)dy_0 + \frac{\partial}{\partial z_0}\left(\frac{1}{r_{P'P_0}}\right)dz_0, \\ &= \left[\hat{x}\frac{\partial}{\partial x_0}\left(\frac{1}{r_{P'P_0}}\right) + \hat{y}\frac{\partial}{\partial y_0}\left(\frac{1}{r_{P'P_0}}\right) + \hat{z}\frac{\partial}{\partial z_0}\left(\frac{1}{r_{P'P_0}}\right)\right] \\ &\quad \cdot [\hat{x}\,dx_0 + \hat{y}\,dy_0 + \hat{z}\,dz_0], \\ &= \nabla_0\left(\frac{1}{r_{P'P_0}}\right) \cdot d\boldsymbol{r}_0. \end{aligned} \tag{2.13}$$

We can obtain the total work W done by the unit charge when it is displaced from the position P_0 to the position P from (2.12) and (2.13) as

$$W = \int_{P_0}^{P}\frac{q}{4\pi\varepsilon}d\left(\frac{1}{r_{P'P_0}}\right) = \frac{q}{4\pi\varepsilon}\left(\frac{1}{r_{P'P}} - \frac{1}{r_{P'P_0}}\right) = \frac{q}{4\pi\varepsilon}\left(\frac{1}{|\boldsymbol{r} - \boldsymbol{r}'|} - \frac{1}{|\boldsymbol{r}_0 - \boldsymbol{r}'|}\right). \tag{2.14}$$

It is important to note that we have made no assumption on the path taken by the unit charge in its displacement from P_0 to P and, consequently, (2.14) is *independent* of the nature of the path from P_0 to P. If we take P_0 to be at infinity, then from (2.14)

$$W = \frac{q}{4\pi\varepsilon\,|\boldsymbol{r} - \boldsymbol{r}'|}, \tag{2.15}$$

which is the work done by the unit charge in being brought from infinity to the location P and is independent of the path. From (2.9) and (2.15), we find that

$$\phi(\boldsymbol{r}) = \frac{q}{4\pi\varepsilon\,|\boldsymbol{r} - \boldsymbol{r}'|} = W. \tag{2.16}$$

We see that a unique scalar function $\phi(\boldsymbol{r})$ can be associated with every point in an electrostatic field and this function merely gives the energy required to move a unit charge from infinity to the point in question. The scalar function $\phi(\boldsymbol{r})$ is called the scalar potential. Once we know the scalar potential, we can evaluate the electric field using (2.8).

From the definition of the electric field, we see that the force acting on a *positive* charge is in the same direction of the electric field. Therefore, a positive charge tends to move in the direction of the electric field and from (2.8) it follows that a positive charge tends to move in the direction of the decreasing potential $\phi(\boldsymbol{r})$. Also, since the unit of potential is the volt, it follows from (2.8) that the unit of electric field is volts/meter.

3. Point Electric Dipole

Let us consider two equal and opposite point charges, q and $-q$, located in free space along the z-axis at $z = h$ and $z = -h$ respectively, as shown in Fig. 4.3. Let P_1 and P_2 denote the positions of the two point charges q and $-q$ respectively. We wish to find the scalar potential $\phi(\boldsymbol{r})$ and the electrostatic field $\boldsymbol{E}(\boldsymbol{r})$ at an observation point P located at a distance $OP = r$ from the origin O of the coordinate system. Let θ be the angle between OP and the positive z-direction, as shown in Fig. 4.3. If r_{P_1P} and r_{P_2P} are the distances between the observation point P and the locations of the charges P_1 and P_2 respectively, then we find with the help of Fig. 4.3 that

$$\frac{r}{r_{P_1P}} = \left[1 - \frac{2h}{r}\cos\theta + \frac{h^2}{r^2}\right]^{-1/2}, \tag{2.17a}$$

$$\frac{r}{r_{P_2P}} = \left[1 + \frac{2h}{r}\cos\theta + \frac{h^2}{r^2}\right]^{-1/2}. \tag{2.17b}$$

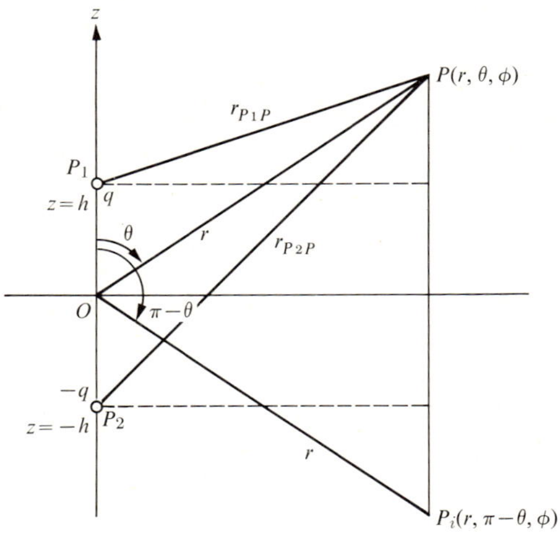

Fig. 4.3 An electric dipole.

We can obtain the scalar potential at P due to both the point charges from (2.16) as

$$\phi(\boldsymbol{r}) = \frac{q}{4\pi\varepsilon_0}\left(\frac{1}{r_{P_1P}} - \frac{1}{r_{P_2P}}\right). \tag{2.18}$$

Let us suppose that the separation distance $2h$ between the two charges is very small in comparison with the distance r to the observation point, that is, $h/r \ll 1$. Then we can expand (2.17a) and (2.17b) into binomial series in powers of h/r and the results of retaining only the first two terms are

$$r/r_{P_1P} = 1 + (h/r)\cos\theta; \qquad r/r_{P_2P} = 1 - (h/r)\cos\theta. \tag{2.19}$$

If we substitute (2.19) into (2.18) we get

$$\phi(\boldsymbol{r}) = q(2h)\cos\theta/4\pi\varepsilon_0 r^2. \tag{2.20}$$

An electric dipole consists of two equal and opposite charges separated by a distance very small in comparison with the other relevant scale lengths such as the distance from the location of the dipole to the point of observation. An electric dipole is characterized by a moment $\boldsymbol{p}$. The magnitude p of $\boldsymbol{p}$ is equal to the product of the magnitude of the individual charge and the separation distance between the charges. The direction $\hat{p}$ of $\boldsymbol{p}$ is along the straight line from the negative to the positive charge. Therefore, we can see that

$$\boldsymbol{p} = \hat{z}q(2h), \tag{2.21}$$

for the electric dipole illustrated in Fig. 4.3.

Let us consider two equal and opposite charges located essentially at the same point. We may assume that this combination of charges represents approximately an atom in which the charges are held together by a binding force. If an electric field is applied in the z-direction, the positive and the negative charges are displaced in the + and the − z-directions respectively so as to acquire an equilibrium position (as shown in Fig. 4.3) in which the binding force on each charge is balanced by the force due to the electric field. We see that the direction of the moment of the electric dipole is the same as that of the electric field that induces it.

If $\boldsymbol{r} = r\hat{r}$ is the position vector of the observation point P, then we can express (2.20) with the help of (2.21) and (2.6) as

$$\phi(\boldsymbol{r}) = \boldsymbol{p}\cdot\hat{r}/4\pi\varepsilon_0 r^2 = -(\boldsymbol{p}/4\pi\varepsilon_0)\cdot\nabla(1/r).. \tag{2.22}$$

We find the spherical components of the electric field due to the electric dipole from (2.8) and (2.20) to be

$$E_r(\boldsymbol{r}) = -\frac{\partial}{\partial r}\phi(\boldsymbol{r}) = \frac{p\cos\theta}{2\pi\varepsilon_0 r^3}, \tag{2.23a}$$

$$E_\theta(\boldsymbol{r}) = -\frac{1}{r}\frac{\partial}{\partial\theta}\phi(\boldsymbol{r}) = \frac{p\sin\theta}{4\pi\varepsilon_0 r^3}, \tag{2.23b}$$

$$E_\phi(\boldsymbol{r}) = -\frac{1}{r\sin\theta}\frac{\partial}{\partial\phi}\phi(\boldsymbol{r}) = 0. \tag{2.23c}$$

4. Conservative Field

Suppose a charge is moved from P_0 to P along one path and is returned to P_0 along another path (Fig. 4.2). Since the scalar potential at P_0 has a unique value, it follows that the work done by the charge in moving from P_0 to P is equal to the work done on the charge in its displacement from P to P_0. Thus, there is neither gain nor loss of energy. Hence the electrostatic field is a *conservative* field. This property of the electrostatic field arises because the electric field is expressible as a gradient of a scalar function of position and because, as a consequence, a unique scalar function can be associated with every point in the electrostatic field. Since the curl of the gradient of a scalar function is zero, it follows from (2.8) that

$$\nabla \times \boldsymbol{E}(\boldsymbol{r}) = 0. \tag{2.24}$$

Thus, on the basis of (2.24), we can define a conservative field as that which has a zero curl. An alternative method of obtaining (2.24) is quite instructive. We have pointed out that the total work done by a unit charge when it is displaced from any point P_0 to any other point P along one path and returned to the original position along another path is zero. In other words, the total work done by a unit charge taken around a closed path is zero and this result is expressed mathematically as

$$\oint_C \boldsymbol{E}(\boldsymbol{r}) \cdot d\boldsymbol{r} = 0. \tag{2.25}$$

In view of Stokes' theorem, (2.25) becomes

$$\int_S \nabla \times \boldsymbol{E}(\boldsymbol{r}) \cdot d\boldsymbol{S} = 0, \tag{2.26}$$

where S is the area bounded by the closed curve C. The vector elemental area $d\boldsymbol{S}$ has a magnitude equal to the area of the element and a direction which is normal to the area of the element. We can determine the direction of the normal to a surface by the following convention. If the fingers of the right hand curl in the direction of integration around the closed path, then the thumb points to the side in which the normal to the surface bounded by the closed path lies. Since (2.25) is true for any closed curve C, it follows that (2.26) is true for any arbitrarily small area S. Consequently, the integrand of (2.26) is equal to zero leading to the previous result given by (2.24).

5. Scalar Potential Due to a Charge Distribution

We can obtain the scalar potential $\phi(\boldsymbol{r})$ at a point P specified by the position vector $\boldsymbol{r}$ with respect to some origin O, due to a point charge q_i located at a point specified by the position vector $\boldsymbol{r}_i$ from (2.9) as

$$\phi(\boldsymbol{r}) = q_i/(4\pi\varepsilon\, |\boldsymbol{r} - \boldsymbol{r}_i|). \tag{2.27}$$

If there are n charges $q_1, q_2, \ldots q_n$ located at points specified by the position vectors $\boldsymbol{r}_1, \boldsymbol{r}_2, \ldots \boldsymbol{r}_n$ respectively, then we can find the resultant scalar potential at P by

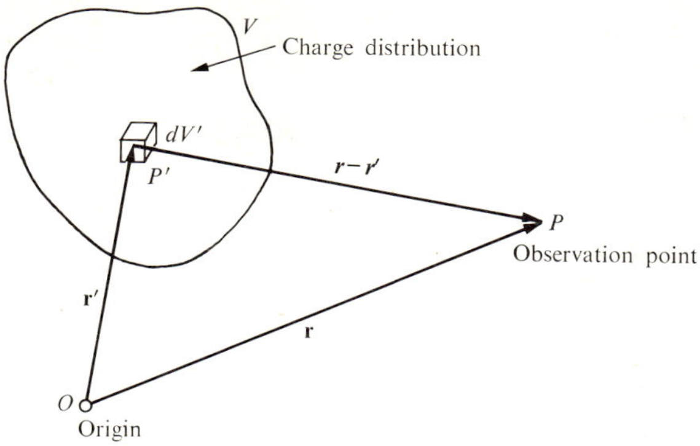

Fig. 4.4 Scalar potential due to a charge distribution.

summing up the individual contributions. Therefore, the resultant scalar potential is

$$\phi(\boldsymbol{r}) = \sum_{i=1}^{i=n} \frac{q_i}{4\pi\varepsilon\, |\boldsymbol{r} - \boldsymbol{r}_i|}. \tag{2.28}$$

Once we know the scalar potential at P, we can obtain the electric field at P by evaluating the negative gradient of the scalar potential in accordance with (2.8).

In a similar manner, we can determine the scalar potential due to a continuous distribution of charge. Let a volume V be occupied by a continuous distribution of charge. Let O be the origin of the coordinate system and let the position vector $\boldsymbol{r}'$ denote a typical point P' inside the volume (Fig. 4.4). Let $\rho(\boldsymbol{r}')$ denote the charge density in coulombs per cubic meter at P'. Let us consider an infinitesimal volume dV' around P'; the charge in dV' is equal to $\rho(\boldsymbol{r}')\, dV'$. Suppose we need the scalar potential at a point P specified by the position vector $\boldsymbol{r}$. We can obtain the contribution $d\phi(\boldsymbol{r})$ to the scalar potential at P due to the charge in dV' from (2.9) to be

$$d\phi(\boldsymbol{r}) = \rho(\boldsymbol{r}')\, dV'/(4\pi\varepsilon\, |\boldsymbol{r} - \boldsymbol{r}'|). \tag{2.29}$$

We can evaluate the total scalar potential $\phi(\boldsymbol{r})$ at P by integrating (2.29) throughout the volume V occupied by the charge distribution. Therefore,

$$\phi(\boldsymbol{r}) = \int_V \frac{\rho(\boldsymbol{r}')\, dV'}{4\pi\varepsilon\, |\boldsymbol{r} - \boldsymbol{r}'|}. \tag{2.30}$$

As before, we deduce the electric field at P from (2.8).

6. Distributed Electric Dipole

It is instructive to deduce the scalar potential due to the following simple linear charge distribution in free space. The charge is distributed along the z-axis from $z = -h$ to $z = h$ as shown in Fig. 4.5. The linear charge density q_L is uniform from $z = 0$ to $z = h$ and has a value q/h coulombs/m. The linear charge density is also

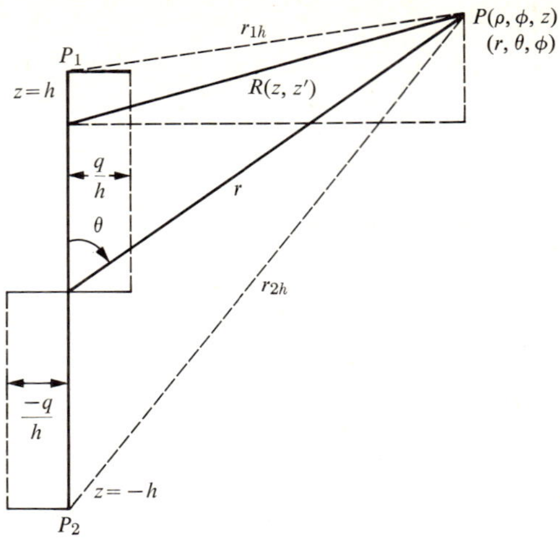

Fig. 4.5 Distributed electric dipole.

uniform from $z = -h$ to $z = 0$ but has a value $-q/h$ coulombs/m. We should note that the linear charge density from $z = -h$ to $z = 0$ is the negative of that from $z = 0$ to $z = h$. This charge distribution bears a close resemblance to the point electric dipole which we treated previously in that the charges q and $-q$, instead of being concentrated at the points $z = h$ and $z = -h$, are now distributed uniformly from $z = 0$ to $z = h$, and from $z = -h$ to $z = 0$ respectively. The linear charge distribution under consideration is called a distributed electric dipole for reasons which we explain subsequently.

It is convenient for us to introduce the cylindrical coordinates ρ, ϕ, and z. The geometry of the linear charge distribution shows that the scalar potential $\phi(\boldsymbol{r})$ and the electric field $\boldsymbol{E}(\boldsymbol{r})$ are independent of the angle ϕ and hence it is sufficient for us to determine the scalar potential at an observation point P in a plane corresponding to one particular value of ϕ, as shown in Fig. 4.5. Let the coordinates of the observation point P be ρ, ϕ, z. Let r_{1h}, r, and r_{2h} be the distances from P to the top end $z = h$, the center $z = 0$, and the bottom end $z = -h$ of the linear charge distribution respectively. Let us consider an element of length dz' of the filament situated at a distance z' from the origin. Let $R(z, z')$ be the distance from the observation point to this element of length at z'. We can obtain from Fig. 4.5 that

$$r_{1h} = \sqrt{(z - h)^2 + \rho^2}; \qquad r_{2h} = \sqrt{(z + h)^2 + \rho^2}; \qquad (2.31\text{a,b})$$

$$r = \sqrt{z^2 + \rho^2}; \qquad R(z, z') = \sqrt{(z - z')^2 + \rho^2}. \qquad (2.31\text{c,d})$$

In the present case, since q_L is the charge per unit length, we find the total charge in an infinitesimal length dz' of the filament as $\rho(\boldsymbol{r}')\, dV' = q_L(z')\, dz'$. Therefore, we obtain from (2.9) that the contribution $d\phi(\boldsymbol{r})$ to the scalar potential at P due to the charge in the infinitesimal length dz' of the filament is

$$d\phi(\mathbf{r}) = q_L(z')\, dz'/4\pi\varepsilon_0 R(z, z'). \tag{2.32}$$

We can determine the total scalar potential $\phi(\mathbf{r})$ at P due to the entire filamentary charge distribution by integrating (2.32) with respect to z' from $z' = -h$ to $z = h$ with the following result

$$\phi(\mathbf{r}) = \frac{1}{4\pi\varepsilon_0}\frac{q}{h}\left[\int_0^h \frac{dz'}{\sqrt{(z - z')^2 + \rho^2}} - \int_{-h}^0 \frac{dz'}{\sqrt{(z - z')^2 + \rho^2}}\right]. \tag{2.33}$$

In obtaining (2.33), $q_L(z)$ has been set equal to q/h for $0 < z < h$ and to $-q/h$ for $-h < z < 0$. We can evaluate the integral in (2.33) to yield

$$\begin{aligned}\phi(\mathbf{r}) &= \frac{-q}{4\pi\varepsilon_0 h}\Big[\ln \{z - z' + \sqrt{(z - z')^2 + \rho^2}\}\Big|_0^h \\ &\quad - \ln \{(z - z') + \sqrt{(z - z')^2 + \rho^2}\}\Big|_{-h}^0\Big], \\ &= \frac{-q}{4\pi\varepsilon_0 h}\left[\ln\left\{\frac{(z/r) - (h/r) + (r_{1h}/r)}{1 + (z/r)}\right\} + \ln\left\{\frac{(z/r) + (h/r) + (r_{2h}/r)}{1 + (z/r)}\right\}\right],\end{aligned} \tag{2.34}$$

where (2.31) has been used. As in the case of the point electric dipole, we assume the linear extent of the filament to be very small in comparison with the distance from the center of the filament to the observation point, that is $h/r \ll 1$.

It is convenient for us to introduce at this stage the spherical coordinates r, θ, and ϕ as is shown in Fig. 4.5. The relationship of r to the cylindrical coordinates is given in (2.31c). Also, we can obtain from Fig. 4.5 that

$$z/r = \cos\theta. \tag{2.35}$$

With the help of (2.31a) and (2.35) we can show that

$$\frac{r_{1h}}{r} = \left(1 - \frac{2h}{r}\cos\theta + \frac{h^2}{r^2}\right)^{1/2}. \tag{2.36}$$

If we expand (2.36) into a binomial series in powers of h/r and retain only the terms up to order h^2/r^2, the result is

$$\frac{r_{1h}}{r} = 1 - \frac{h}{r}\cos\theta + \frac{h^2}{2r^2}(1 - \cos\theta)(1 + \cos\theta). \tag{2.37}$$

We can deduce from (2.35) and (2.37) that

$$\frac{\dfrac{z}{r} - \dfrac{h}{r} + \dfrac{r_{1h}}{r}}{1 + \dfrac{z}{r}} = 1 - \frac{h}{r} + \frac{h^2}{2r^2}(1 - \cos\theta). \tag{2.38}$$

If we use the expansion formula

$$\ln(1 + x) = x - \frac{x^2}{2} + \frac{x^3}{3} \qquad (|x| < 1), \tag{2.39}$$

the first term within the square brackets in (2.34) can be expanded with the help of (2.38) into powers of h/r and if we retain the terms up to order h^2/r^2, we can show that

$$\ln\left[\frac{\dfrac{z}{r} - \dfrac{h}{r} + \dfrac{r_{1h}}{r}}{1 + \dfrac{z}{r}}\right] = -\frac{h}{r} - \frac{h^2}{2r^2}\cos\theta. \tag{2.40}$$

The second term within the square brackets in (2.34) is the same as the first term with h replaced by $-h$. Hence, it follows from (2.40) that

$$\ln\left[\frac{\dfrac{z}{r} + \dfrac{h}{r} + \dfrac{r_{2h}}{r}}{1 + \dfrac{z}{r}}\right] = \frac{h}{r} - \frac{h^2}{2r^2}\cos\theta. \tag{2.41}$$

Since in the sum of (2.40) and (2.41), the terms of order h/r vanish, we find it necessary in all the expansions to retain the terms up to order h^2/r^2. Together with (2.40) and (2.41), (2.34) yields

$$\phi(\boldsymbol{r}) = qh\cos\theta/4\pi\varepsilon_0 r^2. \tag{2.42}$$

If we compare (2.42) with (2.20) we see that the linear filamentary charge distribution under consideration yields for the scalar potential an expression which is very similar to that of the point electric dipole. It is for this reason that the linear charge distribution depicted in Fig. 4.5 is called a distributed electric dipole. In analogy with (2.20), we can use (2.42) to define an equivalent dipole moment $\boldsymbol{p}_e$ for the linear charge distribution of Fig. 4.5 as

$$\boldsymbol{p}_e = \hat{z}qh = \tfrac{1}{2}\boldsymbol{p}. \tag{2.43}$$

It follows that the equivalent dipole moment $\boldsymbol{p}_e$ of the distributed electric dipole is only one half of that due to the equivalent point electric dipole. The point and the distributed electric dipoles are said to be equivalent if their physical extents are the same and if their total charge of each kind is the same. Thus, an electric dipole of given physical extent produces the maximum scalar potential and hence the maximum electric field if the charges are concentrated at the terminal points of the length rather than distributed over the entire length. Other kinds of charge distributions yield different values for the equivalent electric dipole moment.

We derive the scalar potential of another important linear charge distribution in detail in Section 7.3 and discuss its usefulness in obtaining the capacitance C per unit length of a two-wire transmission line.

7. Electric Flux ψ and Electric Flux Density D

The electric field $\boldsymbol{E}(\boldsymbol{r})$ due to a charge is dependent on the properties of the medium in view of the appearance of ε in its defining relation. It is helpful for us to define a physical function associated with a given charge that does not contain ε in its defining relation and is therefore independent of the properties of the medium. The electric flux ψ and the electric flux density $\boldsymbol{D}$ are such physical functions. The meaning of electric flux ψ is best understood by recalling Faraday's experiment. A sphere with charge Q is placed inside, and not touching, a large hollow sphere. The outer sphere is connected to the ground, the connection to the ground is removed and then the inner sphere is removed. The charge on the outer sphere is measured and is found to be equal in magnitude and opposite in sign to the charge on the inner sphere for all dimensions of the inner sphere and for all different types of media between the spheres. Thus, there is an electric displacement from the inner to the outer sphere, the amount of this displacement depending only on the total charge Q on the inner sphere. In the rationalized MKSQ units, the electric displacement or the electric flux is equal to the magnitude of the charge that produces it:

$$\psi = Q \quad \text{C}. \tag{2.44}$$

Let us consider a spherical surface of radius r centered at a point where a point charge q is located. Since the area of the surface is $4\pi r^2$ and since the electric flux through the entire surface is q, we obtain the magnitude $D(\mathbf{r})$ of the flux density as

$$D(\mathbf{r}) = q/4\pi r^2. \tag{2.45}$$

If we compare (2.45) with (2.2) we see that if the magnitudes only are considered, $D(\mathbf{r}) = \varepsilon E(\mathbf{r})$. This suggests the following definition of the electric flux density vector:

$$\mathbf{D}(\mathbf{r}) = \varepsilon \mathbf{E}(\mathbf{r}). \tag{2.46}$$

If ε is a scalar constant as we have implied, then the electric flux density and the electric field are in the same direction. In such a case the medium is said to be isotropic. In an anisotropic medium $\mathbf{D}(\mathbf{r})$ and $\mathbf{E}(\mathbf{r})$ can be in different directions. In this treatment, we consider only isotropic media. In view of (2.46) and (2.2), it follows that

$$\mathbf{D}(\mathbf{r}) = q\hat{\mathbf{r}}/4\pi r^2 \quad \text{C/m}^2. \tag{2.47}$$

8. Gauss's Law

Let us consider an arbitrary surface S enclosing a point charge q (Fig. 4.6). The electric flux $d\psi$ through the elemental area $d\mathbf{S} = \hat{n}\, dS$ is

$$d\psi = \mathbf{D}(\mathbf{r}) \cdot d\mathbf{S} = \mathbf{D}(\mathbf{r}) \cdot \hat{n}\, dS = D(\mathbf{r}) \cos\theta\, dS, \tag{2.48}$$

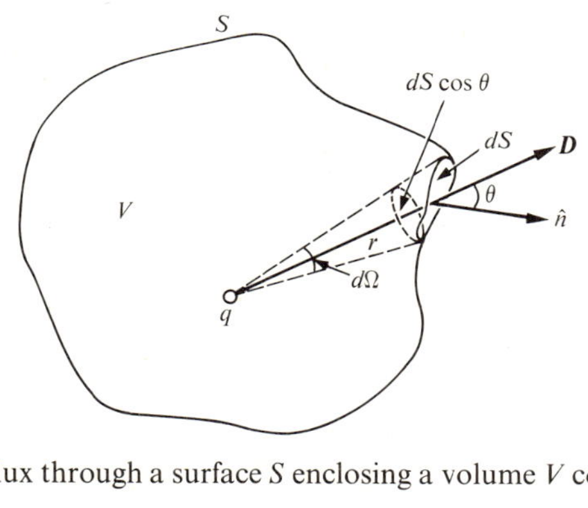

Fig. 4.6 Total electric flux through a surface S enclosing a volume V containing a point charge q.

where θ is the angle between $\boldsymbol{D}(\boldsymbol{r})$ and the outwardly directed unit normal $\hat{n}$ to dS. Since $dS \cos\theta$ is the projection of the element of area normal to the radius vector,

$$d\Omega = dS \cos\theta / r^2, \tag{2.49}$$

where $d\Omega$ is the solid angle subtended at the location of the point charge by the element of area dS. We obtain the total outward electric flux ψ through the entire surface by integrating (2.48) over the entire surface. Together with (2.45) and (2.49), we find that

$$\psi = \int_S D(\boldsymbol{r})\, dS \cos\theta = \int_S \frac{q \cos\theta\, dS}{4\pi r^2} = \frac{q}{4\pi} \int_S d\Omega. \tag{2.50}$$

The total solid angle Ω subtended at the location of q by the surface S completely enclosing the charge is obtained as

$$\int_S d\Omega = \Omega = 4\pi \text{ solid radians.} \tag{2.51}$$

Hence, (2.50) becomes

$$\psi = q, \tag{2.52}$$

which yields Gauss's law. If the charge q is outside the surface S, the total solid angle subtended by the surface S at the location of the charge is zero, with the result that there is no net outward electric flux through a closed surface not containing any charge inside it. Let us suppose that the charge is distributed continuously throughout the volume V enclosed by the surface S. Then, q in (2.52) should be replaced by the total charge inside the volume. Hence

$$\psi = \int_V \rho(\boldsymbol{r})\, dV, \tag{2.53}$$

where $\rho(\boldsymbol{r})$ is the charge density. It is convenient for us to write ψ in vector notation with the help of (2.50) and combine it with (2.53) to yield

$$\int_S \boldsymbol{D}(\boldsymbol{r}) \cdot d\boldsymbol{S} = \int_V \rho(\boldsymbol{r})\, dV. \tag{2.54}$$

Gauss's law expressed mathematically by (2.54) states *that the net electric flux directed out of a surface completely enclosing a volume is numerically equal to the total charge inside the volume.* The use of the divergence theorem enables us to recast (2.54) as

$$\int_V \nabla \cdot \boldsymbol{D}(\boldsymbol{r})\, dV = \int_V \rho(\boldsymbol{r})\, dV. \tag{2.55}$$

Since (2.54) is true for any arbitrarily small surface S and (2.55) is therefore true for any arbitrarily small volume V, it follows that the integrands in (2.55) are equal yielding the following point relation

$$\nabla \cdot \boldsymbol{D}(\boldsymbol{r}) = \rho(\boldsymbol{r}). \tag{2.56}$$

It is usual to refer to (2.54) and (2.56) as the integral and the differential forms respectively of Gauss's law.

9. Poisson's and Laplace's Equation

In view of (2.46) and since ε is a constant independent of position, we may state (2.56) in terms of the electric field as

$$\nabla \cdot \boldsymbol{E}(\boldsymbol{r}) = \rho(\boldsymbol{r})/\varepsilon. \tag{2.57}$$

From (2.8) and (2.57), we can obtain the differential equation satisfied by the scalar potential $\phi(\boldsymbol{r})$ in terms of the charge density as

$$\nabla^2 \phi(\boldsymbol{r}) = -\rho(\boldsymbol{r})/\varepsilon, \tag{2.58}$$

which is known as *Poisson's equation.* In a charge-free region, (2.58) may be specialized to yield

$$\nabla^2 \phi(\boldsymbol{r}) = 0, \tag{2.59}$$

which is known as *Laplace's equation.*

To facilitate a comparison with the magnetostatic field, we collect together the two differential equations (2.24) and (2.57) satisfied by the electrostatic field $\boldsymbol{E}(\boldsymbol{r})$ as:

$$\nabla \times \boldsymbol{E}(\boldsymbol{r}) = 0; \qquad \nabla \cdot \boldsymbol{E}(\boldsymbol{r}) = \rho(\boldsymbol{r})/\varepsilon. \tag{2.60}$$

10. Lines of Electric Force and Electric Flux

The concepts of lines of electric force and electric flux are useful in electrostatic problems. A line of electric force is everywhere parallel to the direction of the electric field. The number of lines per unit area is proportional to the magnitude of the electric field. In a similar manner, a line of electric flux is everywhere parallel to the electric flux density vector and as before, the number of electric flux lines per unit area is proportional to the magnitude of the electric flux density vector. These concepts of lines of electric force and electric flux are useful in the visualization of the nature of the electrostatic field arising due to specified charge distributions. In homogeneous and isotropic media, the lines of electric force and electric flux are in the same direction.

11. Boundary Conditions

The differential equations (2.60) governing the electrostatic field are not valid at the interface between two different media because of the presence of the space derivatives normal to the interface. However, the integral forms (2.25) and (2.54) of the governing equations of the electrostatic field are valid even in a region with boundaries separating two different media and we can use them to deduce the requirements that the field vectors must satisfy at the boundaries. We shall postpone the details of the derivation of the boundary conditions to a later section where we shall give them in connection with the general electromagnetic field quantities. Here, we shall state the boundary conditions satisfied by the electrostatic field without proof.

Let us consider a point on the boundary between two different media. Let us choose the origin of the Cartesian coordinate system x, y, and z to coincide with this point and let the z-axis be normal to the boundary. Moreover, let the coordinate system be conveniently oriented to have its y-axis parallel to the component of the field tangential to the interface. The regions 1 ($z > 0$) and 2 ($z < 0$) are characterized by the permittivities ε_1 and ε_2 respectively. We can show the line integral of the electric field taken around an infinitesimal rectangular path surrounding the origin and lying in the yz-plane to yield

$$E_{y1} = E_{y2}, \qquad \text{that is,} \qquad E_{t1} = E_{t2}, \tag{2.61}$$

where t denotes the components tangential to the interface. In (2.61), the subscripts 1 and 2 indicate the regions corresponding to the field at the interface. If we substitute (2.8) into (2.61) and integrate both sides with respect to the coordinate parallel to the tangential field component we find that

$$\phi_1(\boldsymbol{r}) = \phi_2(\boldsymbol{r}), \tag{2.62}$$

where the constant of integration has been set equal to zero. Therefore, it follows from (2.61) and (2.62) that the tangential electric field and the scalar potential are continuous across the interface separating two different media.

In a similar manner, we can prove by application of the integral form (2.54) of Gauss's law to an infinitesimal pill-box centered around the origin that

$$D_{z1} - D_{z2} = \rho_s, \text{ that is, } D_{n1} - D_{n2} = \rho_s, \tag{2.63}$$

where n denotes the component normal to the interface and ρ_s is the surface charge density. In the absence of free charges on the interface, the normal components of the electric flux density are continuous across the interface.

An important special case occurs when one of the regions is a perfect conductor in which the charges are free to move. In the presence of an electric field, the charges move to new positions of equilibrium in which there are no forces acting on the charges and, therefore, a perfect conductor cannot sustain an electric field. In view of Gauss's law, it follows that for the static case there are no free charges inside a perfect conductor. In electrostatics all the free charges in a perfect conductor are situated on the surface. Since the tangential component of the electric field is continuous across an interface, we see that the tangential component of the electric field is zero on the surface of a perfect conductor since it is zero in the interior. Consequently, the surface of a perfect conductor is an equipotential surface. Also, since the electric field is zero in the interior of a perfect conductor, it follows from (2.62) that the entire perfect conductor is at the same potential. For definiteness, let the region 2 be a perfect conductor. Then, we obtain

$$E_{t1} = 0; \qquad \phi_1(\boldsymbol{r}) = \text{constant}. \tag{2.64}$$

Also, (2.63) yields

$$D_{n1} = D_{z1} = \rho_s, \tag{2.65}$$

that is, the normal component of the electric flux density at the surface of a perfect conductor is numerically equal to the surface charge density and is outwardly directed from a positive charge.

12. Electric Images

The method of images is useful for the evaluation of electrostatic fields and we illustrate it here for a point charge in front of a perfectly conducting plane. Let us consider the point electric dipole treated previously. Let P_i be located at the mirror image of P about the $z = 0$ plane. We ascertain the coordinates of P_i from Fig. 4.3 to be r, $\pi - \theta$, and ϕ. Therefore, we verify from (2.17a and b) that

$$r_{P_1P_i} = r_{P_2P}; \qquad r_{P_2P_i} = r_{P_1P}. \tag{2.66}$$

It is convenient for us to introduce the cylindrical coordinates ρ, ϕ, z. The coordinates of P and P_i are given respectively by (ρ, ϕ, z) and $(\rho, \phi, -z)$. The electrostatic potential $\phi(\rho, \phi, z)$ at P is given by (2.18) and that at P_i we obtain from (2.18) as

$$\phi(\rho, \phi, -z) = \frac{q}{4\pi\varepsilon_0}\left(\frac{1}{r_{P_1P_i}} - \frac{1}{r_{P_2P_i}}\right). \tag{2.67}$$

From (2.18), (2.66), and (2.67) it follows that

$$\phi(\rho, \phi, -z) = -\phi(\rho, \phi, z). \tag{2.68}$$

We obtain from (2.68) that

$$\phi(\rho, \phi, 0) = 0. \tag{2.69}$$

The $z = 0$ plane is an equipotential surface and, therefore, a perfectly conducting plane can be introduced to occupy the plane $z = 0$ without altering the fields. Since the electric fields cannot penetrate a perfect conductor, the fields in the half-spaces $z > 0$ and $z < 0$ are due to the charges q and $-q$ respectively. The charge $-q$ at P_2 can now be removed without affecting the fields in the half-space $z > 0$. Thus, in the half-space $z > 0$ terminated at $z = 0$ by an infinite perfectly conducting plane, the fields due to a point charge situated in that half-space are equal to those of the given charge together with that due to an "image" charge. The "image" charge is located in the half-space $z < 0$ at the mirror image of the location of the actual charge. The "image" charge is equal in magnitude and opposite in sign to that of the actual charge. Therefore, in obtaining the fields due to a point charge in front of a perfectly conducting plane, the effect of the perfectly conducting plane can be reproduced by the introduction of the "image" charge.

We can easily extend the method of images to charge distributions as well as to perfectly conducting cylinders and spheres. It has been found useful in the evaluation of electric fields due to charge distributions in the presence of perfectly conducting boundaries.

Example. Let us consider two perfectly conducting parallel planes of infinite extent situated at $z = 0$ and $z = d$ respectively (Fig. 4.7). The potentials on the two planes are given by $\phi(z = 0) = 0$ and $\phi(z = d) = V$. The regions $0 < z < a$ and $a < z < d$

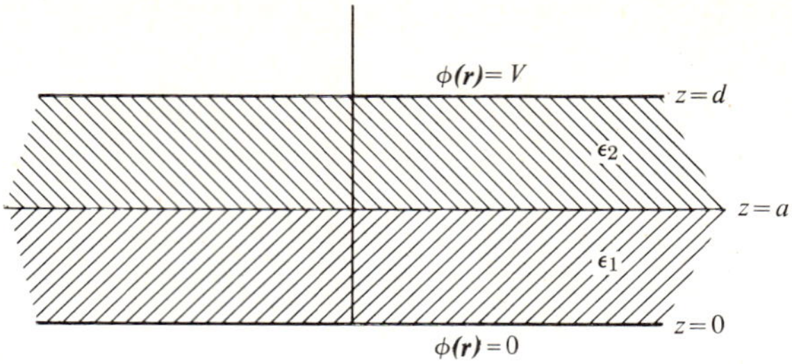

Fig. 4.7 Parallel-plate capacitor with two different dielectrics.

between the parallel planes consist of two different dielectrics characterized by the permittivities ε_1 and ε_2 respectively. We wish to find the potential and the electric field distributions in the region between the parallel planes.

There are no variations in the x- and y-directions with the result that Laplace's equation (2.59) reduces to

$$d^2\phi(\mathbf{r})/dz^2 = 0. \tag{2.70}$$

We obtain the solution of (2.70) in the region $0 < z < a$ subject to the boundary condition $\phi(z = 0) = 0$ as

$$\phi(\mathbf{r}) = Az \qquad (0 < z < a). \tag{2.71}$$

In a similar manner, we deduce the solution of (2.70) in the region $a < z < d$ subject to the boundary condition $\phi(z = d) = V$ to be

$$\phi(z) = B(z - d) + V. \tag{2.72}$$

In (2.71) and (2.72), A and B are constants to be determined.

The electric fields have only z-components and we evaluate them from (2.71) and (2.72) as

$$E_z(z) = -\frac{\partial}{\partial z}\phi(z) = -A \qquad (0 < z < a), \tag{2.73}$$

and

$$E_z(z) = -\frac{\partial}{\partial z}\phi(z) = -B \qquad (a < z < d). \tag{2.74}$$

The electric flux densities also have only z-components and we obtain them from (2.73) and (2.74) as

$$D_z(z) = -A\varepsilon_1 \qquad (0 < z < a), \tag{2.75}$$

$$D_z(z) = -B\varepsilon_2 \qquad (a < z < d). \tag{2.76}$$

If we apply the boundary conditions (2.62) and (2.63) at $z = a$ together with (2.71), (2.72), (2.75), and (2.76) we get

$$Aa + B(d - a) = V, \tag{2.77}$$

$$A\varepsilon_1 - B\varepsilon_2 = 0. \tag{2.78}$$

The solution of (2.77) and (2.78) for A and B gives

$$A = V\varepsilon_2[a\varepsilon_2 + (d - a)\varepsilon_1]^{-1}; \qquad B = V\varepsilon_1[a\varepsilon_2 + (d - a)\varepsilon_1]^{-1}. \tag{2.79}$$

If we substitute (2.79) in (2.71) and (2.72) we obtain the following result for the potential distribution:

$$\phi(\mathbf{r}) = zV\varepsilon_2[a\varepsilon_2 + (d - a)\varepsilon_1]^{-1} \qquad (0 < z < a), \tag{2.80}$$

and

$$\phi(\mathbf{r}) = V[a\varepsilon_2 + (z - a)\varepsilon_1][a\varepsilon_2 + (d - a)\varepsilon_1]^{-1} \qquad (a < z < d). \tag{2.81}$$

In a similar manner, we obtain the electric field and the electric flux density distributions as

$$E_z(z) = -V\varepsilon_2[a\varepsilon_2 + (d - a)\varepsilon_1]^{-1} \qquad (0 < z < a), \tag{2.82}$$

$$E_z(z) = -V\varepsilon_1[a\varepsilon_2 + (d - a)\varepsilon_1]^{-1} \qquad (a < z < d), \tag{2.83}$$

$$D_z(z) = -V\varepsilon_1\varepsilon_2[a\varepsilon_2 + (d - a)\varepsilon_1]^{-1} \qquad (0 < z < d). \tag{2.84}$$

In view of (2.84), we can evaluate the surface charge densities on the two planes with the help of (2.65) as

$$\rho_s(z = 0) = -\rho_s(z = d) = -V\varepsilon_1\varepsilon_2[a\varepsilon_2 + (d - a)\varepsilon_1]^{-1}. \tag{2.85}$$

We should note that the surface charge densities on the two planes are equal in magnitude but opposite in sign.

4.3 POLARIZATION AND THE CONCEPT OF PERMITTIVITY

1. Polarization

The response of material media to an electric field is contained in the macroscopic parameter ε. We wish to examine the introductory aspects of the microscopic theory of matter and develop the concept of permittivity.

Matter consists of atoms and molecules. In atoms, positive and negative charges are held together by the atomic binding forces. When an electric field is applied, the positive charge inside the atom moves in the direction of the electric field and the negative charge moves in the opposite direction, producing a small charge separation. In the presence of an electric field, every atom behaves like an electric dipole with its dipole moment being parallel to the electric field. As we have shown in Fig. 4.8, the electric field due to the induced atomic electric dipoles tends to oppose the applied electric field. A dielectric material in which there is a separation of bound charges is said to be polarized. In order to obtain correctly the response of material media to an electric field, it is necessary for us to take into account the effects of polarization in dielectric materials.

Polarization is of several kinds. The polarization that we have considered so far is on the atomic scale in which the charge separation is due to the displacement of the negative electrons of the atom with respect to the positive nucleus. This type of polarization is called *electronic* polarization. On a molecular scale, the polarization is of two kinds. The polarization resulting from the separation of the positive and

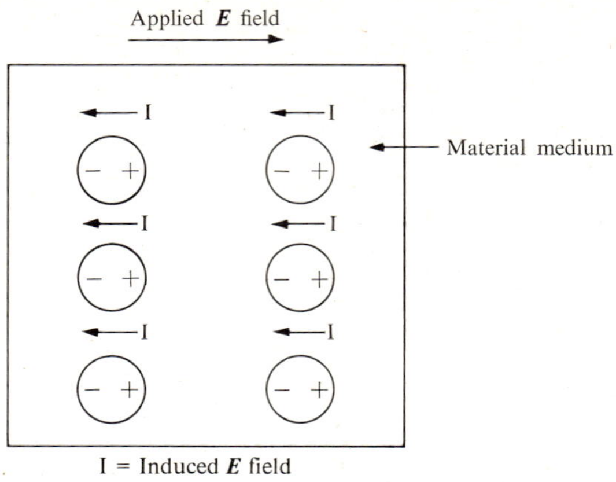

Fig. 4.8 Induced electric dipoles in a material medium.

the negative ions held together by ionic bonds is called *ionic* polarization. Polar molecules are permanently polarized but are randomly oriented with the result that there is no net dipole moment per unit volume. In the presence of an electric field, the molecular dipoles align themselves with the electric field resulting in a net polarization in the material. This type of polarization is called *orientation* polarization. There are also other types of polarization which occur on larger than molecular scale but which we shall not consider here specifically.

We have given the definition of the dipole moment previously. Its magnitude is equal to the product of the individual charge and the separation distance between the charges and its direction is parallel to that of the straight line from the negative to the positive charge. The dipole moment per unit volume is called *polarization* $\boldsymbol{P}(\boldsymbol{r})$. For most materials and for sufficiently small electric fields, polarization $\boldsymbol{P}(\boldsymbol{r})$ is proportional to the electric field, that is,

$$\boldsymbol{P}(\boldsymbol{r}) = \varepsilon_0 \chi \boldsymbol{E}(\boldsymbol{r}), \tag{3.1}$$

where χ is known as the susceptibility of the material.

2. Equivalent Volume and Surface Charge Densities

Let us consider a volume V of a dielectric material enclosed by the surface S as shown in Fig. 4.9. We show now that the macroscopic effect of the polarization of the dielectric is equivalent to a distribution of charges throughout the volume V and the bounding surface S of the dielectric material.

Let O represent the origin of the coordinate system as shown in Fig. 4.9. Let us consider an elementary volume dV' of the material centered around the source point P' whose position vector is given by $\boldsymbol{r}'$. We wish to find the scalar potential at an observation point P which is exterior to the volume V and which is denoted by the position vector $\boldsymbol{r}$. The dipole moment of the elementary volume dV' of the material is given by $\boldsymbol{P}(\boldsymbol{r}')\,dV'$. We obtain the contribution to the scalar potential at the

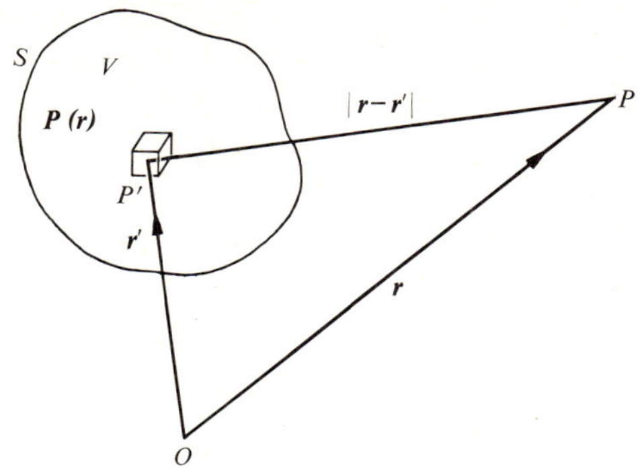

Fig. 4.9 Volume distribution of polarization $\boldsymbol{P}$.

observation point P due to the dipole moment associated with the elementary volume dV' from (2.22) as

$$d\phi(\boldsymbol{r}) = -\frac{\boldsymbol{P}(\boldsymbol{r}')\,dV'}{4\pi\varepsilon_0}\cdot\nabla\left(\frac{1}{|\boldsymbol{r}-\boldsymbol{r}'|}\right) = \frac{\boldsymbol{P}(\boldsymbol{r}')}{4\pi\varepsilon_0}\cdot\nabla'\left(\frac{1}{|\boldsymbol{r}-\boldsymbol{r}'|}\right)dV', \tag{3.2}$$

where ∇' is an operation with respect to the position vector $\boldsymbol{r}'$. We can evaluate the scalar potential $\phi(\boldsymbol{r})$ at the observation point P due to the polarization of the entire volume of the dielectric material by integrating (3.2) throughout the volume V with the following result:

$$\phi(\boldsymbol{r}) = \frac{1}{4\pi\varepsilon_0}\int_V \boldsymbol{P}(\boldsymbol{r}')\cdot\nabla'\left(\frac{1}{|\boldsymbol{r}-\boldsymbol{r}'|}\right)dV'. \tag{3.3}$$

In view of the vector identity

$$\boldsymbol{A}\cdot\nabla a = \nabla\cdot(a\boldsymbol{A}) - a\nabla\cdot\boldsymbol{A}, \tag{3.4}$$

where a is a scalar and $\boldsymbol{A}$ is a vector, we may recast (3.3) into the form

$$\phi(\boldsymbol{r}) = \frac{1}{4\pi\varepsilon_0}\left[\int_V \nabla'\cdot\left(\frac{\boldsymbol{P}(\boldsymbol{r}')}{|\boldsymbol{r}-\boldsymbol{r}'|}\right)dV' - \int_V \frac{\nabla'\cdot\boldsymbol{P}(\boldsymbol{r}')}{|\boldsymbol{r}-\boldsymbol{r}'|}\,dV'\right]. \tag{3.5}$$

With the help of the divergence theorem, we may write the first integral as a surface integral and therefore (3.5) becomes

$$\phi(\boldsymbol{r}) = \frac{1}{4\pi\varepsilon_0}\left[\int_S \frac{\boldsymbol{P}(\boldsymbol{r}')}{|\boldsymbol{r}-\boldsymbol{r}'|}\cdot\hat{n}\,dS' - \int_V \frac{\nabla'\cdot\boldsymbol{P}(\boldsymbol{r}')}{|\boldsymbol{r}-\boldsymbol{r}'|}\,dV'\right], \tag{3.6}$$

where $\hat{n}$ is an outwardly drawn unit normal to the surface. If we compare the form of (3.6) with that of (2.30) we see that the scalar potential at the observation point P is interpretable as that due to the distributions of the following volume and surface charge densities:

$$\rho_b(\boldsymbol{r}) = -\nabla\cdot\boldsymbol{P}(\boldsymbol{r}), \tag{3.7}$$

$$\rho_{bs}(\boldsymbol{r}) = \boldsymbol{P}(\boldsymbol{r})\cdot\hat{n}. \tag{3.8}$$

Therefore, a volume distribution of electric dipoles induced in a dielectric by an electric field, insofar as their macroscopic effects are concerned, can be replaced by the equivalent distributions of volume and surface charge densities as given by (3.7) and (3.8) respectively. We see from (3.7) and (3.8) that bound surface charges appear on the boundaries between dielectrics and on the boundary between a dielectric and free space, whereas bound volume charges appear when the polarization is inhomogeneous, which may arise either due to the inhomogeneity of the dielectric or the inhomogeneity of the electric field.

3. The Permittivity Concept

Let us consider a region of dielectric to consist of a volume distribution of electric dipoles in free space. Then, Gauss's law for an interior point in the dielectric is given by

$$\nabla \cdot \varepsilon_0 \boldsymbol{E}(\boldsymbol{r}) = \rho(\boldsymbol{r}) + \rho_b(\boldsymbol{r}), \tag{3.9}$$

where $\rho(\boldsymbol{r})$ is the free charge density and $\rho_b(\boldsymbol{r})$ is the density of the volume distribution of the equivalent bound charges. Together with (3.7), we can rewrite (3.9) as

$$\nabla \cdot [\varepsilon_0 \boldsymbol{E}(\boldsymbol{r}) + \boldsymbol{P}(\boldsymbol{r})] = \rho(\boldsymbol{r}). \tag{3.10}$$

In analogy with Gauss's law in free space, using (3.10) we can define electric flux density in a dielectric as

$$\boldsymbol{D}(\boldsymbol{r}) = \varepsilon_0 \boldsymbol{E}(\boldsymbol{r}) + \boldsymbol{P}(\boldsymbol{r}). \tag{3.11}$$

In view of (3.1), (3.11) becomes

$$\boldsymbol{D}(\boldsymbol{r}) = \varepsilon_0(1 + \chi)\boldsymbol{E}(\boldsymbol{r}) = \varepsilon_0\varepsilon_r \boldsymbol{E}(\boldsymbol{r}). \tag{3.12}$$

Therefore, in a dielectric medium in which the polarization is linearly related to the electric field, it follows from (3.12) that we can identify $1 + \chi$ as the relative permittivity ε_r and that in linear dielectrics the concepts of polarization and permittivity are equivalent.

Let us consider the interface between a dielectric region and free space as shown in Fig. 4.10. Let E and E_0 be the normal components of the electric field at the interface on the dielectric and the free space sides respectively. Without loss of generality, let us assume that there are no other electric field components. The polarization inside the dielectric is in the same direction as the electric field, namely, normal to the interface as shown in Fig. 4.10; let its magnitude be P. Let $\hat{n}$ be the outwardly drawn unit normal to the dielectric surface. We see from (3.8) and Fig. 4.10 that the density of the surface distribution of the bound charges is given by $\rho_{bs} = P$. If there are no free surface charges, then the application of Gauss's law to an infinitesimal pill-box intersected by the dielectric boundary yields

$$\varepsilon_0 E_0 - \varepsilon_0 E = P \qquad \text{or} \qquad \varepsilon_0 E_0 - (\varepsilon_0 E + P) = 0. \tag{3.13}$$

In view of (3.11) and (3.12), we can conveniently rewrite (3.13) as

$$\varepsilon_0 E_0 - \varepsilon_0 \varepsilon_r E = 0 \qquad \text{or} \qquad \varepsilon_0 E_0 - \varepsilon E = 0, \tag{3.14}$$

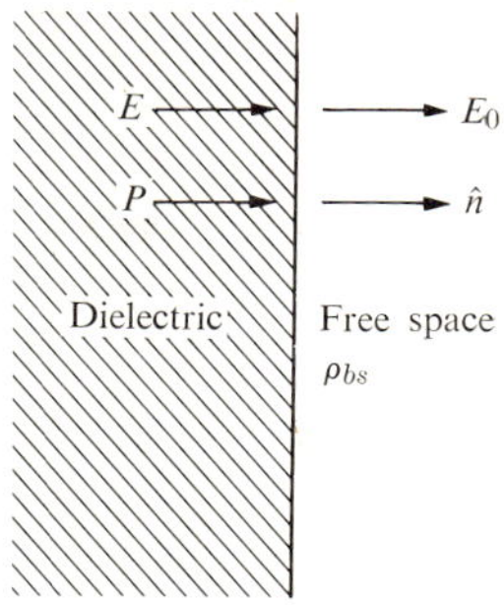

Fig. 4.10 The boundary between a dielectric region and free space.

which states that the normal component of the electric flux density is continuous across the bounding surface of the dielectric if there are no free surface charges. The existence of the bound charges on the dielectric boundary does not affect the boundary condition on the tangential electric field.

If we examine (3.12) and (3.13) we see that the macroscopic effects of a volume distribution of electric dipoles in a polarized dielectric are equivalent to introducing appropriate values for the permittivity and this equivalence holds good inside the dielectric as well as on the boundary of the dielectric. Thus, the permittivity concept is useful in that it incorporates the polarization effects of a dielectric. We can describe the macroscopic behavior of a dielectric either in terms of the permittivity or the polarization with its equivalent volume and surface charges. In linear dielectric media such as those which we are only considering in this treatment, the two methods of description of a dielectric are identical, and therefore we shall use the constitutive parameter ε alone hereafter.

Example. We treated the parallel-plate capacitor with two different dielectrics, as illustrated in Fig. 4.7, in Section 4.2 using the concept of permittivity to describe the dielectrics. It is instructive for us to evaluate the polarizations in the two different dielectric regions and deduce the equivalent volume and surface charge densities.

With the help of (3.11), (2.82), and (2.84) we obtain the polarization in the region $0 < z < a$ as

$$P_z(z) = D_z(z) - \varepsilon_0 E_z(z) = -V\varepsilon_2(\varepsilon_1 - \varepsilon_0)[a\varepsilon_2 + (d - a)\varepsilon_1]^{-1} \quad (\text{for } 0 < z < a). \tag{3.15}$$

In a similar manner, we find from (3.11), (2.83), and (2.84) that

$$P_z(z) = D_z(z) - \varepsilon_0 E_z(z) = -V\varepsilon_1(\varepsilon_2 - \varepsilon_0)[a\varepsilon_2 + (d - a)\varepsilon_1]^{-1} \quad (\text{for } a < z < d). \tag{3.16}$$

The polarizations in the two dielectric regions do not have any components other than those given by (3.15) and (3.16). Since in each dielectric region, the polarization is uniform, it follows from (3.7) that the equivalent volume distribution of bound charges vanishes identically in both the dielectric regions.

We obtain the total surface charge density at $z = 0$ from (2.85), (3.8), and (3.15) as

$$\rho_{ts}(z = 0) = \rho_s(z = 0) - P_z(z) = -V\varepsilon_2\varepsilon_0[a\varepsilon_2 + (d - a)\varepsilon_1]^{-1}. \tag{3.17}$$

Similarly, (3.8), (3.15), and (3.16) yield the total surface charge density at $z = a$ as

$$\rho_{ts}(z = a) = -V\varepsilon_0(\varepsilon_1 - \varepsilon_2)[a\varepsilon_2 + (d - a)\varepsilon_1]^{-1}. \tag{3.18}$$

In the same manner, we find from (2.85), (3.8), and (3.16) that

$$\rho_{ts}(z = d) = \rho_s(z = d) + P_z(z) = V\varepsilon_1\varepsilon_0[a\varepsilon_2 + (d - a)\varepsilon_1]^{-1}. \tag{3.19}$$

Thus, the parallel-plate capacitor is equivalent to three plane surface charges situated at $z = 0$, $z = a$, and $z = d$ in *free space.* Since the sum of the charge densities given by (3.17) to (3.19) is zero, Gauss's law shows that there is no electric field outside the parallel-plate capacitor. We obtain the electric fields in the regions $0 < z < a$ and $a < z < d$ from (2.65), (3.17), and (3.19) as

$$E_z(z) = \rho_{ts}(z = 0)/\varepsilon_0 = -V\varepsilon_2[a\varepsilon_2 + (d - a)\varepsilon_1]^{-1} \qquad (\text{for } 0 < z < a), \tag{3.20}$$

$$E_z(z) = -\rho_{ts}(z = d)/\varepsilon_0 = -V\varepsilon_1[a\varepsilon_2 + (d - a)\varepsilon_1]^{-1} \qquad (\text{for } a < z < d). \tag{3.21}$$

These are identical to those given in (2.82) and (2.83) respectively. Thus, the concepts of permittivity and polarization are two equivalent points of view for the description of the macroscopic behavior of dielectric media.

4.4 MAGNETOSTATICS

1. Steady Current

The principal source of steady magnetic field is the direct current. Permanent magnets also give rise to steady magnetic fields. The electric current is the rate of flow of electric charges. The current density $\boldsymbol{J}(\boldsymbol{r})$ in a conductor is proportional to and in parallel with the applied electric field $\boldsymbol{E}(\boldsymbol{r})$ and hence we may express it as

$$\boldsymbol{J}(\boldsymbol{r}) = \sigma\boldsymbol{E}(\boldsymbol{r}). \tag{4.1}$$

The proportionality factor σ is known as the conductivity and its unit is mhos per meter. The electric field is measured in units of volts per meter. Hence, the unit of the current density $\boldsymbol{J}(\boldsymbol{r})$ is amperes per square meter. Let us consider a small length l of a wire of uniform cross section as shown in Fig. 4.11. Let S be the cross-sectional area of the wire. If $\boldsymbol{J}(\boldsymbol{r})$ is the current density, then the total current I in the wire is

$$I = \int_S \boldsymbol{J}(\boldsymbol{r}) \cdot d\boldsymbol{S} \quad \text{A}. \tag{4.2}$$

If the current density is uniform, it follows from (4.1) and (4.2) that

$$I = SJ(\boldsymbol{r}) = S\sigma E(\boldsymbol{r}). \tag{4.3}$$

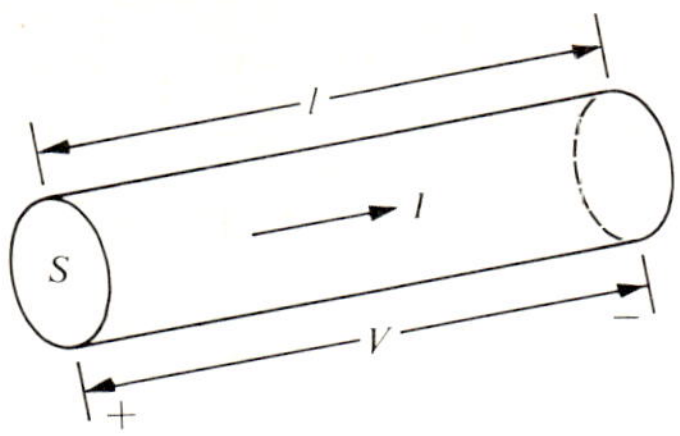

Fig. 4.11 Resistance R of a conducting wire of length l and cross-sectional area S.

Let the two faces of the wire be maintained at different potentials. In the small length of the wire, we may consider the electric field to be conservative with the result that $\boldsymbol{E}(\boldsymbol{r}) = -\nabla\phi(\boldsymbol{r})$. Since the potential varies only along the length of the wire, it is clear that the electric field is along the length of the wire and so also is the current. We can obtain the magnitude of $\boldsymbol{E}(\boldsymbol{r})$ as $E(\boldsymbol{r}) = V/l$ where V is the potential drop along the wire in the direction of the current. Hence, the magnitude of (4.3) becomes

$$I = \frac{S\sigma}{l} V \quad \text{or} \quad V = \frac{l}{S\sigma} I = \rho_{\text{res}} \frac{l}{S} I = RI. \tag{4.4}$$

We should note that the resistance R of a length l of the wire of cross-sectional area S is given by

$$R = \rho_{\text{res}} \frac{l}{S} \quad \text{ohms,} \tag{4.5}$$

where $\rho_{\text{res}} = 1/\sigma$ is the resistivity of the wire in ohms-meter. The current in a conductor arising due to the application of an electric field is called the *conduction* current.

If an insulator containing free electric charges moves as a whole carrying the charges with it, then the resulting flow of charges gives rise to what is known as the *convection current*. An important special case of the convection current is that due to the motion of the charged particles in vacuum under the action of an external force such as that due to the electric field. The electric current due to the motion of the charged particles is dependent on the dynamics of the charged particles. We obtain the trajectories of the charged particles from the equations of motion with the inclusion of the forces of electric origin. The stream of charged particles in a convection current is characterized by the charge density $\rho(\boldsymbol{r})$ and the particle velocity $\boldsymbol{v}(\boldsymbol{r})$. Let us consider an element of area dS normal to the velocity vector $\boldsymbol{v}(\boldsymbol{r})$ in a stream of charged particles as shown in Fig. 4.12. The length dl traveled by the charged particles in the direction of motion in the short time interval dt is $dl = v(\boldsymbol{r})\, dt$. From Fig. 4.12, we can ascertain that the total charge flow across the area dS in time dt is $\rho(\boldsymbol{r})\, dS\, dl$. Hence, we obtain the current which is the rate of flow of charges as

$$I(\boldsymbol{r}) = \rho(\boldsymbol{r}) \frac{dl}{dt} dS = \rho(\boldsymbol{r}) v(\boldsymbol{r})\, dS. \tag{4.6}$$

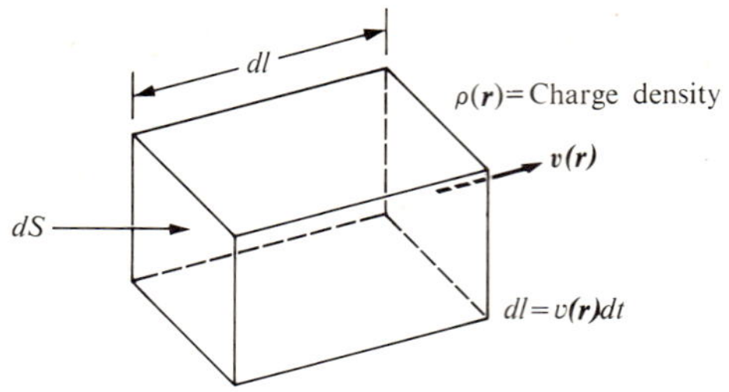

Fig. 4.12 Convection current in a stream of charged particles.

Only the magnitude of the current is given in (4.6) and the direction of the current involving the motion of the positive charged particles is in the direction of the velocity. Together with (4.6), we find the current density vector $\boldsymbol{J}(\boldsymbol{r})$ to be

$$\boldsymbol{J}(\boldsymbol{r}) = \boldsymbol{I}(\boldsymbol{r})/dS = \rho(\boldsymbol{r})\boldsymbol{v}(\boldsymbol{r}). \tag{4.7}$$

Let us consider a surface S enclosing a volume V in a region of current. Since the outwardly directed current through an element of area is $\boldsymbol{J}(\boldsymbol{r}) \cdot d\boldsymbol{S}$, we obtain the total outwardly directed current from the volume V as

$$\int_S \boldsymbol{J}(\boldsymbol{r}) \cdot d\boldsymbol{S} = \int_V \nabla \cdot \boldsymbol{J}(\boldsymbol{r})\, dV. \tag{4.8}$$

We have used the divergence theorem in obtaining the right-hand side of (4.8). In direct currents, there is neither a net accumulation nor a net depletion of current from a volume element in a region of current. Hence, (4.8) is equal to zero. Moreover, the foregoing result is true for any arbitrarily small volume V in a region of current and, therefore, it follows that

$$\nabla \cdot \boldsymbol{J}(\boldsymbol{r}) = 0, \tag{4.9}$$

which is a statement of the continuity of currents. In other words, direct current exists only in closed circuits. Even so, it is convenient for us to deal in terms of current elements. Consider, for example, a wire of cross-sectional area dS and length dl carrying a current $\boldsymbol{I}(\boldsymbol{r})$ whose direction is that of the wire. Then, since the volume dV of the current element is $dV = dS\, dl$, we can specify a current element by one of the three forms:

$$\boldsymbol{J}(\boldsymbol{r})\, dV = \boldsymbol{I}(\boldsymbol{r})\, dl = I(\boldsymbol{r})\, d\boldsymbol{l}. \tag{4.10}$$

2. Ampère's Law of Force

Just as Coulomb's law of force is the basis for the definition of the electric field, so is the law of force of Ampère the basis for the definition of the magnetic field. Let us consider two current elements $\boldsymbol{J}'(\boldsymbol{r}')\, dV'$ and $\boldsymbol{J}(\boldsymbol{r})\, dV$ located at P' and P specified by the radius vectors $\boldsymbol{r}'$ and $\boldsymbol{r}$ respectively as shown in Fig. 4.13. Let $\boldsymbol{r} - \boldsymbol{r}' = \hat{\boldsymbol{r}}_{P'P} r_{P'P}$

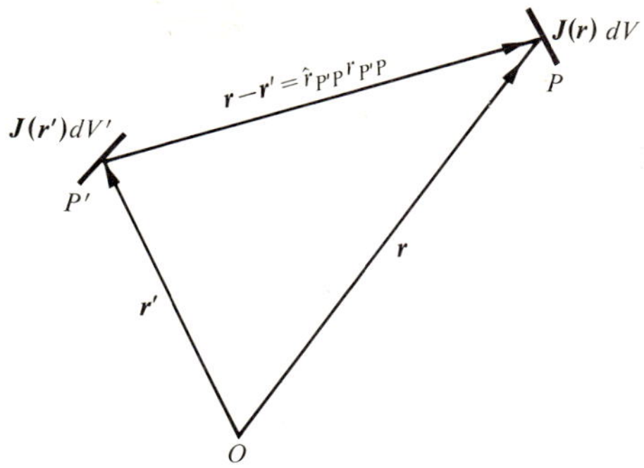

Fig. 4.13 Force between two current elements.

where $\hat{\boldsymbol{r}}_{P'P}$ is a unit vector along the line joining the centers of the two current elements and directed towards the current element at P, and

$$r_{P'P} = [(x - x')^2 + (y - y')^2 + (z - z')^2]^{1/2}$$

is the distance between the centers of the two current elements. The studies by Ampère lead to the important discovery that a force is exerted by one current element on the other and that the element of force $d\boldsymbol{F}_{P'P}$ exerted on the current element $\boldsymbol{J}(\boldsymbol{r})\,dV$ by the current element $\boldsymbol{J}'(\boldsymbol{r}')\,dV'$ is

$$d\boldsymbol{F}_{P'P} = \mu \boldsymbol{J}(\boldsymbol{r})\,dV \times \frac{(\boldsymbol{J}'(\boldsymbol{r}')\,dV' \times \hat{\boldsymbol{r}}_{P'P})}{4\pi r_{P'P}^2}. \tag{4.11}$$

The force is measured in newtons, the current density in amperes per square meter and the distances in meters. The force is different in different media whose property is characterized by the parameter μ, known as the permeability. We shall discuss the concept of permeability of material media in the following section. We denote the permeability of vacuum by μ_0, which is equal to $4\pi \times 10^{-7}$ henry per meter. We express the permeability of other media in terms of that of vacuum as $\mu = \mu_0\mu_r$, where μ_r is known as the relative permeability. As before, we introduce the rationalizing factor 4π into the force equation in order to eliminate its appearance in the field equations. It is worthwhile for us to note that Ampère's law of force is an inverse square law just as Coulomb's law of force is in electrostatics.

We can use the vector relation

$$\boldsymbol{A} \times (\boldsymbol{B} \times \boldsymbol{C}) = \boldsymbol{B}(\boldsymbol{A} \cdot \boldsymbol{C}) - \boldsymbol{C}(\boldsymbol{A} \cdot \boldsymbol{B}) \tag{4.12}$$

to recast (4.11) into the form

$$d\boldsymbol{F}_{P'P} = \frac{\mu}{4\pi r_{P'P}^2}[\boldsymbol{J}'(\boldsymbol{r}')\,dV'\{\boldsymbol{J}(\boldsymbol{r})\,dV \cdot \hat{\boldsymbol{r}}_{P'P}\} - \hat{\boldsymbol{r}}_{P'P}\{\boldsymbol{J}(\boldsymbol{r})\,dV \cdot \boldsymbol{J}'(\boldsymbol{r}')\,dV'\}]. \tag{4.13}$$

Since $\boldsymbol{r} - \boldsymbol{r}' = \hat{r}_{P'P} r_{P'P}$, it follows from (2.6) that $\nabla(1/r_{P'P}) = -\hat{r}_{P'P}/r_{P'P}^2$. Together with (4.10), the first term in (4.13) becomes

$$-\frac{\mu}{4\pi} I'(\boldsymbol{r}')I(\boldsymbol{r})\, d\boldsymbol{l}'[d\boldsymbol{l} \cdot \nabla(1/r_{P'P})]. \tag{4.14}$$

Let C' and C be two closed circuits of which the current elements P' and P respectively form parts. Since the total direct current across any cross-sectional area in a closed circuit should be the same, it follows that $I'(\boldsymbol{r}')$ and $I(\boldsymbol{r})$ are constants independent of the location of the current elements in the circuit. In view of (2.13), we can integrate (4.14) around the closed circuit C to obtain

$$-\frac{\mu}{4\pi} I'I\, d\boldsymbol{l}' \int_C d(1/r_{P'P}) = 0. \tag{4.15}$$

The integral of $d(1/r_{P'P})$ is $1/r_{P'P}$ and since this is a single valued function, it vanishes, as indicated in (4.15), when evaluated around the closed contour C. We obtain the total force $\boldsymbol{F}_{P'P}$ acting on the current loop C due to the current loop C' by integrating (4.13) around the two closed circuits. The result, on using (4.15), is

$$\boldsymbol{F}_{P'P} = -\frac{\mu I'I}{4\pi} \int_C \int_{C'} (d\boldsymbol{l} \cdot d\boldsymbol{l}') \frac{\hat{r}_{P'P}}{r_{P'P}^2}. \tag{4.16}$$

We obtain the force $\boldsymbol{F}_{PP'}$ acting on the current loop C' due to the current loop C from (4.16) by interchanging the primed and the unprimed quantities. Since $\hat{r}_{P'P} = -\hat{r}_{PP'}$, it is found that $\boldsymbol{F}_{PP'} = -\boldsymbol{F}_{P'P}$, which shows that the requirement imposed by Newton's third law is satisfied. However, the element of force $d\boldsymbol{F}_{P'P}$ acting on the current element P due to the current element P' is not required to be equal in magnitude and opposite in direction to the element of force $d\boldsymbol{F}_{PP'}$ acting on the current element P' due to the current element P. Indeed, it follows from (4.13) that $d\boldsymbol{F}_{P'P}$ is not necessarily equal to $-d\boldsymbol{F}_{PP'}$.

Example. A circular loop of radius a carries a current I_0. Find the force exerted on a current element situated along the axis of the loop.

It is convenient for us to use both the rectangular coordinates x, y, and z, and the cylindrical coordinates ρ, ϕ, and z. Let the loop be situated with its center coinciding with the origin of the coordinate system and with its axis coinciding with the z-axis as shown in Fig. 4.14(a). Let us consider an infinitesimal length of the filamentary loop lying between the angles ϕ and $\phi + d\phi$. Hence, the source current element at P' is given by $I_0 a\, d\phi\hat{\phi}$ where $\hat{\phi}$ is the direction of the current element. Let the probe current (I) element be situated along the z-axis at a distance z from the origin and let its length be dz. Therefore, the probe current element at P is given by $I\, dz\hat{z}$. We find from Fig. 4.14(a) that $\boldsymbol{r}_{P'P} = \hat{r}_{P'P} r_{P'P} = -a\hat{\rho} + z\hat{z}$. We see from (4.11) that the element of force $d\mathbf{F}_{P'P}$ exerted by the source current element at P' on the

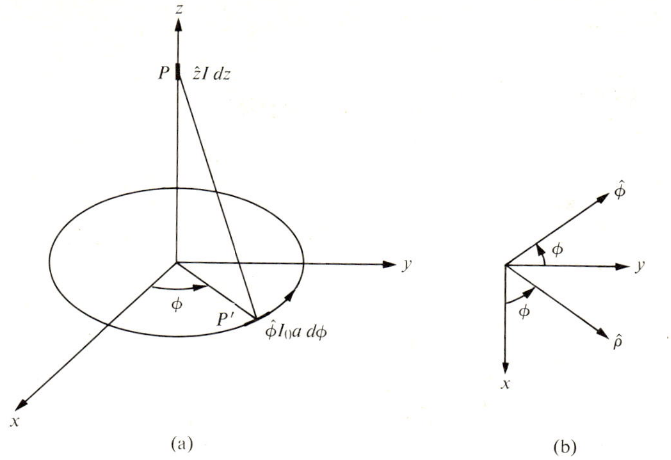

Fig. 4.14 Force exerted by a circular loop of current on a current element situated along its axis.

probe current element at P is given by

$$d\boldsymbol{F}_{P'P} = \frac{\mu}{4\pi r_{P'P}^3}(I\,dz)\hat{z} \times [I_0 a\,d\phi\hat{\phi} \times (-a\hat{\rho} + z\hat{z})],$$

$$= \frac{\mu}{4\pi r_{P'P}^3}(I\,dz)I_0 az\,d\phi\hat{\phi}.$$

Since from Fig. 4.14(b),

$$\hat{\phi} = -\hat{x}\sin\phi + \hat{y}\cos\phi$$

it follows that

$$d\boldsymbol{F}_{P'P} = \frac{\mu}{4\pi r_{P'P}^3}(I\,dz)I_0 az(-\hat{x}\sin\phi + \hat{y}\cos\phi)\,d\phi.$$

We obtain the total force $\boldsymbol{F}_P$ on the probe current element by integrating $d\boldsymbol{F}_{P'P}$ from $\phi = 0$ to $\phi = 2\pi$. In view of the symmetry, $r_{P'P}$ is independent of ϕ. Hence

$$\boldsymbol{F}_P = \int_{\phi=0}^{\phi=2\pi} d\boldsymbol{F}_{P'P} = \frac{\mu(I\,dz)I_0 az}{4\pi r_{P'P}^3}\int_0^{2\pi}(-\hat{x}\sin\phi + \hat{y}\cos\phi)\,d\phi = 0,$$

that is, a circular loop of current exerts no net force on a current element situated along its axis.

3. Biot-Savart's Law

In a region containing a charge, another charge has a force acting on it, as was discovered by Coulomb. This suggested the introduction of the concept of force field in a region containing a charge. Another charge placed in this force field has a force acting on it. The direction of the force is the same as that of the field and the magnitude is equal to the magnitude of the charge times the magnitude of the field.

This force field is by definition the electric field. In the same manner, Ampère's law of force suggests the introduction of another force field in a region containing an electric current. If another current element is placed in this field, it experiences a force whose direction is not the same as that of the force field and Ampère's law of force, in fact, reveals that this force is normal to the plane containing the current element and the direction of the force field. The law of force acting on a current element in a force field due to another current element is known as Biot-Savart's law, whose statement is as follows:

$$d\boldsymbol{F}_{P'P} = \boldsymbol{J}(\boldsymbol{r})\, dV \times d\boldsymbol{B}_{P'P} \tag{4.17}$$

where $\boldsymbol{J}(\boldsymbol{r})\, dV$ is the current element and $d\boldsymbol{B}_{P'P}$ is the element of the force field due to another current element $\boldsymbol{J}'(\boldsymbol{r}')\, dV'$. If we compare (4.17) with (4.11) we find that

$$d\boldsymbol{B}_{P'P} = [\mu \boldsymbol{J}'(\boldsymbol{r}')\, dV' \times \hat{\boldsymbol{r}}_{P'P}]/4\pi r^2_{P'P}. \tag{4.18}$$

From (4.17), we see that the force density $\boldsymbol{f}$, that is the force per unit volume, associated with a current density $\boldsymbol{J}(\boldsymbol{r})$ in this force field $\boldsymbol{B}(\boldsymbol{r})$ is

$$\boldsymbol{f} = \boldsymbol{J}(\boldsymbol{r}) \times \boldsymbol{B}(\boldsymbol{r}), \tag{4.19}$$

where $\boldsymbol{B}(\boldsymbol{r})$ is known as the magnetic flux density. Since $\boldsymbol{B}(\boldsymbol{r})$ is directly related to the force on a current element, it is the fundamental field quantity in a region containing electric currents in the same manner as $\boldsymbol{E}(\boldsymbol{r})$ that is directly related to the force on a charge is the fundamental field quantity in a region containing electric charges.

Since current elements do not exist separately, the only physically meaningful magnetic flux density is that due to a complete current circuit. It is therefore useful for us to obtain on the basis of (4.18) the magnetic flux density due to a volume distribution of electric currents. Let the current distribution be contained inside a volume V. With reference to the origin O, let $\boldsymbol{r}'$ denote a typical point P' inside the volume of current as shown in Fig. 4.15. Let $\boldsymbol{J}(\boldsymbol{r}')$ denote the current density in amperes per square meter at P'. Note that we have omitted the prime on $\boldsymbol{J}$ for convenience since there is no possible confusion. Let us consider an infinitesimal

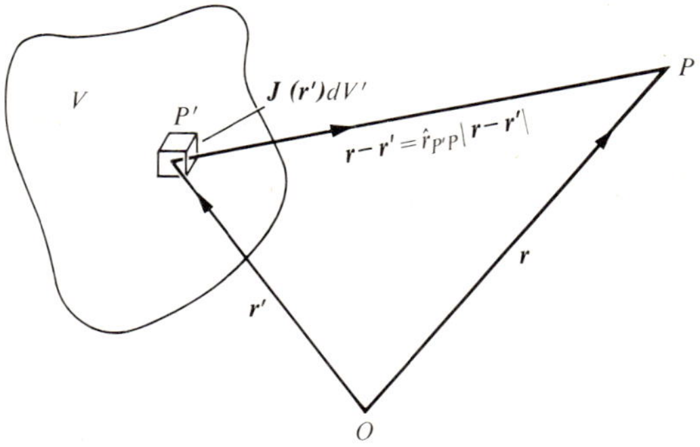

Fig. 4.15 Magnetic flux density due to a volume current distribution.

volume dV' centered around P' leading to the current element $\boldsymbol{J}(\boldsymbol{r}')\,dV'$. We find the contribution $d\boldsymbol{B}(\boldsymbol{r})$ to the magnetic flux density at an observation point P specified by $\boldsymbol{r}$, due to the current element $\boldsymbol{J}(\boldsymbol{r}')\,dV'$, from (4.18) to be

$$d\boldsymbol{B}(\boldsymbol{r}) = [\mu\boldsymbol{J}(\boldsymbol{r}')\,dV' \times \hat{\boldsymbol{r}}_{P'P}]/4\pi\,|\boldsymbol{r} - \boldsymbol{r}'|^2, \tag{4.20}$$

where $\hat{\boldsymbol{r}}_{P'P}$ is the unit vector in the direction of the line joining P' to P and $|\boldsymbol{r} - \boldsymbol{r}'|$ is the distance between P and P'. We obtain the total magnetic flux density $\boldsymbol{B}(\boldsymbol{r})$ at P by integrating (4.20) throughout the volume V occupied by the current distribution. Therefore

$$\boldsymbol{B}(\boldsymbol{r}) = \int_V \frac{\mu\boldsymbol{J}(\boldsymbol{r}')\,dV' \times \hat{\boldsymbol{r}}_{P'P}}{4\pi\,|\boldsymbol{r} - \boldsymbol{r}'|^2}. \tag{4.21}$$

We should note that $\hat{\boldsymbol{r}}_{P'P}$ depends on the variable of integration. In view of (2.6), $\nabla(1/|\boldsymbol{r} - \boldsymbol{r}'|) = -\hat{\boldsymbol{r}}_{P'P}/|\boldsymbol{r} - \boldsymbol{r}'|^2$, where the differential operator ∇ is with respect to the coordinates of P or the unprimed coordinates whereas the integration in (4.21) is with respect to the coordinates of P' or the primed coordinates. Hence (4.21) becomes

$$\boldsymbol{B}(\boldsymbol{r}) = -\int_V \frac{\mu\boldsymbol{J}(\boldsymbol{r}')\,dV'}{4\pi} \times \nabla\left(\frac{1}{|\boldsymbol{r} - \boldsymbol{r}'|}\right). \tag{4.22}$$

With the help of the vector identity

$$\nabla \times (\phi\boldsymbol{A}) = (\nabla\phi) \times \boldsymbol{A} + \phi(\nabla \times \boldsymbol{A}) \tag{4.23}$$

where ϕ is a scalar and $\boldsymbol{A}$ is a vector, we see that

$$\nabla \times \left(\frac{1}{|\boldsymbol{r} - \boldsymbol{r}'|}\boldsymbol{J}(\boldsymbol{r}')\right) = \nabla\left(\frac{1}{|\boldsymbol{r} - \boldsymbol{r}'|}\right) \times \boldsymbol{J}(\boldsymbol{r}') + \frac{1}{|\boldsymbol{r} - \boldsymbol{r}'|}\nabla \times \boldsymbol{J}(\boldsymbol{r}'). \tag{4.24}$$

Since $\boldsymbol{J}(\boldsymbol{r}')$ has no variation with respect to the unprimed coordinates on which ∇ depends, $\nabla \times \boldsymbol{J}(\boldsymbol{r}') = 0$ and therefore (4.24) becomes

$$\nabla \times \left(\frac{\boldsymbol{J}(\boldsymbol{r}')}{|\boldsymbol{r} - \boldsymbol{r}'|}\right) = -\boldsymbol{J}(\boldsymbol{r}') \times \nabla\left(\frac{1}{|\boldsymbol{r} - \boldsymbol{r}'|}\right). \tag{4.25}$$

If we substitute (4.25) in (4.22) and remove the ∇ operator outside the integral sign we obtain

$$\boldsymbol{B}(\boldsymbol{r}) = \nabla \times \int_V \frac{\mu\boldsymbol{J}(\boldsymbol{r}')\,dV'}{4\pi\,|\boldsymbol{r} - \boldsymbol{r}'|}. \tag{4.26}$$

Example. Find the magnetic flux density $\boldsymbol{B}(\boldsymbol{r})$ due to an infinite line current (I_0 A/m) situated along the z-axis.

In view of the translational symmetry in the z-direction, the magnetic flux density $\boldsymbol{B}(\boldsymbol{r})$ is the same at the corresponding points in all the planes transverse to the line current. Therefore, it is sufficient for us to evaluate $\boldsymbol{B}(\boldsymbol{r})$ in one transverse plane. Let the coordinates of the observation point P, as indicated in Fig. 4.16, be given by ρ, ϕ, and 0. Let us consider a current element at P' situated at a distance z from the origin. If its length is dz, we obtain the source current element as $I_0\,dz\hat{\boldsymbol{z}}$. We see

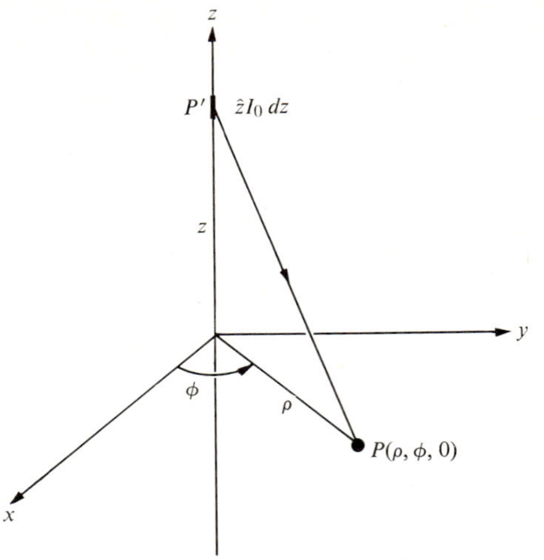

Fig. 4.16 Magnetic flux density $\boldsymbol{B}(\boldsymbol{r})$ due to an infinite line current.

from (4.20) that the contribution to the magnetic flux density at P due to the current element at P' is

$$d\boldsymbol{B}(\boldsymbol{r}) = \frac{\mu I_0\, dz\hat{z} \times [-z\hat{z} + \rho\hat{\rho}]}{4\pi r_{P'P}^3} = \frac{\hat{\phi}\mu I_0 \rho\, dz}{4\pi[\rho^2 + z^2]^{3/2}}.$$

We obtain the magnetic flux density $\boldsymbol{B}(\boldsymbol{r})$ at P due to the infinite line current by integrating $d\boldsymbol{B}(\boldsymbol{r})$ with respect to z from $z = -\infty$ to $z = \infty$. Therefore we find that

$$\boldsymbol{B}(\boldsymbol{r}) = \frac{\hat{\phi}\mu I_0 \rho}{4\pi} \int_{-\infty}^{\infty} \frac{dz}{(\rho^2 + z^2)^{3/2}}.$$

We can evaluate this integral by setting $z = \rho \tan\theta$ to yield the following result:

$$\boldsymbol{B}(\boldsymbol{r}) = \hat{\phi}\mu I_0/2\pi\rho.$$

4. Vector Potential

The form of (4.26) suggests the introduction of an auxiliary vector function $\boldsymbol{A}(\boldsymbol{r})$ as given by

$$\boldsymbol{A}(\boldsymbol{r}) = \int_V \frac{\mu \boldsymbol{J}(\boldsymbol{r}')\, dV'}{4\pi\,|\boldsymbol{r} - \boldsymbol{r}'|}, \tag{4.27}$$

in terms of which we obtain $\boldsymbol{B}(\boldsymbol{r})$ through the relation

$$\boldsymbol{B}(\boldsymbol{r}) = \nabla \times \boldsymbol{A}(\boldsymbol{r}). \tag{4.28}$$

The auxiliary vector function is called the vector potential and is helpful in the determination of the $\boldsymbol{B}$ field due to a given current distribution just as the scalar potential is useful in the evaluation of the electric field due to a given charge distribution.

It is helpful for us to determine the differential equation satisfied by the vector potential. For this purpose, we separate (4.27) into its Cartesian components as follows:

$$A_i(\boldsymbol{r}) = \int_V \frac{\mu J_i(\boldsymbol{r}')\, dV'}{4\pi\, |\boldsymbol{r} - \boldsymbol{r}'|} \qquad (i = x, y, \text{ or } z). \tag{4.29}$$

If we examine (2.58) and its solution (2.30) we see that $A_i(\boldsymbol{r})$ should satisfy the following differential equation:

$$\nabla^2 A_i(\boldsymbol{r}) = -\mu J_i(\boldsymbol{r}) \qquad (i = x, y, \text{ or } z). \tag{4.30}$$

If we then add the three Cartesian components given by (4.30) we get

$$\hat{x}\, \nabla^2 A_x(\boldsymbol{r}) + \hat{y}\, \nabla^2 A_y(\boldsymbol{r}) + \hat{z}\, \nabla^2 A_z(\boldsymbol{r}) = -\mu \boldsymbol{J}(\boldsymbol{r}). \tag{4.31}$$

Since the unit vectors $\hat{x}$, $\hat{y}$, and $\hat{z}$ do not change with the change in the position coordinates x, y, and z, we find that

$$\begin{aligned} \hat{x}\, \nabla^2 A_x(\boldsymbol{r}) + \hat{y}\, \nabla^2 A_y(\boldsymbol{r}) + \hat{z}\, \nabla^2 A_z(\boldsymbol{r}) &= \nabla^2[\hat{x} A_x(\boldsymbol{r}) + \hat{y} A_y(\boldsymbol{r}) + \hat{z} A_z(\boldsymbol{r})] \\ &= \nabla^2 \boldsymbol{A}(\boldsymbol{r}). \end{aligned} \tag{4.32}$$

From (4.31) and (4.32), we see that the vector potential $\boldsymbol{A}(\boldsymbol{r})$ satisfies the following differential equation:

$$\nabla^2 \boldsymbol{A}(\boldsymbol{r}) = -\mu \boldsymbol{J}(\boldsymbol{r}). \tag{4.33}$$

With the help of (4.33), it is possible for us to find the differential equation satisfied by $\boldsymbol{B}(\boldsymbol{r})$. For that purpose, it is first necessary for us to establish the following relation:

$$\nabla \cdot \boldsymbol{A}(\boldsymbol{r}) = \nabla \cdot \int_V \frac{\mu \boldsymbol{J}(\boldsymbol{r}')\, dV'}{4\pi\, |\boldsymbol{r} - \boldsymbol{r}'|} = 0. \tag{4.34}$$

We can take the divergence operator inside the integral sign since it is an operation involving only the unprimed coordinates. Also, in view of the vector identity

$$\nabla \cdot [\phi \boldsymbol{A}] = \boldsymbol{A} \cdot \nabla \phi + \phi\, \nabla \cdot \boldsymbol{A}, \tag{4.35}$$

where ϕ is a scalar and $\boldsymbol{A}$ is a vector, it follows that

$$\begin{aligned} \nabla \cdot \left[\frac{\boldsymbol{J}(\boldsymbol{r}')}{|\boldsymbol{r} - \boldsymbol{r}'|}\right] &= \boldsymbol{J}(\boldsymbol{r}') \cdot \nabla\left(\frac{1}{|\boldsymbol{r} - \boldsymbol{r}'|}\right) + \frac{1}{|\boldsymbol{r} - \boldsymbol{r}'|} \nabla \cdot \boldsymbol{J}(\boldsymbol{r}') \\ &= \boldsymbol{J}(\boldsymbol{r}') \cdot \nabla\left(\frac{1}{|\boldsymbol{r} - \boldsymbol{r}'|}\right) = -\boldsymbol{J}(\boldsymbol{r}') \cdot \nabla'\left(\frac{1}{|\boldsymbol{r} - \boldsymbol{r}'|}\right). \end{aligned} \tag{4.36}$$

In (4.36), ∇' is an operation with respect to the primed coordinates. In the same manner,

$$\begin{aligned} \nabla' \cdot \left[\frac{\boldsymbol{J}(\boldsymbol{r}')}{|\boldsymbol{r} - \boldsymbol{r}'|}\right] &= \boldsymbol{J}(\boldsymbol{r}') \cdot \nabla'\left(\frac{1}{|\boldsymbol{r} - \boldsymbol{r}'|}\right) + \frac{1}{|\boldsymbol{r} - \boldsymbol{r}'|} \nabla' \cdot \boldsymbol{J}(\boldsymbol{r}') \\ &= \boldsymbol{J}(\boldsymbol{r}') \cdot \nabla'\left(\frac{1}{|\boldsymbol{r} - \boldsymbol{r}'|}\right). \end{aligned} \tag{4.37}$$

In (4.36), $\nabla \cdot \boldsymbol{J}(\boldsymbol{r}') = 0$ since $\boldsymbol{J}(\boldsymbol{r}')$ does not vary with respect to the unprimed coordinates and $\nabla' \cdot \boldsymbol{J}(\boldsymbol{r}') = 0$ in (4.37) in view of the equation of continuity (4.9). From (4.36) and (4.37), we see that

$$\nabla \cdot \left[\frac{\boldsymbol{J}(\boldsymbol{r}')}{|\boldsymbol{r} - \boldsymbol{r}'|}\right] = -\nabla' \cdot \left[\frac{\boldsymbol{J}(\boldsymbol{r}')}{|\boldsymbol{r} - \boldsymbol{r}'|}\right]. \tag{4.38}$$

Hence, (4.34) becomes

$$\nabla \cdot \boldsymbol{A}(\boldsymbol{r}) = -\frac{\mu}{4\pi}\int_V \nabla' \cdot \left[\frac{\boldsymbol{J}(\boldsymbol{r}')}{|\boldsymbol{r} - \boldsymbol{r}'|}\right] dV', \tag{4.39}$$

which we can simplify with the help of the divergence theorem to yield

$$\nabla \cdot \boldsymbol{A}(\boldsymbol{r}) = -\frac{\mu}{4\pi}\int_S \frac{\boldsymbol{J}(\boldsymbol{r}')}{|\boldsymbol{r} - \boldsymbol{r}'|} \cdot d\boldsymbol{S}' = 0. \tag{4.40}$$

The surface S in (4.40) completely encloses the volume V which contains entirely all the current distribution. Therefore, there can be no current directed through the surface S and hence, the integral in (4.40) vanishes as indicated.

Since

$$\nabla \times \nabla \times \boldsymbol{A}(\boldsymbol{r}) = \nabla\nabla \cdot \boldsymbol{A}(\boldsymbol{r}) - \nabla^2 \boldsymbol{A}(\boldsymbol{r}), \tag{4.41}$$

we see from (4.33) and (4.40) that

$$\nabla \times \nabla \times \boldsymbol{A}(\boldsymbol{r}) = \mu \boldsymbol{J}(\boldsymbol{r}). \tag{4.42}$$

If we substitute (4.28) in (4.42) we get the following differential equation satisfied by $\boldsymbol{B}(\boldsymbol{r})$:

$$\nabla \times \boldsymbol{B}(\boldsymbol{r}) = \mu \boldsymbol{J}(\boldsymbol{r}). \tag{4.43}$$

5. Magnetic Field

In electrostatics, we derived a field quantity $\boldsymbol{D}(\boldsymbol{r})$ in terms of the fundamental field quantity $\boldsymbol{E}(\boldsymbol{r})$ in order to have a field vector which for a given charge distribution did not show the dependence on the medium parameter, that is, the permittivity ε. In a similar manner, in magnetostatics, it is convenient for us to introduce a field vector $\boldsymbol{H}(\boldsymbol{r})$, known as the magnetic field, which is related to the fundamental field vector $\boldsymbol{B}(\boldsymbol{r})$ and which for a given current distribution does not show the dependence on the medium parameter μ such as for example in the differential equation relating it to the source current. By definition, $\boldsymbol{H}(\boldsymbol{r})$ is related to $\boldsymbol{B}(\boldsymbol{r})$ by the relation

$$\boldsymbol{B}(\boldsymbol{r}) = \mu \boldsymbol{H}(\boldsymbol{r}). \tag{4.44}$$

From (4.43) and (4.44), we find that

$$\nabla \times \boldsymbol{H}(\boldsymbol{r}) = \boldsymbol{J}(\boldsymbol{r}). \tag{4.45}$$

Since the validity of (4.34) is based on the truth of (4.9), we can assert that (4.45) holds good only for direct currents for which (4.9) is valid. In fact, that (4.9) is implied by (4.45) is easily verified by taking the divergence of both sides of (4.45) and noting that the divergence of the curl of any vector is zero. It is helpful for us to anticipate that (4.9) is not valid for time-varying currents and hence we have to modify (4.45) when the current and the fields vary with time. Also, we see from (4.45) that the unit of the magnetic field $\boldsymbol{H}(\boldsymbol{r})$ is amperes per meter.

6. Vector Potential Due to a Magnetic Dipole

Let us consider a filamentary loop of radius a, with its center coinciding with the origin of the coordinate system. Let I_0 be the current in the loop. It is convenient for us to use both the rectangular coordinates x, y, and z, and the spherical coordinates r, θ, and ϕ, as shown in Fig. 4.17(a). Let the axis of the loop coincide with the z-axis. Since it is independent of ϕ, without loss of generality, we can evaluate the vector potential $\boldsymbol{A}(\boldsymbol{r})$ for $\phi = 0$. Let the coordinates of the observation point P which is assumed to lie in the zx-plane be r, θ, and 0. Let us consider an infinitesimal length of the filamentary loop lying between the angles ϕ' and $\phi' + d\phi'$. The length of the current element is $a\,d\phi'$. Let $\hat{\phi}'$ be the direction of the current element. We see from Fig. 4.17(b) that

$$\hat{\phi}' = -\hat{x}\sin\phi' + \hat{y}\cos\phi' = -\hat{x}\sin\phi' + \hat{\phi}\cos\phi', \tag{4.46}$$

since for $\phi = 0$, $\hat{\phi} = \hat{y}$. Let us denote the location of the current element by P'. The coordinates of P' are a, $\pi/2$, and ϕ'.

We obtain the contribution $d\boldsymbol{A}(\boldsymbol{r})$ to the vector potential at P due to the current element at P' from (4.27) as

$$d\boldsymbol{A}(\boldsymbol{r}) = \hat{\phi}'\frac{\mu I_0 a\,d\phi'}{4\pi R(\phi')}, \tag{4.47}$$

where $R(\phi')$ is the distance between the source point P' and the observation point P. In (4.47), $d\boldsymbol{A}(\boldsymbol{r})$ is in the same direction as the current element at P'. We obtain the

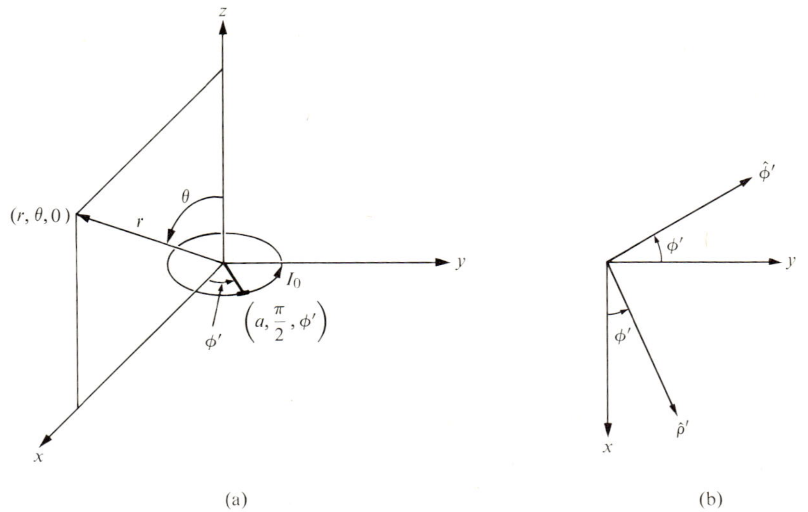

Fig. 4.17 Geometry of a filamentary current loop of radius a, indicating the source and the observation points.

vector potential $\boldsymbol{A}(\boldsymbol{r})$ at P due to the entire filamentary current loop by integrating (4.47) with respect to ϕ' from $\phi' = 0$ to $\phi' = 2\pi$ with the following result:

$$\boldsymbol{A}(\boldsymbol{r}) = \frac{\mu I_0 a}{4\pi} \int_{\phi'=0}^{2\pi} \frac{\hat{\phi}'\, d\phi'}{R(\phi')}, \tag{4.48}$$

where, since it is a function of ϕ', $\hat{\phi}'$ has been retained inside the integral sign. In view of (4.46), we can separate (4.48) as

$$\boldsymbol{A}(\boldsymbol{r}) = \hat{x}A_x(\boldsymbol{r}) + \hat{\phi}A_\phi(\boldsymbol{r}), \tag{4.49}$$

where,

$$A_x(\boldsymbol{r}) = -\frac{\mu I_0 a}{4\pi} \int_0^{2\pi} \frac{d\phi' \sin \phi'}{R(\phi')}, \tag{4.50}$$

and

$$A_\phi(\boldsymbol{r}) = \frac{\mu I_0 a}{4\pi} \int_0^{2\pi} \frac{d\phi' \cos \phi'}{R(\phi')}. \tag{4.51}$$

From Fig. 4.17(a), we see the rectangular coordinates of the observation point P to be $x = r \sin \theta$, $y = 0$, and $z = r \cos \theta$. Similarly, we find the rectangular coordinates of the source point P' to be $x' = a \cos \phi'$, $y' = a \sin \phi'$, and $z' = 0$. We can then evaluate the distance between P and P' as

$$\begin{aligned} R(\phi') &= [(x - x')^2 + (y - y')^2 + (z - z')^2]^{1/2}, \\ &= [(r \sin \theta - a \cos \phi')^2 + a^2 \sin^2 \phi' + r^2 \cos^2 \theta]^{1/2}, \\ &= [r^2 + a^2 - 2ar \sin \theta \cos \phi']^{1/2}. \end{aligned} \tag{4.52}$$

We can verify that $R(2\pi - \phi') = R(\phi')$.

Let us rewrite (4.50) as

$$A_x(\boldsymbol{r}) = -\frac{\mu I_0 a}{4\pi}\left[\int_0^{\pi} \frac{d\phi' \sin \phi'}{R(\phi')} + \int_{\pi}^{2\pi} \frac{d\phi' \sin \phi'}{R(\phi')}\right]. \tag{4.53}$$

The variable of integration of the second integral is changed from ϕ' to $2\pi - \phi'$. Neither $d\phi' \sin \phi'$ nor $R(\phi')$ is changed. But the limits of integration are changed from $(\pi, 2\pi)$ to $(\pi, 0)$. Hence, (4.53) becomes

$$A_x(\boldsymbol{r}) = -\frac{\mu I_0 a}{4\pi}\left[\int_0^{\pi} \frac{d\phi' \sin \phi'}{R(\phi')} + \int_{\pi}^{0} \frac{d\phi' \sin \phi'}{R(\phi')}\right] = 0, \tag{4.54}$$

with the result that $\boldsymbol{A}(\boldsymbol{r})$ due to a filamentary loop of current has only a ϕ-component as given by (4.51).

Suppose that the radius of the loop is very small in comparison with the distance to the observation point, that is, $a/r \ll 1$. Then on writing $1/R(\phi')$ as

$$\frac{1}{R(\phi')} = \frac{1}{r}\left[1 - \frac{2a}{r} \sin \theta \cos \phi' + \frac{a^2}{r^2}\right]^{-1/2}$$

and expanding the square root term into a binomial series in powers of a/r, we obtain the following result if only the first two terms are retained:

$$\frac{1}{R(\phi')} = \frac{1}{r}\left[1 + \frac{a}{r}\sin\theta\cos\phi'\right]. \tag{4.55}$$

Together with (4.55), (4.51) becomes

$$A_\phi(\boldsymbol{r}) = \frac{\mu I_0 a}{4\pi r}\int_0^{2\pi} d\phi'\left[\cos\phi' + \frac{a}{r}\sin\theta\cos^2\phi'\right]. \tag{4.56}$$

The integral of the first term within the square brackets in (4.56) vanishes and that of the second term yields $(\pi a/r)\sin\theta$ with the result that (4.56) reduces to

$$A_\phi(\boldsymbol{r}) = (\mu I_0/4\pi r^2)\pi a^2 \sin\theta. \tag{4.57}$$

We can express (4.57) conveniently in terms of the magnetic moment $\boldsymbol{m}$ of the current loop. We define the magnitude m of the magnetic moment $\boldsymbol{m}$ as the product of the current and the area of the loop, that is, $m = I_0\pi a^2$. We take the direction $\hat{m}$ of $\boldsymbol{m}$ to be normal to the plane of the loop. By convention, if the fingers of the right hand curl in the direction of the current, we assume that $\hat{m}$ is in the direction of the thumb. Therefore, for the current loop of Fig. 4.17(a),

$$\boldsymbol{m} = \hat{z}I_0\pi a^2. \tag{4.58}$$

Since $\hat{\phi}\sin\theta = \hat{z}\times\hat{r}$ where $\hat{r}$ is a unit vector in the direction of the radius vector, we can verify from (4.49), (4.54), and (4.57) that $\boldsymbol{A}(\boldsymbol{r})$ can be expressed as follows:

$$\boldsymbol{A}(\boldsymbol{r}) = \frac{\mu\boldsymbol{m}\times\hat{r}}{4\pi r^2} = -\frac{\mu\boldsymbol{m}}{4\pi}\times\nabla\left(\frac{1}{r}\right). \tag{4.59}$$

A small current loop is called a magnetic dipole. It is characterized by a magnetic dipole moment $\boldsymbol{m}$. In (4.59), we obtain the vector potential due to a magnetic dipole of moment $\boldsymbol{m}$ in a form which is independent of the coordinate system. It is useful for us to express (4.59) in a somewhat more general form in the following manner. Let the magnetic dipole of moment $\boldsymbol{m}$ be located at P' denoted by the position vector $\boldsymbol{r}'$ and let us find the vector potential at the observation point P denoted by the position vector $\boldsymbol{r}$. Then, it follows from (4.59) that

$$\boldsymbol{A}(\boldsymbol{r}) = -\frac{\mu}{4\pi}\boldsymbol{m}\times\nabla(1/|\boldsymbol{r}-\boldsymbol{r}'|). \tag{4.60}$$

In view of (4.28) and (4.44), we can express the magnetic field $\boldsymbol{H}(\boldsymbol{r})$ in terms of the vector potential as

$$\boldsymbol{H}(\boldsymbol{r}) = \frac{1}{\mu}\nabla\times\boldsymbol{A}(\boldsymbol{r}). \tag{4.61}$$

When the vector potential has only a single component as given by (4.57), we can determine the corresponding spherical components of the magnetic field with the help of the spherical components of the curl operator and (4.57). We then obtain the following results:

$$H_r(\boldsymbol{r}) = \frac{1}{\mu r \sin\theta}\frac{\partial}{\partial\theta}(\sin\theta A_\phi(\boldsymbol{r})) = \frac{I_0\pi a^2}{2\pi r^3}\cos\theta, \tag{4.62a}$$

$$H_\theta(\boldsymbol{r}) = -\frac{1}{\mu r}\frac{\partial}{\partial r}(rA_\phi(\boldsymbol{r})) = \frac{I_0\pi a^2}{4\pi r^3}\sin\theta, \tag{4.62b}$$

$$H_\phi(\boldsymbol{r}) = 0. \tag{4.62c}$$

If we compare (4.62) with the fields of an electric dipole as given by (2.23), we see the remarkable similarity between the electric fields of a point electric dipole and the magnetic fields due to a small current loop. It is for this reason that a small current loop is called a magnetic dipole.

We shall determine the vector potential due to two infinite parallel wires carrying equal and opposite currents in detail and discuss its usefulness in deducing the inductance per unit length of a two-wire transmission line in Section 7.3.

7. Magnetostatic Analogue of Gauss's Law

On taking the divergence of both the sides of (4.28), we obtain

$$\nabla \cdot \boldsymbol{B}(\boldsymbol{r}) = 0, \tag{4.63}$$

which is the magnetostatic analogue of Gauss's law. Since $\boldsymbol{D}(\boldsymbol{r})$ and $\boldsymbol{B}(\boldsymbol{r})$ are the electric and magnetic flux densities respectively, if we compare (4.63) with (2.56) we find that there are no isolated magnetic charges. Integrating (4.63) throughout a volume V enclosed by the surface S and using the divergence theorem, we get

$$\int_V \nabla \cdot \boldsymbol{B}(\boldsymbol{r})\, dV = \int_S \boldsymbol{B}(\boldsymbol{r}) \cdot d\boldsymbol{S} = 0, \tag{4.64}$$

which shows that the net outward magnetic flux through a closed surface is zero. Thus, there are no sources of magnetic flux corresponding to the electric charges that produce the electric flux.

8. Ampère's Circuital Law

We derive the circuital law of Ampère from (4.45). In view of its usefulness in subsequent developments, it is advantageous for us to deduce (4.45) directly from (4.26) without having recourse to any analogies with the laws of electrostatics. From (4.28), (4.41), and (4.34) we find that

$$\nabla \times \boldsymbol{B}(\boldsymbol{r}) = \nabla \times \nabla \times \boldsymbol{A}(\boldsymbol{r}) = \nabla\nabla \cdot \boldsymbol{A}(\boldsymbol{r}) - \nabla^2\boldsymbol{A}(\boldsymbol{r}) = -\nabla^2\boldsymbol{A}(\boldsymbol{r}). \tag{4.65}$$

Together with (4.27), (4.65) becomes

$$\nabla \times \boldsymbol{B}(\boldsymbol{r}) = -\nabla^2\boldsymbol{A}(\boldsymbol{r}) = -\frac{\mu}{4\pi}\int_V \boldsymbol{J}(\boldsymbol{r}')\nabla^2\left(\frac{1}{|\boldsymbol{r} - \boldsymbol{r}'|}\right) dV'. \tag{4.66}$$

By differentiating explicitly, we can verify that $\nabla^2(1/|\boldsymbol{r} - \boldsymbol{r}'|) = 0$ if $\boldsymbol{r} \neq \boldsymbol{r}'$, that is,

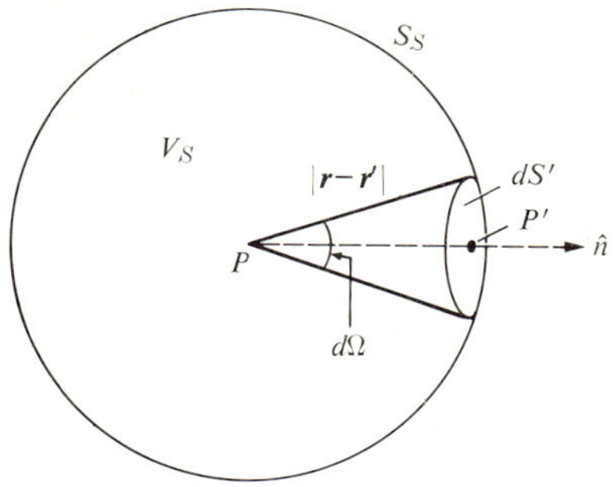

Fig. 4.18 Spherical region of integration around the observation point P when it is inside the source current region.

if the observation point specified by $\boldsymbol{r}$ is outside the volume V containing the source current distribution. Thus, $\nabla \times \boldsymbol{B}(\boldsymbol{r}) = -\nabla^2 \boldsymbol{A}(\boldsymbol{r}) = 0$ outside the source current region. When the observation point is inside the source current region, then it is possible for $\boldsymbol{r}'$ to become equal to $\boldsymbol{r}$ in the process of integration. Therefore, the integrand of (4.66) has a singularity at $\boldsymbol{r} = \boldsymbol{r}'$. We establish now that this singularity can be integrated to yield a finite result. The integrand in (4.66) is zero except in the close neighborhood of $\boldsymbol{r}' = \boldsymbol{r}$. Hence, it is necessary for us to perform the integration in (4.66) only throughout a very small spherical volume V_s centered at the observation point specified by $\boldsymbol{r}$, as shown in Fig. 4.18. Assuming that $\boldsymbol{J}(\boldsymbol{r}')$ is a smooth function, it is legitimate for us to set $\boldsymbol{J}(\boldsymbol{r}') = \boldsymbol{J}(\boldsymbol{r})$ everywhere inside the very small spherical volume V_s, since the radius of the spherical volume V_s can be made as small as desired. Therefore, we can rewrite (4.66) as

$$\nabla \times \boldsymbol{B}(\boldsymbol{r}) = -\nabla^2 \boldsymbol{A}(\boldsymbol{r}) = -\frac{\mu}{4\pi} \boldsymbol{J}(\boldsymbol{r}) \int_{V_s} \nabla^2 \left(\frac{1}{|\boldsymbol{r} - \boldsymbol{r}'|} \right) dV'. \tag{4.67}$$

Since

$$|\boldsymbol{r} - \boldsymbol{r}'| = [(x - x')^2 + (y - y')^2 + (z - z')^2]^{1/2},$$

we obtain the results

$$\nabla \left(\frac{1}{|\boldsymbol{r} - \boldsymbol{r}'|} \right) = -\nabla' \left(\frac{1}{|\boldsymbol{r} - \boldsymbol{r}'|} \right); \qquad \nabla \cdot \nabla \left(\frac{1}{|\boldsymbol{r} - \boldsymbol{r}'|} \right) = \nabla' \cdot \nabla' \left(\frac{1}{|\boldsymbol{r} - \boldsymbol{r}'|} \right).$$

Hence,

$$\int_{V_s} \nabla^2 \left(\frac{1}{|\boldsymbol{r} - \boldsymbol{r}'|} \right) dV' = \int_{V_s} \nabla' \cdot \left[\nabla' \left(\frac{1}{|\boldsymbol{r} - \boldsymbol{r}'|} \right) \right] dV'. \tag{4.68}$$

The use of the divergence theorem and the fact that

$$\nabla' \left(\frac{1}{|\boldsymbol{r} - \boldsymbol{r}'|} \right) = -\nabla \left(\frac{1}{|\boldsymbol{r} - \boldsymbol{r}'|} \right) = \frac{\hat{\boldsymbol{r}}_{P'P}}{|\boldsymbol{r} - \boldsymbol{r}'|^2}$$

enable us to simplify (4.68) as

$$\int_{V_s} \nabla^2\left(\frac{1}{|\boldsymbol{r} - \boldsymbol{r}'|}\right) dV' = \int_{S_s} \frac{\hat{r}_{P'P}}{|\boldsymbol{r} - \boldsymbol{r}'|^2} \cdot \hat{n}\, dS'. \tag{4.69}$$

Since $\hat{n}$ is an outwardly drawn unit normal to the surface S_s at P', it follows that $\hat{r}_{P'P} \cdot \hat{n} = -1$. Also, $dS'/|\boldsymbol{r} - \boldsymbol{r}'|^2 = d\Omega$, the solid angle subtended by the elemental area dS' at the center P of the sphere. The total solid angle subtended by the spherical surface S_s at the center P is equal to 4π and therefore, from (4.69) we obtain

$$\int_{V_s} \nabla^2\left(\frac{1}{|\boldsymbol{r} - \boldsymbol{r}'|}\right) dV' = -4\pi. \tag{4.70}$$

The substitution of (4.70) in (4.67) gives

$$\nabla \times \boldsymbol{B}(\boldsymbol{r}) = -\nabla^2 \boldsymbol{A}(\boldsymbol{r}) = \mu \boldsymbol{J}(\boldsymbol{r}), \tag{4.71}$$

which agrees with (4.33) and (4.43) whose derivation had made use of certain comparisons with the laws of electrostatics. The differential equation (4.45) satisfied by $\boldsymbol{H}(\boldsymbol{r})$ follows from (4.71) and (4.44). We have thus accomplished the alternative derivation of (4.45).

If we integrate (4.45) throughout a surface S bounded by the curve C and use Stokes' theorem we get

$$\int_S \nabla \times \boldsymbol{H}(\boldsymbol{r}) \cdot d\boldsymbol{S} = \int_C \boldsymbol{H}(\boldsymbol{r}) \cdot d\boldsymbol{l} = \int_S \boldsymbol{J}(\boldsymbol{r}) \cdot d\boldsymbol{S}. \tag{4.72}$$

In words, (4.72) states that the line integral of the magnetic field taken around a closed path is equal to the current enclosed. This is Ampère's circuital law. The derivation of (4.72) is based on the truth of (4.9). For the fields and the current densities which vary with time, (4.9) is not valid and (4.72) requires modification.

9. Magnetic Flux

There is another circuital law which is very useful in the evaluation of the magnetic flux Φ. We obtain the magnetic flux Φ through a surface S by the integration of the magnetic flux density $\boldsymbol{B}(\boldsymbol{r})$ over the entire surface. Thus,

$$\Phi = \int_S \boldsymbol{B}(\boldsymbol{r}) \cdot d\boldsymbol{S}. \tag{4.73}$$

Substituting (4.28) in (4.73) and using Stokes' theorem we find that

$$\Phi = \int_S \nabla \times \boldsymbol{A}(\boldsymbol{r}) \cdot d\boldsymbol{S} = \int_C \boldsymbol{A}(\boldsymbol{r}) \cdot d\boldsymbol{l}, \tag{4.74}$$

where C is a closed curve bounding the surface S. Therefore, the line integral of the vector potential taken around a closed path gives the magnetic flux enclosed by the path.

It is instructive for us to collect together the two differential equations (4.43) and (4.63) satisfied by the magnetic flux density $\boldsymbol{B}(\boldsymbol{r})$ as follows:

$$\nabla \times \boldsymbol{B}(\boldsymbol{r}) = \mu \boldsymbol{J}(\boldsymbol{r}), \qquad \nabla \cdot \boldsymbol{B}(\boldsymbol{r}) = 0, \tag{4.75}$$

and compare them with the corresponding equation (2.60) satisfied by the electric field $\boldsymbol{E}(\boldsymbol{r})$. A vector function is completely specified if its curl and divergence are known. Thus, $\boldsymbol{E}(\boldsymbol{r})$ and $\boldsymbol{B}(\boldsymbol{r})$ are uniquely specified by (2.60) and (4.75) respectively. A vector field which has a zero curl is called *irrotational* and that which has zero divergence is called *solenoidal.* Therefore, the electrostatic field is irrotational and the magnetostatic field is solenoidal.

10. Boundary Conditions

The differential equations (4.45) and (4.63) governing the magnetostatic field are not valid at the interface between two different media. However, the integral forms (4.72) and (4.64) of the governing equations of the magnetostatic field are valid even near the interfaces and therefore can be used for the derivation of the boundary conditions. As in electrostatics, we postpone the details of the derivation of the boundary conditions and present them later, in connection with the general electromagnetic field quantities. Here, we give only their statements without proof.

Let us consider a point on the boundary between two different media. Let us choose the origin of the Cartesian coordinate system x, y, and z to coincide with this point and let the z-axis be normal to the boundary. The coordinate system is oriented to have its y-axis parallel to the component of the magnetic field tangential to the interface. The regions 1 ($z > 0$) and 2 ($z < 0$) are characterized by the permeabilities μ_1 and μ_2 respectively. We may show the application of (4.72) around an infinitesimal rectangular path surrounding the origin and lying in the yz-plane to yield

$$H_{y2} - H_{y1} = J_{sx}, \tag{4.76}$$

where J_{sx} is the surface current density in amperes per meter existing at the interface in the x-direction. In (4.76), the subscripts 1 and 2 indicate the regions corresponding to the fields at the interface. In words, (4.76) states that the tangential components of the magnetic field at the boundary between two different media is discontinuous by the amount of the surface current density that might exist on the boundary.

In a similar manner, we may prove by the application of the integral form of Gauss's law given by (4.64) to an infinitesimal pillbox centered around the origin that

$$B_{z1} = B_{z2}, \quad \text{that is,} \quad B_{n1} = B_{n2}, \tag{4.77}$$

where n denotes the component normal to the interface. The normal components of the magnetic flux densities are continuous across any interface separating two different media.

4.5 MAGNETIZATION AND THE CONCEPT OF PERMEABILITY

1. Magnetization

We shall now treat the introductory aspects of the microscopic theory of matter and develop the concept of permeability which describes the macroscopic behavior of materials in the presence of magnetic fields. In view of the presence of charges in motion, atoms have magnetic moments. Some atoms have permanent magnetic moments and the others have magnetic moments induced in them by applied magnetic fields.

When a magnetic field is applied to an atom possessing a permanent magnetic moment, there is a resultant torque which tends to align the magnetic moment of the atom with the external magnetic field whose value is thus enhanced by the magnetic field due to the atom. Materials whose atomic behavior increases the applied magnetic field are called *paramagnetic*. The increasing effect is usually small except in certain class of materials known as *ferromagnets* in which the increase is significantly large.

When a magnetic field is applied to an atom which does not possess a permanent magnetic moment, a magnetic moment is induced so as to oppose the external magnetic field. Such materials whose atomic behavior tends to decrease the applied magnetic field are called *diamagnetic*.

The magnetic moment per unit volume is known as *magnetization* $\boldsymbol{M}(\boldsymbol{r})$. Except for ferromagnets and for sufficiently small magnetic fields, the magnetization is proportional to the magnetic field, that is,

$$\boldsymbol{M}(\boldsymbol{r}) = \chi_m \boldsymbol{H}(\boldsymbol{r}), \tag{5.1}$$

where χ_m is known as the magnetic susceptibility which is positive or negative according to whether the material is paramagnetic or diamagnetic.

2. Equivalent Volume and Surface Current Densities

Let us consider a volume V of a material enclosed by the surface S as shown in Fig. 4.19. We can establish that the macroscopic effect of magnetization is equivalent to a distribution of currents throughout the volume V and the bounding surface S of the material.

Let us consider an elementary volume dV' of the material centered around the source point P' whose position vector is given by $\boldsymbol{r}'$. We wish to find the vector potential at an observation point P which is exterior to the volume V and which is denoted by the position vector $\boldsymbol{r}$. The magnetic moment of the elementary volume dV' of the material is given by $\boldsymbol{M}(\boldsymbol{r}')\, dV'$. We obtain from (4.60) the contribution to the vector potential at the observation point P due to the magnetic moment associated

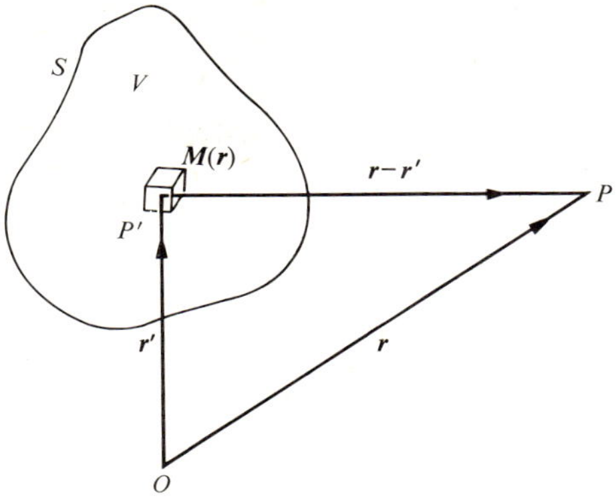

Fig. 4.19 Volume distribution of magnetization $\boldsymbol{M}(\boldsymbol{r})$.

with the elementary volume dV' as

$$d\boldsymbol{A}(\boldsymbol{r}) = \frac{\mu}{4\pi}\boldsymbol{M}(\boldsymbol{r}')\,dV' \times \nabla'\left(\frac{1}{|\boldsymbol{r} - \boldsymbol{r}'|}\right). \tag{5.2}$$

We can determine the vector potential $\boldsymbol{A}(\boldsymbol{r})$ at the observation point P due to the magnetization of the entire volume by integrating (5.2) throughout the volume V with the following result:

$$\boldsymbol{A}(\boldsymbol{r}) = \frac{\mu}{4\pi}\int_V \boldsymbol{M}(\boldsymbol{r}') \times \nabla'\left(\frac{1}{|\boldsymbol{r} - \boldsymbol{r}'|}\right) dV'. \tag{5.3}$$

In view of the vector identity,

$$\boldsymbol{A} \times \nabla a = a\nabla \times \boldsymbol{A} - \nabla \times (a\boldsymbol{A}), \tag{5.4}$$

where a is a scalar and $\boldsymbol{A}$ is a vector, we may rewrite (5.3) as

$$\boldsymbol{A}(\boldsymbol{r}) = \frac{\mu}{4\pi}\left[\int_V \frac{\nabla' \times \boldsymbol{M}(\boldsymbol{r}')}{|\boldsymbol{r} - \boldsymbol{r}'|}\,dV' - \int_V \nabla' \times \left(\frac{\boldsymbol{M}(\boldsymbol{r}')}{|\boldsymbol{r} - \boldsymbol{r}'|}\right) dV'\right]. \tag{5.5}$$

On converting the second volume integral into a surface integral, we find that

$$\boldsymbol{A}(\boldsymbol{r}) = \frac{\mu}{4\pi}\left[\int_V \frac{\nabla' \times \boldsymbol{M}(\boldsymbol{r}')}{|\boldsymbol{r} - \boldsymbol{r}'|}\,dV' + \int_S \frac{\boldsymbol{M}(\boldsymbol{r}') \times \hat{n}}{|\boldsymbol{r} - \boldsymbol{r}'|}\,dS'\right], \tag{5.6}$$

where $\hat{n}$ is an outwardly drawn unit normal to the surface S. If we compare the form of (5.6) with that of (4.27) we find that the vector potential at the observation point P is the same as that due to the distributions of the following volume and surface current densities:

$$\boldsymbol{J}_b(\boldsymbol{r}) = \nabla \times \boldsymbol{M}(\boldsymbol{r}), \tag{5.7}$$

$$\boldsymbol{J}_{bs}(\boldsymbol{r}) = \boldsymbol{M}(\boldsymbol{r}) \times \hat{n}. \tag{5.8}$$

In so far as their macroscopic effects are concerned, we can replace a volume distribution of magnetization induced in a material by a magnetic field by the equivalent distributions of volume and surface current densities given by (5.7) and (5.8) respectively. It follows from (5.7) and (5.8) that equivalent surface currents appear on the boundary between different materials whereas equivalent volume currents appear only when the magnetization is inhomogeneous, which may arise either due to the inhomogeneity of the material or the inhomogeneity of the magnetic field.

3. The Permeability Concept

Let us consider a region to consist of a volume distribution of magnetization in free space. Then, the differential form (4.45) of Ampère's circuital law together with (4.44) for an interior point in the material becomes

$$\nabla \times \frac{\boldsymbol{B}(\boldsymbol{r})}{\mu_0} = \boldsymbol{J}(\boldsymbol{r}) + \boldsymbol{J}_b(\boldsymbol{r}), \tag{5.9}$$

where $\boldsymbol{J}(\boldsymbol{r})$ is the current density and $\boldsymbol{J}_b(\boldsymbol{r})$ is the density of the volume distribution of the equivalent current due to the magnetization. Together with (5.7), we can rewrite (5.9) as

$$\nabla \times \left[\frac{\boldsymbol{B}(\boldsymbol{r})}{\mu_0} - \boldsymbol{M}(\boldsymbol{r})\right] = \boldsymbol{J}(\boldsymbol{r}), \tag{5.10}$$

which with the help of (4.45) enables us to define the magnetic field in a material medium as

$$\boldsymbol{H}(\boldsymbol{r}) = \frac{\boldsymbol{B}(\boldsymbol{r})}{\mu_0} - \boldsymbol{M}(\boldsymbol{r}) \qquad \text{or} \qquad \boldsymbol{B}(\boldsymbol{r}) = \mu_0[\boldsymbol{H}(\boldsymbol{r}) + \boldsymbol{M}(\boldsymbol{r})]. \tag{5.11a,b}$$

With the help of (5.1), we can simplify (5.11b) to

$$\boldsymbol{B}(\boldsymbol{r}) = \mu_0(1 + \chi_m)\boldsymbol{H}(\boldsymbol{r}) = \mu_0\mu_r\boldsymbol{H}(\boldsymbol{r}). \tag{5.12}$$

In a material medium in which the magnetization is linearly related to the magnetic field, it follows from (5.12) that we can identify $1 + \chi_m$ as the relative permeability μ_r and that in linear media the concepts of magnetization and permeability are equivalent.

Let us consider the interface between a magnetized material and free space as shown in Fig. 4.20. Let B and B_0 be the tangential components of the magnetic flux density at the interface on the material and the free space sides respectively. We may assume without loss of generality that there are no other components of magnetic flux density. The magnetization is in the same direction as the magnetic flux density, namely tangential to the interface as shown in Fig. 4.20; let its magnitude be M. Let $\hat{n}$ be the outwardly drawn unit normal to the surface of the material. It follows from (5.8) and Fig. 4.20 that the equivalent surface current density due to the magnetization is given by $\boldsymbol{J}_{bs} = \hat{x}M$. If there are no surface currents other than those due to the magnetization, then we can show that the application of Ampère's circuital law to

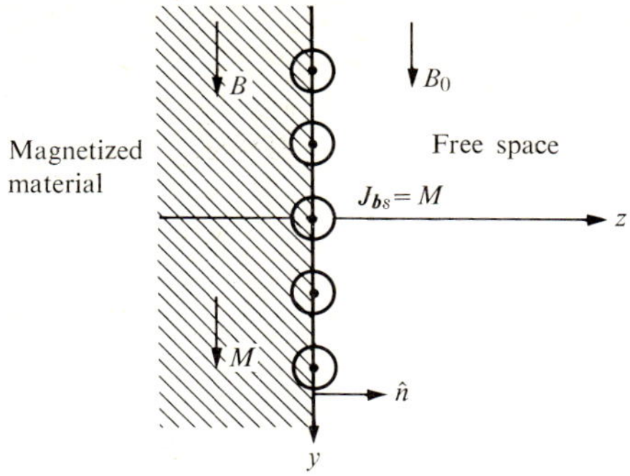

Fig. 4.20 The boundary between a magnetized material and free space.

an infinitesimal rectangular path in the yz-plane intersected by the interface yields

$$\frac{B}{\mu_0} - \frac{B_0}{\mu_0} = M \qquad \text{or} \qquad \left(\frac{B}{\mu_0} - M\right) - \frac{B_0}{\mu_0} = 0. \tag{5.13}$$

In view of (5.11) and (4.44), (5.13) becomes

$$H - H_0 = 0, \tag{5.14}$$

which states that the tangential component of the magnetic field is continuous across the bounding surface of the magnetic material if there are no free surface currents. The existence of the equivalent surface current due to the magnetization does not affect the boundary condition on the normal component of the magnetic field.

We see from (5.12) and (5.13) that the macroscopic effects of a volume distribution of magnetization are equivalent to introducing appropriate values for the permeability and this equivalence holds good inside as well as on the boundary of the material. We can describe the macroscopic behavior of a magnetic material either in terms of the permeability or the magnetization with its equivalent volume and surface currents. In linear materials, the two methods of description are identical and therefore we use the constitutive parameter μ alone.

4.6 MAXWELL'S EQUATIONS FOR TIME-VARYING FIELDS

We summarized the properties of the electric and the magnetic fields in the preceding four sections for the case in which these fields do not vary in time. There is interaction between the electric and the magnetic fields if they are functions of time. The expressions of interrelation between the electric and the magnetic fields are known as Maxwell's equations, whose development we shall present in this section.

1. Divergence Relations

Gauss's law in electrostatics and its magnetostatic analogue which we discussed previously are found to be valid even for the time-varying fields. We can express

Gauss's law for the electric flux density $\boldsymbol{D}(\boldsymbol{r}, t)$ either as the point relation as

$$\nabla \cdot \boldsymbol{D}(\boldsymbol{r}, t) = \rho(\boldsymbol{r}, t), \tag{6.1}$$

or in its large-scale form as

$$\int_S \boldsymbol{D}(\boldsymbol{r}, t) \cdot d\boldsymbol{S} = \int_V \rho(\boldsymbol{r}, t)\, dV. \tag{6.2}$$

We have now added the additional argument t to $\boldsymbol{D}$ and ρ to indicate their dependence on time. As before, the surface S completely encloses the volume V. The mathematical expressions (6.1) and (6.2) state that the net electric flux directed out of a surface completely enclosing a volume is equal to the total charge inside the volume. In a similar manner, we can express the magnetostatic analogue for the magnetic flux density $\boldsymbol{B}(\boldsymbol{r}, t)$ in either of the following two forms:

$$\nabla \cdot \boldsymbol{B}(\boldsymbol{r}, t) = 0, \tag{6.3}$$

or

$$\int_S \boldsymbol{B}(\boldsymbol{r}, t) \cdot d\boldsymbol{S} = 0. \tag{6.4}$$

In the mathematical expressions (6.3) and (6.4) we state that the net magnetic flux directed out of a surface completely enclosing a volume is equal to zero indicating the absence of free magnetic charges.

After the development of Maxwell's equations, we give an alternative interpretation to the divergence relations (6.1) and (6.3).

2. Equation of Continuity

The equation of continuity is the statement of conservation of the total charge. Let us consider a volume V containing a distribution of charge whose density $\rho(\boldsymbol{r}, t)$ varies with time. Let S be the surface completely enclosing the volume V, which we assume not to change with time. We obtain the rate of depletion of the total charge inside the volume V as $-(d/dt)\int_V \rho(\boldsymbol{r}, t)\, dV$ which is equal to $-\int_V (\partial\rho(\boldsymbol{r}, t)/\partial t)\, dV$ since the volume V is finite and does not change with time. The requirement of conservation of charge shows that $-\int_V (\partial\rho(\boldsymbol{r}, t)/\partial t)\, dV$ is the rate of outward flow of charge through the surface S or the outwardly directed current through the surface S. If $\boldsymbol{J}(\boldsymbol{r}, t)$ is the current density, then we obtain the outwardly directed current through the surface S as $\int_S \boldsymbol{J}(\boldsymbol{r}, t) \cdot \hat{n}\, dS$ where $\hat{n}$ is the outwardly directed unit normal to the surface S. On writing the surface integral in the form of an equivalent volume integral $\int_V \nabla \cdot \boldsymbol{J}(\boldsymbol{r}, t)\, dV$ with the help of the divergence theorem, the mathematical statement of the conservation of the total charge becomes

$$\int_V \nabla \cdot \boldsymbol{J}(\boldsymbol{r}, t)\, dV = -\int_V \frac{\partial}{\partial t} \rho(\boldsymbol{r}, t)\, dV. \tag{6.5}$$

Since (6.5) is valid for any arbitrarily small volume V, we obtain the following point relation for the statement of the conservation of the total charge:

$$\nabla \cdot \boldsymbol{J}(\boldsymbol{r}, t) = -\frac{\partial}{\partial t} \rho(\boldsymbol{r}, t). \tag{6.6}$$

Thus, the equation of continuity in the large-scale form and as the point relation are given by (6.5) and (6.6) respectively. When the charge density does not vary with time, we can reproduce the special case of the equation of continuity valid for direct currents as given by (4.9) from (6.6).

3. Faraday's Law of Induction and Maxwell's First Equation

Faraday discovered experimentally that if the magnetic flux linking a closed circuit is changed, an electromotive force (e.m.f.) is induced in that circuit and the induced e.m.f. is proportional to the rate of change of the flux linking the circuit. If the closed circuit is formed by a conducting wire, a current exists in the wire and this current is called the induced current. The principle of operation of electric generators is based on Faraday's law. The induced e.m.f. is defined as the line integral of the electric field $\boldsymbol{E}(\boldsymbol{r}, t)$ taken around the closed path C as given by $\int_C \boldsymbol{E}(\boldsymbol{r}, t) \cdot d\boldsymbol{l}$. The magnetic flux linking the closed curve C is given by $\int_S \boldsymbol{B}(\boldsymbol{r}, t) \cdot d\boldsymbol{S}$ where the surface S is bounded by the curve C. We should remember that $d\boldsymbol{S} = \hat{n}\, dS$ where, if the fingers on the right hand curl in the direction taken along the closed circuit for the evaluation of the e.m.f., the thumb points in the direction of the unit normal $\hat{n}$. The mathematical statement of Faraday's law becomes

$$\int_C \boldsymbol{E}(\boldsymbol{r}, t) \cdot d\boldsymbol{l} = -\frac{d}{dt} \int_S \boldsymbol{B}(\boldsymbol{r}, t) \cdot d\boldsymbol{S}. \tag{6.7}$$

The minus sign in (6.7) is due to the fact that if the magnetic flux is increased, the direction of the induced e.m.f. and therefore the induced current in the closed circuit is such that the magnetic field due to it is in a direction that opposes the increase of the magnetic flux linking the circuit. It is not necessary for the closed circuit to be formed by a conductor in order that an e.m.f. be induced in it. Any closed path in a region of changing magnetic flux has an e.m.f. induced in it. The magnetic flux linking a closed circuit can change with time either (1) due to the change of the magnetic flux density $\boldsymbol{B}(\boldsymbol{r}, t)$ with time, or (2) due to the motion of the closed circuit with time. The induced e.m.f. in the latter case is known as motional e.m.f.

Let us consider first the case in which the closed circuit C and the area S bounded by it are stationary. Then, we can write (6.7) as

$$\int_C \boldsymbol{E}(\boldsymbol{r}, t) \cdot d\boldsymbol{l} = -\int_S \frac{\partial}{\partial t} \boldsymbol{B}(\boldsymbol{r}, t) \cdot d\boldsymbol{S}, \tag{6.8}$$

which shows that a time-varying magnetic flux density $\boldsymbol{B}(\boldsymbol{r}, t)$ gives rise to an electric field $\boldsymbol{E}(\boldsymbol{r}, t)$. The mathematical expression (6.8) of the generalization of Faraday's induction law is the statement of Maxwell's first equation in its large-scale or integral form. The use of Stokes' theorem to convert the line integral in (6.8) into a surface integral yields

$$\int_S \nabla \times \boldsymbol{E}(\boldsymbol{r}, t) \cdot d\boldsymbol{S} = -\int_S \frac{\partial}{\partial t} \boldsymbol{B}(\boldsymbol{r}, t) \cdot d\boldsymbol{S}. \tag{6.9}$$

Since (6.9) is valid for any arbitrarily small closed path C and the associated arbitrarily small area S bounded by C, it follows that the integrands of (6.9) are equal, yielding the point relation

$$\nabla \times \boldsymbol{E}(\boldsymbol{r}, t) = -\frac{\partial}{\partial t} \boldsymbol{B}(\boldsymbol{r}, t), \tag{6.10}$$

which is the differential form of Maxwell's first equation. When $\boldsymbol{B}(\boldsymbol{r}, t)$ is independent of time, (6.10) reduces to (2.24) and therefore we see the limiting form of Faraday's induction law for the static case to be consistent with Coulomb's law.

Example. An electron is accelerated in the presence of an electric field. If the electric field is in the ϕ-direction, then the electron is accelerated in a circular path. The betatron is a device based on this principle and it utilizes the time-varying magnetic flux density in the axial (z) direction to produce the necessary electric field. Suppose that in the spatial region of interest the magnetic flux density is a function of ρ only as given by

$$\boldsymbol{B}(\boldsymbol{r}, t) = \hat{z}\rho t.$$

Then, the z-component of (6.10) together with the fact that there are no variations of the field components with the angular variable ϕ shows that the electric field has only a ϕ-component. The induced e.m.f. in a circular path of radius ρ is therefore obtained from (6.8) as

$$2\pi\rho E_\phi(\boldsymbol{r}) = -\int_0^\rho \rho 2\pi\rho \, d\rho = -2\pi \frac{\rho^3}{3}.$$

Hence,

$$\boldsymbol{E}(\boldsymbol{r}) = -\hat{\phi}\frac{\rho^2}{3},$$

which gives the magnitude and the direction of the electric field at the radial distance ρ.

4. Motional e.m.f.

Let us consider a closed curve C bounding an open surface S in a time-varying and nonuniform magnetic flux density $\boldsymbol{B}(\boldsymbol{r}, t)$. Let the closed curve C move with a uniform velocity $\boldsymbol{v}(\boldsymbol{r})$. We obtained the contribution to the induced e.m.f. due to the time variation of $\boldsymbol{B}(\boldsymbol{r}, t)$ in (6.8). The velocity $\boldsymbol{v}(\boldsymbol{r})$ of the different portions of the curve C need not be the same with the result that the curve C may change in shape as well as undergo translational and rotational motion. The surface integral in (6.8) may still be taken over the original area since we can show that the change in the contribution to the surface integral due to the change in the area with time is negligible.

We now wish to evaluate the contribution representing the motional electromotive force. With reference to Fig. 4.21, the vector element $d\boldsymbol{l}$ of the curve C covers an area $d\boldsymbol{S} = \hat{n}\, dS = \boldsymbol{v}(\boldsymbol{r}) \times d\boldsymbol{l}\, dt$ in a time interval dt, where $\hat{n}$ is a unit normal to

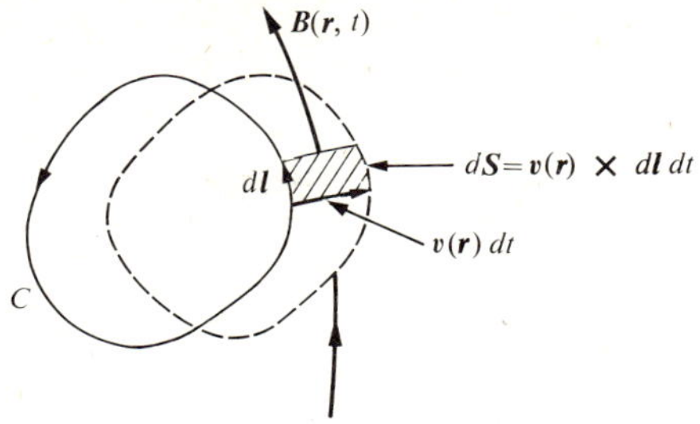

Fig. 4.21 Geometry for the evaluation of the induced motional e.m.f.

the surface related to the direction around the curve by the right-hand rule. The resulting change in the magnetic flux is

$$\boldsymbol{B}(\boldsymbol{r}, t) \cdot \{\boldsymbol{v}(\boldsymbol{r}) \times d\boldsymbol{l}\, dt\} = -\boldsymbol{v}(\boldsymbol{r}) \times \boldsymbol{B}(\boldsymbol{r}, t) \cdot d\boldsymbol{l}\, dt.$$

The contribution to the motional e.m.f. arising from the movement of the vector element of length $d\boldsymbol{l}$ of the curve C is the negative of the associated change in the magnetic flux per unit time, that is $\boldsymbol{v}(\boldsymbol{r}) \times \boldsymbol{B}(\boldsymbol{r}, t) \cdot d\boldsymbol{l}$. If we integrate over the complete closed curve C we obtain the following result for the motional e.m.f. due to the movement of the entire curve C:

$$\text{induced motional e.m.f.} = \int_C \boldsymbol{v}(\boldsymbol{r}) \times \boldsymbol{B}(\boldsymbol{r}, t) \cdot d\boldsymbol{l}. \tag{6.11}$$

Together with (6.8), we obtain the total induced e.m.f. as

$$\int_C \boldsymbol{E}(\boldsymbol{r}, t) \cdot d\boldsymbol{l} = -\int_S \frac{\partial}{\partial t} \boldsymbol{B}(\boldsymbol{r}, t) \cdot d\boldsymbol{S} + \int_C \boldsymbol{v}(\boldsymbol{r}) \times \boldsymbol{B}(\boldsymbol{r}, t) \cdot d\boldsymbol{l}, \tag{6.12}$$

where the first term on the right side of (6.12) is the contribution due to the time variation of $\boldsymbol{B}(\boldsymbol{r}, t)$ and the second term on the right side of (6.12) is the motional induced e.m.f.

Example. Let us consider a uniform magnetic flux density B in the y-direction. Let a single turn circular coil of planar area A be rotated with a uniform angular frequency Ω about a diameter which we assume to coincide with the z-axis. Suppose that the normal to the plane of the loop is in the y-direction at time $t = 0$. Then, the instantaneous value of the area of the loop normal to the magnetic flux density is given by $A \cos \Omega t$. When $\Omega t = 0$, this instantaneous value of the area is equal to A, when $\Omega t = \pi/2$, it is equal to zero and so on. The total magnetic flux linking the plane area bounded by the loop varies with time and hence an electromagnetic force is induced around the periphery of the loop. We obtain its value as

$$\text{induced e.m.f.} = -\frac{d}{dt}[BA \cos \Omega t] = BA\Omega \sin \Omega t.$$

For an N-turn coil, the induced e.m.f. is N times that for a single turn coil.

The result of this example shows that an alternating voltage is generated by the mechanical rotation of a loop in a constant magnetic flux density. This is the principle of an electric generator.

5. Generalization of Ampère's Circuital Law and Maxwell's Second Equation

We deduced Ampère's circuital law (4.45) in magnetostatics with the help of the continuity equation (4.9). For the time-varying fields the equation of continuity is different and is given by (6.6). Consequently, it follows that Ampère's circuital law (4.45) requires modification for the time-varying fields. Though not rigorous, we can use the following simple method to obtain the required modification of Ampère's circuital law for the time-varying fields. If we substitute (6.1) in (6.6), after some rearrangement we get

$$\nabla \cdot \left[\boldsymbol{J}(\boldsymbol{r}, t) + \frac{\partial}{\partial t} \boldsymbol{D}(\boldsymbol{r}, t) \right] = 0. \tag{6.13}$$

Two remarks are now in order. First $(\partial/\partial t)\boldsymbol{D}(\boldsymbol{r}, t)$ has the same dimensions as the current density and is usually called *the displacement current density*. Second, if we compare (6.13) with (4.9) and note the fact that the divergence of a vector quantity represents its net outward flow through a surface enclosing a unit volume, we see that the quantity analogous to $\boldsymbol{J}(\boldsymbol{r})$ in magnetostatics that is continuous in the time-varying case is $\boldsymbol{J}(\boldsymbol{r}, t) + (\partial/\partial t)\boldsymbol{D}(\boldsymbol{r}, t)$. This suggests that we have to modify the circuital law (4.45) by replacing $\boldsymbol{J}(\boldsymbol{r})$ by $\boldsymbol{J}(\boldsymbol{r}, t) + (\partial/\partial t)\boldsymbol{D}(\boldsymbol{r}, t)$. Hence,

$$\nabla \times \boldsymbol{H}(\boldsymbol{r}, t) = \boldsymbol{J}(\boldsymbol{r}, t) + \frac{\partial}{\partial t} \boldsymbol{D}(\boldsymbol{r}, t), \tag{6.14}$$

which is Maxwell's second equation and which embodies the generalization of Ampère's circuital law. Taking the divergence of both sides of (6.14) and noting that $\nabla \cdot \{\nabla \times \boldsymbol{H}(\boldsymbol{r}, t)\} = 0$, we reproduce (6.13). It follows that (6.13) together with the divergence relation (6.1) is consistent with the equation of continuity (6.6). Thus, we have verified (6.14) to be consistent with (6.6). Just as the generalization (6.10) of Faraday's induction law shows that a time-varying magnetic flux density gives rise to an electric field, the generalization (6.14) of Ampère's circuital law shows that a time-varying electric flux density gives rise to a magnetic field. In a similar manner, we can generalize (4.72) to the time-varying case as follows:

$$\int_C \boldsymbol{H}(\boldsymbol{r}, t) \cdot d\boldsymbol{l} = \int_S \left[\boldsymbol{J}(\boldsymbol{r}, t) + \frac{\partial}{\partial t} \boldsymbol{D}(\boldsymbol{r}, t) \right] \cdot d\boldsymbol{S}, \tag{6.15}$$

where as before C is a closed curve bounding the surface S. The use of Stokes' theorem in (6.15) gives (6.14). The large-scale or the integral and the point or the differential forms of Maxwell's second equation are given by (6.15) and (6.14) respectively.

It is instructive for us to consider an example which illustrates the need for the inclusion of the displacement current density $\boldsymbol{J}_d(\boldsymbol{r}, t) = (\partial/\partial t)\boldsymbol{D}(\boldsymbol{r}, t)$ to the current density $\boldsymbol{J}(\boldsymbol{r}, t)$ associated with the motion of the charges. Let us consider a parallel-plate capacitor of capacitance C_0 connected to an a.c. generator of voltage V by means

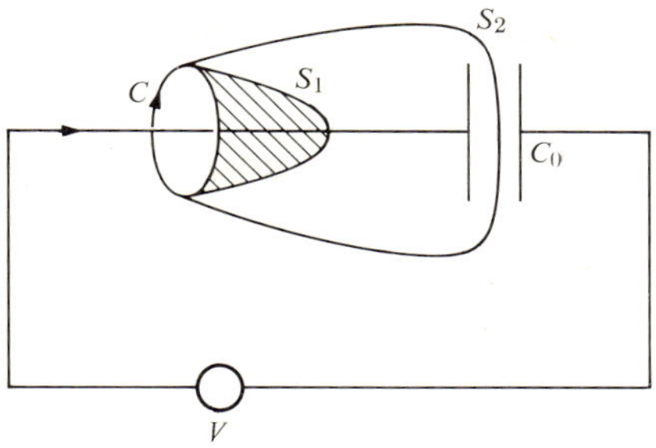

Fig. 4.22 Illustration of the need for the introduction of the displacement current.

of two conducting wires as shown in Fig. 4.22. Let A be the area of the plates, d the distance between them and ε_0 the permittivity of free space between the plates. Then, the electrostatic capacitance of the ideal parallel-plate capacitor is given by $C_0 = \varepsilon_0 A/d$. From the circuit theory, we obtain the charging current in the conducting wires as

$$I_C = C_0 \frac{d}{dt} V. \tag{6.16}$$

This is the conduction current. Since there is no conduction current in the free space between the capacitor plates, it follows that the conduction current is not continuous in the entire circuit.

The electric field inside the capacitor is normal to the plates and has a magnitude given by $E = V/d$. The displacement current is normal to the plates and we find its density to be

$$J_d = \frac{d}{dt} D = \frac{\varepsilon_0}{d} \frac{d}{dt} V. \tag{6.17}$$

The total displacement current I_d existing between the plates is the area of the plates multiplied by the displacement current density. Hence

$$I_d = A \frac{\varepsilon_0}{d} \frac{dV}{dt} = C_0 \frac{dV}{dt}. \tag{6.18}$$

A perfect conductor cannot sustain any electric field. We shall discuss this fact further subsequently. Therefore, there is no displacement current in the conducting wires. Consequently, the displacement current also is not continuous in the entire circuit.

Let us consider a closed path C enclosing the perfectly conducting wire shown in Fig. 4.22. Let us consider two surfaces S_1 and S_2, both of which are bounded by the curve C. Let the surface S_1 intersect the conducting wire and let the surface S_2 pass between the parallel plates of the capacitor. We should expect the current directed through the surfaces S_1 and S_2 to be equal since by Ampère's circuital law both the currents are equal to the same line integral of the magnetic field taken around the path C. If as in magnetostatics we only take the conduction current into account, the current through the surface S_1 as given by (6.16) is not equal to that through the surface S_2 since no conduction current is directed through the surface S_2. If on the other hand we only take the displacement current into account, the current through the surface S_2 as given by (6.18) is not equal to that through the surface S_1 since no displacement current is directed through the surface S_1. If we include both the conduction and the displacement currents, it follows from (6.16) and (6.18) that the total currents through the surfaces S_1 and S_2 are equal, as they should be, since both are equal to the same line integral of the magnetic field taken around the path C. Thus, the requirement of continuity of current in the time-varying case suggests the inclusion of the displacement current to the current due to the flow of charges in Ampère's circuital law.

Example. A uniform electric field of 20 V/m is reduced at a uniform rate to zero in a time $t = 2 \times 10^{-6}$ sec. At time $t = 10^{-6}$ sec, what is the magnetomotive force induced around the periphery of a square loop of 0.1 m on a side if the normal to the plane of the loop is parallel to the electric field? Assume that the medium is free space with the permittivity given by $\varepsilon_0 = 10^{-9}/36\pi$ F/m.

What is the result if we substitute sea water for the free space? The permittivity and the conductivity of sea water are given respectively by $\varepsilon = 9 \times 10^{-9}/4\pi$ F/m and $\sigma = 4$ mhos/m.

We find that dE/dt has a uniform value equal to 10^7. Therefore, we obtain the displacement current density as

$$J_d = \varepsilon_0 \frac{dE}{dt} = \frac{10^{-9}}{36\pi} \times 10^7 = \frac{10^{-2}}{36\pi} \text{ A/m}^2.$$

There is no conduction current in free space. Therefore, the total magnetomotive force induced is equal to the area of the loop multiplied by the displacement current density since the displacement current density is uniform and is normal to the plane of the loop. Hence,

$$\text{induced m.m.f.} = 10^{-4}/36\pi \quad \text{A}.$$

For sea water the displacement current density is given by

$$J_d = \varepsilon(dE/dt) = (9 \times 10^{-9}/4\pi) \times 10^7 = 7.16 \times 10^{-3} \text{ A/m}^2.$$

At $t = 10^{-6}$ sec, the value of the electric field is equal to half the original value, that is, $E = 10$ V/m. Therefore, the conduction current density is $J_{cd} = \sigma E = 40$ A/m^2. As before, the total magnetomotive force induced is equal to the area of the loop multiplied by the sum of the displacement and the conduction current densities. Hence,

$$\text{induced m.m.f.} = 10^{-2}(7.16 \times 10^{-3} + 40) \simeq 0.4 \text{ A}.$$

In sea water, for the time variation of the electric field under consideration, the displacement current density is negligible compared to the conduction current density. Media where this property is satisfied are called good conductors.

6. Summary of Field Equations

The integral forms of Maxwell's first two equations are given by (6.8) and (6.15) respectively and we assemble them here for convenience.

$$\int_C \boldsymbol{E}(\boldsymbol{r}, t) \cdot d\boldsymbol{l} = -\int_S \frac{\partial}{\partial t} \boldsymbol{B}(\boldsymbol{r}, t) \cdot d\boldsymbol{S}, \tag{6.19}$$

and

$$\int_C \boldsymbol{H}(\boldsymbol{r}, t) \cdot d\boldsymbol{l} = \int_S \left[\boldsymbol{J}(\boldsymbol{r}, t) + \frac{\partial}{\partial t} \boldsymbol{D}(\boldsymbol{r}, t) \right] \cdot d\boldsymbol{S}. \tag{6.20}$$

In addition, the conservation of charge gives the equation of continuity (6.5) which we may rewrite as

$$\int_S \boldsymbol{J}(\boldsymbol{r}, t) \cdot d\boldsymbol{S} = -\int_V \frac{\partial}{\partial t} \rho(\boldsymbol{r}, t)\, dV. \tag{6.21}$$

The differential forms of Maxwell's first two equations are given by (6.10) and (6.14) respectively and we collect them together here for convenience:

$$\nabla \times \boldsymbol{E}(\boldsymbol{r}, t) = -\frac{\partial}{\partial t} \boldsymbol{B}(\boldsymbol{r}, t), \tag{6.22}$$

$$\nabla \times \boldsymbol{H}(\boldsymbol{r}, t) = \boldsymbol{J}(\boldsymbol{r}, t) + \frac{\partial}{\partial t} \boldsymbol{D}(\boldsymbol{r}, t). \tag{6.23}$$

In a similar manner, the differential form of the equation of continuity as given by (6.6) is

$$\nabla \cdot \boldsymbol{J}(\boldsymbol{r}, t) = -\frac{\partial}{\partial t} \rho(\boldsymbol{r}, t). \tag{6.24}$$

We now give the alternative interpretation of the divergence laws (6.3) and (6.1). If we take the divergence of both the sides of (6.22) and (6.23) and use (6.24) the following results are obtained:

$$\frac{\partial}{\partial t} [\nabla \cdot \boldsymbol{B}(\boldsymbol{r}, t)] = 0, \tag{6.25}$$

and

$$\frac{\partial}{\partial t} [\nabla \cdot \boldsymbol{D}(\boldsymbol{r}, t) - \rho(\boldsymbol{r}, t)] = 0. \tag{6.26}$$

If the divergence laws

$$\nabla \cdot \boldsymbol{B}(\boldsymbol{r}, t) = 0, \tag{6.27}$$

and

$$\nabla \cdot \boldsymbol{D}(\boldsymbol{r}, t) = \rho(\boldsymbol{r}, t), \tag{6.28}$$

hold good at the origin of time, we see from (6.25) and (6.26) that they are valid for all time. Therefore, we may consider (6.27) and (6.28) as the *initial conditions* to Maxwell's equations (6.22) and (6.23).

In addition, we collect the constitutive relations (2.46) and (4.44) which connect the fundamental field quantities $\boldsymbol{E}(\boldsymbol{r}, t)$ and $\boldsymbol{B}(\boldsymbol{r}, t)$ with the derived field quantities $\boldsymbol{D}(\boldsymbol{r}, t)$ and $\boldsymbol{H}(\boldsymbol{r}, t)$ respectively together as

$$\boldsymbol{D}(\boldsymbol{r}, t) = \varepsilon \boldsymbol{E}(\boldsymbol{r}, t), \tag{6.29}$$

and

$$\boldsymbol{B}(\boldsymbol{r}, t) = \mu \boldsymbol{H}(\boldsymbol{r}, t). \tag{6.30}$$

The permittivity $\varepsilon = \varepsilon_0 \varepsilon_r$ and the permeability $\mu = \mu_0 \mu_r$ characterize the electrical and the magnetic properties of the medium.

Three types of current densities have been introduced and these are

$$\textit{Conduction current:} \qquad \boldsymbol{J}_{cd}(\boldsymbol{r}, t) = \sigma \boldsymbol{E}(\boldsymbol{r}, t), \tag{6.31}$$

$$\textit{Convective current:} \qquad \boldsymbol{J}_{cv}(\boldsymbol{r}, t) = \boldsymbol{v}(\boldsymbol{r}, t)\rho(\boldsymbol{r}, t), \tag{6.32}$$

$$\textit{Displacement current:} \qquad \boldsymbol{J}_{d}(\boldsymbol{r}, t) = \frac{\partial}{\partial t}\boldsymbol{D}(\boldsymbol{r}, t). \tag{6.33}$$

In this treatment, we give attention to the nature of the electromagnetic field only in media where the constitutive parameters μ, ε, and σ are scalar constants independent of location and time.

We should note that Maxwell's equations (6.22) and (6.23), the equation of continuity (6.24), the divergence relations (6.27) and (6.28), and the constitutive relations (6.29) to (6.31) are linear. The relation (6.32) for the convection current can be nonlinear. Either (6.32) is never considered or only its linearized form is treated. Thus, the complete set of field equations that we cover in this treatment is linear.

4.7 BOUNDARY CONDITIONS

In the development of the differential forms of Gauss's laws and Maxwell's equations, although we have not stated so explicitly, we have assumed that the medium is continuous. A vast majority of problems in electromagnetic field theory involve boundaries separating two different media. The differential forms of Maxwell's equations are valid in the interior of each of the media but since they contain space derivatives are not valid and give no information at the interface between two different media. However, the integral forms of Gauss's laws and Maxwell's equations are valid even in a region containing boundaries separating two media with different constitutive parameters and their straightforward application yields certain requirements that the field vectors must satisfy at the boundaries. We shall now take up the derivation of these boundary conditions.

1. Tangential Component of the Electric Field

Let us consider a point on the boundary between two different media. Let us choose the origin of the Cartesian coordinate system x, y, and z to coincide with this point and let the z-axis be normal to the boundary. Without any loss of generality, the coordinate system is oriented in such a way that the component of the field tangential to the interface is parallel to the y-axis. Let us consider an infinitesimal rectangular

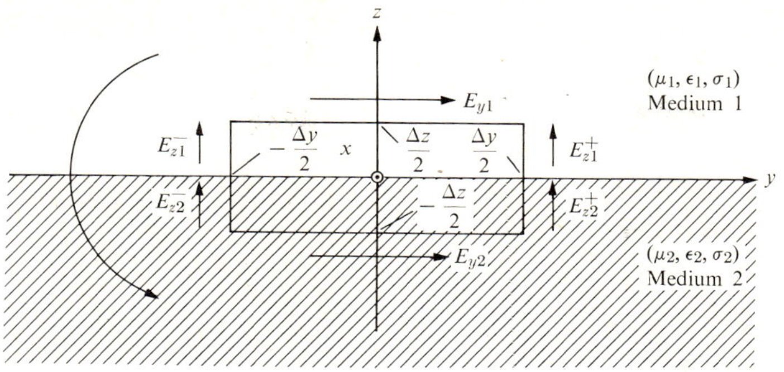

Fig. 4.23 Boundary condition on the tangential component of the electric field.

path in the $x = 0$ plane with the sides specified by $y = \pm\Delta y/2$ and $z = \pm\Delta z/2$. We should note that the interface near the point under consideration is inside the rectangular path. For $z > 0$, the medium has the constitutive parameters μ_1, ε_1, and σ_1 and is designated by 1. Similarly, for $z < 0$, the medium has the constitutive parameters μ_2, ε_2, and σ_2 and is designated by 2. We indicate the field quantities along the sides of the closed rectangular path in Fig. 4.23. By taking the line integral of the electric field along the rectangular path in the direction shown in Fig. 4.23, we obtain with the help of (6.19)

$$-E_{y1}\,\Delta y - E_{z1}^-\frac{\Delta z}{2} - E_{z2}^-\frac{\Delta z}{2} + E_{y2}\,\Delta y + E_{z2}^+\frac{\Delta z}{2} + E_{z1}^+\frac{\Delta z}{2} = -\left(\frac{\partial}{\partial t}B_x\right)\Delta y\,\Delta z, \tag{7.1}$$

where B_x is the average magnetic flux density through the rectangular area. The width Δz of the rectangle is reduced to zero. We should note that the interface is always between the sides of the rectangle. If the electric field is finite and if the rate of change of magnetic flux density is also finite, then all the quantities containing Δz vanish in the limit yielding

$$E_{y1} = E_{y2} \qquad \text{or} \qquad E_{t1} = E_{t2}, \tag{7.2}$$

where the subscript t stands for the tangential component. It follows that the tangential component of the electric field is continuous at the surface of separation of two different media.

2. Tangential Component of the Magnetic Field

If instead of the electric field, we take the line integral of the magnetic field along the rectangular path in the same direction as shown in Fig. 4.24, we obtain from (6.20)

$$-H_{y1}\,\Delta y - H_{z1}^-\frac{\Delta z}{2} - H_{z2}^-\frac{\Delta z}{2} + H_{y2}\,\Delta y + H_{z2}^+\frac{\Delta z}{2} + H_{z1}^+\frac{\Delta z}{2}$$
$$= \left[J_x + \frac{\partial}{\partial t}D_x\right]\Delta y\,\Delta z, \tag{7.3}$$

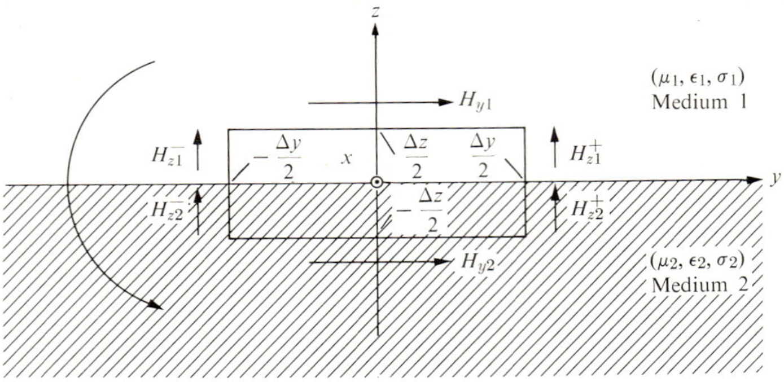

Fig. 4.24 Boundary condition on the tangential component of the magnetic field.

where D_x and J_x are respectively the average electric flux density and the current density through the area $\Delta y \, \Delta z$. As before, we consider the limiting case of Δz going to zero. It is convenient for us to distinguish two cases. In the first case, we assume neither of the two media to be a perfect conductor. Then, the electric current density J_x is finite. Also, if the magnetic field as well as the rate of change of electric flux density are finite, then all the quantities multiplied by Δz in (7.3) vanish in the limit yielding

$$H_{y1} = H_{y2} \qquad \text{or} \qquad H_{t1} = H_{t2}. \tag{7.4}$$

For a very good conductor, σ is large and (6.31) shows that the electric field associated with a finite current density is small. For a *perfect conductor* characterized by $\sigma = \infty$, the electric field is zero. In other words, a perfect conductor cannot sustain any electric field. From Maxwell's first equation (6.22), we see the magnetic field also to be zero inside a perfect conductor. Therefore, the interior of a perfect conductor cannot sustain any electromagnetic field. As we shall establish subsequently, the volume current density $\boldsymbol{J}$ decays from the surface into the interior of a very good conductor and the decay rate increases with the increase in the conductivity. In the limiting case of a perfect conductor, the current density is confined only to the surface of the conductor and is therefore restricted to zero thickness. Consequently the volume current density at the surface of a perfect conductor is infinite since a finite value of current is contained in a volume which goes to zero because the thickness of the current carrying region is zero. It is therefore useful for us to introduce the concept of surface current density $\boldsymbol{J}_s$ as defined by the relation

$$\boldsymbol{J}_s = \lim_{\Delta z \to 0} \boldsymbol{J} \, \Delta z \quad \text{A/m} \tag{7.5}$$

where even though $\boldsymbol{J}$ becomes infinite as Δz goes to zero, their product remains finite in the limit of $\Delta z = 0$.

With this understanding of the response of a perfect conductor to an electromagnetic field, we turn our attention to the second case where Medium 2 in Fig. 4.24 is a perfect conductor. If Δz goes to zero, as in the first case treated previously, all the quantities containing Δz in (7.3) go to zero except $J_x \Delta z$ which yields a finite value J_{xs} as indicated in (7.5). Also, H_{y2} is zero since a perfect conductor cannot sustain any electromagnetic field. Therefore, in the limit of $\Delta z = 0$, (7.3) yields

$$-H_{y1} = J_{xs}. \tag{7.6}$$

The surface current density and the magnetic field are both tangential to the surface of the perfect conductor but are normal to each other. It is therefore convenient for us to express (7.6) as a vector relation as follows:

$$\hat{z} \times (\hat{y}H_{y1}) = \hat{x}J_{xs} \qquad \text{or} \qquad \hat{n} \times \boldsymbol{H}_t = \hat{n} \times \boldsymbol{H} = \boldsymbol{J}_s, \tag{7.7}$$

where $\hat{n}$ is an outwardly directed unit normal to the perfectly conducting surface. If the subscript n denotes the component normal to the surface, it is seen that $\hat{n} \times \hat{n}H_n = 0$; therefore, $\hat{n} \times \boldsymbol{H}_t = \hat{n} \times \boldsymbol{H}$ as indicated in (7.7).

Before proceeding to the consideration of the normal components, it should be noted that if Medium 2 is a perfect conductor $E_{t2} = 0$ and therefore (7.2) simplifies to

$$E_{t1} = 0 \qquad \text{(if Medium 2 is a perfect conductor).} \tag{7.8}$$

3. Normal Component of the Magnetic Flux Density

Let us consider a cylindrical pillbox with its axis coinciding with the z-axis which, as before, we take to be normal to the boundary. Let us assume that the pillbox has an elementary cross-sectional area dS as shown in Fig. 4.25. Let the ends of the pillbox be situated in the planes $z = \Delta z/2$ and $z = -\Delta z/2$. We indicate the field quantities at the ends of the pillbox in Fig. 4.25. If we apply the magnetostatic analogue (6.4) of Gauss's law to the surface of the pillbox we obtain

$$B_{z1}\, dS - B_{z2}\, dS + \Phi_{cyl} = 0, \tag{7.9}$$

where Φ_{cyl} is the total outward magnetic flux through the cylindrical surface of the pillbox. Let us consider the limit of Δz going to zero. If the tangential component

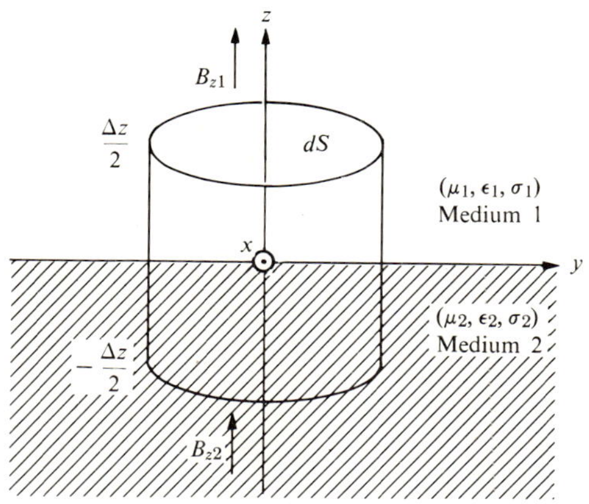

Fig. 4.25 Boundary condition on the normal component of the magnetic flux density.

of the magnetic flux density is finite, Φ_{cyl} vanishes in the limit of $\Delta z = 0$ since the cylindrical surface area vanishes, yielding

$$B_{z1} = B_{z2} \qquad \text{or} \qquad B_{n1} = B_{n2}. \tag{7.10}$$

It follows from (7.10) that the normal component of the magnetic flux density is continuous at the surface of separation of two different media.

4. Normal Component of the Electric Flux Density

If, instead of the magnetic flux density, we consider the electric flux density, the application of Gauss's law (6.2) to the surface of the pillbox as illustrated in Fig. 4.26 gives

$$D_{z1}\, dS - D_{z2}\, dS + \psi_{cyl} = \rho\, \Delta z\, dS, \tag{7.11}$$

where ψ_{cyl} is the total outward electric flux through the cylindrical surface of the pillbox. As with (7.5), we can introduce the concept of the surface charge density ρ_s by the relation

$$\rho_s = \operatorname*{limit}_{\Delta z \to 0} \rho\, \Delta z \ \text{C/m}^2. \tag{7.12}$$

Let us consider the limit of Δz going to zero. If the tangential component of the electric flux density is finite, ψ_{cyl} vanishes in the limit of $\Delta z = 0$ since the cylindrical surface area vanishes. Therefore, in the limit of $\Delta z = 0$, from (7.11) with the help of (7.12) we obtain

$$D_{z1} - D_{z2} = \rho_s \qquad \text{or} \qquad D_{n1} - D_{n2} = \rho_s. \tag{7.13}$$

If there is no surface charge density, we can reduce (7.13) to

$$D_{z1} = D_{z2} \qquad \text{or} \qquad D_{n1} = D_{n2} \qquad (\text{for } \rho_s = 0). \tag{7.14}$$

If Medium 2 is a perfect conductor, $D_{n2} = 0$ and (7.13) becomes

$$D_{z1} = \rho_s \qquad \text{or} \qquad D_{n1} = \rho_s \qquad \text{or} \qquad \hat{n} \cdot \boldsymbol{D} = \rho_s. \tag{7.15}$$

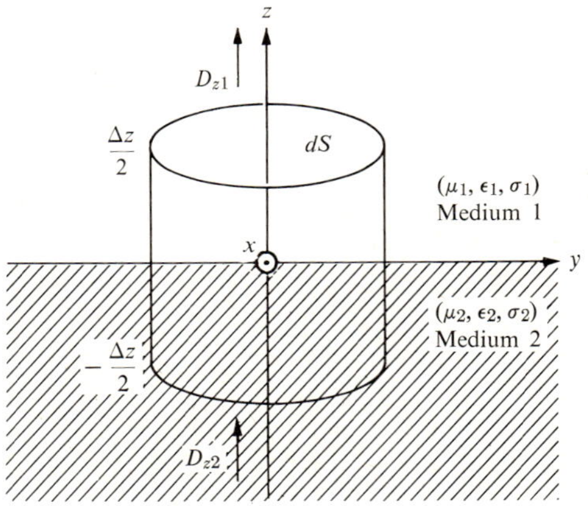

Fig. 4.26 Boundary condition on the normal component of the electric flux density.

It follows from (7.15) that the normal component of the electric flux density at the surface of a perfect conductor is equal to the surface charge density.

5. Summary of the Boundary Conditions

1. The tangential component of the electric field is continuous at the boundary between two different media.
2. The tangential component of the magnetic field is discontinuous at the boundary by the amount of the surface current density.
3. The normal component of the magnetic flux density is continuous at the boundary.
4. The normal component of the electric flux density is discontinuous at the boundary by the amount of the surface charge density.

6. Dependence Among the Boundary Conditions

We obtained the boundary conditions on the tangential components of the field vectors with the help of Maxwell's equations whereas we obtained those on the normal components with the help of Gauss's laws. But Gauss's laws are related to Maxwell's equations. This fact suggests that there is some dependence between the boundary conditions for the tangential and the normal field components. We shall now examine the exact nature of this dependence.

If we refer to Figs. 4.23 and 4.25, since the tangential component of the electric field is in the y-direction and since Maxwell's equation (6.22) is valid in the interior for both the media, we can use the z-component of (6.22) to obtain

$$\frac{\partial}{\partial x} E_{y1} = -\frac{\partial}{\partial t} B_{z1}, \qquad \frac{\partial}{\partial x} E_{y2} = -\frac{\partial}{\partial t} B_{z2}, \tag{7.16}$$

where the subscripts 1 and 2 denote the medium to which the field quantities pertain. From (7.16), we find that

$$\frac{\partial}{\partial x}[E_{y1} - E_{y2}] = -\frac{\partial}{\partial t}[B_{z1} - B_{z2}]. \tag{7.17}$$

In view of (7.2), both E_y as well as its derivative with respect to the coordinate along the boundary surface are continuous. Hence (7.17) becomes

$$\frac{\partial}{\partial t}[B_{z1} - B_{z2}] = 0. \tag{7.18}$$

Therefore, if the boundary condition (7.10) holds at time $t = 0$, then we see from (7.18) that it is valid for all subsequent times. Thus, there is dependence between the boundary condition for the tangential $\boldsymbol{E}$-field and the normal $\boldsymbol{B}$-field.

In the same manner, with reference to Figs. 4.24 and 4.26, we consider two important cases. First, let us assume that Medium 1 and Medium 2 are perfect dielectrics. Thus, the conduction current density is absent in both of the media. Second, let us assume that Medium 1 is a perfect dielectric and Medium 2 is a perfect conductor. Thus, the conduction current density is zero also in the interior of both of the media. Therefore, we can use the z-component of (6.23) together with the fact that the tangential component of the magnetic field is in the y-direction to obtain

$$\frac{\partial H_{y1}}{\partial x} = \frac{\partial}{\partial t} D_{z1}, \qquad \frac{\partial H_{y2}}{\partial x} = \frac{\partial}{\partial t} D_{z2}, \tag{7.19}$$

where, as before, the subscripts 1 and 2 refer to the medium to which the field quantities pertain. From (7.19), we see that

$$\frac{\partial}{\partial x}(H_{y1} - H_{y2}) = \frac{\partial}{\partial t}(D_{z1} - D_{z2}). \tag{7.20}$$

For the first case of two perfect dielectrics, in view of (7.4)

$$\frac{\partial}{\partial t}(D_{z1} - D_{z2}) = 0. \tag{7.21}$$

Therefore, if the boundary condition (7.14) holds initially, we see from (7.21) that it holds good for all time.

In the second case where Medium 2 is a perfect conductor, $H_{y2} = D_{z2} = 0$ and hence with the help of the boundary condition (7.6) for the tangential field component, we can simplify (7.20) to

$$-\frac{\partial}{\partial x} J_{xs} = \frac{\partial D_{z1}}{\partial t}. \tag{7.22}$$

We obtain the equation of continuity for the surface current density in a manner similar to that used in obtaining (6.24) and the result is

$$\nabla_t \cdot \boldsymbol{J}_s = -\frac{\partial}{\partial t} \rho_s, \tag{7.23}$$

where ∇_t is the component of the del operator tangential to the boundary. In the present case, since the surface current density has only an x-component, we can reduce (7.23) to

$$\frac{\partial J_{xs}}{\partial x} = -\frac{\partial}{\partial t} \rho_s. \tag{7.24}$$

Together with (7.24), (7.22) becomes

$$\frac{\partial}{\partial t}[D_{z1} - \rho_s] = 0. \tag{7.25}$$

Hence, if the boundary condition (7.15) is valid at $t = 0$, we see from (7.25) that it is valid for all time. Thus, we have made clear the dependence between the boundary condition for the tangential $\boldsymbol{H}$-field and the normal $\boldsymbol{D}$-field.

Summing up, we should emphasize that subject to the appropriate initial conditions, the boundary conditions on the normal components of the $\boldsymbol{B}$- and the $\boldsymbol{D}$-fields are automatically satisfied, respectively, if the boundary conditions on the tangential components of the $\boldsymbol{E}$-and the $\boldsymbol{H}$-fields are fulfilled. The appropriate initial conditions are that the fields satisfy Gauss's laws on the boundary surfaces. It is pertinent for us to recall that if Maxwell's equations are satisfied, Gauss's laws are automatically satisfied provided Gauss's laws hold good initially and to note that the appropriate initial conditions on the boundaries are the same as those required outside of the boundaries.

4.8 ENERGY DENSITIES AND POWER FLUX DENSITY

Let us consider a conducting wire of cross-sectional area dS and length dl. Let R be its resistance and I be the current in it. The rate of dissipation of energy or the amount of heat Q evolved in unit time is given by

$$Q = I^2 R \quad \text{W}. \tag{8.1}$$

Since $I = J\,dS$ and $R = dl/\sigma\,dS$ where J is the current density and σ is the conductivity, (8.1) becomes

$$Q = J^2\,dS^2\,\frac{dl}{\sigma\,dS} \qquad \text{or} \qquad \frac{Q}{dl\,dS} = \frac{\boldsymbol{J}\cdot\boldsymbol{J}}{\sigma} = \boldsymbol{E}\cdot\boldsymbol{J}. \tag{8.2}$$

From (8.2), it follows that if $\boldsymbol{J}$ is the conduction current density maintained by the electric field $\boldsymbol{E}$, then $\boldsymbol{E}\cdot\boldsymbol{J}$ is the rate of dissipation of energy per unit volume.

The result contained in (8.2) holds good also for the convection currents. If ρ is the charge density and $\boldsymbol{v}$ the velocity of motion of the charge, then $\boldsymbol{J} = \rho\boldsymbol{v}$ is the convection current density. The force density acting on the charge due to the electric field is $\rho\boldsymbol{E}$ and the force density acting on the charge due to the magnetic flux density $\boldsymbol{B}$ we obtain from (4.19) as $\boldsymbol{J}\times\boldsymbol{B}$. Thus, the resultant force density $\boldsymbol{f}$ is

$$\boldsymbol{f} = \rho\boldsymbol{E} + \boldsymbol{J}\times\boldsymbol{B}, \tag{8.3}$$

which is known generally as the *Lorentz force density* equation. If $\boldsymbol{dl}$ is the displacement of the charge in a time interval dt, then the work done on the charge per unit volume in unit time is $\boldsymbol{f}\cdot\boldsymbol{dl}/dt = \boldsymbol{f}\cdot\boldsymbol{v}$. Therefore, from (8.3),

$$\boldsymbol{f}\cdot\boldsymbol{v} = (\rho\boldsymbol{E} + \boldsymbol{J}\times\boldsymbol{B})\cdot\boldsymbol{v} = \rho\boldsymbol{v}\cdot\boldsymbol{E} = \boldsymbol{E}\cdot\boldsymbol{J}. \tag{8.4}$$

We should note that $\boldsymbol{J}\times\boldsymbol{B}$ is normal to $\boldsymbol{J}$ and therefore normal to $\boldsymbol{v}$ also and hence $(\boldsymbol{J}\times\boldsymbol{B})\cdot\boldsymbol{v} = 0$. In (8.4), since $\boldsymbol{E}\cdot\boldsymbol{J}$ is the rate of work done on the charge per unit volume, it represents the rate of dissipation of energy density.

On the other hand, if the current density $\boldsymbol{J}_{\text{ext}}$ is externally maintained and the electric field $\boldsymbol{E}$ is supported by the current density, then $-\boldsymbol{E}\cdot\boldsymbol{J}_{\text{ext}}$ is the rate of work done by the current on the field per unit volume and hence it yields the rate of creation of energy density.

With these preliminaries, we can now derive the expressions for the energy densities and the power flux density. For this purpose, we rewrite Maxwell's equations (6.22) and (6.23) here with the externally impressed current density $\boldsymbol{J}_{\text{ext}}$ separated out from the remainder of the current density as follows:

$$\nabla \times \boldsymbol{E}(\boldsymbol{r}, t) = -\frac{\partial}{\partial t}\boldsymbol{B}(\boldsymbol{r}, t), \tag{8.5}$$

$$\nabla \times \boldsymbol{H}(\boldsymbol{r}, t) = \frac{\partial}{\partial t}\boldsymbol{D}(\boldsymbol{r}, t) + \sigma\boldsymbol{E}(\boldsymbol{r}, t) + \boldsymbol{J}_{\text{ext}}(\boldsymbol{r}, t). \tag{8.6}$$

We should note that in (8.6), apart from the externally impressed current density $\boldsymbol{J}_{\text{ext}}(\boldsymbol{r}, t)$, we have only included the conduction current density $\sigma\boldsymbol{E}(\boldsymbol{r}, t)$. On scalar multiplying (8.5) by $\boldsymbol{H}(\boldsymbol{r}, t)$, (8.6) by $-\boldsymbol{E}(\boldsymbol{r}, t)$, adding the results, and noting that

$$\nabla \cdot \{\boldsymbol{E}(\boldsymbol{r}, t) \times \boldsymbol{H}(\boldsymbol{r}, t)\} = \boldsymbol{H}(\boldsymbol{r}, t) \cdot \{\nabla \times \boldsymbol{E}(\boldsymbol{r}, t)\} - \boldsymbol{E}(\boldsymbol{r}, t) \cdot \{\nabla \times \boldsymbol{H}(\boldsymbol{r}, t)\}, \tag{8.7}$$

we obtain

$$\begin{aligned}
\nabla \cdot \{\boldsymbol{E}(\boldsymbol{r}, t) \times \boldsymbol{H}(\boldsymbol{r}, t)\} &= -\boldsymbol{H}(\boldsymbol{r}, t)\frac{\partial}{\partial t}\boldsymbol{B}(\boldsymbol{r}, t) - \boldsymbol{E}(\boldsymbol{r}, t)\frac{\partial}{\partial t}\boldsymbol{D}(\boldsymbol{r}, t) \\
&\quad - \sigma\boldsymbol{E}(\boldsymbol{r}, t) \cdot \boldsymbol{E}(\boldsymbol{r}, t) - \boldsymbol{E}(\boldsymbol{r}, t) \cdot \boldsymbol{J}_{\text{ext}}(\boldsymbol{r}, t), \\
&= -\frac{\partial}{\partial t}\left[\frac{\mu}{2}\boldsymbol{H}(\boldsymbol{r}, t) \cdot \boldsymbol{H}(\boldsymbol{r}, t) + \frac{\varepsilon}{2}\boldsymbol{E}(\boldsymbol{r}, t) \cdot \boldsymbol{E}(\boldsymbol{r}, t)\right] \\
&\quad - \sigma\boldsymbol{E}(\boldsymbol{r}, t) \cdot \boldsymbol{E}(\boldsymbol{r}, t) - \boldsymbol{E}(\boldsymbol{r}, t) \cdot \boldsymbol{J}_{\text{ext}}(\boldsymbol{r}, t).
\end{aligned} \tag{8.8}$$

It follows from (8.8) that

$$W_E(\boldsymbol{r}, t) = \frac{\varepsilon}{2}\boldsymbol{E}(\boldsymbol{r}, t) \cdot \boldsymbol{E}(\boldsymbol{r}, t), \tag{8.9}$$

and

$$W_H(\boldsymbol{r}, t) = \frac{\mu}{2}\boldsymbol{H}(\boldsymbol{r}, t) \cdot \boldsymbol{H}(\boldsymbol{r}, t), \tag{8.10}$$

have the dimensions of energy density. Therefore, we identify $W_E(\boldsymbol{r}, t)$ and $W_H(\boldsymbol{r}, t)$ as the electric and the magnetic energy densities respectively. If (8.8) is integrated throughout the volume V enclosed by the surface S, the result on using the divergence theorem is

$$\int_V \nabla \cdot \{\boldsymbol{E}(\boldsymbol{r}, t) \times \boldsymbol{H}(\boldsymbol{r}, t)\}\, dV = \int_S \boldsymbol{E}(\boldsymbol{r}, t) \times \boldsymbol{H}(\boldsymbol{r}, t) \cdot d\boldsymbol{S}, \tag{8.11}$$

$$\begin{aligned}
&= -\frac{\partial}{\partial t}\int_V [W_E(\boldsymbol{r}, t) + W_H(\boldsymbol{r}, t)]\, dV \\
&\quad - \int_V \sigma\boldsymbol{E}(\boldsymbol{r}, t) \cdot \boldsymbol{E}(\boldsymbol{r}, t)\, dV - \int_V \boldsymbol{E}(\boldsymbol{r}, t) \cdot \boldsymbol{J}_{\text{ext}}(\boldsymbol{r}, t)\, dV.
\end{aligned} \tag{8.12}$$

The first term on the right side of (8.12) gives the rate of depletion of total energy inside volume V. The term

$$P_D(\mathbf{r}, t) = \sigma \mathbf{E}(\mathbf{r}, t) \cdot \mathbf{E}(\mathbf{r}, t) \tag{8.13}$$

gives the rate of dissipation of energy density and the term

$$P_{\text{source}}(\mathbf{r}, t) = -\mathbf{E}(\mathbf{r}, t) \cdot \mathbf{J}_{\text{ext}}(\mathbf{r}, t) \tag{8.14}$$

gives the rate of creation of energy density. Therefore, the second term on the right side of (8.12) without the minus sign gives the rate of dissipation of energy inside volume V. Similarly, the third term on the right side of (8.12) gives the rate of creation of energy inside volume V. By the principle of conservation of energy, the right side of (8.12) gives the net rate at which energy is leaving the volume V. Hence, we can consider the right side of (8.11) to be the power flux through the surface S enclosing the volume V. It is conventional to identify the integrand

$$\mathbf{S}(\mathbf{r}, t) = \mathbf{E}(\mathbf{r}, t) \times \mathbf{H}(\mathbf{r}, t) \tag{8.15}$$

as the power flux density or the Poynting vector.

It is important for us to realize that (8.15) does not have the unique physical interpretation of power flow per unit area. Let us suppose another vector $\mathbf{S}_0$ for which $\nabla \cdot \mathbf{S}_0 = 0$ is added to the right side of (8.15). Then, it follows from the left side of (8.11) that the addition of the vector $\mathbf{S}_0$ to (8.15) does not result in any change in the integral. Thus, it is legitimate for us to identify $\mathbf{E}(\mathbf{r}, t) \times \mathbf{H}(\mathbf{r}, t) + \mathbf{S}_0$ also as the Poynting vector and give it the physical interpretation of power flow per unit area. However, if $\mathbf{E}(\mathbf{r}, t) \times \mathbf{H}(\mathbf{r}, t) + \mathbf{S}_0$ is integrated over a surface enclosing a volume, the contribution due to $\mathbf{S}_0$ to the surface integral vanishes identically. Thus, the integral of the Poynting vector over a closed surface yields unambiguously the total power flowing through the entire surface.

4.9 TIME-HARMONIC FIELDS

We have placed no restrictions so far on the nature of the time dependence of the field quantities. However, the particular case in which the field quantities have periodic time dependence is important in view of the facts that in a majority of practical applications the field quantities are periodic in time, at least approximately, and that the field quantities with arbitrary time dependence can be synthesized from the time-periodic fields. The periodic phenomenon is also called the steady-state or the time-harmonic problem. For a steady-state problem, it is usual to assume the time dependence of the form

$$f(\mathbf{r}, t) = f(\mathbf{r}, \omega)e^{j\omega t}, \tag{9.1}$$

where $f(\mathbf{r}, t)$ is any scalar field or a component of a vector field. We understand that in (9.1) the real part of the right side is taken before performing any operation with the field quantities, especially in the commonly occurring quadratic forms such as in (8.9), (8.10), and (8.13) to (8.15). The various facts that should be noted in connection with the use of the phasor notation (9.1) we have explained in Section 2.5.

We see from (6.22) and (6.23) that the phasor field quantities $\boldsymbol{E}(\boldsymbol{r})$, $\boldsymbol{B}(\boldsymbol{r})$, $\boldsymbol{D}(\boldsymbol{r})$, and $\boldsymbol{H}(\boldsymbol{r})$ satisfy the following Maxwell's equations

$$\nabla \times \boldsymbol{E}(\boldsymbol{r}) = -j\omega \boldsymbol{B}(\boldsymbol{r}), \tag{9.2}$$

and

$$\nabla \times \boldsymbol{H}(\boldsymbol{r}) = j\omega \boldsymbol{D}(\boldsymbol{r}) + \boldsymbol{J}(\boldsymbol{r}), \tag{9.3}$$

where with the help of (6.29) to (6.31), we obtain the following relations:

$$\boldsymbol{D}(\boldsymbol{r}) = \varepsilon \boldsymbol{E}(\boldsymbol{r}); \qquad \boldsymbol{B}(\boldsymbol{r}) = \mu \boldsymbol{H}(\boldsymbol{r}); \qquad \boldsymbol{J}(\boldsymbol{r}) = \sigma \boldsymbol{E}(\boldsymbol{r}). \tag{9.4}$$

The equation of continuity (6.24) and the divergence laws become in phasor notation

$$\nabla \cdot \boldsymbol{J}(\boldsymbol{r}) = -j\omega\rho(\boldsymbol{r}), \tag{9.5}$$

$$\nabla \cdot \boldsymbol{B}(\boldsymbol{r}) = 0, \tag{9.6}$$

and

$$\nabla \cdot \boldsymbol{D}(\boldsymbol{r}) = \rho(\boldsymbol{r}). \tag{9.7}$$

The boundary conditions which we discussed in Section 4.7 hold good for the time-harmonic fields for the corresponding phasor quantities of the fields. We should note that in (9.2) to (9.7), we have omitted the argument ω for convenience.

To a large extent, we are considering only the time-harmonic fields in this treatment. In the time-harmonic case, since the phenomenon is periodic, we obtain the time average of any physical quantity associated with the fields by taking the average over one time period $2\pi/\omega$. The time-averaged energy and power flux densities are parameters useful in the characterization of the electromagnetic field and we shall now deduce expressions for them

For the time-periodic case of period $2\pi/\omega$, we can write the Poynting vector (8.15) explicitly as follows:

$$\boldsymbol{S}(\boldsymbol{r}, t) = \tfrac{1}{4}[\boldsymbol{E}(\boldsymbol{r})e^{j\omega t} + \boldsymbol{E}^*(\boldsymbol{r})e^{-j\omega t}] \times [\boldsymbol{H}(\boldsymbol{r})e^{j\omega t} + \boldsymbol{H}^*(\boldsymbol{r})e^{-j\omega t}], \tag{9.8}$$

since the real part of $f(\boldsymbol{r})e^{j\omega t} = \tfrac{1}{2}[f(\boldsymbol{r})e^{j\omega t} + f^*(\boldsymbol{r})e^{-j\omega t}]$. If we expand the right side of (9.8) we obtain

$$\begin{aligned} \boldsymbol{S}(\boldsymbol{r}, t) = \tfrac{1}{4}[&\boldsymbol{E}(\boldsymbol{r}) \times \boldsymbol{H}^*(\boldsymbol{r}) + \boldsymbol{E}^*(\boldsymbol{r}) \times \boldsymbol{H}(\boldsymbol{r}) \\ &+ \boldsymbol{E}(\boldsymbol{r}) \times \boldsymbol{H}(\boldsymbol{r})e^{j2\omega t} + \boldsymbol{E}^*(\boldsymbol{r}) \times \boldsymbol{H}^*(\boldsymbol{r})e^{-j2\omega t}]. \end{aligned} \tag{9.9}$$

The time average of $\boldsymbol{S}(\boldsymbol{r}, t)$ over one period is given by

$$\boldsymbol{S}(\boldsymbol{r}) = \frac{\omega}{2\pi}\int_0^{2\pi/\omega} \boldsymbol{S}(\boldsymbol{r}, t)\, dt. \tag{9.10}$$

Since $\int_0^{2\pi/\omega} e^{\pm j2\omega t}\, dt = 0$, we can substitute (9.9) in (9.10) to obtain

$$\begin{aligned} \boldsymbol{S}_{\text{real}}(\boldsymbol{r}) &= \tfrac{1}{4}[\boldsymbol{E}(\boldsymbol{r}) \times \boldsymbol{H}^*(\boldsymbol{r}) + \boldsymbol{E}^*(\boldsymbol{r}) \times \boldsymbol{H}(\boldsymbol{r})], \\ &= \tfrac{1}{2}\,\text{Re}\, \boldsymbol{E}(\boldsymbol{r}) \times \boldsymbol{H}^*(\boldsymbol{r}). \end{aligned} \tag{9.11}$$

The time-averaged Poynting vector, as given by (9.11) is real and is therefore indicated by the subscript real. In analogy with the circuit theory which we discussed in

Section 2.8, it is conventional to introduce the reactive power flow per unit area as

$$\boldsymbol{S}_{\text{reactive}} = \tfrac{1}{2}\,\text{Im}\,\boldsymbol{E}(\boldsymbol{r}) \times \boldsymbol{H}^*(\boldsymbol{r}), \tag{9.12}$$

and the complex Poynting vector

$$\boldsymbol{S}(\boldsymbol{r}) = \boldsymbol{S}_{\text{real}}(\boldsymbol{r}) + j\boldsymbol{S}_{\text{reactive}}(\boldsymbol{r}) = \tfrac{1}{2}\boldsymbol{E}(\boldsymbol{r}) \times \boldsymbol{H}^*(\boldsymbol{r}). \tag{9.13}$$

In a similar manner, we can obtain the time averages of (8.9), (8.10), (8.13), and (8.14) and the results are

$$W_E(\boldsymbol{r}) = \frac{\varepsilon}{4}\,\text{Re}\,\boldsymbol{E}(\boldsymbol{r}) \cdot \boldsymbol{E}^*(\boldsymbol{r}), \tag{9.14}$$

$$W_H(\boldsymbol{r}) = \frac{\mu}{4}\,\text{Re}\,\boldsymbol{H}(\boldsymbol{r}) \cdot \boldsymbol{H}^*(\boldsymbol{r}), \tag{9.15}$$

$$P_D(\boldsymbol{r}) = \tfrac{1}{2}\,\text{Re}\,\sigma\boldsymbol{E}(\boldsymbol{r}) \cdot \boldsymbol{E}^*(\boldsymbol{r}), \tag{9.16}$$

and

$$P_{\text{source}}(\boldsymbol{r}) = -\tfrac{1}{2}\,\text{Re}\,\boldsymbol{E}(\boldsymbol{r}) \cdot \boldsymbol{J}^*_{\text{ext}}(\boldsymbol{r}). \tag{9.17}$$

We should note that in (9.14) to (9.16), Re can be omitted without altering the result. In (9.11) and (9.17), the complex conjugate symbol * can be placed on $\boldsymbol{E}(\boldsymbol{r})$ instead of on $\boldsymbol{H}(\boldsymbol{r})$ and $\boldsymbol{J}_{\text{ext}}(\boldsymbol{r})$ as appears now in (9.11) and (9.17) respectively.

When there is no conduction current density, we can recast (9.3) with the help of (9.4) as

$$\nabla \times \boldsymbol{H}(\boldsymbol{r}) = j\omega\varepsilon\boldsymbol{E}(\boldsymbol{r}). \tag{9.18}$$

With the conduction current density term included (9.3) becomes

$$\begin{aligned}\nabla \times \boldsymbol{H}(\boldsymbol{r}) &= j\omega\varepsilon\boldsymbol{E}(\boldsymbol{r}) + \sigma\boldsymbol{E}(\boldsymbol{r}),\\ &= j\omega\left(\varepsilon + \frac{\sigma}{j\omega}\right)\boldsymbol{E}(\boldsymbol{r}).\end{aligned} \tag{9.19}$$

A comparison of (9.18) and (9.19) shows that in the time-harmonic case, we can take the effects of conduction current conveniently into account by the introduction of an effective permittivity ε_{eff} as given by

$$\varepsilon_{\text{eff}} = \varepsilon + \frac{\sigma}{j\omega}. \tag{9.20}$$

We can see that the effective permittivity (9.20) is complex.

4.10 FREQUENCY RESPONSE OF DIELECTRIC MATERIALS

The polarization of dielectric materials arises because of the displacement of the charged particles. The mass of the charged particles is different for the different types of polarization. For the electrostatic fields, the effects due to the various types of polarization are indistinguishable. But for the electric fields with harmonic time dependence, the effect due to each type of polarization is dominant in certain characteristic frequency regions. These resonance effects in the polarization are due to the characteristic mass and the force constant associated with the charged particles whose displacements give rise to the polarization of the dielectric. The determination of the frequency response of the dielectric materials is difficult if we take all types of polarization into account. Here, we wish to illustrate the resonance behavior by including only the electronic polarization for which we assume a simple model.

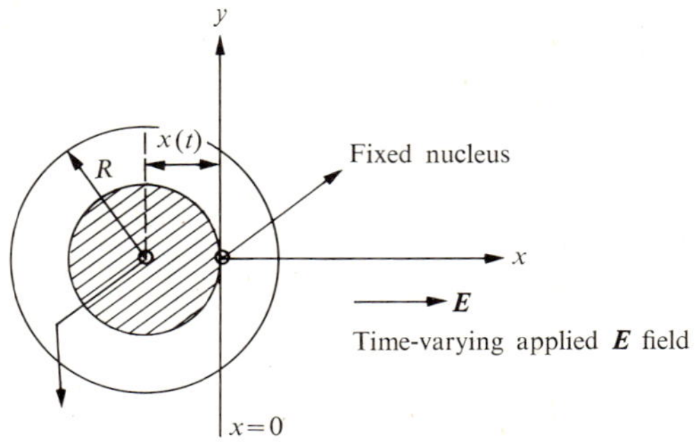

Fig. 4.27 Electronic polarization in a time-varying electric field.

An atom is approximated by a point nucleus of charge $+e$ surrounded by a spherical region in which the electronic charge $-e$ is uniformly distributed. We assume the radius of the sphere representing the electron to be R. In the presence of an electric field, the nucleus and the sphere representing the electron are displaced relative to each other, as shown in Fig. 4.27. It is convenient for us to examine the time-varying displacement of the center of the sphere representing the electron with reference to the position of the nucleus, which may therefore be taken to be stationary. Let the mass of the electron be m_e. The time-varying displacement of the electron is in response to the following forces acting on it: (i) the force due to the applied electric field; (ii) the Coulomb force of attraction between the electron and the positive nucleus; and (iii) the frictional forces which account for the dissipation of the energy.

Without any loss of generality, we assume the applied electric field to be in the x-direction; it has a value $E_x(t)$. The force on the electron due to the applied electric field is

$$-eE_x(t). \tag{10.1}$$

The Coulomb force of attraction is the charge multiplied by the electric field at its location due to the other charge. We can evaluate the attractive force corresponding to the displacement $x(t)$ as follows. The electric field at the location of the nucleus is due to that portion of the electronic charge contained within a sphere of radius $x(t)$ and with its center coinciding with that of the sphere of radius R representing the electron. From Gauss's law, we obtain this electric field as $-ex^3(t)/R^3 4\pi\varepsilon_0 x^2(t)$.

Hence, we determine the Coulomb force of attraction as

$$-kx(t) \qquad (\text{where} \qquad k = e^2/4\pi\varepsilon_0 R^3), \tag{10.2}$$

where k is the force constant. The advantage of assuming a simple model for the electronic polarization is that the force constant k can be evaluated explicitly. We see from (10.2) that the restoring force due to the Coulomb attraction is proportional to the displacement. The frictional force is usually taken to be proportional to the mass of the particle and to its velocity and, therefore, we may assume it to be given by

$$-m_e g\dot{x}(t) \tag{10.3}$$

where g is a positive constant. The sum of the forces given by (10.1) to (10.3) is equal to the mass times the acceleration of the particle with the result that we obtain the following equation of motion for the electron:

$$m_e\ddot{x}(t) = -eE_x(t) - kx(t) - m_e g\dot{x}(t). \tag{10.4}$$

If the electric field $E_x(t)$ has a harmonic time dependence of the form $e^{j\omega t}$, then the displacement also has the same harmonic time dependence. Hence, it is reasonable to assume that

$$E_x(t) = E_{x0}e^{j\omega t}, \qquad x(t) = x_0 e^{j\omega t}, \tag{10.5}$$

where, in accordance with the usual phasor notation, it is implied that $E_x(t)$ and $x(t)$ are equal to the real parts of the expressions on the right side of the equations in (10.5). The substitution of (10.5) into (10.4) enables us to obtain x_0 in terms of E_{x0} with the following result:

$$x_0 = -\frac{e}{m_e}E_{x0}[\omega_0^2 - \omega^2 + j\omega g]^{-1}, \tag{10.6}$$

where $\omega_0 = \sqrt{k/m_e}$. If N_0 is the number of atoms per unit volume, we determine the polarization resulting from the charge displacement (10.6) as

$$P_{x0} = -N_0 e x_0 = \varepsilon_0\chi E_{x0}. \tag{10.7}$$

From (10.7), we find that

$$\chi = -\frac{N_0 e}{\varepsilon_0}\frac{x_0}{E_{x0}}. \tag{10.8}$$

If we substitute the value of x_0/E_{x0} from (10.6) into (10.8) and separate out the resulting expression into its real and imaginary parts, we find that

$$\chi = \frac{N_0 e^2}{m_e\varepsilon_0}\left[\frac{\omega_0^2 - \omega^2}{(\omega_0^2 - \omega^2)^2 + \omega^2 g^2} - j\frac{\omega g}{(\omega_0^2 - \omega^2)^2 + \omega^2 g^2}\right]. \tag{10.9}$$

Together with (3.12), we can use (10.9) to deduce the real and the imaginary parts of the relative permittivity as

$$\varepsilon_r = \varepsilon_r' - j\varepsilon_r'', \tag{10.10}$$

where

$$\varepsilon_r' = 1 + \frac{N_0 e^2}{m_e \varepsilon_0} \frac{(\omega_0^2 - \omega^2)}{(\omega_0^2 - \omega^2)^2 + \omega^2 g^2}, \tag{10.11}$$

and

$$\varepsilon_r'' = \frac{N_0 e^2}{m_e \varepsilon_0} \frac{\omega g}{(\omega_0^2 - \omega^2)^2 + \omega^2 g^2}. \tag{10.12}$$

If we compare (10.10) with (9.20) we see that for the time-harmonic case, the existence of the imaginary part ε_r'' of ε_r has the same effect as the presence of the conductivity σ. In other words, ε_r'' accounts for the dissipation of energy inside the dielectric.

An examination of (10.12) reveals that ε_r'' has a maximum value at $\omega = \omega_0$ and falls off rapidly as ω is changed from the value ω_0 on either side. Therefore, ε_r'' exhibits a resonance at $\omega = \omega_0$. For the electronic polarization, this resonance frequency occurs normally in the ultraviolet region.

If we investigate the frequency dependence of ε_r' as given by (10.11) we see that $(\varepsilon_r' - 1)$ has antiresonance behavior in the neighborhood of $\omega = \omega_0$. At $\omega = \omega_0$, $(\varepsilon_r' - 1) = 0$. Also, $(\varepsilon_r' - 1)$ has a maximum and a minimum respectively for the values of ω slightly less than and slightly greater than ω_0. A detailed study of the antiresonance behavior of $(\varepsilon_r' - 1)$ can be carried out using the procedure employed in Section 2.11 for the treatment of the antiresonance of transmission-line circuits.

The other types of polarization also give rise to resonances in ε_r'' and antiresonances in $(\varepsilon_r' - 1)$ with only the difference that the frequencies of occurrence are much lower. We should note that the resonance frequency is inversely proportional to the square root of the mass of the charged particle whose displacement accounts for the polarization. For ionic and orientational polarization, the effective mass is larger and, therefore, the corresponding resonance frequencies are lower. For ionic and orientational polarizations, these resonance frequencies normally occur in the infrared and the radio frequency regions respectively.

For very large values of ω, we see from (10.11) and (10.12) that $\varepsilon_r' = 1$ and $\varepsilon_r'' = 0$, that is, the polarization effects are negligible. This is to be expected since on account of their finite masses, the charged particles are unable to respond to the applied electric field if the frequency of oscillation is very high. Consequently, in the limit of very high frequencies, the dielectric materials are unable to exhibit polarization effects and hence behave like free space.

PROBLEMS

4.1 Three point charges q, $-2q$, and q are located in free space along the z-axis at $z = 2h$, $z = 0$, and $z = -2h$ respectively. We wish to find the scalar potential $\phi(\mathbf{r})$ at a distance r from the origin in a direction which makes an angle θ with the positive z-direction. If $h/r \ll 1$, show that

$$\phi(\mathbf{r}) = \frac{4qh^2}{4\pi\varepsilon_0 r^3}(3\cos^2\theta - 1).$$

4.2 A distributed electric dipole in free space is in the form of a filament extending from $z = -h$ to $z = h$. The linear charge density is given by $q_L(z) = 2qz/h^2$ C/m. We wish to find the scalar potential $\phi(\boldsymbol{r})$ at a distance r from the origin in a direction which makes an angle θ with the positive z-direction. For $h/r \ll 1$, find the expression for $\phi(\boldsymbol{r})$ and hence prove that the dipole moment of the distributed electric dipole is $\frac{2}{3}$ of that due to the equivalent point electric dipole.

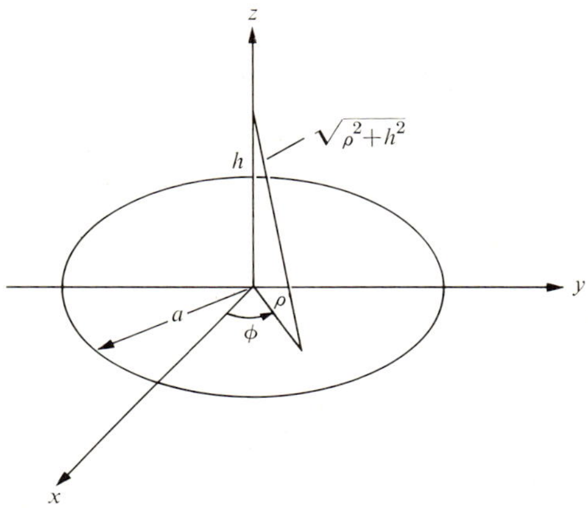

Fig. P4.1 A uniformly charged disc.

4.3 Consider a disc of radius a as illustrated in Fig. P4.1. A charge Q C is distributed uniformly on the disc. Find the electric field on the axis of the disc.

4.4 Find the capacitance of a cylindrical capacitor of length l. The radii of the inner and the outer plates are given by ρ_1 and ρ_2 respectively. The space between the plates is filled with two layers of dielectric. The permittivity of the first layer ($\rho_1 < \rho < a$) is ε_1 and that of the second layer ($a < \rho < \rho_2$) is ε_2.

4.5 Consider a spherical capacitor of inner radius r_1 and outer radius r_2. The space between the two spherical plates is filled with two layers of dielectric. The permittivity of the first layer ($r_1 < r < a$) is ε_1 and that of the second layer ($a < r < r_2$) is ε_2. Find the capacitance.

4.6 For the cylindrical capacitor with two different dielectrics as detailed in Problem 4.4, find the polarizations in the two dielectric regions as well as the equivalent volume and surface charge densities. Evaluate the electric field from the charge densities using Gauss's law and verify that the same results are obtained as in Problem 4.4 for both the dielectric regions as well as for the region exterior to the capacitor.

4.7 Consider two infinitely long parallel filamentary wires 1 and 2 separated by a distance r_{12} and carrying currents I_1 and I_2 respectively. Use Biot-Savart's law to show that the force per unit length along Wire 2 due to the current in Wire 1 is given by

$$\boldsymbol{F}_{12} = -\frac{\mu_0}{2\pi}\frac{I_1 I_2}{r_{12}}\hat{r}_{12} \quad \text{N/m},$$

where μ_0 is the permeability of the medium and $\hat{r}_{12}$ is a unit vector directed from Wire 1 to Wire 2.

4.8 Consider a circular current loop of radius a as shown in Fig. P4.2. Let I be the current in the loop. What is the magnetic field on the axis of the loop?

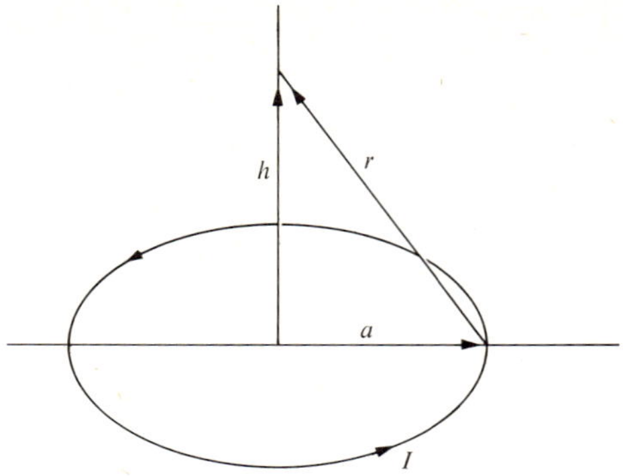

Fig. P4.2 A circular current loop of radius a.

4.9 A coaxial cable is used for transmission of steady current I. The current exists along the central inner conductor of radius ρ_1 and is enclosed by an outer shell of inner and outer radii ρ_2 and ρ_3 respectively. The space between the central conductor and the shell is filled with a dielectric. Find the magnetic field in the following regions:

$$\text{(i) } 0 < \rho < \rho_1, \quad \text{(ii) } \rho_1 < \rho < \rho_2, \quad \text{(iii) } \rho_2 < \rho < \rho_3, \quad \text{and} \quad \text{(iv) } \rho_3 < \rho.$$

4.10 An infinitely long perfectly conducting wire of radius a carries a total axial current of I A. The wire is surrounded by two layers of uniform annular magnetic materials. The materials in the regions $a < \rho < b$ and $b < \rho < c$ are characterized by the permeabilities μ_1 and μ_2 respectively where ρ represents the radial distance from the axis of the wire. The region $\rho > c$ is free space. Find the magnetic field $\boldsymbol{H}(\boldsymbol{r})$ and the magnetic flux density $\boldsymbol{B}(\boldsymbol{r})$ in all the regions of the space.

Determine the magnetization as well as the equivalent volume and surface current densities in the regions (i) $a < \rho < b$ and (ii) $b < \rho < c$ due to the magnetization. With the help of the equivalent volume and surface current densities, evaluate the magnetic field $\boldsymbol{H}(\boldsymbol{r})$ and the magnetic flux density $\boldsymbol{B}(\boldsymbol{r})$ in all the regions of the space and verify that the concept of permeability and magnetization lead to identical results.

4.11 Consider an infinitely long coaxial transmission line consisting of a solid inner conductor of radius a and hollow outer conductor of radius b. Assume a charge distribution of q C/m on the inner conductor and an equal and opposite charge on the inner surface of the outer conductor. With the help of Gauss's law, determine the radial electric field in the coaxial region of permittivity ε. Assume that the line integral of the electric field taken radially from a to b gives the voltage difference between the conductors. Determine this voltage and hence deduce that the capacitance of the coaxial transmission line per unit length is given by

$$C = \frac{2\pi\varepsilon}{\ln(b/a)} \quad \text{F/m.}$$

4.12 A uniform magnetic field in free space has an amplitude of 0.1 A/m and varies in time with the frequency $f = 10^7$ Hz. A square of side 5 cm has its normal inclined at an angle of θ degrees to the magnetic field. What is the amplitude of the voltage induced around the periphery of the square for (i) $\theta = 30°$, (ii) $45°$, (iii) $60°$, and (iv) $90°$? Assume that the permeability of free space is given by $\mu_0 = 4\pi \times 10^{-7}$ H/m.

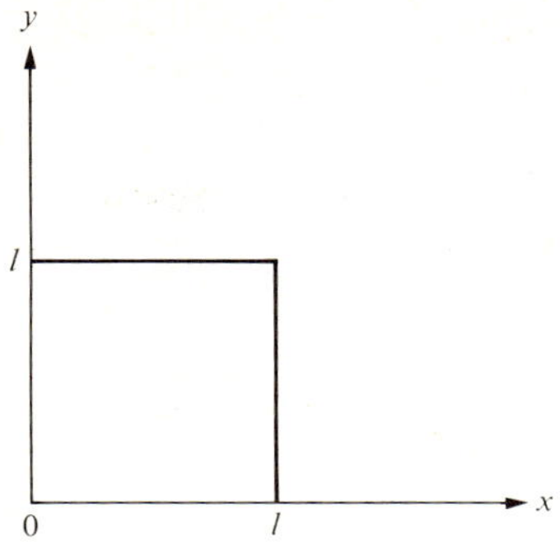

Fig. P4.3 Geometry of the square loop in a time-varying magnetic field.

4.13 A square loop of side l is oriented to lie in the xy-plane as shown in Fig. P4.3 in a loss-free region whose constitutive parameters are given by μ and ε. The magnetic vector in this region is given by

$$\boldsymbol{H}(\boldsymbol{r}, t) = \hat{z} \sin \beta_x x \sin \beta_y y \cos \omega t.$$

(a) Find the voltage induced in the loop by the application of Faraday's law of induction. (b) Evaluate the electric field with the help of the appropriate Maxwell's equation and verify the result obtained in (a) by the calculation of the line integral of the electric field taken around the periphery of the loop. (c) Show that the equality $\beta_x^2 + \beta_y^2 = \omega^2\mu\varepsilon$ holds if Maxwell's equations are to be satisfied. (d) Determine the value of the time-averaged Poynting vector at the center of the loop.

4.14 A square loop of side l is oriented to lie in the yz-plane as shown in Fig. P4.4 in a loss-free region whose constitutive parameters are given by μ and ε. A plane electromagnetic wave of angular frequency ω progressing in the $+x$-direction and having the following magnetic field component

$$\boldsymbol{H}(\boldsymbol{r}, t) = \hat{z} H_0 e^{j(\omega t - \beta x)}, \qquad \beta = \omega\sqrt{\mu\varepsilon},$$

is incident on the loop.

(a) Find the open-circuit voltage induced in the loop. (b) If the loop is reoriented to lie in the xy-plane, what is the open-circuit voltage induced in the loop? (c) Let the loop be rotated about

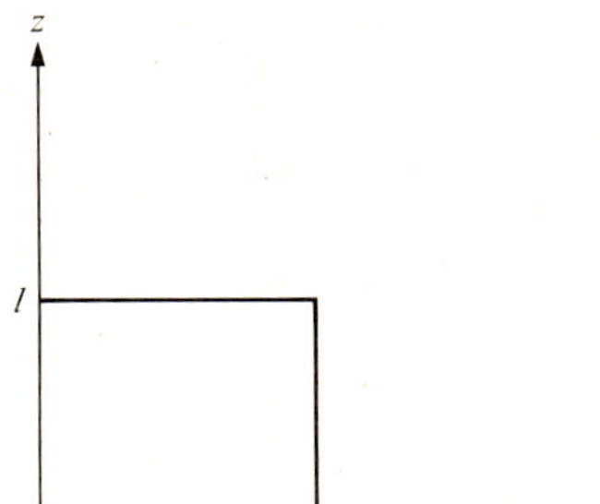

Fig. P4.4 Geometry of the square loop in a plane wave field.

the y-axis and let θ be the angle made by the plane of the loop with respect to the z-axis. Note that cases (a) and (b) correspond to $\theta = 0$ and $\theta = \pi/2$ respectively. Find the open-circuit voltage induced in the loop for an arbitrary angle θ in the range $0 < \theta < \pi/2$. (d) Suggest three methods for increasing the open-circuit voltage induced in the loop.

4.15 A straight wire 0.1 m long is moved at right angles to its length in a region permeated by a uniform magnetic field with the associated magnetic flux density $B = 10^6$ webers/m^2. The magnetic field is perpendicular to both the direction of motion and the length of the wire. The velocity of motion is 10 m/sec. Find the voltage induced in the wire.

4.16 An infinitely long perfectly conducting circular wire of radius a carries a total axial current of I A. With the help of the appropriate Maxwell's equation, find the direction and the magnitude of the magnetic field at a radial distance ρ when (i) $\rho < a$ and (ii) $\rho > a$.

Perform the calculations both for the case of a direct current which is distributed uniformly throughout the cross section of the wire and for the case of an alternating current which is concentrated on the surface of the wire.

4.17 A uniform electric field in free space has an amplitude of 1 V/m and varies in time with the frequency $f = 10^6$ Hz. A square of side 5 cm has its normal in the direction of the electric field. What is the magnetomotive force induced around the periphery of the square? Assume that the permittivity of free space is given by $\varepsilon_0 = 10^{-9}/36\pi$ F/m.

What is the result if sea water is substituted for free space? The permittivity and the conductivity of sea water are given respectively by $\varepsilon = 9 \times 10^{-9}/4\pi$ F/m and $\sigma = 4$ mhos/m.

Repeat the calculations for $f = 10^7$ Hz and find out why the magnetomotive force in sea water is not significantly changed.

4.18 A rectangular dielectric slab of cross-sectional area $A = 0.005$ m^2 and thickness $d = 0.01$ m forms the dielectric of a parallel-plate capacitor. A potential difference of 100 V at a frequency of $f = 5 \times 10^7$ Hz is applied between the two plates of the capacitor. Find the magnetic field inside the parallel-plate capacitor. The permittivity of the dielectric may be assumed to be given by $\varepsilon = 10^{-9}/9\pi$ F/m.

4.19 In the coaxial transmission line of Problem 4.11, let the alternating current on the outer surface of the inner conductor be I A and that on the inner surface of the outer conductor be $-I$ A. With the help of the appropriate Maxwell's equation, determine the azimuthal component of the magnetic field at a radial distance ρ from the axis of the inner conductor in the coaxial dielectric region of permeability μ. Determine the magnetic flux linking the current in one meter of the coaxial transmission line and passing through an area between ρ and $\rho + d\rho$ in the radial direction and one meter in the axial direction. By integrating this flux from $\rho = a$ to $\rho = b$, obtain the total magnetic flux linking the current in one meter of the coaxial transmission line and hence show that the inductance per unit length is given by

$$L = \frac{\mu}{2\pi} \ln\left(\frac{b}{a}\right) \quad \text{H/m.}$$

4.20 In the coaxial transmission line of Problems 4.11 and 4.19, show that the radial electric field and the azimuthal magnetic field vanish identically in the region exterior to the outer conductor.

Verify that there are no other components of the electric and the magnetic fields in the region exterior to the outer conductor.

4.21 With the help of a procedure similar to that employed in Problem 4.19 and making use of the results of Problem 4.16, evaluate the internal inductance L_i per unit length of an infinitely long, perfectly conducting circular wire of radius a carrying a total d.c. axial current of I A.

4.22 A uniform electric field $E = 1$ V/m in free space disappears at a uniform rate in time $t = 10^{-6}$ sec. What is the displacement current through a square of side 15 cm? Assume that the permittivity of free space is given by $\varepsilon_0 = 10^{-9}/36\pi$ F/m.

4.23 The axial current in a copper wire of cross-sectional area $S = 10^{-6}$ m^2 is $I = 2$ A. How many electrons per second pass a unit cross-sectional area in the wire? An electron has a charge -1.6×10^{-19} coulombs. Assuming that the number density of free electrons in copper is 1.68×10^{28} per cubic meter, find the average velocity of the electrons.

4.24 The conduction current in a copper wire of cross-sectional area $S = 10^{-5}$ m^2 is $I = 2$ A. The constitutive parameters of copper are $\mu = 4\pi \times 10^{-7}$ H/m, $\varepsilon = 10^{-9}/36\pi$ F/m and $\sigma = 5.8 \times 10^7$ mhos/m. What is the displacement current in the wire at a frequency $f = 10^5$ Hz?

4.25 Find the current in a cylindrical aluminum wire of radius 0.002 m if the electric field is 0.005 V/m. The constitutive parameters of aluminum are $\varepsilon = 10^{-9}/36\pi$ F/m and $\sigma = 3.72 \times 10^7$ mhos/m.

4.26 Consider a unit length of a coaxial cylindrical capacitor of inner radius a and outer radius b. Assume that the voltage difference between the inner and the outer conductors varies sinusoidally with time. Find the total displacement current directed across any cylindrical surface of radius $\rho(a < \rho < b)$, verify that it is independent of the radius and is equal to the charging current for the capacitor.

4.27 A plane $z = 0$ separates a semi-infinite region $(z < 0)$ occupied by a perfect conductor from another semi-infinite region $(z > 0)$ of free space characterized by the parameters $\varepsilon_0 = 10^{-9}/36\pi$ F/m and $\mu_0 = 4\pi \times 10^{-7}$ H/m. A plane wave is propagating in the $+x$-direction with its magnetic vector given by

$$\mathbf{H}(x) = \hat{y}H_0 f\left(t - \frac{x}{v}\right), \tag{a}$$

where f is any function. In which part of the whole space does (a) exist and what is the numerical value of v? Find the magnitude and the direction of the electric vector. Verify that all the boundary conditions are satisfied. Find the surface charge and the surface current densities. Find the value of the Poynting vector. What does the Poynting vector represent?

4.28 Two deflecting plates of length l are separated by a distance d as illustrated in Fig. P4.5. A uniform electric field $\mathbf{E} = -\hat{x}E_x$ exists between the parallel deflecting plates. An electron of mass m and charge q enters the electric field region at time $t = 0$. The initial conditions at $t = 0$

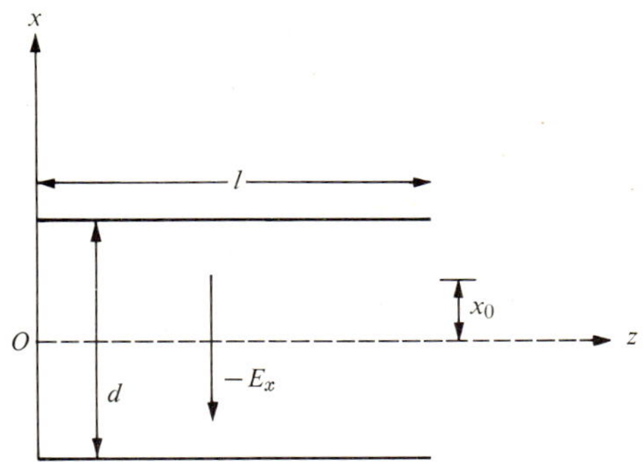

Fig. P4.5 Electrostatic deflection of a uniformly moving electron.

are $x = 0, z = 0, v_x = 0$, and $v_z = v_{z0}$. Write the force equation obtained by equating the Lorentz force to the force given by mass times the acceleration. Integrate this equation twice, and make use of the initial conditions to show that the deflection of the electron in the x-direction when it leaves the electric field region $(0 < z < l)$ is given by

$$x_0 = \frac{q}{m}\frac{E_x}{2}\left(\frac{l}{v_{z0}}\right)^2.$$

If the electric field is $E_x = -5 \times 10^4$ V/m, $v_{z0} = 5 \times 10^7$ m/sec, $l = 0.05$ m and $x_0 = 4.4$ mm, find the ratio of the charge to the mass of an electron.

4.29 Consider the following electromagnetic field:

$$E_y(x, z) = 2jE_{y0} \sin(\beta x \cos\theta) e^{-j\beta z \sin\theta},$$

$$H_x(x, z) = -2j\frac{\sin\theta}{\eta} E_{y0} \sin(\beta x \cos\theta) e^{-j\beta z \sin\theta},$$

and

$$H_z(x, z) = -2\frac{\cos\theta}{\eta} E_{y0} \cos(\beta x \cos\theta) e^{-j\beta z \sin\theta},$$

where E_{y0}, β, θ, and η are real. The other field components are zero. Find the magnitude and the direction of the time-averaged Poynting vector.

4.30 A form of matter known as plasma consists of free space permeated by free charged particles, both negative and positive and equal quantities of them. When a time-varying electric field $\boldsymbol{E}e^{j\omega t}$ acts on a charge, it creates a force and the particle acquires the velocity $\boldsymbol{v}e^{j\omega t}$. Neglect the motion of the heavy positive charged particles. Let N be the number of negative charged particles per unit volume, m the mass of each particle and $-e$ the charge on each particle. Find the force acting on each particle in terms of (i) the electric field and (ii) the velocity. Determine the convection current density $\boldsymbol{J}_{cv}$ and the expression for the conductivity σ. Show that this "conductor" is lossless.

Determine the displacement current density $\boldsymbol{J}_d$ in this conductor and hence show that the effective permittivity of the conducting medium is

$$\varepsilon_0\left(1 - \frac{\omega_p^2}{\omega^2}\right)$$

where $\omega_p^2 = Ne^2/m\varepsilon_0$.

CHAPTER 5

PLANE WAVES IN DIELECTRIC AND CONDUCTING MEDIA

5.1 INTRODUCTION

Many of the features of electromagnetic wave propagation are revealed by a study of the characteristics of plane waves in an unbounded, homogeneous medium, with the result that the analysis of plane waves is a subject of fundamental importance in electromagnetic theory. In this chapter, we shall treat the propagation in a lossless dielectric medium of homogeneous plane waves with arbitrary time dependence both for the case in which the direction of propagation is along one of the axes of the coordinate system and for the case in which it is arbitrarily directed with reference to the coordinate axes of the reference system. We shall examine the response of a highly conducting medium to one-dimensional electromagnetic fields of arbitrary time dependence and emphasize the resulting phenomenon of diffusion and contrast it with the phenomenon of propagation at a finite velocity in a lossless dielectric medium.

We shall then consider the important case of plane waves with harmonic time dependence. We shall discuss for time-harmonic fields, as we did in the case of fields with arbitrary time dependence, the propagation along one of the coordinate axes as well as in an arbitrary direction with reference to the coordinate axes. We shall examine the values of the intrinsic impedance and the propagation coefficient for the various limiting cases, namely, a perfect dielectric, a low-loss dielectric, and a highly conducting medium. We shall study the polarization of plane waves propagating in a given direction and bring out the differences among the linear, the circular, and the elliptic polarizations.

We shall investigate in detail the important problem of the time-harmonic electromagnetic fields in a highly conducting medium and thoroughly examine the concepts of skin depth and surface impedance.

5.2 PROPAGATION OF HOMOGENEOUS PLANE WAVES WITH ARBITRARY TIME DEPENDENCE IN A DIELECTRIC MEDIUM

In a source-free region, the electric $\boldsymbol{E}(\boldsymbol{r}, t)$ and the magnetic $\boldsymbol{H}(\boldsymbol{r}, t)$ fields satisfy the following Maxwell's equations:

$$\nabla \times \boldsymbol{E}(\boldsymbol{r}, t) = -\mu \frac{\partial}{\partial t} \boldsymbol{H}(\boldsymbol{r}, t), \tag{2.1}$$

$$\nabla \times \boldsymbol{H}(\boldsymbol{r}, t) = \varepsilon \frac{\partial}{\partial t} \boldsymbol{E}(\boldsymbol{r}, t) + \sigma \boldsymbol{E}(\boldsymbol{r}, t), \tag{2.2}$$

where μ, ε, and σ are respectively the permeability, the permittivity, and the conductivity of the medium. If we write (2.1) and (2.2) in component form in the Cartesian coordinate system, they become

$$\frac{\partial}{\partial y} E_z(\mathbf{r}, t) - \frac{\partial}{\partial z} E_y(\mathbf{r}, t) = -\mu \frac{\partial}{\partial t} H_x(\mathbf{r}, t), \tag{2.3}$$

$$\frac{\partial}{\partial z} E_x(\mathbf{r}, t) - \frac{\partial}{\partial x} E_z(\mathbf{r}, t) = -\mu \frac{\partial}{\partial t} H_y(\mathbf{r}, t), \tag{2.4}$$

$$\frac{\partial}{\partial x} E_y(\mathbf{r}, t) - \frac{\partial}{\partial y} E_x(\mathbf{r}, t) = -\mu \frac{\partial}{\partial t} H_z(\mathbf{r}, t), \tag{2.5}$$

$$\frac{\partial}{\partial y} H_z(\mathbf{r}, t) - \frac{\partial}{\partial z} H_y(\mathbf{r}, t) = \varepsilon \frac{\partial}{\partial t} E_x(\mathbf{r}, t) + \sigma E_x(\mathbf{r}, t), \tag{2.6}$$

$$\frac{\partial}{\partial z} H_x(\mathbf{r}, t) - \frac{\partial}{\partial x} H_z(\mathbf{r}, t) = \varepsilon \frac{\partial}{\partial t} E_y(\mathbf{r}, t) + \sigma E_y(\mathbf{r}, t), \tag{2.7}$$

$$\frac{\partial}{\partial x} H_y(\mathbf{r}, t) - \frac{\partial}{\partial y} H_x(\mathbf{r}, t) = \varepsilon \frac{\partial}{\partial t} E_z(\mathbf{r}, t) + \sigma E_z(\mathbf{r}, t). \tag{2.8}$$

Let us consider the special case in which the field quantities do not depend on y and z but are functions of only x and t. Then, (2.3) to (2.8) can be reduced to

$$0 = \frac{\partial}{\partial t} H_x(x, t), \tag{2.9}$$

$$-\frac{\partial}{\partial x} E_z(x, t) = -\mu \frac{\partial}{\partial t} H_y(x, t), \tag{2.10}$$

$$\frac{\partial}{\partial x} E_y(x, t) = -\mu \frac{\partial}{\partial t} H_z(x, t), \tag{2.11}$$

$$0 = \varepsilon \frac{\partial}{\partial t} E_x(x, t) + \sigma E_x(x, t), \tag{2.12}$$

$$-\frac{\partial}{\partial x} H_z(x, t) = \varepsilon \frac{\partial}{\partial t} E_y(x, t) + \sigma E_y(x, t), \tag{2.13}$$

$$\frac{\partial}{\partial x} H_y(x, t) = \varepsilon \frac{\partial}{\partial t} E_z(x, t) + \sigma E_z(x, t). \tag{2.14}$$

If $\sigma = 0$, the energy dissipated in the medium is equal to zero and the medium is lossless. We shall treat the properties of the fields in a lossless medium first. In a lossless medium, if we restrict our attention only to the time-varying parts of the fields, it follows from (2.9) and (2.12) that

$$H_x(x, t) = E_x(x, t) = 0. \tag{2.15}$$

Also, (2.11) and (2.13) form one set which contains $E_y(x, t)$ and $H_z(x, t)$. The other independent set containing $E_z(x, t)$ and $H_y(x, t)$ is formed by (2.10) and (2.14). These two independent sets of equations are given by

$$-\frac{\partial}{\partial x}E_y(x,t)=\mu\frac{\partial}{\partial t}H_z(x,t);\qquad -\frac{\partial}{\partial x}H_z(x,t)=\varepsilon\frac{\partial}{\partial t}E_y(x,t);\quad (2.16\text{a,b})$$

$$\frac{\partial}{\partial x}E_z(x,t)=\mu\frac{\partial}{\partial t}H_y(x,t);\qquad \frac{\partial}{\partial x}H_y(x,t)=\varepsilon\frac{\partial}{\partial t}E_z(x,t).\quad (2.17\text{a,b})$$

We can ascertain that (2.16a,b) and (2.17a,b) are similar to the transmission-line equations (1.3.1) and (1.3.2), and we can therefore solve them in the following manner. If we eliminate $H_z(x, t)$ from (2.16a) and (2.16b) we find

$$\frac{\partial^2}{\partial x^2}E_y(x,t)=\frac{1}{v^2}\frac{\partial^2}{\partial t^2}E_y(x,t),\tag{2.18}$$

where

$$v=1/\sqrt{\mu\varepsilon}\tag{2.19}$$

has the dimensions of velocity. We can obtain the time-varying part of the general solution of (2.18) as

$$E_y(x,t)=f_1(x-vt)+f_2(x+vt),\tag{2.20}$$

where f_1 and f_2 are unspecified functions of their respective arguments. From (2.16a) and (2.20), we find that

$$\begin{aligned}\frac{\partial}{\partial t}H_z(x,t)&=-\frac{1}{\mu}\frac{\partial}{\partial x}E_y(x,t),\\&=-\frac{1}{\mu}f'_1(x-vt)-\frac{1}{\mu}f'_2(x+vt),\end{aligned}\tag{2.21}$$

where the primes denote differentiation with respect to the argument. Since

$$\frac{\partial}{\partial t}f_1(x-vt)=-vf'_1(x-vt),\tag{2.22a}$$

and

$$\frac{\partial}{\partial t}f_2(x+vt)=vf'_2(x+vt),\tag{2.22b}$$

(2.21) becomes

$$\frac{\partial}{\partial t}H_z(x,t)=\frac{1}{\eta_0}\frac{\partial}{\partial t}f_1(x-vt)-\frac{1}{\eta_0}\frac{\partial}{\partial t}f_2(x+vt),\tag{2.23}$$

where

$$\eta_0=\mu v=\sqrt{\mu/\varepsilon}\tag{2.24}$$

has the dimensions of impedance. The time-varying part of the solution of (2.23) gives

$$H_z(x,t)=\frac{1}{\eta_0}f_1(x-vt)-\frac{1}{\eta_0}f_2(x+vt).\tag{2.25}$$

We have established in Section 1.4 that $f_1(x - vt)$ represents a wave traveling in the $+x$-direction with a uniform velocity v and without any attenuation or distortion. In a similar manner, $f_2(x + vt)$ represents a wave traveling in the $-x$-direction with the same uniform velocity v and, also, without any attenuation or distortion. The wave fronts of these waves are *planes* perpendicular to the x-axis. The amplitudes of the fields are constants in any wavefront. Thus, (2.20) and (2.25) represent homogeneous plane waves; they propagate either in the $+x$- or the $-x$-direction but with the same uniform velocity. Also, we can see from (2.15) that there are no components of either the electric or the magnetic fields in the direction of propagation; thus, a transverse-electromagnetic (TEM) wave is obtained. Moreover, we find from (2.15), (2.20), and (2.25) that the electric field, the magnetic field, and the direction of propagation, denoted for convenience by $\hat{s}$, form a triad of mutually perpendicular vectors. Also, it follows from (2.20) and (2.25) that

$$\frac{E_y(x, t)}{H_z(x, t)} = \eta_0 \quad (\text{for } \hat{s} = \hat{x}); \qquad \frac{E_y(x, t)}{H_z(x, t)} = -\eta_0 \quad (\text{for } \hat{s} = -\hat{x}). \tag{2.26}$$

Since the electric field has the dimensions of volts per meter and the magnetic field has the dimensions of amperes per meter, the ratio of the electric to the magnetic field has the dimensions of impedance, that is, ohms, and this ratio is called the intrinsic impedance of the medium. We can see that the sign of this ratio depends on the direction of propagation and we shall give a systematic method for ascertaining the sign of the intrinsic impedance later.

We can obtain the time-varying parts of the solution of (2.17a,b) in the same manner as that used for the solution of (2.16a,b) with the following results:

$$E_z(x, t) = g_1(x - vt) + g_2(x + vt), \tag{2.27}$$

$$H_y(x, t) = -\frac{1}{\eta_0} g_1(x - vt) + \frac{1}{\eta_0} g_2(x + vt), \tag{2.28}$$

and

$$\frac{E_z(x, t)}{H_y(x, t)} = -\eta_0 \quad (\text{for } \hat{s} = \hat{x}); \qquad \frac{E_z(x, t)}{H_y(x, t)} = \eta_0 \quad (\text{for } \hat{s} = -\hat{x}). \tag{2.29}$$

As before, (2.27) and (2.28) represent homogeneous plane TEM waves propagating with uniform velocity v in the $+x$- or the $-x$-directions without any attenuation or distortion. Even for the second independent set, the electric field, the magnetic field, and the direction of propagation form a triad of mutually perpendicular vectors. Indeed, it is interesting to observe that we can obtain the second set given by (2.17a,b) from the first set given by (2.16a,b) by a rotation through 90 degrees about the direction of propagation. This rotation transforms $E_y(x, t)$ and $H_z(x, t)$ into $E_z(x, t)$ and $-H_y(x, t)$ respectively. When we apply this transformation to (2.16a,b), we can verify that (2.17a,b) results. In a similar manner, we can apply this transformation to (2.20), (2.25), and (2.26) to obtain the general forms of the solutions given by (2.27), (2.28), and (2.29) respectively. Since (2.16a,b) and (2.17a,b) form two independent sets, the actual functions f_1 and f_2 need not necessarily be equal to g_1 and g_2 respectively.

The whole disturbance together with the associated electric and magnetic fields of the homogeneous plane wave under consideration moves in the $+x$- or the $-x$-directions. Since the disturbance carries with it the associated energy, it follows that the power flux density or the Poynting vector is in the same direction as the direction of motion of the wavefront. We can utilize this observation in obtaining the sign of the intrinsic impedance which is given in (2.26) and (2.29). For example, let us rewrite (2.26) as

$$\frac{E_y(x, t)}{H_z(x, t)} = \eta_0 \text{ sign}, \tag{2.30}$$

where the sign is related to the direction of propagation. For the plane wave with $E_y(x, t)$ and $H_z(x, t)$ as the only field components, we can obtain the Poynting vector $\boldsymbol{S}(x, t)$ with the help of (2.30) as

$$\boldsymbol{S}(x, t) = \hat{s}S(x, t) = \hat{y}E_y(x, t) \times \hat{z}H_z(x, t) = \hat{x}E_y^2(x, t)/\text{sign } \eta_0 \tag{2.31}$$

since $\hat{y} \times \hat{z} = \hat{x}$. We should note that $\hat{s}$ is the direction of propagation and the magnitude $S(x, t)$ of the Poynting vector is equal to $E_y^2(x, t)/\eta_0$, which is positive. Therefore, from (2.31), we see that

$$\text{sign} = \hat{x} \cdot \hat{s}. \tag{2.32}$$

Hence, the sign is seen from (2.32) to be $+$ or $-$ according to whether the direction $\hat{s}$ of propagation is in the $+x(\hat{s} = \hat{x})$- or the $-x(\hat{s} = -\hat{x})$-directions respectively. Thus, (2.30) together with the value of the sign as obtained from the consideration of the Poynting vector is in agreement with the results stated in (2.26).

5.3 PROPAGATION OF GENERAL ONE-DIMENSIONAL FIELDS WITH ARBITRARY TIME DEPENDENCE IN A DIELECTRIC MEDIUM

A field which depends on time and on distance measured along a single axis in space is called a one-dimensional field. The field which we discussed in the previous section is an example of a one-dimensional field in which the preferred direction coincides with an axis of the Cartesian coordinate system. We now wish to treat a general one-dimensional field in which the preferred direction does not necessarily coincide with a coordinate axis of the reference system. Let the field be a function of a coordinate ξ measured along a line whose direction is defined by the unit vector $\hat{n}$. Let n_x, n_y, and n_z be respectively the x, y, and z components of the unit vector $\hat{n}$ with the result that

$$\hat{n} = n_x\hat{x} + n_y\hat{y} + n_z\hat{z}. \tag{3.1}$$

It is conventional to designate n_x, n_y, and n_z as the direction cosines of the new coordinate axis ξ. At each instant of time, the directions and the magnitudes of the vectors $\boldsymbol{E}(\boldsymbol{r}, t)$ and $\boldsymbol{H}(\boldsymbol{r}, t)$ are constants over planes perpendicular to $\hat{n}$. If $\boldsymbol{r}$ is the

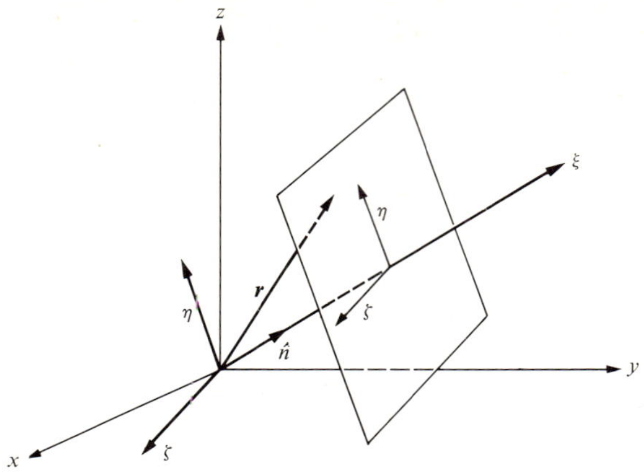

Fig. 5.1 Propagation of homogeneous plane waves in an arbitrary direction specified by the unit vector $\hat{n}$.

radius vector drawn from the origin to any point in space, then

$$\xi = \hat{n} \cdot \boldsymbol{r} = \text{constant} \tag{3.2}$$

defines the planes normal to $\hat{n}$ (Fig. 5.1). The distance ξ along the new coordinate axis is therefore given by

$$\xi = n_x x + n_y y + n_z z, \tag{3.3}$$

where x, y, and z are the Cartesian coordinates of the point in space under consideration. From (3.3), it follows that

$$d\xi = n_x\, dx + n_y\, dy + n_z\, dz, \tag{3.4}$$

where dx, dy, and dz are the changes in x, y, and z respectively and $d\xi$ is the resulting total change in ξ. The field quantities are of the form

$$\boldsymbol{E}(\boldsymbol{r}, t) = \boldsymbol{E}(\xi, t); \qquad \boldsymbol{H}(\boldsymbol{r}, t) = \boldsymbol{H}(\xi, t). \tag{3.5}$$

Hence, if F represents a component of the field, we can write the total change dF in F due to the changes dx, dy, and dz in x, y, and z respectively in the following two forms:

$$dF = \frac{\partial F}{\partial \xi}\, d\xi = \frac{\partial F}{\partial x}\, dx + \frac{\partial F}{\partial y}\, dy + \frac{\partial F}{\partial z}\, dz. \tag{3.6}$$

If we substitute the expression for $d\xi$ given by (3.4) in (3.6) and equate the coefficients of dx, dy, and dz, we find that

$$\frac{\partial}{\partial x} = n_x \frac{\partial}{\partial \xi}; \qquad \frac{\partial}{\partial y} = n_y \frac{\partial}{\partial \xi}; \qquad \frac{\partial}{\partial z} = n_z \frac{\partial}{\partial \xi}. \tag{3.7}$$

The ∇ operator can be simplified with the help of (3.7) as follows:

$$\nabla = \hat{x}\frac{\partial}{\partial x} + \hat{y}\frac{\partial}{\partial y} + \hat{z}\frac{\partial}{\partial z} = (\hat{x}n_x + \hat{y}n_y + \hat{z}n_z)\frac{\partial}{\partial \xi} = \hat{n}\frac{\partial}{\partial \xi}. \tag{3.8}$$

Together with (3.5) and (3.8), Maxwell's equations (2.1) and (2.2) become

$$\hat{n} \times \frac{\partial}{\partial \xi} \boldsymbol{E}(\xi, t) = -\mu \frac{\partial}{\partial t} \boldsymbol{H}(\xi, t), \tag{3.9}$$

and

$$\hat{n} \times \frac{\partial}{\partial \xi} \boldsymbol{H}(\xi, t) = \varepsilon \frac{\partial}{\partial t} \boldsymbol{E}(\xi, t) + \sigma \boldsymbol{E}(\xi, t). \tag{3.10}$$

It is convenient for us to introduce two mutually perpendicular coordinates η and ζ lying in a plane perpendicular to $\hat{n}$ such that ξ, η, and ζ form a triad of mutually perpendicular right-handed system of vectors. Then, we can write $\boldsymbol{E}(\xi, t)$ and $\boldsymbol{H}(\xi, t)$ in component form as follows:

$$\boldsymbol{E}(\xi, t) = \hat{n} E_\xi(\xi, t) + \hat{\eta} E_\eta(\xi, t) + \hat{\zeta} E_\zeta(\xi, t), \tag{3.11}$$

and

$$\boldsymbol{H}(\xi, t) = \hat{n} H_\xi(\xi, t) + \hat{\eta} H_\eta(\xi, t) + \hat{\zeta} H_\zeta(\xi, t), \tag{3.12}$$

where $\hat{\eta}$ and $\hat{\zeta}$ are unit vectors specifying the directions of the coordinates η and ζ respectively. On substituting (3.11) and (3.12) into (3.9) and (3.10) and writing the resulting equations in component form, we find that

$$0 = \frac{\partial}{\partial t} H_\xi(\xi, t), \tag{3.13}$$

$$-\frac{\partial}{\partial \xi} E_\zeta(\xi, t) = -\mu \frac{\partial}{\partial t} H_\eta(\xi, t), \tag{3.14}$$

$$\frac{\partial}{\partial \xi} E_\eta(\xi, t) = -\mu \frac{\partial}{\partial t} H_\zeta(\xi, t), \tag{3.15}$$

$$0 = \varepsilon \frac{\partial}{\partial t} E_\xi(\xi, t) + \sigma E_\xi(\xi, t), \tag{3.16}$$

$$-\frac{\partial}{\partial \xi} H_\zeta(\xi, t) = \varepsilon \frac{\partial}{\partial t} E_\eta(\xi, t) + \sigma E_\eta(\xi, t), \tag{3.17}$$

$$\frac{\partial}{\partial \xi} H_\eta(\xi, t) = \varepsilon \frac{\partial}{\partial t} E_\zeta(\xi, t) + \sigma E_\zeta(\xi, t). \tag{3.18}$$

We should note that (3.13) to (3.18) are exactly identical to (2.9) to (2.14) respectively if x, y, and z are replaced by ξ, η, and ζ respectively. In a lossless medium ($\sigma = 0$), if we restrict our attention to only the time-varying parts of the fields, it follows from (3.13) and (3.16) that

$$H_\xi(\xi, t) = E_\xi(\xi, t) = 0. \tag{3.19}$$

As in the set of equations (2.9) to (2.14), the remaining equations in the set (3.13) to (3.18) are also capable of being decomposed into two independent sets. The first set consists of the field components $E_\eta(\xi, t)$ and $H_\zeta(\xi, t)$. The time-varying parts of the general solution of $E_\eta(\xi, t)$ and $H_\zeta(\xi, t)$ as specified by (3.15) and (3.17) with

$\sigma = 0$ are similar to those given in (2.20) and (2.25) and are easily found to be

$$E_\eta(\xi, t) = f_1(\xi - vt) + f_2(\xi + vt); \tag{3.20}$$

$$H_\zeta(\xi, t) = \frac{f_1}{\eta_0}(\xi - vt) - \frac{f_2}{\eta_0}(\xi + vt). \tag{3.21}$$

Also,

$$\frac{E_\eta(\xi, t)}{H_\zeta(\xi, t)} = \eta_0 \quad (\text{for } \hat{s} = \hat{n}); \qquad \frac{E_\eta(\xi, t)}{H_\zeta(\xi, t)} = -\eta_0 \quad (\text{for } \hat{s} = -\hat{n}). \tag{3.22}$$

It is seen that (3.20) and (3.21) represent homogeneous plane TEM waves propagating with uniform velocity v in the + or $-\xi$-directions without any attenuation or distortion. Note that the + or $-\xi$-directions are the same as the directions specified by $+\hat{n}$ or $-\hat{n}$ respectively. Also, the electric field, the magnetic field, and the direction of propagation form a triad of mutually perpendicular vectors. The ratio $E_\eta(\xi, t)/H_\zeta(\xi, t)$ is, as before, the intrinsic impedance whose exact sign we can determine by using the fact that the direction of the Poynting vector is the same as that of the motion of the wavefront.

The second set consisting of (3.14) and (3.18) with $\sigma = 0$, is formed by the field components $E_\zeta(\xi, t)$ and $H_\eta(\xi, t)$. The time-varying parts of the solution of $E_\zeta(\xi, t)$ and $H_\eta(\xi, t)$ are similar to those in (2.27) and (2.28) and are deduced as

$$E_\zeta(\xi, t) = g_1(\xi - vt) + g_2(\xi + vt), \tag{3.23}$$

and

$$H_\eta(\xi, t) = -\frac{1}{\eta_0}g_1(\xi - vt) + \frac{1}{\eta_0}g_2(\xi + vt). \tag{3.24}$$

Also, we see from (3.23) and (3.24) that

$$\frac{E_\zeta(\xi, t)}{H_\eta(\xi, t)} = -\eta_0 \quad (\text{for } \hat{s} = \hat{n}); \qquad \frac{E_\zeta(\xi, t)}{H_\eta(\xi, t)} = \eta_0 \quad (\text{for } \hat{s} = -\hat{n}). \tag{3.25}$$

The physical interpretations of (3.23) to (3.25) are similar to those we gave for (3.20) to (3.22) respectively.

We have deduced a representation for a plane wave propagating in an arbitrary direction $\hat{n}$ with respect to the Cartesian coordinates x, y, and z to be given by either (3.20) and (3.21) or (3.23) and (3.24). A linear combination of the electric and the magnetic fields given by these two sets also represents a plane wave propagating in the arbitrary direction $\hat{n}$. Since $\hat{n}$ is a unit vector, it follows from (3.1) that

$$n_x^2 + n_y^2 + n_z^2 = 1. \tag{3.26}$$

5.4 DIFFUSION OF ONE-DIMENSIONAL FIELDS IN A HIGHLY CONDUCTING MEDIUM

Let us consider the case of one-dimensional fields which depend only on x and t in a conducting medium characterized by the constitutive parameters μ, ε, and σ. The field components satisfy Maxwell's equations which in component form are the same as those given by (2.9) to (2.14). If we consider the time-varying parts of the fields alone, we see from (2.9) that

$$H_x(x, t) = 0. \tag{4.1}$$

It is convenient for us to designate the direction x of transmission of the disturbance as the longitudinal direction. Thus, the longitudinal component of the magnetic field is zero. From (2.12), we see that if the fields do not vary in time, the longitudinal component of the static electric field is zero. But, if we consider the time-varying parts of the fields, $E_x(x, t) \neq 0$, and we can obtain it as

$$E_x(x, t) = E_x(x, 0)e^{-(\sigma/\varepsilon)t} = E_x(x, 0)e^{-t/\tau}, \tag{4.2}$$

where $E_x(x, 0)$ is the value of $E_x(x, t)$ at time $t = 0$. The factor $\tau = \varepsilon/\sigma$ which controls the rate of decay in time is called the relaxation time. It follows from (4.2) that as the conductivity σ is increased, the relaxation time τ is decreased and the rate of exponential decay of the longitudinal component of the electric field is increased. An important difference between the fields in a lossless dielectric medium treated in Section 5.2 and those in a conducting medium is that the fields in a lossless dielectric medium are of the TEM type whereas those in a conducting medium are of the TM type. As we indicated previously, the TEM wave has no component of either the electric or the magnetic field in the longitudinal direction but a TM wave has no component of the magnetic field alone in the longitudinal direction.

An important special case is that of a highly conducting medium where the conduction current density $\sigma\boldsymbol{E}(x, t)$ is very much larger than the displacement current density $\varepsilon(\partial/\partial t)\boldsymbol{E}(x, t)$ with the result that we can omit the latter in comparison with the former. As before (2.11) and (2.13) containing only $E_y(x, t)$ and $H_z(x, t)$ form one set which, for a highly conducting medium, is approximated by

$$-\frac{\partial}{\partial x}E_y(x, t) = \mu\frac{\partial}{\partial t}H_z(x, t), \tag{4.3}$$

$$-\frac{\partial}{\partial x}H_z(x, t) = \sigma E_y(x, t). \tag{4.4}$$

The other independent set consists of (2.10) and (2.14) and these contain only the field components $E_z(x, t)$ and $H_y(x, t)$. Moreover, we can obtain the second set from the first by replacing $E_y(x, t)$ and $H_z(x, t)$ by $E_z(x, t)$ and $-H_y(x, t)$ respectively. Hence, it is sufficient for us to evaluate the solution of the first set of equations which are reproduced in (4.3) and (4.4) for the special case of a highly conducting medium.

Let a highly conducting medium occupy the half-space $0 < x < \infty$. Suppose $H_z(x, t)$ at $x = 0$ is given by

$$H_z(0, t) = H_0 H(t), \tag{4.5}$$

where H_0 is a constant and $H(t)$ is the Heaviside unit step function. In other words, a constant value H_0 of the z-component of the magnetic field is switched on at time $t = 0$ over the entire $x = 0$ plane. It is instructive for us to examine how the highly conducting medium responds to this excitation. It is helpful for us to note that (4.3),

(4.4), and (4.5) are identical to (2.4.3), (2.4.2), and (2.4.1) respectively with $E_y(x, t)$, $H_z(x, t)$, μ, σ, and H_0 substituted for $I(x, t)$, $V(x, t)$, C, R, and V_0 respectively. Therefore, we obtain $E_y(x, t)$ and $H_z(x, t)$ in the highly conducting medium by comparison with (2.4.6) and (2.4.20) as

$$E_y(x, t) = H_0 H(t) \sqrt{\frac{\mu}{\pi\sigma t}}\, e^{-\mu\sigma x^2/4t}, \tag{4.6}$$

and

$$H_z(x, t) = H_0 H(t)\, \mathrm{erfc}\left(\frac{1}{\sqrt{T}}\right), \qquad T = \frac{4t}{\mu\sigma x^2}, \tag{4.7}$$

where erfc is the complementary error function whose definition, as stated in (2.4.14), is reproduced as:

$$\mathrm{erfc}\,(y_0) = \frac{2}{\sqrt{\pi}} \int_{y_0}^{\infty} e^{-y^2}\, dy. \tag{4.8}$$

The complementary error function is a monotonically decreasing function of its argument. It has the value unity for $y_0 = 0$, decreases as y_0 is increased, and becomes zero for $y_0 = \infty$.

The field components as given in (4.6) and (4.7) reveal two important features. First for $t < 0$, $E_y(x, t)$, and $H_z(x, t)$ are zero everywhere in the conducting medium. We can anticipate this result since there can be no response before the excitation which as given in (4.5) starts only at $t = 0$. Second, as soon as t is greater than zero, no matter by how small a quantity, that is after the source comes into existence, $E_y(x, t)$ and $H_z(x, t)$ are nonvanishing everywhere in the conducting medium. The fields are then said to be *diffused* into the conducting medium. In contrast to a perfect dielectric, the fields are not propagated without distortion or attenuation and with a finite velocity.

Let us consider the limiting case in which the highly conducting medium is a perfect conductor corresponding to $\sigma = \infty$. For $x > 0$, we see from (4.6) and (4.7) that there are no electromagnetic fields in the interior of a perfect conductor. For $x = 0$, (4.6) shows that $E_y(x, t) = 0$ for $\sigma = \infty$ with the result that, in conformity with the usual boundary condition, there is no tangential component of the electric field even on the surface of a perfect conductor. For $x = 0$, (4.7) yields $H_z(0, t) = H_0 H(t)$ and hence, the surface of a perfect conductor can support a nonvanishing tangential component of the magnetic field. There is, therefore, a nonvanishing surface current density $\boldsymbol{J}_s(t)$ which we obtain as

$$\boldsymbol{J}_s(t) = \hat{n} \times \hat{z} H_0 H(t) = -\hat{x} \times \hat{z} H_0 H(t) = \hat{y} H_0 H(t), \tag{4.9}$$

where $\hat{n}$ is an outwardly drawn unit normal to the surface $x = 0$ of the perfect conductor. For a very good but not a perfect conductor, that is, corresponding to σ very large but not infinite, we see from (4.6) that the volume conduction current density $\boldsymbol{J}(x, t) = \hat{y}\sigma E_y(x, t)$ at every instant of time $t > 0$ decays rapidly in an exponential fashion with the distance from the surface inside the good conductor. Since the conduction current density $\boldsymbol{J}(x, t)$ is concentrated near the surface of a good conductor, it is reasonable for us to introduce the concept of *equivalent* surface current density which is the volume current density integrated over the entire depth from $x = 0$ to $x = \infty$. Thus from (4.6), we obtain the equivalent surface current density $\boldsymbol{J}_s^e(t)$ as

$$\boldsymbol{J}_s^e(t) = \hat{y} \int_0^\infty \sigma H_0 H(t) \sqrt{\frac{\mu}{\pi \sigma t}}\, e^{-\mu\sigma x^2/4t}\, dx. \tag{4.10}$$

On changing the variable of integration to $y = \sqrt{\mu\sigma/4t}\, x$, (4.10) becomes

$$\boldsymbol{J}_s^e(t) = \hat{y} H_0 H(t) \operatorname{erfc}(0) = \hat{y} H_0 H(t). \tag{4.11}$$

We can also derive this result directly by integrating both sides of (4.4) with respect to x, from $x = 0$ to $x = \infty$. It is interesting to note that for a specified tangential component of the magnetic field on the bounding plane surface $x = 0$, the actual surface current density we obtained in (4.9) for the case of a perfect conductor and the equivalent surface current density we deduced in (4.11) for the case of a very good conductor are identical. In other words, the surface current density that exists on the surface of a half-space of a perfect conductor is spread throughout the half-space for a very good conductor with the current density concentrated near the surface and falling off exponentially with the distance into the interior of the medium.

We see from (4.7) that for $t > 0$, $H_z(x, t)$ depends only upon T. From Fig. 5.2, where $H_z(x, t)$ is depicted as a function of T, we see $H_z(x, t)$ to increase monotonically

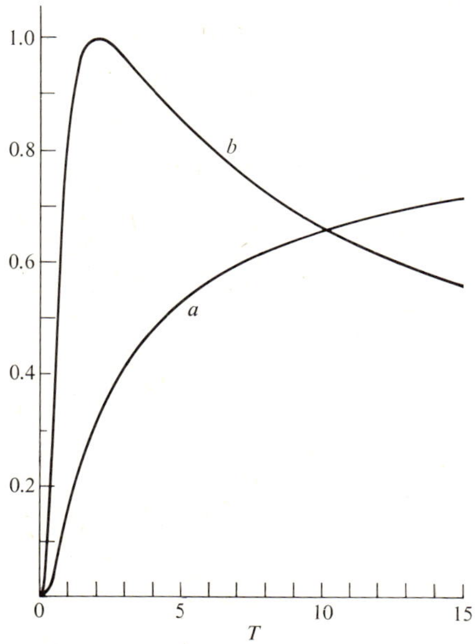

Fig. 5.2 Building-up of magnetic field and volume current density in a half-space of a good conductor. (a) $H_z(x, t)/H_0$; (b) $\tilde{J}(x, t)$; $T = 4t/\mu\sigma x^2$.

from zero and attain the value H_0 for $T = \infty$. If the time t is directly proportional to the square of the distance from the surface we see from (4.7) that T is invariant to the changes in x and hence $H_z(x, t)$ is also not altered. Therefore, the time required for $H_z(x, t)$ at any location to reach a specified value is proportional to the square of the distance of the location from the source position. Thus, the building-up of $H_z(x, t)$ at deeper and deeper locations in the highly conducting medium takes place at progressively smaller rates and thus requires longer time. In the same manner, if the time t is directly proportional to the conductivity σ, we see from (4.7) that the changes in σ do not affect T and therefore, $H_z(x, t)$ either. Hence, the time t required for $H_z(x, t)$ to build up to a specified value at a given location increases proportionately with the increase in the conductivity σ.

The building-up of the volume conduction current density $J_y(x, t) = \sigma E_y(x, t)$ is even more interesting. For $t > 0$ at any location in the highly conducting medium, we can show from (4.6) that $J_y(x, t)$ attains the maximum value given by

$$J_{y,\max}(x, t) = \sqrt{\frac{2}{\pi e}}\frac{H_0}{x}, \tag{4.12}$$

at time

$$t = t_{\max} = \frac{\mu\sigma x^2}{2}. \tag{4.13}$$

With the help of (4.6) and (4.12), we can show that for $t > 0$

$$\tilde{J}_y(x, t) = \frac{J_y(x, t)}{J_{y,\max}(x, t)} = \sqrt{\frac{2e}{T}}\, e^{-1/T}; \qquad T = 4t/\mu\sigma x^2. \tag{4.14}$$

From (4.12) to (4.14), we see that $\tilde{J}_y(x, t)$ reaches the maximum value of unity at $T = 2$. Figure 5.2, which also depicts a plot $\tilde{J}_y(x, t)$ as a function of T, confirms these results. It follows from (4.12) that deeper and deeper in the highly conducting medium, the conduction current density builds up to a smaller and smaller maximum value which is inversely proportional to the distance from the source location. Also, we see from (4.13) that the time required for $J_y(x, t)$ to build up to this smaller maximum increases as the square of the distance from the source location as well as linearly with the conductivity of the medium. Thus, at the locations which are deeper in the highly conducting medium, the conduction current density builds up to a smaller maximum at slower rates, the rate being inversely proportional to the conductivity of the medium. After reaching the maximum value, the conduction current density decreases with the increase in time and ultimately vanishes for $t = \infty$.

5.5 PROPAGATION OF TIME-HARMONIC, PLANE WAVE FIELDS IN A DIELECTRIC MEDIUM

We may assume the time-harmonic electric and magnetic fields to be of the form

$$\boldsymbol{E}(\boldsymbol{r}, t) = \boldsymbol{E}(\boldsymbol{r})e^{j\omega t}, \qquad \boldsymbol{H}(\boldsymbol{r}, t) = \boldsymbol{H}(\boldsymbol{r})e^{j\omega t}, \tag{5.1}$$

where, in a source-free region, $\boldsymbol{E}(\boldsymbol{r})$ and $\boldsymbol{H}(\boldsymbol{r})$ satisfy the following time-harmonic

Maxwell's equations:

$$\nabla \times \boldsymbol{E}(\boldsymbol{r}) = -j\omega\mu \boldsymbol{H}(\boldsymbol{r}), \tag{5.2}$$

$$\nabla \times \boldsymbol{H}(\boldsymbol{r}) = j\omega\varepsilon_{\text{eff}} \boldsymbol{E}(\boldsymbol{r}). \tag{5.3}$$

The effective permittivity ε_{eff} given in (5.3) is defined by

$$\varepsilon_{\text{eff}} = \varepsilon + \frac{\sigma}{j\omega}. \tag{5.4}$$

Let $\boldsymbol{E}(\boldsymbol{r})$ and $\boldsymbol{H}(\boldsymbol{r})$ be assumed to be functions of x only. If we write (5.2) and (5.3) in component form in the Cartesian coordinate system, we obtain the following results:

$$0 = -j\omega\mu H_x(x), \tag{5.5}$$

$$-\frac{\partial}{\partial x} E_z(x) = -j\omega\mu H_y(x), \tag{5.6}$$

$$\frac{\partial}{\partial x} E_y(x) = -j\omega\mu H_z(x), \tag{5.7}$$

$$0 = j\omega\varepsilon_{\text{eff}} E_x(x), \tag{5.8}$$

$$-\frac{\partial}{\partial x} H_z(x) = j\omega\varepsilon_{\text{eff}} E_y(x), \tag{5.9}$$

$$\frac{\partial}{\partial x} H_y(x) = j\omega\varepsilon_{\text{eff}} E_z(x). \tag{5.10}$$

It follows from (5.5) and (5.8) that

$$H_x(x) = E_x(x) = 0. \tag{5.11}$$

Hence, the resulting field structure is of the TEM type. As before, $E_y(x)$ and $H_z(x)$, as specified by (5.7) and (5.9), form one independent set of field quantities. The second set of fields consists of $E_z(x)$ and $H_y(x)$, and is governed by (5.6) and (5.10). Since we can obtain the second set from the first set by replacing $E_y(x)$ and $H_z(x)$ by $E_z(x)$ and $-H_y(x)$ respectively, it is sufficient to consider in detail only the solution of (5.7) and (5.9). If we eliminate $H_z(x)$ from (5.7) and (5.9) we obtain

$$\frac{d^2}{dx^2} E_y(x) - \gamma^2 E_y(x) = 0, \tag{5.12}$$

where,

$$\gamma = \alpha + j\beta = \sqrt{-\omega^2 \mu\varepsilon_{\text{eff}}}. \tag{5.13}$$

The solution of (5.12) is given by

$$E_y(x) = E_{y0}^{+} e^{-\gamma x} + E_{y0}^{-} e^{\gamma x}, \tag{5.14}$$

where E_{y0}^{+} and E_{y0}^{-} are undetermined constants independent of x. If we substitute

(5.14) in (5.9) and integrate with respect to x we get

$$H_z(x) = \frac{E_{y0}^{+}}{\eta_0} e^{-\gamma x} - \frac{E_{y0}^{-}}{\eta_0} e^{\gamma x}, \tag{5.15}$$

where the intrinsic impedance η_0 is given by

$$\eta_0 = \eta_{0r} + j\eta_{0j} = \gamma/j\omega\varepsilon_{\text{eff}} = \sqrt{\mu/\varepsilon_{\text{eff}}}. \tag{5.16}$$

We should note that the integration of (5.9) gives a constant term in (5.15) and when we substitute (5.14) and (5.15) back in (5.7), we see that this constant term is zero.

In (5.13), α is the attenuation coefficient and β is the phase-shift coefficient. The first terms in (5.14) and (5.15) represent fields which are attenuated in the $+x$-direction and whose phase also progresses in the $+x$-direction. Therefore, these terms can be said to represent wave motion progressing in the $+x$-direction. In a similar manner, the second terms in (5.14) and (5.15) represent fields which are attenuated in the $-x$-direction and whose phase also progresses in the $-x$-direction. Thus, the second terms in (5.14) and (5.15) can be said to represent wave motion progressing in the $-x$-direction. These results are conveniently summarized as follows:

Wave Progressing in the $+x$-Direction

$$E_y(x) = E_{y0}^{+} e^{-\gamma x}; \qquad H_z(x) = \frac{E_{y0}^{+}}{\eta_0} e^{-\gamma x}. \tag{5.17}$$

$$\frac{E_y(x)}{H_z(x)} = \eta_0. \tag{5.18}$$

Wave Progressing in the $-x$-Direction

$$E_y(x) = E_{y0}^{-} e^{\gamma x}; \qquad H_z(x) = -\frac{E_{y0}^{-}}{\eta_0} e^{\gamma x}. \tag{5.19}$$

$$\frac{E_y(x)}{H_z(x)} = -\eta_0. \tag{5.20}$$

The superscripts + and − on E_{y0} indicate the directions of progression of the corresponding wave.

We obtain the time-averaged Poynting vector $\mathbf{S}(x)$ for the wave progressing in the $+x$-direction as

$$\begin{aligned} \mathbf{S}(x) &= \tfrac{1}{2} \operatorname{Re} [\hat{y}E_y(x) \times \hat{z}H_z^*(x)] \\ &= \frac{\hat{x}}{2} \operatorname{Re}\left[E_{y0}^{+} e^{-(\alpha+j\beta)x} \frac{E_{y0}^{+*}}{\eta_0^*} e^{-(\alpha-j\beta)x} \right] \\ &= \frac{\hat{x}}{2} \frac{|E_{y0}^{+}|^2 e^{-2\alpha x}}{|\eta_0|^2} \eta_{0r}. \end{aligned} \tag{5.21}$$

In a similar manner, the result for the wave progressing in the $-x$-direction is

$$\mathbf{S}(x) = -\frac{\hat{x}}{2}\frac{|E_{y0}^{-}|^2 e^{2\alpha x}}{|\eta_0|^2}\eta_{0r}. \tag{5.22}$$

It is important for us to note from (5.21) and (5.22) that the directions of the progression of the phase fronts and the attenuation of the amplitude of the wave coincide with that of the time-averaged Poynting vector. This observation is useful in the determination of the sign of the intrinsic impedance. Let

$$E_y(x)/H_z(x) = \text{sign}\ \eta_0. \tag{5.23}$$

Then, we obtain that

$$\mathbf{S}(x) = \tfrac{1}{2}\,\text{Re}\left[\hat{y}E_y(x) \times \hat{z}\frac{E_y^*(x)}{\text{sign}\ \eta_0^*}\right],$$

$$= \frac{\hat{x}\,|E_y(x)|^2}{2\ \text{sign}\ |\eta_0|^2}\eta_{0r}. \tag{5.24}$$

If the wave progresses in the $+x$-direction, then $\mathbf{S}(x)$ is in the $+x$-direction with the result that (5.24) shows that the sign is positive. On the other hand, if the wave progresses in the $-x$-direction, then $\mathbf{S}(x)$ is in the $-x$-direction and (5.24) shows that the sign is negative. These results for the sign of the intrinsic impedance are in agreement with those given by (5.18) and (5.20) respectively.

From (5.17) to (5.20) we see that the phase fronts of the fields under discussion are planes perpendicular to the x-axis. There are no field components in the longitudinal direction. The electric field, the magnetic field, and the direction of propagation form a triad of mutually perpendicular right-handed systems of vectors. The magnitudes and the directions of the field components are constants in any phase front. The magnetic field bears a constant ratio to the electric field and we can determine the sign of this ratio from the consideration that the Poynting vector gives the direction of progression of the phase fronts.

We obtain the solution of (5.6) and (5.10) in a similar fashion as follows:

Wave Progressing in the $+x$-Direction

$$E_z(x) = E_{z0}^{+}e^{-\gamma x}; \qquad H_y(x) = -\frac{E_{z0}^{+}}{\eta_0}e^{-\gamma x}. \tag{5.25}$$

$$\frac{E_z(x)}{H_y(x)} = -\eta_0. \tag{5.26}$$

Wave Progressing in the $-x$-Direction

$$E_z(x) = E_{z0}^{-}e^{\gamma x}; \qquad H_y(x) = \frac{E_{z0}^{-}}{\eta_0}e^{\gamma x}. \tag{5.27}$$

$$\frac{E_z(x)}{H_y(x)} = \eta_0. \tag{5.28}$$

The physical interpretation of (5.25) to (5.28) is very similar to that we gave for (5.17) to (5.20).

Let us consider the important case of a lossless dielectric medium. The absence of loss means that there is no dissipation of energy and this implies that $\sigma = 0$. Therefore, from (5.4), we find that

$$\varepsilon_{\text{eff}} = \varepsilon \qquad \text{(lossless medium).} \tag{5.29}$$

If we substitute (5.29) in (5.13) we obtain

$$\gamma = \alpha + j\beta = \sqrt{-\omega^2\mu\varepsilon} = j\omega\sqrt{\mu\varepsilon}, \tag{5.30}$$

showing that the attenuation coefficient $\alpha = 0$ and the phase-shift coefficient $\beta = \omega\sqrt{\mu\varepsilon} = \omega/v$ where v is the velocity of electromagnetic waves in a medium with the constitutive parameters μ, ε, and $\sigma = 0$. We then determine the phase velocity v_{ph} of the plane waves as

$$v_{ph} = \omega/\beta = v. \tag{5.31}$$

We can deduce the space period or the wavelength λ to be $\lambda = 2\pi/\beta = 2\pi v/\omega = v/f$ where f is the frequency. The intrinsic impedance as given by (5.16) reduces to

$$\eta_0 = \eta_{0r} + j\eta_{0j} = \frac{j\omega\sqrt{\mu\varepsilon}}{j\omega\varepsilon} = \sqrt{\frac{\mu}{\varepsilon}}, \tag{5.32}$$

which shows that the intrinsic impedance in a lossless dielectric is a resistance with the reactive component being equal to zero. As we discussed in Section 2.13, the velocity of propagation of the amplitude maximum of a group of waves with a small spread in the frequency as well as in the phase velocity is the group velocity v_g. For the plane wave in a lossless dielectric, we can deduce from (5.30) that

$$v_g = \partial\omega/\partial\beta = 1/\sqrt{\mu\varepsilon} = v_{ph}, \tag{5.33}$$

that is, the phase and the group velocities are equal.

For the plane wave progressing in the $+x$-direction and having its electric field oriented in the y-direction, we evaluate the time-averaged energy density w_E associated with the electric field from (5.17) and (5.30) as

$$w_E = \frac{\varepsilon}{4}\,\text{Re}\,[E_y(x)E_y^*(x)] = \frac{\varepsilon\,|E_{y0}^+|^2}{4}. \tag{5.34}$$

We obtain the time-averaged energy density w_H associated with the magnetic field from (5.17), (5.30), and (5.32) as

$$w_H = \frac{\mu}{4}\,\text{Re}\,[H_z(x)H_z^*(x)] = \frac{\mu\,|E_{y0}^+|^2}{4\eta_{0r}^2} = \frac{\varepsilon\,|E_{y0}^+|^2}{4}. \tag{5.35}$$

We deduce the time-averaged Poynting vector from (5.21) to be

$$\mathbf{S}(x) = \hat{x}\,\frac{|E_{y0}^+|^2}{2\eta_{0r}} = \frac{\hat{x}}{2}\sqrt{\frac{\varepsilon}{\mu}}\,|E_{y0}^+|^2. \tag{5.36}$$

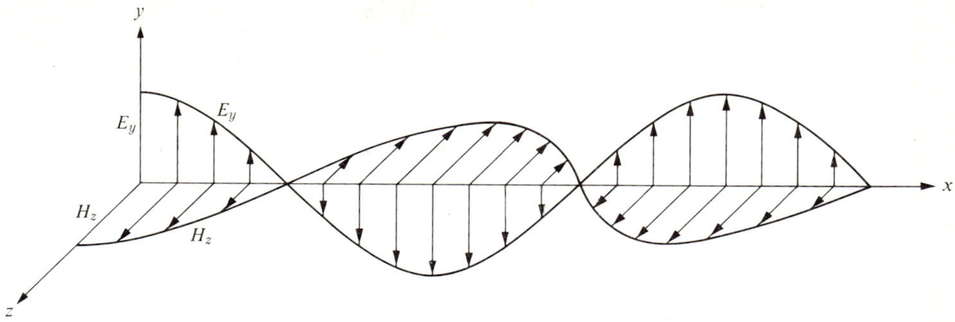

Fig. 5.3 The electric and the magnetic fields of a plane TEM wave propagating in the $+x$-direction in an unbounded lossless dielectric medium.

We can define the energy transport velocity v_E as in Section 3.9 to be the ratio of the time-averaged power flux density $\boldsymbol{S}(x)$ to the time-averaged energy density $w = w_E + w_H$ associated with both the electric and the magnetic fields. Hence, from (5.34) to (5.36),

$$\boldsymbol{v}_E = \frac{\boldsymbol{S}(x)}{w_E + w_H} = \frac{\hat{x}}{2}\sqrt{\frac{\varepsilon}{\mu}}\frac{|E_{y0}^+|^2\, 2}{\varepsilon\, |E_{y0}^+|^2} = \frac{\hat{x}}{\sqrt{\mu\varepsilon}} = \hat{x}v. \tag{5.37}$$

Thus, the energy transport velocity is identical to the phase and the group velocities in a lossless dielectric medium.

For a homogeneous plane wave, the phase fronts are planes over which the direction and the amplitude of the electromagnetic fields do not vary. For the plane wave in an unbounded dielectric medium, the electric and the magnetic fields are perpendicular to the direction of propagation as well as to each other. The plane TEM wave in an unbounded dielectric has an important difference in its characteristic depending on whether the medium is lossless or lossy. For a lossless dielectric medium, we see the intrinsic impedance η_0 from (5.32) to be real and hence, (5.17), (5.19), (5.25), and (5.27) show that the electric and the magnetic fields are in phase. But for a lossy dielectric medium, since η_0 is complex, the electric and the magnetic fields differ in phase. Thus, for the plane TEM wave in a lossless unbounded dielectric, the electric and the magnetic fields are in space quadrature and in time phase, as illustrated in Fig. 5.3.

5.6 INTRINSIC IMPEDANCE AND PROPAGATION COEFFICIENT

The presence of conduction current in a dielectric results in the dissipation of energy. The conduction and the displacement current densities are given respectively by $\boldsymbol{J}_{cd}(\boldsymbol{r}) = \sigma\boldsymbol{E}(\boldsymbol{r})$ and $\boldsymbol{J}_d(\boldsymbol{r}) = j\omega\varepsilon\boldsymbol{E}(\boldsymbol{r})$. If the magnitude of the conduction current density is small compared to the displacement current density, that is, if

$$\frac{|\boldsymbol{J}_{cd}(\boldsymbol{r})|}{|\boldsymbol{J}_d(\boldsymbol{r})|} = \frac{\sigma}{\omega\varepsilon} \ll 1,$$

the dissipation of energy in the dielectric is negligible resulting in a "good dielectric". On the other hand, if the magnitude of the displacement current density is small compared to the conduction current density, that is, if $|\boldsymbol{J}_{cd}(\boldsymbol{r})|/|\boldsymbol{J}_d(\boldsymbol{r})| = \sigma/\omega\varepsilon \gg 1$, the resulting medium constitutes a "good conductor". Thus,

$$\begin{aligned} &\sigma/\omega\varepsilon \ll 1 \qquad \text{(for a "good dielectric")}, \\ &\sigma/\omega\varepsilon \gg 1 \qquad \text{(for a "good conductor")}. \end{aligned} \tag{6.1}$$

We examined the limiting values of the propagation coefficient γ, and the intrinsic impedance η_0, as given by (5.13) and (5.16) respectively in Section 5.5 for the special case of a lossless dielectric. It is instructive for us now to study the limiting values of γ and η_0 for two other important special cases, namely, a "good dielectric" and a "good conductor" as defined in (6.1).

1. Good Dielectric

Let us substitute (5.4) in (5.13) and rearrange the resulting equation to yield

$$\gamma = \alpha + j\beta = \sqrt{-\omega^2\mu\varepsilon\left(1 + \frac{\sigma}{j\omega\varepsilon}\right)} = j\omega\sqrt{\mu\varepsilon}\left[1 + \frac{\sigma}{j\omega\varepsilon}\right]^{1/2}. \tag{6.2}$$

With the use of the binomial expansion formula

$$(1 + y)^{1/2} = 1 + \tfrac{1}{2}y - \tfrac{1}{8}y^2 + \tfrac{1}{16}y^3 \qquad (\text{for } |y| \ll 1), \tag{6.3}$$

if we expand the factor $[1 + \sigma/j\omega\varepsilon]^{1/2}$ for $\sigma/\omega\varepsilon \ll 1$ and retain only the first four terms, (6.2) becomes

$$\gamma = \alpha + j\beta = j\omega\sqrt{\mu\varepsilon}\left[1 + \frac{\sigma}{2j\omega\varepsilon} + \frac{\sigma^2}{8\omega^2\varepsilon^2} - \frac{\sigma^3}{16j\omega^3\varepsilon^3}\right]. \tag{6.4}$$

If we write down the real and the imaginary parts of (6.4) separately, we find that

$$\alpha = \frac{\sigma}{2}\sqrt{\frac{\mu}{\varepsilon}}\left[1 - \frac{\sigma^2}{8\omega^2\varepsilon^2}\right], \tag{6.5}$$

$$\beta = \omega\sqrt{\mu\varepsilon}\left[1 + \frac{\sigma^2}{8\omega^2\varepsilon^2}\right] = \frac{\omega}{v}\left[1 + \frac{\sigma^2}{8\omega^2\varepsilon^2}\right]. \tag{6.6}$$

For a "good dielectric", the attenuation coefficient α as given by (6.5) is nonvanishing and is a function of frequency. The phase-shift coefficient β as given by (6.6) is different from that for the lossless dielectric as given by (5.30). In particular, (6.6) shows that β is not a linear function of frequency.

For a "good dielectric", we obtain the phase velocity from (6.6) as

$$v_{ph} = \frac{\omega}{\beta} = \frac{v}{[1 + \sigma^2/8\omega^2\varepsilon^2]} \approx v\left[1 - \frac{\sigma^2}{8\omega^2\varepsilon^2}\right]. \tag{6.7}$$

Also, we can deduce the group velocity from (6.6) as

$$v_g = \frac{\partial\omega}{\partial\beta} = \frac{1}{\partial\beta/\partial\omega} = \frac{v}{[1 - \sigma^2/8\omega^2\varepsilon^2]} \approx v\left[1 + \frac{\sigma^2}{8\omega^2\varepsilon^2}\right]. \tag{6.8}$$

We can see from (6.7) and (6.8) that in a "good dielectric" the phase velocity increases with an increase in the frequency but the group velocity decreases. Also, the product of the phase and the group velocities is independent of the frequency as is seen from

$$v_{ph}v_g = v^2 \qquad (\text{for } \sigma/\omega\varepsilon \ll 1). \tag{6.9}$$

From (6.7) and (6.8), we also see that an increase in the losses in the dielectric, that is, an increase in $\sigma/\omega\varepsilon$ results in a decrease in the phase velocity and an increase in the group velocity.

As we discussed in detail in Section 2.13, when v_{ph} is a function of the frequency, the medium is said to be dispersive. Since $dv_{ph}/d\omega > 0$ and therefore, $v_g > v_{ph}$, the dispersion in a low-loss dielectric is called *anomalous dispersion*.

From (5.4) and (5.16), we deduce that

$$\eta_0 = \eta_{0r} + j\eta_{0j} = \sqrt{\frac{\mu}{\varepsilon}}\left(1 + \frac{\sigma}{j\omega\varepsilon}\right)^{-1/2}. \tag{6.10}$$

With the use of the binomial expansion formula

$$(1 + y)^{-1/2} = 1 - \tfrac{1}{2}y + \tfrac{3}{8}y^2 - \tfrac{5}{16}y^3 \qquad (\text{for } |y| \ll 1), \tag{6.11}$$

if we expand the factor $(1 + (\sigma/j\omega\varepsilon))^{-1/2}$ for $\sigma/\omega\varepsilon \ll 1$ and retain only the first four terms, (6.10) becomes

$$\eta_0 = \eta_{0r} + j\eta_{0j} = \sqrt{\frac{\mu}{\varepsilon}}\left[1 - \frac{\sigma}{2j\omega\varepsilon} - \frac{3\sigma^2}{8\omega^2\varepsilon^2} + \frac{5}{16}\frac{\sigma^3}{j\omega^3\varepsilon^3}\right]. \tag{6.12}$$

From (6.12), we see that

$$\eta_{0r} = \sqrt{\frac{\mu}{\varepsilon}}\left[1 - \frac{3\sigma^2}{8\omega^2\varepsilon^2}\right], \tag{6.13}$$

and

$$\eta_{0j} = \sqrt{\frac{\mu}{\varepsilon}}\frac{\sigma}{2\omega\varepsilon}\left[1 - \frac{5\sigma^2}{8\omega^2\varepsilon^2}\right]. \tag{6.14}$$

The intrinsic resistance as given by (6.13) is a function of the frequency and is less than that for a perfect dielectric. An increase in the losses in the dielectric or a decrease in the frequency results in the lowering of the intrinsic resistance. Also, as seen from (6.14), the intrinsic reactance in a "good dielectric" is nonvanishing and is inductive.

2. Good Conductor

For a "good conductor," we first rearrange γ as given in (6.2) into the form

$$\gamma = \alpha + j\beta = \sqrt{j\omega\mu\sigma}\left(1 + \frac{j\omega\varepsilon}{\sigma}\right)^{1/2}. \tag{6.15}$$

We should note that $\sqrt{j} = (1 + j)/\sqrt{2}$. As before, we expand the factor $(1 + (j\omega\varepsilon/\sigma))^{1/2}$ with the help of (6.3) for $\omega\varepsilon/\sigma \ll 1$. However, it is sufficient for us to retain only the first two terms for the purpose of obtaining the first two terms in the series for α and β. Then, we can simplify (6.15) to

$$\gamma = \alpha + j\beta = \sqrt{\frac{\omega\mu\sigma}{2}}(1 + j)\left(1 + \frac{j\omega\varepsilon}{2\sigma}\right). \tag{6.16}$$

We can deduce from (6.16) that

$$\alpha = \sqrt{\frac{\omega\mu\sigma}{2}}\left(1 - \frac{\omega\varepsilon}{2\sigma}\right), \tag{6.17}$$

$$\beta = \sqrt{\frac{\omega\mu\sigma}{2}}\left(1 + \frac{\omega\varepsilon}{2\sigma}\right). \tag{6.18}$$

It is usual to retain only the first terms in (6.17) and (6.18). From (6.17) and (6.18), we see that in a "good conductor," the attenuation and the phase-shift coefficients are approximately equal. Also, both α and β increase as the square root of the frequency and as the square root of the conductivity.

In a similar manner, for a good conductor, we can recast the expression for the intrinsic impedance η_0 as given in (6.10) into the following form:

$$\eta_0 = \eta_{0r} + j\eta_{0j} = \sqrt{\frac{j\omega\mu}{\sigma}}\left(1 + \frac{j\omega\varepsilon}{\sigma}\right)^{-1/2}, \tag{6.19}$$

$$= \sqrt{\frac{\omega\mu}{2\sigma}}(1 + j)\left(1 - \frac{j\omega\varepsilon}{2\sigma}\right). \tag{6.20}$$

From (6.20), we see that

$$\eta_{0r} = \sqrt{\frac{\omega\mu}{2\sigma}}\left(1 + \frac{\omega\varepsilon}{2\sigma}\right), \tag{6.21}$$

and

$$\eta_{0j} = \sqrt{\frac{\omega\mu}{2\sigma}}\left(1 - \frac{\omega\varepsilon}{2\sigma}\right). \tag{6.22}$$

As in (6.17) and (6.18), it is conventional to retain only the leading terms in (6.21) and (6.22) also. It follows from (6.21) and (6.22) that the magnitudes of the intrinsic resistance and intrinsic reactance in a "good conductor" are approximately equal. These magnitudes increase as the square root of the frequency and decrease as the inverse square root of the conductivity.

5.7 POLARIZATION OF PLANE WAVES

A uniform plane wave with harmonic time dependence is completely specified by its frequency, its direction of propagation, and the magnitude and the direction of its field vectors. If the magnitude and direction of the electric vector are known, since the direction of propagation is given, the magnitude and direction of the magnetic field vector have been shown to be completely specified. Therefore, it is sufficient to specify the direction and the magnitude of either the electric or the magnetic field vector. It is common to specify the quantities pertaining to the electric vector. For example, a uniform plane wave of angular frequency ω, progressing in the $+x$-direction and having its electric vector in the y-direction as given by (5.17) is completely specified as follows:

$$\begin{aligned} \boldsymbol{E}(x, t) &= \hat{y}\, \text{Re}\, [E_{y0}^{+} e^{-\alpha x} e^{-j\beta x} e^{j\omega t}], \\ &= \hat{y}\, |E_{y0}^{+}|\, e^{-\alpha x} \cos(\omega t + \theta_{E_y} - \beta x), \end{aligned} \tag{7.1}$$

where $E_{y0}^{+} = |E_{y0}^{+}|\, e^{j\theta_{E_y}}$. The argument of the cosine function is called *the phase* of the wave. The magnitude of the coefficient of the cosine function is called *the amplitude* of the wave. In addition to the phase and the amplitude of the wave, it is necessary to specify the direction of the electric vector. The information on the direction of the electric vector is called *the polarization* of the wave. Since it is possible for the direction of the electric vector to change with time, as we shall establish presently, the polarization is usually specified by the locus of the terminal point of the electric vector at any location in space as a function of time for one time period. For any x, for example $x = 0$, the terminal point of the electric vector as given by (7.1) traces a straight line along the y-axis and hence (7.1) represents a wave *linearly* polarized in the y-direction.

In a similar manner, a uniform plane wave of angular frequency ω, progressing in the $+x$-direction and having its electric vector in the z-direction as given by (5.25), is completely specified by

$$\begin{aligned} \boldsymbol{E}(x, t) &= \hat{z}\, \text{Re}\, [E_{z0}^{+} e^{-\alpha x} e^{-j\beta x} e^{j\omega t}], \\ &= \hat{z}\, |E_{z0}^{+}|\, e^{-\alpha x} \cos(\omega t + \theta_{E_z} - \beta x), \end{aligned} \tag{7.2}$$

where $E_{z0}^{+} = |E_{z0}^{+}|\, e^{j\theta_{E_z}}$. We note that (7.2) also represents a linearly polarized wave but one polarized in the z-direction.

A linear combination of (7.1) and (7.2) is also a solution of Maxwell's equations. We obtain the general form of a uniform plane wave progressing in the $+x$-direction from (7.1) and (7.2) as

$$\boldsymbol{E}(x, t) = \text{Re}\, [\hat{y}\, |E_{y0}^{+}|\, e^{j\theta_{E_y}} e^{-\alpha x - j\beta x + j\omega t} + \hat{z}\, |E_{z0}^{+}|\, e^{j\theta_{E_z}} e^{-\alpha x - j\beta x + j\omega t}]. \tag{7.3}$$

We show now that by a suitable combination of the values for the amplitudes and the phases of the two constituent waves, a general uniform plane wave can have in addition to the linear polarization discussed previously, *circular* and *elliptic* polarizations. First, consider the case for which $\theta_{E_y} = \theta_{E_z} = \theta_E$. Then, we can reduce (7.3) to

$$\boldsymbol{E}(x, t) = [\hat{y}\, |E_{y0}^{+}| + \hat{z}\, |E_{z0}^{+}|] e^{-\alpha x} \cos(\omega t + \theta_E - \beta x). \tag{7.4}$$

Let

$$|E_{y0}^{+}| = |E| \cos \phi; \qquad |E_{z0}^{+}| = |E| \sin \phi. \tag{7.5}$$

Then, (7.4) becomes

$$\boldsymbol{E}(x, t) = [\hat{y} \cos \phi + \hat{z} \sin \phi]\, |E|\, e^{-\alpha x} \cos(\omega t + \theta_E - \beta x). \tag{7.6}$$

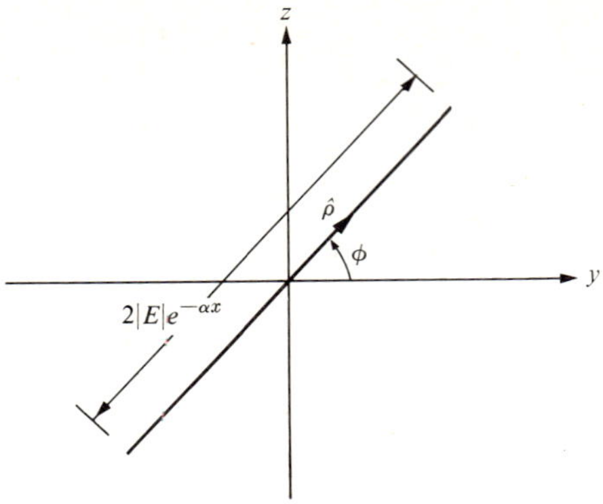

Fig. 5.4 A linearly polarized electric vector.

We see that the terminal point of the electric vector given by (7.6) traces a straight line in the direction $\hat{\rho}$ which makes an angle ϕ with the y-axis, as shown in Fig. 5.4. Therefore, (7.6) represents a wave polarized linearly in the $\hat{\rho}$ direction.

Next, let us consider the case for which $|E_{y0}^{+}| = |E_{z0}^{+}| = |E|$ and $\theta_{E_z} = \theta_{E_y} \pm \pi/2$. For this case we can reduce (7.3) to

$$\boldsymbol{E}(x, t) = |E|\, e^{-\alpha x}[\hat{y} \cos(\omega t + \theta_{E_y} - \beta x) \mp \hat{z} \sin(\omega t + \theta_{E_y} - \beta x)]. \qquad (7.7)$$

From (7.7), we find that

$$E_y^2(x, t) + E_z^2(x, t) = |E|^2\, e^{-2\alpha x}, \qquad (7.8)$$

which shows that the terminal point of the electric vector as given by (7.7) traces a circle in the yz-plane and the radius of the circle is $|E|\, e^{-\alpha x}$. For the wave corresponding to the upper sign in (7.7), the instantaneous position of the electric vector is in the y-direction for $\omega t + \theta_E - \beta x = 0$ and is in the $-z$-direction at a later time corresponding to $\omega t + \theta_E - \beta x = \pi/2$. Thus, the tip of the electric vector moves in the clockwise direction (Fig. 5.5). If the thumb of the left hand points in the direction of propagation, the fingers curl in the direction of rotation of the electric vector. Therefore, the electric field given by the upper sign in (7.7) represents a *left-hand or negative circularly polarized wave*. In a similar manner, we can show that the electric field given by the lower sign in (7.7) represents a *right-hand or positive circularly polarized wave*. In the phasor notation, it follows from (7.3) that we can represent circular polarizations as follows:

$$\boldsymbol{E} = (\hat{y} + j\hat{z})E \qquad \text{(left circular polarization)}, \qquad (7.9)$$

$$\boldsymbol{E} = (\hat{y} - j\hat{z})E \qquad \text{(right circular polarization)}. \qquad (7.10)$$

Finally, let us consider the third case for which $|E_{y0}^{+}| \neq |E_{z0}^{+}|$ but $\theta_{E_z} = \theta_{E_y} \pm \pi/2$. For this case, we can reduce (7.3) to

$$\boldsymbol{E}(x, t) = e^{-\alpha x}[\hat{y}\, |E_{y0}^{+}| \cos(\omega t + \theta_{E_y} - \beta x) \mp \hat{z}\, |E_{z0}^{+}| \sin(\omega t + \theta_{E_y} - \beta x)]. \qquad (7.11)$$

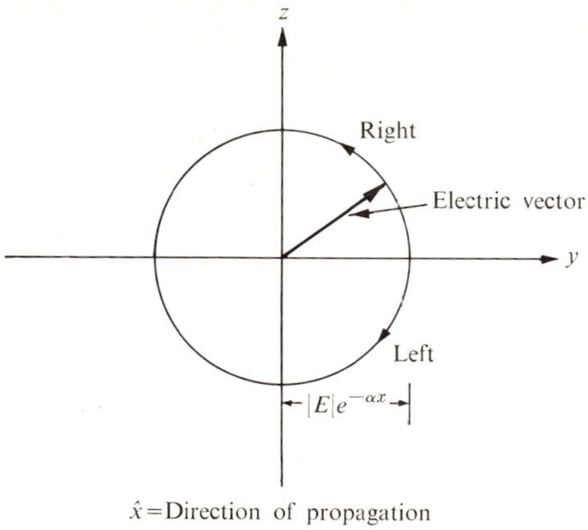

Fig. 5.5 Left and right circularly polarized electric vectors.

From (7.11), we deduce that

$$\left[\frac{E_y(x, t)}{|E_{y0}^+|}\right]^2 + \left[\frac{E_z(x, t)}{|E_{z0}^+|}\right]^2 = e^{-2\alpha x}, \tag{7.12}$$

which shows that the terminal point of the electric vector as given by (7.11) traces an ellipse in the yz-plane. The principal axes of the ellipse are in the y- and z-directions and are of magnitudes $2\,|E_{y0}^+|\,e^{-\alpha x}$ and $2\,|E_{z0}^+|\,e^{-\alpha x}$ respectively (Fig. 5.6). As for the circular polarization, the electric field given by the upper sign in (7.11) represents a *left-hand or negative elliptically polarized wave.* In a similar manner, the electric field given by the lower sign in (7.11) represents a *right-hand or positive elliptically polarized wave.* In the phasor notation, it follows from (7.3) that elliptic polarizations

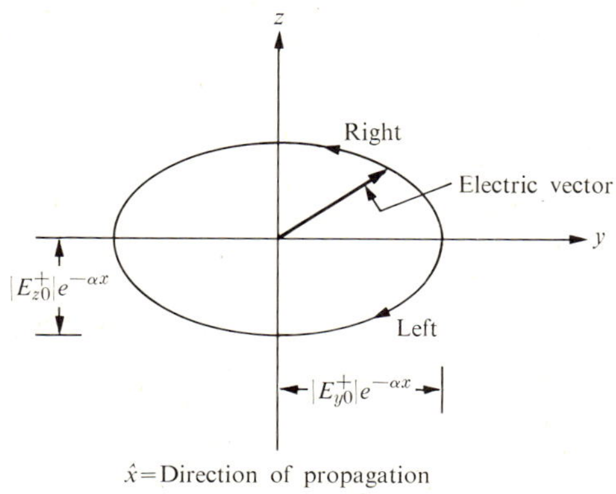

Fig. 5.6 Left and right elliptically polarized electric vectors.

can be represented as follows:

$$\boldsymbol{E} = (\hat{y} + ja\hat{z})E \qquad \text{(left elliptic polarization)}, \tag{7.13}$$

$$\boldsymbol{E} = (\hat{y} - ja\hat{z})E \qquad \text{(right elliptic polarization)}, \tag{7.14}$$

where a is a positive real number.

We have established that a superposition of two linearly polarized waves progressing in the same direction leads to (1) linear, (2) circular, or (3) elliptic polarizations. In all three cases, the constituent polarizations are in space quadrature. We obtain a linear polarization if the constituent polarizations are in time phase. If the constituent polarizations are equal in amplitude and are in phase quadrature, the resulting polarization is circular. An elliptic polarization results if the constituent polarizations are not equal in amplitude but are in phase quadrature. In the circular and the elliptic polarizations, the sense of rotation of the electric vector is reversed if the sign of the phase difference between the two constituent polarizations is reversed.

It is worthwhile for us to remember that, in general, if the magnitudes and/or the phases of the complex amplitudes E_{y0}^+ and E_{z0}^+ of the two constituent polarizations are different, an elliptic polarization results with the principal axes of the ellipse being generally different from the y- and the z-axes. We shall not discuss this general case of elliptic polarization here.

5.8 PROPAGATION OF GENERAL ONE-DIMENSIONAL FIELDS WITH HARMONIC TIME DEPENDENCE

We discussed a general one-dimensional field with arbitrary time dependence in Section 5.3. We can treat the general one-dimensional field with harmonic time dependence in a similar manner. Let the field be a function of a coordinate ξ measured along a line whose direction is defined by the unit vector $\hat{n}$ as given by (3.1). We should note that ξ need not necessarily coincide with a coordinate axis of the reference system. The field quantities are of the form

$$\boldsymbol{E}(\boldsymbol{r}) = \boldsymbol{E}(\xi), \qquad \boldsymbol{H}(\boldsymbol{r}) = \boldsymbol{H}(\xi), \tag{8.1}$$

and ξ is defined by (3.3). Since (3.8) holds, Maxwell's equations (5.2) and (5.3) become

$$\hat{n} \times \frac{\partial}{\partial \xi} \boldsymbol{E}(\xi) = -j\omega\mu \boldsymbol{H}(\xi), \tag{8.2}$$

$$\hat{n} \times \frac{\partial}{\partial \xi} \boldsymbol{H}(\xi) = j\omega\varepsilon_{\text{eff}} \boldsymbol{E}(\xi). \tag{8.3}$$

As before, let us introduce two mutually perpendicular coordinates η and ζ lying in a plane perpendicular to $\hat{n}$ such that ξ, η, and ζ form a triad of mutually perpendicular right-handed system of vectors. Then, we can write $\boldsymbol{E}(\xi)$ and $\boldsymbol{H}(\xi)$ in component form as

$$\boldsymbol{E}(\xi) = \hat{n}E_\xi(\xi) + \hat{\eta}E_\eta(\xi) + \hat{\zeta}E_\zeta(\xi), \tag{8.4}$$

and

$$\boldsymbol{H}(\xi) = \hat{n}H_\xi(\xi) + \hat{\eta}H_\eta(\xi) + \hat{\zeta}H_\zeta(\xi). \tag{8.5}$$

If we substitute (8.4) and (8.5) in (8.2) and (8.3), the result in component form is the same as given in (5.5) to (5.10) with x, y, and z replaced by ξ, η, and ζ respectively.

From (5.11), we obtain

$$H_\xi(\xi) = E_\xi(\xi) = 0. \tag{8.6}$$

Hence, the resulting field structure is of the TEM type with respect to the ξ-direction. From (5.17) through (5.20) and (5.25) through (5.28) we get the following two independent sets of solutions:

First Set

Wave Progressing in the $+\xi$-Direction

$$E_\eta(\xi) = E_{\eta 0}^{+} e^{-\gamma\xi}; \qquad H_\zeta(\xi) = \frac{E_{\eta 0}^{+} e^{-\gamma\xi}}{\eta_0}; \qquad \frac{E_\eta(\xi)}{H_\zeta(\xi)} = \eta_0. \tag{8.7}$$

Wave Progressing in the $-\xi$-Direction

$$E_\eta(\xi) = E_{\eta 0}^{-} e^{\gamma\xi}; \qquad H_\zeta(\xi) = -\frac{E_{\eta 0}^{-} e^{\gamma\xi}}{\eta_0}; \qquad \frac{E_\eta(\xi)}{H_\zeta(\xi)} = -\eta_0. \tag{8.8}$$

Second Set

Wave Progressing in the $+\xi$-Direction

$$E_\zeta(\xi) = E_{\zeta 0}^{+} e^{-\gamma\xi}; \qquad H_\eta(\xi) = -\frac{E_{\zeta 0}^{+} e^{-\gamma\xi}}{\eta_0}; \qquad \frac{E_\zeta(\xi)}{H_\eta(\xi)} = -\eta_0. \tag{8.9}$$

Wave Progressing in the $-\xi$-Direction

$$E_\zeta(\xi) = E_{\zeta 0}^{-} e^{\gamma\xi}; \qquad H_\eta(\xi) = \frac{E_{\zeta 0}^{-} e^{\gamma\xi}}{\eta_0}; \qquad \frac{E_\zeta(\xi)}{H_\eta(\xi)} = \eta_0. \tag{8.10}$$

The propagation coefficient γ and the intrinsic impedance η_0 are the same as those given by (5.13) and (5.16) respectively.

For each of the four constituent waves given by (8.7) to (8.10), we can evaluate the time-averaged Poynting vector $\mathbf{S}(\xi)$ by the method used in obtaining (5.21). We can then verify that the directions of (1) the progression of the phase fronts, (2) the attenuation of the amplitude of the wave and (3) the time-averaged Poynting vector are all identical. This observation is useful in obtaining the correct sign of the intrinsic impedance as we demonstrated in (5.23) and (5.24). It is important to remember that the intrinsic impedance is the ratio of the electric field to the magnetic field and that these fields lie entirely in the plane of the phase fronts. We should note that the phase fronts are the locus of positions in space where the field has the same phase.

The phase fronts of the fields given in (8.7) to (8.10) are planes perpendicular to the ξ-axis or equivalently, the direction specified by $\hat{n}$. The direction of progression of the phase fronts is called the *wave-normal* direction. The wave normals for the fields given by (8.7) and (8.9) are in the $\hat{n}$-direction whereas those for the fields given by (8.8) and (8.10) are in the $-\hat{n}$-direction. It is conventional to call the wave-normal direction also the longitudinal direction. There are no field components in the longitudinal direction and therefore the wave is of the TEM type. The magnitudes and the directions of the field components are constants in any phase front. As we already mentioned, the magnetic field bears a constant ratio to the electric field.

Let us consider a plane wave in an unbounded dielectric medium. Let $\hat{n}$ be the wave-normal direction. The electric and the magnetic fields of this plane wave are given either by (8.7) or (8.9). Since Maxwell's equations (8.2) and (8.3) are linear, a linear combination of the fields given by (8.7) and (8.9) is also an acceptable solution of Maxwell's equations. Consequently, we obtain the general expressions for the fields of a plane wave with the wave normal in the $\hat{n}$-direction by linearly superposing those given by (8.7) and (8.9), with the following results:

$$\boldsymbol{E}(\xi) = [\hat{\eta}E^{+}_{\eta 0} + \hat{\zeta}E^{+}_{\zeta 0}]e^{-\gamma\xi}, \tag{8.11}$$

$$\boldsymbol{H}(\xi) = [\hat{\zeta}E^{+}_{\eta 0} - \hat{\eta}E^{+}_{\zeta 0}]\frac{e^{-\gamma\xi}}{\eta_0}. \tag{8.12}$$

The fact that $\boldsymbol{E}(\xi)$ and $\boldsymbol{H}(\xi)$ have no components in the wave-normal direction is conveniently expressed by the following relations:

$$\hat{n}\cdot\boldsymbol{E}(\xi) = \hat{\xi}\cdot\boldsymbol{E}(\xi) = 0, \tag{8.13}$$

$$\hat{n}\cdot\boldsymbol{H}(\xi) = \hat{\xi}\cdot\boldsymbol{H}(\xi) = 0. \tag{8.14}$$

We can verify that $\boldsymbol{E}(\xi)$ and $\boldsymbol{H}(\xi)$ given by (8.11) and (8.12) satisfy (8.13) and (8.14) respectively.

It is usual to specify a plane wave by giving its wave-normal direction and the magnitude and direction of either the electric or magnetic vector. Of course, the frequency of the wave and the medium parameters are to be specified. Suppose that, in addition to the wave-normal direction, the magnitude and direction of the electric vector are given. The direction of the electric vector must necessarily satisfy (8.13). We can obtain the direction and magnitude of the magnetic vector in the following manner. Let $\hat{n} = \hat{\xi}$ be the direction of propagation and let the magnitude and direction of the electric vector be given by the amplitude term in (8.11), that is,

$$\boldsymbol{E}_0(\xi) = \hat{\eta}E^{+}_{\eta 0} + \hat{\zeta}E^{+}_{\zeta 0}. \tag{8.15}$$

Clearly (8.15) satisfies (8.13). Since the amplitude part of the magnetic field should also satisfy (8.14), we can assume it to be of the form

$$\boldsymbol{H}_0(\xi) = \hat{\eta}H^{+}_{\eta 0} + \hat{\zeta}H^{+}_{\zeta 0}, \tag{8.16}$$

where $H^{+}_{\eta 0}$ and $H^{+}_{\zeta 0}$ are to be determined. We have stated that the electric vector, the magnetic vector, and the wave-normal direction or the Poynting vector form a triad of mutually perpendicular right-handed system of vectors. Therefore

$$\hat{\xi}\times\boldsymbol{E}_0(\xi) = \hat{\xi}\times[\hat{\eta}E^{+}_{\eta 0} + \hat{\zeta}E^{+}_{\zeta 0}] = [\hat{\zeta}E^{+}_{\eta 0} - \hat{\eta}E^{+}_{\zeta 0}] \tag{8.17}$$

should be in the same direction as the magnetic vector. If we compare (8.17) with (8.16) we see that

$$H^{+}_{\eta 0} = -E^{+}_{\zeta 0}A; \qquad H^{+}_{\zeta 0} = E^{+}_{\eta 0}A. \tag{8.18}$$

Thus, $H_{\eta 0}^{+}$ and $H_{\zeta 0}^{+}$ have been determined to within a constant A, which we can evaluate in the following manner. The ratio of the magnitude of the electric field to the magnitude of the magnetic field is equal to the intrinsic impedance η_0. Hence, from (8.18)

$$\frac{\sqrt{E_{\eta 0}^{+2} + E_{\zeta 0}^{+2}}}{\sqrt{H_{\eta 0}^{+2} + H_{\zeta 0}^{+2}}} = 1/A = \eta_0; \qquad \text{that is, } A = 1/\eta_0. \tag{8.19}$$

Together with (8.18) and (8.19), the amplitude term of the magnetic field as given in (8.16) becomes

$$\boldsymbol{H}_0(\xi) = [-\hat{\eta}E_{\zeta 0}^{+} + \hat{\zeta}E_{\eta 0}^{+}]/\eta_0. \tag{8.20}$$

The result obtained in (8.20) is in agreement with the amplitude term of the magnetic field as given in (8.12). We should note that the amplitude term (8.20) should be multiplied by the phase factor $e^{-\gamma\xi}$ to obtain the complete expression for the magnetic vector.

5.9 SKIN DEPTH AND SURFACE IMPEDANCE FOR A GOOD CONDUCTOR

In Section 5.4, we discussed the diffusion of one-dimensional fields in a highly conducting medium. We now wish to examine the interesting features of the time-harmonic, one-dimensional fields supported by a highly conducting medium. Let the half-space $x > 0$ be occupied by a highly conducting medium in which the displacement current density can be neglected in comparison with the conduction current density. We assume the remaining half-space $x < 0$ to be free space. In view of (5.17) and (5.18), we assume a plane wave progressing in the $+x$-direction inside the highly conducting medium to have the following field components:

$$E_y(x) = H_{z0}^{+}\eta_0 e^{-\gamma x} = E_{y0}^{+}e^{-\gamma x}; \qquad H_z(x) = H_{z0}^{+}e^{-\gamma x}. \tag{9.1}$$

It is implied that the source for the fields given by (9.1) is situated in the $x = 0$ plane with the result that the fields given by (5.19) and (5.20) are not applicable since they become infinite at $x = \infty$. Since the displacement current is negligible, it follows that in (5.4) we can omit ε in comparison with $\sigma/j\omega$, as a consequence of which (5.13) and (5.16) reduce to

$$\gamma = \alpha + j\beta = \sqrt{j\omega\mu\sigma}; \qquad \eta_0 = \eta_{0r} + j\eta_{0j} = \sqrt{\frac{j\omega\mu}{\sigma}}. \tag{9.2}$$

From a study of (6.15) to (6.22) or by replacing $\sqrt{j}$ by $(1 + j)/\sqrt{2}$ in (9.2), we find that

$$\alpha = \beta = \sqrt{\frac{\omega\mu\sigma}{2}}; \qquad \eta_{0r} = \eta_{0j} = \sqrt{\frac{\omega\mu}{2\sigma}}. \tag{9.3}$$

Let us first consider the limiting case in which the highly conducting medium is a perfect conductor corresponding to $\sigma = \infty$. For $x > 0$, it follows from (9.3) and (9.1) that there are no electromagnetic fields in the interior of a perfect conductor. But the surface of a perfect conductor can support a nonvanishing tangential component of the magnetic field and hence H_{z0}^{+} is nonvanishing. Since for a perfect conductor $\eta_0 = 0$, it follows that $E_{y0}^+ = 0$ with the result that there is no tangential component of the electric field in the interior or on the surface of a perfect conductor. There is also a nonvanishing surface current density $\boldsymbol{J}_s$ which we obtain as

$$\boldsymbol{J}_s = \hat{n} \times \boldsymbol{H}_0 = -\hat{x} \times \hat{z}H_{z0}^+ = \hat{y}H_{z0}^+ = \hat{y}J_{sy}, \tag{9.4}$$

where $\hat{n}$ is an outwardly drawn unit normal to the surface $x = 0$ of the perfect conductor.

Returning to the case of a very good but not a perfect conductor, that is, corresponding to σ very large but not infinite, we see from (9.3) and (9.1) that the fields decay rapidly in an exponential fashion with the distance from the surface inside the good conductor. In a perfect dielectric, the fields are periodic functions of the distance in the direction of progression of the phase fronts. In such a case, it is conventional to introduce a space scale length for the phenomenon given by the space period, that is, the wavelength. In a highly conducting medium, since the fields decay exponentially with the distance, the concept of wavelength is not physically meaningful, necessitating the introduction of another space scale length. In a highly conducting medium, the space scale length is usually taken to be the distance in which the amplitudes of the fields decay to $1/e$ of their values. This distance is called the depth of penetration or *the skin depth* which we can see from (9.1) and (9.3) to be

$$\delta_{\text{skin}} = 1/\alpha = \sqrt{2/\omega\mu\sigma}. \tag{9.5}$$

The skin depth is inversely proportional both to the square root of the frequency and the square root of the conductivity. Therefore, the skin depth decreases with the increase in the frequency as well as with the increase in the conductivity.

From (9.1), we obtain the volume current density $J_y(x)$ in units of amperes per square meter as

$$J_y(x) = \sigma E_y(x) = \sigma\eta_0 H_{z0}^+ e^{-\gamma x}. \tag{9.6}$$

We should note that the volume current density given by (9.6) also decreases rapidly from the surface of a good conductor. If the skin depth is small and if the volume current density is concentrated near the surface, it is reasonable for us to introduce the concept of equivalent surface current density. We can assume the volume current density integrated over the entire depth from $x = 0$ to $x = \infty$ to be the equivalent current density in amperes per meter concentrated at the surface. Thus, from (9.6) we obtain the *equivalent* surface current density J_{sy} as:

$$J_{sy} = \int_0^\infty J_y(x)\,dx = \sigma\eta_0 H_{z0}^+ \int_0^\infty e^{-\gamma x} = \frac{\sigma\eta_0 H_{z0}^+}{\gamma} = H_{z0}^+. \tag{9.7}$$

We can verify from (9.2) that $\sigma\eta_0 = \gamma$ and we use this result in obtaining (9.7). It is interesting to note that if the tangential component of the magnetic field is maintained to be a constant at the surface, then the actual surface current density as obtained in (9.4) for the case of a perfect conductor, and the equivalent surface current density as obtained in (9.7) for the case of a very good conductor, are identical. We should emphasize that the validity of the concept of equivalent surface current density rests on the fact that the volume current density in a very good conductor is concentrated essentially near the surface.

Let the ratio of the tangential electric field to the tangential magnetic field at the surface of the good conductor be designated as the *surface impedivity* η_s. Then, from (9.1), it follows that

$$\eta_s = E_y(0)/H_z(0) = E_{y0}^+/H_{z0}^+ = \eta_0. \tag{9.8}$$

The surface impedivity has the dimensions of an impedance. It is very important for us to note that (9.8) is valid exactly only when the conducting medium is semi-infinite. On the other hand, if the conducting medium is of finite depth, then there are also waves progressing in the $-x$-direction as given by (5.19) and the ratio of $E_y(0)/H_z(0)$ at the surface is no longer equal to the intrinsic impedance η_0 of the medium. However, if the medium is of finite depth, which is very much greater than the skin depth, the amplitudes of the waves progressing in the $-x$-direction are very much smaller in comparison with those for the waves progressing in the $+x$-direction, with the result that the surface impedivity becomes approximately equal to the intrinsic impedance. Thus, in many practical applications involving good conductors of finite depth, the surface impedivity is taken to be equal to the intrinsic impedance of the medium and, in such cases, it is necessary to ensure that the depth of the conductor is very much larger than the skin depth.

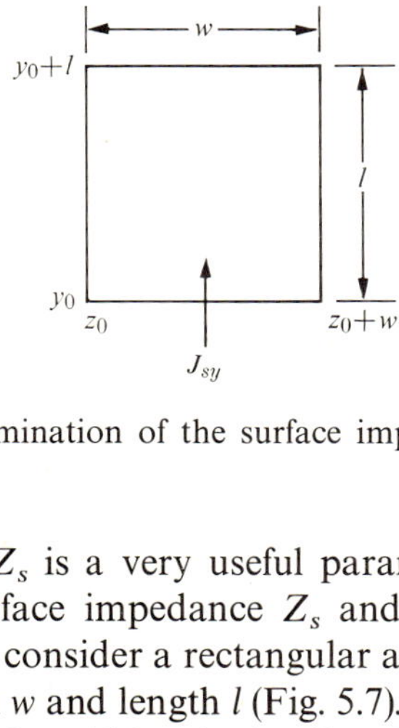

Fig. 5.7 Geometry for the determination of the surface impedance from a knowledge of the surface impedivity.

The surface impedance Z_s is a very useful parameter. We can determine the relationship between the surface impedance Z_s and the surface impedivity η_s in the following manner. Let us consider a rectangular area of the surface $x = 0$ of the very good conductor of width w and length l (Fig. 5.7). Let a uniform surface current density J_{sy} be directed across the width w in the direction of the length l. In the small length l of the surface area, we can consider the electric field to be conservative and, therefore, we obtain the potential drop in the direction of the current across the length of the surface area as $V = E_{y0}^+ l$. The total surface current I_{sy} is given by $I_{sy} = J_{sy}w = H_{z0}^+ w$. Hence, the surface impedance is

$$Z_s = R_s + jX_s = \frac{V}{I_{sy}} = \frac{E_{y0}^+ l}{H_{z0}^+ w} = \eta_s \frac{l}{w}. \tag{9.9}$$

It is apparent from (9.9) that the surface impedance and the surface impedivity have the same dimensions. It is necessary for us, therefore, to be careful not to confuse the surface impedance Z_s with the surface impedivity, although both have the same dimension. If $l = w$, then $Z_s = \eta_s$, and hence for any *square* area of the surface, the surface impedance is equal to the surface impedivity. For this reason, the surface impedance is expressed in *ohms per square*. From (9.3), (9.8), and (9.9) we find that

$$R_s = \eta_{0r}\frac{l}{w} = \sqrt{\frac{\omega\mu}{2\sigma}}\frac{l}{w}. \tag{9.10}$$

We can verify from (9.5) that $\sigma\,\delta_{\text{skin}} = \sqrt{2\sigma/\omega\mu}$ and hence (9.10) becomes

$$R_s = \frac{1}{\sigma}\frac{l}{\delta_{\text{skin}}w}, \tag{9.11}$$

which is amenable to the following simple physical interpretation. Let us consider an elementary volume of the good conductor of width w in the z-direction, length l in the y-direction, that is, the direction of current and of depth δ_{skin} from the surface $x = 0$, as depicted in the Fig. 5.8. Let us assume that the equivalent surface current is distributed uniformly in a depth δ_{skin} of the material with no currents existing at the depths greater than the skin depth. Then, we obtain the resistance R_s to the equivalent current distribution from (4.4.4) to be the same as that given by (9.11). Therefore, the surface resistance can be visualized as arising due to the equivalent surface current being distributed uniformly in a skin depth with no currents at the depths greater than the skin depth. Of course, we should remember that there are currents even at the depths greater than the skin depth and the afore-mentioned picture is helpful only in the interpretation of (9.11) but does not delineate the actual physical process.

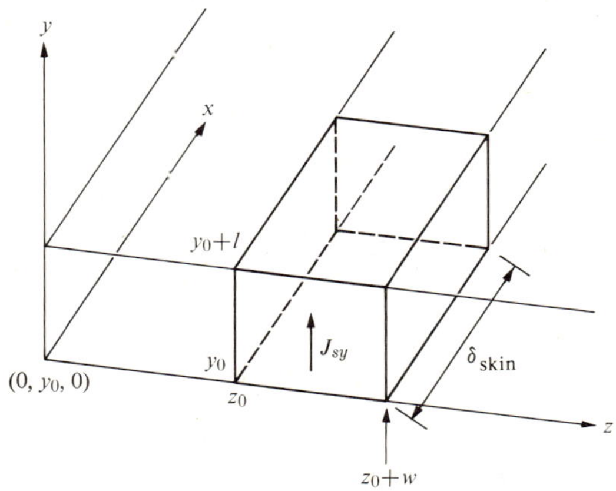

Fig. 5.8 Geometry for the evaluation of the surface resistance R_s.

The surface impedance Z_s in conjunction with the equivalent surface current is useful in obtaining the complex power flow into the conducting medium. Let us consider a volume of the conducting medium bounded by the planes $x = 0$, $x = \infty$, $y = y_0$, $y = y_0 + l$, $z = z_0$, and $z = z_0 + w$ where y_0, z_0, l, and w may have arbitrary values (Fig. 5.8). Since the Poynting vector for the fields given in (9.1) is in the x-direction, the power flow takes place only through planes perpendicular to the x-axis. The fields vanish at $x = \infty$ and therefore there is no power flow through the plane $x = \infty$. Hence, we find the total complex power P flowing into the volume to be

$$P = \int_{y_0}^{y_0+l}\int_{z_0}^{z_0+w} \hat{x}\cdot\tfrac{1}{2}\hat{y}E_y(0)\times\hat{z}H_z^*(0)\,dy\,dz,$$
$$= \tfrac{1}{2}\,|H_{z0}^+|^2\,\eta_0 lw. \qquad (9.12)$$

Since the surface current $I_{sy} = wJ_{sy} = wH_{z0}^+$ and since from (9.8) and (9.9), $Z_s = \eta_0(l/w)$, we may write (9.12) as

$$P = P_{\text{real}} + jP_{\text{reactive}} = \tfrac{1}{2}\,|I_{sy}|^2\,Z_s. \qquad (9.13)$$

Hence,

$$P_{\text{real}} = \tfrac{1}{2}\,|I_{sy}|^2\,R_s, \qquad P_{\text{reactive}} = \tfrac{1}{2}\,|I_{sy}|^2\,X_s, \qquad (9.14)$$

where P_{real} is the time-averaged total power entering the volume. We see from (9.13) that the complex power flow into the volume is obtained from the knowledge of the surface current and the surface impedance.

The principle of energy conservation requires that the total time-averaged power P_{real} flowing into the volume be all dissipated inside it. The expression for the time-averaged dissipated power density is given by (4.9.16). Therefore, we deduce the total power dissipated inside the volume with the help of (9.1) and (9.2) to be

$$P_D^t = \int_{y_0}^{y_0+l} dy \int_{z_0}^{z_0+w} dz \int_0^{\infty} \tfrac{1}{2}\sigma\,|H_{z0}^+|^2\,|\eta_0|^2\,e^{-2\alpha x}\,dx,$$
$$= lw\frac{\sigma}{2}\,|H_{z0}^+|^2\,|\eta_0|^2 \int_0^{\infty} e^{-2\alpha x}\,dx,$$
$$= lw\frac{\sigma}{2}\,|H_{z0}^+|^2\,\frac{|\eta_0|^2}{2\alpha}. \qquad (9.15)$$

From (9.3), we verify that $\sigma\,|\eta_0|^2/2\alpha = \eta_{0r}$. Also, $I_{sy} = wH_{z0}^+$. Hence, (9.15), with the help of (9.8) and (9.9) becomes

$$P_D^t = \tfrac{1}{2}\,|I_{sy}|^2\,\eta_{0r}\frac{l}{w} = \tfrac{1}{2}\,|I_{sy}|^2\,R_s, \qquad (9.16)$$

which, as to be expected, is equal to P_{real} as given in (9.14).

In a similar manner, we can determine an expression for the reactive power density inside the volume and integrate it over the entire volume to obtain a result identical to P_{reactive} as given by (9.14). For this purpose, it is convenient to start from the time-harmonic Maxwell's equations (4.9.2, 4.9.3) and the constituent relations as given in (4.9.4). If the displacement current is neglected as is required for a good conductor, we find that

$$\nabla \times \boldsymbol{E}(\boldsymbol{r}) = -j\omega\mu \boldsymbol{H}(\boldsymbol{r}), \tag{9.17}$$

$$\nabla \times \boldsymbol{H}(\boldsymbol{r}) = \sigma \boldsymbol{E}(\boldsymbol{r}). \tag{9.18}$$

After taking the complex conjugates of both sides of (9.18), we scalar multiply (9.17) and (9.18) by $\frac{1}{2}\boldsymbol{H}^*(\boldsymbol{r})$ and $-\frac{1}{2}\boldsymbol{E}(\boldsymbol{r})$ respectively and add the two terms. The result, on noting that

$$\nabla \cdot [\boldsymbol{E}(\boldsymbol{r}) \times \boldsymbol{H}^*(\boldsymbol{r})] = \boldsymbol{H}^*(\boldsymbol{r}) \cdot [\nabla \times \boldsymbol{E}(\boldsymbol{r})] - \boldsymbol{E}(\boldsymbol{r}) \cdot [\nabla \times \boldsymbol{H}^*(\boldsymbol{r})], \tag{9.19}$$

is

$$\nabla \cdot \left[\frac{\boldsymbol{E}(\boldsymbol{r}) \times \boldsymbol{H}^*(\boldsymbol{r})}{2}\right] = -\frac{j\omega\mu}{2} |\boldsymbol{H}(\boldsymbol{r})|^2 - \frac{\sigma}{2} |\boldsymbol{E}(\boldsymbol{r})|^2. \tag{9.20}$$

When we integrate (9.20) throughout the volume under consideration and use the divergence theorem, we can obtain that

$$\int_S \tfrac{1}{2}\boldsymbol{E}(\boldsymbol{r}) \times \boldsymbol{H}^*(\boldsymbol{r}) \cdot \hat{n}\, dS = -j\frac{\omega\mu}{2} \int_0^\infty dx \int_{y_0}^{y_0+l} dy \int_{z_0}^{z_0+w} dz\, |\boldsymbol{H}(\boldsymbol{r})|^2 - \frac{\sigma}{2} \int_0^\infty dx \int_{y_0}^{y_0+l} dy \int_{z_0}^{z_0+w} dz\, |\boldsymbol{E}(\boldsymbol{r})|^2, \tag{9.21}$$

where $\hat{n}$ is an outwardly directed unit normal to the surface. For the electric and the magnetic fields given in (9.1), $\boldsymbol{E}(\boldsymbol{r}) \times \boldsymbol{H}^*(\boldsymbol{r})$ is in the x-direction with the result that the contribution to the surface integral in (9.21) comes only from the surfaces perpendicular to the x-direction. Moreover, since the fields vanish at $x = \infty$, the contribution to the surface integral in (9.21) comes only from the surface $x = 0$ for which $\hat{n} = -\hat{x}$. Therefore, together with (9.1), we can simplify (9.21) to yield

$$\int_{y_0}^{y_0+l}\int_{z_0}^{z_0+w} \hat{x} \cdot \tfrac{1}{2}\hat{y}E_y(0) \times \hat{z}H_z(0)\, dy\, dz = \frac{j\omega\mu}{2} \int_0^\infty dx \int_{y_0}^{y_0+l} dy \int_{z_0}^{z_0+w} dz\, |H_z(x)|^2 + \frac{\sigma}{2} \int_0^\infty dx \int_{y_0}^{y_0+l} dy \int_{z_0}^{z_0+w} dz\, |E_y(x)|^2. \tag{9.22}$$

We can see from (9.12) that the total complex power P flowing into the volume is given by the left side of (9.22). The real part of P is given by the second term on the right side of (9.22). This second term is equal to the total power dissipated inside the volume and has been evaluated in (9.15) and (9.16). The imaginary part of P is given by the first term on the right side of (9.22). This first term is therefore equal to the total reactive power inside the volume. Hence, the reactive power density in a good conductor is given by

$$P_{\text{Im}} = \frac{\omega\mu}{2} \boldsymbol{H}(\boldsymbol{r}) \cdot \boldsymbol{H}^*(\boldsymbol{r}). \tag{9.23}$$

Explicitly, we have deduced that

$$P_{\text{reactive}} = \frac{\omega\mu}{2} \int_0^{\infty} dx \int_{y_0}^{y_0+l} dy \int_{z_0}^{z_0+w} dz\, |H_z(x)|^2. \tag{9.24}$$

Together with (9.1) to (9.3), we may simplify (9.24) to yield

$$\begin{aligned} P_{\text{reactive}} &= \frac{\omega\mu}{2} \int_{y_0}^{y_0+l} dy \int_{z_0}^{z_0+w} dz \int_0^{\infty} dx\, |H_{z0}^+|^2\, e^{-2\alpha x} \\ &= \frac{\omega\mu}{2}\, lw \frac{|H_{z0}^+|^2}{2\alpha}. \end{aligned} \tag{9.25}$$

From (9.3), we obtain $\omega\mu/2\alpha = \eta_{0j}$. Also, $I_{sy} = wH_{z0}^+$. Hence, (9.25) with the help of (9.8) and (9.9) becomes

$$P_{\text{reactive}} = \tfrac{1}{2}\, |I_{sy}|^2\, \eta_{0j} \frac{l}{w} = \tfrac{1}{2}\, |I_{sy}|^2\, X_s. \tag{9.26}$$

We see that P_{reactive} in (9.26) is equal to that given in (9.14), which was obtained from the evaluation of the surface integral. We have thus established the usefulness of the surface impedance concept since we have shown that both the total power dissipated inside the volume as well as the total reactive power inside the volume can be obtained from a knowledge of the surface impedance and the surface current.

PROBLEMS

5.1 Verify by direct substitution that $E_y(x, t)$ and $H_z(x, t)$ given by (5.2.20) and (5.2.25) respectively satisfy (5.2.16a,b) and (5.2.18).

5.2 Verify that (5.2.17a,b) is obtained from (5.2.16a,b) by a rotation through 90 degrees about the direction of propagation. Show that the same transformation, when applied to (5.2.20), (5.2.25), and (5.2.26) yields (5.2.27), (5.2.28), and (5.2.29) respectively.

5.3 The power flux density is in the same direction as that of motion of the wavefront. Use this *ansatz* to obtain the sign of the intrinsic impedance as given by (5.2.29).

5.4 Show that the charge density ρ in a very good conductor decays to $1/e$ of its original value in time ε/σ sec where ε is the permittivity and σ the conductivity of the medium. Calculate this so-called relaxation time for the sea water for which $\varepsilon = 9 \times 10^{-9}/4\pi$ F/m and $\sigma = 4$ mho/m.

5.5 The half-space $0 < x < \infty$ is occupied by a good conductor (copper) whose constitutive parameters are $\mu = 4\pi \times 10^{-7}$ H/m and $\sigma = 5.8 \times 10^7$ mho/m. A tangential magnetic field of 1 A/m is switched on at time $t = 0$ over the entire $x = 0$ plane. Find the maximum value of the conduction current density at $x = 0.01$ m and the time at which this maximum value is reached. At this time, what is the value of the tangential electric field at the surface $x = 0$ of the good conductor?

Repeat the calculations for iron whose constitutive parameters are $\mu = 4\pi \times 10^{-7}$ H/m and $\sigma = 1.0 \times 10^7$ mho/m.

5.6 A uniform time-harmonic plane wave of frequency $f = 10^{10}$ Hz is progressing in the $+z$-direction in an unbounded medium whose constitutive parameters are $\mu = 4\pi \times 10^{-7}$ H/m, $\varepsilon = 10^{-9}/9\pi$ F/m and $\sigma = 0$ mho/m. The electric vector is in the x-direction and has an amplitude of 1 V/m.

a) Find the phase velocity, the wavelength, the phase-shift coefficient, the group velocity, and the intrinsic impedance of the medium.
b) Find the amplitude and the direction of the magnetic vector.
c) Determine the time-averaged energy densities W_E and W_H associated respectively with the electric and the magnetic fields. Find the value of the time-averaged power flux density and hence deduce the direction and the magnitude of the energy transport velocity.

5.7 In the Problem 5.6, the conductivity instead of being zero is given by $\sigma = 10^{-2}$ mho/m. All the other parameters are unaltered.

a) Find the phase-shift coefficient, the attenuation coefficient, the intrinsic resistance and the intrinsic reactance of the medium.
b) Find the amplitude and the direction of the magnetic vector. Express the complex amplitude in terms of both the real and the imaginary parts, and, the magnitude and the phase angle.

5.8 A good conductor is approximately defined as one for which $\sigma/\omega\varepsilon$ is equal to or greater than 10. The constitutive parameters of sea water are $\sigma = 4$ mho/m and $\varepsilon = 9 \times 10^{-9}/4\pi$ F/m. Find the frequency range for which sea water is a good conductor.

A good dielectric is approximately defined as one for which $\sigma/\omega\varepsilon$ is equal to or less than 0.1. The constitutive parameters of fresh water are $\sigma = 10^{-3}$ mho/m and $\varepsilon = 9 \times 10^{-9}/4\pi$ F/m. Find the frequency range for which fresh water is a good dielectric.

5.9 The constitutive parameters of fused quartz at a frequency $f = 10^6$ Hz are given by $\mu = 4\pi \times 10^{-7}$ H/m, $\varepsilon = 3.78 \times 10^{-9}/36\pi$ F/m, and $\sigma = 5 \times 10^{-6}$ mho/m. Calculate the propagation coefficient, the phase velocity, and the intrinsic impedance using the exact expressions. Find the group velocity with the help of the formula (5.6.9). Also, obtain the results using the approximate expressions which are valid for a good dielectric.

Repeat the computation for ruby mica whose constitutive parameters are $\mu = 4\pi \times 10^{-7}$ H/m, $\varepsilon = 3 \times 10^{-10}/2\pi$ F/m, and $\sigma = 6 \times 10^{-6}$ mho/m. Assume that the frequency is given by $f = 10^8$ Hz.

5.10 The constitutive parameters of sea water are $\mu = 4\pi \times 10^{-7}$ H/m, $\varepsilon = 9 \times 10^{-9}/4\pi$ F/m, and $\sigma = 4$ mho/m. The frequency of operation is $f = 10^6$ Hz. Evaluate the propagation coefficient and the intrinsic impedance using the exact expressions as well as by using the approximate expressions which are valid for a good conductor.

Repeat the calculations for copper whose constitutive parameters may be assumed to be given by $\mu = 4\pi \times 10^{-7}$ H/m, $\varepsilon = 10^{-9}/36\pi$ F/m, and $\sigma = 5.8 \times 10^7$ mho/m. Assume that the frequency of operation is $f = 10^8$ Hz.

5.11 The propagation coefficient $\gamma = \alpha + j\beta$ for plane waves in an unbounded lossy medium is obtained from (5.5.4) and (5.5.13) to be

$$\gamma = \sqrt{-\omega^2\mu\varepsilon + j\omega\mu\sigma}, \tag{a}$$

where μ, ε, and σ are the constitutive parameters of the medium and ω is the wave angular frequency. From (a), deduce that

$$\alpha = \omega\sqrt{\frac{\mu\varepsilon}{2}\left[\sqrt{1 + \frac{\sigma^2}{\omega^2\varepsilon^2}} - 1\right]}, \tag{b}$$

$$\beta = \omega\sqrt{\frac{\mu\varepsilon}{2}\left[\sqrt{1 + \frac{\sigma^2}{\omega^2\varepsilon^2}} + 1\right]}. \tag{c}$$

For $\sigma/\omega\varepsilon \ll 1$, obtain the following approximations from (b) and (c):

$$\alpha = \frac{\sigma}{2}\sqrt{\frac{\mu}{\varepsilon}}\left[1 - \frac{\sigma^2}{8\omega^2\varepsilon^2}\right], \tag{d}$$

$$\beta = \omega\sqrt{\mu\varepsilon}\left[1 + \frac{\sigma^2}{8\omega^2\varepsilon^2}\right]. \tag{e}$$

For $\sigma/\omega\varepsilon \gg 1$, show that (b) and (c) may be approximated as follows:

$$\alpha = \sqrt{\frac{\omega\mu\sigma}{2}}\left(1 - \frac{\omega\varepsilon}{2\sigma}\right), \tag{f}$$

$$\beta = \sqrt{\frac{\omega\mu\sigma}{2}}\left(1 + \frac{\omega\varepsilon}{2\sigma}\right). \tag{g}$$

5.12 From the equation (c) in Problem 5.11, find the expressions for the phase velocity v_{ph}, the wavelength λ, and the group velocity v_g. From these expressions, deduce the following approximations for v_{ph}, λ, and v_g which are valid for a good dielectric for which $\sigma/\omega\varepsilon \ll 1$:

$$v_{ph} = v\left[1 - \frac{\sigma^2}{8\omega^2\varepsilon^2}\right]; \qquad v = \frac{1}{\sqrt{\mu\varepsilon}},$$

$$\lambda = \frac{2\pi v}{\omega}\left[1 - \frac{\sigma^2}{8\omega^2\varepsilon^2}\right],$$

and

$$v_g = v\left[1 + \frac{\sigma^2}{8\omega^2\varepsilon^2}\right].$$

5.13 The intrinsic impedance $\eta_0 = \eta_{0r} + j\eta_{0j}$ for plane waves in an unbounded lossy medium is obtained from (5.5.4) and (5.5.16) to be

$$\eta_0 = \eta_{0r} + j\eta_{0j} = \sqrt{\frac{\mu}{\varepsilon + (\sigma/j\omega)}}, \tag{a}$$

where μ, ε, and σ are the constitutive parameters of the medium and ω is the wave angular frequency. From (a), deduce that

$$\eta_{0r}^2 = \frac{\mu}{2\varepsilon[1 + (\sigma^2/\omega^2\varepsilon^2)]}\left[1 + \sqrt{1 + \frac{\sigma^2}{\omega^2\varepsilon^2}}\right], \tag{b}$$

$$\eta_{0j}^2 = \frac{\mu}{2\varepsilon[1 + (\sigma^2/\omega^2\varepsilon^2)]}\left[-1 + \sqrt{1 + \frac{\sigma^2}{\omega^2\varepsilon^2}}\right]. \tag{c}$$

For $\sigma/\omega\varepsilon \ll 1$, obtain the following approximations from (b) and (c):

$$\eta_{0r} = \sqrt{\frac{\mu}{\varepsilon}}\left(1 - \frac{3\sigma^2}{8\omega^2\varepsilon^2}\right), \tag{d}$$

$$\eta_{0j} = \sqrt{\frac{\mu}{\varepsilon}}\frac{\sigma}{2\omega\varepsilon}\left(1 - \frac{5\sigma^2}{8\omega^2\varepsilon^2}\right). \tag{e}$$

For $\sigma/\omega\varepsilon \gg 1$, show that (b) and (c) may be approximated as follows:

$$\eta_{0r} = \sqrt{\frac{\omega\mu}{2\sigma}}\left(1 + \frac{\omega\varepsilon}{2\sigma}\right), \tag{f}$$

and

$$\eta_{0j} = \sqrt{\frac{\omega\mu}{2\sigma}}\left(1 - \frac{\omega\varepsilon}{2\sigma}\right). \tag{g}$$

5.14 From the treatment indicated in Section 3.10, it is seen that the attenuation coefficient α for the propagation of plane waves in a slightly lossy dielectric medium is given by

$$\alpha = P_L/2P \tag{a}$$

where P is the time-averaged power flux density and P_L is the time-averaged power density dissipated in unit length in the direction of propagation. Consider the plane wave as given by (5.5.17). Show that the attenuation coefficient as deduced from (a) is approximately

$$\alpha = \frac{\sigma}{2}\sqrt{\frac{\varepsilon}{\mu}}. \tag{b}$$

Note that α as given by (b) is equal to the leading term in the expression (5.6.5) which is valid for a good dielectric or a slightly lossy dielectric.

5.15 Two plane waves of angular frequency ω are propagating in the positive z-direction in a medium of constitutive parameters μ and ε. Their electric vectors are given by

$$\text{(i) } E_x = E_{x0}e^{j(\omega t - \beta z)}, \qquad \text{(ii) } E_y = E_{y0}e^{j(\omega t - \beta z)},$$

where $\beta = \omega\sqrt{\mu\varepsilon}$. Obtain the other field components of these two plane waves.

Write the electric vectors associated with these two plane waves as real quantities and examine their polarizations. If these two plane waves are superposed such that $aE_{x0} = E_{y0}$, where a is real, what is the polarization of the resultant plane wave? Obtain the magnetic vector of the resultant plane wave.

5.16 Show that in the phasor notation the circular polarizations can be represented as follows:

$$\mathbf{E} = (\hat{y} + j\hat{z})E \qquad \text{(left circular polarization)}, \tag{a}$$

$$\mathbf{E} = (\hat{y} - j\hat{z})E \qquad \text{(right circular polarization)}. \tag{b}$$

In a similar manner, prove that in the phasor notation the elliptic polarizations can be represented as follows:

$$\mathbf{E} = (\hat{y} + ja\hat{z})E \qquad \text{(left elliptic polarization)}, \tag{c}$$

$$\mathbf{E} = (\hat{y} - ja\hat{z})E \qquad \text{(right elliptic polarization)}, \tag{d}$$

where a is a real constant greater than zero.

Establish that any elliptic polarization such as, for example, that given by (c) can be obtained from a superposition of two oppositely directed circular polarizations as given by (a) and (b).

In a similar manner prove that any circular polarization such as, for example, that given by (b) can be obtained from a superposition of two oppositely directed elliptic polarizations as given by (c) and (d).

5.17 Consider a plane wave propagation in the positive z-direction in an unbounded, nondissipative medium. At $z = 0$, the electric field has only an x-component as given by

$$\boldsymbol{E}\,(z = 0) = \hat{x}E_{x0} = \hat{x}\,|E_{x0}|\,e^{j\theta_{E_0}}. \tag{a}$$

By rewriting (a) as

$$\boldsymbol{E}\,(z = 0) = (\hat{x} + j\hat{y})\frac{E_{x0}}{2} + (\hat{x} - j\hat{y})\frac{E_{x0}}{2}, \tag{b}$$

show that a linear polarization can be decomposed into two oppositely directed circular polarizations. Identify the sense of polarizations in (b).

Let β_z^+ and β_z^- be the phase-shift coefficients associated with the positive and the negative circular polarizations respectively. Assume that the two constituents in (b) propagate in the positive z-direction with the phase-shift coefficients related to their respective polarizations. Obtain the value of the electric field $\boldsymbol{E}(z)$ at an arbitrary position z.

Rewrite $\boldsymbol{E}\,(z = 0)$ as given by (a) and $\boldsymbol{E}(z)$ as real quantities including the time-dependence factor and show that the plane of polarization is rotated by an angle θ_F/z per unit distance:

$$\theta_F/z = (\beta_z^- - \beta_z^+)/2. \tag{c}$$

5.18 The power flux density of a plane wave in free space is 10 W/m^2. Find (i) the amplitude of the electric field, (ii) the amplitude of the magnetic field and (iii) the time-averaged energy densities associated with the electric and the magnetic fields.

How are the results changed if the medium instead of being free space is a dielectric with the following constitutive parameters:

$$\mu = 4\pi \times 10^{-7}\ \text{H/m} \qquad \text{and} \qquad \varepsilon = 10^{-9}/9\pi\ \text{F/m?}$$

5.19 The electric vector of a uniform plane wave in free space is given by

$$\boldsymbol{E}(\boldsymbol{r}) = [3\hat{x} + 4\hat{y} + (3 - j4)\hat{z}]e^{-j2(0.8x - 0.6y) + j\omega t}. \tag{a}$$

Show that the electric field given by (a) is consistent with Maxwell's equations. Find the phase-shift coefficient, the wavelength, and the wave angular frequency ω. Determine the associated magnetic field vector.

Evaluate (i) the time-averaged power flux density, (ii) the time-averaged energy densities associated with the electric and the magnetic fields, and (iii) the energy transport velocity.

Find the direction of progression of the phase fronts and the direction of transport of power.

5.20 Find the frequency for which the skin depth δ_{skin} in copper is 0.01 mm. Assume that the constitutive parameters are given by $\mu = 4\pi \times 10^{-7}$ H/m and $\sigma = 5.8 \times 10^7$ mhos/m.

5.21 Find the value of the skin depth δ_{skin} for the following common materials at a frequency $f = 10^6$ Hz:

silver: $\sigma = 6.17 \times 10^7$ mhos/m;

aluminum: $\sigma = 3.72 \times 10^7$ mhos/m.

Assume that the permeability of these two materials is the same as that of free space.

Repeat the calculations for the frequency $f = 5 \times 10^6$ Hz.

5.22 Show that, for a good conductor, the skin depth is very small in comparison with the wavelength.

5.23 Consider a half-space ($x > 0$) of copper whose constitutive parameters are $\mu = 4\pi \times 10^{-7}$ H/m and $\sigma = 5.8 \times 10^7$ mho/m. A tangential magnetic field of 1 A/m and of frequency

$f = 10^8$ Hz is maintained throughout the surface ($x = 0$) of the conductor. Find the value of the tangential electric field at the surface of the conductor. From the surface impedance concept, evaluate the time-averaged power absorbed by the conductor per unit cross-sectional area. What is the value of the skin depth δ_{skin}?

Repeat the calculations for the wave frequency $f = 10^9$ Hz.

5.24 Consider a half-space ($x > 0$) of sea water whose constitutive parameters are $\mu = 4\pi \times 10^{-7}$ H/m, $\varepsilon = 9 \times 10^{-9}/4\pi$ F/m, and $\sigma = 4$ mho/m. A tangential magnetic field of 1 A/m, and of frequency $f = 10^6$ Hz is maintained throughout the surface $x = 0$. Find the value of the conduction current density at the surface. Evaluate the time-averaged power absorbed by the sea water per unit cross-sectional area.

Determine the skin depth δ_{skin} and the wavelength λ. How do the quantities δ_{skin} and λ compare?

CHAPTER 6

REFLECTION AND TRANSMISSION OF PLANE WAVES

6.1 INTRODUCTION

Many interesting electromagnetic phenomena occur in the presence of boundaries separating different media. Simple cases of such phenomena involve plane boundaries and plane waves. We shall give a treatment of some problems of plane electromagnetic wave propagation in regions with plane boundaries separating media with different constitutive parameters in this chapter. We shall first analyze the problem of reflection and transmission of plane waves normally incident on a plane interface separating two different semi-infinite media for the case in which the two media are lossless and the plane waves are arbitrary functions of time and then extend it to the case in which the two media are lossy and the plane waves are harmonic functions of time. In the remaining part of this chapter we shall consider only the time-harmonic fields. We shall discuss the effect of a resistive sheet of infinitesimal thickness placed at the plane interface between two different media. We shall then examine the problem of plane waves normally incident on a planar dielectric slab in detail and emphasize the usefulness of the techniques, developed first in connection with the transmission-line theory, for the analysis of the dielectric slab problem.

We shall discuss the reflection of plane waves incident obliquely on a perfectly conducting plane surface for both the possible polarizations of the incident field. Then, we shall treat the problem of reflection and transmission of plane waves incident obliquely on a plane interface separating two semi-infinite dielectric media for the two polarizations and investigate the interesting possibilities of total reflection and total transmission. We shall introduce the concept of inhomogeneous plane waves and examine the essential characteristics of such inhomogeneous plane waves supported by a plane reactive surface.

6.2 PLANE WAVES WITH ARBITRARY TIME DEPENDENCE NORMALLY INCIDENT ON A PLANE INTERFACE

Let us consider a plane interface $x = 0$ which separates a half-space ($x < 0$) of Medium 1 characterized by the permeability μ_1, the permittivity ε_1, and zero conductivity from another half-space ($x > 0$) of Medium 2 characterized by the permeability μ_2, the permittivity ε_2, and zero conductivity. There are no losses in either of the two media. A uniform plane wave propagating in the $+x$-direction is incident normally on the interface from the side of Medium 1. We assume the electric vector in the incident wave to be in the y-direction and therefore the magnetic vector in the incident wave is in the z-direction. Hence, we may assume the incident plane wave with arbitrary time dependence to be of the form:

Incident wave ($x < 0$)

$$E_y^i(x, t) = f\left(t - \frac{x}{v_1}\right), \qquad H_z^i(x, t) = g\left(t - \frac{x}{v_1}\right), \tag{2.1}$$

where,

$$g\left(t - \frac{x}{v_1}\right) = f\left(t - \frac{x}{v_1}\right)\bigg/\eta_1, \tag{2.2}$$

and

$$v_1 = 1/\sqrt{\mu_1\varepsilon_1}; \qquad \eta_1 = \sqrt{\mu_1/\varepsilon_1}. \tag{2.3}$$

The plane wave (2.1) when incident on the interface $x = 0$ gives rise to reflected and transmitted waves. The geometry of the problem is invariant to changes in y and z. Therefore, any dependence of the field quantities on y and z should arise only due to the incident wave. Since the incident wave is independent of y and z, the reflected and the transmitted waves are also independent of y and z. Moreover, the reflected and the transmitted waves travel in the $-x$- and $+x$-directions respectively. Consequently, the reflected and the transmitted fields may be sought in the following form:

Reflected wave ($x < 0$)

$$E_y^r(x, t) = f^r\left(t + \frac{x}{v_1}\right), \qquad H_z^r(x, t) = g^r\left(t + \frac{x}{v_1}\right), \tag{2.4}$$

where,

$$g^r\left(t + \frac{x}{v_1}\right) = -f^r\left(t + \frac{x}{v_1}\right)\bigg/\eta_1. \tag{2.5}$$

Transmitted wave ($x > 0$)

$$E_y^t(x, t) = f^t\left(t - \frac{x}{v_2}\right), \qquad H_z^t(x, t) = g^t\left(t - \frac{x}{v_2}\right), \tag{2.6}$$

where,

$$g^t\left(t - \frac{x}{v_2}\right) = f^t\left(t - \frac{x}{v_2}\right)\bigg/\eta_2, \tag{2.7}$$

and

$$v_2 = 1/\sqrt{\mu_2\varepsilon_2}; \qquad \eta_2 = \sqrt{\mu_2/\varepsilon_2}. \tag{2.8}$$

The superscripts i, r, and t on the field quantities indicate the incident, reflected, and transmitted waves respectively.

At the interface $x = 0$, the tangential electric and magnetic fields are required to be continuous. Hence, (2.1), (2.4), and (2.6) yield

$$f(t) + f^r(t) = f^t(t), \tag{2.9}$$

$$g(t) + g^r(t) = g^t(t). \tag{2.10}$$

With the help of (2.2), (2.5), and (2.7), we simplify (2.10) as

$$\frac{f(t)}{\eta_1} - \frac{f^r(t)}{\eta_1} = \frac{f^t(t)}{\eta_2}. \tag{2.11}$$

The successive elimination of $f^t(t)$ and $f^r(t)$ from (2.9) and (2.11) gives us the following results:

$$\frac{f^r(t)}{f(t)} = \rho_E = \frac{\eta_2 - \eta_1}{\eta_2 + \eta_1}, \tag{2.12}$$

$$\frac{f^t(t)}{f(t)} = \tau_E = 1 + \rho_E = \frac{2\eta_2}{\eta_2 + \eta_1}. \tag{2.13}$$

In (2.12) and (2.13), ρ_E and τ_E are called respectively the reflection and the transmission coefficients of the electric field. On using (2.2), (2.5), and (2.7) in (2.12) and (2.13), we find that

$$\frac{g^r(t)}{g(t)} = \rho_H = -\rho_E = -\frac{(\eta_2 - \eta_1)}{(\eta_2 + \eta_1)}, \tag{2.14}$$

$$\frac{g^t(t)}{g(t)} = \tau_H = 1 + \rho_H = 1 - \rho_E = \frac{2\eta_1}{\eta_2 + \eta_1}. \tag{2.15}$$

In (2.14) and (2.15), ρ_H and τ_H are called respectively the reflection and the transmission coefficients of the magnetic field. The reflection coefficient ρ_H for the magnetic field is the negative of the reflection coefficient ρ_E for the electric field. For both the electric and the magnetic fields, the transmission coefficient is equal to one plus the corresponding reflection coefficient. It is usual to refer the reflection coefficient only to the electric field. Unless we specify otherwise, we assume the reflection coefficient to refer to that of the electric field.

From (2.12), changing t to $t + (x/v_1)$, we obtain

$$f^r\left(t + \frac{x}{v_1}\right) = \rho_E f\left(t + \frac{x}{v_1}\right). \tag{2.16}$$

In a similar manner, changing t to $t - (x/v_2)$ in (2.13) gives us

$$f^t\left(t - \frac{x}{v_2}\right) = \tau_E f\left(t - \frac{x}{v_2}\right). \tag{2.17}$$

If we substitute (2.16) and (2.17) into (2.4) and (2.6) respectively and use (2.5) and (2.7), we obtain the following explicit expressions for the reflected and the transmitted fields:

$$E_y^r(x, t) = \rho_E f\left(t + \frac{x}{v_1}\right), \qquad H_z^r(x, t) = \frac{-\rho_E f[t + (x/v_1)]}{\eta_1}, \tag{2.18}$$

and

$$E_y^t(x, t) = \tau_E f\left(t - \frac{x}{v_2}\right), \qquad H_z^t(x, t) = \frac{\tau_E f[t - (x/v_2)]}{\eta_2}. \tag{2.19}$$

The function f gives the shape of the incident wave. We see from (2.18) that the reflected wave is also of the same shape and travels with the same velocity v_1 but in the opposite direction. Also, (2.19) shows us that the transmitted wave is also of the same shape as the incident wave but travels with a different velocity v_2 in the same direction as the incident wave.

It is instructive for us to examine the energy conservation during the process of reflection and transmission of the incident waves. We obtain the incident power flux density $\mathbf{S}^i(0, t)$ at the interface $x = 0$ from (2.1) and (2.2) as

$$\mathbf{S}^i(0, t) = \hat{y}E_y^i(0, t) \times \hat{z}H_z^i(0, t) = \hat{x}\frac{f^2(t)}{\eta_1}. \tag{2.20}$$

The reflected power flux density $\mathbf{S}^r(0, t)$ at the interface $x = 0$ is found from (2.18) to be

$$\mathbf{S}^r(0, t) = \hat{y}E_y^r(0, t) \times \hat{z}H_z^r(0, t) = -\hat{x}\frac{\rho_E^2}{\eta_1}f^2(t). \tag{2.21}$$

The negative sign in (2.21) indicates that the power density is flowing in the $-x$-direction. The sum of (2.20) and (2.21) gives us the net power density flowing into the second medium as:

$$\mathbf{S}^i(0, t) + \mathbf{S}^r(0, t) = \frac{\hat{x}f^2(t)}{\eta_1}(1 - \rho_E^2). \tag{2.22}$$

If we substitute the value of ρ_E from (2.12) into (2.22) we find

$$\mathbf{S}^i(0, t) + \mathbf{S}^r(0, t) = \hat{x}f^2(t)\frac{4\eta_2}{(\eta_2 + \eta_1)^2}. \tag{2.23}$$

The principle of energy conservation requires that (2.23) be equal to the transmitted power flux density. We obtain the transmitted power flux density $\mathbf{S}^t(0, t)$ at the interface $x = 0$ from (2.19) as

$$\mathbf{S}^t(0, t) = \hat{y}E_y^t(0, t) \times \hat{z}H_z^t(0, t) = \hat{x}\tau_E^2\frac{f^2(t)}{\eta_2}. \tag{2.24}$$

If we substitute the value of τ_E from (2.13) into (2.24) we get

$$\mathbf{S}^t(0, t) = \hat{x}f^2(t)\frac{4\eta_2}{(\eta_2 + \eta_1)^2}, \tag{2.25}$$

which we see to be the same as that given in (2.23) in conformity with the principle of energy conservation.

It is important for us to note that in the assumed forms for the reflected and the transmitted fields as given by (2.4) and (2.6), the electric field and the magnetic field were taken to be in the same directions as those for the corresponding fields in the incident wave. Suppose that we assume the reflected and the transmitted fields to have a z-component of the electric field and a y-component of the magnetic field as well. Since the plane waves are of the TEM type, there can be no x-component of either the electric or the magnetic field. On matching the z-components of the electric field and the y-components of the magnetic field at the interface $x = 0$, we see that these field components are identically zero at the interface for all times since similar field components are not present in the incident wave. It follows therefore that in the reflected and the transmitted waves, there is no z-component of the electric or y-component of the magnetic fields. In other words, the polarization of the incident wave is not altered in reflection or transmission at an interface between two isotropic dielectric half-spaces.

Let us consider the special case in which the half-space $x > 0$ is a perfect conductor. In the interior of a perfect conductor, there cannot be any electric or magnetic fields. On the surface of a perfect conductor, the tangential electric field is zero. From (2.1) and (2.4), we see that the boundary condition that the tangential electric field be zero on the surface of a perfect conductor yields the following relation:

$$f(t) + f^r(t) = 0 \qquad \text{or} \qquad \frac{f^r(t)}{f(t)} = -1. \tag{2.26}$$

If we compare (2.26) with (2.12) and use (2.18) we see that the reflected fields for the special case in which the half-space $x > 0$ is a perfect conductor are given by

$$E_y^r(x, t) = -f\left(t + \frac{x}{v_1}\right); \qquad H_z^r(x, t) = \frac{f[t + (x/v_1)]}{\eta_1}. \tag{2.27}$$

We should note that there are no transmitted fields inside Medium 2. We see from (2.1), (2.2), and (2.27) that the resultant tangential component of the magnetic field on the surface $x = 0$ of the perfect conductor is

$$H_z^i(0, t) + H_z^r(0, t) = \frac{f(t)}{\eta_1} + \frac{f(t)}{\eta_1} = \frac{2f(t)}{\eta_1}, \tag{2.28}$$

which is twice the value of the incident tangential magnetic field on the perfectly conducting surface. We then find the surface current density to be

$$\boldsymbol{J}_s(t) = \hat{n} \times \hat{z}\,\frac{2f(t)}{\eta_1} = -\hat{x} \times \hat{z}\,\frac{2f(t)}{\eta_1} = \hat{y}\,\frac{2f(t)}{\eta_1}, \tag{2.29}$$

which is in the same direction as the incident electric field.

We can extend the preceding analysis to the treatment of the phenomenon of reflection and transmission at multiple interfaces. Suppose that we wish to examine the phenomenon of reflection and transmission for a planar dielectric slab with parallel faces coinciding with the planes $x = 0$ and $x = d$. Let the medium external ($-\infty < x < 0$ and $d < x < \infty$) to the slab be characterized by the intrinsic resistance η_1 and the velocity of propagation v_1. We can assume the incident wave to be the same as that given by (2.1) and (2.2). The reflection and the transmission occurring at $x = 0$ are the same as in the preceding analysis. Since the phenomena at the two interfaces take place sequentially in time, the reflection and the transmission occurring at $x = d$ are amenable to a similar treatment. The waves reflected at $x = d$ are again incident on the interface $x = 0$ giving rise to additional reflected and transmitted waves. We should remember that the reflection and the transmission coefficients pertaining to $x = 0$ are different depending on the direction of incidence. Also, we can see that multiple reflections take place between the two parallel faces of the dielectric slab. It is helpful for us to note that this multiple reflection problem can be analyzed in exactly the same fashion as that of the propagation of the transients on a lossless transmission line with multiple discontinuities as discussed in Section 1.11.

6.3 TIME-HARMONIC PLANE WAVES NORMALLY INCIDENT ON A PLANE INTERFACE

We shall now discuss the phenomenon of reflection and transmission of plane waves with harmonic time dependence, normally incident on a plane interface. Let $x = 0$ be the plane interface which separates a half-space ($x < 0$) of Medium 1 from another half-space ($x > 0$) of Medium 2. Let Media 1 and 2 be characterized by the constitutive parameters $(\mu_1, \varepsilon_1, \sigma_1)$ and $(\mu_2, \varepsilon_2, \sigma_2)$ respectively. A uniform plane wave progressing in the $+x$-direction is incident normally on the interface from the side of Medium 1. We assume the incident magnetic vector to be in the z-direction and therefore the incident electric vector is in the y-direction. Hence, we can assume the plane wave with the harmonic time dependence of the form $e^{j\omega t}$ to be:

Incident wave ($x < 0$)

$$E_y^i(x) = \eta_1 H_{z0}^i e^{-\gamma_1 x} = E_{y0}^i e^{-\gamma_1 x}, \qquad H_z^i(x) = H_{z0}^i e^{-\gamma_1 x}, \tag{3.1}$$

where

$$\eta_1 = \eta_{1r} + j\eta_{1j} = \sqrt{\mu_1/\varepsilon_{1,\text{eff}}}, \qquad \gamma_1 = \alpha_1 + j\beta_1 = j\omega\sqrt{\mu_1\varepsilon_{1,\text{eff}}},$$
$$\varepsilon_{1,\text{eff}} = \varepsilon_1 + \sigma_1/j\omega. \tag{3.2}$$

The presence of the interface $x = 0$ gives rise to reflected and transmitted waves. Like the incident wave, the reflected and the transmitted waves are also independent of y and z, because the geometry of the interface has translational invariance with respect to the y- and z-directions. We may assume the reflected wave which progresses in the $-x$-direction and the transmitted wave which progresses in the $+x$-direction to be as follows:

Reflected wave ($x < 0$)

$$E_y^r(x) = -\eta_1 H_{z0}^r e^{\gamma_1 x} = E_{y0}^r e^{\gamma_1 x}; \qquad H_z^r(x) = H_{z0}^r e^{\gamma_1 x}. \tag{3.3}$$

Transmitted wave ($x > 0$)

$$E_y^t(x) = \eta_2 H_{z0}^t e^{-\gamma_2 x} = E_{y0}^t e^{-\gamma_2 x}, \qquad H_z^t(x) = H_{z0}^t e^{-\gamma_2 x}, \tag{3.4}$$

where

$$\eta_2 = \eta_{2r} + j\eta_{2j} = \sqrt{\mu_2/\varepsilon_{2,\text{eff}}}; \qquad \gamma_2 = \alpha_2 + j\beta_2 = j\omega\sqrt{\mu_2\varepsilon_{2,\text{eff}}};$$
$$\varepsilon_{2,\text{eff}} = \varepsilon_2 + \sigma_2/j\omega. \tag{3.5}$$

As before, the superscripts i, r, and t on the field quantities indicate the incident, reflected, and transmitted waves respectively.

At the interface $x = 0$, the tangential electric (E_y) and the tangential magnetic (H_z) fields are required to be continuous. Therefore, we find from (3.1), (3.3), and (3.4) that

$$\eta_1(H_{z0}^i - H_{z0}^r) = \eta_2 H_{z0}^t, \tag{3.6}$$

$$H_{z0}^i + H_{z0}^r = H_{z0}^t. \tag{3.7}$$

From (3.6) and (3.7), we can derive that

$$H_{z0}^r/H_{z0}^i = \rho_H = -(\eta_2 - \eta_1)/(\eta_2 + \eta_1), \tag{3.8}$$

and

$$H_{z0}^t/H_{z0}^i = \tau_H = 1 + \rho_H = 2\eta_1/(\eta_2 + \eta_1). \tag{3.9}$$

We can also deduce with the help of (3.1), (3.3), and (3.4) that

$$E_{y0}^r/E_{y0}^i = -H_{z0}^r/H_{z0}^i = \rho_E = -\rho_H = (\eta_2 - \eta_1)/(\eta_2 + \eta_1), \tag{3.10}$$

and

$$E_{y0}^t/E_{y0}^i = \eta_2 H_{z0}^t/\eta_1 H_{z0}^i = \tau_E = 1 + \rho_E = 2\eta_2/(\eta_2 + \eta_1). \tag{3.11}$$

We should note that the reflection and the transmission coefficients obtained in (3.8) to (3.11) are formally the same as the corresponding quantities given in (2.12) to (2.15) for the fields with arbitrary time dependence. The quantities given in (2.12) to (2.15) are real and are valid only for lossless media whereas those given in (3.8) to (3.11) can be complex and are valid even for lossy media. As before, the reflection coefficient ρ_H for the magnetic field is the negative of the reflection coefficient ρ_E for the electric field. Also, the following relation between the reflection and the transmission coefficients holds good for both the electric and the magnetic fields:

$$\tau = 1 + \rho. \tag{3.12}$$

Since from (3.8) and (3.9) we obtain the undetermined quantities H_{z0}^r and H_{z0}^t in terms of the incident quantity H_{z0}^i, the reflected and the transmitted waves are specified completely by (3.3) and (3.4) respectively.

For $x < 0$, we obtain the total electric and the total magnetic fields from (3.1), (3.3), and (3.10) as

$$E_y^-(x) = E_y^i(x) + E_y^r(x) = \eta_1 H_{z0}^i(e^{-\gamma_1 x} + \rho_E e^{\gamma_1 x}), \tag{3.13}$$

and

$$H_z^-(x) = H_z^i(x) + H_z^r(x) = H_{z0}^i(e^{-\gamma_1 x} - \rho_E e^{\gamma_1 x}), \tag{3.14}$$

where the superscript $-$ denotes that the fields pertain to the region $x < 0$. We can deduce the time-averaged Poynting vector $\boldsymbol{S}^-(x)$ associated with the fields given by (3.13) and (3.14) to be

$$\boldsymbol{S}^-(x) = \tfrac{1}{2}\,\text{Re}\,\hat{y}E_y^-(x) \times \hat{z}H_z^{-*}(x),$$

$$= \frac{\hat{x}}{2}|H_{z0}^i|^2\,[\eta_{1r}e^{-2\alpha_1 x} - \eta_{1r}\,|\rho_E|^2\,e^{2\alpha_1 x} - 2\eta_{1j}\,\text{Im}\,\{\rho_E e^{j2\beta_1 x}\}]. \tag{3.15}$$

If the incident wave as given by (3.1) alone is present, we obtain the corresponding time-averaged Poynting vector $\boldsymbol{S}^i(x)$ as

$$\boldsymbol{S}^i(x) = \tfrac{1}{2}\,\mathrm{Re}\,\hat{y}E_y^i(x) \times \hat{z}H_z^{i*}(x),$$

$$= \frac{\hat{x}}{2}|H_{z0}^i|^2\,\eta_{1r}e^{-2\alpha_1 x}. \tag{3.16}$$

In a similar manner, if the reflected wave as given by (3.3) alone is present, we find the corresponding time-averaged Poynting vector $\boldsymbol{S}^r(x)$ to be

$$\boldsymbol{S}^r(x) = \tfrac{1}{2}\,\mathrm{Re}\,\hat{y}E_y^r(x) \times \hat{z}H_z^{r*}(x),$$

$$= -\frac{\hat{x}}{2}|H_{z0}^i|^2\,\eta_{1r}\,|\rho_E|^2\,e^{2\alpha_1 x}. \tag{3.17}$$

The incident and the reflected power flux densities are given by (3.16) and (3.17) respectively. The minus sign in (3.17) indicates that the reflected power flows in the $-x$-direction. It is important for us to note that the sum of (3.16) and (3.17) is not equal to (3.15). For $x < 0$, as is seen from (3.13) and (3.14), the resultant fields consist of a linear superposition of the corresponding incident and the reflected fields. The time-averaged power flux density $\boldsymbol{S}^-(x)$ associated with $E_y^-(x)$ and $H_z^-(x)$, which consist of a linear superposition of the incident and the reflected waves is not in general equal to the sum of the "incident" and the "reflected" power flux densities as we can see from (3.15) which, in addition to the sum of the terms given in (3.16) and (3.17), contains an additional term. This additional term vanishes for a lossless medium for which $\eta_{1j} = 0$. Therefore, only for a lossless medium is it correct to separate the total power flux density associated with $E_y^-(x)$ and $H_z^-(x)$ as the sum of the "incident" and the "reflected" power densities.

At the interface $x = 0$, we obtain the power flux density associated with the fields in the region $x < 0$ from (3.15) as

$$\boldsymbol{S}^-(0) = \frac{\hat{x}}{2}|H_{z0}^i|^2\,[\eta_{1r}(1 - |\rho_E|^2) - 2\eta_{1j}\,\mathrm{Im}\,\rho_E]. \tag{3.18}$$

For the transmitted wave as given by (3.4), we obtain the time-averaged Poynting vector $\boldsymbol{S}^t(x)$ as

$$\boldsymbol{S}^t(x) = \tfrac{1}{2}\,\mathrm{Re}\,\hat{y}E_y^t(x) \times \hat{z}H_z^{t*}(x),$$

$$= \frac{\hat{x}}{2}\,\mathrm{Re}\,[E_{y0}^t H_{z0}^{t*} e^{-2\alpha_2 x}]. \tag{3.19}$$

From (3.9) to (3.11), it follows that $H_{z0}^t = (1 - \rho_E)H_{z0}^i$ and $E_{y0}^t = (1 + \rho_E)E_{y0}^i$. Also, we see from (3.1) that $E_{y0}^i = \eta_1 H_{z0}^i$. Together with these results, we may simplify (3.19) to yield

$$\begin{aligned} \boldsymbol{S}^t(x) &= \frac{\hat{x}}{2} \operatorname{Re} \left[\eta_1 \, |H^i_{zo}|^2 \, (1 + \rho_E)(1 - \rho_E^*)\right] e^{-2\alpha_2 x}, \\ &= \frac{\hat{x}}{2} e^{-2\alpha_2 x} \, |H^i_{zo}|^2 \left[\eta_{1r}(1 - |\rho_E|^2) - 2\eta_{1j} \operatorname{Im} \rho_E\right]. \end{aligned} \tag{3.20}$$

At the interface $x = 0$, we find the power flux density associated with the fields in the region $x > 0$ from (3.20) to be

$$\boldsymbol{S}^t(0) = \frac{\hat{x}}{2} |H^i_{zo}|^2 \left[\eta_{1r}(1 - |\rho_E|^2) - 2\eta_{1j} \operatorname{Im} \rho_E\right]. \tag{3.21}$$

In verification of the principle of energy conservation, we can see $\boldsymbol{S}^t(0)$ as given in (3.21) to be identical to $\boldsymbol{S}^-(0)$ as given in (3.18).

We should emphasize that in the assumed forms of the reflected and the transmitted waves as given in (3.3) and (3.4) respectively, the electric and the magnetic vectors were taken to be in the same directions as the corresponding vectors of the incident wave. This is due to the fact that, as we have pointed out previously, the polarization of the incident wave is not altered in the process of reflection or transmission at an interface between two isotropic dielectric half-spaces.

Let us consider the special case in which the half-space $x > 0$ is a perfect conductor. Inside a perfect conductor, the electric and the magnetic fields do not exist. On the surface of a perfect conductor, the tangential electric field is zero. Therefore, (3.1) and (3.3) yield the following relation:

$$\eta_1 H^i_{zo} - \eta_1 H^r_{zo} = 0 \quad \text{or} \quad H^r_{zo} = H^i_{zo}. \tag{3.22}$$

The transmitted fields inside Medium 2 vanish and we find from (3.3) and (3.22) that the reflected fields for the special case in which the half-space $x > 0$ is a perfect conductor are given by

$$E^r_y(x) = -\eta_1 H^i_{zo} e^{\gamma_1 x}; \qquad H^r_z(x) = H^i_{zo} e^{\gamma_1 x}. \tag{3.23}$$

We see from (3.1) and (3.23) that the resultant tangential component of the magnetic field on the surface $x = 0$ of the perfect conductor is

$$H^i_z(0) + H^r_z(0) = H^i_{zo} + H^i_{zo} = 2H^i_{zo}, \tag{3.24}$$

which is equal to twice the contribution from the incident field alone. We then derive the surface current density from (3.24) as

$$\boldsymbol{J}_s = \hat{n} \times \hat{z} 2H^i_{zo} = -\hat{x} \times \hat{z} 2H^i_{zo} = \hat{y} 2H^i_{zo}. \tag{3.25}$$

We should note that the surface current density is in the same direction as the incident electric field.

The foregoing analysis is valid even when Medium 2 is a good conductor for which $\sigma_2/\omega\varepsilon_2 \gg 1$. Under this approximation, we may simplify (3.5) to yield

$$\eta_{2r} = \eta_{2j} = \sqrt{\omega\mu_2/2\sigma_2}, \qquad \alpha_2 = \beta_2 = \sqrt{\omega\mu_2\sigma_2/2}, \tag{3.26}$$

which is analogous to that given in (5.9.3). In Section 5.9, we treated the propagation of plane waves in a good conductor and in that treatment the tangential magnetic field H_{z0}^{+} at the surface of the good conductor was assumed to be specified. A comparison of (5.9.1) with (3.4) shows that $H_{z0}^{+} = H_{z0}^{t}$. From (3.9), we obtain H_{z0}^{t} in terms of the amplitude H_{z0}^{i} of the incident magnetic field. Therefore, we can now determine the equivalent surface current density as given in (5.9.4) and the complex power density directed into the good conductor in terms of the amplitude H_{z0}^{i} of the incident magnetic field. Thus, the present analysis together with the discussion contained in Section 5.9 completes the treatment of the phenomenon of plane wave reflection and transmission at a plane interface bounding a half-space of a very good conductor.

6.4 TIME-HARMONIC PLANE WAVES NORMALLY INCIDENT ON A PLANAR RESISTIVE SHEET

Let us consider a plane interface $x = 0$ which separates a half-space ($x < 0$) of Medium 1 from another half-space ($x > 0$) of Medium 2. Let Media 1 and 2 be characterized by the constitutive parameters (μ_1, ε_1) and (μ_2, ε_2) respectively. For simplicity, we shall assume both the media to be lossless. Also, we assume that a thin planar layer of resistive material is sandwiched between the two half-spaces. The thickness of the resistive layer is assumed to be very small in comparison to a wavelength with the result that we can approximate it by a sheet of zero thickness. We assume this resistive sheet of zero thickness to occupy the plane $x = 0$ (Fig. 6.1). The surface current density $\boldsymbol{J}_s$ on the resistive sheet and the electric field tangential to it are related as follows:

$$\boldsymbol{J}_s = \sigma_s \mathbf{E}_t = \mathbf{E}_t/\rho_s \tag{4.1}$$

where σ_s is the surface conductivity and $\rho_s = 1/\sigma_s$ is the surface resistivity. We may

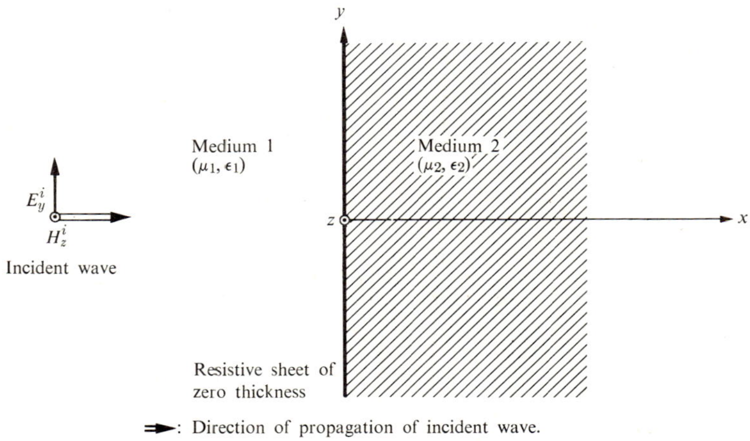

Fig. 6.1 Plane wave normally incident on a planar resistive sheet.

assume a uniform plane wave with harmonic time dependence progressing in the $+x$-direction and incident normally on the resistive sheet from the side of Medium 1 to be of the form:

Incident wave $(x < 0)$

$$E_y^i(x) = E_{y0}^i e^{-j\beta_1 x}, \qquad H_z^i(x) = H_{z0}^i e^{-j\beta_1 x} = \frac{E_{y0}^i}{\eta_{1r}} e^{-j\beta_1 x}, \tag{4.2}$$

where

$$\beta_1 = \omega\sqrt{\mu_1 \varepsilon_1}; \qquad \eta_{1r} = \sqrt{\mu_1/\varepsilon_1}. \tag{4.3}$$

We assume the reflected wave which progresses in the $-x$-direction in Medium 1 and the transmitted wave which progresses in the $+x$-direction in Medium 2 to be as follows:

Reflected wave $(x < 0)$

$$E_y^r(x) = E_{y0}^r e^{j\beta_1 x}; \qquad H_z^r(x) = H_{z0}^r e^{j\beta_1 x} = -\frac{E_{y0}^r}{\eta_{1r}} e^{j\beta_1 x}. \tag{4.4}$$

Transmitted wave $(x > 0)$

$$E_y^t(x) = E_{y0}^t e^{-j\beta_2 x}, \qquad H_z^t(x) = H_{z0}^t e^{-j\beta_2 x} = \frac{E_{y0}^t}{\eta_{2r}} e^{-j\beta_2 x}, \tag{4.5}$$

where

$$\beta_2 = \omega\sqrt{\mu_2 \varepsilon_2}; \qquad \eta_{2r} = \sqrt{\mu_2/\varepsilon_2}. \tag{4.6}$$

At the interface $x = 0$, the tangential electric field (E_y) is continuous and hence from (4.2), (4.4), and (4.5) we obtain that

$$E_{y0}^i + E_{y0}^r = E_{y0}^t. \tag{4.7}$$

There is a surface current density at the interface $x = 0$ due to the presence of the resistive sheet. Since the tangential electric field at the interface is in the y-direction, we see from (4.1) that the surface current density is also in the y-direction and therefore it is appropriate to set $\mathbf{J}_s = \hat{y}J_s$. Since the tangential magnetic field at the interface is discontinuous by the amount of the surface current density, we find the boundary condition on the tangential magnetic field (H_z) from (4.2), (4.4), and (4.5) as

$$H_{z0}^i + H_{z0}^r = J_s + H_{z0}^t. \tag{4.8}$$

Also, it follows from (4.1), (4.2), (4.4), and (4.5) that

$$J_s = \frac{E_{y0}^t}{\rho_s}; \qquad H_{z0}^i = \frac{E_{y0}^i}{\eta_{1r}}; \qquad H_{z0}^r = -\frac{E_{y0}^r}{\eta_{1r}}; \qquad H_{z0}^t = \frac{E_{y0}^t}{\eta_{2r}}. \tag{4.9}$$

Together with (4.9), (4.8) becomes

$$\frac{1}{\eta_{1r}}(E_{y0}^i - E_{y0}^r) = E_{y0}^t\left(\frac{1}{\rho_s} + \frac{1}{\eta_{2r}}\right). \tag{4.10}$$

For convenience in writing the expressions for the reflection and the transmission coefficients, let

$$\eta_{pr} = \frac{\eta_{2r}\rho_s}{\eta_{2r} + \rho_s}. \tag{4.11}$$

By successively eliminating E^t_{y0} and E^r_{y0} from (4.7) and (4.10), we obtain the following expressions for the reflection and the transmission coefficients:

$$\rho_E = \frac{E^r_{y0}}{E^i_{y0}} = \frac{\eta_{pr} - \eta_{1r}}{\eta_{pr} + \eta_{1r}}, \tag{4.12}$$

and

$$\tau_E = 1 + \rho_E = \frac{E^t_{y0}}{E^i_{y0}} = \frac{2\eta_{pr}}{\eta_{pr} + \eta_{1r}}. \tag{4.13}$$

We can separate the total power flux density in the region $x < 0$ into the sum of the incident and the reflected power flux densities since Medium 1 is lossless. We evaluate the incident power flux density from (4.2) as

$$\mathbf{S}^i(x) = \tfrac{1}{2}\,\text{Re}\,\hat{y}E^i_y(x) \times \hat{z}H^{i*}_z(x).$$

$$= \frac{\hat{x}}{2}\frac{|E^i_{y0}|^2}{\eta_{1r}}. \tag{4.14}$$

In a similar manner, we find the reflected power flux density from (4.4) and (4.12) to be

$$\mathbf{S}^r(x) = \tfrac{1}{2}\,\text{Re}\,\hat{y}E^r_y(x) \times \hat{z}H^{r*}_z(x),$$

$$= -\frac{\hat{x}}{2}\frac{|E^i_{y0}|^2}{\eta_{1r}}\rho_E^2. \tag{4.15}$$

We should note that ρ_E and, hence, τ_E are real. It is interesting to observe that (4.14) and (4.15) are independent of x, as is to be expected since the medium is lossless. We can evaluate the transmitted power flux density with the help of (4.5) and (4.13) as

$$\mathbf{S}^t(x) = \tfrac{1}{2}\,\text{Re}\,\hat{y}E^t_y(x) \times \hat{z}H^{t*}_z(x),$$

$$= \frac{\hat{x}}{2}\frac{|E^i_{y0}|^2}{\eta_{2r}}\tau_E^2. \tag{4.16}$$

We find the power dissipated in the resistive sheet per unit square area from (4.1), (4.4), and (4.13) to be

$$S_D = \tfrac{1}{2}\,\text{Re}\,\sigma_s\,|E^t_{y0}|^2,$$

$$= \frac{|E^i_{y0}|^2}{2}\frac{\tau_E^2}{\rho_s}. \tag{4.17}$$

The power density flowing into the interface $x = 0$ from Medium 1 is given by the sum of $\mathbf{S}^i(x)$ and $\mathbf{S}^r(x)$. The power density flowing out of the interface $x = 0$ into Medium 2 is given by $\mathbf{S}^t(x)$. From the principle of energy conservation, it follows that $\hat{x} \cdot [S^i(x) + S^r(x) - S^t(x)]$ is the power dissipated in the resistive sheet per unit square area. From (4.14) to (4.16), we find that

$$\hat{x} \cdot [S^i(x) + S^r(x) - S^t(x)] = \frac{|E^i_{y0}|^2}{2} \tau_E^2 \left[\frac{(1 - \rho_E^2)}{\eta_{1r}\tau_E^2} - \frac{1}{\eta_{2r}} \right]. \tag{4.18}$$

We can show with the help of (4.12) and (4.13) that

$$\frac{(1 - \rho_E^2)}{\eta_{1r}\tau_E^2} = \frac{(1 - \rho_E)}{\eta_{1r}\tau_E} = \frac{2\eta_{1r}(\eta_{pr} + \eta_{1r})}{\eta_{1r}(\eta_{pr} + \eta_{1r})2\eta_{pr}} = \frac{1}{\eta_{pr}}. \tag{4.19}$$

Since from (4.11), $1/\eta_{pr} = 1/\eta_{2r} + 1/\rho_s$, we can substitute (4.19) into (4.18) to obtain

$$\hat{x} \cdot [S^i(x) + S^r(x) - S^t(x)] = \frac{|E^i_{y0}|^2}{2} \frac{\tau_E^2}{\rho_s}. \tag{4.20}$$

In conformity with the principle of energy conservation, we can see (4.20) to be identical with (4.17), which is found to be the power dissipated in the resistive sheet per unit square area.

We can extend the preceding analysis to the case in which Media 1 and 2 are lossy.

6.5 TIME-HARMONIC PLANE WAVES NORMALLY INCIDENT ON A PLANAR DIELECTRIC SLAB

Let us consider a planar dielectric slab bounded by the planes $x = 0$ and $x = d$. Let the dielectric slab region $0 < x < d$, be designated as Medium 2 and be characterized by the constitutive parameters μ_2, ε_2, and σ_2. The half-space $x < 0$ is designated as Medium 1 and its constitutive parameters are taken to be μ_1, ε_1, and 0. In a similar manner the half-space $d < x < \infty$ is called Medium 3, whose constitutive parameters are assumed to be μ_3, ε_3, and 0. We should note that Media 1 and 3 are lossless whereas Medium 2 consisting of the dielectric slab region is lossy. We assume a uniform plane wave with harmonic time dependence progressing in the $+x$-direction and incident normally on the planar dielectric slab from the side of Medium 1, as shown in Fig. 6.2, to be of the form:

$$E^i_y(x) = E^i_{10} e^{-j\beta_1 x}, \qquad H^i_z(x) = \frac{E^i_{10}}{\eta_{1r}} e^{-j\beta_1 x}, \tag{5.1}$$

where

$$\beta_1 = \omega\sqrt{\mu_1 \varepsilon_1}; \qquad \eta_{1r} = \sqrt{\mu_1/\varepsilon_1}. \tag{5.2}$$

We assume the reflected wave which progresses in the $-x$-direction in Medium 1 to be given by

$$E^r_y(x) = E^r_{10} e^{j\beta_1 x}, \qquad H^r_z(x) = -\frac{E^r_{10}}{\eta_{1r}} e^{j\beta_1 x}, \tag{5.3}$$

where the unknown coefficient E^r_{10} depends on the boundary conditions. We obtain

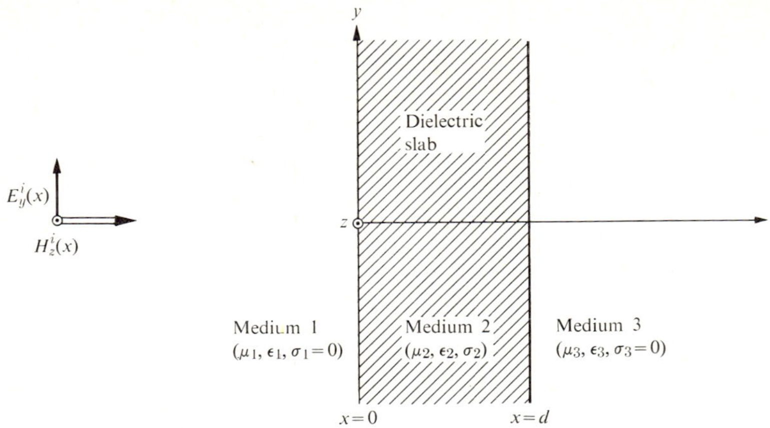

Fig. 6.2 Reflection and transmission of plane waves normally incident on a planar dielectric slab.

the total fields in Medium 1 by superposing (5.1) and (5.3), with the following result:

Medium 1 ($x < 0$)

$$E_{y1}(x) = E_y^i(x) + E_y^r(x) = E_{10}^i e^{-j\beta_1 x} + E_{10}^r e^{j\beta_1 x}, \tag{5.4}$$

$$H_{z1}(x) = H_z^i(x) + H_z^r(x) = \frac{E_{10}^i}{\eta_{1r}} e^{-j\beta_1 x} - \frac{E_{10}^r}{\eta_{1r}} e^{j\beta_1 x}. \tag{5.5}$$

Since Medium 2 is bounded, plane waves progressing in the $+x$- as well as in the $-x$-direction are present. Hence, we may assume the total fields in Medium 2 to be given by:

Medium 2 ($0 < x < d$)

$$E_{y2}(x) = E_{20}^i e^{-\gamma_2 x} + E_{20}^r e^{\gamma_2 x}, \tag{5.6}$$

$$H_{z2}(x) = \frac{E_{20}^i}{\eta_2} e^{-\gamma_2 x} - \frac{E_{20}^r}{\eta_2} e^{\gamma_2 x}, \tag{5.7}$$

where

$$\gamma_2 = \alpha_2 + j\beta_2 = j\omega\sqrt{\mu_2 \varepsilon_{2,\text{eff}}}; \qquad \eta_2 = \eta_{2r} + j\eta_{2j} = \sqrt{\mu_2/\varepsilon_{2,\text{eff}}};$$
$$\varepsilon_{2,\text{eff}} = \varepsilon_2 + \sigma_2/j\omega. \tag{5.8}$$

Medium 3 extends to infinity in the $+x$-direction and, therefore, only the plane waves progressing in the $+x$-direction are present. Consequently, we may assume the total fields in Medium 3 to be of the form:

Medium 3 ($x > d$)

$$E_{y3}(x) = E_{30}^i e^{-j\beta_3 x}, \tag{5.9}$$

$$H_{z3}(x) = \frac{E_{30}^i}{\eta_{3r}} e^{-j\beta_3 x}, \tag{5.10}$$

where

$$\beta_3 = \omega\sqrt{\mu_3\varepsilon_3}; \qquad \eta_{3r} = \sqrt{\mu_3/\varepsilon_3}. \tag{5.11}$$

We have to determine the coefficients E^r_{10}, E^i_{20}, E^r_{20}, and E^i_{30} in terms of E^i_{10} with the help of the boundary conditions that hold at the two interfaces $x = 0$ and $x = d$. In these coefficients, the superscript $i(r)$ denotes that the corresponding waves progress in the $+(-)$ x-direction.

At the interface $x = 0$, the tangential electric field (E_y) and the tangential magnetic field (H_z) are continuous and hence, from (5.4) to (5.7), we find that

$$E^i_{10} + E^r_{10} = E^i_{20} + E^r_{20}, \tag{5.12}$$

and

$$\frac{E^i_{10}}{\eta_{1r}} - \frac{E^r_{10}}{\eta_{1r}} = \frac{E^i_{20}}{\eta_2} - \frac{E^r_{20}}{\eta_2}. \tag{5.13}$$

In a similar manner, from the boundary conditions that the tangential electric (E_y) and the tangential magnetic (H_z) fields are continuous at the interface $x = d$ together with (5.6), (5.7), (5.9), and (5.10) we obtain the following relations:

$$E^i_{20}e^{-\gamma_2 d} + E^r_{20}e^{\gamma_2 d} = E^i_{30}e^{-j\beta_3 d}, \tag{5.14}$$

and

$$\frac{E^i_{20}}{\eta_2}e^{-\gamma_2 d} - \frac{E^r_{20}}{\eta_2}e^{\gamma_2 d} = \frac{E^i_{30}}{\eta_{3r}}e^{-j\beta_3 d}. \tag{5.15}$$

By successively eliminating E^r_{20} and E^i_{20} from (5.14) and (5.15), we can express E^i_{20} and E^r_{20} in terms of E^i_{30} as follows:

$$E^i_{20} = \frac{(\eta_{3r} + \eta_2)}{2\eta_{3r}}e^{\gamma_2 d}E^i_{30}e^{-j\beta_3 d}, \tag{5.16}$$

and

$$E^r_{20} = \frac{(\eta_{3r} - \eta_2)}{2\eta_{3r}}e^{-\gamma_2 d}E^i_{30}e^{-j\beta_3 d}. \tag{5.17}$$

We obtain the following results by substituting (5.16) and (5.17) into (5.12) and (5.13):

$$E^i_{10} + E^r_{10} = [(\eta_{3r} + \eta_2)e^{\gamma_2 d} + (\eta_{3r} - \eta_2)e^{-\gamma_2 d}]\frac{E^i_{30}e^{-j\beta_3 d}}{2\eta_{3r}}, \tag{5.18}$$

$$\frac{E^i_{10}}{\eta_{1r}} - \frac{E^r_{10}}{\eta_{1r}} = [(\eta_{3r} + \eta_2)e^{\gamma_2 d} - (\eta_{3r} - \eta_2)e^{-\gamma_2 d}]\frac{E^i_{30}e^{-j\beta_3 d}}{2\eta_2\eta_{3r}}. \tag{5.19}$$

Again, by successively eliminating E^r_{10} and E^i_{10} from (5.18) and (5.19), we can express E^i_{10} and E^r_{10} in terms of E^i_{30} as follows:

$$E^i_{10} = [(\eta_2 + \eta_{1r})(\eta_{3r} + \eta_2)e^{\gamma_2 d} + (\eta_2 - \eta_{1r})(\eta_{3r} - \eta_2)e^{-\gamma_2 d}]\frac{E^i_{30}e^{-j\beta_3 d}}{4\eta_2\eta_{3r}}, \tag{5.20}$$

$$E^r_{10} = [(\eta_2 - \eta_{1r})(\eta_{3r} + \eta_2)e^{\gamma_2 d} + (\eta_2 + \eta_{1r})(\eta_{3r} - \eta_2)e^{-\gamma_2 d}]\frac{E^i_{30}e^{-j\beta_3 d}}{4\eta_2\eta_{3r}}. \tag{5.21}$$

From (5.20), we find that

$$\frac{E^i_{30}}{E^i_{10}} = T = 4\eta_2\eta_{3r}e^{j\beta_3 d}[(\eta_2 + \eta_{1r})(\eta_{3r} + \eta_2)e^{\gamma_2 d} + (\eta_2 - \eta_{1r})(\eta_{3r} - \eta_2)e^{-\gamma_2 d}]^{-1}. \tag{5.22}$$

If we substitute (5.22) into (5.16) and (5.17) we obtain

$$E^i_{20}/E^i_{10} = 2\eta_2(\eta_{3r} + \eta_2)e^{\gamma_2 d}[(\eta_2 + \eta_{1r})(\eta_{3r} + \eta_2)e^{\gamma_2 d} + (\eta_2 - \eta_{1r})(\eta_{3r} - \eta_2)e^{-\gamma_2 d}]^{-1}, \tag{5.23}$$

$$E^r_{20}/E^i_{10} = 2\eta_2(\eta_{3r} - \eta_2)e^{-\gamma_2 d}[(\eta_2 + \eta_{1r})(\eta_{3r} + \eta_2)e^{\gamma_2 d} + (\eta_2 - \eta_{1r})(\eta_{3r} - \eta_2)e^{-\gamma_2 d}]^{-1}. \tag{5.24}$$

Dividing (5.21) by (5.20) we get

$$\frac{E^r_{10}}{E^i_{10}} = R = \frac{[(\eta_2 - \eta_{1r})(\eta_{3r} + \eta_2)e^{\gamma_2 d} + (\eta_2 + \eta_{1r})(\eta_{3r} - \eta_2)e^{-\gamma_2 d}]}{[(\eta_2 + \eta_{1r})(\eta_{3r} + \eta_2)e^{\gamma_2 d} + (\eta_2 - \eta_{1r})(\eta_{3r} - \eta_2)e^{-\gamma_2 d}]}. \tag{5.25}$$

In (5.22) to (5.25), we have evaluated E^i_{30}, E^i_{20}, E^r_{20}, and E^r_{10} all in terms of the incident field amplitude E^i_{10}. Together with (5.22) to (5.25), (5.3) gives the reflected fields in Medium 1, (5.6) and (5.7) give the fields in the dielectric slab region, and (5.9) and (5.10) give the fields transmitted through the dielectric slab into Medium 3. The ratio E^r_{10}/E^i_{10} of the amplitude of the field reflected by the dielectric slab to the amplitude of the field incident on it is called the reflection coefficient R of the dielectric slab. We should note that R is not the reflection coefficient pertaining to the interface $x = 0$ alone but takes into account also the effect of the interface $x = d$. It is for this reason that R is called the reflection coefficient of the dielectric slab. In a similar manner, the ratio E^i_{30}/E^i_{10} of the amplitude of the field transmitted through the dielectric slab to the amplitude of the field incident on it is called the transmission coefficient T of the dielectric slab.

1. Application of Impedance Transformation Formulas

It is not difficult for us to see that the impedance transformation formulas of transmission-line theory are applicable also to the present case of plane wave propagation in bounded media, since the governing equations for both the cases are similar. We can identify the intrinsic impedance and the surface impedivity with the characteristic impedance and the terminal impedance respectively. It is instructive for us to rederive the expression for the reflection coefficient R of the dielectric slab with the help of the impedance transformation formulas. Since Medium 3 extends to infinity and since in that medium only the waves progressing in the $+x$-direction exist, it follows that the input impedance or the surface impedivity η_{23} at the surface $x = d$ looking into Medium 3 is equal to the intrinsic resistance η_{3r} of Medium 3. We can treat the bounded dielectric slab as a lossy transmission line of length d having the characteristic impedance η_2 and the terminal impedance $\eta_{23} = \eta_{3r}$. We obtain the terminal reflection coefficient ρ_{23} from (3.10) as

$$\rho_{23} = \frac{\eta_{3r} - \eta_2}{\eta_{3r} + \eta_2}. \tag{5.26}$$

From (2.10.2), it follows that the input impedance at $x = 0$ or the surface impedivity η_{12} at the surface $x = 0$ is given by

$$\eta_{12} = \eta_2 \frac{[e^{\gamma_2 d} + \rho_{23} e^{-\gamma_2 d}]}{[e^{\gamma_2 d} - \rho_{23} e^{-\gamma_2 d}]}. \tag{5.27}$$

Together with (5.26), we may simplify (5.27) to

$$\eta_{12} = \eta_2 \frac{[(\eta_{3r} + \eta_2)e^{\gamma_2 d} + (\eta_{3r} - \eta_2)e^{-\gamma_2 d}]}{[(\eta_{3r} + \eta_2)e^{\gamma_2 d} - (\eta_{3r} - \eta_2)e^{-\gamma_2 d}]}. \tag{5.28}$$

Now, we can consider Medium 1 as a lossless transmission line having the characteristic resistance η_{1r} and the terminal impedance η_{12} as given by (5.28). We obtain the terminal reflection coefficient ρ_{12}, as before with the help of (3.10), to be

$$\rho_{12} = \frac{\eta_{12} - \eta_{1r}}{\eta_{12} + \eta_{1r}}. \tag{5.29}$$

Since ρ_{12} gives the ratio of the amplitude of the electric field reflected into Medium 1 from the interface $x = 0$ to the amplitude of the electric field incident on the interface $x = 0$, it follows that ρ_{12} is the same as the reflection coefficient R of the dielectric slab. If we substitute (5.28) into (5.29) and multiply the numerator and the denominator of the resulting equation by the denominator of (5.28), we obtain the following result:

$$\begin{aligned}\rho_{12} &= \frac{\eta_2[(\eta_{3r} + \eta_2)e^{\gamma_2 d} + (\eta_{3r} - \eta_2)e^{-\gamma_2 d}] - \eta_{1r}[(\eta_{3r} + \eta_2)e^{\gamma_2 d} - (\eta_{3r} - \eta_2)e^{-\gamma_2 d}]}{\eta_2[(\eta_{3r} + \eta_2)e^{\gamma_2 d} + (\eta_{3r} - \eta_2)e^{-\gamma_2 d}] + \eta_{1r}[(\eta_{3r} + \eta_2)e^{\gamma_2 d} - (\eta_{3r} - \eta_2)e^{-\gamma_2 d}]},\\ &= \frac{[(\eta_2 - \eta_{1r})(\eta_{3r} + \eta_2)e^{\gamma_2 d} + (\eta_2 + \eta_{1r})(\eta_{3r} - \eta_2)e^{-\gamma_2 d}]}{[(\eta_2 + \eta_{1r})(\eta_{3r} + \eta_2)e^{\gamma_2 d} + (\eta_2 - \eta_{1r})(\eta_{3r} - \eta_2)e^{-\gamma_2 d}]}.\end{aligned} \tag{5.30}$$

If we compare (5.30) with (5.25) we see that the reflection coefficient R of the dielectric slab can also be obtained by the repeated applications of the impedance transformation formulas of transmission-line theory.

2. Electrically Thin Dielectric Slabs

The dielectric slab is said to be electrically thin if $|\gamma_2 d| \ll 1$. It is of interest to us to examine the limiting forms of T and R as given by (5.22) and (5.25) respectively for an electrically thin dielectric slab. For $|\gamma_2 d| \ll 1$, we can approximate $e^{\pm\gamma_2 d}$ by the first two terms $1 \pm \gamma_2 d$ in its power series expansion. When this approximation is used in (5.25) and the resulting expressions for the numerator and the denominator are rearranged, we can deduce that

$$R = \frac{[\eta_2(\eta_{3r} - \eta_{1r}) + \gamma_2 d(\eta_2^2 - \eta_{1r}\eta_{3r})]}{\eta_2(\eta_{3r} + \eta_{1r})}\left[1 + \frac{\gamma_2 d(\eta_2^2 + \eta_{1r}\eta_{3r})}{\eta_2(\eta_{3r} + \eta_{1r})}\right]^{-1}. \tag{5.31}$$

We expand the second factor in (5.31) in binomial series and carry out the multiplication of the two factors explicitly. The result on retaining only the terms up to linear order in $\gamma_2 d$ is

$$R = \frac{(\eta_{3r} - \eta_{1r})}{(\eta_{3r} + \eta_{1r})} + \gamma_2 d 2\eta_{1r} \frac{(\eta_2^2 - \eta_{3r}^2)}{\eta_2(\eta_{3r} + \eta_{1r})^2}. \tag{5.32}$$

In the limit of the thickness of the dielectric slab tending to zero, the dielectric slab problem reduces to that of a plane wave incident normally on a plane interface $x = 0$, separating a semi-infinite region of Medium 3 characterized by the intrinsic resistance η_{3r} from another semi-infinite region of Medium 1 characterized by the intrinsic resistance η_{1r}. For this extreme limiting case, we obtain the reflection coefficient from (3.10) as $(\eta_{3r} - \eta_{1r})/(\eta_{3r} + \eta_{1r})$. We see that in the extreme case of $d = 0$, (5.32) attains the correct limiting form. If the dielectric slab is embedded in the same medium, that is, if Medium 1 and Medium 3 have the same characteristic resistance, say, η_{0r}, the leading term in (5.32) is zero and the remaining term in (5.32) may be simplified to yield

$$R = \gamma_2 d \frac{(\eta_2^2 - \eta_{0r}^2)}{2\eta_2\eta_{0r}} \qquad (\text{if } \eta_{1r} = \eta_{3r} = \eta_{0r}). \tag{5.33}$$

It follows from (5.33) that the reflection coefficient R of an electrically thin dielectric slab embedded in another dielectric medium is proportional to the electrical thickness of the dielectric slab.

If we approximate $e^{\pm\gamma_2 d}$ by $1 \pm \gamma_2 d$, we can simplify (5.22) to yield

$$\begin{aligned} T &= \frac{2\eta_{3r}e^{j\beta_3 d}}{(\eta_{3r} + \eta_{1r})}\left[1 + \gamma_2 d \frac{(\eta_2^2 + \eta_{1r}\eta_{3r})}{\eta_2(\eta_{3r} + \eta_{1r})}\right]^{-1}, \\ &= \frac{2\eta_{3r}e^{j\beta_3 d}}{(\eta_{3r} + \eta_{1r})}\left[1 - \gamma_2 d \frac{(\eta_2^2 + \eta_{1r}\eta_{3r})}{\eta_2(\eta_{3r} + \eta_{1r})}\right]. \end{aligned} \tag{5.34}$$

In the extreme limiting case of zero thickness for the dielectric slab, we obtain the transmission coefficient from (3.11) as $2\eta_{3r}/(\eta_{3r} + \eta_{1r})$. We can verify that in the extreme limiting case of $d = 0$, the transmission coefficient derived in (5.34) attains the correct limiting form. If $\eta_{3r} = \eta_{1r} = \eta_{0r}$, we can simplify (5.34) to

$$T = e^{j\beta_3 d}\left[1 - \gamma_2 d \frac{(\eta_2^2 + \eta_{0r}^2)}{2\eta_2\eta_{0r}}\right] \qquad (\text{if } \eta_{1r} = \eta_{3r} = \eta_{0r}). \tag{5.35}$$

It follows from (5.35) that the transmission coefficient T of an electrically thin dielectric slab embedded in another dielectric medium approaches unity as the slab thickness is reduced to zero.

We wish to consider now specifically the case in which the dielectric slab is lossless. Then, since $\sigma_2 = 0$, it follows from (5.8) that

$$\alpha_2 = 0; \qquad \beta_2 = \omega\sqrt{\mu_2\varepsilon_2}; \qquad \eta_{2j} = 0; \qquad \eta_{2r} = \sqrt{\mu_2/\varepsilon_2}. \tag{5.36}$$

3. Energy Conservation

There is no dissipation of energy inside the dielectric slab. Consequently, the net power flow into the dielectric slab through the interface $x = 0$ must be equal to the net power flow out of the dielectric slab through the interface $x = d$. It is instructive for us to establish that energy is conserved in the process of transmission through the dielectric slab. With the help of (5.8) and (5.36), we may simplify the transmission coefficient T of the dielectric slab as given in (5.22) to yield

$$T = 2\eta_{2r}\eta_{3r}e^{j\beta_3 d}[\eta_{2r}(\eta_{3r} + \eta_{1r})\cos\beta_2 d + j(\eta_{2r}^2 + \eta_{1r}\eta_{3r})\sin\beta_2 d]^{-1}. \tag{5.37}$$

From (5.37), we obtain

$$|T|^2 = \frac{4\eta_{2r}^2\eta_{3r}^2}{[\eta_{2r}^2(\eta_{3r} + \eta_{1r})^2\cos^2\beta_2 d + (\eta_{2r}^2 + \eta_{1r}\eta_{3r})^2\sin^2\beta_2 d]}. \tag{5.38}$$

In a similar manner, for a lossless dielectric slab, we may simplify the reflection coefficient R of the dielectric slab given in (5.25) with the following result:

$$R = \frac{[\eta_{2r}(\eta_{3r} - \eta_{1r})\cos\beta_2 d + j(\eta_{2r}^2 - \eta_{1r}\eta_{3r})\sin\beta_2 d]}{[\eta_{2r}(\eta_{3r} + \eta_{1r})\cos\beta_2 d + j(\eta_{2r}^2 + \eta_{1r}\eta_{3r})\sin\beta_2 d]}. \tag{5.39}$$

Also, from (5.39), we find that

$$|R|^2 = \frac{[\eta_{2r}^2(\eta_{3r} - \eta_{1r})^2\cos^2\beta_2 d + (\eta_{2r}^2 - \eta_{1r}\eta_{3r})^2\sin^2\beta_2 d]}{[\eta_{2r}^2(\eta_{3r} + \eta_{1r})^2\cos^2\beta_2 d + (\eta_{2r}^2 + \eta_{1r}\eta_{3r})^2\sin^2\beta_2 d]}. \tag{5.40}$$

With the help of (5.40), we can deduce that

$$1 - |R|^2 = \frac{4\eta_{1r}\eta_{2r}^2\eta_{3r}}{[\eta_{2r}^2(\eta_{3r} + \eta_{1r})^2\cos^2\beta_2 d + (\eta_{2r}^2 + \eta_{1r}\eta_{3r})^2\sin^2\beta_2 d]}. \tag{5.41}$$

We can verify from (5.38) and (5.41) that

$$(1 - |R|^2)/\eta_{1r} = |T|^2/\eta_{3r}. \tag{5.42}$$

Since Medium 1 is lossless, we can separate the total power flux density in the region $x < 0$ into the sum of the incident and the reflected power flux densities. We obtain the incident power flux density from (5.1) as

$$\mathbf{S}^i(x) = \tfrac{1}{2}\,\mathrm{Re}\,\hat{y}E_y^i(x) \times \hat{z}H_y^{i*}(x) = \frac{\hat{x}}{2}\frac{|E_{10}^i|^2}{\eta_{1r}}. \tag{5.43}$$

We evaluate the reflected power flux density from (5.3) and (5.25) as

$$\mathbf{S}^r(x) = \tfrac{1}{2}\,\mathrm{Re}\,\hat{y}E_y^r(x) \times \hat{z}H_y^{r*}(x) = \frac{-\hat{x}}{2}\frac{|E_{10}^r|^2}{\eta_{1r}} = -\frac{\hat{x}\,|E_{10}^i|^2}{2\eta_{1r}}|R|^2. \tag{5.44}$$

The power density flowing into the dielectric slab through the interface $x = 0$ from Medium 1 is then given by the sum of $\boldsymbol{S}^i(x)$ and $\boldsymbol{S}^r(x)$, which we find with the help of (5.42) to be

$$\boldsymbol{S}^i(x) + \boldsymbol{S}^r(x) = \frac{\hat{x}}{2\eta_{1r}} |E^i_{10}|^2 (1 - |R|^2) = \frac{\hat{x}\, |E^i_{10}|^2 \, |T|^2}{2\eta_{3r}}. \tag{5.45}$$

We evaluate the power density leaving the dielectric slab or the transmitted power flux density from (5.9), (5.10), and (5.22) as

$$\boldsymbol{S}^t(x) = \tfrac{1}{2} \operatorname{Re} \hat{y} E^t_{y3}(x) \times \hat{z} H^{t*}_{y3}(x) = \frac{\hat{x}}{2} \frac{|E^i_{30}|^2}{\eta_{3r}} = \frac{\hat{x}\, |E^i_{10}|^2}{2\eta_{3r}} |T|^2. \tag{5.46}$$

As expected (5.46) is equal to (5.45), verifying the fact that the entire power density flowing into the dielectric slab leaves it since there is no dissipation inside the dielectric slab.

4. Half-a-wavelength Dielectric Slab

The special features of a half-wavelength and a quarter-wavelength thick dielectric slab are worthy of emphasis. If the thickness of the dielectric slab is half of a wavelength in the dielectric Medium 2, then

$$\beta_2 d = \frac{2\pi}{\lambda_2} d = \pi. \tag{5.47}$$

Hence, we see from (5.39) that

$$R = (\eta_{3r} - \eta_{1r})/(\eta_{3r} + \eta_{1r}) \qquad (\text{for } \beta_2 d = \pi). \tag{5.48}$$

The reflection coefficient (5.48) is the same as that for no slab. We can argue that if the thickness of the dielectric slab is an integral multiple of half-a-wavelength in the dielectric Medium 2 and if Media 1 and 3 are the same, there are no reflected waves due to the dielectric slab.

5. Quarter-wavelength Dielectric Slab

If the thickness of the dielectric slab is a quarter of a wavelength in the dielectric Medium 2, then

$$\beta_2 d = \frac{2\pi}{\lambda_2} d = \frac{\pi}{2}. \tag{5.49}$$

Therefore, it follows from (5.39) that

$$R = (\eta_{2r}^2 - \eta_{1r}\eta_{3r})/(\eta_{2r}^2 + \eta_{1r}\eta_{3r}), \qquad (\beta_2 d = \pi/2). \tag{5.50}$$

If

$$\eta_{2r} = \sqrt{\eta_{1r}\eta_{3r}}, \tag{5.51}$$

there are no reflected waves from the dielectric slab. We can nullify the reflection at a plane interface separating two semi-infinite media of characteristic resistances η_{1r} and η_{3r} by interposing a planar dielectric slab having a characteristic resistance η_{2r}, which is the geometric mean of the characteristic resistances of the two media on either side of the slab provided the thickness of the slab is adjusted to be a quarter of a wavelength in the dielectric. This phenomenon is similar to the quarter wavelength transformer in the transmission-line theory discussed in Section 3.8.

Example. A dielectric coating is to be applied at a plane interface between air and glass to eliminate reflections at the boundary. We may assume the constitutive parameters of air to be given by $\mu_0 = 4\pi \times 10^{-7}$ H/m and $\varepsilon_0 = (1/36\pi) \times 10^{-9}$ F/m, and those of glass are given by $\mu_g = \mu_0$ and $\varepsilon_g = 4\varepsilon_0$. If the wave frequency is 3000 MHz, find the thickness of the dielectric coating and its permittivity. Assume that the permeability of the dielectric is the same as that of air and glass. What is the percentage of the incident power flux density that is reflected, if for the designed dielectric coating, the wave frequency is changed to 4000 MHz?

The intrinsic resistance η_0 of air is given by $\eta_0 = \sqrt{\mu_0/\varepsilon_0}$ and that of glass is obtained as $\eta_g = \sqrt{\mu_0/4\varepsilon_0} = \eta_0/2$. We may choose a quarter-wavelength thick dielectric coating to eliminate the reflections at the boundary. In accordance with (5.51), the intrinsic resistance η_d of the dielectric should be $\eta_d = \sqrt{\eta_0\eta_0/2} = \eta_0/\sqrt{2}$. If $\varepsilon_d = \varepsilon_0\varepsilon_r$ is the permittivity of the dielectric, then

$$\eta_d = \sqrt{\mu_0/\varepsilon_0\varepsilon_r} = \eta_0/\sqrt{\varepsilon_r} = \eta_0/\sqrt{2}.$$

Hence, $\varepsilon_r = 2$. The permittivity of the dielectric is therefore given by $\varepsilon_d = 2\varepsilon_0$. The velocity of electromagnetic waves inside the dielectric is

$$v_d = 1/\sqrt{\mu_0 2\varepsilon_0} = 3 \times 10^8/\sqrt{2} \text{ m/sec.}$$

At a frequency of 3000 MHz, we obtain the wavelength as

$$\lambda_d = (3 \times 10^{10})/(\sqrt{2} \times 3 \times 10^9) = 7.07 \text{ cm.}$$

Therefore, the thickness of the dielectric slab should be $d = \lambda_d/4 \simeq 1.77$ cm.

If the phase-shift coefficient inside the dielectric coating is β_2, then at the design frequency of 3000 MHz, $\beta_2 d = \pi/2$. Since $\beta_2 d$ increases linearly with the frequency, we obtain the value of $\beta_2 d$ at the operating frequency of 4000 MHz to be

$$\beta_2 d = (4/3) \times (\pi/2) = 2\pi/3.$$

For the operating frequency, the dielectric coating is no longer quarter-wavelength thick with the result that there are reflections from the dielectric coating. We find the fraction of the incident power density that is reflected back from (5.43) and (5.44) to be $|R|^2$, whose expression is given in (5.40). For $\beta_2 d = 2\pi/3$, $\sin^2 \beta_2 d = 0.75$ and $\cos^2 \beta_2 d = 0.25$. Also, in (5.40); for *air*, $\eta_{1r} = \eta_0$; for *dielectric*, $\eta_{2r} = \eta_0/\sqrt{2}$; for *glass*, $\eta_{3r} = \eta_0/2$. The substitution of these values in (5.40) yields $|R|^2 = \frac{1}{33}$. Hence, we obtain the percentage of the incident power flux density that is reflected as $(\frac{1}{33}) \times 100 = 3.03$.

6.6 PLANE WAVES INCIDENT OBLIQUELY ON A PERFECTLY CONDUCTING PLANE SURFACE

Let us consider a perfectly conducting surface occupying the plane $x = 0$. In the half-space $x > 0$ characterized by the constitutive parameters μ and ε, let a plane wave be incident obliquely on the perfectly conducting plane surface. We assume

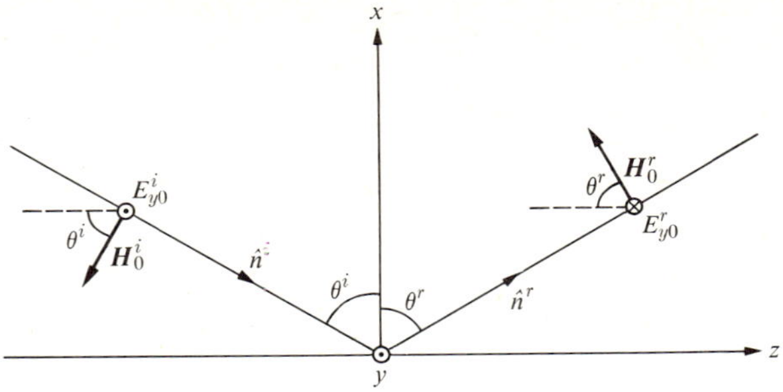

Fig. 6.3 Plane wave reflection by a perfectly conducting plane surface; ***E*** field perpendicular to the plane of incidence.

the wave-normal direction of the incident plane wave to be specified by the unit vector $\hat{n}^i$. Let the plane which is perpendicular to the plane $x = 0$ and which contains the wave-normal direction $\hat{n}^i$ of the incident wave be the $y = 0$ plane as shown in Fig. 6.3. The incident plane wave and the geometry of the reflecting surface have no variations in the y-direction. Consequently, the entire electromagnetic field has no variation in the y-direction, that is, $\partial/\partial y = 0$. Let $\theta^i[0 < \theta^i < \pi/2]$ be the angle between $-\hat{n}^i$ and the $+x$-direction which is normal to the reflecting surface $x = 0$. It is conventional to call θ^i the angle of incidence. Since the angles made by $\hat{n}^i$ with the x, y, and z axes are seen from Fig. 6.3 to be given by $\pi - \theta^i$, $\pi/2$ and $(\pi/2) - \theta^i$ respectively, it follows that

$$\begin{aligned}\hat{n}^i &= \hat{x}\cos(\pi - \theta^i) + \hat{y}\cos\frac{\pi}{2} + \hat{z}\cos\left(\frac{\pi}{2} - \theta^i\right),\\ &= -\hat{x}\cos\theta^i + \hat{z}\sin\theta^i.\end{aligned} \tag{6.1}$$

The zx-plane which is perpendicular to the reflecting surface $x = 0$ and which contains the wave-normal direction $\hat{n}^i$ is called *the plane of incidence.* The electric and the magnetic vectors of the incident plane wave lie in a plane perpendicular to $\hat{n}^i$. We can break down the incident electric vector into two components: one of which is perpendicular to the plane of incidence, and therefore parallel to the reflecting surface, and the other parallel to the plane of incidence but not perpendicular to the reflecting surface. Since the incident electric and the incident magnetic vectors are perpendicular to each other, it follows that for the second case for which the incident electric vector is in the plane of incidence, the corresponding magnetic vector is normal to the plane of incidence and therefore parallel to the reflecting surface. Thus, we can break down the incident wave into two components; namely, one for which the electric vector and the other for which the magnetic vector are perpendicular to the plane of incidence. We find it convenient to treat these two polarizations separately.

1. *E* Field Perpendicular to the Plane of Incidence

We can verify that the polarization of the incident wave is not altered by the reflection at a plane perfectly conducting surface $x = 0$. It follows therefore that for the present polarization

$$E_x(x, z) = E_z(x, z) = H_y(x, z) = 0. \tag{6.2}$$

In view of (6.1), the incident electric vector may be taken as

$$E_y^i(x, z) = E_{y0}^i e^{-j\beta\mathbf{n}\cdot\mathbf{r}} = E_{y0}^i e^{-j\beta(-x\cos\theta^i + z\sin\theta^i)}, \tag{6.3}$$

where $\beta = \omega\sqrt{\mu\varepsilon}$. With the help of (6.2) and the fact that $\partial/\partial y = 0$, we find the nonvanishing rectangular components of the time-harmonic Maxwell's equations to be

$$-j\omega\mu H_x(x, z) = -\frac{\partial}{\partial z} E_y(x, z), \tag{6.4a}$$

$$-j\omega\mu H_z(x, z) = \frac{\partial}{\partial x} E_y(x, z), \tag{6.4b}$$

$$j\omega\varepsilon E_y(x, z) = \frac{\partial}{\partial z} H_x(x, z) - \frac{\partial}{\partial x} H_z(x, z). \tag{6.4c}$$

If we substitute (6.4a) and (6.4b) into (6.4c) we obtain

$$\left[\frac{\partial^2}{\partial x^2} + \frac{\partial^2}{\partial z^2} + \beta^2\right] E_y(x, z) = 0. \tag{6.5}$$

It is an easy matter to verify that $E_y^i(x, z)$ given by (6.3) satisfies the wave equation (6.5). If we substitute (6.3) into (6.4a) and (6.4b) we obtain the following expressions for the rectangular components of the incident magnetic vector:

$$H_x^i(x, z) = -\frac{\sin\theta^i}{\eta_0} E_{y0}^i e^{-j\beta(-x\cos\theta^i + z\sin\theta^i)}, \tag{6.6a}$$

$$H_z^i(x, z) = -\frac{\cos\theta^i}{\eta_0} E_{y0}^i e^{-j\beta(-x\cos\theta^i + z\sin\theta^i)}, \tag{6.6b}$$

where $\eta_0 = \sqrt{\mu/\varepsilon}$.

It is helpful for us to obtain a general vector relation connecting the electric and the magnetic fields of a plane wave. For this purpose, let us assume the electric vector of the plane wave to be of the form

$$\boldsymbol{E}(\boldsymbol{r}) = \boldsymbol{E}_0 e^{-j\beta n\cdot r} = \boldsymbol{E}_0 e^{-j\beta(xn_x + yn_y + zn_z)}, \tag{6.7}$$

where

$$\hat{n} = n_x\hat{x} + n_y\hat{y} + n_z\hat{z}. \tag{6.8}$$

From the first of Maxwell's equations, $\nabla \times \boldsymbol{E}(\boldsymbol{r}) = -j\omega\mu\boldsymbol{H}(\boldsymbol{r})$, it follows that

$$\boldsymbol{H}(\boldsymbol{r}) = -\frac{1}{j\omega\mu}\nabla \times \boldsymbol{E}(\boldsymbol{r}) = -\frac{1}{j\omega\mu}\left[\hat{x}\frac{\partial}{\partial x} + \hat{y}\frac{\partial}{\partial y} + \hat{z}\frac{\partial}{\partial z}\right] \times \boldsymbol{E}_0 e^{-j\beta(n_x x + n_y y + n_z z)},$$

$$= \frac{j\beta(\hat{n} \times \boldsymbol{E}_0)}{j\omega\mu} e^{-j\beta\hat{n}\cdot\boldsymbol{r}} = \frac{(\hat{n} \times \boldsymbol{E}_0)}{\eta_0} e^{-j\beta\hat{n}\cdot\boldsymbol{r}} = \boldsymbol{H}_0 e^{-j\beta\hat{n}\cdot\boldsymbol{r}}. \tag{6.9}$$

In a similar manner, from the second of Maxwell's equations, $\nabla \times \boldsymbol{H}(\boldsymbol{r}) = j\omega\varepsilon \boldsymbol{E}(\boldsymbol{r})$, we can deduce that

$$\boldsymbol{E}(\boldsymbol{r}) = -\eta_0(\hat{n} \times \boldsymbol{H}_0)e^{-j\beta\hat{n}\cdot\boldsymbol{r}} = \boldsymbol{E}_0 e^{-j\beta\hat{n}\cdot\boldsymbol{r}}. \tag{6.10}$$

Thus, we see from (6.9) and (6.10) that the vector amplitudes $\boldsymbol{E}_0$ and $\boldsymbol{H}_0$ of a plane wave characterized by the wave-normal direction $\hat{n}$ satisfy the following relations:

$$\boldsymbol{H}_0 = \frac{(\hat{n} \times \boldsymbol{E}_0)}{\eta_0}; \qquad \boldsymbol{E}_0 = -\eta_0(\hat{n} \times \boldsymbol{H}_0). \tag{6.11a,b}$$

It is a simple exercise to reproduce (6.6a) and (6.6b) from (6.3) and (6.11a).

It follows from (6.3), (6.6a), and (6.6b) that the component of the motion of the phase front of the incident wave normal to the reflecting surface is in the $-x$-direction or towards the reflecting surface. It is intuitively clear that the component of the motion of the phase front of the reflected wave normal to the reflecting surface should be in the $+x$-direction or away from the reflecting surface. Let $\hat{n}^r$ be the wave-normal direction of the reflected wave. Since the reflected field also has no variation in the y-direction, it follows that $\hat{n}^r$ lies in the zx-plane. Let θ^r be the angle between $\hat{n}^r$ and the $+x$-direction as shown in Fig. 6.3. It is usual to call θ^r the angle of reflection. From Fig. 6.3, we can see that:

$$\hat{n}^r = \hat{x}\cos\theta^r + \hat{z}\sin\theta^r. \tag{6.12}$$

Therefore, we may assume the reflected electric vector to be of the form

$$E_y^r(x, z) = E_{y0}^r e^{-j\beta\hat{n}^r\cdot\boldsymbol{r}} = E_{y0}^r e^{-j\beta(x\cos\theta^r + z\sin\theta^r)}, \tag{6.13}$$

where we have to determine E_{y0}^r from the boundary condition. From (6.13) and (6.11a), we can find the magnetic vector of the reflected wave.

The requirement that the direction of progression of the phase of the reflected wave has a component in the $+x$-direction yields the condition

$$-\pi/2 < \theta^r < \pi/2. \tag{6.14}$$

The tangential component of the total electric field is required to vanish on the perfectly conducting surface $x = 0$. Therefore, (6.3) and (6.13) yield the relation

$$E_y^i(0, z) + E_y^r(0, z) = E_{y0}^i e^{-j\beta z\sin\theta^i} + E_{y0}^r e^{-j\beta z\sin\theta^r} = 0. \tag{6.15}$$

The requirement (6.15) can be satisfied for all z only if the phase terms in (6.15) are equal. Hence

$$\sin\theta^i = \sin\theta^r. \tag{6.16}$$

Since $0 < \theta^i < \pi/2$, (6.16) together with (6.14) yields

$$\theta^r = \theta^i, \tag{6.17}$$

which shows that the angles of incidence and reflection are equal. When (6.17) is used in (6.15), we obtain that

$$E_{y0}^r = -E_{y0}^i. \tag{6.18}$$

The substitution of (6.17) and (6.18) in (6.13) yields the following expression for the reflected electric vector:

$$E_y^r(x, z) = -E_{yo}^i e^{-j\beta(x \cos\theta^i + z \sin\theta^i)}. \tag{6.19}$$

With the help of (6.11a), (6.12) (6.17), and (6.18) we may deduce the components of the reflected magnetic vector as

$$H_x^r(x, z) = \frac{\sin\theta^i}{\eta_0} E_{yo}^i e^{-j\beta(x \cos\theta^i + z \sin\theta^i)}, \tag{6.20a}$$

$$H_z^r(x, z) = -\frac{\cos\theta^i}{\eta_0} E_{yo}^i e^{-j\beta(x \cos\theta^i + z \sin\theta^i)}. \tag{6.20b}$$

If we add (6.3), (6.6a), and (6.6b) with (6.19), (6.20a), and (6.20b) respectively we get the following expressions for the total fields in the region $x > 0$:

$$E_y(x, z) = E_y^i(x, z) + E_y^r(x, z) = 2jE_{yo}^i \sin(\beta x \cos\theta^i) e^{-j\beta z \sin\theta^i}, \tag{6.21}$$

$$H_x(x, z) = H_x^i(x, z) + H_x^r(x, z) = -2j\frac{\sin\theta^i}{\eta_0} E_{yo}^i \sin(\beta x \cos\theta^i) e^{-j\beta z \sin\theta^i}, \tag{6.22}$$

$$H_z(x, z) = H_z^i(x, z) + H_z^r(x, z) = -2\frac{\cos\theta^i}{\eta_0} E_{yo}^i \cos(\beta x \cos\theta^i) e^{-j\beta z \sin\theta^i}. \tag{6.23}$$

The foregoing expressions for the total fields together with the discussion contained in Section 3.3 reveal an important feature of these fields. In the z-direction, that is, parallel to the plane reflecting surface, the amplitude of the wave remains constant but there is a linear change in phase. Hence, the fields form a progressive wave in the z-direction. In the x-direction, that is, perpendicular to the reflecting surface, the amplitude of the wave varies periodically reaching the same maximum and minimum and the phase change is not linear. Therefore, the fields form a standing wave in the x-direction. There is a nonvanishing surface current density which we can obtain from (6.23) to be

$$\mathbf{J}_s(z) = \hat{x} \times \hat{z} H_z(0, z) = \hat{y}\frac{2\cos\theta^i}{\eta_0} E_{yo}^i e^{-j\beta z \sin\theta^i}, \tag{6.24}$$

which is also seen to be a progressive wave in the z-direction.

2. *H* Field Perpendicular to the Plane of Incidence

We can treat the phenomenon of reflection of a plane wave with its **H** field normal to the plane of incidence in a manner analogous to that employed for the case in which the ***E*** field is normal to the plane of incidence (Fig. 6.4). For this polarization, it follows that

$$H_x(x, z) = H_z(x, z) = E_y(x, z) = 0. \tag{6.25}$$

If (6.1) gives the incident wave-normal direction, we may assume the incident magnetic vector to be given by

$$H_y^i(x, z) = H_{yo}^i e^{-j\beta\hat{n}^i \cdot \mathbf{r}} = H_{yo}^i e^{-j\beta(-x \cos\theta^i + z \sin\theta^i)}. \tag{6.26}$$

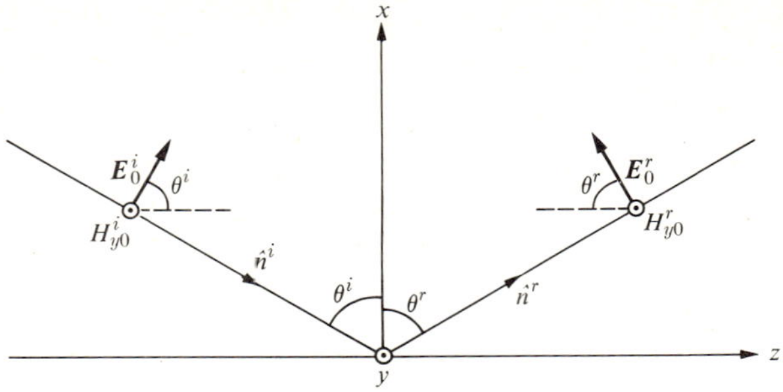

Fig. 6.4 Plane wave reflection by a perfectly conducting plane surface; ***H*** field perpendicular to the plane of incidence.

From (6.25) and the fact that $\partial/\partial y = 0$, we find the nonvanishing rectangular components of the time-harmonic Maxwell's equations to be

$$-j\omega\mu H_y(x, z) = \frac{\partial}{\partial z} E_x(x, z) - \frac{\partial}{\partial x} E_z(x, z), \tag{6.27}$$

$$j\omega\varepsilon E_x(x, z) = -\frac{\partial}{\partial z} H_y(x, z), \tag{6.28a}$$

$$j\omega\varepsilon E_z(x, z) = \frac{\partial}{\partial x} H_y(x, z). \tag{6.28b}$$

If we substitute (6.28a) and (6.28b) in (6.27) we get the wave equation

$$\left[\frac{\partial^2}{\partial x^2} + \frac{\partial^2}{\partial z^2} + \beta^2\right] H_y(x, z) = 0, \tag{6.29}$$

which can be shown to be satisfied by $H_y^i(x, z)$ given in (6.26). Either from (6.28a) and (6.28b) or from (6.11b) together with (6.1) and (6.26), we obtain the rectangular components of the incident electric vector as follows:

$$E_x^i(x, z) = \sin\theta^i \eta_0 H_{y0}^i e^{-j\beta(-x\cos\theta^i + z\sin\theta^i)}, \tag{6.30a}$$

$$E_z^i(x, z) = \cos\theta^i \eta_0 H_{y0}^i e^{-j\beta(-x\cos\theta^i + z\sin\theta^i)}. \tag{6.30b}$$

As before, if the wave normal of the reflected wave is given by (6.12) as illustrated in Fig. 6.4, we may assume the reflected magnetic vector to be of the form

$$H_y^r(x, z) = H_{y0}^r e^{-j\beta\hat{\mathbf{n}}^r\cdot\mathbf{r}} = H_{y0}^r e^{-j\beta(x\cos\theta^r + z\sin\theta^r)}. \tag{6.31}$$

From (6.11b), (6.12), and (6.31), we may deduce the rectangular components of the reflected electric vector to be

$$E_x^r(x, z) = \sin\theta^r \eta_0 H_{y0}^r e^{-j\beta(x\cos\theta^r + z\sin\theta^r)}, \tag{6.32a}$$

$$E_z^r(x, z) = -\cos\theta^r \eta_0 H_{y0}^r e^{-j\beta(x\cos\theta^r + z\sin\theta^r)}. \tag{6.32b}$$

The boundary condition that the tangential component of the total electric field vanish on the perfectly conducting surface $x = 0$ together with (6.30b) and (6.32b) yields

$$E_z^i(0, z) + E_z^r(0, z) = \eta_0[\cos\theta^i H_{y0}^i e^{-j\beta z \sin\theta^i} - \cos\theta^r H_{y0}^r e^{-j\beta z \sin\theta^r}] = 0. \quad (6.33)$$

The requirement (6.33) can be fulfilled for all z only if the phase terms in (6.33) are equal. Hence, we again obtain (6.16). Since (6.14) also is true for the present polarization, (6.17) holds good as before, showing that the angles of incidence and reflection are equal. Therefore, we find from (6.33) that

$$H_{y0}^r = H_{y0}^i. \quad (6.34)$$

Together with (6.17) and (6.34), (6.31) and (6.32) give the complete reflected fields.

We can obtain the total fields in the region $x > 0$ with the help of (6.26), (6.30) to (6.32), (6.17), and (6.34) to be

$$H_y(x, z) = H_y^i(x, z) + H_y^r(x, z) = 2H_{y0}^i \cos(\beta x \cos\theta^i)e^{-j\beta z \sin\theta^i}, \quad (6.35)$$

$$E_x(x, z) = E_x^i(x, z) + E_x^r(x, z) = 2\sin\theta^i \eta_0 H_{y0}^i \cos(\beta x \cos\theta^i)e^{-j\beta z \sin\theta^i}, \quad (6.36)$$

$$E_z(x, z) = E_z^i(x, z) + E_z^r(x, z) = 2j\cos\theta^i \eta_0 H_{y0}^i \sin(\beta x \cos\theta^i)e^{-j\beta z \sin\theta^i}. \quad (6.37)$$

It follows that the total fields given by (6.35) to (6.37) form a progressive wave in the z-direction and a standing wave in the x-direction. We can obtain the surface current density from (6.35) as

$$\mathbf{J}_s(z) = \hat{x} \times \hat{y}H_y(0, z) = \hat{z}2H_{y0}^i e^{-j\beta z \sin\theta^i}, \quad (6.38)$$

which is also a progressive wave in the z-direction.

Let E_{x0}^i, E_{y0}^i, E_{z0}^i, H_{x0}^i, H_{y0}^i, and H_{z0}^i be the components of the amplitude of the incident electric and the incident magnetic vectors. If we compare (6.30a), (6.3), (6.30b), (6.6a), (6.26), and (6.6b) with (6.32a), (6.19), (6.32b), (6.20a), (6.31), and (6.20b) respectively we see that

$$E_{x0}^r = E_{x0}^i; \qquad E_{y0}^r = -E_{y0}^i; \qquad E_{z0}^r = -E_{z0}^i; \quad (6.39)$$

$$H_{x0}^r = -H_{x0}^i; \qquad H_{y0}^r = H_{y0}^i; \qquad H_{z0}^r = H_{z0}^i. \quad (6.40)$$

It is found from (6.39) and (6.40) that the normal component of the incident electric field and the tangential components of the incident magnetic field are not changed by reflection, whereas the tangential components of the incident electric field and the normal component of the incident magnetic field are reversed in direction, but their magnitudes are unchanged, on reflection. Here, the normal and the tangential directions are with reference to the reflecting surface $x = 0$.

It is helpful for us to remember that, in this chapter, we assume the plane interfaces to be parallel to the yz-plane. In Sections 6.2 to 6.5, we only consider the case of normal incidence. If the direction of incidence is changed so as to lie always in the zx-plane, the case of oblique incidence which we treat in the remaining sections of this chapter results. Thus, the normal incidence treated in Sections 6.2 to 6.5 is the special case of the oblique incidence treated in the remaining Sections 6.6 to 6.9 of this chapter.

6.7 PLANE WAVES INCIDENT OBLIQUELY ON A PLANE INTERFACE BETWEEN TWO DIELECTRIC HALF-SPACES

Let the half-spaces $x > 0$ and $x < 0$ be two different dielectric media characterized by the constitutive parameters (μ, ε) and (μ', ε') respectively. In the half-space $x > 0$, we assume a plane wave to be incident obliquely on the interface $x = 0$. As before, let $y = 0$ be the plane of incidence and $\hat{n}^i$ as given by (6.1) be the wave-normal direction of the incident wave. As in the case of reflection by a perfectly conducting plane surface treated in the previous section, it is convenient for us to treat the E-polarization (E-field perpendicular to the plane of incidence) and the H-polarization (H-field perpendicular to the plane of incidence) separately.

1. *E*-polarization

For the E-polarization, $E_y(x, z)$, $H_x(x, z)$, and $H_z(x, z)$ are the only nonvanishing field components and are governed by (6.4a to c). We can assume the incident electric vector to be the same as that given by (6.3). The rectangular components of the incident magnetic vector are given by (6.6a) and (6.6b). In a similar manner, we assume the wave-normal direction $\hat{n}^r$ and the reflected electric vector to be given by (6.12) and (6.13) respectively. As before, we have to fulfill the requirement (6.14) in order that the direction of phase progression of the reflected wave has a component in the $+x$-direction. With the help of (6.11a), (6.12), and (6.13), we evaluate the components of the reflected magnetic field to be

$$H_x^r(x, z) = -\frac{\sin \theta^r}{\eta_0} E_{y0}^r e^{-j\beta(x \cos \theta^r + z \sin \theta^r)}, \tag{7.1a}$$

and

$$H_z^r(x, z) = \frac{\cos \theta^r}{\eta_0} E_{y0}^r e^{-j\beta(x \cos \theta^r + z \sin \theta^r)}. \tag{7.1b}$$

In addition to the reflected fields, there are also transmitted fields in the half-space $x < 0$. The transmitted fields have the same polarization as the incident and the reflected fields. Also, the direction of phase progression of the transmitted wave, like that of the incident wave, should have a component in the $-x$-direction. Let $\hat{n}^t$ be the wave-normal direction of the transmitted wave. Since the transmitted field also has no variation in the y-direction, it follows that $\hat{n}^t$ lies in the zx-plane. Let θ^t be the angle between $-\hat{n}^t$ and the $+x$-direction as shown in Fig. 6.5. As defined in Fig. 6.5, θ^t is called the angle of refraction or transmission. With the help of Fig. 6.5, we can ascertain that

$$\hat{n}^t = -\hat{x} \cos \theta^t + \hat{z} \sin \theta^t. \tag{7.2}$$

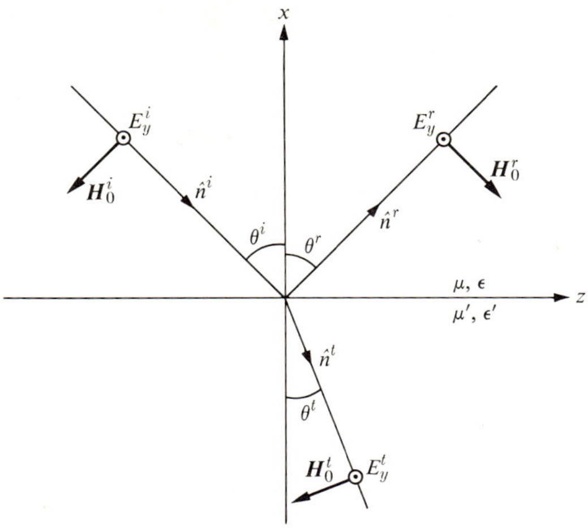

Fig. 6.5 *E*-polarized plane waves incident obliquely on a plane interface separating two different dielectric half-spaces.

Therefore, we may assume the transmitted electric vector in the form

$$E_y^t(x, z) = E_{y0}^t e^{-j\beta' \hat{n}^t \cdot r} = E_{y0}^t e^{-j\beta'(-x \cos\theta^t + z \sin\theta^t)}, \tag{7.3}$$

with

$$\beta' = \omega\sqrt{\mu'\varepsilon'}, \tag{7.4}$$

and we have to determine E_{y0}^t from the boundary conditions at the interface. The fields in the half-space $x < 0$ are also governed by (6.4) and (6.5) with μ, ε, and β replaced by μ', ε', and β' respectively. With the help of (6.11a) and (7.3), we find the components of the transmitted magnetic vector to be

$$H_x^t(x, z) = -\frac{\sin\theta^t}{\eta_0'} E_{y0}^t e^{-j\beta'(-x\cos\theta^t + z\sin\theta^t)}, \tag{7.5a}$$

and

$$H_z^t(x, z) = -\frac{\cos\theta^t}{\eta_0'} E_{y0}^t e^{-j\beta'(-x\cos\theta^t + z\sin\theta^t)}, \tag{7.5b}$$

where

$$\eta_0' = \sqrt{\mu'/\varepsilon'}. \tag{7.6}$$

The requirement that the direction of phase progression of the transmitted wave has a component in the $-x$-direction yields the condition

$$-\pi/2 < \theta^t < \pi/2. \tag{7.7}$$

From (6.3), (6.13), (7.3), (6.6b), (7.1b), and (7.5b) and the boundary conditions that the tangential component of the electric field (E_y) and the tangential component of the magnetic field (H_z) are continuous at the plane interface $x = 0$, we obtain the following relations:

$$E^i_{y0}e^{-j\beta z \sin\theta^i} + E^r_{y0}e^{-j\beta z \sin\theta^r} = E^t_{y0}e^{-j\beta' z \sin\theta^t}, \tag{7.8}$$

$$-\frac{\cos\theta^i}{\eta_0}E^i_{y0}e^{-j\beta z \sin\theta^i} + \frac{\cos\theta^r}{\eta_0}E^r_{y0}e^{-j\beta z \sin\theta^r} = -\frac{\cos\theta^t}{\eta_0'}E^t_{y0}e^{-j\beta' z \sin\theta^t}. \tag{7.9}$$

The relations (7.8) and (7.9) can be satisfied for all z only if the phase terms in (7.8) and (7.9) are equal. Hence

$$\beta \sin\theta^i = \beta \sin\theta^r = \beta' \sin\theta^t. \tag{7.10}$$

From (7.10) and (6.14), (6.16) and (6.17) follow and, as before, the angles of incidence and reflection are equal. Also, from (7.10), we can see that

$$\sin\theta^t/\sin\theta^i = \beta/\beta' = \sqrt{\mu\varepsilon/\mu'\varepsilon'}. \tag{7.11}$$

Since $0 < \theta^i < \pi/2$, it follows from (7.11) and (7.7) that $0 < \theta^t < \pi/2$. The *Snell's law of refraction* (7.11) enables us to evaluate the angle of refraction θ^t from the knowledge of the angle of incidence and the constitutive parameters of the two media. If we substitute (6.17) and (7.11) into (7.8) and (7.9), and successively eliminate E^t_{y0} and E^r_{y0} from the resulting equations, we obtain the reflection coefficient ρ_E and the transmission coefficient τ_E as follows:

$$\frac{E^r_{y0}}{E^i_{y0}} = \frac{\eta_0' \sec\theta^t - \eta_0 \sec\theta^i}{\eta_0' \sec\theta^t + \eta_0 \sec\theta^i} = \rho_E, \tag{7.12}$$

$$\frac{E^t_{y0}}{E^i_{y0}} = \frac{2\eta_0' \sec\theta^t}{\eta_0' \sec\theta^t + \eta_0 \sec\theta^i} = \tau_E. \tag{7.13}$$

Together with (6.17) and (7.11) to (7.13), (6.13), (7.1), (7.3), and (7.5) yield the complete expressions for the reflected and the refracted fields.

We can determine the total electric field in the half-space $x > 0$ where both the incident and the reflected fields exist with the help of (6.3), (6.13), (6.17), and (7.12) as

$$E^+_y(x, z) = E^i_y(x, z) + E^r_y(x, z) = E^i_{y0}[e^{j\beta x \cos\theta^i} + \rho_E e^{-j\beta x \cos\theta^i}]e^{-j\beta z \sin\theta^i}. \tag{7.14}$$

The z-dependence of $E^+_y(x, z)$ is controlled by the factor $e^{-j\beta z \sin\theta^i}$ whose amplitude is unity and whose phase has a linear variation in z. Hence, (7.14) represents a progressive wave in the z-direction. The x-dependence of $E^+_y(x, z)$ is controlled by the factor inside the square brackets in (7.14) and we can conveniently rewrite this factor as

$$F_x = e^{j\beta x \cos\theta^i}[1 + |\rho_E|\, e^{j\phi}], \qquad \phi = \theta_{\rho_E} - 2\beta x \cos\theta^i, \tag{7.15}$$

where $|\rho_E|$ and θ_{ρ_E} are the magnitude and the phase of the reflection coefficient ρ_E. We can evaluate the magnitude of (7.15) as

$$|F_x| = [1 + |\rho_E|^2 + 2\,|\rho_E| \cos\phi]^{1/2}. \tag{7.16}$$

We can verify that the phase of F_x does not vary linearly with x. The amplitude term (7.16) varies periodically with x and its maximum and minimum values are given by $1 + |\rho_E|$ and $1 - |\rho_E|$ respectively. As with the concepts developed in the transmission-line theory, the ratio $(1 + |\rho_E|)/(1 - |\rho_E|)$ of the maximum to the minimum of the amplitude of the electric field in the half-space $x > 0$ is called the standing-wave ratio S. Since the amplitude varies periodically and the phase nonlinearly with x, the electric field given in (7.14) forms a standing wave in the x-direction, that is, normal to the interface separating the two half-spaces. In the half-space $x < 0$, the transmitted fields given by (7.3) and (7.5) form progressive waves both in the x- and in the z-directions. We can evaluate the components of the total magnetic field in the half-space $x > 0$ from (6.6), (7.1), (6.17), and (7.12) as

$$H_x^+(x, z) = H_x^i(x, z) + H_x^r(x, z) = -\frac{\sin \theta^i}{\eta_0} E_{y0}^i\left[e^{j\beta x \cos \theta^i} + \rho_E e^{-j\beta x \cos \theta^i}\right]e^{-j\beta z \sin \theta^i}, \tag{7.17}$$

$$H_z^+(x, z) = H_z^i(x, z) + H_z^r(x, z) = -\frac{\cos \theta^i}{\eta_0} E_{y0}^i\left[e^{j\beta x \cos \theta^i} - \rho_E e^{-j\beta x \cos \theta^i}\right]e^{-j\beta z \sin \theta^i}. \tag{7.18}$$

We can see from (7.17) and (7.18) that the magnetic field components also, in the half-space $x > 0$, form standing waves in the x-direction and progressive waves in the z-direction.

It is helpful for us to relate the problem of plane waves incident obliquely on a plane interface between two dielectric half-spaces to the corresponding problem of normal incidence, for such an identification enables us to obtain the reflection and the transmission coefficients given respectively by (7.12) and (7.13) in a somewhat simpler manner by suitable modifications of the corresponding expressions for the normal incidence. The requirement that the boundary conditions be fulfilled everywhere along the interface has necessitated all the field components to have the same phase variation along the interface. For example, it is required that

$$E_y^{i,r}(x, z) = E_y^{i,r}(x)e^{-j\beta z \sin \theta^i}, \qquad E_y^t(x, z) = E_y^t(x)e^{-j\beta' z \sin \theta^t}, \tag{7.19}$$

where, in view of (7.11) we obtain the same phase variation. We should remember that (6.4a to c) govern the fields in both the dielectric half-spaces if the appropriate values are used for the parameters μ, ε, β, and θ^i in the case of the half-space $x < 0$. If we substitute (6.4a) into (6.4c) in order to eliminate $H_x(x, z)$ we obtain

$$j\omega\varepsilon E_y(x, z) = \frac{1}{j\omega\mu}\frac{\partial^2}{\partial z^2} E_y(x, z) - \frac{\partial}{\partial x} H_z(x, z). \tag{7.20}$$

With the help of the relations similar to (7.19), we can reduce (7.20) to yield

$$j\omega\varepsilon \cos^2 \theta^i E_y(x, z) = -\frac{\partial}{\partial x} H_z(x, z). \tag{7.21}$$

Also, we reproduce (6.4b) here for convenience:

$$j\omega\mu H_z(x, z) = -\frac{\partial}{\partial x} E_y(x, z). \tag{7.22}$$

For normal incidence $\theta^i = 0$, (7.21) and (7.22) become

$$j\omega\varepsilon E_y(x, z) = -\frac{\partial}{\partial x} H_z(x, z), \tag{7.23}$$

$$j\omega\mu H_z(x, z) = -\frac{\partial}{\partial x} E_y(x, z). \tag{7.24}$$

For normal incidence, even though the fields are independent of z, we have retained the argument z for simplicity in (7.23) and (7.24). We should note that (7.23) and (7.24) are identical to (5.5.9) with $\sigma = 0$ and (5.5.7) respectively. The generalization to oblique incidence for the E-polarization has resulted in the change of ε to $\varepsilon \cos^2 \theta^i$. Along the x-axis, the incident wave is progressing in the $-x$-direction. Hence, for normal incidence, (5.5.20) yields the value $-\sqrt{\mu/\varepsilon}$ for the intrinsic impedance. Since ε changes to $\varepsilon \cos^2 \theta^i$ for the case of oblique incidence, the *effective* intrinsic impedance becomes $\tilde{\eta}_0 = -\eta_0 \sec \theta^i$. In a similar manner, we find the *effective* intrinsic impedance for the half-space $x < 0$ to be $\tilde{\eta}'_0 = -\eta'_0 \sec \theta^t$. If instead of the actual intrinsic impedances, we use the effective values in (3.10) and (3.11), we obtain the reflection and the transmission coefficients for the obliquely incident E-polarized wave as

$$\rho_E = \frac{\tilde{\eta}'_0 - \tilde{\eta}_0}{\tilde{\eta}'_0 + \tilde{\eta}_0} = \frac{\eta'_0 \sec \theta^t - \eta_0 \sec \theta^i}{\eta'_0 \sec \theta^t + \eta_0 \sec \theta^i}, \tag{7.25}$$

$$\tau_E = \frac{2\tilde{\eta}'_0}{\tilde{\eta}'_0 + \tilde{\eta}_0} = \frac{2\eta'_0 \sec \theta^t}{\eta'_0 \sec \theta^t + \eta_0 \sec \theta^i}. \tag{7.26}$$

These expressions for the reflection and the transmission coefficients are the same as those in (7.12) and (7.13). Thus, insofar as the determination of the reflection and the transmission coefficients is concerned, in the corresponding expressions for the normal incidence, ε has to be changed to $\varepsilon \cos^2 \theta$ where θ is the angle of incidence or the angle of refraction according to whether ε refers to the half-space $x > 0$ or $x < 0$.

In the phenomenon of reflection and transmission of E-polarized plane waves obliquely incident on a dielectric half-space, we can establish that the energy is conserved. Since there are no losses, we can obtain the power flux density in the half-space $x > 0$ as the sum of the incident and the reflected power flux densities. In view of the fact that the electric vector has only a y-component and the magnetic vector has both x- and z-components it follows that the Poynting vector has both x- and z-components. But only the component of the Poynting vector in the $-x$-direction contributes to the power flux into the half-space $x < 0$. We obtain the

$-x$-component of the incident power flux density from (6.3) and (6.6b) as

$$S^i_{-x} = -\hat{x} \cdot \tfrac{1}{2} \operatorname{Re} \hat{y}E^i_y(x, z) \times \hat{z}H^{i*}_z(x, z),$$
$$= \frac{|E^i_{y0}|^2}{2\eta_0} \cos \theta^i. \tag{7.27}$$

In a similar manner, we show the $-x$-component of the reflected power flux density with the help of (6.13), (7.1b), (6.17), and (7.12) to be

$$S^r_{-x} = -\hat{x} \cdot \tfrac{1}{2} \operatorname{Re} \hat{y}E^r_y(x, z) \times \hat{z}H^{r*}_z(x, z),$$
$$= -\frac{|E^i_{y0}|^2}{2\eta_0} \cos \theta^i \, |\rho_E|^2. \tag{7.28}$$

Thus, the sum of (7.27) and (7.28) gives the power density flowing into the half-space $x < 0$ in the $-x$-direction. We can show with the help of (7.12) that

$$S^i_{-x} + S^r_{-x} = \frac{|E^i_{y0}|^2}{2\eta_0} \cos \theta^i [1 - |\rho_E|^2],$$
$$= |E^i_{y0}|^2 \, 2\eta'_0 \sec \theta^t [\eta'_0 \sec \theta^t + \eta_0 \sec \theta^i]^{-2}. \tag{7.29}$$

From (7.3) and (7.5b), we evaluate the $-x$-component of the transmitted power flux density as

$$S^t_{-x} = -\hat{x} \cdot \tfrac{1}{2} \operatorname{Re} \hat{y}E^t_y(x, z) \times \hat{z}H^{t*}_z(x, z),$$
$$= \frac{|E^t_{y0}|^2 \cos \theta^t}{2\eta'_0}. \tag{7.30}$$

If we substitute (7.13) in (7.30) we get

$$S^t_{-x} = |E^i_{y0}|^2 \, 2\eta'_0 \sec \theta^t [\eta'_0 \sec \theta^t + \eta_0 \sec \theta^i]^{-2}. \tag{7.31}$$

It follows from (7.29) and (7.31) that the power flux density leaving the half-space $x > 0$ through the interface $x = 0$ is equal to that entering the half-space $x < 0$ through the same interface, in confirmation of the fact that the energy is conserved in the process of reflection and transmission of E-polarized plane waves obliquely incident on a dielectric half-space.

If the permeabilities of the two half-spaces are the same, that is, if $\mu = \mu'$, we can recast the expressions for ρ_E and τ_E given by (7.12) and (7.13) respectively into compact forms. If $\mu = \mu'$, (7.11) gives

$$\sqrt{\varepsilon} = \sqrt{\varepsilon'} \frac{\sin \theta^t}{\sin \theta^i}. \tag{7.32}$$

Also, we can rewrite (7.12) as

$$\rho_E = \frac{\sqrt{\varepsilon} \cos \theta^i - \sqrt{\varepsilon'} \cos \theta^t}{\sqrt{\varepsilon} \cos \theta^i + \sqrt{\varepsilon'} \cos \theta^t}. \tag{7.33}$$

If we replace $\sqrt{\varepsilon}$ in (7.33) by its equivalent given by (7.32) and if we multiply the numerator and the denominator of the resulting expression by $\sin \theta^i$, (7.33) can be simplified to yield

$$\rho_E = \frac{\sin(\theta^t - \theta^i)}{\sin(\theta^t + \theta^i)}. \tag{7.34}$$

It is possible to manipulate (7.13) into the form

$$\tau_E = \frac{2\sqrt{\varepsilon}\cos\theta^i}{\sqrt{\varepsilon}\cos\theta^i + \sqrt{\varepsilon'}\cos\theta^t}, \tag{7.35}$$

which with the help of (7.32) we can simplify to give

$$\tau_E = \frac{2\sin\theta^t\cos\theta^i}{\sin(\theta^t + \theta^i)}. \tag{7.36}$$

It is helpful to consider the following numerical problem.

Example. The half-space $x > 0$ is characterized by the constitutive parameters $\mu_0 = 4\pi \times 10^{-7}$ H/m and $\varepsilon_0 = (1/36\pi) \times 10^{-9}$ F/m. The other half-space $x < 0$ consists of a dielectric whose constitutive parameters are given by $\mu' = \mu_0$ and $\varepsilon' = 3\varepsilon_0$. A plane wave with the electric field normal to the plane of incidence is incident obliquely on the interface from the half-space $x > 0$. If the angle of incidence $\theta^i = 60$ degrees, find the angle of reflection θ^r and the angle of refraction θ^t. Find the values of the reflection and the transmission coefficients.

Since the permeabilities of the two half-spaces are equal, the simplified expressions given in (7.32), (7.34), and (7.36) can be used. The angle of reflection $\theta^r = \theta^i = 60$ degrees. From (7.32), we see that $\sin\theta^t = \sin 60°(\sqrt{\varepsilon_0}/\sqrt{3\varepsilon_0}) = (\sqrt{3}/2)(1/\sqrt{3}) = \frac{1}{2}$. Hence, the angle of refraction $\theta^t = 30$ degrees. We find from (7.34) that

$$\rho_E = \frac{\sin(30° - 60°)}{\sin(30° + 60°)} = -\frac{1}{2}.$$

The reflection coefficient $\rho_E = -\frac{1}{2}$, and since it is negative, the phase of the electric field is changed by 180 degrees on reflection. We can see from (7.36) that

$$\tau_E = \frac{2\sin 30°\cos 60°}{\sin(30° + 60°)} = \frac{1}{2}.$$

We can verify that $1 + \rho_E = \tau_E$, as it should be.

We obtain the fraction of the incident power flux density that is reflected back from (7.27) and (7.28) as $|\rho_E|^2 = \frac{1}{4}$. We find from (7.27) and (7.30) that the fraction of the incident power flux density that is transmitted is

$$|\tau_E|^2 \frac{\eta_0}{\eta_0'}\frac{\cos\theta^t}{\cos\theta^i} = \frac{1}{4}\sqrt{\frac{\mu_0}{\varepsilon_0}}\sqrt{\frac{3\varepsilon_0}{\mu_0}}\frac{\cos 30°}{\cos 60°} = \frac{3}{4}.$$

Note that the sum of the fraction of the incident power flux density that is reflected and that fraction which is transmitted is equal to unity, in verification of the principle of energy conservation.

2. *H*-polarization

We can treat the phenomenon of reflection and transmission of H-polarized plane waves incident obliquely on a plane interface between two dielectric half-spaces by techniques analogous to those used for the E-polarization. For the H-polarization (Fig. 6.6), $H_y(x, z)$, $E_x(x, z)$ and $E_z(x, z)$ are the only nonvanishing field components and are governed by (6.27) and (6.28). We may assume the incident magnetic vector to be the same as that given by (6.26). Then, the components of the incident electric vector are given as in (6.30). We assume that the reflected magnetic vector and the components of the reflected electric vector are, as before, given by (6.31) and (6.32) respectively. The angle of reflection θ^r has to satisfy the condition (6.14) in order that the direction of progression of the phase of the reflected wave has a component in the $+x$-direction. The wave normal $\hat{n}^t$ of the transmitted fields that exist in the half-space $x < 0$ is expressed in (7.2) in terms of the angle of refraction θ^t. We may assume that the transmitted magnetic field is of the form

$$H_y^t(x, z) = H_{yo}^t e^{-j\beta'\hat{n}^t \cdot \mathbf{r}} = H_{yo}^t e^{-j\beta'(-x \cos \theta^t + z \sin \theta^t)}, \tag{7.37}$$

where the constant H_{y0}^t has to be determined from the boundary conditions. The fields in the half-space $x < 0$ are also governed by (6.27) to (6.29) with μ', ε', and β' substituted for μ, ε, and β respectively. From (6.11b) and (7.37), we can determine the components of the transmitted electric vector as

$$E_x^t(x, z) = \sin \theta^t \eta_0' H_{yo}^t e^{-j\beta'(-x \cos \theta^t + z \sin \theta^t)}, \tag{7.38a}$$

$$E_z^t(x, z) = \cos \theta^t \eta_0' H_{yo}^t e^{-j\beta'(-x \cos \theta^t + z \sin \theta^t)}. \tag{7.38b}$$

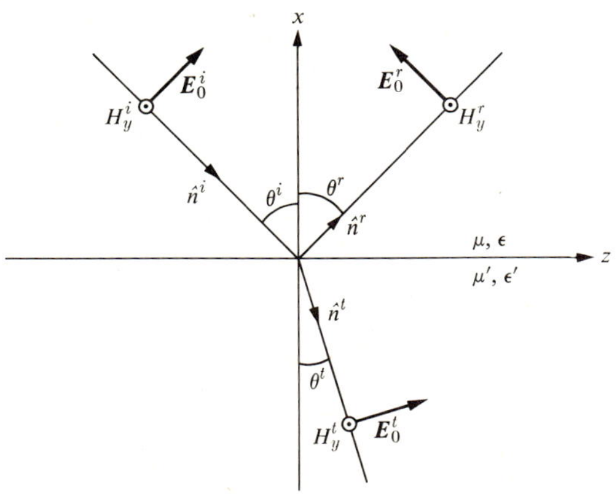

Fig. 6.6 H-polarized plane waves incident obliquely on a plane interface separating two different dielectric half-spaces.

The angle of refraction θ^t needs to satisfy the condition (7.7) in order that the direction of phase progression of the transmitted wave has a component in the $-x$-direction.

The boundary conditions that the tangential component of the magnetic field (H_y) and the tangential component of the electric field (E_z) are continuous at the plane interface $x = 0$, together with (6.26), (6.31), (7.37), (6.30b), (6.32b), and (7.38b), give rise to the following relations:

$$H^i_{y0}e^{-j\beta z \sin\theta^i} + H^r_{y0}e^{-j\beta z \sin\theta^r} = H^t_{y0}e^{-j\beta' z \sin\theta^t}, \tag{7.39}$$

$$\cos\theta^i\eta_0 H^i_{y0}e^{-j\beta z \sin\theta^i} - \cos\theta^r\eta_0 H^r_{y0}e^{-j\beta z \sin\theta^r} = \cos\theta^t\eta'_0 H^t_{y0}e^{-j\beta' z \sin\theta^t}. \tag{7.40}$$

The phase terms in (7.39) and (7.40) should be equal in order that (7.39) and (7.40) be satisfied for all z. Hence, (7.10), (6.16), and (6.17) follow leading to the general result that the angles of incidence and reflection are equal. Also, we find the Snell's law of refraction (7.11) to be valid even for the H-polarization. If we substitute (6.17) and (7.11) in (7.39) and (7.40), and successively eliminate H^t_{y0} and H^r_{y0} from the resulting equations we obtain the reflection coefficient ρ_H and the transmission coefficient τ_H as follows:

$$\frac{H^r_{y0}}{H^i_{y0}} = -\frac{[\eta'_0 \cos\theta^t - \eta_0 \cos\theta^i]}{[\eta'_0 \cos\theta^t + \eta_0 \cos\theta^i]} = \rho_H, \tag{7.41}$$

$$\frac{H^t_{y0}}{H^i_{y0}} = \frac{2\eta_0 \cos\theta^i}{[\eta'_0 \cos\theta^t + \eta_0 \cos\theta^i]} = \tau_H. \tag{7.42}$$

Together with (6.17), (7.11), (7.41), and (7.42), (6.31), (6.32), (7.37), and (7.38) yield the complete expressions for the reflected and the transmitted fields.

We find the total fields in the half-space $x > 0$, where both the incident and the reflected fields exist, with the help of (6.26), (6.30) to (6.32), and (7.41) to be

$$H^+_y(x, z) = H^i_y(x, z) + H^r_y(x, z) = H^i_{y0}[e^{j\beta x \cos\theta^i} + \rho_H e^{-j\beta x \cos\theta^i}]e^{-j\beta z \sin\theta^i} \tag{7.43}$$

$$E^+_x(x, z) = E^i_x(x, z) + E^r_x(x, z) = \sin\theta^i\eta_0 H^i_{y0}[e^{j\beta x \cos\theta^i} + \rho_H e^{-j\beta x \cos\theta^i}]e^{-j\beta z \sin\theta^i}, \tag{7.44}$$

$$E^+_z(x, z) = E^i_z(x, z) + E^r_z(x, z) = \cos\theta^i\eta_0 H^i_{y0}[e^{j\beta x \cos\theta^i} - \rho_H e^{-j\beta x \cos\theta^i}]e^{-j\beta z \sin\theta^i}. \tag{7.45}$$

If we compare (7.43) to (7.45) with (7.14), (7.17), and (7.18), it follows that the fields given by (7.43) to (7.45) form progressive waves in the z-direction and standing waves in the x-direction. The standing-wave ratio is obtained as $(1 + |\rho_H|)/(1 - |\rho_H|)$. The fields in the half-space $x < 0$ as given by (7.37) and (7.38), on the other hand, form progressive waves both in the x- and the z-directions.

In the case of H-polarization as in the case of E-polarization, we can relate the problem of oblique incidence to the corresponding problem of normal incidence. For the case of H-polarization, the fields in both the dielectric half-spaces are governed by (6.27) and (6.28); however, for the half-space $x < 0$, μ', ε', β', and θ^t are to be substituted for μ, ε, β, and θ^i respectively. If $E_x(x, z)$ is eliminated from (6.27) with the help of (6.28a), we see that

$$-j\omega\mu H_y(x, z) = -\frac{1}{j\omega\varepsilon}\frac{\partial^2}{\partial z^2} H_y(x, z) - \frac{\partial}{\partial x} E_z(x, z). \tag{7.46}$$

Since relations similar to (7.19) are found to be true for the previously stated reasons, we can reduce (7.46) to yield

$$-j\omega\mu \cos^2 \theta^i H_y(x, z) = -\frac{\partial}{\partial x} E_z(x, z). \tag{7.47}$$

For convenience, we reproduce (6.28b) here:

$$j\omega\varepsilon E_z(x, z) = \frac{\partial}{\partial x} H_y(x, z). \tag{7.48}$$

For the special case of normal incidence $\theta^i = 0$, (7.47) and (7.48) are identical to (5.5.6) and (5.5.10) with $\sigma = 0$ respectively. The generalization to oblique incidence for the H-polarization has resulted in the change of μ to $\mu \cos^2 \theta^i$. Along the x-axis, the incident wave is progressing in the $-x$-direction. Hence, for the normal incidence we obtain the intrinsic impedance from (5.5.28) as $\sqrt{\mu/\varepsilon}$. Since μ changes to $\mu \cos^2 \theta^i$ for the case of oblique incidence, the *effective* intrinsic impedance becomes $\tilde{\eta}_0 = \eta_0 \cos \theta^i$. In a similar manner, we find the *effective* intrinsic impedance for the half-space $x < 0$ to be $\tilde{\eta}_0' = \eta_0' \cos \theta^t$. If instead of the actual intrinsic impedances, we use the effective values in (3.8) and (3.9), we obtain the reflection and the transmission coefficients for the obliquely incident H-polarized wave as

$$\rho_H = -\frac{(\tilde{\eta}_0' - \tilde{\eta}_0)}{(\tilde{\eta}_0' + \tilde{\eta}_0)} = -\frac{(\eta_0' \cos \theta^t - \eta_0 \cos \theta^i)}{(\eta_0' \cos \theta^t + \eta_0 \cos \theta^i)}, \tag{7.49}$$

$$\tau_H = \frac{2\tilde{\eta}_0}{(\tilde{\eta}_0' + \tilde{\eta}_0)} = \frac{2\eta_0 \cos \theta^i}{(\eta_0' \cos \theta^t + \eta_0 \cos \theta^i)}. \tag{7.50}$$

These expressions for the reflection and the transmission coefficients are the same as those given by (7.41) and (7.42). It follows therefore that insofar as the determination of the reflection and the transmission coefficients is concerned, in the corresponding expressions for the normal incidence, μ has to be changed to $\mu \cos^2 \theta$ where $\theta = \theta^i$ or $\theta = \theta^t$ according to whether μ refers to the half-space $x > 0$ or $x < 0$.

Just as in the case of E-polarization, we can verify that the principle of energy conservation is obeyed in the case of H-polarization also. We obtain the $-x$-component of the incident power flux density from (6.26) and (6.30) as

$$S^i_{-x} = -\hat{x} \cdot \tfrac{1}{2} \operatorname{Re} \hat{z}E^i_z(x, z) \times \hat{y}H^{i*}_y(x, z) = \tfrac{1}{2} |H^i_{y0}|^2 \eta_0 \cos \theta^i. \tag{7.51}$$

With the help of (6.31), (6.32b), (6.17), and (7.41), we find the $-x$-component of the reflected power flux density to be

$$S^r_{-x} = -\hat{x} \cdot \tfrac{1}{2} \operatorname{Re} \hat{z}E^r_z(x, z) \times \hat{y}H^{r*}_y(x, z) = -\tfrac{1}{2} |H^i_{y0}|^2 \eta_0 \cos \theta^i |\rho_H|^2. \tag{7.52}$$

We can show from (7.51), (7.52), and (7.41) that

$$S^i_{-x} + S^r_{-x} = \tfrac{1}{2}\,|H^i_{y0}|^2\,\eta_0\cos\theta^i[1 - |\rho_H|^2],$$
$$= |H^i_{y0}|^2\,2\eta_0^2\eta_0'\cos^2\theta^i\cos\theta^t[\eta_0'\cos\theta^t + \eta_0\cos\theta^i]^{-2}. \qquad (7.53)$$

Note that (7.53) gives the net power density flowing into the half-space $x < 0$. From (7.37), (7.38b), and (7.42), we evaluate the $-x$- component of the transmitted power flux density as

$$S^t_{-x} = -\hat{x}\cdot\tfrac{1}{2}\,\mathrm{Re}\,\hat{z}E^t_z(x, z)\times\hat{y}H^{t*}_y(x, z) = \tfrac{1}{2}\,|H^i_{y0}|^2\,\eta_0'\cos\theta^t\,|\tau_H|^2,$$
$$= |H^i_{y0}|^2\,2\eta_0^2\eta_0'\cos^2\theta^i\cos\theta^t[\eta_0'\cos\theta^t + \eta_0\cos\theta^i]^{-2}. \qquad (7.54)$$

It follows that the principle of energy conservation is obeyed since the power flux density as given by (7.53), leaving the half-space $x > 0$ through the interface $x = 0$, is equal to the power flux density as given by (7.54), entering the half-space $x < 0$ through the same interface.

We can write the expressions for the reflection and the transmission coefficients as given by (7.41) and (7.42) respectively in compact forms if the permeabilities of the two half-spaces are the same. From (7.41) and (7.32), we can show in this case that

$$\rho_H = -\frac{[(\cos\theta^t/\sqrt{\varepsilon'}) - (\cos\theta^i/\sqrt{\varepsilon})]}{[(\cos\theta^t/\sqrt{\varepsilon'}) + (\cos\theta^i/\sqrt{\varepsilon})]},$$
$$= -\frac{[\sin 2\theta^t - \sin 2\theta^i]}{[\sin 2\theta^t + \sin 2\theta^i]} = -\frac{2\sin(\theta^t - \theta^i)\cos(\theta^t + \theta^i)}{2\sin(\theta^t + \theta^i)\cos(\theta^t - \theta^i)},$$
$$= \frac{\tan(\theta^i - \theta^t)}{\tan(\theta^i + \theta^t)}. \qquad (7.55)$$

In a similar manner, with the help of (7.32), we can manipulate (7.42) to yield

$$\tau_H = \frac{2\cos\theta^i/\sqrt{\varepsilon}}{[(\cos\theta^t/\sqrt{\varepsilon'}) + (\cos\theta^i/\sqrt{\varepsilon})]},$$
$$= \frac{\sin 2\theta^i}{\sin(\theta^t + \theta^i)\cos(\theta^t - \theta^i)}. \qquad (7.56)$$

Example. The half-space $x > 0$ is characterized by the constitutive parameters $\mu_0 = 4\pi\times 10^{-7}$ H/m and $\varepsilon_0 = (1/36\pi)\times 10^{-9}$ F/m. The other half-space $x < 0$ consists of a dielectric whose constitutive parameters are given by $\mu' = \mu_0$ and $\varepsilon' = 3.732\varepsilon_0$. A plane wave with its magnetic vector normal to the plane of incidence is incident obliquely on the interface from the half-space $x > 0$. If the angle of incidence $\theta^i = 30$ degrees, find the angle of reflection θ^r and the angle of refraction θ^t. Find the values of the reflection and the transmission coefficients.

Since the permeabilities of the two half-spaces are equal, we can use the simplified expressions given in (7.32), (7.55), and (7.56). The angle of reflection $\theta^r = \theta^i = 30$

degrees. From (7.32), we see that

$$\sin \theta^t = \sqrt{\frac{\varepsilon}{\varepsilon'}} \sin \theta^i = \sqrt{\frac{\varepsilon_0}{3.732\varepsilon_0}} \sin 30^\circ = 0.2588.$$

Hence the angle of refraction $\theta^t = 15$ degrees. From (7.55) we obtain that

$$\rho_H = \frac{\tan (30^\circ - 15^\circ)}{\tan (30^\circ + 15^\circ)} = 0.268.$$

Also, from (7.56), we can show that

$$\tau_H = \frac{\sin 60^\circ}{\sin (30^\circ + 15^\circ) \cos (15^\circ - 30^\circ)} = 1.268.$$

It is observed that $1 + \rho_H = \tau_H$, as it should be.

We obtain the fraction of the incident power flux density that is reflected back from (7.51) and (7.52) as $|\rho_H|^2 = 0.072$. We find from (7.51) and (7.54) that the fraction of the incident power flux density that is transmitted is

$$|\tau_H|^2 \frac{\eta_0' \cos \theta^t}{\eta_0 \cos \theta^i} = (1.268)^2 \sqrt{\frac{\varepsilon_0}{3.732\varepsilon_0}} \frac{\cos 15^\circ}{\cos 30^\circ} = 0.928.$$

We should note that the sum of the fraction of the incident power flux density that is reflected and that fraction which is transmitted is equal to unity, as it should be.

The expressions (7.12), (7.13), (7.41), and (7.42) are all called Fresnel coefficients for reflection and transmission.

6.8 TOTAL REFLECTION AND TOTAL TRANSMISSION

It is convenient for us to begin by introducing the concept of vector phase-shift coefficient. For example, for the transmitted field as given in (7.3), the vector phase-shift coefficient is defined as

$$\boldsymbol{\beta}' = \beta' \hat{n}^t, \tag{8.1}$$

whose magnitude is as given in (7.4) and whose direction coincides with that of the wave normal. With the help of (7.2), we obtain the x- and z-components of (8.1) as

$$\beta_x' = -\beta' \cos \theta^t; \qquad \beta_z' = \beta' \sin \theta^t. \tag{8.2}$$

We deduce the phase velocities and the wavelengths in the x- and z-directions from (8.2) as

$$v_{ph,x}' = \omega/\beta_x' = -v_{ph}'/\cos \theta^t; \qquad v_{ph,z} = \omega/\beta_z' = v_{ph}'/\sin \theta^t; \qquad v_{ph}' = 1/\sqrt{\mu'\varepsilon'}. \tag{8.3}$$

$$\lambda_x' = 2\pi/|\beta_x'| = \lambda'/\cos \theta^t; \qquad \lambda_z' = 2\pi/|\beta_z'| = \lambda'/\sin \theta^t; \qquad \lambda' = 2\pi v_{ph}'/\omega. \tag{8.4}$$

In (8.3) and (8.4), v'_{ph} and λ' are respectively the phase velocity and the wavelength of the transmitted plane wave in the wave-normal ($\hat{n}^t$) direction. For θ^t real, the phase velocities as well as the wavelengths in the x- and z-directions are larger than their corresponding values in the wave-normal direction. In particular, the transmitted wave along the interface has a phase velocity which is greater than that in the wave-normal direction and hence for θ^t real, the transmitted wave along the interface is a "fast wave."

1. Total Reflection

The character of the transmitted wave changes significantly if the angle of refraction θ^t becomes complex. We can now show that under certain circumstances θ^t can become complex. Let us consider the case of transmission from a "slower" medium to a "faster" medium. Since $1/\sqrt{\mu\varepsilon} < 1/\sqrt{\mu'\varepsilon'}$, it follows from (7.11) that $\sin\theta^t > \sin\theta^i$ and, therefore, $\theta^t > \theta^i$, since both the angles are in the range from 0 to $\pi/2$. There is a particular value of $\theta^i = \theta^i_c$ for which $\theta^t = \pi/2$. This value of θ^i is called the *critical angle of incidence.* We can show from (7.11) that θ^i_c is defined by the relation

$$\sin\theta^i_c = \sqrt{\mu'\varepsilon'/\mu\varepsilon}\,\sin(\pi/2) = \sqrt{\mu'\varepsilon'/\mu\varepsilon}. \tag{8.5}$$

Consequently, we can rewrite Snell's law of refraction (7.11) conveniently in the form

$$\sin\theta^t = \sin\theta^i/\sin\theta^i_c. \tag{8.6}$$

For $\theta^i < \theta^i_c$, $\sin\theta^t < 1$ and we can find a real value of θ^t. If $\theta^i > \theta^i_c$, $\sin\theta^t > 1$ and no real value of θ^t can satisfy (8.6). For $\theta^i > \theta^i_c$, let $\theta^t = (\pi/2) + j\theta^t_j$ and therefore,

$$\sin\theta^t = \cosh\theta^t_j; \qquad \cos\theta^t = -j\sinh\theta^t_j. \tag{8.7}$$

For $\theta^i > \theta^i_c$, if we assume $\theta^t = (\pi/2) + j\theta^t_j$ to be the solution for θ^t, then it follows from (8.6) and (8.7) that

$$\cosh\theta^t_j = \sin\theta^i/\sin\theta^i_c. \tag{8.8}$$

Since the cosh of a real angle is always greater than unity, (8.8) shows that for $\theta^i > \theta^i_c$ a real angle θ^t_j exists. Hence, it follows that for $\theta^i > \theta^i_c$, the value of θ^t is complex and is of the form $\theta^t = (\pi/2) + j\theta^t_j$ where θ^t_j is defined by (8.8).

For $\theta^i > \theta^i_c$, we see the phase velocity $v'_{ph,z}$ of the transmitted wave along the interface from (8.3) and (8.7) to be

$$v_{ph,z} = v'_{ph}/\cosh\theta^t_j. \tag{8.9}$$

Since $\cosh\theta^t_j$ is greater than unity, the transmitted wave becomes a "slow wave" along the interface for $\theta^i > \theta^i_c$. We obtain the phase term of the transmitted wave from (7.3) and (8.7) to be

$$e^{-j\beta'(-x\cos\theta^t + z\sin\theta^t)} = e^{x\beta'\sinh\theta^t_j - jz\beta'\cosh\theta^t_j}. \tag{8.10}$$

It follows from (8.10) that as x increases negatively, that is, as the distance in the second medium from the interface increases, the field is attenuated exponentially and $\beta'\sinh\theta^t_j$ is the attenuation factor. We see from (8.10) that the equiphase surfaces of the transmitted field are (z = constant) planes which are perpendicular to the interface. We find from (8.10) that the surfaces where the transmitted field has a constant amplitude are (x = constant) planes which are parallel to the interface. Thus, the transmitted wave for $\theta^i > \theta^i_c$ becomes a *nonuniform* plane wave since on any of its plane phase fronts, the amplitude decreases exponentially away from the interface. We have to contrast this situation with the character of the transmitted wave for $\theta^i < \theta^i_c$. The transmitted wave is a "fast wave" along the interface since the phase velocity along the interface is greater than that in the wave-normal direction. Also, the constant phase surfaces are planes perpendicular to the wave-normal $\hat{n}^t$-direction and the amplitude of the fields does not vary on the plane phase fronts with the result that for $\theta^i < \theta^i_c$, the transmitted field is a *uniform* plane wave.

For the E-polarization, for $\theta^i > \theta^i_c$, we find the transmitted fields from (7.3), (7.5b), and (8.7) to be

$$E^t_y(x, z) = E^t_{y0} e^{x\beta' \sinh \theta^t_j - jz\beta' \cosh \theta^t_j}, \tag{8.11}$$

$$H^t_z(x, z) = \frac{j \sinh \theta^t_j}{\eta'_0} E^t_{y0} e^{x\beta' \sinh \theta^t_j - jz\beta' \cosh \theta^t_j}. \tag{8.12}$$

From (8.11) and (8.12), we evaluate the $-x$-component of the transmitted power flux density as

$$\begin{aligned} S^t_{-x} &= -\hat{x} \cdot \tfrac{1}{2} \operatorname{Re} \hat{y} E^t_y(x, z) \times \hat{z} H^{t*}_z(x, z), \\ &= \tfrac{1}{2} \operatorname{Re} j \sinh \theta^t_j \frac{|E^t_{y0}|^2}{\eta'_0} e^{2x\beta' \sinh \theta^t_j} = 0. \end{aligned} \tag{8.13}$$

For the H-polarization, for $\theta^i > \theta^i_c$, we find the transmitted fields from (7.37), (7.38b), and (8.7) to be

$$H^t_y(x, z) = H^t_{y0} e^{x\beta' \sinh \theta^t_j - jz\beta' \cosh \theta^t_j}, \tag{8.14}$$

$$E^t_z(x, z) = -j \sinh \theta^t_j \eta'_0 H^t_{y0} e^{x\beta' \sinh \theta^t_j - jz\beta' \cosh \theta^t_j}. \tag{8.15}$$

We deduce the $-x$-component of the transmitted power flux density from (8.14) and (8.15) for the H-polarization case as

$$\begin{aligned} S^t_{-x} &= -\hat{x} \cdot \tfrac{1}{2} \operatorname{Re} \hat{z} E^t_z(x, z) \times \hat{y} H^{t*}_y(x, z) \\ &= \tfrac{1}{2} \operatorname{Re} \left[-j \sinh \theta^t_j \eta'_0 \, |H^t_{y0}|^2 \, e^{2x\beta' \sinh \theta^t_j} \right] = 0. \end{aligned} \tag{8.16}$$

Therefore, for an angle of incidence greater than the critical angle of incidence, no time-averaged power is transmitted into the lower half-space $x < 0$. For $\theta^i > \theta^i_c$, we obtain the reflection coefficients for the two polarizations from (7.12), (7.41), and (8.7) as

$$\rho_E = \frac{j\eta'_0 \operatorname{cosech} \theta^t_j - \eta_0 \sec \theta^i}{j\eta'_0 \operatorname{cosech} \theta^t_j + \eta_0 \sec \theta^i}, \tag{8.17}$$

$$\rho_H = -\frac{[-j\eta'_0 \sinh \theta^t_j - \eta_0 \cos \theta^i]}{[-j\eta'_0 \sinh \theta^t_j + \eta_0 \cos \theta^i]}, \tag{8.18}$$

where $\operatorname{cosech} \theta_j^t = 1/\sinh \theta_j^t$. We can verify from (8.17) and (8.18) that $|\rho_E| = 1$ and $|\rho_H| = 1$. Thus, for $\theta^i > \theta_c^i$, the entire time-averaged incident power is reflected back.

The intrinsic resistance of the half-space $x < 0$ is $\sqrt{\mu'/\varepsilon'}$. For the case of oblique incidence, we find the effective value of the intrinsic resistance to be $\sqrt{\mu'/\varepsilon'}$ with ε' changed to $\varepsilon' \cos^2 \theta^t$ for the E-polarization and μ' changed to $\mu' \cos^2 \theta^t$ for the H-polarization. Therefore, we see that for the oblique incidence the intrinsic resistances are

$$\begin{aligned} \text{Intrinsic resistance} &= \sqrt{\mu'/\varepsilon'} \sec \theta^t \qquad (E\text{-polarization}), \\ &= \sqrt{\mu'/\varepsilon'} \cos \theta^t \qquad (H\text{-polarization}). \end{aligned} \tag{8.19}$$

For $\theta^i > \theta_c^i$, since $\cos \theta^t$ becomes imaginary as we can see from (8.7), the intrinsic impedance of the half-space $x < 0$ becomes reactive. We can easily ascertain from (8.19) and (8.7) that

$$\begin{aligned} \text{Intrinsic reactance} &= j\sqrt{\mu'/\varepsilon'} \operatorname{cosech} \theta_j^t \qquad (E\text{-polarization}), \\ &= -j\sqrt{\mu'/\varepsilon'} \sinh \theta_j^t \qquad (H\text{-polarization}). \end{aligned} \tag{8.20}$$

For $\theta^i < \theta_c^i$, the intrinsic impedance of the half-space $x < 0$ is a resistance but for $\theta^i > \theta_c^i$, the intrinsic impedance of the half-space $x < 0$ we can see from (8.20) to be a reactance which is inductive for the E-polarization and is capacitive for the H-polarization.

We shall now summarize the phenomenon of the so-called *total internal reflection.* For the case of propagation of plane waves from a "slower" to a "faster" medium, there is an angle of incidence for which the transmitted waves just graze the interface. If $\theta^i > \theta_c^i$, the total time-averaged power in the incident wave is reflected back with no transmission of the time-averaged power into the second medium. The transmitted wave is a "slow wave", has phase propagation along the interface and is attenuated exponentially perpendicular to the interface. The second medium presents only a reactance for $\theta^i > \theta_c^i$.

When $\theta^i < \theta_c^i$, we can see from (7.12) and (7.41) that ρ_E and ρ_H can be either positive or negative with the result that at the interface the reflected fields can be either in phase or 180 degrees out of phase with the corresponding incident fields. The magnitudes of ρ_E and ρ_H are both less than unity and we can obtain them from (7.12) and (7.41) respectively. However, if $\theta^i > \theta_c^i$, the magnitudes of both ρ_E and ρ_H have been shown to be unity. We still have to determine the phase angles. It is convenient for us to rewrite ρ_E given in (8.17) as follows:

$$\rho_E = -\frac{[\eta_0 \sinh \theta_j^t - j\eta_0' \cos \theta^i]}{[\eta_0 \sinh \theta_j^t + j\eta_0' \cos \theta^i]}. \tag{8.21}$$

If

$$e^{-j(\tilde{\theta}_E/2)} = \eta_0 \sinh \theta_j^t - j\eta_0' \cos \theta^i, \tag{8.22}$$

(8.21) becomes

$$\rho_E = e^{j(\pi - \tilde{\theta}_E)} = e^{j\theta_E}. \tag{8.23}$$

From (8.22), (8.5), and (8.8), it follows that

$$\tan(\tilde{\theta}_E/2) = (\eta_0' \cos\theta^i)/(\eta_0 \sinh\theta_j^t)$$
$$= \frac{\eta_0' \cos\theta^i}{\eta_0\sqrt{(\mu\varepsilon/\mu'\varepsilon')\sin^2\theta^i - 1}}. \tag{8.24}$$

Thus, for the E-polarization case, the phase difference at the interface between the reflected and the incident electric fields normal to the plane of incidence is given by $\theta_E = \pi - \tilde{\theta}_E$ [where $\tilde{\theta}_E$ is defined in (8.24)] in terms of the constitutive parameters of the two media and the angle of incidence. In a similar manner, if

$$e^{j(\theta_H/2)} = \eta_0 \cos\theta^i + j\eta_0' \sinh\theta_j^t, \tag{8.25}$$

then we can simplify ρ_H given in (8.18) to

$$\rho_H = e^{j\theta_H}. \tag{8.26}$$

From (8.5), (8.8), and (8.25), we can see that

$$\tan(\theta_H/2) = (\eta_0' \sinh\theta_j^t)/(\eta_0 \cos\theta^i)$$
$$= \frac{\eta_0'\sqrt{(\mu\varepsilon/\mu'\varepsilon')\sin^2\theta^i - 1}}{\eta_0 \cos\theta^i}. \tag{8.27}$$

Therefore, for the H-polarization case, the phase difference at the interface between the reflected and the incident magnetic fields normal to the plane of incidence is given by θ_H where θ_H is defined in (8.27).

2. Total Transmission

We have investigated the possibility of obtaining total reflection. It is of interest for us to examine if total transmission is possible under some circumstances. For the E-polarized plane waves, we see from (7.12) and (7.11) that total transmission ($\rho_E = 0$) is possible only if

$$\eta_0 \cos\theta^t = \eta_0' \cos\theta^i \quad \text{or} \quad \sqrt{\mu/\varepsilon}\sqrt{1 - (\mu\varepsilon/\mu'\varepsilon')\sin^2\theta^i} = \sqrt{\mu'/\varepsilon'}\cos\theta^i. \tag{8.28}$$

Let us consider the case in which the two permeabilities are equal, that is, $\mu = \mu'$. Then, we can recast the condition (8.28) into the form

$$\sqrt{1 - (\varepsilon/\varepsilon')\sin^2\theta^i} = \sqrt{(\varepsilon/\varepsilon') - (\varepsilon/\varepsilon')\sin^2\theta^i}, \tag{8.29}$$

which equality can never be satisfied since ε and ε' are different. Thus, if $\mu = \mu'$, for the E-polarization, total transmission is not possible.

From (7.11) and (7.41), we see total transmission ($\rho_H = 0$) to be possible for the H-polarized waves if

$$\eta_0' \cos\theta^t = \eta_0 \cos\theta^i \quad \text{or} \quad \sqrt{\mu'/\varepsilon'}\sqrt{1 - (\mu\varepsilon/\mu'\varepsilon')\sin^2\theta^i} = \sqrt{\mu/\varepsilon}\cos\theta^i. \tag{8.30}$$

Again, if $\mu = \mu'$, we can recast the condition (8.30) into the form

$$1 - (\varepsilon/\varepsilon')\sin^2\theta^i = (\varepsilon'/\varepsilon)\cos^2\theta^i. \tag{8.31}$$

Let the value of θ^i which satisfies (8.31) be denoted by θ^i_B. We can show that by dividing both sides of (8.31) by $\cos^2\theta^i$ and noting that

$$\sec^2\theta^i = 1/\cos^2\theta^i = 1 + \tan^2\theta^i,$$

$$\tan\theta^i_B = \sqrt{\varepsilon'/\varepsilon}. \tag{8.32}$$

We should note that irrespective of whether ε' is greater or less than ε, a real value of θ^i_B can always be found. For $\theta^i = \theta^i_B$ and $\mu = \mu'$, there is no reflected wave for the case of H-polarization.

If a plane wave with an arbitrary polarization is incident at the angle $\theta^i = \theta^i_B$, a part of the E-polarized wave is reflected but there is no H-polarized wave in the reflected fields. Thus, the reflected wave is polarized with its electric vector lying normal to the plane of incidence. Since a pure E-polarization can be obtained on reflection at the angle $\theta^i = \theta^i_B$, θ^i_B is called *the polarizing or the Brewster angle.*

There is a simple relation between the angles of incidence and refraction when the angle of incidence is equal to the Brewster angle. Let the value of θ^t when $\theta^i = \theta^i_B$ be denoted by θ^t_B. Since $\mu = \mu'$, (7.11) and (8.32) give

$$\sin\theta^t_B/\sin\theta^i_B = \sqrt{\varepsilon/\varepsilon'} = 1/\tan\theta^i_B = \cos\theta^i_B/\sin\theta^i_B. \tag{8.33}$$

Hence,

$$\sin\theta^t_B = \cos\theta^i_B = \sin\left(\frac{\pi}{2} - \theta^i_B\right) \quad \text{or} \quad \theta^t_B = \frac{\pi}{2} - \theta^i_B. \tag{8.34}$$

An interesting dielectric device making use of the principles of total reflection and total transmission has been reported by Rosenbaum and Coleman [*IEEE Transactions on Microwave Theory and Techniques*, **MTT-11,** 302–311, September 1963]. The cross section of the dielectric device is shown schematically in Fig. 6.7. For experimental convenience, it is necessary to have a plane wave which is propagating at an angle θ to the face AB of a dielectric region emerging in a direction parallel to the face AB. The parameters of the shaped dielectric region used for this purpose are also indicated in Fig. 6.7. The dielectric has a relative permittivity, $\varepsilon_r = 2.04$. Therefore $\varepsilon = 2.04\,\varepsilon_0$, $\varepsilon' = \varepsilon_0$, and $\mu = \mu'$. Then, we can evaluate the critical angle

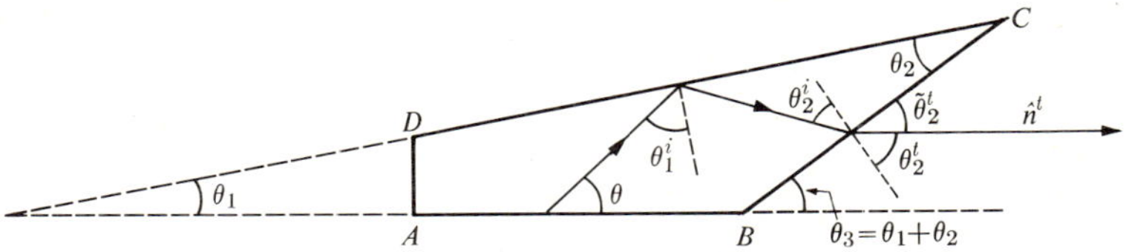

Fig. 6.7 Cross section of the dielectric coupler.

$\theta_1 = 10.5°$; $\theta_2 = 24.5°$; $\theta_3 = 35°$;
$\theta = 41°$; $\theta^i_1 = 59.5°$; $\theta^i_2 = 35°$.
$\theta^t_2 = 55°$; $\tilde{\theta}^t_2 = 35°$;

of incidence θ_c^i and the Brewster angle θ_B^i with the help of (8.5) and (8.32) as: $\sin \theta_c^i = 1/\sqrt{2.04}$ or $\theta_c^i = 44.4°$; $\tan \theta_B^i = 1/\sqrt{2.04}$ or $\theta_B^i = 35°$. Also, from (8.34), $\theta_B^t = 55°$. The wave normal of the plane wave makes an angle $\theta = 41$ degrees to the face AB. From Fig. 6.7 we can ascertain that the angle of incidence at the face CD is given by $\theta_1^i = 59.5°$. Since $\theta_1^i > \theta_c^i$, there is total reflection. Again, from Fig. 6.7, we find that at the face BC, the angle of incidence $\theta_2^i = 35°$. Since $\theta_2^i = \theta_B^i$, there is complete transmission for the H-polarized waves. Also, since $\theta_2^t = \theta_B^t = 55°$, we see that in Fig. 6.7, $\tilde{\theta}_2^t = 90° - 55° = 35° = \theta_3$. Hence, it follows that the wave-normal direction $\hat{n}^t$ of the emerging wave is parallel to the face AB. This device works only for the H-polarized waves since for the angle of incidence equal to the Brewster angle, only the H-polarized waves are totally transmitted at the face BC.

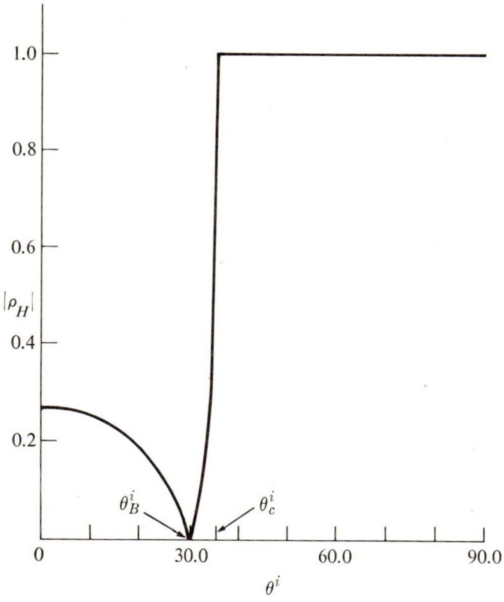

Fig. 6.8 The Brewster angle and the critical angle of incidence for the H-polarized wave. $\mu = \mu'$; $\varepsilon = 3\varepsilon_0$; $\varepsilon' = \varepsilon_0$.

The existence of the Brewster angle and the critical angle of incidence for the H-polarized waves is portrayed in Fig. 6.8, where $|\rho_H|$ as given by (7.55) and (8.18) is shown as a function of θ^i for $0 < \theta^i < 90°$. We have used the values of $\mu = \mu'$, $\varepsilon = 3\varepsilon_0$, and $\varepsilon' = \varepsilon_0$ as the data for Fig. 6.8. From (8.32), we find that $\theta_B^i = 30°$. We see from Fig. 6.8 that $|\rho_H| = 0.268$ for $\theta^i = 0°$, decreases monotonically as θ^i is increased, and reaches the value of zero at the Brewster angle $\theta^i = \theta_B^i$. For the further increase in the value of θ^i beyond θ_B^i, we see $|\rho_H|$ to increase monotonically and attain the value of unity at $\theta^i = 35.3°$ approximately. From (8.5), we can verify that the critical angle of incidence is given by $\theta_c^i = 35.3°$. It is therefore to be expected that $|\rho_H| = 1$ for $\theta^i = \theta_c^i$. For $\theta_c^i < \theta^i < 90°$, we find from (8.18) that $|\rho_H| = 1$, as is shown in Fig. 6.8.

6.9 INHOMOGENEOUS PLANE WAVES SUPPORTED BY A REACTIVE SURFACE

For a plane wave incident from a "slower" medium at an angle greater than the critical angle of incidence, we have shown that the transmitted plane wave in a "faster" medium is an inhomogeneous plane wave. We shall establish here that such an inhomogeneous plane wave is capable of being supported also by a reactive surface. Let $x = 0$ be a plane reactive surface and let z be the preferred direction. We shall only consider the two-dimensional fields which have no dependence on y. Then, the total electromagnetic fields are capable of being separated into two independent components one of which has the E-field and the other the H-field in the y-direction. The first case which has no component of the electric field in the preferred z-direction is denoted as the transverse electric (TE) mode and the second case which has no component of the magnetic field in the preferred z-direction is denoted as the transverse magnetic (TM) mode.

For the TE mode, $E_y(x, z)$, $H_x(x, z)$, and $H_z(x, z)$ are the only nonvanishing field components and are governed by (6.4) and (6.5). We obtain the impedance looking into the plane surface, that is, the impedance in the $-x$-direction at the plane surface from (5.5.20) to be $-E_y(0, z)/H_z(0, z)$ and this impedance is assumed to be reactive. The plane surface has a translational invariance in the z-direction and, therefore, it is reasonable for us to look for the electromagnetic field which has the z-dependence as given by

$$E_y(x, z) = E_y(x)e^{-j\beta_z z}. \tag{9.1}$$

Substituting (9.1) in (6.5) we obtain

$$\left[\frac{d^2}{dx^2} + \beta^2 - \beta_z^2\right]E_y(x) = 0, \tag{9.2}$$

where $\beta = \omega\sqrt{\mu\varepsilon}$ and, μ and ε are the permeability and the permittivity of the medium occupying the half-space $x > 0$. We can express the solution of (9.2) as

$$E_y(x) = E_{y0}e^{-\sqrt{\beta_z^2 - \beta^2}\,x} + \tilde{E}_{y0}e^{\sqrt{\beta_z^2 - \beta^2}\,x}. \tag{9.3}$$

We have to rule out the second term in (9.3) since x extends up to infinity. The solution expressed in (9.3) is convenient if $\beta_z > \beta$. Otherwise, it is necessary to set $\sqrt{\beta_z^2 - \beta^2} = j\sqrt{\beta^2 - \beta_z^2}$ for $\beta_z < \beta$. With this change in (9.3), we have again to rule out the second term in (9.3) since it represents a wave with its phase progressing in the $-x$-direction and also since such a wave is not possible unless there are some sources at $x = \infty$. From (9.1) and (9.3), we may seek $E_y(x, z)$ in the form

$$E_y(x, z) = E_{y0}e^{-\sqrt{\beta_z^2 - \beta^2}\,x - j\beta_z z}. \tag{9.4}$$

Substituting (9.4) into (6.4a) and (6.4b) we obtain the following expressions for the magnetic fields:

$$H_x(x, z) = -\frac{\beta_z}{\omega\mu} E_{yo} e^{-\sqrt{\beta_z^2-\beta^2}\,x - j\beta_z z}, \tag{9.5a}$$

and

$$H_z(x, z) = \frac{\sqrt{\beta_z^2 - \beta^2}}{j\omega\mu} E_{yo} e^{-\sqrt{\beta_z^2-\beta^2}\,x - j\beta_z z}. \tag{9.5b}$$

We find the impedance in the $-x$-direction at the plane surface $x = 0$ from (9.4) and (9.5b) to be $-j\omega\mu/\sqrt{\beta_z^2 - \beta^2}$, and if the impedance looking into the plane surface is a *capacitive* reactance $-jX_s(X_s > 0)$, it follows that

$$\omega\mu/\sqrt{\beta_z^2 - \beta^2} = X_s. \tag{9.6}$$

It is clear from (9.6) that $\sqrt{\beta_z^2 - \beta^2} = \omega\mu/X_s$ is positive real. Also,

$$\beta_z = \sqrt{(\omega\mu/X_s)^2 + \beta^2}$$

is greater than β and real.

The wave represented by (9.4) is an inhomogeneous wave. Its phase progresses in the z-direction and the phase fronts are planes (z = constant) perpendicular to the plane reactive surface. The amplitude of the fields decays in the x-direction away from the plane reactive surface and, therefore, the amplitude fronts are planes (x = constant) parallel to the plane reactive surface. Since $\beta_z > \beta$, the wave is a "slow wave" since its phase velocity along the interface is less than that of the plane electromagnetic waves in an unbounded dielectric medium with the same constitutive parameters μ and ε. We obtain the time-averaged Poynting vector in the x-direction from (9.4) and (9.5b) as

$$\begin{aligned} S_x &= \hat{x} \cdot \tfrac{1}{2} \operatorname{Re} \hat{y}E_y(x, z) \times \hat{z}H_z^*(x, z), \\ &= \tfrac{1}{2} \operatorname{Re}\left[\frac{\sqrt{\beta_z^2 - \beta^2}}{-j\omega\mu} |E_{yo}|^2 e^{-2\sqrt{\beta_z^2-\beta^2}\,x}\right] = 0. \end{aligned} \tag{9.7}$$

Similarly, we find the time-averaged Poynting vector in the z-direction from (9.4) and (9.5a) to be

$$\begin{aligned} S_z &= \hat{z} \cdot \tfrac{1}{2} \operatorname{Re} \hat{y}E_y(x, z) \times \hat{x}H_x^*(x, z), \\ &= \tfrac{1}{2} \operatorname{Re}\left[\frac{\beta_z}{\omega\mu} |E_{yo}|^2 e^{-2\sqrt{\beta_z^2-\beta^2}\,x}\right], \\ &= \frac{\beta_z}{2\omega\mu} |E_{yo}|^2 e^{-2\sqrt{\beta_z^2-\beta^2}\,x}. \end{aligned} \tag{9.8}$$

We observe from (9.7) and (9.8) that the inhomogeneous plane wave does not transport power normal to the plane reactive surface but transports power only along the reactive surface.

The inhomogeneous plane wave given by (9.4) and (9.5) is called a *plane surface wave*. This wave transports power only along the reactive surface. Since the fields decay exponentially with distance away from the reactive surface, the fields are essentially concentrated near the surface. We should note that *the TE plane surface waves* can be supported only by a plane reactive surface having a *capacitive* reactance $-jX_s$. We can ascertain that as the magnitude of the reactance X_s is decreased, the phase velocity of the wave along the surface is decreased and since the attenuation increases, the wave becomes more tightly bound to the reactive surface.

For the TM mode, $H_y(x, z)$, $E_x(x, z)$, and $E_z(x, z)$ are the only nonvanishing field components and are governed by (6.27) to (6.29). We obtain the impedance looking into the plane surface, that is, the impedance in the $-x$-direction at the plane surface from (5.5.28) to be $E_z(0, z)/H_y(0, z)$ and this impedance is, as before, assumed to be reactive. For the TM mode, as in the case of the TE mode, the solution for $H_y(x, z)$ as specified by (6.29) can be expressed in the form

$$H_y(x, z) = H_{y0}e^{-\sqrt{\beta_z^2-\beta^2}\,x - j\beta_z z}. \tag{9.9}$$

Substituting (9.9) into (6.28a) and (6.28b), we can determine the electric field components as

$$E_x(x, z) = \frac{\beta_z}{\omega\varepsilon} H_{y0}e^{-\sqrt{\beta_z^2-\beta^2}\,x - j\beta_z z}, \tag{9.10a}$$

and

$$E_z(x, z) = -\frac{\sqrt{\beta_z^2 - \beta^2}}{j\omega\varepsilon} H_{y0}e^{-\sqrt{\beta_z^2-\beta^2}\,x - j\beta_z z}. \tag{9.10b}$$

We find the impedance in the $-x$-direction at the plane surface $x = 0$ from (9.9) and (9.10b) to be $j\sqrt{\beta_z^2 - \beta^2}/\omega\varepsilon$ and if the impedance looking into the plane surface is an *inductive* reactance $jX_s(X_s > 0)$, we can see that

$$\frac{\sqrt{\beta_z^2 - \beta^2}}{\omega\varepsilon} = X_s \tag{9.11}$$

It follows from (9.11) that $\sqrt{\beta_z^2 - \beta^2} = \omega\varepsilon X_s$ is positive real and $\beta_z = \sqrt{(\omega\varepsilon X_s)^2 + \beta^2}$ is greater than β and real.

The field given by (9.9) has the same properties as that of the TE mode. It is an inhomogeneous wave having its plane phase and plane amplitude fronts respectively perpendicular and parallel to the plane reactive surface. It is also a "slow wave". We obtain the time-averaged Poynting vector in the x-direction from (9.9) and (9.10b) as

$$\begin{aligned} S_x &= \hat{x} \cdot \tfrac{1}{2}\,\mathrm{Re}\,\hat{z}E_z(x, z) \times \hat{y}H_y^*(x, z), \\ &= -\tfrac{1}{2}\,\mathrm{Re}\,[|H_{y0}|^2 e^{-2\sqrt{\beta_z^2-\beta^2}\,x} j\sqrt{\beta_z^2 - \beta^2}/\omega\varepsilon] = 0. \end{aligned} \tag{9.12}$$

In a similar manner, we can evaluate the time-averaged Poynting vector in the z-direction with the help of (9.9) and (9.10a) as

$$\begin{aligned} S_z &= \hat{z} \cdot \tfrac{1}{2}\,\mathrm{Re}\,\hat{x}E_x(x, z) \times \hat{y}H_y^*(x, z), \\ &= \frac{\beta_z}{2\omega\varepsilon} |H_{y0}|^2 e^{-2\sqrt{\beta_z^2-\beta^2}\,x}. \end{aligned} \tag{9.13}$$

We should note from (9.12) and (9.13) that the inhomogeneous plane wave (9.9) transports power only along the plane reactive surface.

As in the case of the TE mode, the TM mode with the field components as given in (9.9) and (9.10) is also a *plane surface wave.* But, *the TM plane surface waves* can be supported only by a plane reactive surface having an *inductive* reactance jX_s. We can argue that as the magnitude X_s of the inductive reactance is increased, the phase velocity of the wave along the surface is decreased, the attenuation increases and therefore the wave becomes more tightly bound to the reactive surface.

There are a number of ways in which a reactive surface can be realized in practice and we shall discuss some of them in Chapter 8. One of the simplest methods to obtain a reactive surface is by coating a perfectly conducting plane surface with a lossless dielectric layer. We examine the electromagnetic fields supported by a dielectric coated plane perfect conductor thoroughly in Chapter 8.

PROBLEMS

6.1 Consider a plane interface $x = 0$ which separates a half-space ($x < 0$) of free space (Medium 1) characterized by the permeability μ_0, the permittivity ε_0, and zero conductivity from another half-space ($x > 0$) of dielectric (Medium 2) characterized by the permeability μ_0, the permittivity $4\varepsilon_0$, and zero conductivity. A uniform plane wave propagating in the $+x$-direction is incident normally on the interface from the side of Medium 1. The incident wave has its electric vector in the y-direction and is in the form of a rectangular pulse of amplitude 10 V/m and time duration $T = 10^{-5}$ sec.

Determine (i) the incident magnetic field, (ii) the reflected electric and magnetic fields, and (iii) the transmitted electric and magnetic fields.

Find the energies in (i) the incident, (ii) the reflected, and (iii) the transmitted waves and verify that the total energy is conserved.

6.2 Suppose that Medium 2 of Problem 6.1 extends only from $x = 0$ to $x = d = 150$ m and that the half-space $d < x < \infty$ consists of free space (Medium 3). A uniform plane wave in the form of a d.c. pulse with the amplitude of the electric field equal to 1 V/m is incident normally on the interface $x = 0$ at time $t = 0$ from the region $x < 0$.

Plot the values of the electric and the magnetic fields on the interfaces (i) $x = 0$ and (ii) $x = d$, as a function of time from $t = 0$ to $t = 5 \times 10^{-6}$ sec.

Plot the values of the electric and the magnetic fields at time $t = 3.5 \times 10^{-6}$ sec as a function of distance from $x = 0$ to $x = d$.

6.3 Let $x = 0$ be the plane interface which separates a half-space ($x < 0$) of free space (Medium 1) from a half-space ($x > 0$) of dielectric (Medium 2). Let Media 1 and 2 be characterized by the constitutive parameters ($\mu_1 = \mu_0, \varepsilon_1 = \varepsilon_0, \sigma_1 = 0$) and ($\mu_2 = \mu_0, \varepsilon_2 = 2.25\varepsilon_0, \sigma_2 = 10^{-3}$ mho/m) respectively. A uniform plane wave progressing in the $+x$-direction is incident normally on the interface from the side of Medium 1. The incident magnetic vector of frequency $f = 10^8$ Hz and of amplitude $H_0 = 1$ A/m is in the z-direction.

Determine (i) the incident electric vector, (ii) the reflected electric and magnetic vectors and (iii) the transmitted electric and magnetic vectors.

Evaluate (i) the incident, (ii) the reflected and (iii) the transmitted time-averaged Poynting vectors and verify that the principle of energy conservation is satisfied.

6.4 Let $x = 0$ be the plane interface which separates a half-space ($x < 0$) of free space (Medium 1) from a half-space ($x > 0$) of conductor (Medium 2). Let Media 1 and 2 be characterized by the constitutive parameters ($\mu_1 = \mu_0$, $\varepsilon_1 = \varepsilon_0$, $\sigma_1 = 0$) and ($\mu_2 = \mu_0$, $\varepsilon_2 = \varepsilon_0$, $\sigma_2 = 5.8 \times 10^7$ mhos/m) respectively. A uniform plane wave progressing in the $+x$-direction is incident normally on the interface from the side of Medium 1. The incident magnetic vector of frequency $f = 10^8$ Hz and of amplitude $H_0 = 1$ A/m is in the z-direction.

Evaluate the time-averaged power per unit cross-sectional area that is absorbed by the conducting medium by calculating the time-averaged powers per unit cross-sectional area in the incident and in the reflected waves.

At what distance from the interface in the conducting medium does the electric field fall to $1/e$ of its value at the interface?

6.5 The plane surface of a large (may be taken as semi-infinite) glass block is coated with a very thin layer of silver. The constitutive parameters of glass are $\varepsilon = 2.25\varepsilon_0$ and $\mu = \mu_0$. The silver film of negligible thickness, has a surface resistivity of R ohms. If a plane wave is normally incident on the surface, find the ratios of reflected and transmitted electric field amplitudes to the amplitude of the incident electric field for two cases where: (a) the wave is incident from the free space side and (b) the wave is incident from the dielectric side.

From these results, determine whether or not (c) the fraction of the incident power which is transmitted into the second medium is the same in both cases, (d) the fraction of the incident power which is reflected is the same in both cases, and (e) it is possible by a proper choice of R to eliminate the reflection in one case or the other.

6.6 A plane wave of angular frequency ω in free space (μ_0, ε_0) is incident normally on a half-space of a very good conductor (μ_0, ε_0, σ). Show that the ratio of the reflected to the incident time-averaged Poynting vectors is approximately: $R_s = 1 - 2\beta\delta$, where $\beta = \omega\sqrt{\mu_0\varepsilon_0}$ and δ is the skin depth.

6.7 A planar slab of dielectric ($\mu = \mu_0$, $\varepsilon = 4\varepsilon_0$, $\sigma = 0$) is in free space (μ_0, ε_0). The thickness of the dielectric slab is $d = 7.5$ cm. A plane electromagnetic wave is normally incident on the slab. Show that the entire incident power is transmitted through the slab if the wave frequency is $f = 10^9$ Hz. What is the fraction of the incident power that is transmitted through the slab if the wave frequency is changed to $f = 5 \times 10^8$ Hz?

6.8 A planar slab of dielectric ($\mu = \mu_0$, $\varepsilon = (1 + \Delta)\varepsilon_0$, $\sigma = 0$) is in free space (μ_0, ε_0). The thickness of the dielectric slab is d. A plane electromagnetic wave of angular frequency ω is normally incident on the slab. Find the expressions for the reflection and the transmission coefficients, R and T, as defined in Section 6.5 in the limiting case of $\Delta \ll 1$.

6.9 A planar slab of dielectric ($\mu = \mu_0$, $\varepsilon = 4\varepsilon_0$, $\sigma = 10^{-2}$ mho/m) is in free space (μ_0, ε_0). The thickness of the dielectric slab is $d = 10$ cm. A plane electromagnetic wave of frequency $f = 10^9$ Hz is incident normally on the slab. The time-averaged incident Poynting vector is 1 W/m^2. Evaluate the time-averaged power per unit cross-sectional area that is absorbed in the dielectric slab.

6.10 A plane electromagnetic wave of frequency $f = 4 \times 10^8$ Hz is incident normally on a half-space of dielectric ($\mu = \mu_0$, $\varepsilon = 4.84\varepsilon_0$) from a semi-infinite region of free space (μ_0, ε_0). What is the fraction of the incident power flux density that is reflected? Find the thickness and the relative permittivity required for a plate placed on the interface between the free space and the dielectric in order to eliminate the reflection of a wave incident normally on the plate. Assume that the permeability of the plate is given by μ_0. What is the percentage of the incident power flux density that is reflected, if for the designed plate, the wave frequency is changed to $f = 6 \times 10^8$ Hz?

6.11 For a plane interface between a semi-infinite dielectric material and a semi-infinite free space, reflections of an incident plane wave from free space are observed to produce a standing-wave ratio of 3 in the free space. The interface is found to be an electric minimum. Find the reflection coefficient and the permittivity of the material. Assume that the permeability of the dielectric material is the same as that of free space. What is the standing-wave ratio if the plane wave is incident from the dielectric side?

6.12 A plane electromagnetic wave is propagating in the x-direction of a right-handed Cartesian coordinate system. The magnetic field component is in the y-direction and has a magnitude of 1 A/m. This plane wave is incident normally on a plane interface separating two media whose characteristics are as follows: Medium 1, relative permittivity 25/9; Medium 2, relative permittivity 1. The relative permeability is unity for both the media. Assume that the permeability and the permittivity of vacuum are given respectively by $\mu_0 = 4\pi \times 10^{-7}$ H/m and $\varepsilon_0 = 10^{-9}/36\pi$ F/m. It is found that the wave is slowed down after transmission through the interface.

(a) Draw a schematic diagram for this problem. (b) Find the direction and the magnitude of the incident electric field as well as the incident power flux density. (c) What are the directions and the magnitudes of the transmitted magnetic and electric fields as well as the transmitted power flux density? (d) From the knowledge of the incident and the transmitted electric and magnetic vectors, evaluate the reflected electric and magnetic vectors. Compute the reflected power flux density and examine, if possible, its correctness from the results of (b) and (c).

6.13 A perfect *magnetic conductor* occupies the plane $x = 0$. On the surface of a perfect magnetic conductor, the tangential magnetic field is zero. The half-space $x > 0$ is characterized by the constitutive parameters μ_0 and ε_0. A plane wave with its (a) electric and (b) magnetic vectors perpendicular to the plane of incidence is incident on the conductor at an angle of incidence equal to θ^i. Note that the field quantities do not vary in the y-direction.

Obtain all the components of the incident and the reflected fields. Find the phase velocity and the wavelength along the surface of the conductor.

6.14 In the geometry of Problem 6.13, suppose that another perfect magnetic conductor is introduced to occupy the plane $x = d$. For each of the two polarizations, deduce the set of discrete values of $\theta^i = \theta^i_g$ for which the sum of the incident and the reflected fields obtained in Problem 6.13 satisfy also the boundary condition on the plane $x = d$ and hence determine all the field components of the TE and the TM modes supported in the dielectric region between two parallel planes of perfect magnetic conductors.

6.15 A planar dielectric slab is bounded by the planes $x = 0$ and $x = d$. The semi-infinite regions $x < 0$ and $x > d$ consist of free space (μ_0, ε_0). The constitutive parameters of the dielectric are given by μ_0 and $\varepsilon_0\varepsilon_r$. A uniform plane wave which is independent of y and which has its magnetic vector perpendicular to the plane of incidence is incident on the interface $x = 0$ of the dielectric slab at the Brewster angle θ^i_B. From first principles, prove that the incident wave is transmitted without any reflection at both the interfaces.

If the slab is made of quartz for which $\varepsilon_r = 2.1$, what is the value of the Brewster angle θ^i_B for transmission (a) from free space to dielectric and (b) from dielectric to free space?

Note that the "Brewster angle" quartz windows are used in gas lasers to minimize reflection losses.

6.16 The medium in front of a perfectly conducting plane is free space whose constitutive parameters are $\mu_0 = 4\pi \times 10^{-7}$ H/m and $\varepsilon_0 = 10^{-9}/36\pi$ F/m. In front of the conducting plane and at a distance $d = 12.5$ cm from it, is a resistive sheet of negligible thickness and of surface resistivity $\eta_{sr} = 120\pi$ ohms. A plane electromagnetic wave of frequency $f = 6 \times 10^8$ Hz is incident normally on the resistive sheet. What is the fraction of the incident power flux density that is reflected?

If the frequency is changed to $f = 10^9$ Hz, then what is the fraction of the incident power flux density that is reflected?

6.17 Consider a planar slab of dielectric in free space (μ_0, ε_0). The constitutive parameters of the dielectric are μ_0 and $4\varepsilon_0$. The thickness of the dielectric slab is $d = 5$ cm. If a plane electromagnetic wave of frequency $f = 1.5 \times 10^9$ Hz is incident normally on the dielectric slab, what is the percentage of the incident power flux density that is reflected if the losses in the dielectric are neglected?

Suppose that the conductivity of the dielectric is $\sigma = 10^{-3}$ mho/m. Find the percentage of the incident power flux density that is reflected if the losses in the dielectric are taken into account.

6.18 Consider a semi-infinite region of free space (μ_0, ε_0) and a semi-infinite region of dielectric $(\mu_0, 4\varepsilon_0)$. Design (a) one quarter-wavelength matching slab and (b) two quarter-wavelength matching slabs in series which when interposed between the free space and the dielectric regions eliminates the reflections of a plane wave of frequency $f = 5 \times 10^8$ Hz incident normally on the interfaces from the free space side. Assume that the permeability of the material of the matching slabs is μ_0.

If the frequency is changed to $f = 6 \times 10^8$ Hz, what is the percentage of the incident power flux density that is reflected?

6.19 Evaluate the critical angle of incidence for transmission of a plane electromagnetic wave from the following dielectrics into free space (μ_0, ε_0): (a) mica $(\mu_0, 6\varepsilon_0)$; (b) water $(\mu_0, 81\varepsilon_0)$.

6.20 Let the half-spaces $x > 0$ and $x < 0$ be two different dielectric media characterized by the constitutive parameters (μ, ε) and (μ, ε') respectively as shown in Fig. 6.5. In the half-space $x > 0$, a plane wave is assumed to be incident obliquely on the interface $x = 0$. Let $y = 0$ be the plane of incidence. Let θ^i, θ^r, and θ^t be the angles of incidence, reflection, and transmission respectively.

Consider the E-polarization for which $E_y(x, z)$, $H_x(x, z)$, and $H_z(x, z)$ are the only nonvanishing field components. Let E^i_{y0}, H^i_{x0}, and H^i_{z0} be the amplitudes of the incident field components at the interface. Let E^r_{y0}, H^r_{x0}, and H^r_{z0} be the corresponding reflected quantities and E^t_{y0}, H^t_{x0}, and H^t_{z0} be the corresponding transmitted quantities. Verify the following results:

$$E^r_{y0}/E^i_{y0} = H^r_{x0}/H^i_{x0} = -H^r_{z0}/H^i_{z0} = \rho_E = \sin(\theta^t - \theta^i)/\sin(\theta^t + \theta^i), \tag{a}$$

$$H^t_{x0}/H^i_{x0} = E^t_{y0}/E^i_{y0} = \tau_E = 2\sin\theta^t\cos\theta^i/\sin(\theta^t + \theta^i), \tag{b}$$

and

$$H^t_{z0}/H^i_{z0} = 2\cos\theta^t\sin\theta^i/\sin(\theta^t + \theta^i). \tag{c}$$

If H^i_0, H^r_0, and H^t_0 are respectively the amplitudes of the incident, the reflected, and the transmitted magnetic fields at the interface, find the values of the ratios H^r_0/H^i_0 and H^t_0/H^i_0.

6.21 In Problem 6.20, consider the H-polarization for which $H_y(x, z)$, $E_x(x, z)$, and $E_z(x, z)$ are the only nonvanishing field components. Let H^i_{y0}, E^i_{x0}, and E^i_{z0} be the amplitudes of the incident field components at the interface. Let H^r_{y0}, E^r_{x0}, and E^r_{z0} be the corresponding reflected quantities and H^t_{y0}, E^t_{x0}, and E^t_{z0} be the corresponding transmitted quantities. Verify the following results:

$$H^r_{y0}/H^i_{y0} = E^r_{x0}/E^i_{x0} = -E^r_{z0}/E^i_{z0} = \rho_H = \tan(\theta^i - \theta^t)/\tan(\theta^i + \theta^t), \tag{a}$$

$$\frac{E^t_{x0}}{E^i_{x0}} = \frac{\sin^2\theta^t\, H^t_{y0}}{\sin^2\theta^i\, H^i_{y0}} = \frac{\sin^2\theta^t}{\sin^2\theta^i}\tau_H = \frac{\sin^2\theta^t}{\sin^2\theta^i}\,\frac{\sin 2\theta^i}{\sin(\theta^t + \theta^i)\cos(\theta^t - \theta^i)}, \tag{b}$$

$$E^t_{z0}/E^i_{z0} = \sin 2\theta^t/[\sin(\theta^t + \theta^i)\cos(\theta^t - \theta^i)]. \tag{c}$$

If E_0^i, E_0^r, and E_0^t are respectively the amplitudes of the incident, the reflected, and the transmitted electric fields at the interface, find the values of the ratios E_0^r/E_0^i and E_0^t/E_0^i.

6.22 For (i) the TE and (ii) the TM surface waves treated in Section 6.9 evaluate the phase, the group, and the energy transport velocities in the neighborhood of frequencies where the reactance of the surface is a constant with respect to the frequency.

CHAPTER 7

ELECTROMAGNETIC THEORY OF TRANSMISSION LINES

7.1 INTRODUCTION

In the previous treatment, we considered the transmission line as a distributed parameter electric circuit. We pointed out that the circuit parameters L, C, R, and G depend on the geometry of the transmission line and have to be obtained from the considerations of electromagnetic field theory. In this chapter, we shall discuss the electromagnetic theory of the transmission lines.

Certain arrangements of perfect conductors which form the boundary of dielectric regions and which are uniform in a given direction are able to support the so-called TEM guided waves. For the case of TEM guided waves, these conductors behave like ordinary transmission lines and can therefore be treated as distributed parameter electric circuits. We can determine the values of these circuit parameters in terms of the geometry of the transmission line by analyzing these structures from the point of view of electromagnetic theory. In Section 7.2, we shall discuss the general characteristics of the TEM guided waves and apply them in Sections 7.3 and 7.4 to a two-wire transmission line. We shall examine the TEM guided wave in a coaxial transmission line in Sections 7.5 to 7.7 where we derive the circuit parameters L,C, R, and G of a coaxial transmission line. In Sections 7.8 to 7.10 we give the treatment of the TEM guided wave in a parallel-plate transmission line and also evaluate the circuit parameters of a parallel-plate transmission line. Even though there are other arrangements of conductors which support TEM guided waves, the two-wire, the coaxial, and the parallel-plate are the transmission lines most commonly used.

7.2 GENERAL CHARACTERISTICS OF TRANSVERSE ELECTROMAGNETIC GUIDED MODES

In a right-handed rectangular coordinate system x, y, and z, let z be the direction of propagation. A wave having no components of the electric and the magnetic fields in the direction of propagation is called a transverse electromagnetic or TEM wave. Since z is the direction of propagation, for the TEM wave, $E_z = H_z = 0$. Let the medium be an unbounded dielectric characterized by the permeability μ_d and the permittivity ε_d, where the subscript d denotes that the corresponding quantities pertain to the dielectric region. For a plane wave, the magnitudes and the directions of the field components are constants in any transverse plane, that is, a plane perpendicular to the direction of propagation. A plane wave propagating in an unbounded dielectric region was shown in Section 5.5 to be a TEM wave. The time-averaged power flow per unit area in the direction of propagation is independent of the location in the transverse plane. This plane wave is called a free wave or an unguided wave. Under certain circumstances, it is possible to have a guided wave that also has no electric or magnetic field components in the direction of propagation. The following three bounded structures support TEM guided waves which are also sometimes called transmission-line modes:

1. Two-wire transmission line.
2. Coaxial transmission line.
3. Parallel-plate transmission line.

A two-wire transmission line consists of a pair of parallel wires usually of circular cross section. Let a be the radius of each of the wires and b the separation distance between the axes of the wires. These wires are in an otherwise unbounded dielectric region with the constitutive parameters μ_d and ε_d. For definiteness, we assume the axes of the wires to coincide with the lines $x = \pm b/2$, $y = 0$. A coaxial transmission line consists of a coaxial dielectric region (μ_d, ε_d) between an inner cylinder of radius a and a coaxial outer cylinder of inner radius b and outer radius b_e. We assume the common axis of the two cylinders to coincide with the z-axis. A parallel-plate transmission line consists of a dielectric slab region (μ_d, ε_d) sandwiched between two parallel plates. The lower and the upper surfaces of the lower plate coincide with the planes $x = -t$ and $x = 0$ respectively. Similarly, the lower and the upper surfaces of the upper plate coincide with the planes $x = a$ and $x = a_e$ respectively. The thickness of the dielectric slab situated between the two parallel plates is a. The dielectric region is usually free space. We assume all the conductors which form the boundaries of the dielectric region to be perfect conductors. All the three structures are uniform in the z-direction, which is assumed to be the direction of propagation. The cross sections of the two-wire, the coaxial, and the parallel-plate transmission lines by a transverse plane are shown respectively in Figs. 7.1, 7.2, and 7.3.

The conductors are perfect and therefore the electromagnetic fields cannot penetrate them. If electromagnetic fields are set up in the coaxial dielectric region of a coaxial transmission line or in the dielectric slab region of a parallel-plate transmission line, they remain confined to the respective dielectric regions without

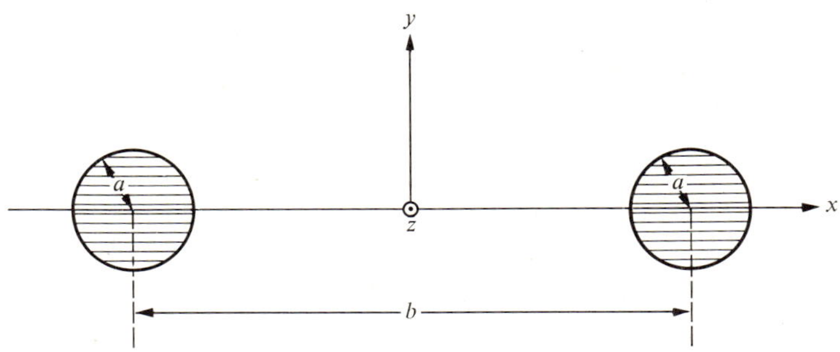

Fig. 7.1 Schematic diagram of a two-wire transmission line.

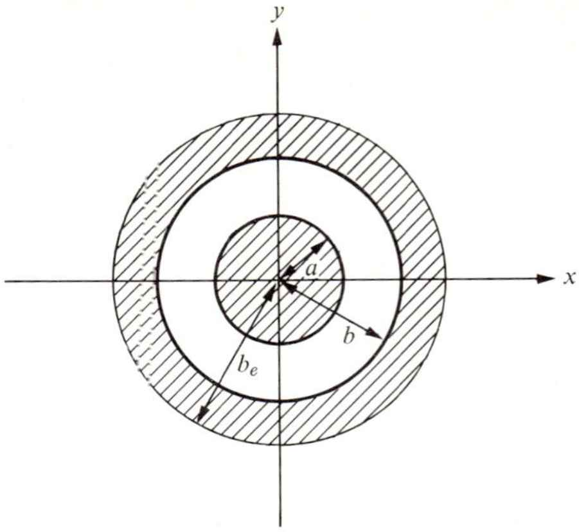

Fig. 7.2 Schematic diagram of a coaxial transmission line.

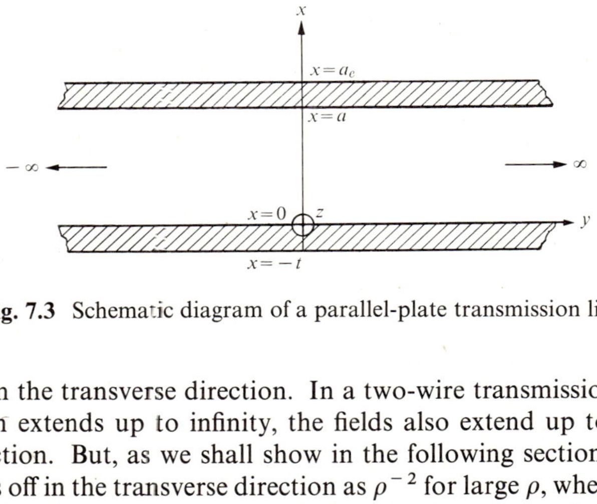

Fig. 7.3 Schematic diagram of a parallel-plate transmission line.

spreading out in the transverse direction. In a two-wire transmission line, since the dielectric region extends up to infinity, the fields also extend up to infinity in the transverse direction. But, as we shall show in the following section, the amplitude of the fields falls off in the transverse direction as ρ^{-2} for large ρ, where ρ is the radial distance from the observation point to the z-axis. The power flux density in the direction of propagation falls off in the transverse direction as ρ^{-4} with the result that even though the fields extend up to infinity, the energy of the wave is concentrated near the two wires. Thus, the two-wire, the coaxial, and the parallel-plate transmission lines serve either to concentrate or confine the energy of the wave in their close neighborhood and, hence, play the role of guiding structures for the electromagnetic waves.

There are other structures which guide TEM guided waves but the three structures we have previously discussed are the most commonly used transmission lines for supporting TEM waves. A feature common to all the structures supporting TEM guided waves is that they have at least two distinct conductors.

The TEM guided wave has the following three important general characteristics:

1. The propagation coefficient is the same as that for a uniform plane wave in the corresponding unbounded dielectric region without any guiding structures.
2. The transverse electric field $\boldsymbol{E}$, the transverse magnetic field $\boldsymbol{H}$, and the direction of propagation $\hat{z}$ form a triad of mutually perpendicular vectors. In this respect, the TEM guided wave is the same as the uniform plane wave.
3. In any transverse plane the distributions of the electric and the magnetic fields are the same as in the corresponding static case with direct currents and time-independent charge distributions.

It is instructive to deduce these results from Maxwell's equations.

The electric and the magnetic fields satisfy the following time-harmonic Maxwell's equations in the dielectric region:

$$\nabla \times \boldsymbol{E}(\boldsymbol{r}) = -j\omega\mu_d \boldsymbol{H}(\boldsymbol{r}), \tag{2.1}$$

and

$$\nabla \times \boldsymbol{H}(\boldsymbol{r}) = j\omega\varepsilon_d \boldsymbol{E}(\boldsymbol{r}), \tag{2.2}$$

where the position vector $\boldsymbol{r}$ specifies the observation point. Let us assume that the dependence of all the field quantities on z, which is the direction of propagation, is of the form

$$\boldsymbol{E}(\boldsymbol{r}) = \boldsymbol{E}(x, y)e^{-\gamma z}, \qquad \boldsymbol{H}(\boldsymbol{r}) = \boldsymbol{H}(x, y)e^{-\gamma z}, \tag{2.3}$$

where γ is the propagation coefficient. If we substitute (2.3) in (2.1) and (2.2) and if the resulting equations are written in component form, we find that

$$\frac{\partial}{\partial y} E_z(x, y) + \gamma E_y(x, y) = -j\omega\mu_d H_x(x, y), \tag{2.4}$$

$$-\gamma E_x(x, y) - \frac{\partial}{\partial x} E_z(x, y) = -j\omega\mu_d H_y(x, y), \tag{2.5}$$

$$\frac{\partial}{\partial x} E_y(x, y) - \frac{\partial}{\partial y} E_x(x, y) = -j\omega\mu_d H_z(x, y), \tag{2.6}$$

$$\frac{\partial}{\partial y} H_z(x, y) + \gamma H_y(x, y) = j\omega\varepsilon_d E_x(x, y), \tag{2.7}$$

$$-\gamma H_x(x, y) - \frac{\partial}{\partial x} H_z(x, y) = j\omega\varepsilon_d E_y(x, y), \tag{2.8}$$

$$\frac{\partial}{\partial x} H_y(x, y) - \frac{\partial}{\partial y} H_x(x, y) = j\omega\varepsilon_d E_z(x, y). \tag{2.9}$$

We omit the arguments (x, y) of the field components wherever possible for the sake of convenience. For a TEM wave

$$E_z = H_z = 0. \tag{2.10}$$

Then, from (2.4), (2.5), (2.7), and (2.8), we obtain the following results:

$$E_x = \frac{j\omega\mu_d}{\gamma} H_y; \qquad H_y = \frac{j\omega\varepsilon_d}{\gamma} E_x; \tag{2.11a,b}$$

$$E_y = \frac{-j\omega\mu_d}{\gamma} H_x; \qquad H_x = \frac{-j\omega\varepsilon_d}{\gamma} E_y. \tag{2.12a,b}$$

Eliminating H_y and E_x successively from (2.11a) and (2.11b) we get

$$(\gamma^2 + \omega^2\mu_d\varepsilon_d)E_x = 0; \qquad (\gamma^2 + \omega^2\mu_d\varepsilon_d)H_y = 0. \tag{2.13}$$

In a similar manner, we can obtain from (2.12a,b) that

$$(\gamma^2 + \omega^2\mu_d\varepsilon_d)E_y = 0; \qquad (\gamma^2 + \omega^2\mu_d\varepsilon_d)H_x = 0. \tag{2.14}$$

If E_x and H_y are to be nonvanishing, (2.13) is satisfied only if

$$\gamma = \alpha + j\beta = \pm j\omega\sqrt{\mu_d\varepsilon_d}. \tag{2.15}$$

In a similar manner, if E_y and H_x are to be nonvanishing, (2.14) is satisfied only if again (2.15) is fulfilled. If the dielectric has a small loss, then ε_d has to be replaced by its effective value $[\varepsilon_d - j(\sigma_d/\omega)]$ where σ_d is the conductivity of the dielectric and is assumed to be small in comparison with $\omega\varepsilon_d$. The binomial expansion of the square root in (2.15) yields a term which has a negative imaginary part and, therefore, the upper sign in (2.15) yields a positive α. From (2.3), we can see that the field components decay with increasing z. If we use the lower sign in (2.15), α is negative and we see the field components from (2.3) to grow with increasing z and this is physically unacceptable. Hence, we only choose the upper sign in (2.15). Then we find the propagation coefficient γ to be the same as that for a uniform plane wave in the corresponding unbounded dielectric region. We should note that the foregoing derivation is essentially the same as that used for the uniform plane waves. Whereas in the uniform plane wave, the transverse field components are not subjected to any boundary conditions, the transverse field components in a guided TEM wave must satisfy boundary conditions in view of the presence of the boundaries. Consequently, in general, the transverse field components of a guided TEM wave, unlike in a uniform plane wave, become functions of the location in the transverse plane.

With the use of the upper sign in (2.15) for γ, we can show that

$$j\omega\mu_d/\gamma = j\omega\mu_d/j\omega\sqrt{\mu_d\varepsilon_d} = \sqrt{\mu_d/\varepsilon_d} = \eta_d, \tag{2.16a}$$

and

$$\gamma/j\omega\varepsilon_d = j\omega\sqrt{\mu_d\varepsilon_d}/j\omega\varepsilon_d = \sqrt{\mu_d/\varepsilon_d} = \eta_d. \tag{2.16b}$$

It is convenient for us to rewrite (2.11) and (2.12) with the help of (2.16) as follows:

$$E_x = \eta_d H_y, \qquad E_y = -\eta_d H_x, \tag{2.17}$$

where η_d is the intrinsic impedance of the dielectric.

Let the electric and the magnetic vectors at a point in space be denoted by

$$\boldsymbol{E} = \hat{x}E_x + \hat{y}E_y, \tag{2.18}$$

$$\boldsymbol{H} = \hat{x}H_x + \hat{y}H_y. \tag{2.19}$$

Substituting (2.17) in (2.19) we obtain

$$\boldsymbol{H} = -\hat{x}\frac{E_y}{\eta_d} + \hat{y}\frac{E_x}{\eta_d}. \tag{2.20}$$

From (2.18) and (2.20), we find that

$$\boldsymbol{E}\cdot\boldsymbol{H} = (\hat{x}E_x + \hat{y}E_y)\cdot\left(-\hat{x}\frac{E_y}{\eta_d} + \hat{y}\frac{E_x}{\eta_d}\right) = 0. \tag{2.21}$$

Therefore, the electric and the magnetic vectors are everywhere normal to each other. Also, from (2.3), (2.15), (2.18), and (2.20), we obtain

$$\begin{aligned}\tfrac{1}{2}\operatorname{Re}\boldsymbol{E}(\boldsymbol{r})\times\boldsymbol{H}^*(\boldsymbol{r}) &= \tfrac{1}{2}\operatorname{Re}\boldsymbol{E}\times\boldsymbol{H}^*e^{-2\alpha z},\\ &= \tfrac{1}{2}e^{-2\alpha z}\operatorname{Re}\left[\hat{x}E_x + \hat{y}E_y\right]\times\left[-\hat{x}\frac{E_y^*}{\eta_d^*} + \hat{y}\frac{E_x^*}{\eta_d^*}\right],\\ &= \frac{\hat{z}}{2}e^{-2\alpha z}(|E_x|^2 + |E_y|^2)\operatorname{Re}\left(\frac{1}{\eta_d^*}\right).\end{aligned} \tag{2.22}$$

We can see the direction of flow of the time-averaged power from (2.22) to coincide with the direction (z) of propagation. Also, the electric and the magnetic vectors are perpendicular to the direction of flow of the time-averaged power. Thus, $\boldsymbol{E}$, $\boldsymbol{H}$, and $\hat{z}$ form a triad of mutually perpendicular vectors.

From (2.17), (2.9), and (2.10), we obtain

$$\frac{\partial}{\partial x}E_x = \eta_d\frac{\partial}{\partial x}H_y; \qquad \frac{\partial H_y}{\partial x} = \frac{\partial H_x}{\partial y}. \tag{2.23}$$

From (2.17), we find that

$$\frac{\partial H_x}{\partial y} = -\frac{1}{\eta_d}\frac{\partial E_y}{\partial y}. \tag{2.24}$$

Combining (2.23) and (2.24) we get

$$\frac{\partial}{\partial x}E_x = -\frac{\partial}{\partial y}E_y. \tag{2.25}$$

On differentiating both sides of (2.25) with respect to x, we obtain

$$\frac{\partial^2}{\partial x^2}E_x = -\frac{\partial}{\partial y}\left(\frac{\partial E_y}{\partial x}\right). \tag{2.26}$$

Since from (2.6) and (2.10)

$$\frac{\partial E_y}{\partial x} = \frac{\partial E_x}{\partial y}, \tag{2.27}$$

(2.26) becomes

$$\frac{\partial^2}{\partial x^2} E_x + \frac{\partial^2}{\partial y^2} E_x = 0. \tag{2.28}$$

Differentiating both sides of (2.25) with respect to y, we find that

$$\frac{\partial^2}{\partial y^2} E_y = -\frac{\partial}{\partial x}\left(\frac{\partial E_x}{\partial y}\right). \tag{2.29}$$

If we replace $\partial E_x/\partial y$ in (2.29) by its equivalent expression in terms of E_y as given in (2.27), we can show (2.29) to yield

$$\frac{\partial^2}{\partial x^2} E_y + \frac{\partial^2}{\partial y^2} E_y = 0. \tag{2.30}$$

Substituting $E_x = \eta_d H_y$ and $E_y = -\eta_d H_x$ respectively in (2.28) and (2.30) we get the following results:

$$\frac{\partial^2}{\partial x^2} H_x + \frac{\partial^2}{\partial y^2} H_x = 0, \tag{2.31}$$

$$\frac{\partial^2}{\partial x^2} H_y + \frac{\partial^2}{\partial y^2} H_y = 0. \tag{2.32}$$

In a source-free region, we see from (4.2.60) that the electrostatic fields are governed by the equations

$$\nabla \times \mathbf{E}(\mathbf{r}) = 0; \qquad \nabla \cdot \mathbf{E}(\mathbf{r}) = 0. \tag{2.33}$$

If the fields are functions of only the coordinates x and y, then we can reduce (2.33) to

$$E_z = 0; \tag{2.34}$$

$$\frac{\partial}{\partial x} E_y - \frac{\partial}{\partial y} E_x = 0; \qquad \frac{\partial}{\partial x} E_x + \frac{\partial}{\partial y} E_y = 0. \tag{2.35a,b}$$

Since (2.35a) and (2.35b) are identical to (2.27) and (2.25) respectively, they lead to (2.28) and (2.30). In a similar manner, we find from (4.4.75) that the magnetostatic fields in a source-free region satisfy the following equations

$$\nabla \times \mathbf{H}(\mathbf{r}) = 0; \qquad \nabla \cdot \mathbf{H}(\mathbf{r}) = 0. \tag{2.36}$$

As before, if the fields are functions of only the coordinates x and y, then we can simplify (2.36) to

$$H_z = 0; \tag{2.37}$$

$$\frac{\partial}{\partial x} H_y - \frac{\partial}{\partial y} H_x = 0; \qquad \frac{\partial}{\partial x} H_x + \frac{\partial}{\partial y} H_y = 0. \tag{2.38a,b}$$

We can show that (2.38a,b) lead to (2.31) and (2.32). Moreover, we note that (2.34) and (2.37) are identical to the defining relation (2.10) for the TEM wave. Thus, (2.28) and (2.30) to (2.32) show that in the transverse plane the electric and the magnetic fields of a TEM wave satisfy the same two-dimensional Laplace's equation as the corresponding electrostatic and magnetostatic fields. Also, for the TEM guided wave, the following boundary conditions should be satisfied on the surface of a perfect conductor:

$$E_t = 0 \qquad \text{and} \qquad H_n = 0, \tag{2.39}$$

where t and n denote respectively the components tangential and normal to the surface of the perfect conductor. The electrostatic and the magnetostatic fields also satisfy the same boundary conditions as in (2.39). Since in the transverse plane the differential equations and the boundary conditions satisfied by the electric and the magnetic fields for the TEM guided wave are the same as the corresponding electrostatic and magnetostatic fields, it follows that in the transverse plane, the distributions of the electric and the magnetic fields associated with a TEM guided wave are the same as the corresponding static field distributions. Consequently, we can determine the capacitance C per unit length and the inductance L per unit length of a uniform transmission line supporting a TEM guided wave by using the same methods as used in electrostatics and magnetostatics respectively.

7.3 EVALUATION OF C AND L OF A TWO-WIRE TRANSMISSION LINE

We used the capacitance C and the inductance L of unit length of the transmission line in the course of the treatment of wave propagation along a lossless two-wire transmission line. We indicated that the expressions for C and L in terms of the geometry of the transmission line have to be deduced from the considerations of electromagnetic field theory. Here, we shall derive the values of C and L for a two-wire transmission line using the methods of electrostatics and magnetostatics.

For a balanced transmission line, in any transverse plane, the total axial (z) current in one wire is equal in magnitude and opposite in direction to that in the other wire. There are no other components of the electric current in the two wires since H_z is zero. Consequently, in any transverse plane, the charge per unit length in one wire is equal in magnitude and opposite in sign to that in the other wire. Only a balanced transmission line can support a TEM guided wave. Since C the capacitance and L the inductance of unit length of a transmission line can be meaningfully defined only for a TEM guided wave, we assume *a priori* that the two-wire transmission line is balanced and therefore, in any transverse plane, the axial current and the charge per unit length in one wire are equal in magnitude and opposite in sign respectively to those in the other wire.

It is convenient for analytical reasons for us to begin with the simpler case in which the radius of each of the wires is zero so that the wires degenerate to the case of filaments. Let us consider two parallel, infinitely long filaments A and B situated

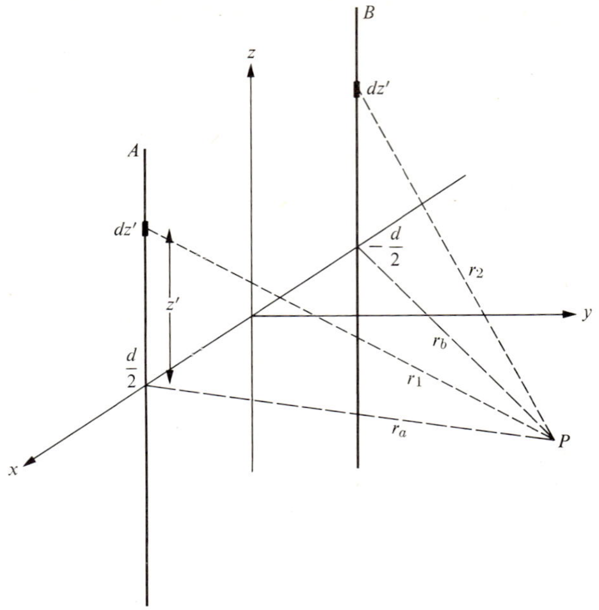

Fig. 7.4 Two infinitely long, parallel filaments.

along the lines $x = d/2$, $y = 0$ and $x = -d/2$, $y = 0$ respectively (Fig. 7.4). Let us assume that a linear charge density q C/m is distributed along wire A and $-q$ C/m along wire B. We know from (4.2.30) that the scalar potential ϕ can be obtained from the knowledge of the charge distribution with the help of the relation

$$\phi = \frac{1}{4\pi\varepsilon_d} \int \frac{\rho \, dV}{r}, \tag{3.1}$$

where ε_d is the permittivity of the medium, ρ is the charge per unit volume or charge density, r is the distance of the observation point to the location of the charge, and the integration extends over all regions where there are charges. In the present case, since q is the charge per unit length, we obtain the total charge in an infinitesimal length dz of the filament as $q\,dz = \rho\,dV$. Let P be the observation point in the transverse plane $z = 0$ and let r_a and r_b be the distances of the observation point to the points of intersection of the transverse plane $z = 0$ with filaments A and B respectively. Let us consider infinitesimal lengths dz' of the two filaments located near the transverse plane $z = z'$. Let r_1 and r_2 be the distances to the observation point from the locations of the infinitesimal lengths dz' of filaments A and B respectively. From Fig. 7.4, we see that

$$r_1 = \sqrt{r_a^2 + z'^2}; \qquad r_2 = \sqrt{r_b^2 + z'^2}. \tag{3.2}$$

We therefore determine the contribution $d\phi$ to the scalar potential at P due to the charges in the infinitesimal lengths dz' of the two filaments from (3.1) and (3.2) as

$$d\phi = \frac{q}{4\pi\varepsilon_d}\left(\frac{1}{\sqrt{r_a^2 + z'^2}} - \frac{1}{\sqrt{r_b^2 + z'^2}}\right) dz'. \tag{3.3}$$

We obtain the total scalar potential ϕ at P due to the entire charge distributions in the two filaments by integrating (3.3) with respect to z' from $z' = -\infty$ to $z' = \infty$ with the following results:

$$\begin{aligned}\phi &= \frac{q}{4\pi\varepsilon_d}\int_{-\infty}^{\infty}\left[\frac{1}{\sqrt{r_a^2 + z'^2}} - \frac{1}{\sqrt{r_b^2 + z'^2}}\right] dz', \\ &= \frac{q}{2\pi\varepsilon_d}\int_{0}^{\infty}\left[\frac{1}{\sqrt{r_a^2 + z'^2}} - \frac{1}{\sqrt{r_b^2 + z'^2}}\right] dz', \\ &= \frac{q}{2\pi\varepsilon_d}\left[\ln\{z' + \sqrt{r_a^2 + z'^2}\} - \ln\{z' + \sqrt{r_b^2 + z'^2}\}\right]_0^\infty \\ &= \frac{q}{2\pi\varepsilon_d}\left[\ln\left\{\frac{z' + \sqrt{r_a^2 + z'^2}}{z' + \sqrt{r_b^2 + z'^2}}\right\}\right]_0^\infty.\end{aligned} \tag{3.4}$$

As z' approaches infinity the ratio $(z' + \sqrt{r_a^2 + z'^2})/(z' + \sqrt{r_b^2 + z'^2})$ approaches unity and the logarithm of the ratio tends to zero. Therefore, when the limits are inserted, (3.4) becomes

$$\phi = -\frac{q}{2\pi\varepsilon_d}\ln\left(\frac{r_a}{r_b}\right) = \frac{q}{2\pi\varepsilon_d}\ln\left(\frac{r_b}{r_a}\right). \tag{3.5}$$

In the plane of the symmetry between the wires, $r_a = r_b$ and then we see from (3.5) the potential ϕ to be zero.

From (3.5), we can find the equation for the *equipotential surface* on which the potential is everywhere the same. Since any two points on the equipotential surface have zero potential difference between them, everywhere, the electric field tangential to the equipotential surface is zero. Therefore, it follows that (a) the movement of a charge over such a surface requires no work and (b) the electric field is always normal to equipotential surfaces. We obtain the locus of a constant potential from (3.5) as

$$\frac{q}{2\pi\varepsilon_d}\ln\left(\frac{r_b}{r_a}\right) = \text{constant} \qquad \text{or} \qquad \frac{r_b}{r_a} = K. \tag{3.6}$$

From Fig. 7.4, we obtain

$$r_a^2 = \left(x - \frac{d}{2}\right)^2 + y^2; \qquad r_b^2 = \left(x + \frac{d}{2}\right)^2 + y^2. \tag{3.7}$$

Together with (3.7), (3.6) yields

$$\left(x + \frac{d}{2}\right)^2 + y^2 = K^2\left[\left(x - \frac{d}{2}\right)^2 + y^2\right], \tag{3.8}$$

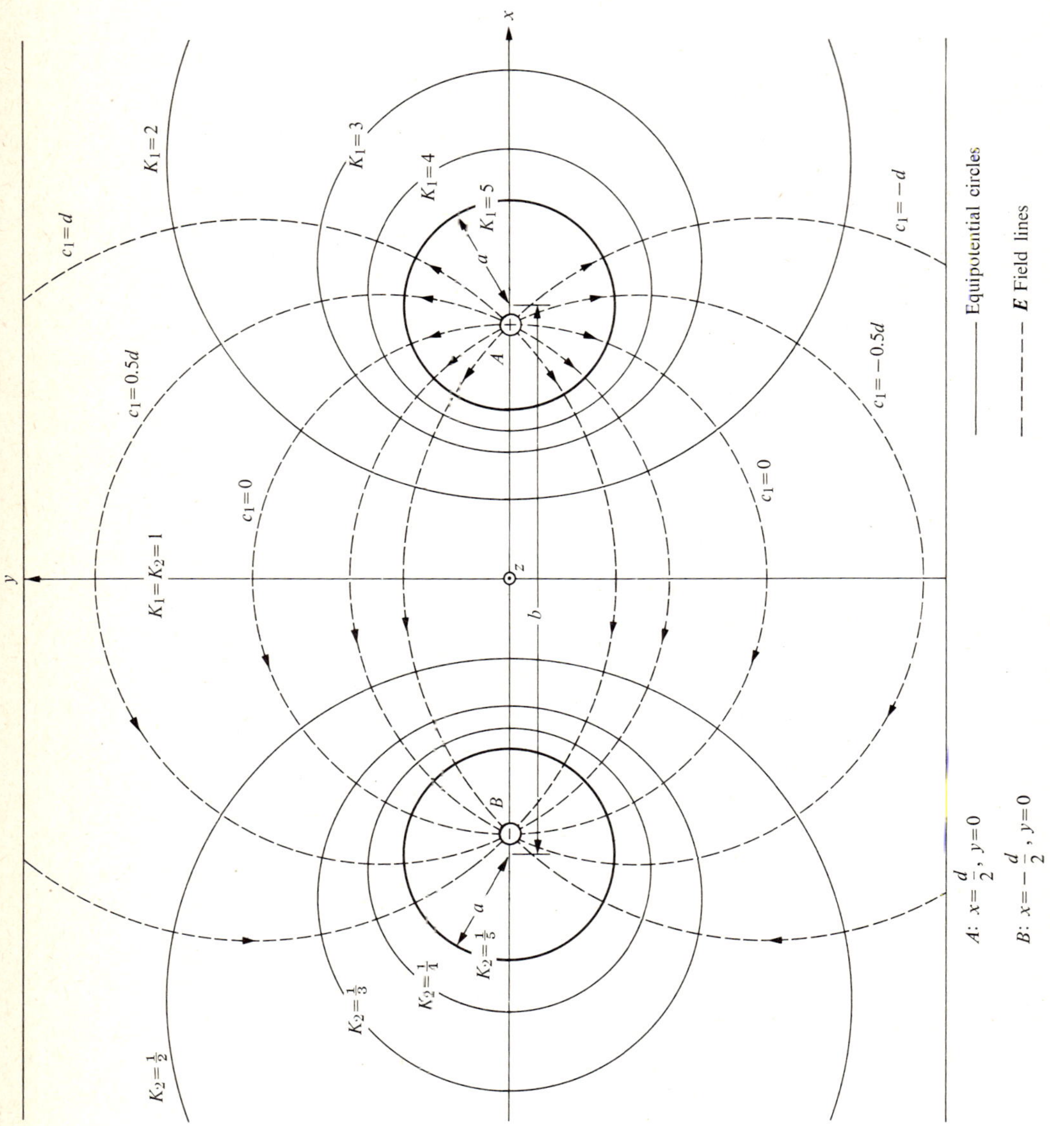

Fig. 7.5 Equipotential circles and ***E***-field lines about two infinitely long parallel filaments *A* and *B*, carrying equal and opposite charges uniformly throughout their lengths.

which we can rearrange as follows:

$$x^2(1 - K^2) + xd(1 + K^2) + y^2(1 - K^2) + \frac{d^2}{4}(1 - K^2) = 0,$$

$$x^2 + xd\frac{(1 + K^2)}{(1 - K^2)} + y^2 + \frac{d^2}{4} = 0. \tag{3.9}$$

On completing the square in x, we find that

$$\left[x + \frac{d}{2}\frac{(1 + K^2)}{(1 - K^2)}\right]^2 + y^2 = \frac{d^2}{4}\left[\frac{(1 + K^2)^2}{(1 - K^2)^2} - 1\right] = \frac{d^2K^2}{(1 - K^2)^2}, \tag{3.10}$$

which defines two sets of circles. For the first set, let $K = K_1 > 1$ and for the second set, let $K = K_2 < 1$. The centers of the two sets of circles are at

$$x = x_1 = \frac{d}{2}\frac{(K_1^2 + 1)}{(K_1^2 - 1)}; \qquad y = y_1 = 0 \qquad (\text{with } K_1 > 1), \tag{3.11a}$$

and

$$x = x_2 = -\frac{d}{2}\frac{(1 + K_2^2)}{(1 - K_2^2)}; \qquad y = y_2 = 0 \qquad (\text{with } K_2 < 1). \tag{3.11b}$$

Their radii are given by

$$a_1 = \frac{dK_1}{K_1^2 - 1}; \qquad K_1 > 1, \tag{3.12a}$$

and

$$a_2 = \frac{dK_2}{1 - K_2^2}; \qquad K_2 < 1. \tag{3.12b}$$

Note that $x_1 > 0$ and $x_2 < 0$. From (3.11a) and (3.12a), we can show that

$$x_1 - a_1 = \frac{d}{2}\frac{1}{(K_1^2 - 1)}[K_1^2 + 1 - 2K_1] = \frac{d}{2}\frac{(K_1 - 1)}{(K_1 + 1)} > 0. \tag{3.13a}$$

In a similar manner, we can deduce that

$$|x_2| - a_2 = \frac{d}{2}\frac{1}{(1 - K_2^2)}(1 + K_2^2 - 2K_2) = \frac{d}{2}\frac{(1 - K_2)}{(1 + K_2)} > 0. \tag{3.13b}$$

From (3.13a,b), we can argue that the first set of circles centered at $(x_1, 0)$ and radius equal to a_1 lie completely in the region $x > 0$ and that the second set of circles centered at $(x_2, 0)$ and radius equal to a_2 lie completely in the region $x < 0$. The circles in both the sets corresponding to $K_1 = K_2 = 1$ degenerate into the same straight line which coincides with the y-axis. Thus, the two sets of circles are separated by the y-axis. The two sets of circles and the limiting straight line are shown in Fig. 7.5.

Let us consider a pair of circles of equal radii as given by

$$a_1 = a_2 = a. \tag{3.14}$$

From (3.12a,b) and (3.14), we can prove that

$$K_1 = 1/K_2. \tag{3.15}$$

If ϕ_1 is the potential corresponding to a circle of a given radius in the first set and if ϕ_2 is the potential corresponding to a circle in the second set and having the same given radius, then it follows from (3.15), (3.5), and (3.6) that

$$\phi_1 = -\phi_2. \tag{3.16}$$

We should remember that the foregoing development is true for all transverse planes with the result that the equipotential surfaces are cylindrical surfaces whose cross section by a transverse plane is as shown in Fig. 7.5.

Let b be the distance between the centers of the two circles with equal radii. Then from (3.11a), (3.11b), and (3.15), we find that

$$b = x_1 - x_2 = d\frac{(K_1^2 + 1)}{(K_1^2 - 1)}. \tag{3.17}$$

If $K_1 = \infty$, $K_2 = 0$ and (3.12a,b) shows that $a_1 = a_2 = 0$. For $K_1 = \infty$, (3.17) yields $b = d$, as it should be since $a_1 = a_2 = 0$ shows that the cylinders degenerate into two parallel filaments whose separation distance we denoted by d to start with. From (3.12a), we see that

$$a = \frac{dK_1}{K_1^2 - 1}. \tag{3.18}$$

Eliminating d from (3.17) and (3.18) we get

$$\frac{b}{2a} = \frac{K_1^2 + 1}{2K_1} = \cosh u; \qquad u \text{ is real and positive.} \tag{3.19}$$

Since $b \geq 2a$, we have set $b/2a$ equal to $\cosh u$ since $\cosh u$ is always greater than unity. From (3.19), we can find K_1 in terms of u in the following manner. We can obtain a quadratic equation in K_1 from (3.19) as

$$K_1^2 - 2K_1 \cosh u + 1 = 0, \tag{3.20}$$

which we can solve to yield the following two solutions:

$$K_1 = e^u, e^{-u}. \tag{3.21}$$

Since $K_1 > 1$, we have to rule out the latter solution in (3.21). With the help of (3.5), (3.6), (3.19), and (3.21), we can deduce that

$$\phi_1 = -\phi_2 = \frac{q}{2\pi\varepsilon_d} \ln K_1 = \frac{q}{2\pi\varepsilon_d} \cosh^{-1}\left(\frac{b}{2a}\right). \tag{3.22}$$

Writing $\cosh u = (e^u + e^{-u})/2$, we can obtain a quadratic equation in e^u in terms of b/a from (3.19) as follows:

$$e^{2u} - \frac{b}{a} e^u + 1 = 0, \tag{3.23}$$

whose two solutions are given by

$$e^u = \frac{b}{2a} \pm \sqrt{(b/2a)^2 - 1},$$

$$= \frac{b}{2a}[1 \pm \sqrt{1 - (2a/b)^2}]. \qquad (3.24)$$

In the limit when $2a/b$ tends to zero, (3.24) becomes

$$e^u = \frac{b}{2a}\left[1 \pm \left\{1 - \frac{1}{2}\left(\frac{2a}{b}\right)^2\right\}\right] \simeq \frac{b}{a}, \frac{a}{b}. \qquad (3.25)$$

The second solution for e^u yields a value less than unity and we therefore rule it out. Then, from (3.19) and (3.24), we see that

$$\cosh^{-1}\left(\frac{b}{2a}\right) = \ln\left[\frac{b}{2a}\left\{1 + \sqrt{1 - \left(\frac{2a}{b}\right)^2}\right\}\right]. \qquad (3.26)$$

When $a^2/b^2 \ll 1$, we obtain an approximation for $\cosh^{-1}(b/2a)$ from (3.26) as

$$\cosh^{-1}(b/2a) \approx \ln(b/a) \qquad \text{if} \qquad (a/b)^2 \ll 1. \qquad (3.27)$$

Under this approximation, (3.22) reduces to

$$\phi_1 = -\phi_2 = \frac{q}{2\pi\varepsilon_d}\ln\left(\frac{b}{a}\right) \qquad \text{if} \qquad \left(\frac{a}{b}\right)^2 \ll 1. \qquad (3.28)$$

If we use (3.7) in (3.5) we get the following expression for the scalar potential:

$$\phi = \frac{q}{4\pi\varepsilon_d}\ln\left[\frac{[x + (d/2)]^2 + y^2}{[x - (d/2)]^2 + y^2}\right]. \qquad (3.29)$$

With the help of the relation

$$\boldsymbol{E} = -\nabla\phi = -\hat{x}\frac{\partial\phi}{\partial x} - \hat{y}\frac{\partial\phi}{\partial y}, \qquad (3.30)$$

and (3.29), we can show E_x and E_y to be

$$E_x = -\frac{\partial\phi}{\partial x} = -\frac{q}{2\pi\varepsilon_d}\left[\frac{x + (d/2)}{[x + (d/2)]^2 + y^2} - \frac{x - (d/2)}{[x - (d/2)]^2 + y^2}\right], \qquad (3.31)$$

$$E_y = -\frac{\partial\phi}{\partial y} = -\frac{qy}{2\pi\varepsilon_d}\left[\frac{1}{[x + (d/2)]^2 + y^2} - \frac{1}{[x - (d/2)]^2 + y^2}\right]. \qquad (3.32)$$

In the description of a vector field such as that given by (3.31) and (3.32), it is convenient for us to introduce field lines whose tangent at every point gives the direction of the field at that point. Thus, at any point x, y in the transverse plane, the directional tangent dy/dx to the field line gives the ratio of the y- to the x-component of the field at that point:

$$dy/dx = E_y/E_x. \qquad (3.33)$$

Substituting (3.31) and (3.32) in (3.33) we get after some simplification

$$\frac{dy}{dx} = \frac{2xy}{x^2 - y^2 - (d/2)^2}. \tag{3.34}$$

The solution of the nonlinear differential equation (3.34) results in the equation of the electric field lines. Rearranging (3.34) as

$$2xy\,dx - \left\{x^2 - y^2 - \left(\frac{d}{2}\right)^2\right\} dy = 0 \tag{3.35}$$

and then dividing throughout by y^2 we get

$$\frac{2x}{y}\,dx + \left\{1 + \left(\frac{d}{2}\right)^2 \frac{1}{y^2} - \frac{x^2}{y^2}\right\} dy = 0. \tag{3.36}$$

Let

$$f(x, y) = y - \left(\frac{d}{2}\right)^2 \frac{1}{y} + \frac{x^2}{y}. \tag{3.37}$$

We can see from (3.37) that

$$\frac{\partial f(x, y)}{\partial x} = \frac{2x}{y}; \qquad \frac{\partial f(x, y)}{\partial y} = 1 + \left(\frac{d}{2}\right)^2 \frac{1}{y^2} - \frac{x^2}{y^2}. \tag{3.38}$$

Substituting (3.38) in (3.36) we obtain

$$\frac{\partial f(x, y)}{\partial x}\,dx + \frac{\partial f(x, y)}{\partial y}\,dy = df(x, y) = 0. \tag{3.39}$$

Therefore, we see from (3.39) that

$$f(x, y) = \text{constant} = c_1 \tag{3.40}$$

is the equation for the electric field lines. The constant c_1 is arbitrary. Together with (3.37), we can rearrange (3.40) to give

$$x^2 + \left(y - \frac{c_1}{2}\right)^2 = \left(\frac{d}{2}\right)^2 + \left(\frac{c_1}{2}\right)^2, \tag{3.41}$$

which represents a family of circles centered at $x = 0$; $y = c_1/2$ and radius $a_E = \sqrt{(d/2)^2 + (c_1/2)^2}$. We can verify that $x = \pm d/2$, $y = 0$ satisfies (3.41) for all d and c_1. Therefore, the family of circles as given in (3.41) all pass through the locations ($x = d/2$, $y = 0$) and ($x = -d/2$, $y = 0$) where the two parallel filaments are situated. The arbitrary constant c_1 can be assigned any value from $-\infty$ to ∞ and corresponding to each value of c_1, (3.41) defines a field line for the $\boldsymbol{E}$ field. A few field lines for the $\boldsymbol{E}$ field are shown by dashed lines in Fig. 7.5.

We find the sense of the electric field along the field lines in the following manner. With the help of (3.41), we may simplify (3.31) and (3.32) to give

$$E_x = -\frac{q\,dy}{\pi\varepsilon_d}\frac{[y - (c_1/2)]}{\{[x - (d/2)]^2 + y^2\}\{[x + (d/2)]^2 + y^2\}}, \tag{3.42}$$

$$E_y = \frac{q\,d}{\pi\varepsilon_d}\frac{xy}{\{[x - (d/2)]^2 + y^2\}\{[x + (d/2)]^2 + y^2\}}. \tag{3.43}$$

From (3.42), it follows that E_x is negative or positive according to whether y and $[y - (c_1/2)]$ are of the same or opposite sign respectively. Similarly, (3.43) shows that E_y is positive or negative according to whether x and y are of the same or opposite sign. With these results, it is possible for us to determine the sense of the electric field along the field lines. Near the location of filament A, since $x > 0$, for $y > 0$, E_y is positive. Therefore, for $y > 0$, the electric vector is outwardly directed along the field lines with respect to the location of filament A. For $y < 0$, E_y becomes negative near the location of filament A and hence for $y < 0$ also, the electric vector is outwardly directed from the location of filament A. Near the location of filament B, since $x < 0$, we can argue with the help of (3.43) that $E_y \lessgtr 0$ according to whether $y \gtrless 0$. These facts lead us to the conclusion that the electric vector is inwardly directed along the field lines with respect to the location of filament B. For $x = 0$, we find from (3.37) that $y(y - (c_1/2)) > 0$. Hence, (3.42) and (3.43) show that for $x = 0$, $E_y = 0$ and $E_x < 0$. Therefore, for $x = 0$, the electric vector along all the field lines is directed from filament A to filament B. Following the foregoing procedure, that the electric vector is outwardly (or inwardly) directed along the field lines with respect to the location of filament A (or B) can be proved, with the help of (3.42) and (3.43), to be true generally at any location on the field lines. The directions of the electric vector along the field lines are also shown in Fig. 7.5. Since along the equipotential circles the tangential electric field is zero, it follows that the electric field lines are everywhere normal to equipotential circles.

It is now possible for us to replace the filaments by perfectly conducting cylinders of finite radius. Since the tangential electric field is zero on it, a perfectly conducting surface introduced so as to coincide with an equipotential surface does not alter the fields. Suppose that a perfectly conducting, infinitely long cylindrical surface A of radius a is introduced so as to coincide with the equipotential (positive) cylindrical surface of the same radius a. The existing electric and magnetic fields are unaffected. Since electromagnetic fields cannot penetrate a perfect conductor, it follows that the fields inside the cylindrical surface A are due to filament A and those outside it are due to filament B. Now, if we remove filament A, the fields inside the cylindrical surface A vanish and those outside it are unchanged. All the electric field lines which emanated from filament A now emanate from the cylindrical surface A with the result that the charge per unit length on the cylindrical surface A is the same, namely q, as that on filament A. The region inside the cylindrical surface A can be filled with a perfect conductor without affecting the fields exterior to it. In the same manner, filament B can be replaced by another perfectly conducting, infinitely long cylinder B

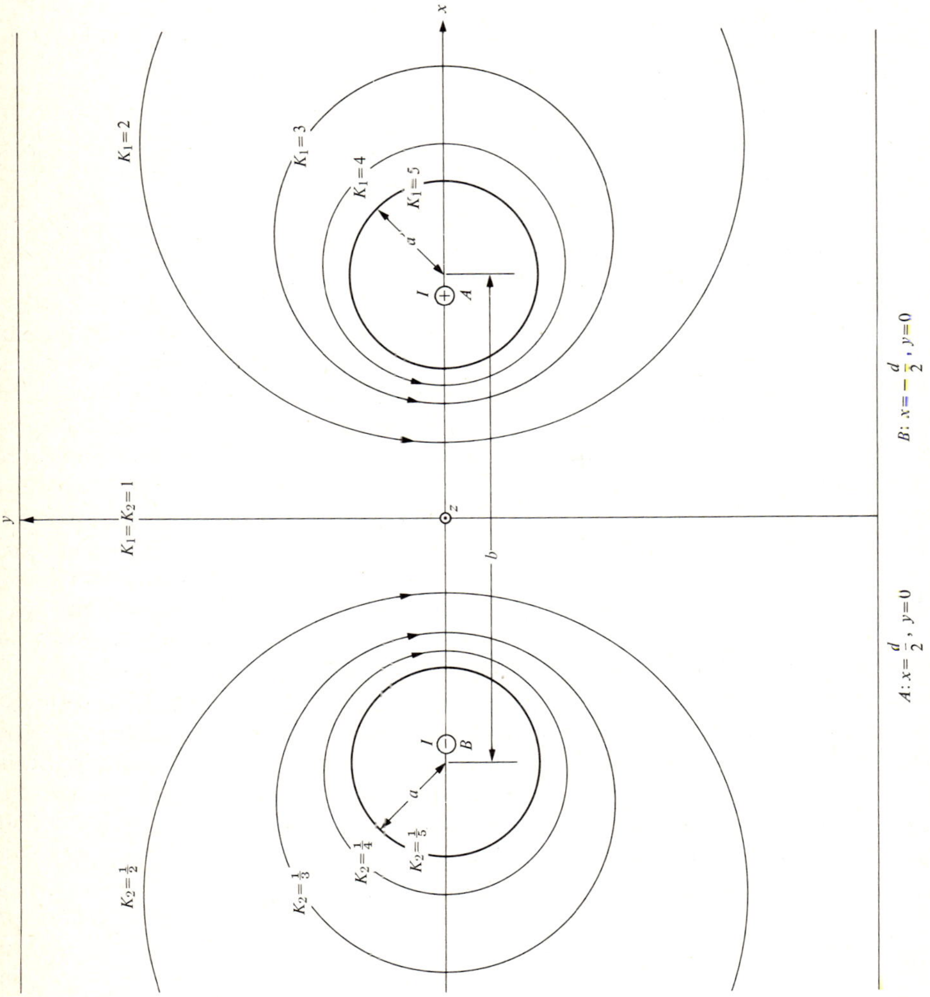

Fig. 7.6 ***H*-field lines about two infinitely long parallel filaments *A* and *B* carrying equal and oppositely directed currents. Note that the equipotential (vector potential) circles are the same as the *H*-field lines.**

of radius a so that its surface coincides with the equipotential (negative) cylindrical surface of the same radius. All the electric field lines which terminated on filament B now terminate on cylinder B with the result that the charge per unit length on cylinder B is the same, namely $-q$, as that on filament B. There is no electric field inside cylinder B and the electric field outside it is unaffected. Thus, we have obtained a two-wire transmission line with a for the radius of each of the two wires, as also the electric field external to the two wires. If b is the distance between the axes of the two wires, we can obtain the potential difference V between them from (3.22) to be

$$V = \phi_1 - \phi_2 = \frac{q}{\pi\varepsilon_d} \cosh^{-1}\left(\frac{b}{2a}\right). \tag{3.44}$$

We can derive the capacitance C per unit length of the two-wire transmission line from (3.44) to be

$$C = \frac{q}{V} = \frac{\pi\varepsilon_d}{\cosh^{-1}(b/2a)}, \tag{3.45a}$$

$$= \frac{\pi\varepsilon_d}{\ln(b/a)} \qquad (\text{if } a^2/b^2 \ll 1). \tag{3.45b}$$

For the evaluation of the inductance L per unit length of the two-wire transmission line, it is again convenient for us to start with the same two parallel filaments now carrying currents. We assume the current in the $+z$-direction to be positive. Let the currents in filaments A and B be I and $-I$ A, respectively. We know from (4.4.27) that the vector potential $\boldsymbol{A}$ can be obtained from the knowledge of the current distribution with the help of the relation

$$\boldsymbol{A} = \frac{\mu_d}{4\pi} \int \frac{\boldsymbol{J}\, dV}{r}, \tag{3.46}$$

where $\boldsymbol{J}$ is the current density, r is the distance of the observation point to the location of the current, and the integration extends over all current carrying regions. Since the vector potential is in the same direction as the current, in the present case, the vector potential has only a z-component. The current in an infinitesimal length dz of the filament is given by $\hat{z}I\, dz = \boldsymbol{J}\, dV$. As before, we assume the observation point to be P in the transverse plane $z = 0$. We can obtain the vector potential A_z at P due to the two parallel current filaments in the same manner as the scalar potential ϕ due to the two parallel charge filaments. The result is

$$A_z = \frac{\mu_d I}{2\pi} \ln\left(\frac{r_b}{r_a}\right). \tag{3.47}$$

If we compare (3.47) and (3.5) we see that the *equipotential surfaces* for the vector potential are the same as those for the scalar potential. The equipotential surfaces are two sets of cylindrical surfaces with the generators parallel to the z-axis. The cross section of these cylindrical surfaces by the transverse plane $z = 0$ yields two sets of circles whose properties are as we described in (3.11a,b) and (3.12a,b). These circles are the same as in Fig. 7.5 and are reproduced again in Fig. 7.6.

The ratio r_b/r_a for the first set of circles occupying the region $x > 0$ is denoted by K_1 and the same ratio corresponding to the second set of circles occupying the region $x < 0$ is denoted by K_2. Let us consider a pair of circles of equal radii as given by (3.14). We have proved the K values of these circles to be inverses of each other, as indicated by (3.15). If A_{z1} is the vector potential corresponding to a circle of a given radius in the first set and if A_{z2} is the vector potential corresponding to a circle in the second set and having the same given radius, then it follows from (3.15), (3.19), (3.21), and (3.47) that

$$A_{z1} = -A_{z2} = \frac{\mu_d I}{2\pi} \cosh^{-1}\left(\frac{b}{2a}\right), \tag{3.48}$$

where a is the radius of each of the two circles and b is the distance between their centers.

If we use (3.7) in (3.47) we obtain the following expression for the vector potential

$$A_z = \frac{\mu_d I}{4\pi} \ln\left\{\frac{[x + (d/2)]^2 + y^2}{[x - (d/2)]^2 + y^2}\right\}. \tag{3.49}$$

With the help of the relation

$$\boldsymbol{H} = \frac{1}{\mu_d} \nabla \times \hat{z} A_z = \left[\hat{x}\frac{\partial A_z}{\partial y} - \hat{y}\frac{\partial A_z}{\partial x}\right]\Big/\mu_d, \tag{3.50}$$

and (3.49), we obtain H_x and H_y as

$$H_x = \frac{1}{\mu_d}\frac{\partial A_z}{\partial y} = \frac{Iy}{2\pi}\left[\frac{1}{[x + (d/2)]^2 + y^2} - \frac{1}{[x - (d/2)]^2 + y^2}\right], \tag{3.51}$$

$$H_y = -\frac{1}{\mu_d}\frac{\partial A_z}{\partial x} = -\frac{I}{2\pi}\left[\frac{[x + (d/2)]}{[x + (d/2)]^2 + y^2} - \frac{[x - (d/2)]}{[x - (d/2)]^2 + y^2}\right]. \tag{3.52}$$

As in the case of the electric field, we can conveniently introduce the magnetic field lines. We deduce the directional tangent dy/dx to the magnetic field lines from (3.51) and (3.52) as

$$dy/dx = H_y/H_x = [y^2 - x^2 + (d/2)^2]/2xy. \tag{3.53}$$

Rearranging (3.53) as

$$\left[y^2 - x^2 + \left(\frac{d}{2}\right)^2\right] dx - 2xy\, dy = 0, \tag{3.54}$$

and then dividing throughout by x^2, we get

$$\left[\frac{y^2}{x^2} - 1 + \left(\frac{d}{2}\right)^2 \frac{1}{x^2}\right] dx - \frac{2y}{x}\, dy = 0. \tag{3.55}$$

Let

$$g(x, y) = -\frac{y^2}{x} - x - \left(\frac{d}{2}\right)^2 \frac{1}{x}. \tag{3.56}$$

We can see from (3.56) that

$$\frac{\partial g(x, y)}{\partial x} = \frac{y^2}{x^2} - 1 + \left(\frac{d}{2}\right)^2 \frac{1}{x^2}; \qquad \frac{\partial g(x, y)}{\partial y} = -\frac{2y}{x}. \tag{3.57}$$

Substituting (3.57) in (3.55) we obtain

$$\frac{\partial}{\partial x} g(x, y)\, dx + \frac{\partial}{\partial y} g(x, y)\, dy = dg(x, y) = 0, \tag{3.58}$$

and therefore

$$g(x, y) = \text{constant} = c_2 \tag{3.59}$$

is the equation for the magnetic field lines. Together with (3.56), we can rearrange (3.59) to give

$$\left(x + \frac{c_2}{2}\right)^2 + y^2 = \left(\frac{c_2}{2}\right)^2 - \left(\frac{d}{2}\right)^2, \tag{3.60}$$

which represents a set of circles of center at $x = -c_2/2$, $y = 0$ and radius equal to $\sqrt{(c_2/2)^2 - (d/2)^2}$. The arbitrary constant c_2 can take any value from $-\infty$ to ∞. But, in order for the circles to be real, it is necessary that $|c_2| \geq d$. For K varying from 0 to ∞, $(1 + K^2)/(1 - K^2)$ assumes all values from $-\infty$ to -1 and 1 to ∞. Insofar as obtaining all physically real magnetic field lines, there is no loss of generality in setting

$$c_2 = d\frac{(1 + K^2)}{(1 - K^2)} \qquad (\text{where } 0 < K < \infty). \tag{3.61}$$

Together with (3.61), (3.60) becomes

$$\left[x + \frac{d}{2}\frac{(1 + K^2)}{(1 - K^2)}\right]^2 + y^2 = \frac{d^2K^2}{(1 - K^2)^2}. \tag{3.62}$$

If we compare (3.62) with (3.10) we see that the magnetic field lines are identical with the equipotential circles. This result is to be expected since we know that **E** and **H** are everywhere normal to each other and that the **E**-field lines are everywhere normal to the equipotential circles.

We find the sense of the magnetic field along the field lines in the following manner. With the help of (3.56) and (3.59), we may simplify (3.51) and (3.52) to give

$$H_x = -\frac{I\, dxy}{\pi\{[x - (d/2)]^2 + y^2\}\{[x + (d/2)]^2 + y^2\}}, \tag{3.63}$$

$$H_y = \frac{I\, dx[x + (c_2/2)]}{\pi\{[x - (d/2)]^2 + y^2\}\{[x + (d/2)]^2 + y^2\}}. \tag{3.64}$$

First, let $c_2 < 0$. Then, from (3.60), we see that the field lines are the circles occupying the region $x > 0$. For $x > 0$, we find from (3.63) that $H_x \lesseqgtr 0$ according to whether $y \gtrless 0$. Similarly, from (3.64), it follows that $H_y \gtrless 0$ according as $x \gtrless |c_2|/2$. From the foregoing results, we can argue that for the **H**-field lines which are circles centered at $x = |c_2|/2$, $y = 0$, the magnetic vector is in the counter-clockwise direction. Now, let $c_2 > 0$. The associated field lines are circles occupying the region $x < 0$. For $x < 0$, we see from (3.63) that $H_x \gtrless 0$ according to whether $y \gtrless 0$. Similarly, we find from (3.64) that $H_y \gtrless 0$ according as $x \lessgtr -c_2/2$. These results show that for the $\boldsymbol{H}$-field lines which are circles centered at $x = -|c_2|/2$, $y = 0$, the magnetic vector is in the clockwise direction. Consequently, we can obtain the direction of the magnetic field around each current filament by the right-hand rule. If the thumb of the right hand points in the direction of the current, the fingers curl in the direction of the magnetic field. In Fig. 7.6, the directions of the magnetic field are also indicated along the field lines.

We can now replace the current filaments by perfectly conducting cylinders of finite radius. Since the boundary condition is that the normal component of the magnetic field is zero on a perfectly conducting surface, a perfectly conducting surface introduced so as to coincide with an equipotential surface does not alter the fields. Suppose that a perfectly conducting, infinitely long cylindrical surface A of radius a is introduced so as to coincide with the equipotential (positive) cylindrical surface of the same radius a. The existing electric and magnetic fields are unchanged. The electromagnetic fields cannot penetrate a perfect conductor. Therefore, all the fields inside the cylindrical surface A are due to current filament A and those outside it are due to current filament B. If filament A is removed, the fields inside the cylindrical surface vanish and those outside it are unaffected. Since the line integral of the magnetic field taken around a circle is equal to the total current enclosed, it follows that the total axial current on the cylindrical surface A is the same, namely I, as that on filament A. We can fill the region inside the cylindrical surface A with a perfect conductor without affecting the fields both on the exterior and on the interior to it, since the time-varying electric current exists only on the surface of a perfect conductor. In a similar manner, we replace filament B by another perfectly conducting, infinitely long cylinder B of radius a so that its surface coincides with the equipotential (negative) cylindrical surface of the same radius. Again, the current exists only on the surface of the cylinder, with the result that there is no magnetic field inside cylinder B also. The total axial current on the surface of cylinder B is the same as that on filament B and the $\boldsymbol{H}$-field distribution external to cylinder B is also unaffected. Thus, we have obtained a two-wire transmission line as well as the proper distribution of the magnetic field around it. We assume the wires to be perfectly conducting and the current to exist only on their surfaces, as is the case if the current is time-varying. The vector-potential difference W between the two wires is found from (3.48) as

$$W = A_{z1} - A_{z2} = \frac{\mu_d I}{\pi} \cosh^{-1}\left(\frac{b}{2a}\right). \tag{3.65}$$

Let us consider a unit length of the two-wire transmission line lying between the transverse planes $z = z$ and $z = z + 1$. We can consider the conduction current along wire A, the displacement current from wire A to wire B, the conduction current in the reverse direction along wire B and the displacement current directed back from wire B to wire A to form an electric circuit. We wish to find the total magnetic

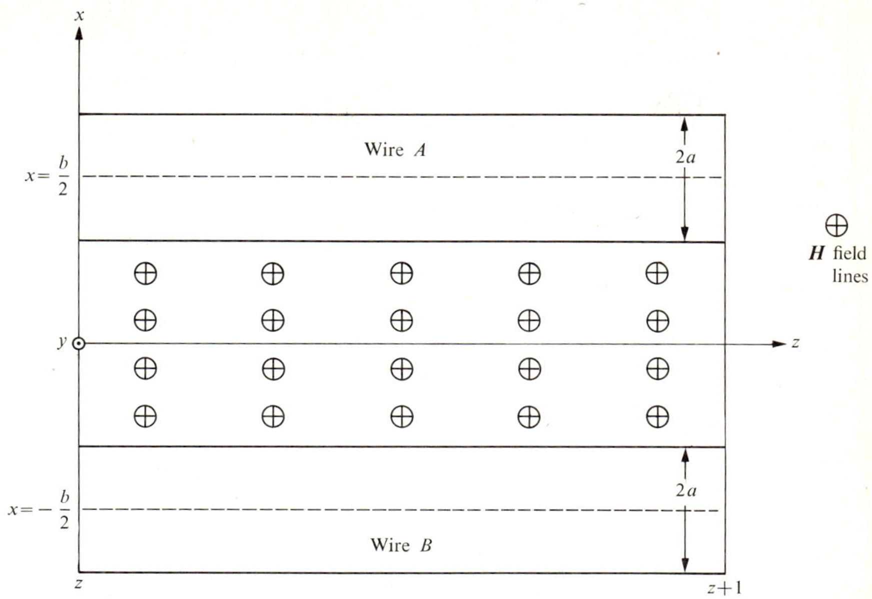

Fig. 7.7 Schematic diagram of a two-wire transmission line for the purpose of evaluating the magnetic flux linkage with unit length of the transmission line.

flux linking with this circuit. For this purpose, we need a surface area bounded by the previously mentioned electric circuit. The simplest of such surfaces is that which lies in the $y = 0$ plane, as shown in Fig. 7.7. We can show from (3.56) that

$$x[x + (c_2/2)] = \tfrac{1}{2}[x^2 - (d/2)^2] < 0$$

for the region between the wires since $|x| < d/2$. Hence, we see the magnetic field penetrating the bounded surface area in the $y = 0$ plane from (3.63) and (3.64) to be in the $-y$-direction. Therefore, we deduce the total magnetic flux linking the electric circuit as

$$\Phi = \int_B^A - \mu_d H_y \, dx, \tag{3.66}$$

where the integration extends from the surface of wire B to that of wire A. Since from (3.50), $H_y = -(1/\mu_d)(\partial A_z/\partial x)$, it follows from (3.66) or (4.4.74) that

$$\Phi = A_{z1} - A_{z2}. \tag{3.67}$$

Therefore, we derive the external inductance L per unit length of the two-wire transmission line from (3.65), (3.67), and (3.27) to be

$$L = \frac{\Phi}{I} = \frac{\mu_d}{\pi} \cosh^{-1}\left(\frac{b}{2a}\right), \tag{3.68a}$$

$$= \frac{\mu_d}{\pi} \ln\left(\frac{b}{a}\right) \qquad (\text{if } a^2/b^2 \ll 1). \tag{3.68b}$$

In the course of the derivation of the capacitance C and the inductance L of unit length of the transmission line, we have obtained the electric and the magnetic field distributions in a plane transverse to the two wires. The field distributions from one transverse plane to the other differ only by the phase term $e^{-j\beta z}$ where $\beta = \omega/v$ and v is the velocity of the plane waves in the medium. From the circuit theory of the transmission line, we obtained the velocity v as

$$v = 1/\sqrt{LC}. \tag{3.69}$$

Then, from (3.45) and (3.68), (3.69) becomes

$$v = 1/\sqrt{\mu_d \varepsilon_d}, \tag{3.70}$$

which verifies the fact that the velocity v of the TEM mode on a transmission line is the same as that of the homogeneous plane wave in the corresponding unbounded medium.

From (3.45), (3.68), and (3.27) we find the characteristic resistance R_0 of the two-wire transmission line to be

$$R_0 = \frac{\eta_d}{\pi} \cosh^{-1}\left(\frac{b}{2a}\right), \tag{3.71a}$$

$$= \frac{\eta_d}{\pi} \ln\left(\frac{b}{a}\right) \qquad (\text{for } a^2/b^2 \ll 1), \tag{3.71b}$$

where $\eta_d = \sqrt{\mu_d/\varepsilon_d}$ is the intrinsic resistance of the dielectric medium.

We can ascertain from (3.42), (3.43), (3.63), and (3.64) that the amplitudes of the electromagnetic fields generated by the two-wire transmission line fall off in the transverse direction as ρ^{-2} for large ρ, where ρ is the radial distance from the observation point to the central axis of the transmission line. Thus as has been alluded to in the previous section, the energy of the TEM wave supported by the two-wire transmission line is concentrated in its close neighborhood.

7.4 EVALUATION OF G AND R OF A TWO-WIRE TRANSMISSION LINE

Let us suppose that the dielectric medium in which the two-wire transmission line is situated, instead of being perfect, has slight losses. This results in a conduction current leaking through the dielectric medium from wire A to wire B. Consequently, in the distributed parameter representation of the two-wire transmission line, there is a shunt conductance to account for the leakage currents. We can deduce the shunt conductance G per unit length of the two-wire transmission line in a simple manner. In a perfect dielectric, the shunt element consists only of a capacitance and therefore we obtain the shunt admittance Y per unit length of a two-wire transmission line from (3.45a) as

$$Y = \frac{j\omega\pi\varepsilon_d}{\cosh^{-1}(b/2a)}. \tag{4.1}$$

When the dielectric becomes lossy, we have to replace the real permittivity ε_d by the complex effective permittivity $(\varepsilon_d + (\sigma_d/j\omega))$ where σ_d is the conductivity of the dielectric. For a lossy dielectric (4.1) becomes

$$Y = \frac{j\omega\pi[\varepsilon_d + (\sigma_d/j\omega)]}{\cosh^{-1}(b/2a)}, \tag{4.2}$$

which has both real and imaginary parts. The real part of (4.2) is the conductance G per unit length of a two-wire transmission line. Hence,

$$G = \frac{\pi\sigma_d}{\cosh^{-1}(b/2a)}, \tag{4.3a}$$

$$\approx \frac{\pi\sigma_d}{\ln(b/a)} \qquad (\text{if } a^2/b^2 \ll 1). \tag{4.3b}$$

On account of the dielectric losses, the field distribution from one transverse plane to the other in the $+z$-direction differs now by the factor $e^{-(\alpha+j\beta)z}$ where α is the attenuation coefficient. In general, for very small losses, the following approximation is valid:

$$\alpha = \alpha_d + \alpha_c, \tag{4.4}$$

where α_d is the contribution to the attenuation coefficient due to the dielectric losses and α_c is that due to the imperfect conductivity of the wires. From the leading term in (2.7.14), we can see that

$$\alpha_d = \frac{G}{2}R_0; \qquad \alpha_c = \frac{R}{2R_0}, \tag{4.5a,b}$$

where R_0 is the characteristic resistance and R, the resistance per unit length of the transmission line, arises due to the imperfect conductivity of the wires. We shall derive an approximate expression for R later in this section. From (4.3) and (3.71), we find that

$$\alpha_d = \sigma_d\eta_d/2. \tag{4.6}$$

It is worthwhile to note that to a first approximation, the contribution α_d to the attenuation coefficient arising from the dielectric losses is independent of the geometrical dimensions of the transmission line.

Let us suppose that the two wires of the transmission line instead of being perfect conductors, are extremely good conductors. The conductivity σ_c of the two wires is extremely large but not infinite. Therefore, the electromagnetic field penetrates into the conductor. The time-varying current instead of being on the surface of the wires, as is the case if the wires are perfect conductors, penetrates into the conductor. Associated with the electromagnetic fields inside the conductor, there is complex power whose value depends on the total axial current along the wire. Consequently, in the distributed parameter representation of the two-wire transmission line, there is a series impedance to account for the complex power inside the conductors. The real part of the complex power represents the power actually dissipated inside the conductors and, therefore, gives rise to R, the series resistance per unit length of the transmission line. Since there is current inside the wire and since the magnetic field penetrates into the conductor, there are magnetic flux linkages with the current inside the conductor giving rise to L_i, the series internal inductance per unit length of the transmission line. Together R and L_i constitute the series impedance which accounts for the complex power inside the conductor. We wish to find the approximate expressions for R and L_i for the two-wire transmission line.

In order for power to flow into the conductors, it is necessary to have an axial component of the electric field, namely E_z, so that E_z together with the component of the transverse magnetic field. H_t, tangential to the surface of the conductors, gives rise to a Poynting vector which is directed into the surface of the conductors. Therefore, when the wires of the two-wire transmission line become imperfectly conducting, a TEM guided wave is no longer possible. The guided wave becomes a transverse magnetic (TM) wave which has a component of $\boldsymbol{E}$ field but not that of $\boldsymbol{H}$ field in the propagation direction. However, we assume the conductivity of the wires to be extremely large though not infinite with the result that it is reasonable to expect that the field distributions around a transmission line with imperfect conductors differ only by a small extent from the corresponding field distributions obtained with perfect conductors. The axial component of the electric field, E_z, is very small compared to the transverse field components. The field distributions from one transverse plane to the other in the $+z$-direction differs by the factor $e^{-(\alpha+j\beta)z}$ where the attenuation coefficient α is due to the power loss in the conductors. Since the inductance L per unit length of the transmission line is increased by the internal inductance L_i also per unit length, it follows that the phase-shift coefficient β is changed slightly when the effect of the imperfect conductivity of the wires is included.

The intensity or the magnitude of the magnetic field is proportional to the density of the field lines, that is the number of field lines passing through a unit area. Since the current density is equal to the tangential component of the magnetic field, it follows from Fig. 7.6 that the axial current density on the surface of a wire of a two-wire transmission line is largest at the location nearest to the second wire and decreases as the distance from the second wire increases. In other words, the axial current density is not uniform around the circumference of the wires. For an isolated cylindrical wire, the existence of the cylindrical symmetry shows that the axial current density is uniform around the circumference of the wire. The nonuniformity in the axial current distribution around the circumference of the wires in a two-wire transmission line arises due to the close presence of the second wire or *the proximity effect.* When we evaluated C and L in the last section, the proximity effect was taken into account. In the approximate evaluation of R and L_i, we neglect the proximity effect. Hence the axial current distribution around the circumference of the wires is uniform, as is the case if the other wire is not present. Consequently, we can evaluate R and L_i for a unit length of a single wire of radius a and since there are two wires in a two-wire transmission line, the actual values of R and L_i are twice those for a single wire.

Let us consider the cross section by a transverse plane of a single wire of radius a as shown in Fig. 7.8 together with the polar coordinates ρ, ϕ, and z. Also, let a unit

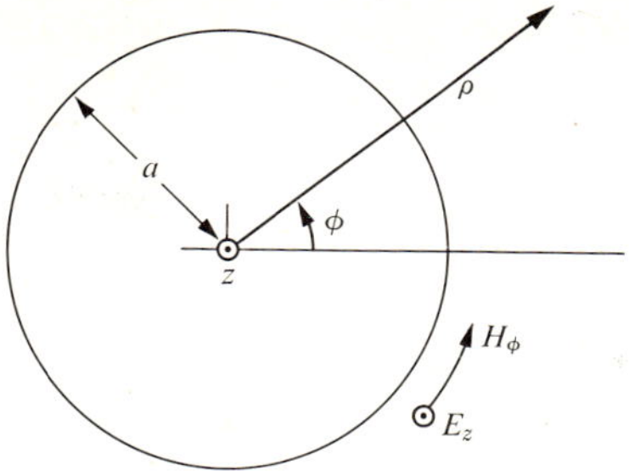

Fig. 7.8 Cross section of a single, imperfectly conducting wire of radius a.

length of the single wire be assumed for the purposes of the present discussion. Let H_ϕ be the tangential component of the magnetic field at the surface of the wire, when the wire is perfectly conducting. If the conductivity of the wire is very large but not infinite, there is a small electric field in the z-direction but H_ϕ is not altered appreciably with the result that the following approximate relation holds

$$E_z/H_\phi = \eta_s, \tag{4.7}$$

where η_s is the surface impedivity. The component of the complex Poynting vector directed into the surface of the wire is

$$-S_\rho = \tfrac{1}{2}E_z H_\phi^* = \tfrac{1}{2}\eta_s H_\phi H_\phi^*. \tag{4.8}$$

If I is the total axial current, we find that

$$2\pi a H_\phi = I. \tag{4.9}$$

The total complex power P_L associated with a unit length of the wire is equal to the total surface area of a unit length of the wire times $-S_\rho$, that is

$$P_L = -2\pi a S_\rho = (2\pi a)\tfrac{1}{2}\eta_s H_\phi H_\phi^*. \tag{4.10}$$

Together with (4.9), (4.10) becomes

$$P_L = \tfrac{1}{2}(\eta_s/2\pi a)II^*. \tag{4.11}$$

From (4.11), we obtain the series impedance $(R + j\omega L_i)$ which accounts for the complex power flow into the wire of radius a as

$$R + j\omega L_i = \eta_s/2\pi a \qquad \text{(single wire)}. \tag{4.12}$$

We assume the radius a of the wire to be very large in comparison with the skin depth that the surface impedivity of the curved surface of the wire can be approximated by that of a plane surface. Hence, (4.12) becomes

$$R + j\omega L_i = \frac{(1 + j)}{2\pi a}\sqrt{\frac{\omega\mu_c}{2\sigma_c}} \qquad \text{(single wire)}, \tag{4.13}$$

where μ_c and σ_c are the permeability and the conductivity of the wires respectively. The separation of the real and the imaginary parts of (4.13) yields

$$R = \frac{1}{2\pi a}\sqrt{\frac{\omega\mu_c}{2\sigma_c}} \qquad \text{(single wire)}, \tag{4.14}$$

and

$$L_i = \frac{1}{\omega 2\pi a}\sqrt{\frac{\omega\mu_c}{2\sigma_c}} \qquad \text{(single wire)}. \tag{4.15}$$

Since a two-wire transmission line consists of two wires of the same radius a, the values of R and L_i pertaining to a unit length of a two-wire transmission line are twice those given in (4.14) and (4.15). Therefore, we obtain the series resistance R and the series internal inductance L_i per unit length of a two-wire transmission line as

$$R = \frac{1}{\pi a}\sqrt{\frac{\omega\mu_c}{2\sigma_c}}, \tag{4.16}$$

and

$$L_i = \frac{1}{\omega\pi a}\sqrt{\frac{\omega\mu_c}{2\sigma_c}}. \tag{4.17}$$

We deduce the attenuation coefficient α_c arising from the imperfect conductivity of the wires from (4.5b), (3.71), and (4.16) as

$$\alpha_c = \frac{1}{2a}\sqrt{\frac{\omega\mu_c}{2\sigma_c}}\frac{1}{\eta_d \cosh^{-1}(b/2a)}, \tag{4.18a}$$

$$= \frac{1}{2a}\sqrt{\frac{\omega\mu_c}{2\sigma_c}}\frac{1}{\eta_d \ln(b/a)} \qquad (\text{if } a^2/b^2 \ll 1). \tag{4.18b}$$

Note that in contrast to α_d, α_c depends on the geometrical parameters of the two-wire transmission line.

7.5 TEM MODE IN A COAXIAL TRANSMISSION LINE

Let us consider an infinitely long coaxial transmission line. Its cross section by a transverse plane is shown in Fig. 7.2. We assume both the inner and the outer walls to be perfect conductors. The coaxial dielectric region is able to support a TEM guided mode whose properties are investigated in this section. The electric $\boldsymbol{E}(\boldsymbol{r})$ and the magnetic $\boldsymbol{H}(\boldsymbol{r})$ vectors satisfy the time-harmonic Maxwell's equations (2.1) and (2.2). We assume that the field components do not vary in the azimuthal direction, that is, $\partial/\partial\phi = 0$. If the field components have azimuthal variations, then we can show that a TEM guided wave cannot be obtained. When the ρ, ϕ, and z components of (2.1) and (2.2) are written down and $\partial/\partial\phi$ is set equal to zero, we obtain the following results:

$$-j\omega\mu_d H_\rho(\rho, z) = -\frac{\partial}{\partial z} E_\phi(\rho, z), \tag{5.1}$$

$$-j\omega\mu_d H_\phi(\rho, z) = \frac{\partial}{\partial z} E_\rho(\rho, z) - \frac{\partial}{\partial \rho} E_z(\rho, z), \tag{5.2}$$

$$-j\omega\mu_d H_z(\rho, z) = \frac{1}{\rho}\frac{\partial}{\partial \rho}(\rho E_\phi(\rho, z)), \tag{5.3}$$

$$j\omega\varepsilon_d E_\rho(\rho, z) = -\frac{\partial}{\partial z} H_\phi(\rho, z), \tag{5.4}$$

$$j\omega\varepsilon_d E_\phi(\rho, z) = \frac{\partial}{\partial z} H_\rho(\rho, z) - \frac{\partial}{\partial \rho} H_z(\rho, z), \tag{5.5}$$

$$j\omega\varepsilon_d E_z(\rho, z) = \frac{1}{\rho}\frac{\partial}{\partial \rho}(\rho H_\phi(\rho, z)). \tag{5.6}$$

For a TEM wave

$$E_z(\rho, z) = H_z(\rho, z) = 0. \tag{5.7}$$

From (5.7) and (5.3), we can show that

$$E_\phi(\rho, z) = E_{\phi 0}(z)/\rho, \tag{5.8}$$

where $E_{\phi 0}(z)$ is a function of z only. For $\rho = a$ and $\rho = b$, since it is tangential to the surface of a perfect conductor, $E_\phi(\rho, z)$ should vanish. This boundary condition can be fulfilled by (5.8) only if $E_{\phi 0}(z)$ and therefore $E_\phi(\rho, z)$ vanish. Then, from (5.1), it follows that $H_\rho(\rho, z)$ also vanishes. Thus, for a TEM guided wave, in addition to (5.7), we have proved that

$$E_\phi(\rho, z) = H_\rho(\rho, z) = 0. \tag{5.9}$$

The only nonvanishing field components are $H_\phi(\rho, z)$ and $E_\rho(\rho, z)$. It is convenient for us to rewrite (5.2) together with (5.7), and (5.4) as follows:

$$-\frac{\partial}{\partial z} E_\rho(\rho, z) = j\omega\mu_d H_\phi(\rho, z), \tag{5.10}$$

$$-\frac{\partial}{\partial z} H_\phi(\rho, z) = j\omega\varepsilon_d E_\rho(\rho, z). \tag{5.11}$$

We can see with the help of (5.6) and (5.7) that

$$H_\phi(\rho, z) = H_{\phi 0}(z)/\rho, \tag{5.12}$$

where $H_{\phi 0}(z)$ is a function of z only. From (5.10) and (5.12), it follows that

$$E_\rho(\rho, z) = E_{\rho 0}(z)/\rho, \tag{5.13}$$

where $E_{\rho 0}(z)$ is also a function of z only. Substituting (5.12) and (5.13) in (5.10) and (5.11) we obtain

$$-\frac{d}{dz}E_{\rho 0}(z) = j\omega\mu_d H_{\phi 0}(z), \tag{5.14}$$

$$-\frac{d}{dz}H_{\phi 0}(z) = j\omega\varepsilon_d E_{\rho 0}(z). \tag{5.15}$$

It is worth noting that (5.14) and (5.15) are very similar to the transmission-line equations for the current and the voltage.

On differentiating (5.15) with respect to z and then making use of (5.14), we find that

$$\frac{d^2}{dz^2}H_{\phi 0}(z) + \beta^2 H_{\phi 0}(z) = 0, \tag{5.16}$$

where

$$\beta = \omega\sqrt{\mu_d\varepsilon_d}. \tag{5.17}$$

The general solution of (5.16) is given by

$$H_{\phi 0}(z) = H_1 e^{-j\beta z} + H_2 e^{j\beta z}, \tag{5.18}$$

where H_1 and H_2 are constants. The first term on the right-hand side of (5.18) is a wave progressing in the $+z$-direction and the second term represents a wave progressing in the $-z$-direction. Considering only the wave progressing in the $+z$-direction, we find from (5.17), (5.18), and (5.15) that

$$E_{\rho 0}(z) = \eta_d H_{\phi 0}(z). \tag{5.19}$$

With the help of (5.12), (5.13), (5.18), and (5.19), for a wave progressing in the $+z$-direction, we obtain the field components of the TEM guided wave in a coaxial transmission line as

$$H_\phi(\rho, z) = \frac{H_1}{\rho}e^{-j\beta z}, \tag{5.20}$$

$$E_\rho(\rho, z) = \frac{\eta_d H_1}{\rho}e^{-j\beta z} = \frac{E_1}{\rho}e^{-j\beta z}. \tag{5.21}$$

We see from (5.17) that the propagation coefficient of a TEM guided wave in a coaxial transmission line is the same as that of a homogeneous plane wave in the corresponding unbounded dielectric. Also, it is clear from (5.20) and (5.21) that the electric field $E_\rho(\rho, z)$, the magnetic field $H_\phi(\rho, z)$, and the direction of propagation z form a mutually perpendicular triad. We have thus shown that the coaxial transmission line supports a TEM guided wave and the properties of this wave are the same as those for a general TEM guided wave, as we have already enumerated in Section 7.2.

7.6 EVALUATION OF C AND L OF A COAXIAL TRANSMISSION LINE

We obtain the surface charge density ρ_s on the inner surface of the outer wall $\rho = b$ from (5.21) as

$$\rho_s = -\hat{\rho} \cdot \hat{\rho}\varepsilon_d \frac{E_1}{b} e^{-j\beta z} = -\frac{\varepsilon_d E_1}{b} e^{-j\beta z}. \tag{6.1}$$

Therefore, the charge per unit length of the outer wall is

$$q(z) = -2\pi b \frac{\varepsilon_d E_1}{b} e^{-j\beta z} = -qe^{-j\beta z} \qquad (\text{for } \rho = b), \tag{6.2}$$

where

$$q = 2\pi\varepsilon_d E_1. \tag{6.3}$$

In a similar manner, we can deduce the charge per unit length of the inner wall as

$$q(z) = qe^{-j\beta z} \qquad (\text{for } \rho = a). \tag{6.4}$$

In any transverse plane, we see the amplitudes of the charge per unit length on the outer and the inner conductors from (6.2) and (6.4) to be equal in magnitude and opposite in sign as on the two wires of a two-wire transmission line.

We can deduce the surface current density $\boldsymbol{J}_s$ on the inner surface of the outer wall $\rho = b$ from (5.20) as

$$\boldsymbol{J}_s = -\hat{\rho} \times \hat{\phi} \frac{H_1}{b} e^{-j\beta z} = -\hat{z} \frac{H_1}{b} e^{-j\beta z}. \tag{6.5}$$

Therefore, the total axial current on the outer wall is

$$I(z) = -2\pi b \frac{H_1}{b} e^{-j\beta z} = -Ie^{-j\beta z} \qquad (\text{for } \rho = b), \tag{6.6}$$

where

$$I = 2\pi H_1. \tag{6.7}$$

Similarly, we can obtain the total axial current on the inner wall as

$$I(z) = Ie^{-j\beta z} \qquad (\text{for } \rho = a). \tag{6.8}$$

We find the total axial currents on the outer and the inner conductors in any transverse plane from (6.6) and (6.8) to be equal in magnitude and opposite in direction. In terms of $I(z)$ in (6.7) and (6.8), we can rewrite (5.20) as

$$H_\phi(\rho, z) = I(z)/2\pi\rho, \tag{6.9}$$

where $I(z)$ is the current on the inner conductor.

We shall show now that for the TEM guided wave, a unique voltage difference between the two conductors can be defined. Let us consider any closed path $ABCDA$

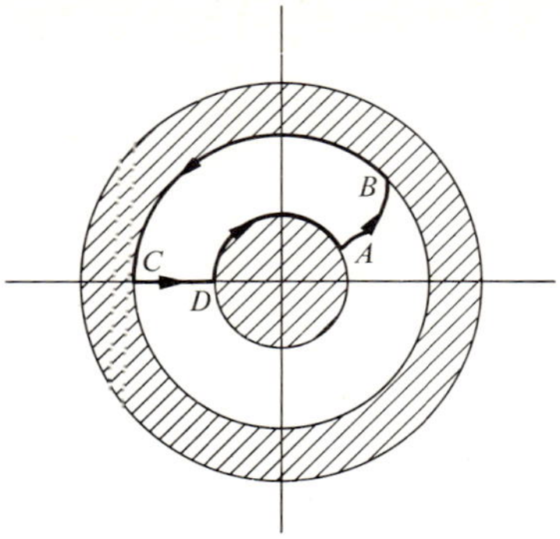

Fig. 7.9 Cross section of a coaxial transmission line showing the path *ABCDA* for the purpose of evaluating the voltage difference between the inner and the outer conductors.

in the transverse plane joining the inner and the outer conductors as shown in Fig. 7.9. The line integral of the electric field taken around the closed contour *ABCDA* is equal to the negative of the time rate of change of the total magnetic flux through the surface area enclosed by the contour. Since $H_z = 0$ for a TEM guided wave, there is no magnetic flux through the surface area enclosed by the path *ABCDA* and, hence, the line integral of the electric field taken around the closed contour *ABCDA* is zero:

$$\int_A^B \mathbf{E} \cdot d\mathbf{s} + \int_B^C \mathbf{E} \cdot d\mathbf{s} + \int_C^D \mathbf{E} \cdot d\mathbf{s} + \int_D^A \mathbf{E} \cdot d\mathbf{s} = 0. \tag{6.10}$$

The second and the fourth integrals in (6.10) vanish, since the electric field tangential to the surface of a perfect conductor is zero. Therefore, (6.10) yields

$$\int_A^B \mathbf{E} \cdot d\mathbf{s} = \int_D^C \mathbf{E} \cdot d\mathbf{s}. \tag{6.11}$$

Since the path *ABCDA* is arbitrarily chosen, it follows from (6.11) that the line integral of the electric field taken from any point on the inner conductor to another point on the outer conductor in the same transverse plane has a unique value independent of the path from the inner to the outer conductor. This unique value of the line integral of the electric field is the voltage difference between the two conductors. For the purpose of ensuring proper sign conventions for the current and the voltage, we may identify schematically the inner and the outer conductors of a coaxial transmission line with the upper and the lower conductors in a two-wire transmission line shown in Fig. 1.2. Since the lower conductor in a two-wire transmission line is taken as the reference for the voltage, it follows that the line integral of the electric field taken from the *inner* to the *outer* conductor in a coaxial transmission line gives the voltage difference $V(z)$ with the proper sign convention. Hence, from (5.21)

$$V(z) = \int_a^b \frac{E_1}{\rho} e^{-j\beta z} = E_1 \ln\left(\frac{b}{a}\right) e^{-j\beta z}. \tag{6.12}$$

In terms of $V(z)$ as given in (6.12), we can write $E_\rho(\rho, z)$ in (5.21) as

$$E_\rho(\rho, z) = \frac{V(z)}{\rho \ln (b/a)}. \tag{6.13}$$

Eliminating $E_\rho(\rho, z)$ and $H_\phi(\rho, z)$ from (5.10) and (5.11) by their equivalent expressions in terms of the voltage $V(z)$ and the current $I(z)$ as given in (6.13) and (6.9) respectively, we get after some rearrangement

$$-\frac{\partial}{\partial z} V(z) = j\omega\left[\frac{\mu_d \ln (b/a)}{2\pi}\right] I(z), \tag{6.14}$$

$$-\frac{\partial}{\partial z} I(z) = j\omega\left[\frac{2\pi\varepsilon_d}{\ln (b/a)}\right] V(z). \tag{6.15}$$

If we compare (6.14) and (6.15) with the equations for the voltage and the current on a lossless transmission line in terms of the inductance L and the capacitance C per unit length we obtain the following expressions for L and C of a coaxial transmission line in terms of the geometrical parameters:

$$L = \frac{\mu_d \ln (b/a)}{2\pi}, \tag{6.16}$$

$$C = \frac{2\pi\varepsilon_d}{\ln (b/a)}. \tag{6.17}$$

From (6.16) and (6.17), we find the characteristic resistance of a coaxial transmission line to be

$$R_0 = \sqrt{\frac{L}{C}} = \frac{\eta_d \ln (b/a)}{2\pi}. \tag{6.18}$$

Also, we obtain the velocity of propagation v as

$$v = 1/\sqrt{LC} = 1/\sqrt{\mu_d \varepsilon_d}, \tag{6.19}$$

which is, as it should be, the same as that of the homogeneous plane wave in the corresponding unbounded dielectric medium.

7.7 EVALUATION OF G AND R OF A COAXIAL TRANSMISSION LINE

If the coaxial dielectric is lossy, there is conduction current leaking from one conductor to the other with the result that in the distributed parameter representation of the coaxial transmission line, there is a shunt conductance whose value G per unit length we can determine using the same technique as was employed in the evaluation of G of a two-wire transmission line. We obtain the shunt admittance per unit length of a coaxial transmission line with a perfect dielectric from (6.17), then replace the real permittivity ε_d by the complex effective permittivity $[\varepsilon_d + (\sigma_d/j\omega)]$ to account for the dielectric losses, and identify the real part of the resulting shunt admittance as the conductance G per unit length of the coaxial transmission line. We then obtain the following result:

$$G = \frac{2\pi\sigma_d}{\ln(b/a)}. \tag{7.1}$$

We derive an expression for α_d, the contribution to the attenuation coefficient due to the dielectric losses with the help of (4.5a), (6.18), and (7.1) and the result is the same as that given in (4.6) which, we should note, is independent of the geometrical parameters of the transmission line.

Let us suppose that the conductors of the coaxial transmission line are not perfect, that is, the conductivity of the inner and the outer conductors is extremely large but not infinite. Then, in the equivalent circuit representation of the coaxial transmission line, there arise the series resistance R, and the series internal inductance L_i, both per unit length. We should remember that as soon as the walls of the coaxial transmission line become imperfectly conducting, a TEM guided wave is no longer possible since an axial component E_z of the electric field is present to account for the power transport into the conductors. We can derive approximate expressions for R and L_i in terms of the geometrical parameters of the coaxial transmission line using the same techniques as were employed in connection with the two-wire transmission line. On account of the cylindrical symmetry, the axial current density in each conductor is uniform around the circumference with the result that there is no need to neglect the proximity effect as was done for the case of a two-wire transmission line. We assume that the radius of the inner conductor as well as the thickness of the outer conductor are very large in comparison with the skin depth. Consequently, we can use the development contained in Section 7.4 for the evaluation of R and L_i for a single wire to find the corresponding quantities of a coaxial transmission line. The contributions to R and L_i due to the imperfect conductivity of the inner conductor (radius a) are the same as those given in (4.14) and (4.15) respectively, whereas we evaluate the contributions arising from the imperfect conductivity of the outer conductor (radius b) from (4.14) and (4.15) respectively by replacing a by b. Therefore, for a coaxial transmission line, we obtain the following results:

$$R = \frac{1}{2\pi}\sqrt{\frac{\omega\mu_c}{2\sigma_c}}\left(\frac{1}{a} + \frac{1}{b}\right), \tag{7.2}$$

and

$$L_i = \frac{1}{\omega 2\pi}\sqrt{\frac{\omega\mu_c}{2\sigma_c}}\left(\frac{1}{a} + \frac{1}{b}\right). \tag{7.3}$$

We deduce the attenuation coefficient α_c arising from the imperfect conductivity of the conductors from (4.5b), (6.18), and (7.2) as

$$\alpha_c = \frac{1}{2}\sqrt{\frac{\omega\mu_c}{2\sigma_c}}\left(\frac{1}{a} + \frac{1}{b}\right)\frac{1}{\eta_d \ln(b/a)}, \tag{7.4}$$

which unlike α_d depends on the geometrical parameters of the coaxial transmission line.

7.8 TEM MODE IN A PARALLEL-PLATE TRANSMISSION LINE

The cross section by a transverse plane of an infinitely long parallel-plate transmission line is shown in Fig. 7.3. The parallel plates extend from $-\infty$ to $+\infty$ both in the y- and in the z-directions. We assume the propagation to take place in the z-direction. We assume both the lower and the upper plates to be perfect conductors. The dielectric slab region is able to support a TEM guided wave whose electric $\boldsymbol{E}(\boldsymbol{r})$ and magnetic $\boldsymbol{H}(\boldsymbol{r})$ vectors satisfy the time-harmonic Maxwell's equations (2.1) and (2.2). We assume that the field components do not vary in the y-direction, that is, $\partial/\partial y = 0$. This condition is necessary in order to be able to obtain a TEM guided wave. When we write down the x, y, and z components of (2.1) and (2.2) and set $\partial/\partial y$ equal to zero, we obtain the following results:

$$-j\omega\mu_d H_x(x, z) = -\frac{\partial}{\partial z} E_y(x, z), \tag{8.1}$$

$$-j\omega\mu_d H_y(x, z) = \frac{\partial}{\partial z} E_x(x, z) - \frac{\partial}{\partial x} E_z(x, z), \tag{8.2}$$

$$-j\omega\mu_d H_z(x, z) = \frac{\partial}{\partial x} E_y(x, z), \tag{8.3}$$

$$j\omega\varepsilon_d E_x(x, z) = -\frac{\partial}{\partial z} H_y(x, z), \tag{8.4}$$

$$j\omega\varepsilon_d E_y(x, z) = \frac{\partial}{\partial z} H_x(x, z) - \frac{\partial}{\partial x} H_z(x, z), \tag{8.5}$$

$$j\omega\varepsilon_d E_z(x, z) = \frac{\partial}{\partial x} H_y(x, z). \tag{8.6}$$

For a TEM wave

$$E_z(x, z) = H_z(x, z) = 0. \tag{8.7}$$

From (8.7) and (8.3), we can derive that

$$E_y(x, z) = E_y(z), \tag{8.8}$$

which is independent of x. For $x = 0$ and $x = a$, since it is tangential to the surface of a perfect conductor, $E_y(x, z)$ should vanish. This boundary condition can be fulfilled by (8.8) only if $E_y(z)$ and $E_y(x, z)$ vanish. Then, from (8.1), we see that $H_x(x, z)$ also vanishes. Thus, for a TEM guided wave, in addition to (8.7), we have deduced that

$$E_y(x, z) = H_x(x, z) = 0. \tag{8.9}$$

The only nonvanishing field components are $E_x(x, z)$ and $H_y(x, z)$. It is convenient for us to rewrite (8.2) together with (8.7), and (8.4) as follows:

$$-\frac{\partial}{\partial z} E_x(x, z) = j\omega\mu_d H_y(x, z), \tag{8.10}$$

$$-\frac{\partial}{\partial z} H_y(x, z) = j\omega\varepsilon_d E_x(x, z). \tag{8.11}$$

We can verify with the help of (8.6) and (8.7) that

$$H_y(x, z) = H_y(z), \tag{8.12}$$

which is independent of x. From (8.10) and (8.12), we see that

$$E_x(x, z) = E_x(z), \tag{8.13}$$

which is also independent of x. If we substitute (8.12) and (8.13) in (8.10) and (8.11) we obtain

$$-\frac{d}{dz} E_x(z) = j\omega\mu_d H_y(z), \tag{8.14}$$

and

$$-\frac{d}{dz} H_y(z) = j\omega\varepsilon_d E_x(z), \tag{8.15}$$

which are very similar to the transmission-line equations for the voltage and the current.

On differentiating (8.15) with respect to z and then making use of (8.14), we obtain

$$\frac{d^2}{dz^2} H_y(z) + \beta^2 H_y(z) = 0, \tag{8.16}$$

where,

$$\beta = \omega\sqrt{\mu_d\varepsilon_d}. \tag{8.17}$$

The general solution of (8.16), as in (5.18), consists of two waves, one progressing in the $+z$- and the other in the $-z$-directions respectively. Considering only the wave progressing in the $+z$-direction, we determine the solution of (8.16) as

$$H_y(z) = H_y(x, z) = H_1 e^{-j\beta z}. \tag{8.18}$$

From (8.17), (8.18), and (8.15), we can show that

$$E_x(z) = E_x(x, z) = \eta_d H_1 e^{-j\beta z} = E_1 e^{-j\beta z}. \tag{8.19}$$

Thus, (8.18) and (8.19) are the field components of the TEM mode in a parallel-plate transmission line.

The propagation coefficient of the TEM guided wave in a parallel-plate transmission line is seen from (8.17) to be the same as that of a homogeneous plane wave in the corresponding unbounded dielectric. Also, it follows from (8.18) and (8.19) that the electric field $E_x(x, z)$, the magnetic field $H_y(x, z)$ and the direction of propagation z form a mutually perpendicular triad. Hence, we have verified the TEM guided wave supported by a parallel-plate transmission line as having all the general properties of any TEM guided wave.

7.9 DETERMINATION OF C AND L OF A PARALLEL-PLATE TRANSMISSION LINE

The parallel-plate transmission line is uniform in the y-direction. For the purpose of determining the parameters of the equivalent circuit representation of the parallel-plate transmission line, we only consider a width b of the plates in the y-direction. We obtain the surface charge density ρ_s on the lower surface of the upper plate from (8.19) to be

$$\rho_s = -\hat{x} \cdot \hat{x}\varepsilon_d E_1 e^{-j\beta z} = -\varepsilon_d E_1 e^{-j\beta z} \qquad (\text{for } x = a). \tag{9.1}$$

Therefore, for a width b and for unit length in the z-direction, the charge on the lower surface of the upper plate is

$$q(z) = -b\varepsilon_d E_1 e^{-j\beta z} = -qe^{-j\beta z} \qquad (\text{for } x = a), \tag{9.2}$$

where

$$q = b\varepsilon_d E_1. \tag{9.3}$$

In a similar manner, we can derive the charge per unit length on the upper surface of the lower plate as

$$q(z) = qe^{-j\beta z} \qquad (\text{for } x = 0). \tag{9.4}$$

In any transverse plane, we see the amplitudes of the charge per unit length on the upper and the lower plates from (9.2) and (9.4) to be equal in magnitude and opposite in sign as on the two wires of a two-wire transmission line or the two conductors of a coaxial transmission line.

We can deduce the surface current density $\boldsymbol{J}_s$ on the lower surface of the upper plate from (8.18) to be

$$\boldsymbol{J}_s = -\hat{x} \times \hat{y}H_1 e^{-j\beta z} = -\hat{z}H_1 e^{-j\beta z} \qquad (\text{for } x = a). \tag{9.5}$$

Therefore, the total axial current on the upper plate is

$$I(z) = -bH_1 e^{-j\beta z} = -Ie^{-j\beta z} \qquad (\text{for } x = a), \tag{9.6}$$

where

$$I = bH_1. \tag{9.7}$$

In a similar manner, we obtain the total axial current on the upper surface of the lower plate as

$$I(z) = Ie^{-j\beta z} \qquad (\text{for } x = 0). \tag{9.8}$$

As in the case of the charge, we also find the total axial currents on the upper and the lower conductors in any transverse plane from (9.6) and (9.8) to be equal in magnitude and opposite in direction. With the help of (9.7) and (9.8), we can rewrite (8.18) in terms of the axial current in the lower plate as

$$H_y(x, z) = I(z)/b. \tag{9.9}$$

As in the case of a coaxial transmission line, we can prove that there is a unique voltage difference between the upper and the lower plates of a parallel-plate transmission line supporting the TEM guided wave. For the purpose of obtaining the proper sign conventions, for the current and the voltage, we may identify the lower and the upper plates with the upper and the lower conductors of a two-wire transmission line shown in Fig. 1.2. Since the lower conductor in a two-wire transmission line is taken as the reference for the voltage, it follows that the line integral of the electric field taken from the lower to the upper plate by a path in the transverse plane gives the voltage difference $V(z)$ with the proper sign convention. Hence, from (8.19)

$$V(z) = \int_0^a E_1 e^{-j\beta z}\, dx = aE_1 e^{-j\beta z}. \tag{9.10}$$

In terms of $V(z)$ as given in (9.10), we can rewrite $E_x(x, z)$ in (8.19) as

$$E_x(x, z) = V(z)/a. \tag{9.11}$$

When (9.9) and (9.11) are substituted in (8.10) and (8.11), we obtain the following results after some rearrangement:

$$-\frac{\partial}{\partial z} V(z) = j\omega\left(\frac{\mu_d a}{b}\right) I(z), \tag{9.12}$$

and

$$-\frac{\partial}{\partial z} I(z) = j\omega\left(\frac{\varepsilon_d b}{a}\right) V(z). \tag{9.13}$$

If we compare (9.12) and (9.13) with the transmission-line equations we find the following expressions for L and C of a parallel-plate transmission line in terms of the geometrical parameters:

$$L = \mu_d \frac{a}{b}, \tag{9.14}$$

and

$$C = \varepsilon_d \frac{b}{a}. \tag{9.15}$$

From (9.14) and (9.15), we can see that the characteristic resistance of a parallel-plate transmission line is given by

$$R_0 = \sqrt{\frac{L}{C}} = \eta_d \frac{a}{b}. \tag{9.16}$$

Also, we obtain the velocity of propagation v as

$$v = 1/\sqrt{LC} = 1/\sqrt{\mu_d \varepsilon_d}. \tag{9.17}$$

We should note that (9.14) to (9.16) correspond only to a width b of the parallel-plate transmission line. In order for (9.14) to (9.16) to be strictly correct, the parallel plates should extend from $-\infty$ to $+\infty$ in the y-direction, even though we evaluate L, C, and R_0 only for a width b. For a finite width of the parallel plates, the fields in the y-direction are no longer uniform. In particular, near the ends of the plates, the fields are quite different from those given in (8.18) and (8.19), resulting in what is known as fringe effects. Consequently, (9.14) to (9.16) are not strictly correct for a finite-width parallel-plate transmission line. However, if b is very much larger in comparison with a, the overall effect of the fringing fields is small and (9.14) to (9.16) are very good approximations for the corresponding parameters of a parallel-plate transmission line of finite width b.

7.10 DETERMINATION OF G AND R OF A PARALLEL-PLATE TRANSMISSION LINE

If the dielectric slab is lossy, there is conduction current leaking from one parallel plate to the other. Therefore, in the equivalent circuit representation of the parallel-plate transmission line, there is a shunt conductance whose value G per unit length we can determine using the same technique as we employed in the evaluation of G for two-wire and coaxial transmission lines. With the help of (9.15), we can find G for a parallel-plate transmission line to be

$$G = \sigma_d \frac{b}{a}. \tag{10.1}$$

We can then derive an expression for α_d with the help of (4.5a), (9.16), and (10.1) and the result is the same as that given in (4.6).

Let us suppose that the conductivity of the upper and the lower plates is extremely large but not infinite. Then, the series resistance R and the series internal inductance L_i arise. As in the case of the two-wire and the coaxial transmission lines, so in the case of the parallel-plate transmission line a TEM guided wave is not possible when the plates are imperfectly conducting, since an axial component E_z of the electric field is present to account for the power transport into the conductors. We can derive approximate expressions for R and L_i corresponding to a parallel-plate transmission line in the following manner.

Let us consider a unit length of the parallel-plate transmission line of width b. Let H_y be the tangential component of the magnetic field at the upper surface of the lower plate when the plate is perfectly conducting. If the conductivity of the plate is very large but not infinite, there is a small electric field in the z-direction but H_y is not changed appreciably. If η_s is the surface impedivity, approximately

$$E_z/H_y = \eta_s. \tag{10.2}$$

The component of the complex Poynting vector directed into the surface of the lower plate is

$$-S_x = \tfrac{1}{2}E_z H_y^* = \tfrac{1}{2}\eta_s H_y H_y^*. \tag{10.3}$$

If I is the total axial current in the lower plate, we see that

$$bH_y = I. \tag{10.4}$$

The total complex power P_L associated with a unit length of the plate is equal to the total surface area per unit length of the lower plate multiplied by $-S_x$, that is

$$P_L = -bS_x = \frac{b}{2}\eta_s H_y H_y^*. \tag{10.5}$$

Together with (10.4), (10.5) becomes

$$P_L = \frac{1}{2}\frac{\eta_s}{b} II^*. \tag{10.6}$$

From (10.6), we see the series impedance that accounts for the complex power flow into the lower plate of width b and a unit length to be given by η_s/b. In a similar manner, we can show the series impedance that accounts for the complex power flow into the upper plate of width b and a unit length to be equal to the same quantity η_s/b with the result that the series impedance per unit length of the parallel-plate transmission line of width b is given by

$$R + j\omega L_i = 2\frac{\eta_s}{b}. \tag{10.7}$$

Since

$$\eta_s = (1 + j)\sqrt{\frac{\omega\mu_c}{2\sigma_c}},$$

it follows from (10.7) that

$$R = \frac{2}{b}\sqrt{\frac{\omega\mu_c}{2\sigma_c}}, \tag{10.8}$$

$$L_i = \frac{2}{\omega b}\sqrt{\frac{\omega\mu_c}{2\sigma_c}}. \tag{10.9}$$

We can deduce the attenuation coefficient α_c arising from the imperfect conductivity of the parallel plates from (4.5b), (9.16), and (10.8) as

$$\alpha_c = \frac{1}{\eta_d a}\sqrt{\frac{\omega\mu_c}{2\sigma_c}}. \tag{10.10}$$

We should note the interesting fact that α_c for a parallel-plate transmission line is independent of the width b to within the approximation under which (9.16) and (10.8) are valid.

PROBLEMS

7.1 A two-wire transmission line consists of a pair of parallel wires A_1 and A_2 of circular cross section. The radii of wires A_1 and A_2 are given by a_1 and a_2 respectively and b is the separation distance between the axes of the wires. These wires are in an otherwise unbounded dielectric region with the constitutive parameters μ_d and ε_d. Find the capacitance C of unit length of the transmission line. Assume that the wires are perfect conductors.

For this purpose, consider the two parallel, infinitely long filaments A and B situated along the lines $x = d/2$, $y = 0$, $-\infty < z < \infty$ and $x = -d/2$, $y = 0$, $-\infty < z < \infty$ as treated in Section 7.3. The equipotential cylindrical surface with the axis along the line

$$x = x_1 = \frac{d\,(K_1^2 + 1)}{2\,(K_1^2 - 1)},$$

$$y = y_1 = 0,$$

$$-\infty < z < \infty$$

and

$$\text{radius } a_1 = \frac{dK_1}{K_1^2 - 1},$$

where $K_1 > 1$ is assumed to coincide with the surface of wire A_1 and the equipotential cylindrical surface with the axis along the line

$$x = x_2 = -\frac{d\,(1 + K_2^2)}{2\,(1 - K_2^2)},$$

$$y = y_2 = 0,$$

$$-\infty < z < \infty$$

and

$$\text{radius } a_2 = \frac{dK_2}{1 - K_2^2},$$

where $K_2 < 1$ is assumed to coincide with the surface of wire A_2. By setting

$$x_1/a_1 = \cosh u_1 \quad \text{and} \quad -x_2/a_2 = \cosh u_2, \tag{a}$$

show that the potentials on wires A_1 and A_2 are given respectively by

$$\phi_1 = \frac{q}{2\pi\varepsilon_d} u_1; \qquad \phi_2 = -\frac{q}{2\pi\varepsilon_d} u_2. \tag{b}$$

Verify the following relations connecting a_1, a_2, and b to u_1 and u_2:

$$a_1 \sinh u_1 - a_2 \sinh u_2 = 0, \tag{c}$$

$$a_1 \cosh u_1 + a_2 \cosh u_2 = b. \tag{d}$$

With the help of (c) and (d) show that

$$u_1 = \cosh^{-1} \psi_1 \quad \text{and} \quad u_2 = \cosh^{-1} \psi_2 \tag{e}$$

where

$$\psi_1 = \frac{b^2 + a_1^2 - a_2^2}{2a_1 b}; \qquad \psi_2 = \frac{b^2 + a_2^2 - a_1^2}{2a_2 b}. \tag{f}$$

Deduce that the capacitance C of unit length of the transmission line is given by

$$C = 2\pi\varepsilon_d/[\cosh^{-1} \psi_1 + \cosh^{-1} \psi_2]. \tag{g}$$

For widely separated wires for which the conditions $b^2 \gg a_1^2$ and $b^2 \gg a_2^2$ are satisfied, show that

$$C = \frac{\pi \varepsilon_d}{\ln (b/\sqrt{a_1 a_2})}. \tag{h}$$

7.2 For the two-wire transmission line of Problem 7.1, show that the inductance L of unit length of the transmission line is given by

$$L = \frac{\mu_d}{2\pi} [\cosh^{-1} \psi_1 + \cosh^{-1} \psi_2].$$

If the wires are widely separated, for which the conditions $b^2 \gg a_1^2$ and $b^2 \gg a_2^2$ are satisfied, show that

$$L = \frac{\mu_d}{\pi} \ln \left(\frac{b}{\sqrt{a_1 a_2}} \right).$$

7.3 For the two-wire transmission line of Problem 7.1, deduce that the conductance G of unit length of the transmission line is given by

$$G = 2\pi\sigma_d/[\cosh^{-1} \psi_1 + \cosh^{-1} \psi_2],$$

where σ_d is the conductivity of the lossy dielectric region.

For the widely spaced wires, show that

$$G = \frac{\pi \sigma_d}{\ln (b/\sqrt{a_1 a_2})}.$$

7.4 Suppose that the wires of the two-wire transmission line of Problem 7.1 are imperfectly conducting. Let μ_c and σ_c be the permeability and the conductivity of the wires respectively. Deduce the expressions for the resistance R and the internal inductance L_i of unit length of the transmission line.

7.5 Prove that the capacitance C, the inductance L, and the conductance G of unit length of the two-wire transmission line of Problem 7.1 can be expressed as follows:

$$C = 2\pi\varepsilon_d/\cosh^{-1} \psi, \tag{a}$$

$$L = \frac{\mu_d}{2\pi} \cosh^{-1} \psi, \tag{b}$$

$$G = 2\pi\sigma_d/\cosh^{-1} \psi, \tag{c}$$

where,

$$\psi = \frac{b^2 - a_1^2 - a_2^2}{2a_1 a_2}.$$

Suppose that the two wires are of equal radii, that is $a_1 = a_2 = a$. Then, show that the expressions for C, L, and G as given by (a), (b), and (c) respectively reduce to the corresponding expressions appropriate to a two-wire transmission line with the two wires having equal radii, as treated in Sections 7.3 and 7.4.

7.6 A wire P of circular cross section of radius a is situated at a distance $b/2$ from a perfectly conducting infinite plane. The wire P and its image Q form a two-wire transmission line. We wish to find the capacitance C and the inductance L of unit length of the transmission line. Let μ_d and ε_d be the constitutive parameters of the medium in front of the perfectly conducting plane.

For this purpose, consider the two parallel, infinitely long filaments A and B situated along the lines $x = d/2$, $y = 0$, $-\infty < z < \infty$, and $x = -d/2$, $y = 0$, $-\infty < z < \infty$, as treated in Section 7.3. The equipotential cylindrical surface with the axis along the line

$$x = x_1 = \frac{d\,(K_1^2 + 1)}{2\,(K_1^2 - 1)},$$

$$y = y_1 = 0,$$

$$-\infty < z < \infty$$

and

$$\text{radius } a = \frac{dK_1}{K_1^2 - 1},$$

where $K_1 > 1$, is assumed to coincide with the surface of the wire P and the equipotential surface $x = 0$ is assumed to coincide with the perfectly conducting plane. The potential on the perfectly conducting plane is zero. Find the potential on the wire P in terms of the charge q coulombs per unit length, b and a. Hence, show that the capacitance C per unit length of the wire is

$$C = \frac{2\pi\varepsilon_d}{\cosh^{-1}(b/2a)}. \tag{a}$$

In a similar manner, show that the inductance L per unit length of the wire is

$$L = \frac{\mu_d}{2\pi}\cosh^{-1}\left(\frac{b}{2a}\right). \tag{b}$$

If the dielectric is lossy and σ_d is the conductivity of the dielectric, prove that the conductance G per unit length of the wire is

$$G = \frac{2\pi\sigma_d}{\cosh^{-1}(b/2a)}. \tag{c}$$

Obtain the approximations to C, L, and G as given by (a) to (c) respectively for $a^2/b^2 \ll 1$.

7.7 A shielded transmission line has an eccentric inner conductor A_1 which is a solid cylinder of radius a_1. The outer conductor is a hollow cylinder A_2 of inner radius a_2 and D is the distance between the axes of the two cylinders. Let μ_d and ε_d be the constitutive parameters of the medium between the two cylinders. Find the expressions for the transmission line parameters C and L.

For this purpose, consider two parallel infinitely long filaments A and B situated along the lines $x = d/2$, $y = 0$, $-\infty < z < \infty$ and $x = -d/2$, $y = 0$, $-\infty < z < \infty$ as treated in Section 7.3. The equipotential cylindrical surface with the axis along the line

$$x = x_1 = \frac{d\,(K_1^2 + 1)}{2\,(K_1^2 - 1)},$$

$$y = y_1 = 0,$$

$$-\infty < z < \infty$$

and

$$\text{radius } a_1 = \frac{dK_1}{K_1^2 - 1},$$

where $K_1 > 1$ is assumed to coincide with the surface of wire A_1 and the equipotential surface with the axis along the line

$$x = x_2 = \frac{d\,(K_1'^2 + 1)}{2\,(K_1'^2 - 1)},$$

$$y = y_2 = 0,$$

$$-\infty < z < \infty$$

and

$$\text{radius } a_2 = \frac{dK_1'}{K_1'^2 - 1},$$

where $K_1 > K_1' > 1$ is assumed to coincide with the surface of the hollow cylinder A_2. With the help of a procedure similar to that indicated in Problem 7.1, deduce that the capacitance C of unit length of the shielded transmission line is given by

$$C = 2\pi\varepsilon_d/[\cosh^{-1}\psi_1 - \cosh^{-1}\psi_2],$$
$$= 2\pi\varepsilon_d/\cosh^{-1}\psi,$$

where,

$$\psi_1 = \frac{a_2^2 - a_1^2 - D^2}{2a_1 D}; \qquad \psi_2 = \frac{a_2^2 - a_1^2 + D^2}{2a_2 D}; \qquad \psi = \frac{a_2^2 + a_1^2 - D^2}{2a_1 a_2}.$$

In a similar manner, show that the inductance L of unit length of the shielded transmission line is given by

$$L = \frac{\mu_d}{2\pi}\cosh^{-1}\psi.$$

If the dielectric is lossy and σ_d is the conductivity of the dielectric, prove that the conductance G of unit length of the shielded transmission line is given by

$$G = 2\pi\sigma_d/\cosh^{-1}\psi.$$

7.8 Show that the expressions for C, L, and G given in Problem 7.7 for a shielded transmission line with an eccentric inner conductor reduce to the appropriate expressions pertaining to a coaxial transmission line if the separation distance D between the axes of the two cylinders vanishes.

7.9 Show that for a coaxial transmission line with an outer conductor of fixed inside radius r_0, the attenuation coefficient due to the conductor losses is a minimum when $r_0/r_i = 3.6$, where r_i is the outside radius of the inner conductor. What is the characteristic impedance of such a coaxial transmission line with the minimum attenuation? Investigate the minimum, if any, of the attenuation coefficient of the coaxial transmission line in which the outside radius r_i of the inner conductor is fixed.

7.10 A two-wire transmission line is situated in a medium which has the following constitutive parameters: $\mu_d = 4\pi \times 10^{-7}$ H/m and $\varepsilon_d = 10^{-9}/36\pi$ F/m. The wires are of equal radius: $a = 0.15$ cm and the separation distance between the axes of the two wires $b = 2.5$ cm. Determine the capacitance C and the inductance L of unit length of the transmission line by using the exact hyperbolic expressions. Find the values of C and L by using the approximate logarithmic expressions. Find the percentage error in the values of C and L if the approximate expressions are used.

If the wires are made of copper which has the following constitutive parameters:

$$\mu_c = 4\pi \times 10^{-7}\ \text{H/m} \qquad \text{and} \qquad \sigma_c = 5.8 \times 10^7\ \text{mho/m},$$

evaluate the resistance R and the internal inductance L_i per unit length of the transmission line as a function of the frequency f.

7.11 A two-wire transmission line is situated in a medium which has the following constitutive parameters: $\mu_d = 4\pi \times 10^{-7}$ H/m, $\varepsilon_d = (5 \times 10^{-9})/36\pi$ F/m and $\sigma_d = 10^{-5}$ mho/m. The wires are of unequal radius; the radii of the two wires are given by $a_1 = 0.10$ cm and $a_2 = 0.20$ cm respectively. The separation distance between the axes of the two wires $b = 2.5$ cm. Determine the capacitance C, the inductance L, and the conductance G of unit length of the transmission line by using (i) the exact hyperbolic expressions and (ii) the approximate logarithmic expressions.

7.12 A wire of circular cross section of radius $a = 0.20$ cm is situated at a distance $b/2 = 1.5$ cm from a perfectly conducting infinite plane. The constitutive parameters of the medium in front of the perfectly conducting plane are given by $\mu_d = 4\pi \times 10^{-7}$ H/m, $\varepsilon_d = 10^{-9}/36\pi$ F/m, and $\sigma_d = 10^{-3}$ mho/m. Evaluate the capacitance C, the inductance L, and the conductance G of unit length of the wire.

7.13 A coaxial transmission line is composed of copper conductors and has the following dimensions: inside radius of outer tube = 1.00 cm; radius of inner conductor = 0.32 cm. The constitutive parameters of copper are given by $\mu_c = 4\pi \times 10^{-7}$ H/m and $\sigma_c = 5.8 \times 10^7$ mho/m. The constitutive parameters of the dielectric region are as follows: $\mu_d = 4\pi \times 10^{-7}$ H/m and $\varepsilon_d = (2.25 \times 10^{-9})/36\pi$ F/m.

For a frequency of 10^8 Hz, compute the capacitance C, the inductance L, and the resistance R of unit length of the coaxial transmission line. Find the value of the attenuation coefficient α in nepers per meter.

7.14 A shielded transmission line has an eccentric inner conductor whose radius is $a_1 = 0.32$ cm. The inside radius of the hollow outer cylinder is $a_2 = 1.00$ cm. The separation distance between the axes of the two cylinders is $D = 0.30$ cm. The constitutive parameters of the dielectric region are given by $\mu_d = 4\pi \times 10^{-7}$ H/m and $\varepsilon_d = 10^{-9}/36\pi$ F/m. Assume that the conductors are made of silver whose constitutive parameters are given by $\mu_c = 4\pi \times 10^{-7}$ H/m and $\sigma_c = 6.2 \times 10^7$ mho/m.

For a frequency of 10^7 Hz, find the capacitance C, the inductance L, and the resistance R of unit length of the transmission line. Determine the values of the characteristic impedance and the propagation coefficient.

Repeat the calculations for the corresponding coaxial transmission line for which the separation distance D between the axes of the two cylinders is zero.

7.15 A parallel-plate transmission line is made of copper strips of width $b = 6.0$ cm. The spacing between the strips is $a = 0.5$ cm. The constitutive parameters of the dielectric region between the plates are given by $\mu_d = 4\pi \times 10^{-7}$ H/m and $\varepsilon_d = 10^{-9}/36\pi$ F/m. The constitutive parameters of copper are given by $\mu_c = 4\pi \times 10^{-7}$ H/m and $\sigma_c = 5.8 \times 10^7$ mho/m. Find the capacitance C, the inductance L, and the resistance R of unit length of the transmission line. Evaluate the characteristic impedance and the propagation coefficient. Assume that the frequency of operation is $f = 10^8$ Hz.

7.16 An infinitely long cylindrical wire of radius a carries direct current. The permeability and the conductivity of the material of the wire are given by μ_c and σ_c respectively. Find the value of the resistance R and the internal inductance L_i of unit length of the wire.

7.17 A two-wire transmission line is situated in a medium which has the following constitutive parameters: $\mu_d = 4\pi \times 10^{-7}$ H/m and $\varepsilon_d = (2.25 \times 10^{-9})/36\pi$ F/m. The wires are of equal radius: $a = 0.06$ cm and the separation distance between the axes of the two wires $b = 2.0$ cm. Determine the capacitance C and the inductance L of unit length of the transmission line.

If the wires are made of copper which has the following constitutive parameters: $\mu_c = 4\pi \times 10^{-7}$ H/m and $\sigma_c = 5.8 \times 10^7$ mho/m, evaluate the resistance R and the internal inductance L_i of unit length of the transmission line. Assume that the frequency of operation is given by $f = 10^9$ Hz.

Evaluate the characteristic impedance and the propagation coefficient.

7.18 A coaxial transmission line is composed of brass conductors and has the following dimensions: inside radius of outer tube = 2.00 cm; radius of inner conductor = 0.85 cm. The constitutive parameters of brass are given by $\mu_c = 4\pi \times 10^{-7}$ H/m and $\sigma_c = 1 \times 10^7$ mho/m. The constitutive parameters of the dielectric region are as follows: $\mu_d = 4\pi \times 10^{-7}$ H/m and $\varepsilon_d = 10^{-9}/9\pi$ F/m.

For a frequency of 2×10^8 Hz, determine the capacitance C, the inductance L, and the resistance R of unit length of the coaxial transmission line. Find the values of the propagation coefficient and the characteristic impedance.

7.19 A parallel-plate transmission line is made of brass strips of width $b = 8.0$ cm. The spacing between the strips $a = 0.6$ cm. The constitutive parameters of the dielectric region between the parallel plates are given by $\mu_d = 4\pi \times 10^{-7}$ H/m, $\varepsilon_d = 10^{-9}/12\pi$ F/m, and $\sigma_d = 10^{-3}$ mho/m. The constitutive parameters of brass are given by $\mu_c = 4\pi \times 10^{-7}$ H/m and $\sigma_c = 1 \times 10^7$ mho/m. The frequency of operation is 3×10^8 Hz. Find the capacitance C, the inductance L, the conductance G, and the resistance R of unit length of the transmission line. Determine the values of the characteristic impedance and the propagation coefficient.

7.20 Consider a two-wire transmission line as depicted in Fig. 7.1. Show that the amplitudes of the fields fall off in the transverse direction as ρ^{-2} for large ρ, where ρ is the distance from the observation point to the z-axis.

7.21 For the TEM mode in a coaxial transmission line as treated in Section 7.5, determine the time-averaged energy densities $W_E(\rho)$ and $W_H(\rho)$ associated with the electric and the magnetic fields respectively. Obtain the spatial averages $\overline{W}_E$ and $\overline{W}_H$ of $W_E(\rho)$ and $W_H(\rho)$ over the cross section of the coaxial transmission line. Determine the time-averaged power flux density $S_z(\rho)$ and its spatial average $\bar{S}_z$ over the cross section of the coaxial transmission line. Show that the ratio $\bar{S}_z/(\overline{W}_E + \overline{W}_H)$ has the dimensions of a velocity and find its value.

7.22 Consider a two-wire transmission line as illustrated in Fig. 7.1. Assume that the separation distance b between the axes of the two wires is fixed and is large enough compared to the radius a of the wires to justify the use of the approximate logarithmic expressions for the capacitance C and the inductance L of unit length of the transmission line. Find the value of the radius a of the wires for which the attenuation coefficient α_c arising from the imperfect conductivity of the wires is a minimum. What is the characteristic impedance of the transmission line with the minimum attenuation, if the transmission line is situated in free space?

7.23 The inside radius b of the outer conductor of a coaxial transmission line is a fixed constant. If V is the potential difference between the inner and the outer conductors and a is the outside radius of the inner conductor, find the expression for the electric field. Determine the maximum value of the electric field and show that it occurs at the surface of the inner conductor. Find the value of the outside radius of the inner conductor for which the strength of the electric field at the surface of the inner conductor is a minimum. Note that the design of the inner conductor in the foregoing manner leads to the highest voltage for the occurrence of the breakdown of the dielectric.

7.24 The exact expressions for the capacitance C and the inductance L per unit length of the two-wire transmission line are given by (3.45a) and (3.68a) respectively. Note that a is the radius of the wires and b is the separation distance between the axes of the wires. The corresponding approximate expressions valid for $a^2/b^2 \ll 1$ are given by (3.45b) and (3.68b) respectively. Examine the difference in the physical process corresponding to the exact and the approximate expressions for C and L.

7.25 Consider an infinitely long, perfectly conducting cylinder of radius a and an infinitely long filamentary charge of q C/m situated parallel to the axis of the conducting cylinder, and at a radial distance ρ from the axis. From the developments contained in Section 7.3, show that the image of the filamentary charge is also a filamentary charge of $-q$ C/m situated parallel to the axis of the conducting cylinder, and at a radial distance a^2/ρ from the axis.

CHAPTER 8

GUIDED ELECTROMAGNETIC WAVES

8.1 INTRODUCTION

We have examined the characteristics of plane electromagnetic waves in an unbounded medium and in the presence of plane interfaces. We have also analyzed the electromagnetic theory of some common types of transmission lines for the transverse electromagnetic mode of propagation. Only the two-wire and the coaxial transmission line have been considered in addition to the parallel-plate waveguide. For the TEM mode of propagation, the characteristics of the fields propagated along these transmission lines bear a close resemblance to those of the plane electromagnetic waves in an unbounded medium. In this chapter, we shall develop the general theory of the guided waves.

In addition to the TEM mode, the parallel-plate waveguide can support the transverse electric (TE) and the transverse magnetic (TM) modes of propagation. We shall synthesize the field components associated with these modes of propagation with the use of the properly chosen plane wave fields and discuss their characteristics. We shall then determine the attenuation of the waves resulting from the dielectric losses and the imperfect conductivity of the walls of the waveguide.

We shall treat the theory of the rectangular waveguide in detail. We shall show that a rectangular waveguide cannot support a TEM mode of propagation. We shall then deduce the field components associated with the TE and the TM modes of propagation in a rectangular waveguide directly from the solutions of Maxwell's equations subject to the appropriate boundary conditions on the walls of the waveguide. We shall examine the phase, the group, and the energy transport velocities of propagation as well as the surface charge and the surface current densities on the walls of the waveguide. As we did in the case of the parallel-plate waveguide, so for the case of the rectangular waveguide we shall obtain the expressions for the attenuation coefficient due to the dielectric and the conductor losses.

We shall treat a simple case of an "open" waveguide in the form of a planar dielectric slab giving consideration to both the TE and the TM surface waves. We shall analyze the characteristics of the TE surface waves directly from Maxwell's equations as well as the appropriate boundary conditions at the interfaces, and synthesize those of the TM surface waves from the pertinent plane wave fields.

With a view to bringing out the essential characteristics of the fields sustained by electromagnetic cavity resonators, we shall give a detailed theory of the rectangular cavity resonator and obtain the expressions for the resonant frequencies and the associated field patterns for the case of the TE mode. Finally, we shall consider the damping of the modes due to the imperfect conductivity of the walls of the cavity resonator and deduce an expression for the resulting selectivity of the resonator.

8.2 PARALLEL-PLATE WAVEGUIDE: TE MODE

Let us consider two-dimensional fields which are independent of y. Then, we can separate the electromagnetic fields into two independent sets. The first set consists of $E_y(x, z)$, $H_x(x, z)$, and $H_z(x, z)$, which are governed by (6.6.4) and (6.6.5). The second set comprises of $H_y(x, z)$, $E_x(x, z)$, and $E_z(x, z)$ and these are governed by (6.6.27) to (6.6.29). Let z be the preferred direction. As we shall clarify later, z is the direction of propagation. Since the electric field is entirely transverse to the direction of propagation, we designate the first set as the TE mode. In a similar manner, since the magnetic field is entirely transverse to the direction of propagation, we call the second set the TM mode. Now, let us examine the characteristics of the TE mode propagating in a lossless dielectric region between two parallel, perfectly conducting infinite plates. We should point out that the characteristics of the modes sustained by a parallel-plate waveguide are obtained here by a synthesis of the constituent plane wave fields.

Let $x = 0$ be a perfectly conducting, infinite plane. We deduced in Section 6.6 the TE wave which can exist in the region $x > 0$ and which is a solution of the time-harmonic Maxwell's equations subject to the boundary condition at $x = 0$. The field components of such a TE wave were found as

$$E_y(x, z) = 2jE_{y0}^i \sin(\beta x \cos\theta^i)e^{-j\beta z \sin\theta^i}, \tag{2.1}$$

$$H_x(x, z) = -2j\frac{\sin\theta^i}{\eta_0}E_{y0}^i \sin(\beta x \cos\theta^i)e^{-j\beta z \sin\theta^i}, \tag{2.2}$$

$$H_z(x, z) = -2\frac{\cos\theta^i}{\eta_0}E_{y0}^i \cos(\beta x \cos\theta^i)e^{-j\beta z \sin\theta^i}. \tag{2.3}$$

In (2.1) to (2.3), $\beta = \omega\sqrt{\mu\varepsilon} = \omega/v$, $\eta_0 = \sqrt{\mu/\varepsilon}$ and these parameters depend on the frequency and the macroscopic properties of the dielectric in the region $x > 0$. We should note that E_{y0}^i is an arbitrary constant. For the manner in which it is defined (see Fig. 6.3), θ^i can have any value from $-\pi/2$ to $\pi/2$. Since $\sin\theta^i > 0$ for $0 < \theta^i < \pi/2$, it follows that if the fields (2.1) to (2.3) are restricted to represent waves progressing in the $+z$-direction, θ^i is limited to have any value only in the range $0 < \theta^i < \pi/2$. We should note that θ^i can assume *arbitrarily* any value in the range $0 < \theta^i < \pi/2$.

1. Synthesis of the Waveguide Fields

The factor $e^{-j\beta z \sin\theta^i}$ represents the z-dependence of the fields. In the z-direction, the fields have constant amplitudes but have the same linear variations in phase with the result that the fields form a progressive wave in the z-direction. The factor $\sin(\beta x \cos\theta^i)$ represents the x-dependence of $E_y(x, z)$. In the x-direction, the magnitude of $E_y(x, z)$ varies periodically from 0 to 1 and the phase does not vary linearly with x but remains constant except at some discrete values of x where the phase changes abruptly by π radians. For the values of x such that $\sin(\beta x \cos\theta^i) = 0$, that is, for $x = m\pi/\beta\cos\theta^i$ where m is a positive integer, the magnitude of $E_y(x, z)$ is zero. We should note that at these values of x, $E_y(x, z) = 0$ for all times. Therefore, at these values of x, we can introduce perfectly conducting planes without disturbing the fields. For a given frequency, β is fixed and, therefore, the planes where $E_y(x, z) = 0$ can be changed by varying θ^i. Suppose that at $x = d$ we introduce another perfectly conducting, infinite plane. Therefore, at $x = d$, $E_y(x, z)$ is required to vanish. This requirement cannot be satisfied for all values of θ^i in the range $0 < \theta^i < \pi/2$. Consequently, the admissible values of θ^i are limited. If θ_g^i denotes the admissible values of θ^i, then it follows that

$$\beta d \cos\theta_g^i = m\pi \qquad (m = 1, 2, \ldots). \tag{2.4}$$

We can obtain from (2.4) that

$$\beta_z = \beta\sin\theta_g^i = \sqrt{\frac{\omega^2}{v^2} - \left(\frac{m\pi}{d}\right)^2}. \tag{2.5}$$

We should note that β_z is the component of the phase-shift coefficient in the z-direction.

The field components given by (2.1) to (2.3) with the additional requirement that $\theta^i = \theta_g^i$ as specified by (2.4), are the solutions of Maxwell's equations in the region $0 < x < d$ and they also satisfy the boundary conditions at the planes $x = 0$ and $x = d$. Consequently, (2.1) to (2.3) with $\theta^i = \theta_g^i$ are the complete expressions for the field components of the TE mode sustained in the dielectric region between the two perfectly conducting, parallel planes. Moreover, if the sources of the fields are situated within the parallel-plate waveguide, since no electromagnetic fields can penetrate a perfect conductor, there are no electromagnetic fields exterior to ($x < 0$ and $x > d$) the two parallel planes.

With the help of (2.4) and (2.5), we can express the field components of the TE mode in a parallel-plate waveguide as given by (2.1) to (2.3) conveniently as follows:

$$E_y(x, z) = 2jE_{y0}^i \sin\left(\frac{m\pi x}{d}\right)e^{-j\beta_z z}, \tag{2.6}$$

$$H_x(x, z) = -\frac{2j\beta_z}{\eta_0\beta}E_{y0}^i \sin\left(\frac{m\pi x}{d}\right)e^{-j\beta_z z}, \tag{2.7}$$

$$H_z(x, z) = -\frac{2m\pi}{\eta_0\beta d}E_{y0}^i \cos\left(\frac{m\pi x}{d}\right)e^{-j\beta_z z}. \tag{2.8}$$

For a given frequency, if $\omega/v > m\pi/d$, β_z is real and we see that (2.6) to (2.8) form a progressive wave in the z-direction. If $\omega/v < m\pi/d$, β_z is imaginary and hence we can recast β_2 into the form

$$\beta_z = -j\sqrt{\left(\frac{m\pi}{d}\right)^2 - \frac{\omega^2}{v^2}} = -j\alpha_z, \tag{2.9}$$

where α_z is real. Then, the z-dependence of the fields in (2.6) to (2.8) becomes of the form $e^{-\alpha_z z}$, which shows that the fields are attenuated with increasing z. In (2.9), if we use $+j$ in front of the radical, the z-dependence of the fields becomes of the form $e^{\alpha_z z}$. These fields grow with increasing z and since they become infinite for $z = \infty$, they are physically unacceptable. Consequently, $+j$ should not be used in front of the radical in (2.9). The character of the wave changes as β_z changes from real to imaginary and therefore the value of ω at which $\beta_z = 0$ is called the critical or the cut-off angular frequency ω_c, which we can see from (2.5) to be

$$\omega_{c,m} = m\pi v/d. \tag{2.10}$$

2. Brillouin Diagram (ω *versus* β_z *diagram*)

For a given ω, β_z determines the character of the wave and it is therefore useful for us to depict an ω versus β_z diagram. This diagram is also called the Brillouin diagram. In Fig. 8.1, β_z and α_z as defined by (2.5) and (2.9) are shown as functions of the angular frequency ω for three values of m. The curve β is also shown. We should note that the ω versus β curve is a straight line whose slope $\omega/\beta = v$ is the velocity of the electromagnetic waves in an unbounded dielectric having the same constitutive parameters as the medium between the parallel plates. For a straight line joining any point to the right of the ω-β straight line, the slope is less than v and therefore the corresponding phase velocity is less than v. Hence, the region to the right of the

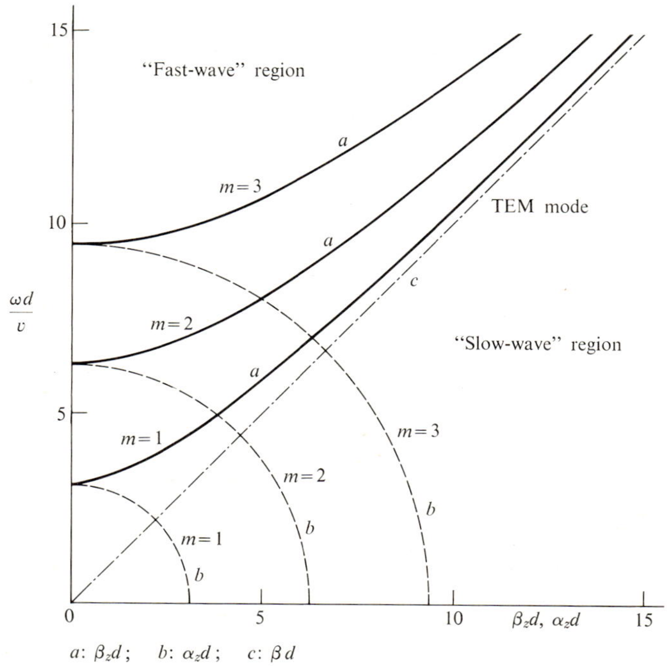

Fig. 8.1 Brillouin diagram for the first three modes guided by a parallel-plate waveguide.

ω-β straight line is called the "slow-wave" region. In a similar manner, the slopes of straight lines joining the origin to points on the left of the ω-β straight line are greater than v and hence, the region to the left of the ω-β straight line is called the "fast-wave" region. We can see that the ω-β_z curves all lie in the fast-wave region, emphasizing the fact that the modes of propagation inside a parallel-plate waveguide are all fast waves. We also observe that all the ω-β_z curves approach the ω-β curve at very high frequencies, leading to the fact that the characteristics of propagation inside a waveguide approach those in an unbounded dielectric in the limit of very high frequencies. Also Fig. 8.1 shows that the attenuation coefficient α_z increases monotonically as ω is decreased below the respective critical angular frequency $\omega_{c,m}$. In Fig. 8.1 β_z, β, and α_z are normalized by multiplying them by d and ω is normalized by multiplying it by d/v.

3. Evanescent and Propagating Modes

We obtain the x-component of the time-averaged Poynting vector from (2.6) and (2.8) as

$$S_x(x) = \tfrac{1}{2} \operatorname{Re} E_y(x, z) H_z^*(x, z),$$
$$= \tfrac{1}{2} \operatorname{Re}\left[-\frac{4jm\pi}{\eta_0 \beta d} \sin\left(\frac{m\pi x}{d}\right) \cos\left(\frac{m\pi x}{d}\right) |E_{yo}^i|^2 \, e^{-j(\beta_z - \beta_z^*)z}\right] = 0, \quad (2.11)$$

which is true irrespective of whether β_z is real or imaginary. Similarly, we find the z-component of the time-averaged Poynting vector from (2.6) and (2.7) to be

$$S_z(x) = -\tfrac{1}{2} \operatorname{Re} E_y(x, z) H_x^*(x, z),$$
$$= \frac{2\,|E_{yo}^i|^2}{\eta_c \beta} \sin^2\left(\frac{m\pi x}{d}\right) \operatorname{Re}\left[\beta_z^* e^{-j(\beta_z - \beta_z^*)z}\right]. \quad (2.12)$$

It is convenient for us to distinguish the following two cases: (1) $\omega < \omega_{c,m}$ and (2) $\omega > \omega_{c,m}$. For $\omega < \omega_{c,m}$, β_z is imaginary as given by (2.9) and $S_z(x) = 0$. For $\omega > \omega_{c,m}$, β_z is real and (2.12) becomes

$$S_z(x) = \frac{2\beta_z\,|E_{yo}^i|^2}{\eta_0 \beta} \sin^2\left(\frac{m\pi x}{d}\right). \quad (2.13)$$

By examining the Poynting vector we have revealed the following important feature of the waveguide mode. For $\omega < \omega_{c,m}$, the fields decay exponentially in the z-direction and therefore do not transport any time-averaged power in the z-direction. There are no losses in the dielectric region or on the bounding walls of the waveguide. Yet, the fields decay exponentially in the z-direction. Consequently, the exponential decay of the fields arises due to kinematic reasons and not due to dissipation. Such exponentially decaying fields are called *evanescent* modes. On the other hand, for $\omega > \omega_{c,m}$, the fields have phase progression in the z-direction and therefore yield transport of time-averaged power in the z-direction. Consequently, for $\omega > \omega_{c,m}$

the fields are said to belong to *propagating* modes. Thus, the critical angular frequency marks the transition from evanescence to propagation.

4. Wavelength, Phase and Group Velocities

We can obtain the phase velocity $v_{ph,z}$ and the wavelength λ_z in the direction of propagation from (2.5) and (2.10) as

$$v_{ph,z} = \frac{\omega}{\beta_z} = \frac{\omega}{\sqrt{(\omega/v)^2 - (m\pi/d)^2}} = \frac{v}{\sqrt{1 - (\omega_{c,m}/\omega)^2}}, \tag{2.14}$$

$$\lambda_z = \frac{2\pi}{\beta_z} = \frac{2\pi}{\sqrt{(\omega/v)^2 - (m\pi/d)^2}} = \frac{\lambda}{\sqrt{1 - (\omega_{c,m}/\omega)^2}}, \qquad \lambda = \frac{2\pi v}{\omega}, \tag{2.15}$$

where v and λ are the phase velocity and the wavelength in the corresponding unbounded dielectric. The expressions (2.14) and (2.15) are valid for $\omega > \omega_{c,m}$ since only for that frequency range is there propagation. Both $v_{ph,z}$ and λ_z are infinite at the cut-off or the critical angular frequency $\omega = \omega_{c,m}$, and decrease monotonically to their respective values in a corresponding unbounded dielectric as ω is increased beyond $\omega_{c,m}$. The group velocity v_g as defined in Section 2.13 is also in the z-direction and we can evaluate its magnitude as

$$v_{g,z} = \frac{\partial\omega}{\partial\beta_z} = 1\bigg/\frac{d\beta_z}{d\omega} = v\sqrt{1 - \left(\frac{\omega_{c,m}}{\omega}\right)^2}. \tag{2.16}$$

The group velocity $v_{g,z} = 0$ for $\omega = \omega_{c,m}$ and increases monotonically to the value v as ω is increased beyond $\omega_{c,m}$. We should note from (2.14) and (2.16) that $v_{ph,z}v_{g,z} = v^2$, which is a constant. In Fig. 8.2, $v_{ph,z}/v$ and $v_{g,z}/v$ are depicted as functions of $\omega/\omega_{c,m}$. We see that $(d/d\omega)v_{ph,z} < 0$ and therefore, $v_{g,z} < v_{ph,z}$ with the result that inside the waveguide *normal dispersion* is obtained. A medium is said to be dispersive if the phase velocity is a function of frequency. A lossless *unbounded* dielectric is nondispersive as we may verify from (5.5.31). It is interesting to note that, in general, the same lossless dielectric when bounded by parallel plates becomes dispersive.

5. Energy Transport Velocity

The power flux density $S_z(x)$ in (2.13) varies in the cross section of the parallel-plate waveguide. It is useful for us to introduce the spatial average of $S_z(x)$ as defined by

$$\bar{S}_z = \frac{1}{d}\int_0^d S_z(x)\, dx. \tag{2.17}$$

If (2.13) is substituted in (2.17), we obtain

$$\bar{S}_z = \frac{2\beta_z\,|E_{y0}^i|^2}{\eta_0\beta}\,\frac{1}{d}\int_0^d \sin^2\left(\frac{m\pi x}{d}\right) dx = \frac{\beta_z\,|E_{y0}^i|^2}{\eta_0\beta}. \tag{2.18}$$

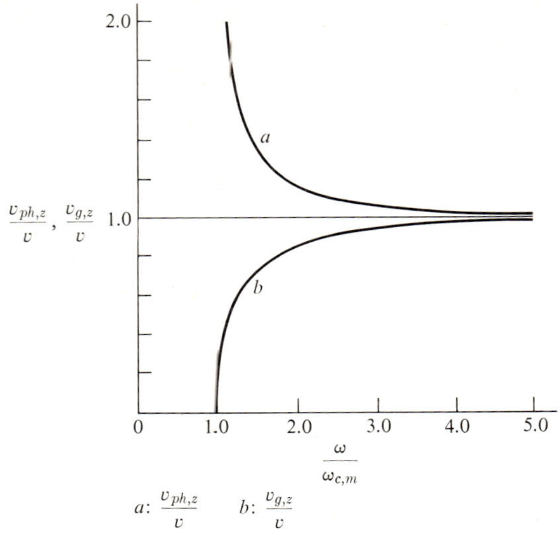

Fig. 8.2 Variation of the normalized phase velocity and the normalized group velocity as a function of the normalized frequency.

We can show from (2.6) that the time-averaged energy density $W_E(x)$ associated with the electric field is

$$W_E(x) = \frac{\varepsilon}{4} |E_y(x, z)|^2 = \varepsilon |E^i_{y0}|^2 \sin^2 (m\pi x/d). \tag{2.19}$$

As in (2.17), we find the spatial average of (2.19) to be

$$\overline{W_E} = \frac{1}{d} \int_0^d W_E(x)\, dx = \frac{\varepsilon |E^i_{y0}|^2}{2}. \tag{2.20}$$

We obtain the time-averaged energy density $W_H(x)$ associated with the magnetic field from (2.7) and (2.8) as

$$\begin{aligned} W_H(x) &= \frac{\mu}{4} |H(x, z)|^2 = \frac{\mu}{4} [|H_x(x, z)|^2 + |H_z(x, z)|^2], \\ &= \frac{\mu |E^i_{y0}|^2}{\eta_0^2 \beta^2} \left[\beta_z^2 \sin^2 \left(\frac{m\pi x}{d}\right) + \left(\frac{m\pi}{d}\right)^2 \cos^2 \left(\frac{m\pi x}{d}\right)\right]. \end{aligned} \tag{2.21}$$

With the help of (2.5), we obtain the spatial average of (2.21) as

$$\begin{aligned} \overline{W_H} &= \frac{1}{d} \int_0^d W_H(x)\, dx = \frac{\mu}{2\eta_0^2 \beta^2} |E^i_{y0}|^2 \left[\beta_z^2 + \left(\frac{m\pi}{d}\right)^2 \right], \\ &= \frac{\varepsilon}{2} |E^i_{y0}|^2, \end{aligned} \tag{2.22}$$

since $\eta_0^2 = \mu/\varepsilon$ and $\beta^2 = \omega^2/v^2 = \beta_z^2 + (m\pi/d)^2$. It is interesting to note that

$\overline{W}_E = \overline{W}_H$. The energy transport velocity v_E is defined as the ratio of the averaged (over both time and space) power flux density $\bar{S}_z$ to the averaged total energy density $\overline{W} = \overline{W}_E + \overline{W}_H$. Hence, from (2.18), (2.20), and (2.22), we can show that

$$v_E = \frac{\bar{S}_z}{\overline{W}_E + \overline{W}_H} = \frac{\beta_z}{\eta_0 \beta \varepsilon} = v\sqrt{1 - \left(\frac{\omega_{c,m}}{\omega}\right)^2}. \tag{2.23}$$

We can see the energy transport velocity as given in (2.23) to be identical with the group velocity obtained in (2.16). In Section 5.5, the energy transport velocity was defined as the ratio of the time-averaged power flux density to the time-averaged total energy density. Since the densities obtained in Section 5.5 are independent of position, the spatial averages of the densities are the same as the densities themselves. Consequently, the definition of the energy transport velocity we used in (2.23) is general and that which we used in Section 5.5 is a specialization that is valid for the case of uniform densities.

6. Surface Charge and Surface Current Densities

We can see from (2.6) that the electric field of the TE mode in a parallel-plate waveguide is parallel to the bounding walls. Since the surface charge density ρ_s is equal to the normal component of the electric flux density, it follows that there are no surface charges on the bounding walls. We obtain the surface current density $\boldsymbol{J}_s\,(x = 0)$ on the lower plate as

$$\begin{aligned}\boldsymbol{J}_s\,(x = 0) &= \hat{n} \times \boldsymbol{H}(0, z) = \hat{x} \times [\hat{x}H_x(0, z) + \hat{z}H_z(0, z)],\\ &= -\hat{y}H_z(0, z) = \hat{y}\frac{2}{\eta_0\beta}\frac{m\pi}{d}E_{y0}^i e^{-j\beta_z z}.\end{aligned} \tag{2.24}$$

Similarly, we find the surface current density $\boldsymbol{J}_s\,(x = d)$ on the upper plate to be

$$\begin{aligned}\boldsymbol{J}_s\,(x = d) &= \hat{y}H_z(d, z) = -\hat{y}\frac{2}{\eta_0\beta}\frac{m\pi}{d}E_{y0}^i \cos(m\pi)e^{-j\beta_z z},\\ &= \hat{y}\frac{2}{\eta_0\beta}\frac{m\pi}{d}E_{y0}^i(-1)^{m-1}e^{-j\beta_z z}.\end{aligned} \tag{2.25}$$

For m odd, the surface current densities on the two plates are in the same direction, but for m even, they are in opposite directions. However, the magnitudes of the surface current densities on the two plates are equal.

7. Designation Procedure for Modes

We can see from (2.6) and (2.7) that $E_y(x, z)$ and $H_x(x, z)$ are both zero at $x = 0$ and $x = d$. For $m = 1$, the amplitude of these fields have one maximum in the cross section of the waveguide. For $m = 2$, there are two maxima and so on. Similarly, we can see from (2.8) that the amplitude of $H_z(x, z)$ has a maximum at $x = 0$ and $x = d$. For $m = 1$, the amplitude of $H_z(x, z)$ has one zero in the cross section of the waveguide. For $m = 2$, there are two zeros and so on (Figs. 8.3a and 8.3b). In other

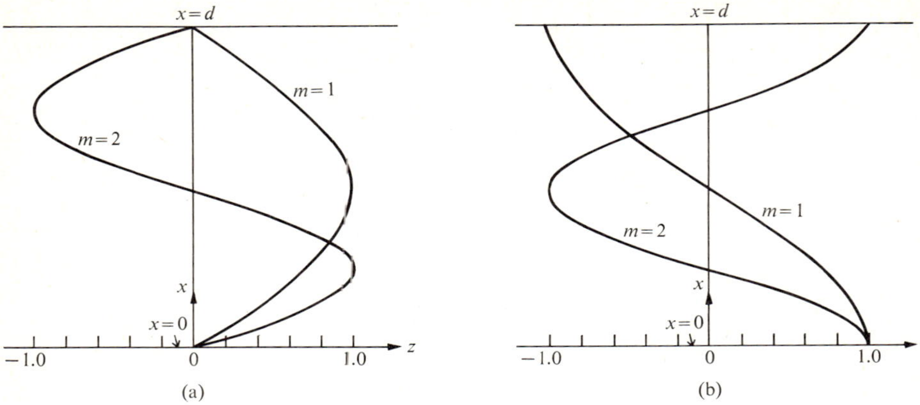

Fig. 8.3 Relative distribution of the amplitude of the fields in the x-direction between the parallel plates: (a) $E_y(x, z)$, $H_x(x, z)$; (b) $H_z(x, z)$.

words, the field components vary periodically with the distance in the region between the parallel plates and m gives the number of cycles of variation of the fields in the cross section of the waveguide. The foregoing discussion makes it possible for us to explain physically the designation procedure used for the waveguide modes. The mode of propagation under discussion is designated as the TE_{m0} mode where TE stands for transverse electric and m stands for the number of cycles of variation of the fields in the x-direction. The second subscript 0 indicates that the fields do not vary in the y-direction. As we shall see later, in a rectangular waveguide, the fields vary in the y-direction and in that case, the second subscript is a nonvanishing integer, usually n, which, in turn, gives the number of cycles of variation of the fields in the y-direction.

8. Electric and Magnetic Field Lines

It is beyond the scope of this treatment for us to give an account of the methods of excitation of the waveguide modes. Since a picture of the electric and the magnetic field lines is useful in the choice of the proper method of excitation of the waveguide modes, we shall include a brief discussion of the field lines. From (2.6), it follows that the electric field lines are in the y-direction and the density of the field lines varies as $\sin(m\pi x/d)$. For $m = 1$, as depicted in Fig. 8.4(a), the density of the field lines is maximum at the center of the waveguide, falls off on either side, and becomes zero at the two bounding walls.

The magnetic vector has both x- and z-components, with the result that the magnetic field lines are more difficult to evaluate. From (2.7) and (2.8), we obtain the expressions for the magnetic field components in the time domain as

$$H_x(x, z, t) = -\frac{2\beta_z}{\eta_0\beta} E^i_{y0} \sin\left(\frac{m\pi x}{d}\right) \sin(\beta_z z - \omega t), \tag{2.26}$$

$$H_z(x, z, t) = -\frac{2}{\eta_0\beta}\frac{m\pi}{d} E^i_{y0} \cos\left(\frac{m\pi x}{d}\right) \cos(\beta_z z - \omega t). \tag{2.27}$$

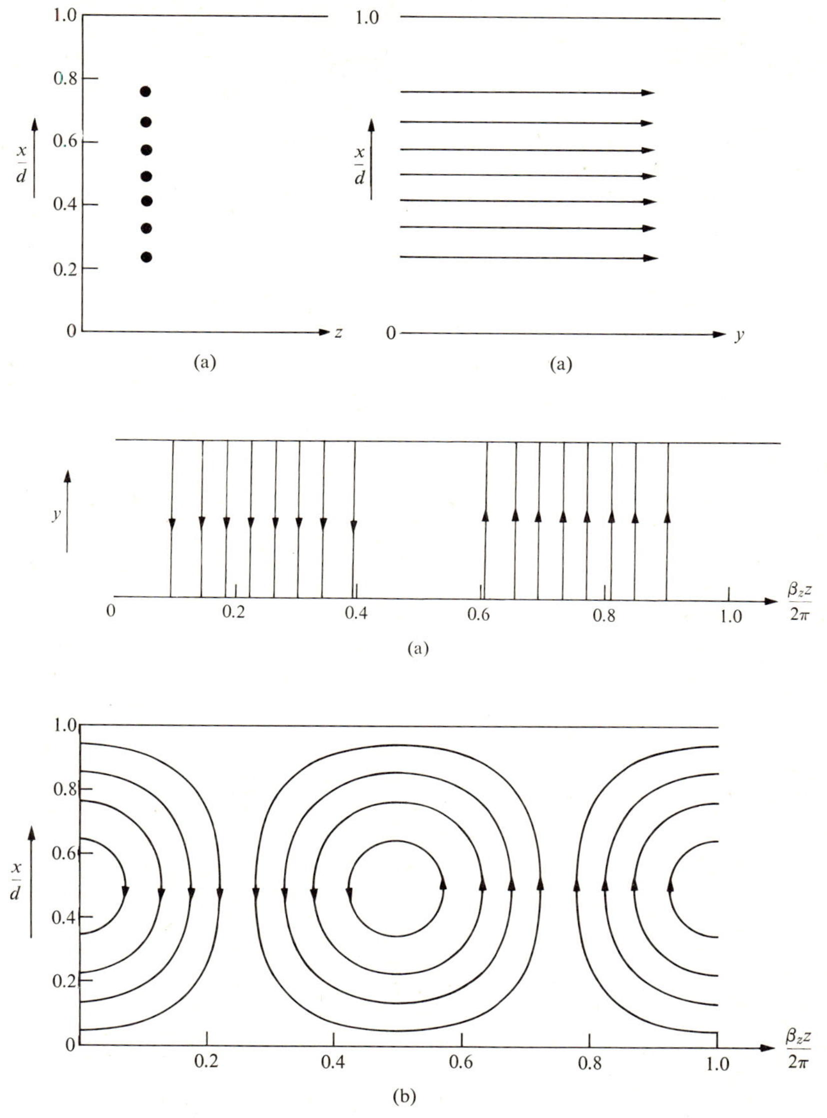

Fig. 8.4 Field lines of the TE_{10} mode of a parallel-plate waveguide. (a) Electric field lines; (b) magnetic field lines.

In obtaining (2.26) and (2.27), we have assumed E_{y0}^i to be real without any loss of generality. In the xz-plane, the directional tangent dx/dz to the field line gives the ratio of the x- to the z-component of the field at that point:

$$\frac{H_x(x, z, t)}{H_z(x, z, t)} = \frac{dx}{dz} = \frac{\beta_z}{(m\pi/d)} \frac{\sin(m\pi x/d)\sin(\beta_z z - \omega t)}{\cos(m\pi x/d)\cos(\beta_z z - \omega t)}. \tag{2.28}$$

Therefore, on rearrangement, we find that

$$\frac{(m\pi/d)\cos(m\pi x/d)\,dx}{\sin(m\pi x/d)} = \frac{\beta_z \sin(\beta_z z - \omega t)\,dz}{\cos(\beta_z z - \omega t)},$$

or

$$\frac{d[\sin(m\pi x/d)]}{\sin(m\pi x/d)} = -\frac{d[\cos(\beta_z z - \omega t)]}{\cos(\beta_z z - \omega t)}. \tag{2.29}$$

If we integrate (2.29) we obtain

$$\ln[\sin(m\pi x/d)\cos(\beta_z z - \omega t)] = \text{constant}, \tag{2.30}$$

which can be rewritten as

$$\sin(m\pi x/d)\cos(\beta_z z - \omega t) = \text{constant} = \pm\sin(m\pi x_0/d). \tag{2.31}$$

In view of the expression on the left-hand side, the constant in (2.31) can vary only from -1 to $+1$ and hence has been set equal to $\pm\sin(m\pi x_0/d)$, where the upper or the lower sign is chosen according to whether $\cos(\beta_z z - \omega t)$ is positive or negative. If $\beta_z z - \omega t = n\pi$ where n is an integer, it is convenient for us to choose the solution $x = x_0$ and, therefore, x_0 is the value of x for a particular curve corresponding to $\beta_z z - \omega t = n\pi$. The equation for the magnetic field lines is given by (2.31). Since z and t appear only in the combination $\beta_z z - \omega t$, it follows that as time progresses, the field lines move in the positive z-direction with a velocity equal to the phase velocity $\omega/\beta_z = v_{ph,z}$. Consequently, it is sufficient for us to examine the field lines at one value of time t which is conveniently taken to be equal to zero.

We illustrate the construction of the magnetic field lines in detail for the fundamental TE mode corresponding to $m = 1$. For this case, (2.31) becomes

$$\pm\cos\beta_z z = \frac{\sin(\pi x_0/d)}{\sin(\pi x/d)}. \tag{2.32}$$

Since (2.32) is not changed when $\beta_z z$ increases by 2π, it follows that the field lines repeat themselves for every increase of $\beta_z z$ by 2π and that a complete cycle of the field lines is contained in the interval $0 < \beta_z z < 2\pi$. Therefore, the field lines are plotted only for the interval $0 < \beta_z z < 2\pi$. For $\beta_z z = 0$, we can see from (2.26) that $H_x(x, z, 0) = 0$ and $H_z(x, z, 0)$ varies as $\cos(\pi x/d)$. We choose a set of values of x_0/d in the interval from 0 to 1 such that the density of points varies as $|\cos(\pi x/d)|$. This choice of values of x_0/d is needed since the density of the field lines should be proportional to the magnitude of the field. The value of $\beta_z z$ is changed in steps of $\pi/20$ from $\beta_z z = 0$ to $\beta_z z = 2\pi$. For each value of $\beta_z z$ and for each of the chosen set of values of x_0/d, we determine the value of x/d which satisfies (2.32). The points corresponding to the same values of x_0/d are connected together and the resulting field lines are depicted in Fig. 8.4(b). We find that the magnetic field lines form closed loops.

We can find the sense of the magnetic field along the field lines in the following manner. For $t = 0$ and $m = 1$, (2.26) and (2.27) become

$$H_x(x, z, 0) = -\frac{2\beta_z}{\eta_0\beta} E^i_{y0} \sin(\pi x/d) \sin(\beta_z z), \tag{2.33}$$

$$H_z(x, z, 0) = -\frac{2}{\eta_0\beta}\frac{m\pi}{d} E^i_{y0} \cos(\pi x/d) \cos(\beta_z z). \tag{2.34}$$

Let E^i_{y0} be positive. We can verify that

$$\sin\beta_z z > 0, \qquad \cos\beta_z z > 0 \text{ (for } 0 < \beta_z z < \pi/2); \tag{2.35}$$

$$\sin\pi x/d > 0, \qquad \cos\pi x/d > 0 \text{ (for } 0 < x/d < 1/2); \tag{2.36}$$

$$\sin\pi x/d > 0, \qquad \cos\pi x/d < 0 \text{ (for } 1/2 < x/d < 1). \tag{2.37}$$

From (2.33) to (2.36), it follows that for $0 < \beta_z z < \pi/2$ and $0 < x/d < 1/2$, $H_x(x, z, 0) < 0$ and $H_z(x, z, 0) < 0$. Also, from (2.33) to (2.35) and (2.37), we can ascertain that for $0 < \beta_z z < \pi/2$ and $1/2 < x/d < 1$, $H_x(x, z, 0) < 0$ and $H_z(x, z, 0) > 0$. Consequently, as indicated in Fig. 8.4(b), the magnetic vector is in the clockwise direction for $0 < \beta_z z < \pi/2$. Since

$$\sin\beta_z z > 0, \qquad \cos\beta_z z < 0, \qquad (\text{for } \pi/2 < \beta_z z < \pi), \tag{2.38}$$

we can verify with the help of (2.33), (2.34), (2.36), and (2.38) that for $\pi/2 < \beta_z z < \pi$ and $0 < x/d < 1/2$, $H_x(x, z, 0) < 0$ and $H_z(x, z, 0) > 0$. Also, it follows that for $\pi/2 < \beta_z z < \pi$ and $1/2 < x/d < 1$, $H_x(x, z, 0) < 0$ and $H_z(x, z, 0) < 0$. Therefore, the magnetic vector is in the counterclockwise direction for $\pi/2 < \beta_z z < \pi$. In a similar manner, we can establish that the magnetic vector is in the counterclockwise direction for $\pi < \beta_z z < 3\pi/2$ and is in the clockwise direction for $3\pi/2 < \beta_z z < 2\pi$. Thus, the magnetic field lines form closed loops which are alternatively in the clockwise and in the counterclockwise directions.

8.3 PARALLEL-PLATE WAVEGUIDE: TM MODE

As before, let $x = 0$ be a perfectly conducting, infinite plane. We deduced the field components of the TM wave existing in the region $x > 0$ in Section 6.6 with the following results:

$$H_y(x, z) = 2H_{y0}^i \cos(\beta x \cos\theta^i)e^{-j\beta z \sin\theta^i}, \tag{3.1}$$

$$E_x(x, z) = 2\sin\theta^i\eta_0 H_{y0}^i \cos(\beta x \cos\theta^i)e^{-j\beta z \sin\theta^i}, \tag{3.2}$$

$$E_z(x, z) = 2j\cos\theta^i\eta_0 H_{y0}^i \sin(\beta x \cos\theta^i)e^{-j\beta z \sin\theta^i}. \tag{3.3}$$

We assume that θ^i has arbitrarily any value in the range $0 < \theta^i < \pi/2$.

1. Synthesis of the Waveguide Fields

We can synthesize the field components of the TM mode inside a parallel-plate waveguide consisting of parallel, perfectly conducting, infinite planes situated at $x = 0$ and $x = d$ from (3.1) to (3.3) in the following manner. At $x = d$, the tangential electric field component $E_z(x, z)$ is required to vanish. This requirement cannot be satisfied by the fields given by (3.1) to (3.3) for all values of θ^i in the range $0 < \theta^i < \pi/2$. If we restrict θ^i to have only the values given by θ_g^i as defined in (2.4), $E_z(x, z) = 0$. Consequently, (3.1) to (3.3) with $\theta^i = \theta_g^i$ are the complete expressions of the field components of the TM mode sustained inside the parallel-plate region, since these field components satisfy Maxwell's equations and the appropriate boundary conditions at $x = 0$ and $x = d$.

With the help of (2.4) and (2.5), we can conveniently express as follows the field components of the TM mode in a parallel-plate waveguide, as given by (3.1) to (3.3):

$$H_y(x, z) = 2H_{y0}^i \cos\left(\frac{m\pi x}{d}\right)e^{-j\beta_z z}, \tag{3.4}$$

$$E_x(x, z) = 2\frac{\beta_z}{\beta}\eta_0 H_{y0}^i \cos\left(\frac{m\pi x}{d}\right)e^{-j\beta_z z}, \tag{3.5}$$

$$E_z(x, z) = 2j\frac{m\pi}{\beta d}\eta_0 H_{y0}^i \sin\left(\frac{m\pi x}{d}\right)e^{-j\beta_z z}, \tag{3.6}$$

where β_z is the same as that defined by (2.5) for $\omega > \omega_{c,m}$ and (2.9) for $\omega < \omega_{c,m}$. The critical angular frequency $\omega_{c,m}$ is given by (2.10).

2. TEM Mode

When $m = 0$, the field components (2.6) to (2.8) of the TE mode inside a parallel-plate waveguide vanish leading to a trivial solution. On the other hand, for the TM mode, we can see from (3.4) to (3.6) that only $E_z(x, z) = 0$ but $H_y(x, z)$ and $E_x(x, z)$ are nonvanishing. Also, (2.4) shows us that for $m = 0$, $\theta_g^i = \pi/2$ and hence

$$\beta_z = \beta = \omega/v = \omega\sqrt{\mu\varepsilon}. \tag{3.7}$$

For $m = 0$, there is neither an electric nor a magnetic field component in the direction of propagation and the propagation coefficient becomes the same as that in the corresponding unbounded dielectric. Therefore, for $m = 0$, the TM_{m0} mode inside a parallel-plate waveguide, reduces to the TEM mode which we have treated previously in Section 7.8.

The dispersion diagram, that is, the ω versus β_z diagram for the TM_{m0} mode is the same as that given in Fig. 8.1. We can now identify the ω-β curve in Fig. 8.1 as the dispersion curve of the TEM mode.

3. Evanescent and Propagating Regions

We can prove, as before, that the evanescence and the propagation of the waveguide modes correspond to $\omega < \omega_{c,m}$ and $\omega > \omega_{c,m}$ respectively. We obtain the x-component of the time-averaged Poynting vector from (3.4) and (3.6) as

$$\begin{aligned} S_x(x) &= -\tfrac{1}{2}\,\text{Re}\, E_z(x,z)H_y^*(x,z), \\ &= -\tfrac{1}{2}\,\text{Re}\left[\frac{4jm\pi\eta_0}{\beta d}\cos\left(\frac{m\pi x}{d}\right)\sin\left(\frac{m\pi x}{d}\right)|H_{y0}^i|^2\, e^{-j(\beta_z-\beta_z^*)z}\right] = 0. \end{aligned} \qquad (3.8)$$

The result given by (3.8) is true irrespective of whether β_z is real or imaginary. Similarly, we find the z-component of the time-averaged Poynting vector from (3.4) and (3.5) as

$$\begin{aligned} S_z(x) &= \tfrac{1}{2}\,\text{Re}\, E_x(x,z)H_y^*(x,z), \\ &= \frac{2\,|H_{y0}^i|^2\,\eta_0}{\beta}\cos^2\left(\frac{m\pi x}{d}\right)\text{Re}\,[\beta_z e^{j(\beta_z^*-\beta_z)z}]. \end{aligned} \qquad (3.9)$$

For $\omega < \omega_{c,m}$, β_z is imaginary as given by (2.9) and then (3.9) yields $S_z(x) = 0$. For $\omega > \omega_{c,m}$, β_z is real and (3.9) becomes

$$S_z(x) = \frac{2\beta_z\eta_0}{\beta}|H_{y0}^i|^2\cos^2\left(\frac{m\pi x}{d}\right). \qquad (3.10)$$

Thus, there is no transport of energy inside the waveguide for $\omega < \omega_{c,m}$ which, therefore, defines the evanescent region. On the other hand, there is transport of energy for $\omega > \omega_{c,m}$, which frequency range is, therefore, the propagation region.

Since β_z for the TM_{m0} mode is the same as that for the TE_{m0} mode, it follows that the phase velocity $v_{ph,z}$, the wavelength λ_z, and the group velocity $v_{g,z}$ in the direction of propagation for the TM_{m0} mode are the same as the corresponding quantities for the TE_{m0} mode and the expressions of these quantities are given by (2.14), (2.15), and (2.16) respectively. We should remember that there is no TE_{m0} mode corresponding to $m = 0$ and the TM_{m0} mode corresponding to $m = 0$ is the TEM mode. In Fig. 8.2, the straight line yields the normalized phase and the normalized group velocities of the TEM mode. We should note that the TEM mode inside the parallel-plate waveguide has no dispersion.

4. Energy Transport Velocity

We can obtain the spatial average of the power flux density $S_z(x)$ defined by (3.10) as

$$\bar{S}_z = \frac{1}{d}\int_0^d S_z(x)\,dx = \frac{2\beta_z\eta_0}{\beta}|H_{y0}^i|^2\frac{1}{d}\int_0^d \cos^2\left(\frac{m\pi x}{d}\right)dx = \frac{\beta_z\eta_0\,|H_{y0}^i|^2}{\beta}. \qquad (3.11)$$

We can show from (3.5) and (3.6) that the time-averaged energy density $W_E(x)$ associated with the electric field is

$$W_E(x) = \frac{\varepsilon}{4}|E(x, z)|^2 = \frac{\varepsilon}{4}[|E_x(x, z)|^2 + |E_z(x, z)|^2],$$
$$= \frac{\varepsilon\,|H_{y0}^i|^2}{\beta^2}\eta_0^2\left[\beta_z^2\cos^2\left(\frac{m\pi x}{d}\right) + \left(\frac{m\pi}{d}\right)^2\sin^2\left(\frac{m\pi x}{d}\right)\right]. \tag{3.12}$$

With the help of (2.5), we obtain the spatial average of (3.12) as

$$\overline{W}_E = \frac{1}{d}\int_0^d W_E(x)\,dx = \frac{\varepsilon\,|H_{y0}^i|^2\,\eta_0^2}{2\beta^2}\left[\beta_z^2 + \left(\frac{m\pi}{d}\right)^2\right] = \frac{\mu}{2}|H_{y0}^i|^2, \tag{3.13}$$

since $\eta_0^2 = \mu/\varepsilon$ and $\beta^2 = \omega^2/v^2 = \beta_z^2 + (m\pi/d)^2$. We deduce the time-averaged energy density $W_H(x)$ associated with the magnetic field from (3.4) as

$$W_H(x) = \frac{\mu}{4}|H(x, z)|^2 = \mu\,|H_{y0}^i|^2\cos^2\left(\frac{m\pi x}{d}\right). \tag{3.14}$$

We find the spatial average of (3.14) to be

$$\overline{W}_H = \frac{1}{d}\int_0^d W_H(x)\,dx = \frac{\mu}{2}|H_{y0}^i|^2. \tag{3.15}$$

We should note, as before, that $\overline{W}_E = \overline{W}_H$. We evaluate the energy transport velocity v_E with the help of (3.11), (3.13), and (3.15) as

$$v_E = \frac{\bar{S}_z}{\overline{W}_E + \overline{W}_H} = \frac{\beta_z\eta_0\,|H_{y0}^i|^2}{\beta\mu\,|H_{y0}^i|^2} = v\sqrt{1 - \left(\frac{\omega_{c,m}}{\omega}\right)^2}. \tag{3.16}$$

Since the time-averaged Poynting vector is in the z-direction, the energy transport velocity is also in the z-direction. If we compare (3.16) with (2.16) we see that the energy transport velocity and the group velocity are identical.

5. Surface Charge and Surface Current Densities

We can obtain the surface charge density $\rho_s\,(x = 0)$ on the lower plate from (3.5) as

$$\rho_s\,(x = 0) = \hat{x}\cdot\hat{x}\varepsilon E_x(0, z) = \varepsilon 2(\beta_z/\beta)\eta_0 H_{y0}^i e^{-j\beta_z z}. \tag{3.17}$$

Similarly, we find from (3.5) the surface charge density $\rho_s\,(x = d)$ on the upper plate to be

$$\rho_s\,(x = d) = -\hat{x}\cdot\hat{x}\varepsilon E_x(d, z) = \varepsilon 2(\beta_z/\beta)\eta_0(-1)^{m+1}H_{y0}^i e^{-j\beta_z z}. \tag{3.18}$$

For m even, the surface charge densities on the two plates are of opposite signs but for m odd, they are of the same sign. The magnitudes of the surface charge densities are equal on both the plates.

We can deduce the surface current density $\boldsymbol{J}_s\,(x = 0)$ on the lower plate to be

$$\boldsymbol{J}_s\,(x = 0) = \hat{n}\times\boldsymbol{H}(0, z) = \hat{x}\times\hat{y}H_y(0, z) = \hat{z}2H_{y0}^i e^{-j\beta_z z}. \tag{3.19}$$

Similarly, we find the surface current density $\boldsymbol{J}_s\,(x = d)$ on the upper plate to be

$$\boldsymbol{J}_s\,(x = d) = -\hat{x} \times \hat{y}H_y(d, z) = \hat{z}2H^i_{y0}(-1)^{m+1}e^{-j\beta_z z}. \qquad (3.20)$$

We can see from (3.19) and (3.20) that for m odd the surface current densities on the two plates are in the same direction but for m even they are in the opposite directions. But, the magnitudes of the surface current densities on the two plates are equal.

The designation procedure for the transverse magnetic modes is the same as that for the transverse electric modes. However, we should note that there is no TE_{m0} mode for $m = 0$ and the TM_{m0} mode for $m = 0$ is the TEM mode.

6. Electric and Magnetic Field Lines

We can obtain the electric and the magnetic field lines for the TM_{m0} mode in the same manner as those for the TE_{m0} mode. From (3.4), it follows that the magnetic field lines are in the y-direction and the density of the field lines varies as $\cos\,(m\pi x/d)$. For $m = 1$, as depicted in Fig. 8.5(a), the density of the field lines is zero at the center of the waveguide, increases on either side, the increase being in the $+y$-direction on one side and in the $-y$-direction on the other, and becomes a maximum at the two bounding walls.

The electric vector has both x- and z-components and, therefore, the electric field lines are somewhat more difficult to determine. As in the case of the TE mode, in the TM mode the field lines move in the $+z$-direction with a uniform velocity $v_{ph,z} = \omega/\beta_z$. It is therefore sufficient for us to examine the field lines at one value of time t which, as before, is taken to be equal to zero. Without any loss of generality, we assume H^i_{y0} to be real. We find from (3.5) and (3.6) that the expressions for the electric field components, in the time domain, at $t = 0$ are

$$E_x(x, z, t = 0) = 2(\beta_z/\beta)\eta_0 H^i_{y0} \cos\,(m\pi x/d) \cos\,(\beta_z z), \qquad (3.21)$$

$$E_z(x, z, t = 0) = 2(m\pi/\beta d)\eta_0 H^i_{y0} \sin\,(m\pi x/d) \sin\,(\beta_z z). \qquad (3.22)$$

In the xz-plane, the directional tangent dx/dz to the field lines gives the ratio of the x- to the z-component of the field at that point. Hence,

$$\frac{E_x(x, z, t = 0)}{E_z(x, z, t = 0)} = \frac{dx}{dz} = \frac{\beta_z}{(m\pi/d)} \frac{\cos\,(m\pi x/d) \cos\,(\beta_z z)}{\sin\,(m\pi x/d) \sin\,(\beta_z z)}. \qquad (3.23)$$

It is possible for us to integrate (3.23) to yield the following result:

$$\cos\,(m\pi x/d) \sin\,(\beta_z z) = \text{constant} = \pm\cos\,(m\pi x_0/d). \qquad (3.24)$$

In view of the expression on the left-hand side of (3.24), the constant in (3.24) can vary only from -1 to $+1$ and hence has been set equal to $\pm\cos\,(m\pi x_0/d)$, where the upper or the lower sign is chosen according to whether $\sin\,\beta_z z$ is positive or negative. If $\beta_z z = (n - \frac{1}{2})\pi$ where n is an integer, it is convenient for us to choose the solution $x = x_0$ and therefore x_0 is the value of x for a particular curve corresponding to $\beta_z z = (n - \frac{1}{2})\pi$.

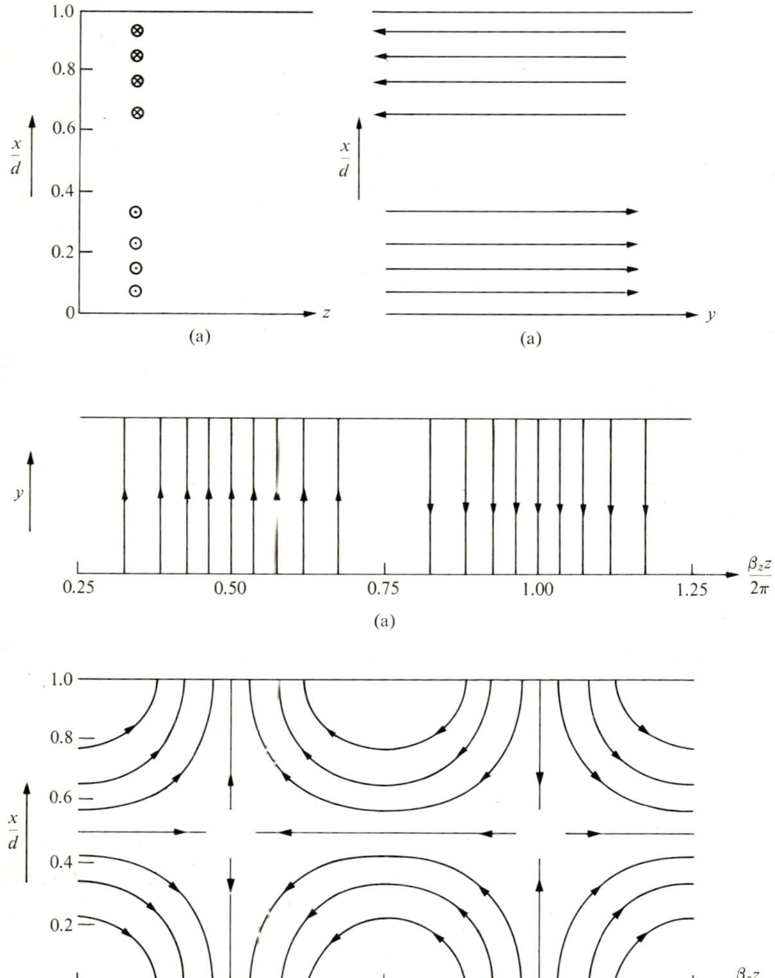

Fig. 8.5 Field lines of the TM_{10} mode of a parallel-plate waveguide. (a) Magnetic field lines; (b) electric field lines.

We shall illustrate the construction of the electric field lines for the TM_{10} mode. For this case, we can rewrite (3.24) as

$$\pm \sin(\beta_z z) = \frac{\cos(\pi x_0/d)}{\cos(\pi x/d)}. \tag{3.25}$$

We can argue, as before, from (3.25) that a complete cycle of the field lines is contained in the interval $\pi/2 < \beta_z z < 5\pi/2$. Therefore, the field lines are plotted only for the interval $\pi/2 < \beta_z z < 5\pi/2$. For $\beta_z z = \pi/2$ we can see from (3.21) and (3.22) that $E_x(x, z, 0) = 0$ and $E_z(x, z, 0)$ varies as $\sin(\pi x/d)$. For the reasons mentioned previously, a set of values of x_0/d is chosen in the interval from 0 to 1 such that the density of the points varies as $\sin(\pi x/d)$. The value of $\beta_z z$ is changed in steps of $\pi/20$ from $\beta_z z = \pi/2$ to $\beta_z z = 5\pi/2$. For each value of $\beta_z z$ and for each of the chosen set of values of x_0/d, we determine the value of x/d which satisfies (3.25). The points corresponding to the same values of x_0/d are connected together and the resulting field lines are shown in Fig. 8.5(b). We see that the electric field lines terminate on the two parallel plates.

We can find the sense of the electric field along the field lines in the following manner. Let H^i_{y0} be positive. We can ascertain that

$$\sin \beta_z z > 0; \qquad \cos \beta_z z < 0; \qquad \text{for } \pi/2 < \beta_z z < \pi. \tag{3.26}$$

From (3.21), (3.22), (3.26), and (2.36), it follows that for $\pi/2 < \beta_z z < \pi$ and $0 < x/d < \frac{1}{2}$, $E_x(x, z, 0) < 0$ and $E_z(x, z, 0) > 0$. Also, from (3.21), (3.22), (3.26), and (2.37), we can verify that for $\pi/2 < \beta_z z < \pi$ and $\frac{1}{2} < x/d < 1$, $E_x(x, z, 0) > 0$ and $E_z(x, z, 0) > 0$. Therefore, we can argue that, as indicated in Fig. 8.5(b), the electric vector bends in the clockwise direction for $0 < x/d < \frac{1}{2}$ and in the counterclockwise direction for $\frac{1}{2} < x/d < 1$ in the range of $\beta_z z$ given by $\pi/2 < \beta_z z < \pi$. Since

$$\sin \beta_z z < 0, \qquad \cos \beta_z z < 0, \qquad (\text{for } \pi < \beta_z z < 3\pi/2), \tag{3.27}$$

we can verify with the help of (3.21), (3.22), (3.27), and (2.36), that for $\pi < \beta_z z < 3\pi/2$ and $0 < x/d < \frac{1}{2}$, $E_x(x, z, 0) < 0$ and $E_z(x, z, 0) > 0$. Also we may obtain for $\pi < \beta_z z < 3\pi/2$ and $\frac{1}{2} < x/d < 1$, $E_x(x, z, 0) > 0$ and $E_z(x, z, 0) < 0$. Therefore, the electric vector bends in the counterclockwise direction for $0 < x/d < \frac{1}{2}$ and in the clockwise direction for $\frac{1}{2} < x/d < 1$ in the range of $\beta_z z$ given by $\pi < \beta_z z < 3\pi/2$. In a similar manner, we can obtain the direction of the electric vector along the field lines for the ranges $3\pi/2 < \beta_z z < 2\pi$ and $2\pi < \beta_z z < 5\pi/2$ to yield the results as shown in Fig. 8.5(b). The magnetic field lines of the TE mode form closed loops but the electric field lines of the TM mode terminate on the two parallel plates, accounting for the surface charges on the bounding walls of the parallel-plate waveguide.

8.4 EFFECT OF LOSSES ON THE MODES OF A PARALLEL-PLATE WAVEGUIDE

We discussed the characteristics of waves propagating between two infinite, conducting, parallel planes in the preceding two sections under the assumption that the dielectric region between the planes is perfect, that is, has no losses, and the conductivity of the bounding walls is infinite. In practice, these ideal conditions are not obtained. The dielectric between the parallel planes has losses and the bounding walls are not perfect conductors. The exact solution of the problem of wave propagation in a lossy dielectric region bounded by two infinite, imperfectly conducting, parallel planes is quite difficult and we shall not attempt it here. However, we shall take advantage of the fact that the losses are usually small to obtain a simple physical picture of the effect of the losses on the modes guided by a parallel-plate waveguide. The approximate procedure used is valid only when the losses in the dielectric are small and when the conductivity of the bounding walls, though not infinite, is very large.

It is intuitively obvious that when the losses are very small, the modes of an ideal parallel-plate waveguide are subject to some perturbation. An examination of the field components (2.6) to (2.8) of the TE mode and (3.4) to (3.6) of the TM mode reveals to us an important general feature of the field components of the guided waves. These field components consist of two parts. The so-called "longitudinal" part is given by $e^{-j\beta_z z}$ which represents, in general, a progressive wave or propagation in the "longitudinal" or the z-direction. The remaining part is called the "transverse" part which gives the field pattern or the mode pattern in the cross section of the waveguide. We assume that the very small losses do not affect the mode pattern of the guided waves but merely perturb the longitudinal part so as to introduce attenuation in the direction of propagation. We propose to determine the attenuation coefficient separately for the dielectric and the conductor losses.

1. Dielectric Losses

For both the TE and the TM modes, we can obtain from (2.5)

$$j\beta_z = j\sqrt{\omega^2\mu\varepsilon - (m\pi/d)^2} = j\omega\sqrt{\mu\varepsilon}[1 - (\omega_{c,m}/\omega)^2]^{1/2}. \tag{4.1}$$

If the dielectric is lossless ε is real and, also, if $\omega > \omega_{c,m}$, it follows from (4.1) that β_z is real. Since the z-dependence of the field components is of the form $e^{-j\beta_z z}$, there is propagation without attenuation for ε real and $\omega > \omega_{c,m}$. We wish to find the attenuation coefficient $\alpha_{d,z}$ due to the dielectric losses in the frequency range $\omega > \omega_{c,m}$ where there is propagation without attenuation in the absence of losses. If there are losses in the dielectric, real ε changes to complex $\varepsilon_{\text{eff}} = \varepsilon + (\sigma/j\omega)$ where σ is the conductivity of the dielectric. As a consequence (4.1), which is imaginary, becomes a complex quantity γ. Therefore, it follows from (4.1) that

$$\begin{aligned}\gamma &= j\sqrt{\omega^2\mu\left(\varepsilon + \frac{\sigma}{j\omega}\right) - \left(\frac{m\pi}{d}\right)^2},\\ &= j\sqrt{\omega^2\mu\varepsilon - \left(\frac{m\pi}{d}\right)^2 - j\omega\mu\sigma}.\end{aligned} \tag{4.2}$$

We should note that the z-dependence of the field components is now of the form $e^{-\gamma z}$. We can find the real and the imaginary parts of γ from (4.2) and the real part gives the attenuation coefficient. Under certain conditions, we can obtain an explicit expression for $\alpha_{d,z}$. On rewriting (4.2) as

$$\gamma = j\sqrt{\omega^2\mu\varepsilon - (m\pi/d)^2}\,[1 - j\omega\mu\sigma\{\omega^2\mu\varepsilon - (m\pi/d)^2\}^{-1}]^{1/2}, \tag{4.3}$$

and using the condition

$$\omega\mu\sigma \ll |\{\omega^2\mu\varepsilon - (m\pi/d)^2\}| = \omega^2\mu\varepsilon\,|\{1 - (\omega_{c,m}/\omega)^2\}|, \tag{4.4}$$

the square root term in (4.3) can be expanded into a binomial series. The result of retaining only the first two terms of the series is

$$\gamma = \alpha_{d,z} + j\beta_z = \frac{\sigma}{2}\sqrt{\frac{\mu}{\varepsilon}}\left\{1 - \left(\frac{\omega_{c,m}}{\omega}\right)^2\right\}^{-1/2} + j\omega\sqrt{\mu\varepsilon}\left\{1 - \left(\frac{\omega_{c,m}}{\omega}\right)^2\right\}^{1/2}. \tag{4.5}$$

First, within the approximation specified by (4.4), the phase-shift coefficient β_z with a lossy dielectric is the same as that for a lossless dielectric. For $\omega > \omega_{c,m}$, the attenuation coefficient $\alpha_{d,z}$ decreases monotonically with the increase in the frequency. The attenuation coefficient $\alpha_{d,z}$ appears to become infinite at the critical angular frequency $\omega = \omega_{c,m}$ but, in actuality, this is not so since as ω approaches $\omega_{c,m}$ the condition (4.4) no longer holds and the approximation (4.5) is then not valid. However, we can find the correct value of $\alpha_{d,z}$ at $\omega = \omega_{c,m}$ directly from (4.2).

2. Conductor Losses

Let $\alpha_{c,z}$ be the attenuation coefficient resulting from the imperfect conductivity of the walls of the parallel-plate waveguide. The attenuation coefficient $\alpha_{c,z}$ is different for the TE and the TM modes. We shall discuss an approximate method for the determination of $\alpha_{c,z}$ for the TE modes first.

We should emphasize that we have assumed that for very small losses, the field components given by (2.6) to (2.8) are unaltered except for the multiplicative factor $e^{-\alpha_{c,z}z}$. Consequently, for $\omega > \omega_{c,m}$, the time-averaged Poynting vector $S_z(x, z)$ in the direction ($\hat{z}$) of propagation, as given by (2.13), has to be modified to

$$S_z(x, z) = \frac{2\beta_z\,|E^i_{y0}|^2}{\eta_0\beta}\sin^2\left(\frac{m\pi x}{d}\right)e^{-2\alpha_{c,z}z}. \tag{4.6}$$

Let us consider a parallel-plate waveguide of unit width in the y-direction. Then, we can obtain the total time-averaged power flowing through the cross section of the waveguide from (4.6) as

$$P(z) = \int_0^1 dy\int_0^d dx S_z(x, z) = \frac{d\beta_z\,|E^i_{y0}|^2}{\eta_0\beta}e^{-2\alpha_{c,z}z} = P_0e^{-2\alpha_{c,z}z}, \tag{4.7}$$

where

$$P_0 = \frac{d\beta_z\,|E^i_{y0}|^2}{\eta_0\beta}. \tag{4.8}$$

The form of $P(z)$, though obtained in (4.7) for the TE mode in a parallel-plate waveguide, is generally true. We can evaluate the decrease of total time-averaged power flow through the cross section of the waveguide per unit distance in the direction of propagation from (4.7) as

$$-dP(z)/dz = 2\alpha_{c,z}P(z) = P_L(z), \tag{4.9}$$

and this must necessarily be equal to the total power $P_L(z)$ dissipated per unit distance in the direction of propagation. From (4.9), it follows that

$$\alpha_{c,z} = P_L(z)/2P(z), \tag{4.10}$$

which we can use to determine the expression for $\alpha_{c,z}$ for any waveguide mode.

3. TE Mode

The total power $P_L(z)$ dissipated per unit distance in the direction of propagation is actually the power flowing into the bounding walls per unit width in the y-direction and unit length in the z-direction. The implication is that the tangential component of the electric field does not vanish at the surface of the walls of the waveguide but has a small non-zero value which we can determine from the knowledge of the tangential magnetic field and the concept of surface impedance developed in Section 5.9. For the purposes of obtaining $P_L(z)$, it is not necessary for us to evaluate this non-zero tangential electric field explicitly since $P_L(z)$ can be determined directly from the knowledge of the surface current and the surface resistivity. From (5.9.8) and (5.9.10), we find that the surface resistivity is

$$\eta_{sr} = \sqrt{\omega\mu_c/2\sigma_c}, \tag{4.11}$$

which is equal to the surface resistance of a square of equal width and length. In (4.11), μ_c is the permeability and σ_c is the conductivity of the material of the walls of the waveguide. We find the magnitudes of the surface current densities on the lower and the upper walls of the waveguide to be the same from (2.24) and (2.25), and are given by

$$|\boldsymbol{I}_s| = |\boldsymbol{J}_s| = \frac{2}{\eta_0\beta}\frac{m\pi}{d}|E^i_{y0}|\,e^{-\alpha_{c,z}z}. \tag{4.12}$$

The total surface current $\boldsymbol{I}_s$, as indicated in (4.12), is equal to the surface current density since the length is equal to unity. The exponential factor has to be included in (4.12) since in the presence of the conductor losses, all the field components have to be multiplied by the factor $e^{-\alpha_{c,z}z}$. From (5.9.16) and (4.12), we can determine the power $P_{LL}(z)$ flowing into the lower wall of the waveguide per unit width and per unit length as

$$P_{LL}(z) = \tfrac{1}{2}|\boldsymbol{I}_s|^2\eta_{sr} = \frac{\eta_{sr}}{2}\frac{4}{\eta_0^2\beta^2}\left(\frac{m\pi}{d}\right)^2|E^i_{y0}|^2\,e^{-2\alpha_{c,z}z}. \tag{4.13}$$

We can show the power $P_{LU}(z)$ flowing into the upper wall of the waveguide per unit width and per unit length to be equal to $P_{LL}(z)$. Hence,

$$P_L(z) = 2P_{LL}(z) = \frac{4\eta_{sr}}{\eta_0^2\beta^2}\left(\frac{m\pi}{d}\right)^2 |E_{y0}^i|^2 e^{-2\alpha_{c,z}z}. \tag{4.14}$$

With the help of (4.10), (4.7), (4.14), (2.5), and (2.10), we can deduce in a straightforward manner that

$$\alpha_{c,z} = \frac{2\eta_{sr}(\omega_{c,m}/\omega)^2}{\eta_0 d\sqrt{1-(\omega_{c,m}/\omega)^2}} \qquad \text{(TE mode)}, \tag{4.15}$$

which is the attenuation coefficient due to the imperfect conductivity of the walls for the TE mode in a parallel-plate waveguide.

4. TM Mode

As we mentioned earlier, for the TM mode the attenuation coefficient $\alpha_{c,z}$ due to the imperfect conductivity of the bounding walls is different from that for the TE mode and we can deduce its expression in the following manner. With the conductor losses, for $\omega > \omega_{c,m}$, we have to alter the time-averaged Poynting vector $S_z(x, z)$ in the direction ($\hat{z}$) of propagation, as given by (3.10), to

$$S_z(x, z) = \frac{2\beta_z\eta_0}{\beta}|H_{y0}^i|^2 \cos^2\left(\frac{m\pi x}{d}\right)e^{-2\alpha_{c,z}z}. \tag{4.16}$$

As before, let us consider a parallel-plate waveguide of unit width in the y-direction. Then, we obtain the total time-averaged power flowing through the cross section of the waveguide from (4.16) as

$$P(z) = \int_0^1 dy \int_0^d dx S_z(x, z) = \frac{d\beta_z\eta_0}{\beta}|H_{y0}^i|^2 e^{-2\alpha_{c,z}z} \qquad (\text{for } m \neq 0). \tag{4.17}$$

We can see from (3.19) and (3.20) that the magnitudes of the surface current densities on the lower and the upper walls of the waveguide are the same, and are given by

$$|\boldsymbol{I}_s| = |\boldsymbol{J}_s| = 2\,|H_{y0}^i|\, e^{-\alpha_{c,z}z}. \tag{4.18}$$

We have included the exponential factor in (4.18) to take into account the conductor losses. As before, the surface current is equal to the surface current density since only unit width of the waveguide is under consideration. The powers $P_{LL}(z)$ and $P_{LU}(z)$ flowing respectively into the lower and the upper walls of the waveguide per unit width and per unit length we can deduce from (5.9.16) and (4.18) to be

$$P_{LL}(z) = P_{LU}(z) = \tfrac{1}{2}\,|\boldsymbol{I}_s|^2\,\eta_{sr} = \frac{\eta_{sr}}{2}\,4\,|H_{y0}^i|^2 e^{-2\alpha_{c,z}z}. \tag{4.19}$$

Hence,

$$P_L(z) = 2P_{LL}(z) = 4\eta_{sr}\,|H_{y0}^i|^2 e^{-2\alpha_{c,z}z}. \tag{4.20}$$

If we substitute (4.17) and (4.20) into (4.10) and use (2.5) and (2.10) we obtain

$$\alpha_{c,z} = \frac{2\eta_{sr}}{\eta_0 d\sqrt{1-(\omega_{c,m}/\omega)^2}} \qquad (m \neq 0;\ \text{TM mode}), \tag{4.21}$$

which is the attenuation coefficient due to the imperfect conductivity of the walls, for the TM mode in a parallel-plate waveguide.

5. TEM Mode

We have pointed out earlier that the TM_{m0} mode of a parallel-plate waveguide for the special case of $m = 0$ degenerates to the case of the TEM mode, whose attenuation coefficient can also be found. For $m = 0$, (4.16) and (4.17) yield

$$P(z) = \frac{2d\beta_z\eta_0}{\beta} |H^i_{y0}|^2 e^{-2\alpha_{c,z}z}. \tag{4.22}$$

It follows from (4.20) that $P_L(z)$ for $m = 0$ is the same as that given in (4.20). As before, we may prove the substitution of (4.22) and (4.20) into (4.10) to give the following result for the attenuation coefficient due to the conductor losses for the TEM mode in a parallel-plate waveguide:

$$\alpha_{c,z} = \eta_{sr}/\eta_0 d \qquad (m = 0;\ \text{TEM mode}). \tag{4.23}$$

It is necessary for us to exercise caution against substituting $m = 0 = \omega_{c,m}$ in (4.21) to obtain the attenuation coefficient of the TEM mode. We have determined the

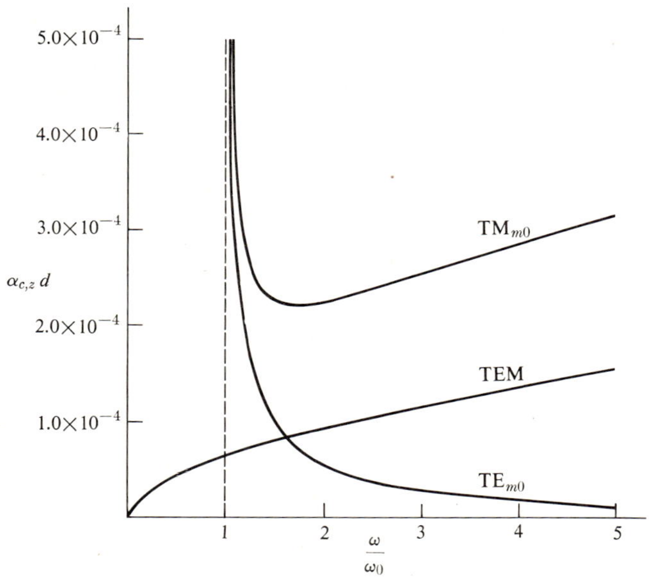

$\frac{\omega_0}{2\pi} = 10^{10}$ Hz; For $\omega=\omega_0$, $\eta_{sr}=2.608\times10^{-2}$ ohm; $\eta_0=377$ ohm

$v=3\times10^8$ m/sec. For the TE_{m0} and the TM_{m0} modes $d=1.5m$ cm.

Fig. 8.6 Normalized attenuation coefficients of the modes of a parallel-plate waveguide with imperfectly conducting walls as a function of the normalized frequency.

result given by (4.23) in (7.10.10) in the course of the treatment of the electromagnetic theory of the transmission lines.

For the TE_{m0}, the TM_{m0}, and the TEM modes, the normalized attenuation coefficient $\alpha_{c,z}d$ is shown in Fig. 8.6 as a function of the normalized frequency ω/ω_0 where the normalizing frequency has been assumed to be equal to $\omega_0/2\pi = 10^{10}$ Hz. At this frequency, we have assumed the surface resistivity to be $\bar{\eta}_{sr} = 2.608 \times 10^{-2}$ ohm. This value of $\bar{\eta}_{sr}$ corresponds to the case in which the bounding walls of the waveguide are made of copper. We assume the dielectric region between the parallel plates to be free space characterized by $\eta_0 = 377$ ohms and $v = 3 \times 10^8$ m/sec. For the TE_{m0} and the TM_{m0} modes, we take the critical frequency to be equal to the normalizing frequency with the result that $d = (\frac{3}{2}m)$ cm. For example, for the TE_{10} and the TM_{10} modes, the separation distance between the parallel plates is equal to $d = 1.5$ cm. We can see from Fig. 8.6 that the attenuation coefficient of the TEM mode increases as the square root of the frequency. For the TE_{m0} mode, the attenuation coefficient is infinite at the cut-off frequency and decreases monotonically as the frequency is increased. But for the TM_{m0} mode, the attenuation coefficient is infinite at the cut-off frequency, decreases and reaches a minimum as the frequency is increased, and then increases for further increases in the frequency.

8.5 RECTANGULAR WAVEGUIDE: TE MODE

In any cross section of a parallel-plate waveguide, the electromagnetic energy is confined to a finite region in one direction but not in the perpendicular direction. In practice, the plates are finite in width. At the edges of the plates, the field pattern is disturbed and there is a leakage of electromagnetic energy through the sides. For this reason, only completely enclosed waveguides are preferred in practice. The geometry of the waveguides has translational invariance in the direction of propagation and the cross section of the waveguide is either rectangular or circular. The difficulty of fabrication prevents the usuage of waveguides having other than rectangular or circular cross sections. We shall now only discuss rectangular waveguides since the theory of wave propagation in rectangular waveguides involves only simple functions, making the investigation of all the general features of the guided waves possible without at the same time introducing complexities in the detail.

We obtained the field components of a parallel-plate waveguide in terms of the constituent plane wave fields by using a *synthesis* procedure. In contrast, we evaluate the field components of a rectangular waveguide with the help of an *analysis* procedure, starting from Maxwell's equations. To begin with, we manipulate the time-harmonic Maxwell's equations into a form suitable for the analysis of the guided waves.

We assume the geometry of the waveguide to have translational invariance in the z-direction, which is called the "longitudinal" direction. The dependence of the field quantities on z is of the form $e^{-\gamma z}$ where γ, in general, is a complex quantity known as the propagation coefficient. If the real part of γ is positive, it follows that $e^{-\gamma z}$ represents a wave propagating in the $+z$-direction for the general case of the nonvanishing imaginary part of γ. The dependence of the field quantities on z can also be of the form $e^{\gamma z}$ which represents a wave propagating in the $-z$-direction. Since the essential characteristics of the guided waves are independent of the direction of propagation, we consider only the wave propagation in the $+z$-direction and we assume the dependence of the field quantities to be of the form $e^{-\gamma z}$. Hence, we can write any field quantity $f(x, y, z)$ as

$$f(x, y, z) = f(x, y)e^{-\gamma z}. \tag{5.1}$$

Since it is invariant to changes in z, the geometry of the waveguide cannot introduce any z-dependence into the field quantities, with the result that the z-dependence of the field quantities is the same throughout the length of the waveguide, providing the justification for seeking the field components in the form (5.1). In view of (5.1), it is evident that

$$\frac{\partial}{\partial z} f(x, y, z) = -\gamma f(x, y)e^{-\gamma z}. \tag{5.2}$$

We write the time-harmonic Maxwell's equations in component form in the rectangular coordinate system x, y, and z and make use of (5.1) and (5.2) for all the field components. We can then factor out the term $e^{-\gamma z}$ yielding the following set of equations which govern the behavior of the field components in any transverse (x, y) plane:

$$-j\omega\mu H_x(x, y) = \frac{\partial}{\partial y} E_z(x, y) + \gamma E_y(x, y), \tag{5.3}$$

$$-j\omega\mu H_y(x, y) = -\gamma E_x(x, y) - \frac{\partial}{\partial x} E_z(x, y), \tag{5.4}$$

$$-j\omega\mu H_z(x, y) = \frac{\partial}{\partial x} E_y(x, y) - \frac{\partial}{\partial y} E_x(x, y), \tag{5.5}$$

$$j\omega\varepsilon E_x(x, y) = \frac{\partial}{\partial y} H_z(x, y) + \gamma H_y(x, y), \tag{5.6}$$

$$j\omega\varepsilon E_y(x, y) = -\gamma H_x(x, y) - \frac{\partial}{\partial x} H_z(x, y), \tag{5.7}$$

$$j\omega\varepsilon E_z(x, y) = \frac{\partial}{\partial x} H_y(x, y) - \frac{\partial}{\partial y} H_x(x, y). \tag{5.8}$$

The field components $E_z(x, y)$ and $H_z(x, y)$ are called the longitudinal field components; also, $E_x(x, y)$, $E_y(x, y)$, $H_x(x, y)$, and $H_y(x, y)$ are called the transverse field components. First, we shall express all the four transverse field components in terms of the two longitudinal field components. Such a procedure is useful only when $E_z(x, y)$ and $H_z(x, y)$ are not *both* identically zero.

1. Non-existence of TEM Mode

We can argue now that $E_z(x, y)$ and $H_z(x, y)$ cannot both be identically zero in a hollow waveguide such as a rectangular waveguide. We can establish this result by showing that the assumption $E_z(x, y) = H_z(x, y) = 0$ leads to a contradiction.

Since $H_z(x, y) = 0$ and since $\nabla_t \cdot \boldsymbol{H}_t(x, y) = 0$, where the subscript t denotes the components transverse to the z-direction, it follows that the magnetic field lines form closed loops in the transverse plane. The second of Maxwell's equations shows that the line integral of the magnetic field taken around any of the above-mentioned closed paths is equal to the total axial current enclosed. The absence of conductors inside the rectangular waveguide leads to the absence of conduction current in the axial direction. Also, the displacement current in the axial direction is zero since we have assumed that $E_z(x, y) = 0$. Thus, the total axial current is zero and hence there cannot be any closed loop of magnetic field lines in the transverse plane. This is a contradiction. Hence, the original assumption that $E_z(x, y) = H_z(x, y) = 0$ is not valid. Therefore, $E_z(x, y)$ and $H_z(x, y)$ cannot be identically zero in a hollow waveguide. In other words, there is no TEM mode in a rectangular waveguide.

2. Governing Equations for TE and TM Modes

We may show the solution of the simultaneous equations (5.3) and (5.7) for $H_x(x, y)$ and $E_y(x, y)$ to yield the following results:

$$H_x(x, y) = \frac{j\omega\varepsilon}{h^2}\frac{\partial}{\partial y} E_z(x, y) - \frac{\gamma}{h^2}\frac{\partial}{\partial x} H_z(x, y), \tag{5.9}$$

$$E_y(x, y) = -\frac{\gamma}{h^2}\frac{\partial}{\partial y} E_z(x, y) + \frac{j\omega\mu}{h^2}\frac{\partial}{\partial x} H_z(x, y), \tag{5.10}$$

where,

$$h^2 = \gamma^2 + \omega^2\mu\varepsilon = \gamma^2 + \omega^2/v^2. \tag{5.11}$$

In a similar manner, we can obtain from (5.4) and (5.6):

$$H_y(x, y) = -\frac{j\omega\varepsilon}{h^2}\frac{\partial}{\partial x} E_z(x, y) - \frac{\gamma}{h^2}\frac{\partial}{\partial y} H_z(x, y), \tag{5.12}$$

$$E_x(x, y) = -\frac{\gamma}{h^2}\frac{\partial}{\partial x} E_z(x, y) - \frac{j\omega\mu}{h^2}\frac{\partial}{\partial y} H_z(x, y). \tag{5.13}$$

If we substitute (5.13) and (5.10) into (5.5), as well as (5.9) and (5.12) into (5.8), we find the differential equations for $E_z(x, y)$ and $H_z(x, y)$ to be

$$\left(\frac{\partial^2}{\partial x^2} + \frac{\partial^2}{\partial y^2} + h^2\right)E_z(x, y) = 0, \tag{5.14}$$

and

$$\left(\frac{\partial^2}{\partial x^2} + \frac{\partial^2}{\partial y^2} + h^2\right)H_z(x, y) = 0. \tag{5.15}$$

Once we have obtained the longitudinal field components from the solutions of (5.14) and (5.15), the transverse field components can be evaluated with the help of (5.9), (5.10), (5.12), and (5.13).

As in the case of the parallel-plate waveguide, we can conveniently decompose the field components supported by a rectangular waveguide into the following two sets: (1) the transverse electric (TE) mode for which $E_z(x, y) \equiv 0$, and (2) the transverse magnetic (TM) mode for which $H_z(x, y) \equiv 0$. We should note, however, that as we have established previously, the TEM mode is not possible inside a rectangular waveguide. From (5.9), (5.10), and (5.12) to (5.15), we may collect the governing equations for the TE and the TM modes together as follows:

TE Mode $(E_z(x, y) \equiv 0)$

$$E_x(x, y) = -\frac{j\omega\mu}{h^2}\frac{\partial}{\partial y}H_z(x, y); \qquad E_y(x, y) = \frac{j\omega\mu}{h^2}\frac{\partial}{\partial x}H_z(x, y); \tag{5.16}$$

$$H_x(x, y) = -\frac{\gamma}{h^2}\frac{\partial}{\partial x}H_z(x, y); \qquad H_y(x, y) = -\frac{\gamma}{h^2}\frac{\partial}{\partial y}H_z(x, y); \tag{5.17}$$

$$\left(\frac{\partial^2}{\partial x^2} + \frac{\partial^2}{\partial y^2} + h^2\right)H_z(x, y) = 0. \tag{5.18}$$

TM Mode $(H_z(x, y) \equiv 0)$

$$E_x(x, y) = -\frac{\gamma}{h^2}\frac{\partial}{\partial x}E_z(x, y); \qquad E_y(x, y) = -\frac{\gamma}{h^2}\frac{\partial}{\partial y}E_z(x, y); \tag{5.19}$$

$$H_x(x, y) = \frac{j\omega\varepsilon}{h^2}\frac{\partial}{\partial y}E_z(x, y); \qquad H_y(x, y) = -\frac{j\omega\varepsilon}{h^2}\frac{\partial}{\partial x}E_z(x, y); \tag{5.20}$$

$$\left(\frac{\partial^2}{\partial x^2} + \frac{\partial^2}{\partial y^2} + h^2\right)E_z(x, y) = 0. \tag{5.21}$$

We shall study the characteristics of the TE mode first.

3. TE Mode

We can solve the partial differential equation (5.18) by the method of separation of variables. The principle of this method is to seek a solution of $H_z(x, y)$ in the form of a product of two functions one of which is a function of one of the variables and the other is a function of the other variable. Thus,

$$H_z(x, y) = X(x)Y(y). \tag{5.22}$$

If we substitute (5.22) into (5.18) and divide the resulting equation by (5.22) we obtain

$$\frac{X''(x)}{X(x)} + \frac{Y''(y)}{Y(y)} + h^2 = 0, \tag{5.23}$$

where each prime denotes a single differentiation with respect to the argument. On rearranging (5.23) as

$$\frac{X''(x)}{X(x)} = -\frac{Y''(y)}{Y(y)} - h^2, \tag{5.24}$$

we note that the left side is a function of x only and the right side is a function of y only. Also, the equality given by (5.24) is true for all values of x and y. We can obtain such a situation only when each side is equal to the same constant. If we denote this constant as $-\beta_x^2$, (5.24) may be seen to yield the following two ordinary differential equations:

$$X''(x) + \beta_x^2 X(x) = 0, \tag{5.25}$$

$$Y''(y) + \beta_y^2 Y(y) = 0, \tag{5.26}$$

where,

$$h^2 = \beta_x^2 + \beta_y^2. \tag{5.27}$$

We should note that β_x and β_y are arbitrary constants subject only to the stipulation given by (5.27). The partial differential equation (5.18) alone does not lead to any restrictions on the values of β_x and β_y. We shall show later that the boundary conditions on the walls of the waveguide stipulate that β_x and β_y can assume only certain specified values.

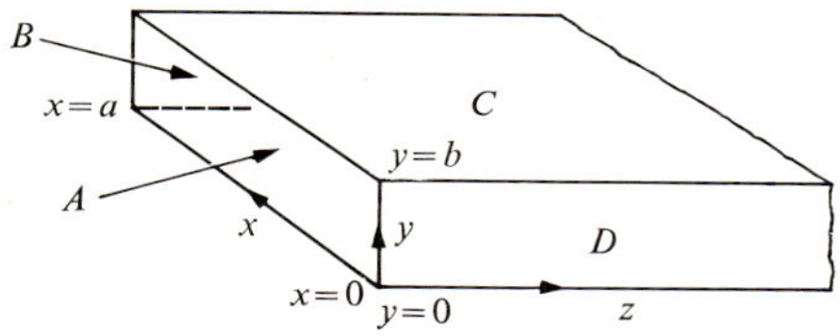

Fig. 8.7 Geometry of a rectangular waveguide.

We evaluate the general solutions of (5.25) and (5.26) as

$$X(x) = A_1 \cos \beta_x x + A_2 \sin \beta_x x, \tag{5.28}$$

$$Y(y) = B_1 \cos \beta_y y + B_2 \sin \beta_y y. \tag{5.29}$$

From (5.22), (5.28), and (5.29), we can obtain

$$H_z(x, y) = [A_1 \cos \beta_x x + A_2 \sin \beta_x x][B_1 \cos \beta_y y + B_2 \sin \beta_y y]. \tag{5.30}$$

We should note that A_1, A_2, B_1, and B_2 are arbitrary constants.

Before proceeding to apply the boundary conditions, we must specify the geometry of the rectangular waveguide. The waveguide extends from $x = 0$ to $x = a$ and from $y = 0$ to $y = b$, as shown in Fig. 8.7. In the z-direction, the waveguide extends from $z = -\infty$ to $z = \infty$ and, therefore, is uniform in the z-direction. We first assume that the walls of the waveguide are perfect conductors and subsequently discuss the effect of imperfect conductivity. In accordance with the usual convention, we assume the transverse dimension of the waveguide, parallel to the x-axis, to be greater than that parallel to the y-axis.

The appropriate boundary conditions are

$$E_y(x, y) = 0 \qquad (\text{for } x = 0, x = a;\ \text{for } 0 < y < b), \tag{5.31}$$

$$E_x(x, y) = 0 \qquad (\text{for } y = 0, y = b;\ \text{for } 0 < x < a). \tag{5.32}$$

We observe from (5.16) that (5.31) and (5.32) are respectively equivalent to the following:

$$\frac{\partial}{\partial x} H_z(x, y) = 0 \qquad (\text{for } x = 0, x = a; \text{ for } 0 < y < b), \tag{5.33}$$

$$\frac{\partial}{\partial y} H_z(x, y) = 0 \qquad (\text{for } y = 0, y = b; \text{ for } 0 < x < a). \tag{5.34}$$

From (5.29) and (5.30), we find that

$$\frac{\partial}{\partial x} H_z(x, y) = \beta_x[-A_1 \sin \beta_x x + A_2 \cos \beta_x x]Y(y). \tag{5.35}$$

At $x = 0$, (5.35) becomes

$$\frac{\partial}{\partial x} H_z(x, y) = \beta_x A_2 Y(y) \qquad (\text{for } x = 0). \tag{5.36}$$

In order for (5.36) to be equal to zero for any y in the range $0 < y < b$, in accordance with (5.33), it is necessary that $A_2 = 0$. Even though $\beta_x = 0$ also satisfies the required boundary condition, we shall show later that $\beta_x = 0$ is a special case of the general result that is derived starting from the fact that $A_2 = 0$. Since $A_2 = 0$, we can easily deduce from (5.35) that for $x = a$

$$\frac{\partial}{\partial x} H_z(x, y) = -\beta_x A_1 \sin(\beta_x a) Y(y) \qquad (\text{for } x = a). \tag{5.37}$$

We can verify that (5.37) is equal to zero for any y in the range $0 < y < b$, in accordance with (5.33), if either $A_1 = 0$ or $\beta_x = 0$ or $\sin \beta_x a = 0$. If $A_1 = 0$, since $A_2 = 0$ also, we can see from (5.30) that $H_z(x, y) = 0$, which is a trivial solution. In order to obtain a nontrivial solution for $H_z(x, y)$, it follows that A_1 should not be equal to zero. Also, as we mentioned before, $\beta_x = 0$ is a special case which is a part of the general result and hence the result $\beta_x = 0$ need not be considered further. Consequently, $\sin \beta_x a = 0$ or

$$\beta_x = m\pi/a \qquad (m \text{ is an integer}), \tag{5.38}$$

is the general requirement for (5.37) to be equal to zero for any y in the range $0 < y < b$. As we indicated earlier, the special result $\beta_x = 0$ is contained in (5.38) since for $m = 0$, $\beta_x = 0$. Therefore, the fulfillment of the boundary condition (5.33) requires that $X(x)$ given by (5.28) should be still further restricted to be of the form:

$$X(x) = A_1 \cos(m\pi x/a). \tag{5.39}$$

In a similar manner, we can show that the boundary condition (5.34) can be fulfilled if $B_2 = 0$ and

$$\beta_y = n\pi/b \qquad (n \text{ is an integer}), \tag{5.40}$$

or equivalently, $Y(y)$ given by (5.29) should be stipulated to be of the form:

$$Y(y) = B_1 \cos(n\pi y/b). \tag{5.41}$$

Hence, from (5.39) and (5.41), we have deduced $H_z(x, y)$ given by (5.30) to be of the form

$$H_z(x, y) = H_0 \cos(m\pi x/a) \cos(n\pi y/b), \tag{5.42}$$

where the arbitrary constant A_1B_1 has been denoted by H_0. We should emphasize that $H_z(x, y)$ given by (5.42) satisfies the partial differential equation (5.18) and the boundary conditions (5.33) and (5.34) and, therefore, (5.42) gives the *proper* form of the solution of $H_z(x, y)$ inside the rectangular waveguide.

We have not determined the constant H_0 and cannot evaluate it with the information that has been supplied. It is intuitively obvious that if there are no sources inside the waveguide, there can be no electromagnetic fields either. The constant H_0 is nonvanishing only in the presence of the source which, therefore, completely determines H_0. Consequently, H_0 may be called the excitation coefficient. As we have pointed out earlier, we do not propose to give a treatment of the excitation of the waveguide modes. Insofar as the development of the theory of the waveguide modes as contained here is concerned, H_0 is an arbitrary constant.

4. Critical or Cut-off Frequency

We can obtain from (5.11), (5.27), (5.38), and (5.40) that

$$\gamma = \sqrt{(m\pi/a)^2 + (n\pi/b)^2 - (\omega^2/v^2)}. \tag{5.43}$$

Let

$$\omega_{m,n} = v\sqrt{(m\pi/a)^2 + (n\pi/b)^2}. \tag{5.44}$$

From (5.43) and (5.44), we see that

$$\gamma = \alpha_z = \frac{\omega}{v}\sqrt{\left(\frac{\omega_{m,n}}{\omega}\right)^2 - 1}; \qquad \beta_z = 0 \qquad (\text{for } \omega < \omega_{m,n}); \tag{5.45a}$$

$$\gamma = j\beta_z = j\frac{\omega}{v}\sqrt{1 - \left(\frac{\omega_{m,n}}{\omega}\right)^2}; \qquad \alpha_z = 0 \qquad (\text{for } \omega > \omega_{m,n}). \tag{5.45b}$$

It follows from (5.45) that γ is imaginary for $\omega > \omega_{m,n}$ and γ is real for $\omega < \omega_{m,n}$. Consequently, (5.1) shows us that for $\omega < \omega_{m,n}$, the fields are exponentially attenuated in the z-direction and for $\omega > \omega_{m,n}$ the fields have phase progression or, therefore, propagation in the $+z$-direction. Since $\omega = \omega_{m,n}$ demarcates the boundary between the attenuation and the propagation regions, $\omega_{m,n}$ is called the cut-off or the critical angular frequency. In the remainder of the treatment of the characteristics of the TE mode of a rectangular waveguide, unless stated otherwise, we shall assume that $\omega > \omega_{m,n}$ and that the fields are propagating in the $+z$-direction.

5. Fundamental Mode

The field components obtained from (5.42), (5.16), and (5.17) are said to belong to the TE_{mn} mode, where TE denotes that the electric field is entirely in the transverse $(E_z(x, y) \equiv 0)$ direction. The subscripts m and n denote the number of periodic variations of the field components in the x- and y-directions respectively. If $m = n = 0$, $H_z(x, y)$ is a constant and (5.16) and (5.17) show that all the transverse field components vanish leading to no transport of electromagnetic energy. In other words, there is no TE_{00} mode. But, we can see that TE_{m0} and TE_{0n} modes are possible.

We can see from (5.44) that the critical angular frequency $\omega_{m,n}$ increases with the increase in the order m or n of the mode. Therefore, $\omega_{1,n} < \omega_{m,n}$ for $m > 1$ and any n. Also, $\omega_{m,1} < \omega_{m,n}$ for any m and $n > 1$. Thus, the critical angular frequencies of the two of the lower order modes are given by $\omega_{1,0} = \pi v/a$ and $\omega_{0,1} = \pi v/b$. Since by convention $a > b$, it follows that $\omega_{1,0}$ is the lowest critical angular frequency associated with the TE mode. We shall show later that the lowest critical angular frequency associated with the TM mode is greater than that associated with the TE mode. Consequently, TE_{10} mode has the lowest critical angular frequency. Since the propagation inside the waveguide takes place only at frequencies greater than the critical frequency, we see that the TE_{10} mode allows the propagation at frequencies lower than that permitted by any other mode. For this reason, TE_{10} mode in a rectangular waveguide is called *the fundamental mode.*

6. Brillouin Diagram

The Brillouin diagram, that is the plot of ω versus β_z, portrays clearly the propagation and the attenuation regions of the modes. The Brillouin diagram for the first few lower order TE modes is shown in Fig. 8.8. The normalized frequency $\omega/\omega_{1,0}$ is used, where $\omega_{1,0}$ is the critical angular frequency of the fundamental mode. For Fig. 8.8, we assume that $a = 2b = \pi$ cm. We observe that all the dispersion curves lie

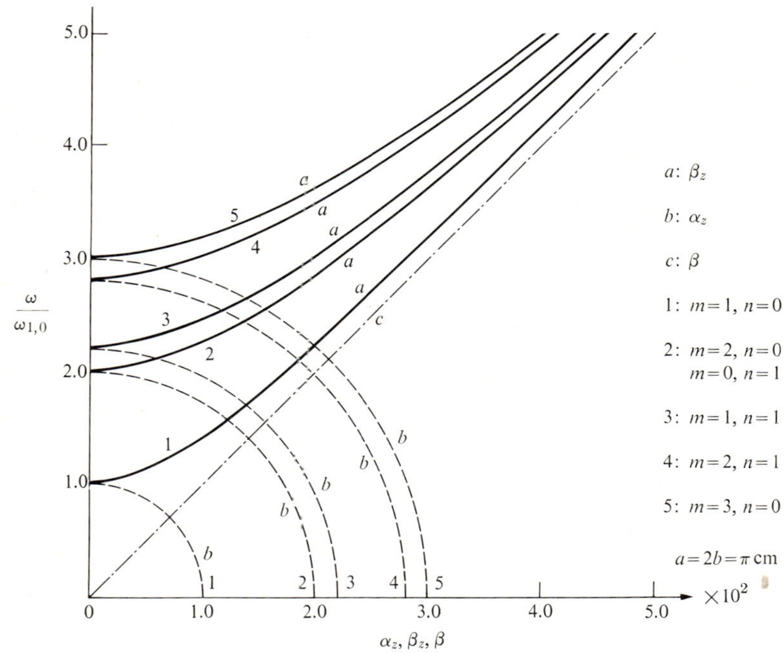

Fig. 8.8 Brillouin diagram for the first few lowest order modes guided by a rectangular waveguide.

in the fast-wave region. In the limit of very high frequencies, all the ω-β_z curves approach the ω-β straight line showing that the characteristics of all the waveguide modes approach those of the plane wave in the corresponding unbounded medium. Other features of the dispersion curves are evident from Fig. 8.8.

7. Degenerate Modes

For $a = 2b$, we can verify from (5.44) that $\omega_{2i,0} = \omega_{0,i}$, where i is an integer. Thus, the modes $TE_{2i,0}$ and $TE_{0,i}$ have identical critical (or cut-off) angular frequencies. Since the values of m and n are different for the two sets of modes, it follows from (5.42) that the field distributions inside the waveguide for the above-mentioned two sets of modes are different. The modes such as these, which have the same critical angular frequencies but different field distributions, are known as *degenerate modes.*

8. Phase Velocity, Wavelength and Group Velocity

For $\omega > \omega_{m,n}$, that is, in the propagation region, we can evaluate from (5.45b) the phase velocity $v_{ph,z}$, the wavelength λ_z, and the group velocity $v_{g,z}$, all in the direction ($\hat{z}$) of propagation, with the following results:

$$v_{ph,z} = v\{1 - (\omega_{m,n}/\omega)^2\}^{-1/2}, \tag{5.46}$$

$$\lambda_z = \lambda\{1 - (\omega_{m,n}/\omega)^2\}^{-1/2}, \tag{5.47}$$

$$v_{g,z} = \partial\omega/\partial\beta_z = v\{1 - (\omega_{m,n}/\omega)^2\}^{1/2}, \tag{5.48}$$

where v is the velocity and λ is the wavelength of the electromagnetic wave in the corresponding unbounded region. We can see from (5.46) that the guided wave propagation is dispersive since the phase velocity is a function of the frequency. The following feature of the waves in a dielectric is worthy of emphasis. The wave propagation in an *unbounded*, ordinary dielectric is *nondispersive.* But, the same dielectric when *bounded* by the perfectly conducting walls of a waveguide becomes *dispersive.*

From (5.1), (5.16), (5.17), and (5.42), we may deduce the field components of the TE_{mn} mode to be

$$E_x(x, y, z) = \frac{j\omega\mu}{h^2}\left(\frac{n\pi}{b}\right)H_0 \cos\left(\frac{m\pi x}{a}\right) \sin\left(\frac{n\pi y}{b}\right)e^{-\gamma z}, \tag{5.49}$$

$$E_y(x, y, z) = -\frac{j\omega\mu}{h^2}\left(\frac{m\pi}{a}\right)H_0 \sin\left(\frac{m\pi x}{a}\right) \cos\left(\frac{n\pi y}{b}\right)e^{-\gamma z}, \tag{5.50}$$

$$E_z(x, y, z) = 0, \tag{5.51}$$

$$H_x(x, y, z) = \frac{\gamma}{h^2}\left(\frac{m\pi}{a}\right)H_0 \sin\left(\frac{m\pi x}{a}\right) \cos\left(\frac{n\pi y}{b}\right)e^{-\gamma z}, \tag{5.52}$$

$$H_y(x, y, z) = \frac{\gamma}{h^2}\left(\frac{n\pi}{b}\right)H_0 \cos\left(\frac{m\pi x}{a}\right) \sin\left(\frac{n\pi y}{b}\right)e^{-\gamma z}, \tag{5.53}$$

$$H_z(x, y, z) = H_0 \cos\left(\frac{m\pi x}{a}\right) \cos\left(\frac{n\pi y}{b}\right)e^{-\gamma z}. \tag{5.54}$$

From (5.27), (5.38), and (5.40) we obtain

$$h^2 = (m\pi/a)^2 + (n\pi/b)^2, \tag{5.55}$$

which is real.

9. Propagation and Evanescence

We can evaluate the x-component of the time-averaged Poynting vector from (5.50) and (5.54) as

$$\begin{aligned} S_x(x, y, z) &= \tfrac{1}{2} \operatorname{Re} \left[E_y(x, y, z)H_z^*(x, y, z)\right], \\ &= \tfrac{1}{2} \operatorname{Re} \left[-\frac{j\omega\mu}{h^2}\left(\frac{m\pi}{a}\right)|H_0|^2 \sin\left(\frac{m\pi x}{a}\right)\cos\left(\frac{m\pi x}{a}\right)\cos^2\left(\frac{n\pi y}{b}\right)e^{-(\gamma-\gamma^*)z}\right]. \end{aligned} \tag{5.56}$$

We can see from (5.56) that $S_x(x, y, z) = 0$ irrespective of whether γ is real or imaginary corresponding to $\omega < \omega_{m,n}$ or $\omega > \omega_{m,n}$ respectively. Similarly, we find the y-component of the time-averaged Poynting vector from (5.49) and (5.54) to be

$$\begin{aligned} S_y(x, y, z) &= -\tfrac{1}{2} \operatorname{Re} \left[E_x(x, y, z)H_z^*(x, y, z)\right], \\ &= -\tfrac{1}{2} \operatorname{Re} \left[\frac{j\omega\mu}{h^2}\left(\frac{n\pi}{b}\right)|H_0|^2 \cos^2\left(\frac{m\pi x}{a}\right)\sin\left(\frac{n\pi y}{b}\right)\cos\left(\frac{n\pi y}{b}\right)e^{-(\gamma+\gamma^*)z}\right]. \end{aligned} \tag{5.57}$$

From (5.57), we find that $S_y(x, y, z) = 0$ both for $\omega < \omega_{m,n}$ and for $\omega > \omega_{m,n}$. We obtain the z-component of the time-averaged Poynting vector $S_z(x, y, z)$ from (5.49), (5.50), (5.52), and (5.53) as

$$\begin{aligned} S_z(x, y, z) &= \tfrac{1}{2} \operatorname{Re} \left[E_x(x, y, z)H_y^*(x, y, z) - E_y(x, y, z)H_x^*(x, y, z)\right], \\ &= \tfrac{1}{2} \operatorname{Re} \left[\frac{j\omega\mu\gamma^*}{h^4}\left(\frac{n\pi}{b}\right)^2|H_0|^2 \cos^2\left(\frac{m\pi x}{a}\right)\sin^2\left(\frac{n\pi y}{b}\right)e^{-(\gamma+\gamma^*)z}\right. \\ &\quad \left. + \frac{j\omega\mu\gamma^*}{h^4}\left(\frac{m\pi}{a}\right)^2|H_0|^2 \sin^2\left(\frac{m\pi x}{a}\right)\cos^2\left(\frac{n\pi y}{b}\right)e^{-(\gamma+\gamma^*)z}\right]. \end{aligned} \tag{5.58}$$

For $\omega < \omega_{m,n}$, $\gamma = \alpha_z$, which is real with the result that we can verify from (5.58) that $S_z(x, y, z) = 0$. But for $\omega > \omega_{m,n}$, $\gamma = j\beta_z$, which is imaginary; as a consequence, (5.58) is a nonvanishing quantity as given by

$$\begin{aligned} S_z(x, y, z) = \frac{\omega\mu\beta_z}{2h^4}|H_0|^2 &\left[\left(\frac{n\pi}{b}\right)^2\cos^2\left(\frac{m\pi x}{a}\right)\sin^2\left(\frac{n\pi y}{b}\right)\right. \\ &\left. + \left(\frac{m\pi}{a}\right)^2\sin^2\left(\frac{m\pi x}{a}\right)\cos^2\left(\frac{n\pi y}{b}\right)\right]. \end{aligned} \tag{5.59}$$

Thus, for $\omega < \omega_{m,n}$, there is no transport of electromagnetic energy, but for $\omega > \omega_{m,n}$ there is propagation of electromagnetic energy. Consequently, $\omega < \omega_{m,n}$ and $\omega > \omega_{m,n}$ correspond respectively to the evanescent and the propagation regions.

This general result, valid for ideal waveguides, has been proved before for the parallel-plate waveguide and is established here for the TE_{mn} modes in an ideal rectangular waveguide.

10. Energy Transport Velocity

We obtain the total time-averaged power flow through the cross section of the waveguide by integrating (5.59) from $x = 0$ to $x = a$ and from $y = 0$ to $y = b$. Hence,

$$P = \int_0^a dx \int_0^b dy S_z(x, y, z). \tag{5.60}$$

Since

$$\int_0^a \left[\begin{matrix}\cos\\ \sin\end{matrix}\left(\frac{m\pi x}{a}\right)\right]^2 dx = \frac{a}{\varepsilon_{0m}}, \qquad \int_0^b \left[\begin{matrix}\cos\\ \sin\end{matrix}\left(\frac{n\pi y}{b}\right)\right]^2 dy = \frac{b}{\varepsilon_{0m}}; \tag{5.61a}$$

where,

$$\varepsilon_{0i} = 1 \quad (\text{for } i = 0); \qquad \varepsilon_{0i} = 2 \quad (\text{for } i \neq 0); \qquad i = m, n, \tag{5.61b}$$

(5.60) together with (5.55) and (5.59) yields

$$P = \frac{\omega\mu\beta_z\, |H_0|^2\, ab}{2h^2\varepsilon_{0m}\varepsilon_{0n}}. \tag{5.62}$$

Hence, we find the spatial average of the time-averaged power flow through the cross section of the waveguide from (5.62) to be

$$\bar{S}_z = \frac{\omega\mu\beta_z\, |H_0|^2}{2h^2\varepsilon_{0m}\varepsilon_{0n}}. \tag{5.63}$$

For $\omega > \omega_{m,n}$, $\gamma = j\beta_z$ and we can evaluate the time-averaged energy density $W_E(x, y)$ associated with the electric field from (5.49) and (5.50) as

$$\begin{aligned} W_E(x, y) &= \frac{\varepsilon}{4}\,|\boldsymbol{E}(x, y, z)|^2 = \frac{\varepsilon}{4}\,[|E_x(x, y, z)|^2 + |E_y(x, y, z)|^2], \\ &= \frac{\varepsilon}{4}\frac{\omega^2\mu^2}{h^4}\,|H_0|^2 \left[\left(\frac{n\pi}{b}\right)^2 \cos^2\left(\frac{m\pi x}{a}\right)\sin^2\left(\frac{n\pi y}{b}\right)\right. \\ &\quad \left. + \left(\frac{m\pi}{a}\right)^2 \sin^2\left(\frac{m\pi x}{a}\right)\cos^2\left(\frac{n\pi y}{b}\right)\right]. \end{aligned} \tag{5.64}$$

We can calculate the spatial average of $W_E(x, y)$ from (5.64) to yield

$$\overline{W}_E = \frac{1}{ab}\int_0^a dx \int_0^b dy W_E(x, y) = \frac{\varepsilon}{4}\frac{\omega^2\mu^2}{h^2}\frac{|H_0|^2}{\varepsilon_{0m}\varepsilon_{0n}}. \tag{5.65}$$

By using (5.52) to (5.54) we can determine the time-averaged energy density $W_H(x, y)$ associated with the magnetic field as

$$\begin{aligned} W_H(x, y) &= \frac{\mu}{4}\,|\boldsymbol{H}(x, y, z)|^2 = \frac{\mu}{4}\,[|H_x(x, y, z)|^2 + |H_y(x, y, z)|^2 + |H_z(x, y, z)|^2], \\ &= \frac{\mu}{4}\,|H_0|^2 \left[\frac{\beta_z^2}{h^4}\left(\frac{m\pi}{a}\right)^2 \sin^2\left(\frac{m\pi x}{a}\right)\cos^2\left(\frac{n\pi y}{b}\right)\right. \\ &\quad \left. + \frac{\beta_z^2}{h^4}\left(\frac{n\pi}{b}\right)^2 \cos^2\left(\frac{m\pi x}{a}\right)\sin^2\left(\frac{n\pi y}{b}\right) + \cos^2\left(\frac{m\pi x}{a}\right)\cos^2\left(\frac{n\pi y}{b}\right)\right]. \end{aligned} \tag{5.66}$$

We can deduce the spatial average of $W_H(x, y)$ from (5.66) as

$$\overline{W_H} = \frac{1}{ab}\int_0^a dx \int_0^b dy W_H(x, y) = \frac{\mu}{4}\frac{|H_0|^2}{\varepsilon_{0m}\varepsilon_{0n}}\left[\frac{\beta_z^2\{(m\pi/a)^2 + (n\pi/b)^2\}}{h^4} + 1\right]. \quad (5.67)$$

With the help of (5.11) and (5.55), we can simplify (5.67) to yield

$$\overline{W_H} = \frac{\mu}{4}\frac{|H_0|^2}{\varepsilon_{0m}\varepsilon_{0n}}\frac{\omega^2\mu\varepsilon}{h^2}. \quad (5.68)$$

We observe from (5.65) and (5.68) that $\overline{W_E} = \overline{W_H}$, as is to be expected. We obtain the energy transport velocity v_E from (5.45b), (5.63), (5.65), and (5.68) as

$$v_E = \frac{\bar{S}_z}{\overline{W_E} + \overline{W_H}} = \frac{\omega\mu\beta_z |H_0|^2 \varepsilon_{0m}\varepsilon_{0n} 2h^2}{\varepsilon_{0m}\varepsilon_{0n} 2h^2\omega^2\mu^2\varepsilon |H_0|^2} = \frac{v^2}{\omega}\beta_z = v\left\{1 - \left(\frac{\omega_{m,n}}{\omega}\right)^2\right\}^{1/2}. \quad (5.69)$$

The energy transport velocity, like the time-averaged Poynting vector, is in the z-direction. If we compare (5.69) with (5.48) we see that the group and the energy transport velocities are identical. In actuality, the result that the group and the energy transport velocities are identical is generally valid for any nondissipative system.

11. Surface Charge and Surface Current Densities

We now propose to determine the surface charge and the surface current densities on the four walls of the waveguide. For the sake of convenience, the walls along the planes $y = 0$ and $y = b$ are designated as A and C respectively. Similarly, the walls along the planes $x = 0$ and $x = a$ are designated as D and B respectively. These designations are indicated in Fig. 8.7. We obtain the surface charge density $\rho_s\ (y = 0)$ on the wall A from (5.50) as

$$\rho_s\,(y = 0) = \hat{y}\cdot\varepsilon\hat{y}E_y(x, 0, z) = -\frac{j\omega\mu\varepsilon}{h^2}\left(\frac{m\pi}{a}\right)H_0 \sin\left(\frac{m\pi x}{a}\right)e^{-j\beta_z z}. \quad (5.70)$$

We find the surface charge density $\rho_s\ (y = b)$ on the wall C from (5.50) to be

$$\rho_s\,(y = b) = -\hat{y}\cdot\varepsilon\hat{y}E_y(x, b, z) = -\frac{j\omega\mu\varepsilon}{h^2}\left(\frac{m\pi}{a}\right)H_0 \sin\left(\frac{m\pi x}{a}\right)(-1)^{n+1}e^{-j\beta_z z}. \quad (5.71)$$

Similarly, we evaluate the surface charge densities $\rho_s\ (x = 0)$ and $\rho_s\ (x = a)$ respectively on the walls D and B from (5.49) as

$$\rho_s\,(x = 0) = \hat{x}\cdot\varepsilon\hat{x}E_x(0, y, z) = \frac{j\omega\mu\varepsilon}{h^2}\left(\frac{n\pi}{b}\right)H_0 \sin\left(\frac{n\pi y}{b}\right)e^{-j\beta_z z}, \quad (5.72)$$

and

$$\rho_s\,(x = a) = -\hat{x}\cdot\varepsilon\hat{x}E_x(a, y, z) = \frac{j\omega\mu\varepsilon}{h^2}\left(\frac{n\pi}{b}\right)H_0(-1)^{m+1} \sin\left(\frac{n\pi y}{b}\right)e^{-j\beta_z z}. \quad (5.73)$$

We can see from (5.70) and (5.71) that the magnitudes of the surface charge densities on the walls A and C are equal. Also, we can verify from (5.72) and (5.73) that the magnitudes of the surface charge densities on the walls D and B are equal.

We obtain the surface current density $\boldsymbol{J}_s\,(y = 0)$ on the wall A from (5.52) and (5.54) as

$$\begin{aligned}\boldsymbol{J}_s\,(y = 0) &= \hat{y} \times [\hat{x}H_x(x, 0, z) + \hat{z}H_z(x, 0, z)],\\ &= \hat{x}H_z(x, 0, z) - \hat{z}H_x(x, 0, z),\\ &= \hat{x}H_0 \cos\left(\frac{m\pi x}{a}\right)e^{-j\beta_z z} - \hat{z}\frac{j\beta_z}{h^2}\left(\frac{m\pi}{a}\right)H_0 \sin\left(\frac{m\pi x}{a}\right)e^{-j\beta_z z}.\end{aligned} \tag{5.74}$$

We also find the surface current density $\boldsymbol{J}_s\,(y = b)$ on the wall C from (5.52) and (5.54) as

$$\begin{aligned}\boldsymbol{J}_s\,(y = b) &= -\hat{y} \times [\hat{x}H_x(x, b, z) + \hat{z}H_z(x, b, z)],\\ &= -\hat{x}H_z(x, b, z) + \hat{z}H_x(x, b, z),\\ &= \hat{x}H_0 \cos\left(\frac{m\pi x}{a}\right)(-1)^{n+1}e^{-j\beta_z z}\\ &\quad - \hat{z}\frac{j\beta_z}{h^2}\left(\frac{m\pi}{a}\right)H_0 \sin\left(\frac{m\pi x}{a}\right)(-1)^{n+1}e^{-j\beta_z z}.\end{aligned} \tag{5.75}$$

In a similar manner, we can evaluate the surface current densities $\boldsymbol{J}_s\,(x = 0)$ and $\boldsymbol{J}_s\,(x = a)$ respectively on the walls D and B from (5.53) and (5.54) as

$$\begin{aligned}\boldsymbol{J}_s\,(x = 0) &= \hat{x} \times [\hat{y}H_y(0, y, z) + \hat{z}H_z(0, y, z)],\\ &= -\hat{y}H_0 \cos\left(\frac{n\pi y}{b}\right)e^{-j\beta_z z} + \hat{z}\frac{j\beta_z}{h^2}\left(\frac{n\pi}{b}\right)H_0 \sin\left(\frac{n\pi y}{b}\right)e^{-j\beta_z z},\end{aligned} \tag{5.76}$$

and

$$\begin{aligned}\boldsymbol{J}_s\,(x = a) &= -\hat{x} \times [\hat{y}H_y(a, y, z) + \hat{z}H_z(a, y, z)],\\ &= -\hat{y}H_0(-1)^{m+1} \cos\left(\frac{n\pi y}{b}\right)e^{-j\beta_z z}\\ &\quad + \hat{z}\frac{j\beta_z}{h^2}\left(\frac{n\pi}{b}\right)H_0(-1)^{m+1} \sin\left(\frac{n\pi y}{b}\right)e^{-j\beta_z z}.\end{aligned} \tag{5.77}$$

From (5.74) and (5.11), we deduce that

$$\begin{aligned}\nabla \cdot \{\boldsymbol{J}_s\,(y = 0)\} &= -\left(\frac{m\pi}{a}\right)H_0 \sin\left(\frac{m\pi x}{a}\right)e^{-j\beta_z z}\\ &\quad - \frac{\beta_z^2}{h^2}\left(\frac{m\pi}{a}\right)H_0 \sin\left(\frac{m\pi x}{a}\right)e^{-j\beta_z z},\\ &= -\frac{\omega^2\mu\varepsilon}{h^2}\left(\frac{m\pi}{a}\right)H_0 \sin\left(\frac{m\pi x}{a}\right)e^{-j\beta_z z}.\end{aligned} \tag{5.78}$$

We can show that the substitution of (5.70) in (5.78) gives

$$\nabla \cdot \{\boldsymbol{J}_s\,(y = 0)\} + j\omega\rho_s\,(y = 0) = 0, \tag{5.79}$$

which is nothing but the equation of continuity for the surface current. In a similar manner, we can prove that the surface current densities given by (5.75), (5.76), and (5.77) are related respectively to the surface charge densities given by (5.71), (5.72), and (5.73) through the equation of continuity similar to that given by (5.79). Therefore, we can evaluate the surface charge densities alternatively from a knowledge of the surface current densities.

The surface current densities given by (5.74) to (5.77), in general, have two components. It is helpful for us to define, in a manner similar to the electric and the magnetic field lines, the lines of surface current density as being everywhere parallel to the surface current density; also, the density of the lines is proportional to the magnitude of the surface current density. We can employ a procedure similar to that used in determining the field lines of the TE and the TM modes in a parallel-plate waveguide to evaluate the lines of the surface current densities on the four walls of the rectangular waveguide. Such a mapping of the lines of the surface current density on the waveguide walls is useful in assessing the effect of the slots on the waveguide walls.

12. Wave Impedance

For the case of plane wave propagation in an unbounded medium, we found the concept of intrinsic impedance to be very helpful. In a similar manner, we can introduce the concept of wave impedance for the field components propagating in a bounded medium such as a waveguide. We observe from (5.49), (5.50), (5.52), and (5.53), together with (5.45b), that

$$\frac{E_x(x, y, z)}{H_y(x, y, z)} = -\frac{E_y(x, y, z)}{H_x(x, y, z)} = \eta_0 \left\{1 - \left(\frac{\omega_{m,n}}{\omega}\right)^2\right\}^{-1/2} = Z_{\text{TE}}. \tag{5.80}$$

We can see the electric and the magnetic field components in the transverse plane from (5.80) to be related by the wave impedance Z_{TE}, where the subscript indicates that the wave impedance pertains to the TE mode. We should note that Z_{TE} is reactive below the cut-off frequency, is infinite at the cut-off frequency, and is resistive above the cut-off frequency. We should also note that Z_{TE} approaches η_0 in the limit of very high frequencies, as should be the case since in that limit, as mentioned earlier, the characteristics of wave propagation inside a waveguide approach those of the waves in the corresponding unbounded medium.

13. Characteristics of the Fundamental Mode

The fundamental mode (TE_{10}) is used extensively in practice and, hence, it is appropriate for us to summarize its characteristics. From (5.49) to (5.55), we obtain the field components of the TE_{10} mode as

$$E_y(x, y, z) = -\tilde{Z}_{\text{TE}} H_x(x, y, z) = -\frac{j\omega\mu}{(\pi/a)} H_0 \sin\left(\frac{\pi x}{a}\right) e^{-j\tilde{\beta}_z z}, \tag{5.81}$$

$$H_z(x, y, z) = H_0 \cos\left(\frac{\pi x}{a}\right) e^{-j\tilde{\beta}_z z}, \tag{5.82}$$

$$E_x(x, y, z) = E_z(x, y, z) = H_y(x, y, z) = 0, \tag{5.83}$$

where

$$\tilde{Z}_{TE} = \eta_0 \left\{1 - \left(\frac{\omega_{1,0}}{\omega}\right)^2\right\}^{-1/2}; \qquad \beta_z = \frac{\omega}{v}\sqrt{1 - \left(\frac{\omega_{1,0}}{\omega}\right)^2}; \qquad \omega_{1,0} = \frac{\pi v}{a}. \quad (5.84)$$

We can verify that the field components as given by (5.81) to (5.83) are of the same form as those of the TE_{10} mode in a parallel-plate waveguide with the walls occupying the planes $x = 0$ and $x = a$. We treated the TE_{10} mode in a parallel-plate waveguide in Section 8.2. There, we find that there are no electric field components parallel to the xz-plane. Consequently, perfectly conducting planes parallel to the xz-plane can be introduced without disturbing the existing fields, since the boundary conditions on these newly introduced planes are automatically satisfied by the original fields. If the newly introduced planes coincide with the planes $y = 0$ and $y = b$, the parallel-plate waveguide of Section 8.2 becomes the rectangular waveguide shown in Fig. 8.7. Therefore, as we have obtained in (5.81) and (5.82), the field components of the TE_{10} mode of a parallel-plate waveguide are of the same form as those of the TE_{10} mode in a rectangular waveguide. There is no stipulation on the exact locations of the newly introduced planes except that these perfectly conducting planes be parallel to the xz-plane. Therefore, we expect the characteristics of the TE_{10} mode to be independent of the dimension b. If we examine (5.84) we see that such is the case.

For the TE_{10} mode in a rectangular waveguide, the electric field lines in the yz-plane and the magnetic field lines in the xz-plane are identical to those of the TE_{10} mode in a parallel-plate waveguide and have already been shown in Fig. 8.4. The electric field lines in the cross section of the rectangular waveguide, that is, in the xy-plane, are shown in Fig. 8.9 for the TE_{10} mode.

The surface charge densities on the walls A and C are given respectively by (5.70) and (5.71) with $m = 1$ and $n = 0$. For the TE_{10} mode, the surface charge densities on the walls D and B we can see from (5.72) and (5.73) to be identically zero. Consequently, the surface current densities on the walls D and B should have no divergence. We see from (5.76) and (5.77) that the surface current densities on the walls D and B are only in the y-direction but are independent of y. Hence, the divergences of these surface current densities vanish in conformity with the results we obtained from (5.72) and (5.73). The surface current densities on the walls A and C are given respectively

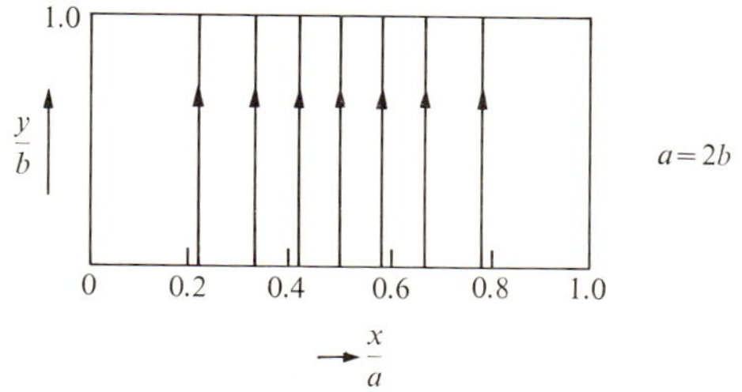

Fig. 8.9 Electric field lines of the TE_{10} mode guided by a rectangular waveguide.

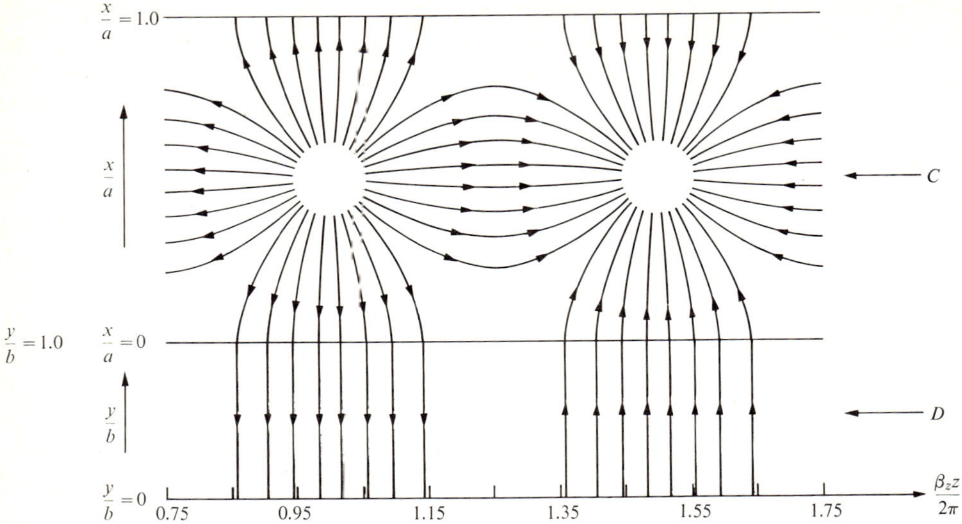

Fig. 8.10 Lines of electric current density on the walls of a rectangular waveguide propagating the TE_{10} mode.

by (5.74) and (5.75) with $m = 1$ and $n = 0$, and these current densities have two components.

For the TE_{10} mode of the rectangular waveguide, we obtain the surface current density on the wall D from (5.76) as

$$\mathbf{J}_s\,(x = 0) = -\hat{y}H_0 e^{-j\tilde{\beta}_z z}. \tag{5.85}$$

Similarly, we find the surface current density on the wall C from (5.75) as

$$\mathbf{J}_s\,(y = b) = -\hat{x}H_0 \cos\left(\frac{\pi x}{a}\right)e^{-j\tilde{\beta}_z z} + \hat{z}\,\frac{j\tilde{\beta}_z}{\pi/a}\,H_0 \sin\left(\frac{\pi x}{a}\right)e^{-j\tilde{\beta}_z z}. \tag{5.86}$$

We can verify that, in general, the terminal point of the current density vector as given by (5.86) traces an ellipse. The lines of surface current densities at a particular time ($t = 0$), as evaluated from (5.85) and (5.86), are depicted in Fig. 8.10. We should note that the lines of surface current densities on walls A and B are essentially the same as those on walls C and D respectively.

8.6 RECTANGULAR WAVEGUIDE: TM MODE

The governing equations of the TM mode are given by (5.19) to (5.21). We can solve the partial differential equation (5.21) in the same manner as we did (5.18). The appropriate boundary conditions for the TM mode are as follows:

$$E_z(x, y) = 0 \qquad (\text{for } x = 0, x = a;\ \text{for } 0 < y < b), \tag{6.1}$$

$$E_z(x, y) = 0 \qquad (\text{for } y = 0, y = b;\ \text{for } 0 < x < a), \tag{6.2}$$

$$E_y(x, y) = 0 \qquad (\text{for } x = 0, x = a;\ \text{for } 0 < y < b), \tag{6.3}$$

$$E_x(x, y) = 0 \qquad (\text{for } y = 0, y = b;\ \text{for } 0 < x < a). \tag{6.4}$$

We see from (5.19) that if (6.1) is satisfied, (6.3) is also automatically satisfied; also, if (6.2) is fulfilled, (6.4) is also automatically fulfilled. Hence, the necessary and the sufficient boundary conditions for the TM mode are those given by (6.1) and (6.2). By employing a procedure similar to that used for the TE mode, we can prove that the solution of (5.21) subject to the boundary conditions (6.1) and (6.2) is

$$E_z(x, y) = E_0 \sin \beta_x x \sin \beta_y y = E_0 \sin (m\pi x/a) \sin (n\pi y/b), \tag{6.5}$$

where β_x and β_y are identical to those given by (5.38) and (5.40) respectively. Also, E_0 is an arbitrary constant known as the excitation coefficient and its determination is not possible without the specification of the source which excites the guided waves. As for the case of the TE mode, we can ascertain that for the TM mode also the critical angular frequency $\omega_{m,n}$, the attenuation coefficient α_z for $\omega < \omega_{m,n}$, and the phase-shift coefficient β_z for $\omega > \omega_{m,n}$ are given by (5.44), (5.45a), and (5.45b) respectively. The TM modes are also exponentially attenuated for $\omega < \omega_{m,n}$ and are propagated without attenuation for $\omega > \omega_{m,n}$.

If either m or n is equal to zero, we can see from (6.5) together with (5.19) and (5.20) that all the field components vanish leading to a trivial solution. Hence, the lowest TM mode corresponds to $m = n = 1$. Since $\omega_{1,0} < \omega_{1,1}$, it follows that the critical angular frequency of the TM_{11} mode is greater than that of the TE_{10} mode. Hence, as we pointed out earlier, taking into account both the TE and the TM modes, TE_{10} is the fundamental mode in a rectangular waveguide.

Since β_z for the TM mode is the same as that for the TE mode, the Brillouin diagrams of the TM and the TE modes are the same for identical mode orders m and n. In Fig. 8.8, the Brillouin diagrams for the first few lower-order TM modes are also indicated for the waveguide of the same dimensions as used for the case of the TE mode. For identical mode orders, the dispersion characteristics of the TE and the TM modes are the same. Note that *the degenerate modes* also arise for the case of the TM modes. In addition, in the direction of propagation, the phase velocity $v_{ph,z}$, the wavelength λ_z, and the group velocity $v_{g,z}$ of the TM mode are the same as those given by (5.46) to (5.48) respectively.

From (6.5), (5.19), and (5.20), we may deduce the field components of the TM_{mn} mode to be

$$E_x(x, y, z) = -\frac{j\beta_z}{h^2}\left(\frac{m\pi}{a}\right)E_0 \cos\left(\frac{m\pi x}{a}\right) \sin\left(\frac{n\pi y}{b}\right)e^{-j\beta_z z}, \tag{6.6}$$

$$E_y(x, y, z) = -\frac{j\beta_z}{h^2}\left(\frac{n\pi}{b}\right)E_0 \sin\left(\frac{m\pi x}{a}\right) \cos\left(\frac{n\pi y}{b}\right)e^{-j\beta_z z}, \tag{6.7}$$

$$E_z(x, y, z) = E_0 \sin\left(\frac{m\pi x}{a}\right) \sin\left(\frac{n\pi y}{b}\right)e^{-j\beta_z z}, \tag{6.8}$$

$$H_x(x, y, z) = \frac{j\omega\varepsilon}{h^2}\left(\frac{n\pi}{b}\right)E_0 \sin\left(\frac{m\pi x}{a}\right) \cos\left(\frac{n\pi y}{b}\right)e^{-j\beta_z z}, \tag{6.9}$$

$$H_y(x, y, z) = -\frac{j\omega\varepsilon}{h^2}\left(\frac{m\pi}{a}\right)E_0 \cos\left(\frac{m\pi x}{a}\right) \sin\left(\frac{n\pi y}{b}\right)e^{-j\beta_z z}, \tag{6.10}$$

$$H_z(x, y, z) = 0. \tag{6.11}$$

With the help of (6.6) to (6.11), we can show the x- and the y-components of the time-averaged Poynting vectors, $S_x(x, y, z)$ and $S_y(x, y, z)$, to vanish both for $\omega < \omega_{m,n}$ and for $\omega > \omega_{m,n}$. In a similar manner, we can prove the z-component of the time-averaged Poynting vector to be equal to zero for $\omega < \omega_{m,n}$ and non-zero for $\omega > \omega_{m,n}$. There is no transport of electromagnetic energy for $\omega < \omega_{m,n}$ but the electromagnetic energy is propagated for $\omega > \omega_{m,n}$. Therefore, as before, $\omega < \omega_{m,n}$ and $\omega > \omega_{m,n}$ correspond to the evanescent and the propagation regions respectively.

We can evaluate the energy transport velocity for the TM_{mn} mode in the same manner as for the TE_{mn} mode and then verify that it is equal to the group velocity as given by (5.48), in conformity with the generally valid result for any nondissipative system.

We can determine the surface charge densities on the four walls of the waveguide from (6.6) and (6.7); the results are not included here. We can evaluate the surface current densities on the four walls of the waveguide with the help of (6.9) to (6.11) with the following results:

$$\textit{Wall A:}\ \mathbf{J}_s(y = 0) = \hat{y} \times \hat{x}H_x(x, 0, z) = -\hat{z}\frac{j\omega\varepsilon}{h^2}\left(\frac{n\pi}{b}\right)E_0 \sin\left(\frac{m\pi x}{a}\right)e^{-j\beta_z z}, \tag{6.12}$$

$$\textit{Wall C:}\ \mathbf{J}_s(y = b) = -\hat{y} \times \hat{x}H_x(x, b, z)$$

$$= -\hat{z}\frac{j\omega\varepsilon}{h^2}\left(\frac{n\pi}{b}\right)E_0 \sin\left(\frac{m\pi x}{a}\right)(-1)^{n+1}e^{-j\beta_z z}, \tag{6.13}$$

$$\textit{Wall D:}\ \mathbf{J}_s(x = 0) = \hat{x} \times \hat{y}H_y(0, y, z) = -\hat{z}\frac{j\omega\varepsilon}{h^2}\left(\frac{m\pi}{a}\right)E_0 \sin\left(\frac{n\pi y}{b}\right)e^{-j\beta_z z}, \tag{6.14}$$

$$\textit{Wall B:}\ \mathbf{J}_s(x = a) = -\hat{x} \times \hat{y}H_y(a, y, z)$$

$$= -\hat{z}\frac{j\omega\varepsilon}{h^2}\left(\frac{m\pi}{a}\right)E_0(-1)^{m+1} \sin\left(\frac{n\pi y}{b}\right)e^{-j\beta_z z}. \tag{6.15}$$

In contrast to the case of the TE_{mn} mode, the surface current densities on the walls of the waveguide propagating the TM_{mn} mode have only one component which is in the "longitudinal" direction. Consequently, we can see that the lines of surface current densities on the four walls of a rectangular waveguide supporting a TM_{mn} mode are a set of straight lines parallel to the z-direction.

We can deduce the wave impedance Z_{TM} for the TM_{mn} mode in a rectangular waveguide from (6.6), (6.7), (6.9), and (6.10) as

$$\frac{E_x(x, y, z)}{H_y(x, y, z)} = -\frac{E_y(x, y, z)}{H_x(x, y, z)} = \eta_0\left\{1 - \left(\frac{\omega_{m,n}}{\omega}\right)^2\right\}^{1/2} = Z_{\text{TM}}. \tag{6.16}$$

If we compare (6.16) with (5.80) we see that Z_{TM} and Z_{TE} are different. The wave impedance Z_{TM} of the TM_{mn} mode is reactive below the cut-off frequency, is zero at the cut-off frequency, and is resistive above the cut-off frequency. We should note that Z_{TM}, like Z_{TE}, approaches η_0 in the limit of very high frequencies.

8.7 EFFECT OF LOSSES ON THE MODES OF A RECTANGULAR WAVEGUIDE

We have treated the characteristics of the guided modes of an ideal rectangular waveguide in the preceding two sections. We can evaluate the effect of the losses in the dielectric and that due to the imperfect conductivity of the walls of the waveguide using the same procedure as employed for the parallel-plate waveguide. As before, we assume that these losses are very small with the result that they do not perturb the field pattern significantly but merely introduce attenuation in the direction of propagation. We can determine the attenuation coefficient separately for the dielectric and the conductor losses.

1. Dielectric Losses

For both the TE and the TM modes, we find from (5.43) that

$$j\beta_z = j\sqrt{\omega^2\mu\varepsilon - (m\pi/a)^2 - (n\pi/b)^2}. \tag{7.1}$$

We should note that for ε real and $\omega > \omega_{m,n}$, β_z is real, resulting in the propagation of the modes without any attenuation. If there are losses in the dielectric, real ε changes to complex $\varepsilon_{\mathrm{eff}} = \varepsilon + (\sigma/j\omega)$, where σ is the conductivity of the dielectric. Therefore, (7.1) becomes

$$\gamma = j\sqrt{\omega^2\mu\varepsilon - (m\pi/a)^2 - (n\pi/b)^2 - j\omega\mu\sigma}. \tag{7.2}$$

The real part $\alpha_{d,z}$ of γ as given in (7.2) yields the attenuation due to the dielectric losses. If

$$\omega\mu\sigma \ll |\{\omega^2\mu\varepsilon - (m\pi/a)^2 - (n\pi/b)^2\}| = \omega^2\mu\varepsilon\,|\{1 - (\omega_{m,n}/\omega)^2\}|, \tag{7.3}$$

we can expand (7.2) into a binomial series and the result of retaining only the first two terms of the series is found to yield

$$\gamma = \alpha_{d,z} + j\beta_z = \frac{\sigma}{2}\sqrt{\frac{\mu}{\varepsilon}}\left\{1 - \left(\frac{\omega_{m,n}}{\omega}\right)^2\right\}^{-1/2} + j\omega\sqrt{\mu\varepsilon}\left\{1 - \left(\frac{\omega_{m,n}}{\omega}\right)^2\right\}^{1/2}. \tag{7.4}$$

We should note that, in terms of the cut-off angular frequency $\omega_{m,n}$, (7.4) is the same as (4.5). Therefore, the remarks given in connection with (4.5) are also applicable to (7.4).

2. Conductor Losses: TE Mode

Let $\alpha_{c,z}$ be the attenuation coefficient due to the imperfect conductivity of the walls of the rectangular waveguide. All the field components of the TE mode supported by a rectangular waveguide with perfectly conducting walls now have to include the multiplicative factor $e^{-\alpha_{c,z}z}$. Therefore, for the TE mode, we obtain the total time-averaged power flowing through the cross section of the waveguide from (5.62) as

$$P(z) = \frac{\omega\mu\beta_z\,|H_0|^2\,ab}{2h^2\varepsilon_{0m}\varepsilon_{0n}}\,e^{-2\alpha_{c,z}z}. \tag{7.5}$$

It is now necessary for us to compute the total power $P_L(z)$ dissipated per unit distance in the direction of propagation. This power is equal to that flowing into the bounding walls per unit distance in the z-direction. Let us consider first the bounding wall A as shown in Fig. 8.7. We obtain the surface current density $\boldsymbol{J}_s\,(y = 0)$ on wall A from (5.74) as:

$$\boldsymbol{J}_s\,(y = 0) = \hat{x}H_0 \cos\left(\frac{m\pi x}{a}\right)e^{-j\beta_z z - \alpha_{c,z}z} - \hat{z}\frac{j\beta_z}{h^2}\left(\frac{m\pi}{a}\right)H_0 \sin\left(\frac{m\pi x}{a}\right)e^{-j\beta_z z - \alpha_{c,z}z}. \tag{7.6}$$

Let us consider an elemental area $dx\,dz$ of an infinitesimal length dx in the x-direction and dz in the z-direction. Let

$$\eta_{sr} = \sqrt{\omega\mu_c/2\sigma_c}, \tag{7.7}$$

be the surface resistivity of the walls of the waveguide. In (7.7), μ_c is the permeability and σ_c is the conductivity of the material of the walls of the waveguide. Then, we find from (7.6) that the time-averaged power $dP_{La}(z)$ flowing through the elemental area $dx\,dz$ in wall A of the waveguide is

$$\begin{aligned} dP_{La}(z) &= \tfrac{1}{2}\eta_{sr}\,|\boldsymbol{J}_s\,(y = 0)|^2\,dx\,dz, \\ &= \frac{\eta_{sr}}{2}|H_0|^2\,e^{-2\alpha_{c,z}z}\left[\cos^2\left(\frac{m\pi x}{a}\right) + \frac{\beta_z^2}{h^4}\left(\frac{m\pi}{a}\right)^2\sin^2\left(\frac{m\pi x}{a}\right)\right]dx\,dz. \end{aligned} \tag{7.8}$$

If we integrate $dP_{La}(z)$ given by (7.8) with respect to x from $x = 0$ to $x = a$ and divide the resulting expression by dz, we obtain the expression for the time-averaged power $P_{La}(z)$ flowing into wall A of the waveguide *per* unit distance in the direction ($\hat{z}$) of propagation. Hence,

$$\begin{aligned} P_{La}(z) &= \frac{\eta_{sr}}{2}|H_0|^2\,e^{-2\alpha_{c,z}z}\int_0^a\left[\cos^2\left(\frac{m\pi x}{a}\right) + \frac{\beta_z^2}{h^4}\left(\frac{m\pi}{a}\right)^2\sin^2\left(\frac{m\pi x}{a}\right)\right]dx, \\ &= \frac{\eta_{sr}}{2}|H_0|^2\,e^{-2\alpha_{c,z}z}\,\frac{a}{\varepsilon_{0m}}\left[1 + \frac{\beta_z^2}{h^4}\left(\frac{m\pi}{a}\right)^2\right]. \end{aligned} \tag{7.9}$$

In evaluating the integral in (7.9), we have used (5.61a). In a similar manner, we can determine the time-averaged power $P_{Lc}(z)$ flowing into wall C of the waveguide *per* unit distance in the direction ($\hat{z}$) of propagation by using (5.75), with the following result:

$$P_{Lc}(z) = P_{La}(z). \tag{7.10}$$

We can evaluate the time-averaged power $P_{Ld}(z)$ flowing through an elemental area $dy\,dz$ in wall D of the waveguide from (5.76) as

$$dP_{Ld}(z) = \frac{\eta_{sr}}{2}|H_0|^2\,e^{-2\alpha_{c,z}z}\left[\cos^2\left(\frac{n\pi y}{b}\right) + \frac{\beta_z^2}{h^4}\left(\frac{n\pi}{b}\right)^2\sin^2\left(\frac{n\pi y}{b}\right)\right]dy\,dz. \tag{7.11}$$

We should note that the current density given by (5.76) is multiplied by the factor $e^{-\alpha_{c,z}z}$ to take into account the attenuation due to the conductor losses. With the help of (7.11) and (5.61a), we can evaluate the time-averaged power $P_{Ld}(z)$ flowing into wall D of the waveguide *per* unit distance in the direction of propagation as

$$P_{Ld}(z) = \frac{\eta_{sr}}{2}|H_0|^2 e^{-2\alpha_{c,z}z}\frac{b}{\varepsilon_{0n}}\left[1 + \frac{\beta_z^2}{h^4}\left(\frac{n\pi}{b}\right)^2\right]. \tag{7.12}$$

Similarly, we can deduce the time-averaged power $P_{Lb}(z)$ flowing into wall B of the waveguide *per* unit distance in the direction of propagation from (5.77) as

$$P_{Lb}(z) = P_{Ld}(z). \tag{7.13}$$

From (7.9), (7.10), (7.12), and (7.13), we obtain the total time-averaged power $P_L(z)$ dissipated per unit distance in the direction of propagation as

$$\begin{aligned} P_L(z) &= P_{La}(z) + P_{Lb}(z) + P_{Lc}(z) + P_{Ld}(z), \\ &= \frac{\eta_{sr}|H_0|^2 e^{-2\alpha_{c,z}z}}{\varepsilon_{0m}\varepsilon_{0n}}\left[\varepsilon_{0n}a + \varepsilon_{0m}b + \frac{\beta_z^2}{h^4}\left\{\varepsilon_{0n}a\left(\frac{m\pi}{a}\right)^2 + \varepsilon_{0m}b\left(\frac{n\pi}{b}\right)^2\right\}\right]. \end{aligned} \tag{7.14}$$

We can show that the substitution of (7.5) and (7.14) into (4.10) together with the use of (5.44) and (5.45b) yields the following expression for the attenuation coefficient $\alpha_{c,z}$:

TE$_{mn}$ *Mode*

$$\begin{aligned} \alpha_{c,z} = \frac{\eta_{sr}}{\eta_0 ab}\left[1 - \left(\frac{\omega_{m,n}}{\omega}\right)^2\right]^{-1/2}&\left[\left(\frac{\omega_{m,n}}{\omega}\right)^2(\varepsilon_{0n}a + \varepsilon_{0m}b)\right. \\ &\left. + \frac{\{1 - (\omega_{m,n}/\omega)^2\}}{h^2}\left\{\varepsilon_{0n}a\left(\frac{m\pi}{a}\right)^2 + \varepsilon_{0m}b\left(\frac{n\pi}{b}\right)^2\right\}\right], \end{aligned} \tag{7.15}$$

where the expression for h^2 is given by (5.55). For the dominant or the fundamental mode, that is the TE$_{10}$ mode, $m = 1, n = 0, \varepsilon_{0m} = 2$, and $\varepsilon_{0n} = 1$. Therefore, from (7.15), we obtain the following expression for the attenuation coefficient of the dominant mode:

TE$_{10}$ *Mode*

$$\alpha_{c,z} = \frac{\eta_{sr}}{\eta_0 b}\left\{1 - \left(\frac{\omega_{1,0}}{\omega}\right)^2\right\}^{-1/2}\left[1 + \frac{2b}{a}\left(\frac{\omega_{1,0}}{\omega}\right)^2\right]. \tag{7.16}$$

3. Conductor Losses: TM Mode

We can deduce the attenuation coefficient $\alpha_{c,z}$ due to the imperfect conductivity of the walls of a rectangular waveguide supporting a TM mode with the help of (6.6), (6.7), (6.9), (6.10), and (6.12) to (6.15) by employing a procedure similar to that used for the TE mode. We omit the details and the result is:

TM *Mode*

$$\alpha_{c,z} = \frac{2\eta_{sr}}{\eta_0 ab}\left\{1 - \left(\frac{\omega_{m,n}}{\omega}\right)^2\right\}^{-1/2}\frac{[b(m\pi/a)^2 + a(n\pi/b)^2]}{h^2}. \tag{7.17}$$

We should note that the lowest TM mode corresponds to $m = n = 1$.

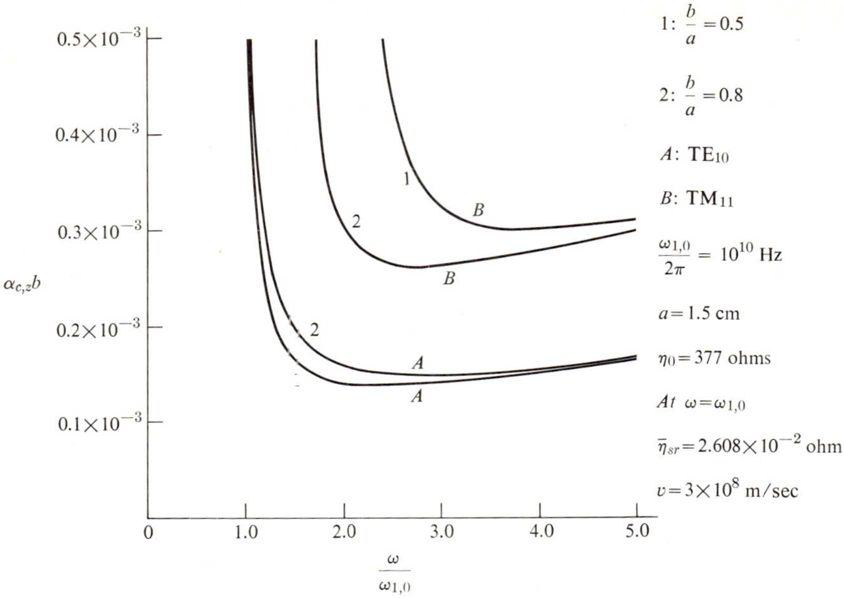

Fig. 8.11 Normalized attenuation coefficients of the TE_{10} and the TM_{11} modes of a rectangular waveguide with imperfectly conducting walls as a function of the normalized frequency.

The attenuation coefficient $\alpha_{c,z}$ due to the conductor losses are depicted in Fig. 8.11 for the TE_{10} and the TM_{11} modes. We observe that the attenuation coefficient is infinite at the critical angular frequency $\omega_{m,n}$ decreases as ω is increased beyond $\omega_{m,n}$, reaches a minimum and then increases for the further increases in ω.

We should remember that the use of the concept of the surface resistivity in the evaluation of the attenuation coefficient of the guided modes due to the imperfect conductivity of the walls of the waveguide implies that the thickness of the waveguide walls is very much larger than the skin depth. This requirement is usually met in practice.

For the evaluation of the effect of the dielectric and the conductor losses on the modes of a waveguide, we assume that these losses are very small and that they do not perturb the field patterns significantly but merely introduce attenuation in the direction of propagation. It is worthwhile for us to point out that the field patterns are affected by the losses in the waveguide. When there are losses in the waveguide, it is reasonable for us to expect the fields to be attenuated in the direction of propagation since part of the wave energy is dissipated by the dielectric and the conductor losses. Hence, the propagation coefficient $\gamma = \alpha_z + j\beta_z$ is complex and therefore γ^2 is also complex. In the absence of losses, γ is real for frequencies below, and imaginary for frequencies above the cut-off frequency and, therefore, γ^2 is real. It follows from (5.11) and (5.27) that, in general, β_x and β_y are real in the absence of losses but become complex with the inclusion of the losses. Let us suppose that $\beta_x = \beta_{xr} + j\beta_{xj}$, where

β_{xr} and β_{xj} are real. In the presence of losses, we cannot find a solution of the general form $H_z(x, y, z) = H_0 Y(y) \cos \beta_x x$ to satisfy the boundary condition (5.33) for $x = a$. Consequently, (5.42) is not a true solution. The implication is that the losses affect the field patterns also. But, if the losses are so small that the imaginary parts of β_x and β_y are negligible compared to their real parts, then (5.42) is a good approximation to the actual solution and the field patterns are then not affected significantly. The foregoing discussion shows that the procedure used for the evaluation of the attenuation coefficient due to the dielectric and the conductor losses is an approximation valid only when the losses are so small that the resulting imaginary parts of β_x and β_y are negligible compared to their real parts, leading to field patterns that are essentially unaffected by the inclusion of the losses. We should emphasize that a rigorous solution of the problem of the waveguide with losses is quite difficult.

8.8 DIELECTRIC SLAB WAVEGUIDE: TE MODE

For the waveguides with perfectly conducting walls, in the transverse plane, the electromagnetic energy is completely confined within the waveguide but is propagated in the longitudinal direction, that is, in the direction of the axis of the waveguide. It is not necessary to have completely enclosed structures to localize the electromagnetic energy in a small region in the transverse plane. A dielectric region which has translational invariance in the longitudinal direction can support electromagnetic fields which, in the ideal case, are propagated in the longitudinal direction and which, in the transverse plane, decay very rapidly away from the surface of the dielectric region. Even though the guiding structure is open, there is no transport of electromagnetic energy in the transverse plane. The energy density falls off very rapidly from the surface of the dielectric region with the result that the electromagnetic energy, though in principle extending throughout in the transverse plane, is for all essential purposes confined to the very close neighborhood of the surface of the dielectric region. Consequently, the guided electromagnetic waves supported by the "open" dielectric structures are called the surface waves and the guiding structures are called the surface waveguides. The simplest of all the possible surface waveguides is an infinite, planar dielectric slab. We propose to treat the characteristics of the guided waves supported by a dielectric slab waveguide with a view to bringing out all the essential features of the surface waveguides.

Let us consider an infinite planar dielectric slab bounded by the parallel planes at $x = d$ and $x = -d$ respectively, as shown in Fig. 8.12. The slab thickness is $2d$. Let μ_d and ε_d be the constitutive parameters of the dielectric slab. The dielectric slab is surrounded by free space whose permeability and permittivity are given by μ_0 and ε_0 respectively. We shall only consider the two-dimensional fields which have no dependence on y. We showed in Section 6.6 that the two-dimensional electromagnetic fields are capable of being decomposed into two independent sets, one of which has the E-field and the other the H-field in the y-direction. The first case which has no component of the electric field in the propagation direction ($\hat{z}$) is called the TE mode and the second case which has no component of the magnetic field in the

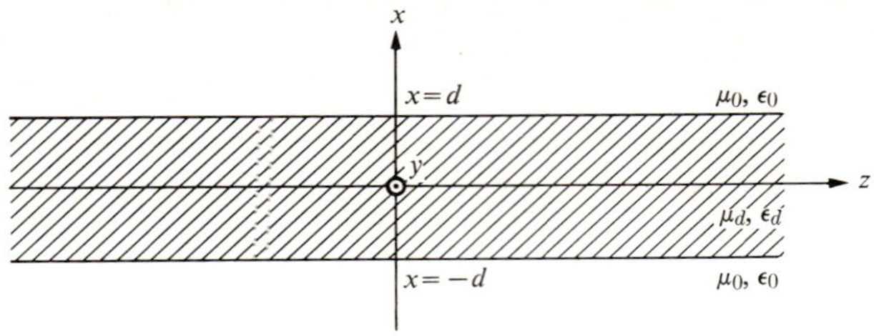

Fig. 8.12 Geometry of the dielectric slab waveguide.

propagation direction is called the TM mode. In this section, we shall analyze the characteristics of the TE surface wave guided by a planar dielectric slab.

For the TE mode, $E_y(x, z)$, $H_x(x, z)$, and $H_z(x, z)$ are the only nonvanishing field components and we can reproduce the appropriate governing equations from (6.6.4) and (6.6.5) as

$$-j\omega\mu H_x(x, z) = -\frac{\partial}{\partial z} E_y(x, z), \tag{8.1}$$

$$-j\omega\mu H_z(x, z) = \frac{\partial}{\partial x} E_y(x, z), \tag{8.2}$$

$$\left[\frac{\partial^2}{\partial x^2} + \frac{\partial^2}{\partial z^2} + \beta^2\right] E_y(x, z) = 0; \qquad \beta = \omega\sqrt{\mu\varepsilon}. \tag{8.3}$$

In the dielectric region $|x| < d$, $\mu = \mu_d$, $\varepsilon = \varepsilon_d$, and the field components and β are distinguished by the additional subscript 1. In the free space region $|x| > d$, $\mu = \mu_0$, $\varepsilon = \varepsilon_0$ and the field components and β are denoted by the additional subscript 2. It is convenient for us to subdivide $E_y(x, z)$ as

$$E_y(x, z) = E_y^o(x, z) + E_y^e(x, z), \tag{8.4}$$

where $E_y^o(x, z)$ and $E_y^e(x, z)$ are respectively odd and even functions of x. The TE mode obtained from $E_y^o(x, z)$ is called the *odd* TE mode. Similarly, the TE mode obtained from $E_y^e(x, z)$ is called the *even* TE mode. If $E_y(x, z)$ is an odd function of x, we see from (8.1) and (8.2) that $H_x(x, z)$ and $H_z(x, z)$ are respectively odd and even functions of x. On the other hand, if $E_y(x, z)$ is an even function of x, we see from (8.1) and (8.2) that $H_x(x, z)$ and $H_z(x, z)$ are respectively even and odd functions of x. Thus, for the odd (even) TE mode, $E_y(x, z)$ and $H_x(x, z)$ are odd (even) functions and $H_z(x, z)$ is an even (odd) function of x. We should note that the presence of $\partial/\partial x$ changes the type of symmetry of the function on which it operates. The advantage of subdividing the TE mode into odd and even sets is that it is sufficient for us to consider only the field components for the half-space $x > 0$, since those for $x < 0$ can be obtained from the symmetry consideration. Consequently, the boundary conditions need be imposed on the interface $x = d$ only.

In order that the boundary conditions be fulfilled for all z, it is necessary for the field components in the dielectric and in the free space regions to have the same

z-dependence. Since the geometry of the slab has translational invariance in the z-direction, it follows that the z-dependence of the field components should be the same throughout the length of the slab. If we are interested in ascertaining the characteristics of the surface waves guided in the $+z$-direction, we may assume the z-dependence of all the field quantities to be of the form

$$f(x, z) = f(x)e^{-j\beta_z z}. \tag{8.5}$$

On the other hand, if we assume the z-dependence of the form $e^{j\beta_z z}$, we obtain the characteristics of the surface waves propagating in the $-z$-direction. The essential characteristics of the surface waves are independent of the direction and hence we use the z-dependence of the form (8.5) alone.

We shall examine the odd TE mode first. For the dielectric region, we obtain from (8.3) and (8.5):

$$\left[\frac{d^2}{dx^2} + \beta_{1x}^2\right] E_{1y}(x) = 0, \tag{8.6}$$

where

$$\beta_{1x}^2 = \beta_1^2 - \beta_z^2 = \omega^2 \mu_d \varepsilon_d - \beta_z^2. \tag{8.7}$$

We find the solution of (8.6) which is an odd function of x as

$$E_{1y}(x) = E_1^o \sin \beta_{1x} x. \tag{8.8}$$

From (8.1), (8.2), (8.5), and (8.8), we obtain the magnetic field components in the dielectric region as

$$H_{1x}(x) = -\frac{\beta_z}{\omega \mu_d} E_1^o \sin \beta_{1x} x; \qquad H_{1z}(x) = -\frac{\beta_{1x}}{j\omega \mu_d} E_1^o \cos \beta_{1x} x. \tag{8.9}$$

In the free space region $x > d$, (8.3) and (8.5) yield

$$\left[\frac{d^2}{dx^2} - \alpha_{2x}^2\right] E_{2y}(x) = 0, \tag{8.10}$$

where

$$\alpha_{2x}^2 = \beta_z^2 - \beta_2^2 = \beta_z^2 - \omega^2 \mu_0 \varepsilon_0. \tag{8.11}$$

We can express the solution of (8.10) as

$$E_{2y}(x) = E_2^o e^{-\alpha_{2x} x} + \tilde{E}_2^o e^{\alpha_{2x} x}. \tag{8.12}$$

The form of the solution given by (8.12) is convenient for the case in which $\beta_z > \beta_2$ or α_{2x} is real. It becomes clear subsequently that α_{2x} is real and that is the reason for expressing the solution of (8.10) in the form given by (8.12). For the region in which (8.12) applies, x extends up to infinity. Hence, the second term in (8.12) becomes infinite for x tending to infinity and, therefore, has to be ruled out as being physically unacceptable. Hence, we obtain the proper solution of (8.10) from (8.12) as

$$E_{2y}(x) = E_2^o e^{-\alpha_{2x} x}. \tag{8.13}$$

From (8.1), (8.2), (8.5), and (8.13), we can deduce the magnetic field components in the free space region $x > d$ as

$$H_{2x}(x) = -\frac{\beta_z}{j\omega\mu_0} E_2^o e^{-\alpha_{2x}x}; \qquad H_{2z}(x) = \frac{\alpha_{2x}}{j\omega\mu_0} E_2^o e^{-\alpha_{2x}x}. \tag{8.14}$$

At the plane interface $x = d$, the tangential field components $E_y(x)$ and $H_z(x)$ should be continuous. If we apply these boundary conditions, with the help of (8.8), (8.9), (8.13), and (8.14), we obtain the following relations:

$$E_1^o \sin \beta_{1x}d - E_2^o e^{-\alpha_{2x}d} = 0, \tag{8.15}$$

$$-E_1^o \frac{\beta_{1x}}{j\omega\mu_d} \cos \beta_{1x}d - E_2^o \frac{\alpha_{2x}}{j\omega\mu_0} e^{-\alpha_{2x}d} = 0. \tag{8.16}$$

In order for (8.15) and (8.16) to yield a nontrivial solution, the determinant of the coefficients of E_1^o and E_2^o should vanish, resulting in the following requirement:

$$-\frac{\mu_0}{\mu_d} \beta_{1x}d \cot (\beta_{1x}d) = \alpha_{2x}d \qquad [\text{odd TE mode}]. \tag{8.17}$$

As in the case of the waveguide with perfectly conducting walls, E_1^o and E_2^o are called the excitation coefficients, which we can determine only if the sources are specified. Together with (8.7) and (8.11), (8.17) gives the equation for the determination of the phase-shift coefficient β_z. We should note that (8.17) is known as *the dispersion equation*. Only for the values of β_z which are solutions of (8.17) together with (8.7) and (8.11) do the fields given by (8.8), (8.9), (8.13), and (8.14) satisfy the appropriate Maxwell's equations and the required boundary conditions at the interface $x = d$. In contrast to a rectangular waveguide, for which we obtained β_z explicitly as a function of ω and the waveguide dimensions, for the dielectric slab waveguide we have found β_z only as an implicit function of ω and the slab thickness d, as expressed in the transcendental equation (8.17).

With the help of (8.7), (8.11), and (8.17), we can evaluate the value of β_z graphically or numerically by using a computer. Before proceeding to determine the exact values of β_z, it is helpful for us to deduce analytically certain bounds on the values of β_z. If β_{1x} is imaginary, that is, $\beta_{1x} = \pm j\alpha_{1x}$, where α_{1x} is real, (8.17) becomes

$$-\frac{\mu_0}{\mu_d} (\alpha_{1x}d) \coth (\alpha_{1x}d) = \alpha_{2x}d. \tag{8.18}$$

From (8.7), we observe that β_{1x} is either real or imaginary. In either case, we see from (8.18) that the left side of (8.17) is always real. Hence, α_{2x} is real, as we had expected earlier. We should note that in (8.13) we have assumed α_{2x} to be positive so as to yield a solution that remains bounded at infinity. The left side of (8.18) is negative whereas the right side is positive. Hence, (8.18) has no real solution for β_z. In other words, β_{1x} cannot be imaginary. The left side of (8.17) is positive for certain ranges of values of $\beta_{1x}d$ and therefore it is possible for (8.17) to have a solution. Consequently, the real solutions of β_z are possible only in the region in which β_{1x} is also real. Since β_{1x} is real, we see from (8.7) that $\beta_1 > \beta_z$. Similarly, since α_{2x} is real, (8.11) shows that

$\beta_z > \beta_2$. Hence

$$\beta_2 < \beta_z < \beta_1. \tag{8.19}$$

We find from (8.7) and (8.11) that

$$\beta_{1x}^2 + \alpha_{2x}^2 = \beta_1^2 - \beta_2^2 = \omega^2(\mu_d\varepsilon_d - \mu_0\varepsilon_0). \tag{8.20}$$

Therefore,

$$\alpha_{2x}d = [\omega^2 d^2(\mu_d\varepsilon_d - \mu_0\varepsilon_0) - \beta_{1x}^2 d^2]^{1/2}. \tag{8.21}$$

Together with (8.21), (8.17) becomes

$$-\frac{\mu_0}{\mu_d}\beta_{1x}d \cot(\beta_{1x}d) = [(\omega d)^2(\mu_d\varepsilon_d - \mu_0\varepsilon_0) - (\beta_{1x}d)^2]^{1/2} \quad \text{[odd TE mode]}. \tag{8.22}$$

By a simple graphical procedure, we can solve (8.22) to yield the value of $\beta_{1x}d$. The left side of (8.22) is plotted as a function of $\beta_{1x}d$ as shown in Fig. 8.13. Since the dispersion equation (8.22) can have solutions only when the left side of (8.22) is positive, only the positive values of the left side of (8.22) are depicted. We can see that these positive values occur periodically in the intervals $(n - \frac{1}{2})\pi < \beta_{1x}d < n\pi$ where n is a nonzero integer. On the same graph the right side of (8.22), that is the value of $\alpha_{2x}d$ as given by (8.21), is plotted as a function of $\beta_{1x}d$. We can verify that the curve of $\alpha_{2x}d$

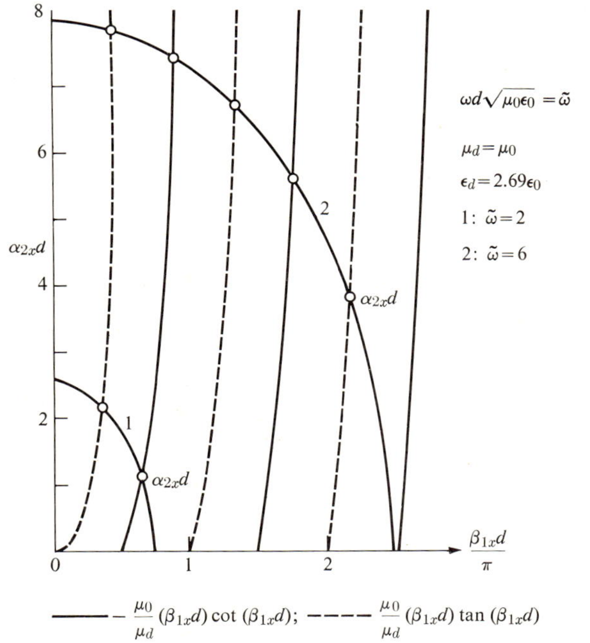

Fig. 8.13 Graphical solution of the dispersion equations of the odd and the even TE modes of a dielectric slab waveguide.

as a function of $\beta_{1x}d$ is a quadrant of a circle of radius equal to $[(\omega d)^2(\mu_d\varepsilon_d - \mu_0\varepsilon_0)]^{1/2}$. We should note that as we deduced earlier, only the positive values of $\alpha_{2x}d$ and $\beta_{1x}d$ are relevant. The points of intersection give the values of $\beta_{1x}d$ that are solutions of the dispersion equation (8.22). Once $\beta_{1x}d$ is known, we can obtain $\beta_z d$ from (8.7) as

$$\beta_z d = \sqrt{\omega^2 d^2 \mu_d \varepsilon_d - (\beta_{1x}d)^2}. \tag{8.23}$$

In portraying the propagation characteristics of the guided waves, it is advantageous for us to use the following normalized parameters: the normalized frequency $\tilde{\omega} = \omega\sqrt{\mu_0\varepsilon_0}\, d$ and the normalized propagation coefficient $\tilde{\beta}_z = \beta_z d$.

We can verify from Fig. 8.13 that for sufficiently large $\tilde{\omega}$ there are a number of possible solutions of $\beta_{1x}d$ and hence also of the normalized propagation coefficient $\tilde{\beta}_z$. Thus, there are a number of possible modes of guided wave propagation. The order of the mode is indicated by n and we assume it to increase in the direction of the increasing values of $\beta_{1x}d$. Hence, from (8.23), we see that for the lowest order ($n = 1$) mode, the value of $\tilde{\beta}_z$ has the largest value and decreases with the increase in the mode order n.

For each order of the mode, we can see from Fig. 8.13 that the decrease of $\tilde{\omega}$ results in the decrease of both $\alpha_{2x}d$ and $\beta_{1x}d$. Since $\alpha_{2x}d$ gives the normalized attenuation factor of the surface wave in the x-direction, it follows that the decrease of $\tilde{\omega}$ results in the guided wave being more and more loosely bound to the interface. Let us consider the limiting values of $\tilde{\omega}$ for which

$$\beta_{1x}d = (n - \tfrac{1}{2})\pi. \tag{8.24}$$

It follows from Fig. 8.13 or from (8.21) and (8.22) that corresponding to (8.24),

$$\alpha_{2x}d = 0. \tag{8.25}$$

For these limiting frequencies, the waves become homogeneous in the transverse plane. We can see from (8.11) and (8.25) that the propagation coefficient of the surface waves approaches that of the unbounded medium, which in the present case is free space. The corresponding modes disappear and do not propagate if the frequencies are lowered below these limiting values. Hence, these limiting frequencies are called the cut-off frequencies. We can obtain their values from (8.21), (8.24), and (8.25) as

$$\omega_{c,n} = \frac{(n - \frac{1}{2})\pi}{d\sqrt{\mu_d\varepsilon_d - \mu_0\varepsilon_0}} \qquad (n = 1, 2, 3, \ldots) \qquad [\text{odd TE mode}]. \tag{8.26}$$

The mode of order n propagates only for $\omega > \omega_{c,n}$.

When $\tilde{\omega}$ becomes very large, we notice from Fig. 8.13 that at least for the lower order modes, the following result obtains approximately

$$\beta_{1x}d \approx n\pi \qquad (n = 1, 2, 3, \ldots). \tag{8.27}$$

For small n, (8.7) shows that β_z approaches β_1 since β_{1x}^2 becomes negligible compared with $\omega^2\mu_d\varepsilon_d$. Hence, for very large $\tilde{\omega}$, that is, for very high frequencies or very large

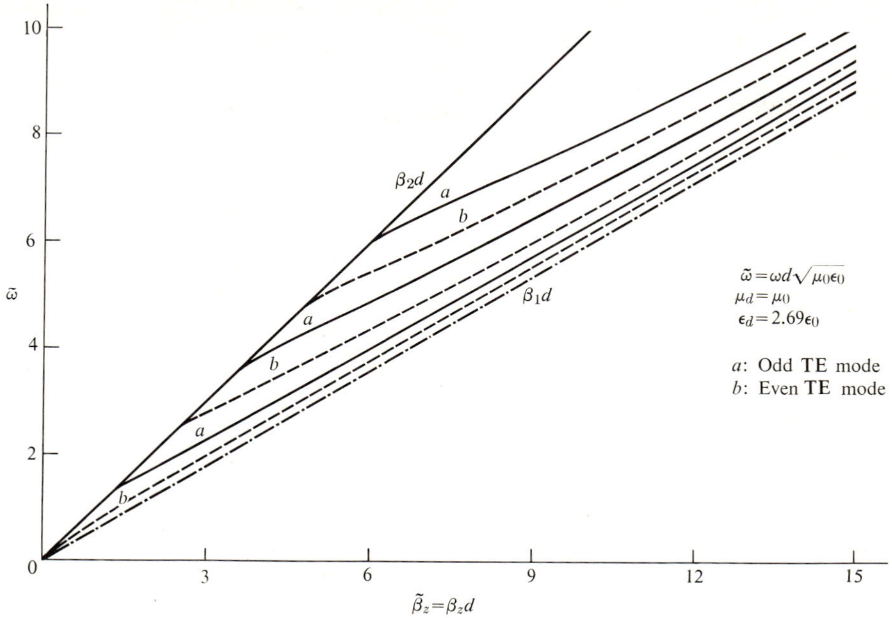

Fig. 8.14 Brillouin diagram for the first three lowest order odd and even TE modes guided by a dielectric slab waveguide.

thicknesses of the dielectric slab, the propagation coefficient approaches that of the unbounded region having the same constitutive parameters as the dielectric slab.

From the solutions of (8.22), the normalized Brillouin diagram, that is, the plot of $\tilde{\omega}$ versus $\tilde{\beta}_z$, is constructed and is shown in Fig. 8.14 for the first three lowest order modes. We see from Fig. 8.14 that the increase of $\tilde{\omega}$ results in the increase of $\tilde{\beta}_z$ for every mode.

For the odd TE mode, we see from (8.8) that $E_y(x, z) = 0$ for $x = 0$. Therefore, a perfect conductor can be introduced to coincide with the plane $x = 0$ without affecting the existing fields, since the required boundary condition on the perfectly conducting plane is automatically satisfied. Consequently, the characteristics of the odd TE surface wave supported by a planar dielectric slab backed by a perfectly conducting plane are the same as those of the corresponding mode supported by a planar dielectric slab of double the thickness but without the backing of the perfectly conducting plane. For a given frequency, in order that at least one odd TE mode may propagate, we see from (8.26) that the thickness of the dielectric slab should be greater than the minimum value given by

$$d_{\min} = \frac{\pi}{2\omega\sqrt{\mu_d\varepsilon_d - \mu_0\varepsilon_0}} \qquad \text{(for odd TE mode).} \tag{8.28}$$

With the help of (8.5), (8.8), (8.9), (8.13), and (8.14), we find the field components of the odd TE mode to be as follows:

Dielectric Region $(0 < x < d)$

$$E_{1y}(x, z) = E_1^o \sin(\beta_{1x}x)e^{-j\beta_z z}, \tag{8.29}$$

$$H_{1x}(x, z) = -\frac{\beta_z}{\omega\mu_d} E_1^o \sin(\beta_{1x}x)e^{-j\beta_z z}, \tag{8.30}$$

$$H_{1z}(x, z) = -\frac{\beta_{1x}}{j\omega\mu_d} E_1^o \cos(\beta_{1x}x)e^{-j\beta_z z}. \tag{8.31}$$

Free Space Region $(x > d)$

$$E_{2y}(x, z) = E_2^o e^{-\alpha_{2x}x}e^{-j\beta_z z}, \tag{8.32}$$

$$H_{2x}(x, z) = -\frac{\beta_z}{j\omega\mu_0} E_2^o e^{-\alpha_{2x}x}e^{-j\beta_z z}, \tag{8.33}$$

$$H_{2z}(x, z) = \frac{\alpha_{2x}}{j\omega\mu_0} E_2^o e^{-\alpha_{2x}x}e^{-j\beta_z z}. \tag{8.34}$$

The ratio E_1^o/E_2^o is fixed either by (8.15) or (8.16). We should note that β_z is determined from (8.22) and (8.23), and α_{2x} is evaluated from (8.21) and (8.22).

We can evaluate the impedance in the $-x$-direction at the plane interface $x = d$ from (8.32) and (8.34) as

$$Z_s = -E_{2y}(d, z)/H_{2z}(d, z) = -j\omega\mu_0/\alpha_{2x}, \tag{8.35}$$

which is a *capacitive* reactance. We discussed the characteristics of the TE surface waves supported by a plane surface with a capacitive reactance previously in Section 6.9. There, we did not specify how a reactance surface is realized in practice. We establish here that a perfectly conducting plane coated with a dielectric layer of sufficient thickness provides the necessary capacitive reactance to sustain a TE surface wave.

So far, we have discussed only the characteristics of the odd TE mode. Since we can obtain the characteristics of the even TE mode in a similar manner, we shall only briefly summarize the properties of the even TE mode here. For the even TE mode, we should change $E_{1y}(x)$ given by (8.8) to

$$E_{1y}(x) = E_1^e \cos \beta_{1x}x. \tag{8.36}$$

Then, we can deduce the magnetic field components from (8.1), (8.2), (8.5), and (8.36) as

$$H_{1x}(x) = -\frac{\beta_z}{\omega\mu_d} E_1^e \cos \beta_{1x}x; \qquad H_{1z}(x) = \frac{\beta_{1x}}{j\omega\mu_d} E_1^e \sin \beta_{1x}x. \tag{8.37}$$

We may verify the field components in the free space region $x > d$ to be given by (8.13) and (8.14). For the sake of consistency of notation, we replace E_2^o by E_2^e where the superscript e stands for even. If we apply the boundary conditions at the interface

$x = d$ the following dispersion equation results:

$$\frac{\mu_0}{\mu_d}(\beta_{1x}d)\tan(\beta_{1x}d) = \alpha_{2x}d \qquad [\text{even TE mode}], \tag{8.38}$$

where $\alpha_{2x}d$ is defined in (8.21). After finding $\beta_{1x}d$ from (8.21) and (8.38), we can determine $\beta_z d$ from (8.23). As before, we can prove that α_{2x} and β_{1x} have to be real and the bounds on β_z given by (8.19) continue to hold.

We can solve the dispersion equation (8.38) together with (8.21) by the same graphical procedure, as before. The graph of the left side of (8.38) is also included in Fig. 8.13 and is shown by dashed lines. Since only the positive values of the left side of (8.38) are needed, only those portions of the graph are depicted in Fig. 8.13. We can see that these positive values occur periodically in the intervals

$$(n-1)\pi < \beta_{1x}d < (n-\tfrac{1}{2})\pi$$

where n is a nonzero integer. The graph of the right side of (8.38) is the same as that of (8.22). The points of intersection of these two graphs yield the values of $\beta_{1x}d$ which satisfy the dispersion equation (8.38) and we can then evaluate the corresponding values of $\beta_z d$ from (8.23). As in the case of the odd TE mode, several orders of the even TE mode are also possible. In general, the dependence on the normalized frequency of the characteristics of the even TE mode is similar to those of the odd TE mode. We can determine the cut-off angular frequencies of the even TE modes as

$$\omega_{c,n} = \frac{(n-1)\pi}{d\sqrt{\mu_d\varepsilon_d - \mu_0\varepsilon_0}} \qquad (n = 1, 2, 3, \ldots) \qquad [\text{even TE mode}]. \tag{8.39}$$

The cut-off angular frequency $\omega_{c,1}$ of the lowest order even TE mode is zero; in other words, the lowest order even TE mode propagates down to zero frequencies. As before, the mode of order n propagates only for $\omega > \omega_{c,n}$. The propagation coefficient β_z approaches either that of an unbounded free space or that of an unbounded region having the same constitutive parameters as the dielectric slab according to whether ω is near $\omega_{c,n}$ or is very large. From the results of the solutions of (8.38), we can construct the normalized Brillouin diagrams. The normalized Brillouin diagrams for the first three lowest order even TE modes are depicted by dashed lines in Fig. 8.14. We should note from Fig. 8.14 that the dispersion curves of the even and the odd TE modes are interspersed. As in the case of the odd TE mode, so in the case of the even TE mode the increase of $\tilde{\omega}$ results in the increase of $\tilde{\beta}_z$ for every mode. We should note in particular that the lowest order even TE mode extends down to zero frequencies.

There are some interesting differences between the odd and the even TE modes. For the even TE mode, since $E_y(x, z) \neq 0$ for $x = 0$, we cannot introduce a perfectly conducting plane to coincide with the plane $x = 0$. Therefore, it follows that a perfectly conducting plane coated with a dielectric layer cannot support even TE modes. Since the cut-off angular frequency of the lowest order even TE mode is zero, we can verify that for a given frequency, a dielectric slab in free space can support the lowest order even TE mode even for extremely small thicknesses of the slab. On the other hand, the thickness of the dielectric slab has to be larger than the minimum value given by (8.28) in order that it may be able to sustain the lowest order odd TE mode.

With the help of (8.5), (8.36), (8.37), (8.13), and (8.14), we can evaluate the detailed expressions for the field components of the even TE mode but we do not include them here. In particular, we can determine the impedance in the $-x$-direction at the plane interface $x = d$ and we find it to be the same as in (8.35). It follows, therefore, that the reactance has to be capacitive for the plane surface to be able to support a TE mode irrespective of whether it is an odd or an even mode.

8.9 DIELECTRIC SLAB WAVEGUIDE: TM MODE

We *analyzed* the characteristics of the TE surface waves guided by a planar dielectric slab starting from Maxwell's equations and the appropriate boundary conditions. In this section, we shall *synthesize* the electromagnetic fields associated with the TM surface waves guided by a planar dielectric slab starting from the uniform plane wave fields. In Section 6.8, we showed that for the plane wave incidence at a plane interface separating two different media, there is a critical angle of incidence if the wave is incident from the slower (denser) to the faster (rarer) medium. Let us consider a planar dielectric slab as shown in Fig. 8.15. For the sake of convenience in referring to them, we shall designate the plane interfaces at $x = d$ and $x = -d$ as the interfaces 1 and 2 respectively. The geometry of the dielectric slab as well as the constitutive parameters of the dielectric and the free space regions are the same as in the preceding section. Let us consider a uniform plane wave incident on the *first* interface from the dielectric side. Let us suppose that the angle of incidence is greater than the critical angle. Then, as we discussed in Section 6.8, in the free space region $x > d$, the transmitted wave is exponentially attenuated away from the interface and has linearly varying phase progression along the interface. The wave reflected at the first interface is incident on the second interface. Since the two interfaces are parallel to each other, the angle of incidence at the second interface is the same as that at the first interface and, hence, is also greater than the critical angle. Consequently, in the free space

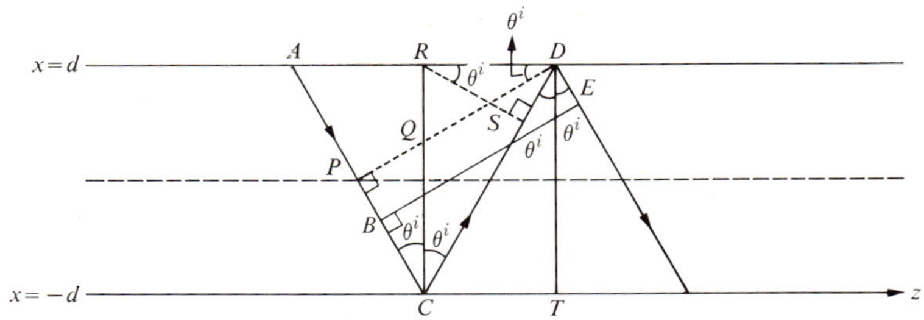

Fig. 8.15 Geometry of the dielectric slab waveguide for the determination of the condition under which the fields inside the dielectric are describable in terms of two uniform plane waves.

region $x < -d$ also, the transmitted wave is exponentially attenuated away from the interface and has linearly varying phase progression along the interface. The wave reflected at the second interface is incident again on the first interface, and these transmission and reflection processes continue along the dielectric slab in the longitudinal direction. Thus, inside the dielectric slab, there are two sets of multiply reflected waves, one set emanating from the first interface and the other from the second interface. If a particular condition is satisfied, the set of reflected waves arising from the first interface becomes a single uniform plane wave. In view of the symmetry, the set of waves arising from the second interface also becomes another single uniform plane wave under the same condition. The first and the second uniform plane waves are just the reflected and the corresponding incident waves at the first interface and vice versa at the second interface. We pointed out in Section 6.7, that the incident, the reflected, and the transmitted waves all have the same propagation coefficient in the longitudinal direction. Thus, the following total picture of the surface waves guided by a planar dielectric slab emerges. The electromagnetic fields in the dielectric and in the free space regions are guided in the longitudinal direction with the same propagation coefficient. In the transverse plane, the fields in the free space region decay exponentially from the interfaces but the fields in the dielectric slab form a standing wave due to the superposition of the two constituent uniform plane waves. The number of periodic variations of the fields in the transverse plane can have any value and is precisely the basis for the ordering of the modes. We find that the general characteristics of the fields guided by a planar dielectric slab, as built up from the superposition of plane waves, are identical to those we obtained in the preceding section from a straightforward analysis starting from Maxwell's equations.

The condition under which the set of waves arising from the two interfaces become two uniform plane waves is the dispersion equation which we shall now derive for the TM surface waves. Let us consider a plane front whose cross section by the plane $y = 0$ is shown by BE in Fig. 8.15. Since the problem is two-dimensional, it is sufficient for us to consider the various phenomena occurring only in the xz-plane. The points B and E on the plane front are respectively due to the reflected waves from A and D on interface 1. The plane front forms the phase front of a uniform plane wave only when the phase difference between the waves progressing away from interface 1 at B and E is an integral multiple of 2π. The wave phenomenon at B travels along BC, gets reflected at C, travels along CD, gets reflected again at D, and travels along DE to reach E. The phase difference between the waves at B and E is the phase change introduced as the uniform plane wave travels in the dielectric the path length $BCDE$, which for the sake of convenience we denote by l, together with the phase changes introduced by the reflections at C and D. The angles of incidence and reflection at C and D are all equal and are denoted by θ^i. Therefore, the phase changes introduced by the reflection at C and D are equal; we have deduced its value θ_H in (6.8.26) and (6.8.27) for the TM waves for which the magnetic field is perpendicular to the plane of incidence (xz-plane). With the appropriate changes in the notation, we can reproduce (6.8.27) here as

$$\tan\left(\frac{\theta_H}{2}\right) = \frac{\sqrt{\mu_0/\varepsilon_0}[(\mu_d\varepsilon_d/\mu_0\varepsilon_0)\sin^2\theta^i - 1]^{1/2}}{\sqrt{\mu_d/\varepsilon_d}\cos\theta^i}. \tag{9.1}$$

Since $\beta_1 = \omega\sqrt{\mu_d\varepsilon_d}$ is the phase-shift coefficient of the uniform plane waves in a medium of the constitutive parameters μ_d and ε_d, we obtain the phase difference ϕ_{ph} between the waves at B and E as

$$\phi_{\mathrm{ph}} = -\beta_1 l + 2\theta_H. \tag{9.2}$$

It is convenient for us to distinguish two cases, namely,

Case 1. $$\phi_{ph} = -\beta_1 l + 2\theta_H = 2\pi(2n-1) \qquad (n = 1, 2, 3, \ldots), \tag{9.3}$$

Case 2. $$\phi_{ph} = -\beta_1 l + 2\theta_H = 2\pi(2n) \qquad (n = 1, 2, 3, \ldots). \tag{9.4}$$

In both the cases 1 and 2, the phase of the waves at B and E are the same and therefore BE is a phase front. In other words, the set of waves reflected from interface 1 becomes a uniform plane wave if (9.3) or (9.4) is satisfied. By symmetry, for the same conditions as expressed by (9.3) and (9.4), the set of waves reflected from interface 2 also becomes a uniform plane wave. Thus, (9.3) and (9.4) are the desired dispersion relations for the TM surface waves.

Let β_z be the propagation coefficient in the longitudinal direction. We should note that β_z is the same in the dielectric and in the free space regions. Let β_{1x} be the propagation coefficient inside the dielectric region in the x-direction. Since β_1 is the propagation coefficient in the direction CD which gives the wave normal direction, it follows from Fig. 8.15 that

$$\beta_{1x} = \beta_1\cos\theta^i; \qquad \beta_z = \beta_1\sin\theta^i. \tag{9.5}$$

In view of (9.5),

$$\beta_{1x}^2 + \beta_z^2 = \beta_1^2 = \omega^2\mu_d\varepsilon_d, \tag{9.6}$$

which we see to be identical to that in (8.7). In the free space, let α_{2x} be the attenuation coefficient in the x-direction. As indicated before, β_z is the propagation coefficient in the z-direction. After the required changes in the notation, it follows from (6.8.10) that

$$\alpha_{2x} = \beta_2\sinh\theta_j^t, \qquad \beta_z = \beta_2\cosh\theta_j^t, \tag{9.7}$$

where θ_j^t is defined by (6.8.8). We should note that θ_c^i is the critical angle of incidence as defined by

$$\sin\theta_c^i = \sqrt{\mu_0\varepsilon_0/\mu_d\varepsilon_d}. \tag{9.8}$$

We can verify from (9.7) that

$$-\alpha_{2x}^2 + \beta_z^2 = \beta_2^2 = \omega^2\mu_0\varepsilon_0, \tag{9.9}$$

which we see to be identical with that in (8.11). From (9.9), we can show that

$$\frac{\beta_z^2}{\beta_2^2} - 1 = \frac{\beta_z^2}{\omega^2\mu_0\varepsilon_0} - 1 = \frac{\alpha_{2x}^2}{\beta_2^2} = \frac{\alpha_{2x}^2}{\omega^2\mu_0\varepsilon_0}. \tag{9.10}$$

With the use of the value of $\sin\theta^i$ from (9.5), we find that

$$\left(\frac{\mu_d\varepsilon_d}{\mu_0\varepsilon_0}\sin^2\theta^i - 1\right)^{1/2} = \left(\frac{\mu_d\varepsilon_d}{\mu_0\varepsilon_0}\frac{\beta_z^2}{\beta_1^2} - 1\right)^{1/2}. \tag{9.11}$$

Since $\beta_1^2/\beta_2^2 = \mu_d\varepsilon_d/\mu_0\varepsilon_0$, it follows from (9.10) and (9.11) that

$$\left(\frac{\mu_d\varepsilon_d}{\mu_0\varepsilon_0}\sin^2\theta^i - 1\right)^{1/2} = \frac{\alpha_{2x}}{\omega\sqrt{\mu_0\varepsilon_0}}. \tag{9.12}$$

If we substitute (9.12) and the value of $\cos\theta^i$ from (9.5) into (9.1) we can prove that

$$\tan\left(\frac{\theta_H}{2}\right) = \sqrt{\frac{\mu_0}{\varepsilon_0}}\frac{\alpha_{2x}}{\omega\sqrt{\mu_0\varepsilon_0}}\sqrt{\frac{\varepsilon_d}{\mu_d}}\frac{\omega\sqrt{\mu_d\varepsilon_d}}{\beta_{1x}} = \frac{\varepsilon_d}{\varepsilon_0}\frac{\alpha_{2x}}{\beta_{1x}}. \tag{9.13}$$

We can express the path length l in terms of the thickness $2d$ of the dielectric slab and the angle of incidence θ^i. With reference to Fig. 8.15, the path length l is given by $BCDE$. From D, we draw a straight line parallel to EB so as to intersect the straight line CA at P. The path $BCDE$ is equal to the path length PCD since $PB = DE$. The straight line CQR is normal to the lines defined by $x = d$ and $x = -d$. A perpendicular is dropped from R to CD so as to intersect CD at S. We can verify that $\measuredangle QDR = \measuredangle DRS = \theta^i$. From the triangles PQC and RCS, we obtain

$$PC = CQ\cos\theta^i; \qquad CS = CR\cos\theta^i. \tag{9.14}$$

From the triangle QRD, we see that

$$QR\cos\theta^i = RD\sin\theta^i. \tag{9.15}$$

In a similar manner, we find from the triangle SDR that

$$RD\sin\theta^i = SD. \tag{9.16}$$

From (9.15) and (9.16), it is found that

$$SD = QR\cos\theta^i. \tag{9.17}$$

We can verify from (9.14) and (9.17) that

$$PC + CS + SD = l = \cos\theta^i[CQ + QR + CR] = 4d\cos\theta^i. \tag{9.18}$$

Hence, with the help of (9.5) and (9.18), we may recast (9.3) and (9.4) as

Case 1. $$\frac{\theta_H}{2} = \beta_{1x}d + \frac{\pi}{2}(2n-1) \qquad (n = 1, 2, 3, \ldots), \tag{9.19}$$

Case 2. $$\frac{\theta_H}{2} = \beta_{1x}d + \pi n \qquad (n = 1, 2, 3, \ldots). \tag{9.20}$$

On taking the tangent of both the sides of (9.19) and (9.20) and replacing the value of tan $(\theta_H/2)$ from (9.13), we deduce the following dispersion relations:

Case 1.
$$-(\beta_{1x}d)\cot(\beta_{1x}d) = \frac{\varepsilon_d}{\varepsilon_0}(\alpha_{2x}d), \tag{9.21}$$

Case 2.
$$(\beta_{1x}d)\tan(\beta_{1x}d) = \frac{\varepsilon_d}{\varepsilon_0}(\alpha_{2x}d). \tag{9.22}$$

As in the preceding section, we can show that the left sides of (9.21) and (9.22) are real irrespective of whether β_{1x} is real or imaginary. We should note from (9.6) that β_{1x} can be only either real or imaginary. Therefore, for the fulfillment of (9.21) and (9.22) α_{2x} has to be real. Hence, from (9.12) we see that

$$\frac{\mu_d\varepsilon_d}{\mu_0\varepsilon_0}\sin^2\theta^i > 1. \tag{9.23}$$

From (9.8), and (9.23), we obtain

$$\sin^2\theta^i > \sin^2\theta^i_c \qquad \text{or} \qquad \theta^i > \theta^i_c. \tag{9.24}$$

The implication of the result (9.24) is quite important. We showed that the multiply reflected waves from the two parallel interfaces bounding the dielectric slab can be described in terms of two uniform plane waves only when the dispersion relations are satisfied. It follows from (9.24) that the dispersion relations are satisfied only if the angle of incidence at the two interfaces exceeds the critical angle. For this case, as we explained previously, the fields in the free space region decay exponentially away from the interface, but are propagated in the longitudinal direction with the same propagation coefficient as in the dielectric. Thus, the planar dielectric slab is able to support a surface wave.

The magnetic field in the dielectric is of the form

$$H_{1y}(x, z) = H_{1+}e^{-j\beta_{1x}x}e^{-j\beta_z z} + H_{1-}e^{j\beta_{1x}x}e^{-j\beta_z z}, \tag{9.25}$$

where the first and the second terms respectively give the uniform plane waves incident and reflected at interface 1. We can obtain the values of the incident and the reflected fields at interface 1 from (9.25) as

$$\textit{incident: } H_{1+}e^{-j\beta_{1x}d}e^{-j\beta_z z}; \qquad \textit{reflected: } H_{1-}e^{j\beta_{1x}d}e^{-j\beta_z z}. \tag{9.26}$$

From the definition of the reflection coefficient as given by (6.8.26) and (9.26), we find that

$$\frac{H_{1-}e^{j\beta_{1x}d}e^{-j\beta_z z}}{H_{1+}e^{-j\beta_{1x}d}e^{-j\beta_z z}} = e^{j\theta_H} \qquad \text{or} \qquad H_{1-} = H_{1+}e^{j\{-2\beta_{1x}d+\theta_H\}}. \tag{9.27}$$

It is convenient for us to distinguish the two cases at this stage. For Case 1, we see from (9.19) and (9.27) that

$$H_{1-} = H_{1+}e^{j\pi(2n-1)} = -H_{1+}. \tag{9.28}$$

Hence, (9.25) becomes

$$H_{1y}(x, z) = H_1^o \sin(\beta_{1x}x)e^{-j\beta_z z}, \tag{9.29}$$

where the constant $2jH_{1-} = -2jH_{1+}$ has been replaced by another constant H_1^o. For the TM mode, the nonvanishing electric field components $E_x(x, z)$ and $E_z(x, z)$ are related to the magnetic field $H_y(x, z)$ by the expressions given by (6.6.28), which we reproduce here for convenience:

$$E_x(x, z) = -\frac{1}{j\omega\varepsilon}\frac{\partial}{\partial z}H_y(x, z); \qquad E_z(x, z) = \frac{1}{j\omega\varepsilon}\frac{\partial}{\partial x}H_y(x, z). \tag{9.30}$$

We use the same procedure employed in the preceding section for distinguishing the field components and the constitutive relations in the dielectric and in the free space regions. In the dielectric region, we can evaluate the electric field components from (9.29) and (9.30) as

$$E_{1x}(x, z) = \frac{\beta_z}{\omega\varepsilon_d}H_1^o \sin(\beta_{1x}x)e^{-j\beta_z z}, \tag{9.31}$$

$$E_{1z}(x, z) = \frac{\beta_{1x}}{j\omega\varepsilon_d}H_1^o \cos(\beta_{1x}x)e^{-j\beta_z z}. \tag{9.32}$$

In the free space region $x > d$ and $x < -d$, the magnetic fields are of the form

$$H_{2y}(x, z) = H_2^o e^{-\alpha_{2x}x}e^{-j\beta_z z} \qquad (x > d), \tag{9.33}$$

$$H_{2y}(x, z) = H_{2-}^o e^{\alpha_{2x}x}e^{-j\beta_z z} \qquad (x < -d). \tag{9.34}$$

We obtain the value of the transmitted field at $x = d$ from (9.33) as

$$\textit{transmitted:}\ H_2^o e^{-\alpha_{2x}d}e^{-j\beta_z z}. \tag{9.35}$$

From the definition of the transmission coefficient τ_H as defined by (6.7.42), (9.26) and (9.35), we see that

$$H_2^o e^{-\alpha_{2x}d}/H_{1+}e^{-j\beta_{1x}d} = \tau_H. \tag{9.36}$$

We find the incident and the transmitted fields at $x = -d$ from (9.25) and (9.34) to be

$$\textit{incident:}\ H_{1-}e^{-j\beta_{1x}d}e^{-j\beta_z z}; \qquad \textit{transmitted:}\ H_{2-}^o e^{-\alpha_{2x}d}e^{-j\beta_z z}. \tag{9.37}$$

It is clear from (9.37) that

$$H_{2-}^o e^{-\alpha_{2x}d}/H_{1-}e^{-j\beta_{1x}d} = \tau_H. \tag{9.38}$$

From (9.28), (9.36), and (9.38) we obtain

$$H_{2-}^o = -H_2^o. \tag{9.39}$$

We can determine the electric field components in the free space region from (9.30), (9.33), (9.34), and (9.39) as

$$E_{2x}(x, z) = \frac{\beta_z}{\omega\varepsilon_0} H_2^o e^{-\alpha_{2x}x} e^{-j\beta_z z} \qquad (x > d), \tag{9.40}$$

$$E_{2z}(x, z) = -\frac{\alpha_{2x}}{j\omega\varepsilon_0} H_2^o e^{-\alpha_{2x}x} e^{-j\beta_z z} \qquad (x > d), \tag{9.41}$$

$$E_{2x}(x, z) = -\frac{\beta_z}{\omega\varepsilon_0} H_2^o e^{\alpha_{2x}x} e^{-j\beta_z z} \qquad (x < -d), \tag{9.42}$$

$$E_{2z}(x, z) = -\frac{\alpha_{2x}}{j\omega\varepsilon_0} H_2^o e^{\alpha_{2x}x} e^{-j\beta_z z} \qquad (x < -d). \tag{9.43}$$

We observe from (9.29), (9.33), (9.34), and (9.39) that $H_y(x, z)$ is an odd function of x and therefore, Case 1 is identified as the odd TM mode. For the odd TM mode, (9.31), (9.32), and (9.40) to (9.43) show that $H_y(x, z)$ and $E_x(x, z)$ are odd functions of x whereas $E_z(x, z)$ is an even function of x.

In a similar manner, we can deduce that Case 2 corresponds to the even TM mode for which $H_y(x, z)$ and $E_x(x, z)$ are even functions of x whereas $E_x(x, z)$ is an odd function of x. In view of the symmetry, it is sufficient for us to give the expressions for the field components only for $x > 0$. The results are as follows:

Dielectric Region: ($|x| < d$)

$$H_{1y}(x, z) = H_1^e \cos(\beta_{1x}x) e^{-j\beta_z z} \qquad \text{[even TM mode]}, \tag{9.44}$$

$$E_{1x}(x, z) = \frac{\beta_z}{\omega\varepsilon_d} H_1^e \cos(\beta_{1x}x) e^{-j\beta_z z} \qquad \text{[even TM mode]}, \tag{9.45}$$

$$E_{1z}(x, z) = -\frac{\beta_{1x}}{j\omega\varepsilon_d} H_1^e \sin(\beta_{1x}x) e^{-j\beta_z z} \qquad \text{[even TM mode]}. \tag{9.46}$$

Free Space Region: ($x > d$)

$$H_{2y}(x, z) = H_2^e e^{-\alpha_{2x}x} e^{-j\beta_z z} \qquad \text{[even TM mode]}, \tag{9.47}$$

$$E_{2x}(x, z) = \frac{\beta_z}{\omega\varepsilon_0} H_2^e e^{-\alpha_{2x}x} e^{-j\beta_z z} \qquad \text{[even TM mode]}, \tag{9.48}$$

$$E_{2z}(x, z) = \frac{-\alpha_{2x}}{j\omega\varepsilon_0} H_2^e e^{-\alpha_{2x}x} e^{-j\beta_z z} \qquad \text{[even TM mode]}. \tag{9.49}$$

For the even TM mode, the tangential electric field component $E_z(x, z) = 0$ for $x = 0$. Hence, a perfect conductor can be introduced to coincide with the plane $x = 0$ without affecting the existing field components. It follows therefore that the characteristics of the TM surface wave mode supported by a perfectly conducting plane coated with a dielectric layer of appropriate thickness are identical to those of the even TM surface wave mode sustained by an ungrounded dielectric slab of double the thickness.

As in the preceding section, we can show that the cut-off angular frequencies $\omega_{c,n}$ correspond to $\alpha_{2x} = 0$. It is therefore possible for us to deduce the expressions for the cut-off angular frequencies from (9.6), (9.9), (9.21), and (9.22). The results are:

$$\omega_{c,n} = \frac{(n - \frac{1}{2})\pi}{d\sqrt{\mu_d \varepsilon_d - \mu_0 \varepsilon_0}} \qquad (n = 1, 2, 3, \ldots) \qquad \text{[odd TM mode]}, \tag{9.50}$$

$$\omega_{c,n} = \frac{(n - 1)\pi}{d\sqrt{\mu_d \varepsilon_d - \mu_0 \varepsilon_0}} \qquad (n = 1, 2, 3, \ldots) \qquad \text{[even TM mode]}. \tag{9.51}$$

We observe the lowest order even TM mode from (9.51) to extend down to zero frequencies. In other words, for a given frequency, the lowest order even TM mode exists no matter how small the thickness of the dielectric slab is. Consequently, a perfectly conducting plane coated with a dielectric layer is able to maintain the lowest order even TM mode of any frequency even for extremely small thicknesses of the dielectric coating. But, as we pointed out in the preceding section, the lowest order odd TE mode of any frequency can be supported by a dielectric-coated perfectly conducting plane only if the thickness is sufficiently large.

For the TM mode, we obtain the impedance at the interface, looking into the dielectric region from the free space side, from (9.33), (9.41), (9.47), and (9.49) as

$$Z_s = E_{2z}(d, z)/H_{2y}(d, z) = j\alpha_{2x}/\omega\varepsilon_0. \tag{9.52}$$

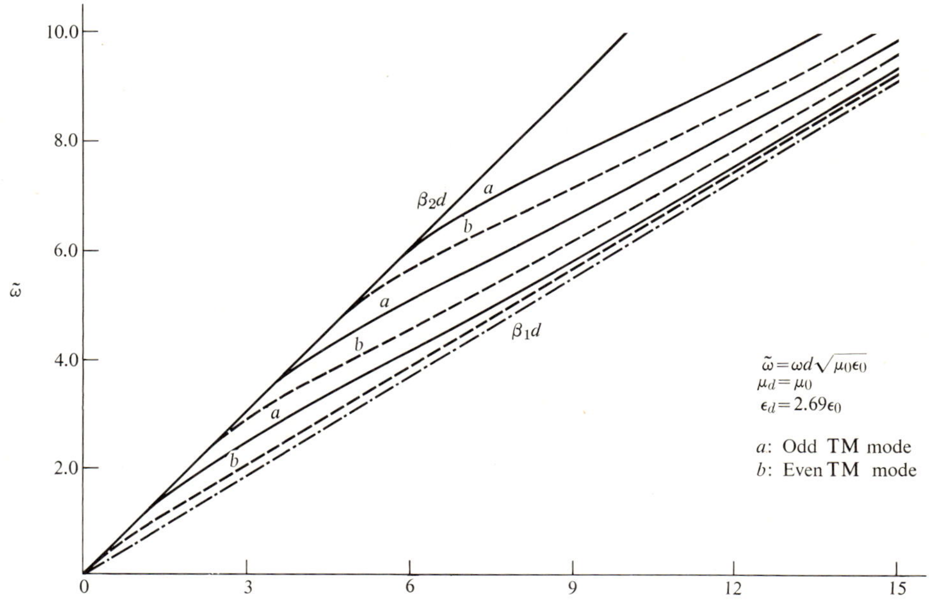

Fig. 8.16 Brillouin diagram for the first three lowest order odd and even TM modes guided by a dielectric slab waveguide.

We should note that the surface impedance (9.52) has been evaluated for the interface $x = d$ but the same result is also found to be true for the interface $x = -d$. We can see from (9.52) that the surface impedance is an inductive reactance. In Section 6.9, we deduced that a plane reactive surface can support a TM surface wave provided the reactance is inductive without considering the methods of realizing such an inductive reactance surface. We establish here that a perfectly conducting plane coated with a dielectric layer of even very small thickness presents an inductive reactance so as to be able to sustain a TM surface wave mode of propagation.

We can solve the dispersion relations (9.21) and (9.22) either by the graphical method as enumerated in the preceding section or by the numerical techniques using a computer. After we determine $\beta_{1x}d$ from (9.21) and (9.22) we can evaluate $\tilde{\beta}_z = \beta_z d$ from (9.6) as a function of $\tilde{\omega} = \omega\sqrt{\mu_0\varepsilon_0}\, d$. The normalized dispersion curves obtained in this manner are depicted in Fig. 8.16, for the first three odd and even TM modes. If we examine Fig. 8.16 we see that the dependence of the normalized propagation coefficient $\tilde{\beta}_z$ on the normalized frequency $\tilde{\omega}$ for the TM modes has the same general features as those for the TE modes.

8.10 RECTANGULAR CAVITY RESONATOR

We pointed out in Section 2.11 that a two-wire transmission line with short-circuit terminations at both the ends has resonance at certain discrete frequencies for which the length of the transmission line is an integral multiple of half of a wavelength. Since there are an infinite number of integral multiples of half of a wavelength in a given length of the transmission line, it follows that the resonant frequencies form an infinite set. The arrangement of a transmission line with short-circuit terminations at both the ends can be used as a resonant circuit element at high frequencies. In a similar manner, we should expect that a waveguide closed at both the ends by perfectly conducting walls can sustain electromagnetic fields at certain discrete frequencies. We shall establish the validity of this expectation now. In actuality, any dielectric region completely enclosed by conducting walls forms an *electromagnetic cavity resonator* which is capable of sustaining electromagnetic fields at certain discrete frequencies with a very small expenditure of power. The only power needed is that required to compensate for the dissipation due to the dielectric losses and the imperfect conductivity of the walls of the cavity resonator as well as that transported outside of the resonator by the coupling devices. At very high frequencies, the electromagnetic cavity resonators are used as circuit elements. Although an electromagnetic cavity resonator can be of any arbitrary shape, an important class of the cavity resonators consists of enclosures formed by waveguides of finite length, terminated at both the ends by plane perfectly conducting walls situated so as to be perpendicular to the axial direction. We shall restrict our attention to only this important class of resonators which, however, contain all the general characteristics of a cavity resonator of any arbitrary shape. Moreover, we shall assume the waveguides to be rectangular in cross section since the theory of rectangular waveguides

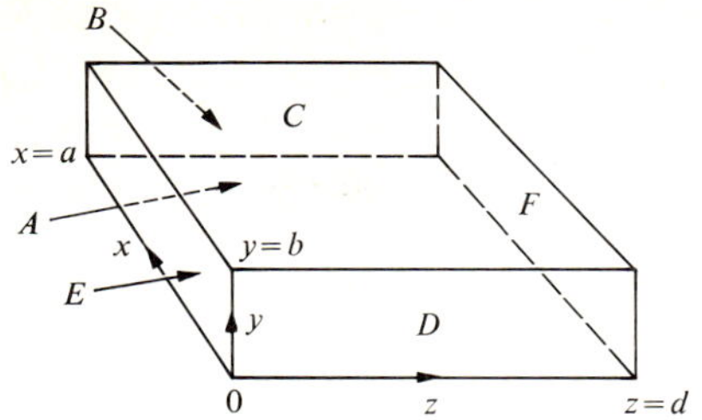

Fig. 8.17 Geometry of a rectangular cavity resonator.

which we developed in Sections 8.5 to 8.7 can be extended and applied to the analysis of the rectangular cavity resonator.

1. TE_{mnp} Mode Pattern

Let us consider an ideal rectangular waveguide as shown in Fig. 8.7, supporting a TE_{mn} mode. Let us introduce two perfectly conducting planes to coincide with the planes $z = 0$ and $z = d$, with the result that the rectangular waveguide depicted in Fig. 8.7 becomes a rectangular cavity as illustrated in Fig. 8.17. For the sake of convenience in referring to them, we designate the planes at $z = 0$ and $z = d$ as E and F respectively. Since the waveguide is finite in the longitudinal ($\hat{z}$) direction, there are waves progressing both in the $+$ and in the $-z$-directions. We assume that the angular frequency of the electromagnetic wave is such that $\omega > \omega_{m,n}$. Therefore, the constituent waves progressing in the $+$ and the $-z$-directions propagate without attenuation. Hence, from (5.1), (5.42) and (5.45b), we can deduce the longitudinal component of the magnetic field, which satisfies Maxwell's equations and the appropriate boundary conditions on all the walls excepting E and F, as

$$H_z(x, y, z) = \cos(m\pi x/a)\cos(n\pi y/b)[H_{0+}e^{-j\beta_z z} + H_{0-}e^{j\beta_z z}]. \tag{10.1}$$

We should note that only one of the two integers m and n can be zero. In (10.1) β_z is a continuous variable and is a function of ω which can assume any value in the range $\omega > \omega_{m,n}$. We shall show now that in order for the boundary conditions on the walls E and F to be fulfilled, only certain discrete values of β_z are admissible and, in turn, the admissible values of ω are also restricted to a set of discrete values. The constants H_{0+} and H_{0-} are arbitrary; their precise values depend on the exciting source. But, the boundary conditions on the walls E and F impose also a certain relationship between H_{0+} and H_{0-}.

From (5.16), it follows that in order for the tangential electric field components $E_x(x, y, z)$ and $E_y(x, y, z)$ to vanish on the walls E and F, it is necessary and sufficient for $H_z(x, y, z)$ to vanish on the walls E and F. It follows from (10.1) that $H_z(x, y, z) = 0$ for $z = 0$ and for $0 < x < a$ and $0 < y < b$ only if

$$H_{0+} = -H_{0-} = jH_0/2, \tag{10.2}$$

where H_0 is a new arbitrary constant. Substituting (10.2) into (10.1) we obtain

$$H_z(x, y, z) = H_0 \cos(m\pi x/a) \cos(n\pi y/b) \sin \beta_z z. \tag{10.3}$$

We can see from (10.3) that $H_z(x, y, z) = 0$ for $z = d$ and for $0 < x < a$ and $0 < y < b$ only if

$$\text{(i) } H_0 = 0 \quad \text{or} \quad \text{(ii) } \sin \beta_z d = 0. \tag{10.4}$$

The condition $H_0 = 0$ leads to a trivial solution since $H_z(x, y, z) \equiv 0$. Therefore, only the second condition in (10.4) is acceptable which stipulates that β_z can assume only the following discrete values

$$\beta_z = p\pi/d \qquad [p \text{ is an integer greater than zero}]. \tag{10.5}$$

If $p = 0$, $\beta_z = 0$, and $H_z(x, y, z) \equiv 0$, again a trivial solution results. Hence, p is restricted to be an integer greater than zero. We obtain the longitudinal magnetic field, which satisfies Maxwell's equations and the boundary conditions on all the six walls of the rectangular cavity, from (10.3) and (10.5) to be of the form

$$H_z(x, y, z) = H_0 \cos(m\pi x/a) \cos(n\pi y/b) \sin(p\pi z/d), \tag{10.6}$$

where the excitation coefficient H_0 is undetermined.

We see from (5.43) that

$$\omega^2/v^2 = (m\pi/a)^2 + (n\pi/b)^2 + \beta_z^2. \tag{10.7}$$

Since β_z in (10.7) is restricted to have only the values given by (10.5), the admissible values of ω are given by

$$\omega_{m,n,p} = v\sqrt{(m\pi/a)^2 + (n\pi/b)^2 + (p\pi/d)^2}. \tag{10.8}$$

The values of ω given by (10.8) are called the resonant angular frequencies. The subscripts m, n, and p indicate the number of periodic variations of the electromagnetic fields in the x-, y-, and z-directions.

For the purpose of deducing the electromagnetic fields associated with the resonant angular frequencies given by (10.8), it is convenient for us to rewrite (10.3), with the help of (10.1) and (10.2), as the sum of the following two constituent parts:

$$H_z(x, y, z) = H_{z+}(x, y, z) + H_{z-}(x, y, z), \tag{10.9}$$

where

$$H_{z+}(x, y, z) = \frac{jH_0}{2} \cos\left(\frac{m\pi x}{a}\right) \cos\left(\frac{n\pi y}{b}\right) e^{-j(p\pi/d)z}, \tag{10.10}$$

and

$$H_{z-}(x, y, z) = -\frac{jH_0}{2} \cos\left(\frac{m\pi x}{a}\right) \cos\left(\frac{n\pi y}{b}\right) e^{j(p\pi/d)z}. \tag{10.11}$$

Evidently, (10.10) and (10.11) represent the waves progressing in the + and the $-z$-directions respectively. With the help of (5.16), (5.17), (10.10), and (10.11), we can obtain the constituent parts of all the field components. We should note that in (5.17), $\gamma = j\beta_z = j(p\pi/d)$ for the waves progressing in the $+z$-direction and for the waves progressing in the $-z$-direction, γ should be changed to

$$-\gamma = -j\beta_z = -j(p\pi/d).$$

From (5.16), (5.17), and (10.10), we can evaluate those parts of all the field components which represent the waves progressing in the $+z$-direction as

$$E_{x+}(x, y, z) = \frac{j\omega\mu}{h^2}\left(\frac{n\pi}{b}\right)\frac{jH_0}{2}\cos\left(\frac{m\pi x}{a}\right)\sin\left(\frac{n\pi y}{b}\right)e^{-j(p\pi/d)z}, \tag{10.12}$$

$$E_{y+}(x, y, z) = -\frac{j\omega\mu}{h^2}\left(\frac{m\pi}{a}\right)\frac{jH_0}{2}\sin\left(\frac{m\pi x}{a}\right)\cos\left(\frac{n\pi y}{b}\right)e^{-j(p\pi/d)z}, \tag{10.13}$$

$$E_{z+}(x, y, z) = 0, \tag{10.14}$$

$$H_{x+}(x, y, z) = \frac{j}{h^2}\left(\frac{p\pi}{d}\right)\left(\frac{m\pi}{a}\right)\frac{jH_0}{2}\sin\left(\frac{m\pi x}{a}\right)\cos\left(\frac{n\pi y}{b}\right)e^{-j(p\pi/d)z}, \tag{10.15}$$

$$H_{y+}(x, y, z) = \frac{j}{h^2}\left(\frac{p\pi}{d}\right)\left(\frac{n\pi}{b}\right)\frac{jH_0}{2}\cos\left(\frac{m\pi x}{a}\right)\sin\left(\frac{n\pi y}{b}\right)e^{-j(p\pi/d)z}. \tag{10.16}$$

In a similar manner, from (5.16), (5.17), and (10.11), we can deduce the parts of all the field components which represent the waves progressing in the $-z$-direction as

$$E_{x-}(x, y, z) = \frac{j\omega\mu}{h^2}\left(\frac{n\pi}{b}\right)\left(-\frac{jH_0}{2}\right)\cos\left(\frac{m\pi x}{a}\right)\sin\left(\frac{n\pi y}{b}\right)e^{j(p\pi/d)z}, \tag{10.17}$$

$$E_{y-}(x, y, z) = -\frac{j\omega\mu}{h^2}\left(\frac{m\pi}{a}\right)\left(-\frac{jH_0}{2}\right)\sin\left(\frac{m\pi x}{a}\right)\cos\left(\frac{n\pi y}{b}\right)e^{j(p\pi/d)z}, \tag{10.18}$$

$$E_{z-}(x, y, z) = 0, \tag{10.19}$$

$$H_{x-}(x, y, z) = -\frac{j}{h^2}\left(\frac{p\pi}{d}\right)\left(\frac{m\pi}{a}\right)\left(-\frac{jH_0}{2}\right)\sin\left(\frac{m\pi x}{a}\right)\cos\left(\frac{n\pi y}{b}\right)e^{j(p\pi/d)z}, \tag{10.20}$$

$$H_{y-}(x, y, z) = -\frac{j}{h^2}\left(\frac{p\pi}{d}\right)\left(\frac{n\pi}{b}\right)\left(-\frac{jH_0}{2}\right)\cos\left(\frac{m\pi x}{a}\right)\sin\left(\frac{n\pi y}{b}\right)e^{j(p\pi/d)z}. \tag{10.21}$$

If we add the field components given by (10.12) to (10.16) respectively with those given by (10.17) to (10.21), we obtain the following expressions for the components of the electromagnetic fields associated with the resonant frequency given by (10.8):

$$E_x(x, y, z) = \frac{j\omega\mu}{h^2}\left(\frac{n\pi}{b}\right)H_0\cos\left(\frac{m\pi x}{a}\right)\sin\left(\frac{n\pi y}{b}\right)\sin\left(\frac{p\pi z}{d}\right), \tag{10.22}$$

$$E_y(x, y, z) = -\frac{j\omega\mu}{h^2}\left(\frac{m\pi}{a}\right)H_0\sin\left(\frac{m\pi x}{a}\right)\cos\left(\frac{n\pi y}{b}\right)\sin\left(\frac{p\pi z}{d}\right), \tag{10.23}$$

$$E_z(x, y, z) = 0, \tag{10.24}$$

$$H_x(x, y, z) = -\frac{1}{h^2}\left(\frac{p\pi}{d}\right)\left(\frac{m\pi}{a}\right)H_0\sin\left(\frac{m\pi x}{a}\right)\cos\left(\frac{n\pi y}{b}\right)\cos\left(\frac{p\pi z}{d}\right), \tag{10.25}$$

$$H_y(x, y, z) = -\frac{1}{h^2}\left(\frac{p\pi}{d}\right)\left(\frac{n\pi}{b}\right)H_0\cos\left(\frac{m\pi x}{a}\right)\sin\left(\frac{n\pi y}{b}\right)\cos\left(\frac{p\pi z}{d}\right), \tag{10.26}$$

$$H_z(x, y, z) = H_0\cos\left(\frac{m\pi x}{a}\right)\cos\left(\frac{n\pi y}{b}\right)\sin\left(\frac{p\pi z}{d}\right), \tag{10.27}$$

where (10.27) is reproduced from (10.6).

We can determine the time-averaged energy associated with the electric field inside the cavity from (10.22) to (10.24) as

$$W_E = \int_0^a\int_0^b\int_0^d \frac{\varepsilon}{4}|\boldsymbol{E}(x, y, z)|^2\,dx\,dy\,dz = \frac{\varepsilon}{4}\frac{\omega^2\mu^2}{h^4}|H_0|^2\int_0^a\int_0^b\int_0^d$$
$$\times\left[\left(\frac{n\pi}{b}\right)^2\cos^2\left(\frac{m\pi x}{a}\right)\sin^2\left(\frac{n\pi y}{b}\right)\sin^2\left(\frac{p\pi z}{d}\right)\right.$$
$$\left.+\left(\frac{m\pi}{a}\right)^2\sin^2\left(\frac{m\pi x}{a}\right)\cos^2\left(\frac{n\pi y}{b}\right)\sin^2\left(\frac{p\pi z}{d}\right)\right],$$
$$= \frac{\varepsilon}{8}\frac{\omega^2\mu^2}{h^2}\frac{|H_0|^2}{\varepsilon_{0m}\varepsilon_{0n}}abd, \tag{10.28}$$

where ε_{0m} and ε_{0n} are defined in (5.61b). Note that p can never be equal to zero. We can deduce the time-averaged energy associated with the magnetic field inside the cavity from (10.25) to (10.27) as

$$W_H = \int_0^a\int_0^b\int_0^d \frac{\mu}{4}|\boldsymbol{H}(x, y, z)|^2\,dx\,dy\,dz = |H_0|^2\frac{\mu}{4}\int_0^a\int_0^b\int_0^d$$
$$\times\left[\frac{1}{h^4}\left(\frac{p\pi}{d}\right)^2\left(\frac{m\pi}{a}\right)^2\sin^2\left(\frac{m\pi x}{a}\right)\cos^2\left(\frac{n\pi y}{b}\right)\cos^2\left(\frac{p\pi z}{d}\right)\right.$$
$$+\frac{1}{h^4}\left(\frac{p\pi}{d}\right)^2\left(\frac{n\pi}{b}\right)^2\cos^2\left(\frac{m\pi x}{a}\right)\sin^2\left(\frac{n\pi y}{b}\right)\cos^2\left(\frac{p\pi z}{d}\right)$$
$$\left.+\cos^2\left(\frac{m\pi x}{a}\right)\cos^2\left(\frac{n\pi y}{b}\right)\sin^2\left(\frac{p\pi z}{d}\right)\right],$$
$$= \frac{|H_0|^2\,\mu abd}{8\varepsilon_{0m}\varepsilon_{0n}}\left[\frac{1}{h^4}\left(\frac{p\pi}{d}\right)^2\left\{\left(\frac{m\pi}{a}\right)^2+\left(\frac{n\pi}{b}\right)^2\right\}+1\right] = \frac{|H_0|^2\,\mu\omega_{m,n,p}^2\mu\varepsilon abd}{8h^2\varepsilon_{0m}\varepsilon_{0n}}. \tag{10.29}$$

For $\omega = \omega_{m,n,p}$, we can see from (10.28) and (10.29) that $W_E = W_H$ and hence we find the total energy inside the cavity to be

$$W = W_E + W_H = \frac{|H_0|^2\,\omega_{m,n,p}^2\mu^2\varepsilon abd}{4h^2\varepsilon_{0m}\varepsilon_{0n}} \qquad (\text{for } \omega = \omega_{m,n,p}). \tag{10.30}$$

2. Attenuation Coefficient

We have established that a cavity resonator supports electromagnetic oscillations at certain discrete frequencies with a definite mode pattern for each resonant frequency. If we wish to excite a particular mode of oscillation, the desired field pattern cannot be set up unless the exciting frequency is made identical to the appropriate resonant frequency. However, in practice there is a narrow band of frequencies around the resonant frequency over which significant excitation occurs. The spreading out of the resonant frequency of oscillation is due to the loss of energy from the cavity resulting from the dielectric and the conductor losses or the coupling of energy to the external devices. In the actual design of the cavity resonator, the shape and the dimensions are adjusted so that in the neighborhood of the frequency of operation, the resonant frequencies are well separated with the result that the instability or the tendency to switch to a different mode of oscillation due to the possible drifts of the resonant frequency can be avoided. In the present analysis, we have assumed the cavity to be completely closed and we have not treated the difficult problem of coupling of energy to the external devices. Moreover, since the principal losses arise due to the imperfect conductivity of the walls of the cavity, we have determined the attenuation resulting from the conductor losses alone here. We assume that the conductor losses are sufficiently small for the field patterns not to be significantly perturbed.

In the time domain a typical field component, such as the longitudinal magnetic field as given by (10.6), is of the form

$$H_z(x, y, z, t) = H_z(x, y, z) \cos \omega t, \tag{10.31}$$

where for the sake of simplicity we have assumed H_0 to be real. If ω coincides with one of the resonant frequencies, we see from (10.31) that the amplitude of the mode pattern as a function of time is unaltered. This result applies to all the field components. Consequently, the energy associated with the electromagnetic fields, after it is averaged over a period, remains a constant independent of time. If the walls of the cavity are imperfectly conducting, some energy is dissipated in the conductor walls, as a consequence of which the energy associated with the electromagnetic fields, after it is averaged over a period, decays exponentially as a function of time. Therefore, in the presence of the conductor losses, the field components, such as that given by (10.31), become of the form

$$H_z(x, y, z, t) = H_z(x, y, z)e^{-\alpha t} \cos \omega t, \tag{10.32}$$

where α is known as the attenuation coefficient. We assume that the conductor losses are sufficiently small and therefore α also is sufficiently small for the decay of the amplitude in *a* period of oscillation to be very small. In the evaluation of the time-averaged energies, there occur integrals with respect to time over a period $(0 < t < 2\pi/\omega)$ of quantities which are similar to the square of (10.32). Since $e^{-2\alpha t}$ varies only very slightly in a period, it can be taken outside the integral. Hence, it follows that the time-averaged energy associated with the electromagnetic fields inside the cavity decays exponentially with respect to time as

$$W(t) = We^{-2\alpha t}, \tag{10.33}$$

where W is the time-averaged energy inside the cavity at $t = 0$. The negative rate of change of $W(t)$ with respect to time must equal the power loss in the walls and therefore

$$-dW(t)/dt = 2\alpha W(t) = P_L(t) \qquad \text{or} \qquad \alpha = P_L(t)/2W(t) = P_L/2W. \tag{10.34}$$

In the presence of the losses, all the field components have the multiplicative factor $e^{-\alpha t}$ with the result that both $W(t)$ and $P_L(t)$ contain the same time factor $e^{-2\alpha t}$, which cancels out in the expression for α. Consequently, we can use P_L and W in obtaining the expression for α. We have already evaluated the value of W and it is given in (10.30).

For the purpose of obtaining the power lost in the walls of the cavity, it is necessary for us to determine the surface current densities on all the walls of the cavity. With the help of (10.25) to (10.27), we can obtain the following results:

$$\textit{Wall A:}\quad \mathbf{J}_s\,(y=0) = \hat{y}\times[\hat{x}H_x(x,0,z)+\hat{z}H_z(x,0,z)]$$
$$= \hat{x}H_0\cos\left(\frac{m\pi x}{a}\right)\sin\left(\frac{p\pi z}{d}\right) + \hat{z}\frac{1}{h^2}\left(\frac{p\pi}{d}\right)\left(\frac{m\pi}{a}\right)H_0\sin\left(\frac{m\pi x}{a}\right)\cos\left(\frac{p\pi z}{d}\right), \tag{10.35}$$

$$\textit{Wall C:}\quad \mathbf{J}_s\,(y=b) = -\hat{y}\times[\hat{x}H_x(x,b,z)+\hat{z}H_z(x,b,z)] = (-1)^{n+1}\mathbf{J}_s\,(y=0), \tag{10.36}$$

$$\textit{Wall D:}\quad \mathbf{J}_s\,(x=0) = \hat{x}\times[\hat{y}H_y(0,y,z)+\hat{z}H_z(0,y,z)]$$
$$= -\hat{y}H_0\cos\left(\frac{n\pi y}{b}\right)\sin\left(\frac{p\pi z}{d}\right) - \hat{z}\frac{1}{h^2}\left(\frac{p\pi}{d}\right)\left(\frac{n\pi}{b}\right)H_0\sin\left(\frac{n\pi y}{b}\right)\cos\left(\frac{p\pi z}{d}\right), \tag{10.37}$$

$$\textit{Wall B:}\quad \mathbf{J}_s\,(x=a) = -\hat{x}\times[\hat{y}H_y(a,y,z)+\hat{z}H_z(a,y,z)] = (-1)^{m+1}\mathbf{J}_s\,(x=0), \tag{10.38}$$

$$\textit{Wall E:}\quad \mathbf{J}_s\,(z=0) = \hat{z}\times[\hat{x}H_x(x,y,0)+\hat{y}H_y(x,y,0)]$$
$$= \hat{x}\frac{1}{h^2}\left(\frac{p\pi}{d}\right)\left(\frac{n\pi}{b}\right)H_0\cos\left(\frac{m\pi x}{a}\right)\sin\left(\frac{n\pi y}{b}\right) - \hat{y}\frac{1}{h^2}\left(\frac{p\pi}{d}\right)\left(\frac{m\pi}{a}\right)H_0\sin\left(\frac{m\pi x}{a}\right)\cos\left(\frac{n\pi y}{b}\right), \tag{10.39}$$

$$\textit{Wall F:}\quad \mathbf{J}_s\,(z=d) = -\hat{z}\times[\hat{x}H_x(x,y,d)+\hat{y}H_y(x,y,d)] = (-1)^{p+1}\mathbf{J}_s\,(z=0). \tag{10.40}$$

Since we have assumed the conductor losses to be small, it is legitimate for us to assume also that the surface currents on the walls of a cavity resonator with imperfectly conducting walls are not appreciably different from those of the corresponding ideal cavity resonator and hence are the same as those given by (10.35) to (10.40). Let η_{sr} be the surface resistivity of the walls of the cavity resonator. Then, we obtain the powers P_{La} and P_{Lc} dissipated respectively in walls A and C from (10.35) and (10.36) as

$$P_{La} = P_{Lc} = \frac{\eta_{sr}}{2}\int_0^a dx\int_0^d dz\,|\mathbf{J}_s\,(y=0)|^2 = \frac{\eta_{sr}}{4}\frac{|H_0|^2}{\varepsilon_{0m}}\left[1+\frac{(p\pi/d)^2(m\pi/a)^2}{h^4}\right]ad. \tag{10.41}$$

In a similar manner, we may deduce the powers P_{Ld} and P_{Lb} dissipated respectively in walls D and B with the help of (10.37) and (10.38) as

$$P_{Ld} = P_{Lb} = \frac{\eta_{sr}}{2}\int_0^b dy \int_0^d dz\, |\mathbf{J}_s\,(x = 0)|^2 = \frac{\eta_{sr}}{4}\frac{|H_0|^2}{\varepsilon_{0n}}\left[1 + \frac{(p\pi/d)^2(n\pi/b)^2}{h^4}\right]bd. \quad (10.42)$$

Also, we can determine the powers P_{Le} and P_{Lf} dissipated respectively in walls E and F from (10.39) and (10.40) with the following result:

$$P_{Le} = P_{Lf} = \frac{\eta_{sr}}{2}\int_0^a dx \int_0^b dy\, |\mathbf{J}_s\,(z = 0)|^2 = \frac{\eta_{sr}}{2}\frac{|H_0|^2}{\varepsilon_{0m}\varepsilon_{0n}h^2}\left(\frac{p\pi}{d}\right)^2 ab. \quad (10.43)$$

We find from (10.41) to (10.43) that

$$\begin{aligned} P_L &= P_{La} + P_{Lb} + P_{Lc} + P_{Ld} + P_{Le} + P_{Lf}, \\ &= \frac{\eta_{sr}}{2}\frac{|H_0|^2}{\varepsilon_{0m}\varepsilon_{0n}h^2}\left[\varepsilon_{0n}\left\{h^2 + \frac{(p\pi/d)^2(m\pi/a)^2}{h^2}\right\}ad \right. \\ &\quad \left. + \varepsilon_{0m}\left\{h^2 + \frac{(p\pi/d)^2(n\pi/b)^2}{h^2}\right\}bd + 2\left(\frac{p\pi}{d}\right)^2 ab\right]. \end{aligned} \quad (10.44)$$

Substituting (10.30) and (10.44) into (10.34) we get

$$\begin{aligned} \alpha = \frac{\eta_{sr}}{\omega_{m,n,p}^2\mu^2\varepsilon abd}&\left[\varepsilon_{0n}\left\{h^2 + h^{-2}\left(\frac{p\pi}{d}\right)^2\left(\frac{m\pi}{a}\right)^2\right\}ad \right. \\ &\left. + \varepsilon_{0m}\left\{h^2 + h^{-2}\left(\frac{p\pi}{d}\right)^2\left(\frac{n\pi}{b}\right)^2\right\}bd + 2\left(\frac{p\pi}{d}\right)^2 ab\right]. \end{aligned} \quad (10.45)$$

3. Selectivity

In the presence of the losses, we pointed out that there is an appreciable excitation of the electromagnetic fields inside the cavity over a narrow band of frequencies centered on the resonant frequency. As is to be expected, the excitation is a maximum at the resonant frequency and falls off as the frequency deviates from the resonant frequency on either side. It is useful for us to introduce a parameter which is a measure of the sharpness of the response of the cavity resonator to a given excitation. This parameter is called the selectivity Q which is defined as in (3.10.8) by

$$Q = \omega_{m,n,p}\frac{W}{P_L}, \quad (10.46)$$

where $\omega_{m,n,p}$ is the resonant frequency, W is the time-averaged energy inside the cavity and P_L is the power dissipated within the cavity. We should note that the definition (10.46) is the same as that used in connection with the transmission-line resonant circuits. From (10.34) and (10.46), it follows that

$$Q = \omega_{m,n,p}/2\alpha. \quad (10.47)$$

Substituting (10.45) into (10.47) gives us the following expression for the selectivity:

$$Q = \frac{\eta_0 \beta^3_{m,n,p} abd}{2\eta_{sr}} \left[\varepsilon_{0n} \left\{ h^2 + h^{-2} \left(\frac{p\pi}{d}\right)^2 \left(\frac{m\pi}{a}\right)^2 \right\} ad \right.$$
$$\left. + \varepsilon_{0m} \left\{ h^2 - h^{-2} \left(\frac{p\pi}{d}\right)^2 \left(\frac{n\pi}{b}\right)^2 \right\} bd + 2\left(\frac{p\pi}{d}\right)^2 ab \right]^{-1}, \tag{10.48}$$

where

$$\eta_0 = \sqrt{\mu/\varepsilon} \quad \text{and} \quad \beta_{m,n,p} = \omega_{m,n,p}/v. \tag{10.49}$$

The selectivity given by (10.48) pertains to the TE_{mnp} mode where the subscripts m, n, and p represent the number of periodic variations of the electromagnetic fields in the x-, y-, and z-directions respectively. We can obtain the characteristics of the TM_{mnp} mode of a rectangular cavity resonator in a manner analogous to that used for the TE_{mnp} mode, but we do not include them here. We should note that for the TM_{mnp} mode neither m nor n can be zero, but p can be zero.

4. TE_{101} Mode

For the TE_{101} mode, we can obtain Q from (10.48) to be given by

$$TE_{101}\ \textit{Mode:} \quad Q = \frac{\eta_0(\beta_{m,n,p}\, da)^3 b}{2\eta_{sr}\pi^2} [ad(a^2 + d^2) + 2b(a^3 + d^3)]^{-1}. \tag{10.50}$$

From (10.8), we find the resonant angular frequency of the TE_{101} mode to be

$$\omega_{1,0,1} = v\pi \left[\frac{1}{a^2} + \frac{1}{d^2}\right]^{1/2}. \tag{10.51}$$

For the case of the cavity resonator in the form of a cube, $a = b = d$, with the result that we can simplify (10.50) and (10.51) to yield the following results:

$$TE_{101}\ \textit{Mode:} \quad Q = \eta_0 \pi \sqrt{2}/6\eta_{sr} \quad (\text{for } a = b = d), \tag{10.52}$$

and

$$f_{1,0,1} = v/\sqrt{2}\,a \quad (\text{for } a = b = d). \tag{10.53}$$

If the cavity resonator is filled with air, $v = 3 \times 10^8$ m/sec. Also, if $a = 2.121$ cm, we can evaluate the resonant frequency as $f_{1,0,1} = 10^{10}$ Hz. The intrinsic impedance of free space is given by $\eta_0 = 377$ ohms. If the walls of the cavity are made of copper, at the frequency of 10^{10} Hz, the surface resistivity is approximately $\eta_{sr} = 0.028$ ohm. Hence, we can show Q given by (10.52) to be approximately 10^4. In contrast, Q of an ordinary LC circuit at low frequencies is only of the order of a few hundred. In view of the very high selectivity obtained with a cavity resonator at very high frequencies, it is used as a resonant circuit in high-frequency tubes and in wavemeters for the frequency measurements.

PROBLEMS

8.1 Consider a parallel-plate waveguide with the perfectly conducting surfaces occupying the planes $x = 0$ and $x = d$ respectively. The fields are independent of y but propagate in the z-direction. Deduce the field components of the TE mode as given in (8.2.6) to (8.2.8) directly from the solution of Maxwell's equations subject to the appropriate boundary conditions. Determine the expression for the wave impedance.

8.2 For the parallel-plate waveguide of Problem 8.1, find the expressions for the field components of the TE mode if the fields propagate in the $-z$-direction. What is the value of the wave impedance?

8.3 For the parallel-plate waveguide of Problem 8.1, deduce the field components of the TM mode as given in (8.3.4) to (8.3.6) directly from the solution of Maxwell's equations subject to the appropriate boundary conditions for the case of propagation in the $+z$-direction. Determine the expression for the wave impedance.

If the propagation takes place in the $-z$-direction, what are the expressions for the field components of the TM mode and what is the value of the wave impedance?

8.4 A parallel-plate waveguide consists of two parallel, perfectly conducting infinite planes spaced $d = 10$ cm apart. The constitutive parameters of the dielectric region between the parallel planes are given by $\mu_d = 4\pi \times 10^{-7}$ H/m and $\varepsilon_d = 10^{-9}/36\pi$ F/m. The wave frequency is $f = 10^9$ Hz. For the *dominant* mode of propagation, evaluate (a) the propagation coefficient; (b) the cut-off frequency; and (c) the phase velocity, the wavelength, and the group velocity inside the waveguide.

Repeat the calculations for the next higher order mode of propagation. Clearly designate the modes whose characteristics are under consideration.

8.5 A waveguide is formed by two parallel, perfectly conducting infinite planes spaced $d = 5$ cm apart in free space. The frequency of operation is $f = 10^{10}$ Hz. Determine the effects on (a) the cut-off frequency; (b) the propagation coefficient; and (c) the phase velocity, the wavelength, and the group velocity inside the waveguide, if a loss-free dielectric of relative permittivity $\varepsilon_r = 2.25$ is substituted for the free space region between the parallel plates. Assume that the propagation takes place in the TE_{10} mode.

8.6 A waveguide is formed by two parallel sheets of copper spaced $d = 20$ cm apart in free space. The constitutive parameters of copper are $\mu_c = 4\pi \times 10^{-7}$ H/m and $\sigma_c = 5.8 \times 10^7$ mho/m. Note that the constitutive parameters of free space are $\mu_d = 4\pi \times 10^{-7}$ H/m. $\varepsilon_d = 10^{-9}/36\pi$ F/m. The frequency of operation is $f = 10^9$ Hz. Evaluate (a) the phase-shift coefficient and (b) the attenuation coefficient due to the conductor losses for (i) the TEM, (ii) the TE_{10}, and (iii) the TM_{10} modes of operation.

8.7 A waveguide is formed by two parallel, perfectly conducting infinite planes spaced $d = 20$ cm apart. The constitutive parameters of the dielectric region between the parallel planes are given by $\mu_d = 4\pi \times 10^{-7}$ H/m, $\varepsilon_d = 10^{-9}/9\pi$ F/m, and $\sigma_d = 10^{-3}$ mho/m. The frequency of operation is $f = 10^9$ Hz. Evaluate (a) the phase-shift coefficient and (b) the attenuation coefficient due to the losses in the dielectric for (i) the TEM, (ii) the TE_{10}, and (iii) the TM_{10} modes of propagation.

8.8 Show that the attenuation coefficient α_c due to the conductor losses of the TM_{m0} ($m \neq 0$) mode in a parallel-plate waveguide is a minimum at a frequency which is $\sqrt{3}$ times the cut-off frequency.

8.9 Consider a parallel-plate waveguide with the perfectly conducting surfaces occupying the planes $x = 0$ and $x = d$ respectively. The fields are independent of y but propagate in the z-direction. The dielectric region between the planes is assumed to be slightly lossy. Deduce the value of the attenuation coefficient α_d due to the losses in the dielectric with the help of the relation (8.4.10) for (a) the TE_{m0} and (b) the TM_{m0} modes and compare it with the result given in (8.4.5).

8.10 A waveguide is formed by two parallel perfectly conducting infinite planes spaced $d = 2.5$ cm apart. The constitutive parameters of the dielectric region between the parallel planes are given by $\mu_d = 4\pi \times 10^{-7}$ H/m and $\varepsilon_d = 10^{-9}/36\pi$ F/m. The frequency of operation is $f = 10^{10}$ Hz. Assume that a field strength of 10^6 V/m causes the voltage breakdown of the dielectric. Find the maximum time-averaged power per unit width of the waveguide that can be propagated without causing the voltage breakdown of the dielectric. Determine the result for the following three modes of propagation: (i) TEM mode, (ii) TE_{10} mode, and (iii) TM_{10} mode. For each case, evaluate the surface charge and the surface current densities on the two walls of the waveguide.

8.11 Consider the surface current density $\boldsymbol{J}_s\ (y = b)$ as given by (8.5.86) on wall C of a rectangular waveguide supporting the dominant TE_{10} mode. Show that the lines of surface current density at a particular time $t = 0$ are specified by the equation

$$[\cos(\pi x/a)]^p = \text{constant} \sin \tilde{\beta}_z z, \tag{a}$$

where

$$p = (\omega/\omega_{1,0})^2 - 1. \tag{b}$$

Note that ω is the angular frequency of operation and $\omega_{1,0}$ is the cut-off angular frequency of the TE_{10} mode. The current density lines obtained from (a) are shown in Fig. 8.10 for $\omega/\omega_{1,0} = \sqrt{3}$. The current density lines are determined in the following manner.

It is seen from Fig. 8.10 that the current density lines consist of two sets: one set is essentially in the longitudinal direction and the other set is essentially in the transverse direction. First, the current density lines which are essentially in the longitudinal direction are considered. For $\tilde{\beta}_z z = 3\pi/2$, it is convenient to take the constant in (a) as $-[\cos(\pi x_0/a)]^p$ with the result that the current density lines pass through $x/a = x_0/a$ for $\tilde{\beta}_z z = 3\pi/2$. It is seen from (8.5.86) that for $\tilde{\beta}_z z = 3\pi/2$, $J_{sx}\ (y = b) = 0$ and $J_{sz}\ (y = b)$ varies as $\sin(\pi x/a)$ in the interval $0 < x/a < 1$. A set of values of x_0/a is chosen in the interval $0 < x/a < 1$ such that the density of points varies as $\sin(\pi x/a)$. This choice of values of x_0/a is needed since the density of the current lines should be proportional to the magnitude of the current. The equation (a) specifying the current lines is conveniently rewritten as follows:

$$\left[\frac{\cos(\pi x/a)}{\cos(\pi x_0/a)}\right]^p = -\sin \tilde{\beta}_z z. \tag{c}$$

The value of $\tilde{\beta}_z z$ is changed in small intervals from $\tilde{\beta}_z z = 3\pi/2$ to $\tilde{\beta}_z z = 7\pi/2$. For each value of $\tilde{\beta}_z z$ and for each of the chosen set of values of x_0/a, the value of x/a which satisfies (c) is determined. The points corresponding to the same values of x_0/a are connected together to yield the first set of current lines. Note that a complete cycle of current lines is contained in the interval $3\pi/2 < \tilde{\beta}_z z < 7\pi/2$.

Choose eight points in the interval $0 < x/a < 1$ such that the density of the points varies as $\sin(\pi x/a)$. Obtain the first set of current lines in the foregoing manner and plot them.

The current density lines which are essentially in the transverse direction are now considered. For $x/a = 0$, it is convenient to take the constant in (a) as $1/\sin \tilde{\beta}_z z_0$ with the result that the current density lines pass through $\tilde{\beta}_z z = \tilde{\beta}_z z_0$ for $x/a = 0$. It is seen from (8.5.86) that for $x/a = 0$, $J_{sz}\ (y = b) = 0$ and $J_{sx}\ (y = b)$ varies as $\cos \tilde{\beta}_z z$. A set of values of $\tilde{\beta}_z z_0/2\pi$ is chosen in the interval $3/4 < \tilde{\beta}_z z/2\pi < 7/4$ such that the density of points varies as $\cos \tilde{\beta}_z z$. As before, this choice of values of $\tilde{\beta}_z z_0$ is needed since the density of the current lines should be proportional to the magnitude of the current. The equation (a) specifying the current lines is conveniently rewritten as

$$[\cos(\pi x/a)]^p = \sin(\tilde{\beta}_z z)/\sin(\tilde{\beta}_z z_0). \tag{d}$$

The second set of current lines which are essentially in the transverse direction are obtained from (d) by following a procedure similar to that used for (c).

Choose sixteen points in the interval $3/4 < \tilde{\beta}_z z/2\pi < 7/4$ such that the density of the points varies as $\cos \tilde{\beta}_z z$. Determine the second set of current lines, plot them also in the same figure which contains the first set of current lines, and thus reproduce Fig. 8.10.

8.12 For the TM mode of a rectangular waveguide treated in Section 8.6, prove that $E_z(x, y)$ given by (8.6.5) is the solution of the differential equation (8.5.21) subject to the boundary conditions given by (8.6.1) and (8.6.2).

8.13 The field components of the TM mode of a rectangular waveguide are given by (8.6.6) to (8.6.11). With the help of (8.6.6) to (8.6.11), show that the x- and y-components of the time-averaged Poynting vectors, $S_x(x, y, z)$ and $S_y(x, y, z)$, vanish both for $\omega < \omega_{m,n}$ and for $\omega > \omega_{m,n}$, where $\omega_{m,n}$ is the critical angular frequency of the TM_{mn} mode. Also, prove that the z-component of the time-averaged Poynting vector $S_z(x, y, z)$ is equal to zero for $\omega < \omega_{m,n}$. Find the value of $S_z(x, y, z)$ for $\omega > \omega_{m,n}$.

Deduce the expression for the energy transport velocity of the TM_{mn} mode and verify that it is equal to the group velocity as given by (8.5.48).

Obtain the surface charge densities on the four walls of the rectangular waveguide for the TM_{mn} mode of propagation.

8.14 Deduce the expression (8.7.17) for the attenuation coefficient α_c due to the imperfect conductivity of the walls of a rectangular waveguide supporting a TM mode.

8.15 Let f_c be the cut-off frequency of the propagating mode inside a rectangular waveguide and let $f(>f_c)$ be the frequency of operation. The wavelengths corresponding to f and f_c in an unbounded medium having the same constitutive parameters as that enclosed by the walls of the waveguide are denoted by λ and λ_c respectively. Let λ_g be the wavelength inside the waveguide. Then, show that

$$\frac{1}{\lambda_g^2} + \frac{1}{\lambda_c^2} = \frac{1}{\lambda^2}.$$

8.16 Show that the attenuation coefficient α_c of the TE_{mn} mode due to the imperfect conductivity of the walls of a rectangular waveguide as given by (8.7.15) has a minimum and find the value of the frequency for which the minimum attenuation occurs.

Show that the attenuation coefficient α_c of the TM_{mn} mode as given by (8.7.17) also has a minimum. Determine the value of the frequency for which the minimum attenuation occurs.

8.17 Consider a rectangular waveguide with air as a dielectric. The dimensions of the waveguide are given by $a = 5$ cm and $b = 3$ cm. Determine the critical frequency for the dominant mode. If the wavelength in unbounded air is 4 cm, find the phase velocity, the wavelength, and the group velocity inside the waveguide. Determine the wave impedance corresponding to the TE_{10} mode of the waveguide. Find the angle that the constituent plane waves of the TE_{10} mode make with the axis of the waveguide.

8.18 An air-filled rectangular waveguide has the following cross-sectional dimensions: $a = 10$ cm and $b = 7.5$ cm. How many TE modes can be transmitted by this waveguide at frequencies below $f = 5 \times 10^9$ Hz? State the designations of these modes and find their cut-off frequencies.

8.19 An air-filled rectangular waveguide has the following cross-sectional dimensions: $a = 10$ cm; $b = 7$ cm. Find the phase velocity of the TE_{10} mode at a frequency which is twice the cut-off frequency. What is the angle made by the constituent plane wave with the axis of the waveguide?

8.20 A rectangular waveguide is to be designed to operate in the TE_{10} mode at a frequency of $f = 10^{10}$ Hz. It is desired that the frequency of operation be at least 15% *above* the cut-off frequency of the propagating mode and 20% *below* the cut-off frequency of the next higher order mode. Find the dimensions of the rectangular waveguide and determine the cut-off frequencies of the first three lowest order modes.

8.21 An air-filled rectangular waveguide has the following cross-sectional dimensions: $a = 2.5$ cm and $b = 1.0$ cm. The constitutive parameters of the air dielectric are $\mu_d = 4\pi \times 10^{-7}$ H/m and $\varepsilon_d = 10^{-9}/36\pi$ F/m. Assume that the voltage breakdown of the air dielectric occurs if the electric field intensity exceeds 5×10^6 V/m. The frequency of operation is $f = 10^{10}$ Hz. Find the maximum time-averaged power that can be transmitted in the TE_{10} mode without causing the voltage breakdown of the dielectric. What are the surface charge and the surface current densities on all the walls of the waveguide under the condition of maximum transport of power?

8.22 A rectangular waveguide with copper walls operates in the TE_{10} mode at a frequency of $f = 10^{10}$ Hz. The constitutive parameters of copper are $\mu_c = 4\pi \times 10^{-7}$ H/m and $\sigma_c = 5.8 \times 10^7$ mhos/m. The constitutive parameters of the air dielectric are $\mu_d = 4\pi \times 10^{-7}$ H/m and $\varepsilon_d = 10^{-9}/36\pi$ F/m. The wavelength inside the waveguide is 1.5 times the corresponding wavelength in an unbounded air medium. The waveguide has a height-to-width ratio of $b/a = 0.5$. The waveguide is required to transmit 600 W of time-averaged power.

(a) Evaluate the maximum values of the electric and the magnetic fields, and find the locations where these maximum values occur. (b) Find the surface charge and the surface current densities on all the four walls of the waveguide. (c) Determine the attenuation coefficient α_c due to the conductor losses.

8.23 An air-filled rectangular waveguide transmits 500 W of time-averaged power at a frequency of $f = 10^9$ Hz. The energy transport velocity is 0.75 times the corresponding velocity in an unbounded air medium. The maximum value of the electric field is 10^5 V/m.

Find the dimensions of the waveguide. Determine the amplitudes of all the field components. Evaluate the surface charge and the surface current densities on all the walls of the waveguide. Assume that the propagation takes place in the TE_{10} mode. Repeat the calculations for the TM_{11} mode of propagation.

8.24 An air-filled rectangular waveguide has the following cross-sectional dimensions: $a = 3.2$ cm and $b = 1.6$ cm. The TE_{10} mode propagates in the waveguide at a frequency which is 20% above the cut-off frequency of the propagating mode. The maximum value of the electric field is 10^5 V/m. Find the time-averaged power transmitted through the waveguide.

8.25 An air-filled ideal rectangular waveguide extends from $x = 0$ to $x = 0.02$ m and from $y = 0$ to $y = 0.01$ m. The longitudinal electric field in the waveguide is given by

$$E_z(\mathbf{r}, t) = 2 \sin(50\pi x) \sin(100\pi y) \cos(\omega t - 200\pi z) \text{ V/m}.$$

where x, y, and z are in meters. Evaluate (a) the frequency of operation, (b) the remaining field components, (c) the surface charge and the surface current densities on all the walls of the waveguide, and (d) the time-averaged power transported by the waveguide.

8.26 An ideal rectangular waveguide extends from $x = 0$ to $x = a$ and from $y = 0$ to $y = b$. In the region $0 \le z \le d$, the waveguide is filled with a dielectric of constitutive parameters μ and ε. The remaining part ($-\infty < z < 0$ and $d < z < \infty$) of the waveguide is filled with air of constitutive parameters μ_0 and ε_0. A wave is incident on the dielectric from the region $z < 0$. Determine the amplitude of the wave transmitted past the dielectric block in terms of the wave incident on the block if (a) the incident wave is the TE_{10} mode, (b) the incident wave is the TM_{11} mode.

8.27 A rectangular waveguide with copper walls has the following cross-sectional dimensions: $a = 4$ cm and $b = 2$ cm. The constitutive parameters of copper are $\mu_c = 4\pi \times 10^{-7}$ H/m and $\sigma_c = 5.8 \times 10^7$ mhos/m. The constitutive parameters of the dielectric region enclosed by the waveguide are $\mu_d = 4\pi \times 10^{-7}$ H/m, $\varepsilon_d = 10^{-9}/36\pi$ F/m, and $\sigma_d = 10^{-4}$ mho/m. The frequency of operation is $f = 10^{10}$ Hz. Evaluate (a) the phase-shift coefficient, (b) the attenuation coefficient due to the losses in the dielectric, and (c) the attenuation coefficient due to the imperfect conductivity of the walls of the waveguide. Obtain the results both for (1) the TE_{10} and (2) the TM_{11} modes of propagation. (d) Determine the frequency for which the attenuation coefficient of the TE_{10} mode due to the imperfect conductivity of the walls of the waveguide is a minimum and find the value of the minimum attenuation. Repeat the calculations for the TM_{11} mode of propagation.

8.28 Consider a rectangular waveguide with perfectly conducting walls extending from $x = 0$ to $x = a$ and from $y = 0$ to $y = b$. The fields propagate in the z-direction. The dielectric region enclosed by the walls of the waveguide is assumed to be slightly lossy. Deduce the value of the attenuation coefficient α_d due to the losses in the dielectric with the help of the relation (8.4.10) for (a) the TE_{mn} and (b) the TM_{mn} modes and compare it with the result given in (8.7.4).

8.29 Consider a rectangular waveguide as shown in Fig. 8.7. The walls of the waveguide occupying the planes $x = 0$, $y = 0$, and $y = b$ are perfect electric conductors on which the tangential component of the electric field vanishes. The wall occupying the plane $x = a$ is a fictitious magnetic conductor on which the tangential component of the magnetic field vanishes. Assume that the dielectric has no losses.

Evaluate the longitudinal component of the magnetic field which satisfies Maxwell's equations and the appropriate boundary conditions on the walls of the waveguide. Obtain the transverse components of the magnetic and the electric fields and thus deduce the field components of the TE_{mn} mode supported by the rectangular waveguide under consideration.

In a similar manner, deduce all the field components of the TM_{mn} mode also.

Designate the dominant mode of propagation and examine the field configuration of this mode.

8.30 An ideal air-filled rectangular waveguide has the following cross-sectional dimensions: $a = 3$ cm and $b = 1.5$ cm. Assume the TE_{10} mode of propagation. If the frequency of operation is $f = 10^{10}$ Hz, what is the value of the phase-shift coefficient?

If the frequency is below the cut-off frequency, the wave is attenuated. Hence, a waveguide operating below cut-off can be used as an attenuator. What is the attenuation coefficient for $f = 10^8$ Hz and what is the length of the waveguide in which the amplitude of the field is reduced by a factor of 1000?

If the frequency of operation is very much below the cut-off frequency, show that the attenuation coefficient becomes independent of the frequency and find its value. For this case, determine the length of the waveguide in which the amplitude of the field is reduced by a factor of 1000.

8.31 Consider a dielectric slab waveguide as shown in Fig. 8.12. Evaluate the characteristics of both (i) the odd TE and (ii) the even TE modes using the *synthesis* procedure starting from the uniform plane wave fields. In particular, obtain the dispersion equation and the field components both inside and outside the dielectric region. Compare the results with those given in Section 8.8.

8.32 Consider a dielectric slab waveguide as shown in Fig. 8.12. Evaluate the characteristics of both (i) the odd TM and (ii) the even TM modes using the *analysis* procedure starting from Maxwell's equations and the appropriate boundary conditions. In particular, obtain the dispersion equation and the field components both inside and outside the dielectric region. Compare the results with those obtained in Section 8.9 using the synthesis procedure.

8.33 A dielectric slab waveguide is in free space whose constitutive parameters are μ_0 and ε_0. The constitutive parameters of the dielectric are $\mu_d = \mu_0$ and $\varepsilon_d = 2.69\varepsilon_0$. The frequency of operation is $f = 10^9$ Hz. Find the minimum value d_{min} of the thickness of the dielectric slab for which an odd TE mode can be supported by the dielectric slab waveguide.

8.34 A dielectric slab waveguide is in free space whose constitutive parameters are μ_0 and ε_0. The constitutive parameters of the dielectric are $\mu_d = \mu_0$ and $\varepsilon_d = 2.69\varepsilon_0$. The normalized frequency is given by $\tilde{\omega} = \omega d\sqrt{\mu_0\varepsilon_0} = 2$ where $2d$ is the thickness of the dielectric slab. Find the value of the normalized propagation coefficient $\tilde{\beta}_z = \beta_z d$ of (i) the odd TE mode, (ii) the even TE mode, (iii) the odd TM mode, and (iv) the even TM mode by solving graphically the appropriate dispersion equation. Compare the results with those given in Figs. 8.14 and 8.16. How many orders of modes of each kind propagate?

8.35 An ideal air-filled rectangular cavity has the following dimensions: $a = 4$ cm, $b = 2$ cm, and $d = 5$ cm. Designate the first three (i) TE and (ii) TM modes of oscillation. Find their resonant frequencies.

8.36 Consider a rectangular waveguide as shown in Fig. 8.7 supporting a TM_{mn} mode. Let two conducting planes be introduced to coincide with the planes $z = 0$ and $z = d$ with the result that the rectangular waveguide becomes a rectangular cavity as illustrated in Fig. 8.17. Evaluate the resonant frequency and the field components of the TM_{mnp} mode of oscillation. Prove that for the TM_{mnp} mode neither m nor n can be zero but p can be zero.

Deduce the expression for the Q of the rectangular cavity resonator for the TM_{mnp} mode due to the imperfect conductivity of the walls of the cavity.

8.37 A rectangular cavity has the following dimensions: $a = 4$ cm, $b = 2$ cm, and $d = 5$ cm. The cavity is filled with air having the following constitutive parameters: $\mu_d = 4\pi \times 10^{-7}$ H/m and $\varepsilon_d = 10^{-9}/36\pi$ F/m. The walls are made of copper whose constitutive parameters are $\mu_c = 4\pi \times 10^{-7}$ H/m and $\sigma_c = 5.8 \times 10^7$ mhos/m. Evaluate the resonant frequency and the Q of the cavity for the following modes of oscillation: (i) TE_{101} mode, and (ii) TM_{110} mode.

8.38 A rectangular cavity resonator is filled with a lossy dielectric. Show that for the TE_{mnp} mode of oscillation, the Q of the cavity due to the losses in the dielectric is given by $Q = \omega\varepsilon/\sigma$, where ε and σ are respectively the permittivity and the conductivity of the dielectric. Also, $\omega/2\pi$ is the resonant frequency of oscillation.

8.39 A cavity resonator has the following dimensions: $a = b = d = 10$ cm. The cavity is filled with air having the following constitutive parameters: $\mu_d = 4\pi \times 10^{-7}$ H/m and $\varepsilon_d = 10^{-9}/36\pi$ F/m. The walls are made of brass whose constitutive parameters are $\mu_c = 4\pi \times 10^{-7}$ H/m and $\sigma_c = 1.5 \times 10^7$ mhos/m. Evaluate the resonant frequency and the Q of the cavity for the *dominant* mode of oscillation.

If the cavity is filled with a lossless dielectric material with a relative permittivity of 2.25, what are the new values of the resonant frequency and the Q of the cavity?

8.40 An ideal air-filled rectangular cavity extends from $x = 0$ to $x = 4$ cm, from $y = 0$ to $y = 2$ cm, and from $z = 0$ to $z = 5$ cm. The walls occupying the planes $z = 0$ and $z = 5$ cm are designated as E and F respectively. Find the total charge on the walls E and F in terms of the amplitude of the longitudinal electric field. Determine the potential difference between the walls

E and F at their centers and hence obtain the value of the equivalent capacitance of the cavity. Assume the lowest order mode for which the longitudinal electric field does not vary in the longitudinal direction.

8.41 If $H_z(x, y) = 0$ and $\nabla_t \cdot \boldsymbol{H}_t(x, y) = 0$ where the subscript t denotes the components transverse to the z-direction, show that the magnetic field lines form closed loops in the transverse (x, y) plane.

If the total current in the z-direction is zero, prove that there cannot be any closed loop of magnetic field lines in the transverse plane.

CHAPTER 9

ELECTROMAGNETIC RADIATION

9.1 INTRODUCTION

In the previous chapters, we treated the characteristics of electromagnetic waves in unbounded and bounded media without taking into account the sources which give rise to these electromagnetic fields. In this chapter, we shall specifically consider the effect of time-varying charges and currents. We shall introduce certain auxiliary functions which are known as the vector and the scalar potentials and which facilitate the determination of the electromagnetic fields due to the time-varying charges and currents. We shall deduce the partial differential equations satisfied by the potentials and indicate their solutions in terms of the charge and the current distributions. Then, we shall give a treatment of the method of obtaining the electromagnetic fields from the knowledge of the potentials.

We shall consider the radiation from some simple current distributions. We shall examine in detail the fields radiated by the Hertzian dipole and the radiation characteristics of a finite length current filament with prescribed current distributions in the form of both a standing wave and a progressive wave. Such physical parameters as the radiation resistance, the directive gain and the effective height which are useful in the characterization of the transmitting properties of a radiating structure are introduced. We shall then analyze the imaging properties of an infinite, perfectly conducting plane systematically.

We shall next discuss the radiation characteristics of a small electric current loop and point out the dual relationship that exists between the fields associated with an electric current element and a small electric current loop. Finally, we shall introduce the concept of the equivalent magnetic charges and magnetic currents and illustrate their usefulness by the evaluation of the electromagnetic fields radiated by a magnetic current element.

9.2 VECTOR AND SCALAR POTENTIALS

Let us consider a volume distribution of charge density $\rho(\boldsymbol{r}, t)$ which varies with time. From the equation of continuity,

$$\nabla \cdot \boldsymbol{J}(\boldsymbol{r}, t) + \frac{\partial}{\partial t}\rho(\boldsymbol{r}, t) = 0, \tag{2.1}$$

it follows that there is an accompanying volume distribution of current density, $\boldsymbol{J}(\boldsymbol{r}, t)$. The time-varying charge and current distributions give rise to the electric and the magnetic fields, $\boldsymbol{E}(\boldsymbol{r}, t)$ and $\boldsymbol{H}(\boldsymbol{r}, t)$, which satisfy the following Maxwell's equations:

$$\nabla \times \boldsymbol{E}(\boldsymbol{r}, t) = -\frac{\partial}{\partial t}\boldsymbol{B}(\boldsymbol{r}, t), \tag{2.2}$$

$$\nabla \times \boldsymbol{H}(\boldsymbol{r}, t) = \frac{\partial}{\partial t}\boldsymbol{D}(\boldsymbol{r}, t) + \boldsymbol{J}(\boldsymbol{r}, t). \tag{2.3}$$

We assume the medium to be lossless and therefore the following constitutive relations hold:

$$\boldsymbol{B}(\boldsymbol{r}, t) = \mu\boldsymbol{H}(\boldsymbol{r}, t); \qquad \boldsymbol{D}(\boldsymbol{r}, t) = \varepsilon\boldsymbol{E}(\boldsymbol{r}, t). \tag{2.4}$$

In addition, the fields satisfy the following two divergence relations:

$$\nabla \cdot \boldsymbol{D}(\boldsymbol{r}, t) = \rho(\boldsymbol{r}, t), \tag{2.5}$$

$$\nabla \cdot \boldsymbol{B}(\boldsymbol{r}, t) = 0. \tag{2.6}$$

We wish to relate $\boldsymbol{E}(\boldsymbol{r}, t)$ and $\boldsymbol{H}(\boldsymbol{r}, t)$ to the sources $\rho(\boldsymbol{r}, t)$ and $\boldsymbol{J}(\boldsymbol{r}, t)$. The divergence relations (2.5) and (2.6) contain the three components of $\boldsymbol{E}(\boldsymbol{r}, t)$ and the three components of $\boldsymbol{H}(\boldsymbol{r}, t)$ respectively with the result that it is not possible for us to relate any of the field components to the sources with the help of (2.5) and (2.6) alone. Also, the six scalar equations represented by (2.2) and (2.3) contain the three components of $\boldsymbol{E}(\boldsymbol{r}, t)$ and the three components of $\boldsymbol{H}(\boldsymbol{r}, t)$ coupled in a complex fashion, with the result that it is difficult for us to relate $\boldsymbol{E}(\boldsymbol{r}, t)$ and $\boldsymbol{H}(\boldsymbol{r}, t)$ to $\rho(\boldsymbol{r}, t)$ and $\boldsymbol{J}(\boldsymbol{r}, t)$ directly from the six equations represented by (2.2) and (2.3).

It is possible for us to manipulate (2.2) and (2.3) with the help of (2.4) to (2.6) to obtain six separate equations each of which contains only one of the field components. For this purpose, we take the curl of both sides of (2.2) and make use of (2.3) and (2.4) to obtain:

$$\nabla \times \nabla \times \boldsymbol{E}(\boldsymbol{r}, t) = -\mu\frac{\partial}{\partial t}[\nabla \times \boldsymbol{H}(\boldsymbol{r}, t)] = -\mu\varepsilon\frac{\partial^2}{\partial t^2}\boldsymbol{E}(\boldsymbol{r}, t) - \mu\frac{\partial}{\partial t}\boldsymbol{J}(\boldsymbol{r}, t). \tag{2.7}$$

From (2.4) and (2.5), we find that

$$\nabla \times \nabla \times \boldsymbol{E}(\boldsymbol{r}, t) = \nabla[\nabla \cdot \boldsymbol{E}(\boldsymbol{r}, t)] - \nabla^2\boldsymbol{E}(\boldsymbol{r}, t) = \nabla\left(\frac{\rho(\boldsymbol{r}, t)}{\varepsilon}\right) - \nabla^2\boldsymbol{E}(\boldsymbol{r}, t). \tag{2.8}$$

Together with (2.8), we may simplify (2.7) to yield

$$\nabla^2\boldsymbol{E}(\boldsymbol{r}, t) - \mu\varepsilon\frac{\partial^2}{\partial t^2}\boldsymbol{E}(\boldsymbol{r}, t) = \frac{1}{\varepsilon}\nabla\rho(\boldsymbol{r}, t) + \mu\frac{\partial}{\partial t}\boldsymbol{J}(\boldsymbol{r}, t). \tag{2.9}$$

Since $\nabla^2\boldsymbol{E}(\boldsymbol{r}, t) = \hat{x}\nabla^2E_x(\boldsymbol{r}, t) + \hat{y}\nabla^2E_y(\boldsymbol{r}, t) + \hat{z}\nabla^2E_z(\boldsymbol{r}, t)$, it follows that (2.9) represents three separate equations for the three *rectangular* components of $\boldsymbol{E}(\boldsymbol{r}, t)$. In a similar manner, taking the curl of both sides of (2.3) together with (2.2) and (2.4) gives

$$\begin{aligned}\nabla \times \nabla \times \boldsymbol{H}(\boldsymbol{r}, t) &= \varepsilon\frac{\partial}{\partial t}[\nabla \times \boldsymbol{E}(\boldsymbol{r}, t)] + \nabla \times \boldsymbol{J}(\boldsymbol{r}, t),\\ &= -\mu\varepsilon\frac{\partial^2}{\partial t^2}\boldsymbol{H}(\boldsymbol{r}, t) + \nabla \times \boldsymbol{J}(\boldsymbol{r}, t).\end{aligned} \tag{2.10}$$

Also, from (2.6), we find that

$$\nabla \times \nabla \times \boldsymbol{H}(\boldsymbol{r}, t) = \nabla[\nabla \cdot \boldsymbol{H}(\boldsymbol{r}, t)] - \nabla^2 \boldsymbol{H}(\boldsymbol{r}, t) = -\nabla^2 \boldsymbol{H}(\boldsymbol{r}, t). \tag{2.11}$$

From (2.10) and (2.11) we obtain

$$\nabla^2 \boldsymbol{H}(\boldsymbol{r}, t) - \mu\varepsilon \frac{\partial^2}{\partial t^2} \boldsymbol{H}(\boldsymbol{r}, t) = -\nabla \times \boldsymbol{J}(\boldsymbol{r}, t). \tag{2.12}$$

In the same manner as (2.9), (2.12) also represents three separate equations for the three *rectangular* components of $\boldsymbol{H}(\boldsymbol{r}, t)$. The source terms in the separate differential equations (2.9) and (2.12) for $\boldsymbol{E}(\boldsymbol{r}, t)$ and $\boldsymbol{H}(\boldsymbol{r}, t)$ respectively occur in a fairly complicated manner. Consequently, we find it simpler to introduce certain auxiliary functions known as the vector and the scalar potentials, in whose differential equations the source terms occur in a fairly straightforward manner. The field vectors $\boldsymbol{E}(\boldsymbol{r}, t)$ and $\boldsymbol{H}(\boldsymbol{r}, t)$ are obtainable from the vector and the scalar potentials.

For any vector $\boldsymbol{F}$, $\nabla \cdot (\nabla \times \boldsymbol{F}) = 0$ and therefore from (2.6) we can see that it is reasonable to set

$$\boldsymbol{B}(\boldsymbol{r}, t) = \nabla \times \boldsymbol{A}(\boldsymbol{r}, t), \tag{2.13}$$

which defines the vector potential $\boldsymbol{A}(\boldsymbol{r}, t)$. If we substitute (2.13) in (2.2), we can rearrange the resulting equation to yield

$$\nabla \times \left[\boldsymbol{E}(\boldsymbol{r}, t) + \frac{\partial}{\partial t} \boldsymbol{A}(\boldsymbol{r}, t)\right] = 0. \tag{2.14}$$

Since for any scalar f, $\nabla \times [\nabla f] = 0$, it follows from (2.14) that

$$\boldsymbol{E}(\boldsymbol{r}, t) + \frac{\partial}{\partial t} \boldsymbol{A}(\boldsymbol{r}, t) = -\nabla \phi(\boldsymbol{r}, t),$$

or

$$\boldsymbol{E}(\boldsymbol{r}, t) = -\frac{\partial}{\partial t} \boldsymbol{A}(\boldsymbol{r}, t) - \nabla \phi(\boldsymbol{r}, t), \tag{2.15}$$

where $\phi(\boldsymbol{r}, t)$ is any scalar function. We should note that $\phi(\boldsymbol{r}, t)$ is the scalar potential. The definition of $\boldsymbol{B}(\boldsymbol{r}, t)$ and $\boldsymbol{E}(\boldsymbol{r}, t)$ in terms of $\boldsymbol{A}(\boldsymbol{r}, t)$ and $\phi(\boldsymbol{r}, t)$ as given by (2.13) and (2.15) respectively satisfies the two homogeneous Maxwell's equations (2.2) and (2.6). The two inhomogeneous Maxwell's equations (2.3) and (2.5) enable us to determine the relationships of $\boldsymbol{A}(\boldsymbol{r}, t)$ and $\phi(\boldsymbol{r}, t)$ to the sources $\rho(\boldsymbol{r}, t)$ and $\boldsymbol{J}(\boldsymbol{r}, t)$.

Substituting (2.4), (2.13), and (2.15) in (2.3) and (2.5) we obtain respectively the following two equations:

$$\frac{1}{\mu}[\nabla\{\nabla \cdot \boldsymbol{A}(\boldsymbol{r}, t)\} - \nabla^2 \boldsymbol{A}(\boldsymbol{r}, t)] = -\varepsilon \frac{\partial^2}{\partial t^2} \boldsymbol{A}(\boldsymbol{r}, t) - \varepsilon \nabla \frac{\partial}{\partial t} \phi(\boldsymbol{r}, t) + \boldsymbol{J}(\boldsymbol{r}, t), \tag{2.16}$$

$$-\frac{\partial}{\partial t}\{\nabla \cdot \boldsymbol{A}(\boldsymbol{r}, t)\} - \nabla^2 \phi(\boldsymbol{r}, t) = \frac{\rho(\boldsymbol{r}, t)}{\varepsilon}. \tag{2.17}$$

The partial differential equations (2.16) and (2.17) are coupled since each of them contains $\boldsymbol{A}(\boldsymbol{r}, t)$ and $\phi(\boldsymbol{r}, t)$. A vector field is completely specified only if its curl and its divergence are defined. Since only the curl of $\boldsymbol{A}(\boldsymbol{r}, t)$ is defined in (2.13), $\boldsymbol{A}(\boldsymbol{r}, t)$ is not uniquely specified. We can remove this arbitrariness in the definition of $\boldsymbol{A}(\boldsymbol{r}, t)$ by defining the divergence of $\boldsymbol{A}(\boldsymbol{r}, t)$ through the so-called Lorentz condition:

$$\nabla \cdot \boldsymbol{A}(\boldsymbol{r}, t) + \mu\varepsilon \frac{\partial}{\partial t} \phi(\boldsymbol{r}, t) = 0. \tag{2.18}$$

With the help of (2.18), we can simplify (2.16) to yield

$$\nabla^2 \boldsymbol{A}(\boldsymbol{r}, t) - \mu\varepsilon \frac{\partial^2}{\partial t^2} \boldsymbol{A}(\boldsymbol{r}, t) = -\mu \boldsymbol{J}(\boldsymbol{r}, t). \tag{2.19}$$

Also, if we substitute (2.18) into (2.17), we find that

$$\nabla^2 \phi(\boldsymbol{r}, t) - \mu\varepsilon \frac{\partial^2}{\partial t^2} \phi(\boldsymbol{r}, t) = -\frac{\rho(\boldsymbol{r}, t)}{\varepsilon}. \tag{2.20}$$

The Lorentz condition (2.18) has enabled us to decouple the coupled equations (2.16) and (2.17) into two separate differential equations (2.19) and (2.20), respectively, for the vector and the scalar potentials and has therefore facilitated the determination of $\boldsymbol{A}(\boldsymbol{r}, t)$ and $\phi(\boldsymbol{r}, t)$ in terms of $\boldsymbol{J}(\boldsymbol{r}, t)$ and $\rho(\boldsymbol{r}, t)$ respectively. We should note that (2.19) represents three separate equations for the rectangular components of the vector potential. In contrast to (2.9) and (2.12), $\boldsymbol{J}(\boldsymbol{r}, t)$ and $\rho(\boldsymbol{r}, t)$ occur separately and directly in (2.19) and (2.20). Consequently, we can relate $\boldsymbol{A}(\boldsymbol{r}, t)$ and $\phi(\boldsymbol{r}, t)$ in a simple manner to $\boldsymbol{J}(\boldsymbol{r}, t)$ and $\rho(\boldsymbol{r}, t)$ respectively. Once $\boldsymbol{A}(\boldsymbol{r}, t)$ and $\phi(\boldsymbol{r}, t)$ are determined, we can determine all the field vectors from (2.13) and (2.15).

In obtaining (2.13), (2.15), (2.19), and (2.20), we have used Maxwell's equations (2.2), (2.3), (2.5), and (2.6). We shall now show that the Lorentz condition (2.18) which was used to simplify the differential equations (2.16) and (2.17) for $\boldsymbol{A}(\boldsymbol{r}, t)$ and $\phi(\boldsymbol{r}, t)$ is consistent with the equation of continuity (2.1). For any vector $\boldsymbol{F}$, since

$$\nabla^2 \boldsymbol{F} = \hat{x}\nabla^2 F_x + \hat{y}\nabla^2 F_y + \hat{z}\nabla^2 F_z,$$

it is seen that

$$\nabla \cdot [\nabla^2 \boldsymbol{F}] = \frac{\partial}{\partial x}\nabla^2 F_x + \frac{\partial}{\partial y}\nabla^2 F_y + \frac{\partial}{\partial z}\nabla^2 F_z = \nabla^2 \frac{\partial F_x}{\partial x} + \nabla^2 \frac{\partial F_y}{\partial y} + \nabla^2 \frac{\partial F_z}{\partial z}.$$

Also, since

$$\nabla^2[\nabla \cdot \boldsymbol{F}] = \nabla^2\left[\frac{\partial F_x}{\partial x} + \frac{\partial F_y}{\partial y} + \frac{\partial F_z}{\partial z}\right] = \nabla^2 \frac{\partial F_x}{\partial x} + \nabla^2 \frac{\partial F_y}{\partial y} + \nabla^2 \frac{\partial F_z}{\partial z},$$

it follows that

$$\nabla \cdot \nabla^2 \boldsymbol{F} = \nabla^2[\nabla \cdot \boldsymbol{F}]. \tag{2.21}$$

From (2.19) and (2.21), we see that

$$\nabla \cdot \boldsymbol{J}(\boldsymbol{r}, t) = -\frac{1}{\mu}\left[\nabla^2\{\nabla \cdot \boldsymbol{A}(\boldsymbol{r}, t)\} - \mu\varepsilon \frac{\partial^2}{\partial t^2}\{\nabla \cdot \boldsymbol{A}(\boldsymbol{r}, t)\}\right]. \tag{2.22}$$

Similarly, from (2.20), we obtain

$$\frac{\partial}{\partial t}\rho(\boldsymbol{r}, t) = -\varepsilon\left[\nabla^2\left\{\frac{\partial}{\partial t}\phi(\boldsymbol{r}, t)\right\} - \mu\varepsilon \frac{\partial^2}{\partial t^2}\left\{\frac{\partial}{\partial t}\phi(\boldsymbol{r}, t)\right\}\right]. \tag{2.23}$$

The addition of (2.22) and (2.23) yields

$$\begin{aligned}\nabla \cdot \boldsymbol{J}(\boldsymbol{r}, t) + \frac{\partial}{\partial t}\rho(\boldsymbol{r}, t) &= -\frac{1}{\mu}\left[\nabla^2\left\{\nabla \cdot \boldsymbol{A}(\boldsymbol{r}, t) + \mu\varepsilon\frac{\partial}{\partial t}\phi(\boldsymbol{r}, t)\right\}\right. \\ &\qquad \left. - \mu\varepsilon\frac{\partial^2}{\partial t^2}\left\{\nabla \cdot \boldsymbol{A}(\boldsymbol{r}, t) + \mu\varepsilon\frac{\partial}{\partial t}\phi(\boldsymbol{r}, t)\right\}\right], \\ &= -\frac{1}{\mu}\left[\nabla^2 - \mu\varepsilon\frac{\partial^2}{\partial t^2}\right]\left\{\nabla \cdot \boldsymbol{A}(\boldsymbol{r}, t) + \mu\varepsilon\frac{\partial}{\partial t}\phi(\boldsymbol{r}, t)\right\}. \qquad (2.24)\end{aligned}$$

If (2.18) holds, it follows from (2.24) that (2.1) is reproduced. In other words, the Lorentz condition (2.18) is consistent with the equation of continuity (2.1). On the other hand, if we assume the truth of (2.1), (2.18) does not follow uniquely. For the purpose of exhibiting this result clearly, let

$$f(\boldsymbol{r}, t) = \nabla \cdot \boldsymbol{A}(\boldsymbol{r}, t) + \mu\varepsilon\frac{\partial}{\partial t}\phi(\boldsymbol{r}, t).$$

Then, from (2.1) and (2.24), we find that

$$\left[\nabla^2 - \mu\varepsilon\frac{\partial^2}{\partial t^2}\right]f(\boldsymbol{r}, t) = 0. \qquad (2.25)$$

We should note that $f(\boldsymbol{r}, t) = 0$ is not the only solution of (2.25). Therefore, we should not misconstrue that the Lorentz condition (2.18) *follows* from the equation of continuity (2.1).

9.3 RETARDED AND ADVANCED POTENTIALS

The solutions of the differential equations (2.19) and (2.20) give $\boldsymbol{A}(\boldsymbol{r}, t)$ and $\phi(\boldsymbol{r}, t)$ in terms of $\boldsymbol{J}(\boldsymbol{r}, t)$ and $\rho(\boldsymbol{r}, t)$ respectively. A systematic deduction of the solutions of (2.19) and (2.20) is beyond the scope of this treatment. Therefore, we state the expressions of $\boldsymbol{A}(\boldsymbol{r}, t)$ and $\phi(\boldsymbol{r}, t)$ in terms of $\boldsymbol{J}(\boldsymbol{r}, t)$ and $\rho(\boldsymbol{r}, t)$ respectively. Then, we verify these expressions of $\boldsymbol{A}(\boldsymbol{r}, t)$ and $\phi(\boldsymbol{r}, t)$ to be the solutions of (2.19) and (2.20) respectively.

When $\phi(\boldsymbol{r}, t)$ is independent of time, we can reduce (2.20) to

$$\nabla^2\phi(\boldsymbol{r}, t) = -\rho(\boldsymbol{r}, t)/\varepsilon, \qquad (3.1)$$

which, as it should be, is Poisson's equation in electrostatics as stated in (4.2.58). The solution of Poisson's equation is given by (4.2.30), which suggests the following solution of the differential equation (2.20) for the time-dependent scalar potential $\phi(\boldsymbol{r}, t)$:

$$\phi(\boldsymbol{r}, t) = \int_V \frac{\rho(\boldsymbol{r}', t - |\boldsymbol{r} - \boldsymbol{r}'|/v)\, dV'}{4\pi\varepsilon\,|\boldsymbol{r} - \boldsymbol{r}'|}, \qquad (3.2)$$

where $v = 1/\sqrt{\mu\varepsilon}$. The scalar potential at P denoted by the position vector $\boldsymbol{r}$ with reference to some origin and at time t due to the charge contained in an element of volume dV' at P' denoted by the position vector $\boldsymbol{r}'$ is given by the integrand in (3.2) and is dependent only on the scalar distance $|\boldsymbol{r} - \boldsymbol{r}'|$ between the source and the observation points. Since ∇^2 is an operation with respect to the unprimed coordinates, it follows that

$$\nabla^2\phi(\boldsymbol{r}, t) = \int_V \nabla^2\left[\frac{\rho(\boldsymbol{r}', t - |\boldsymbol{r} - \boldsymbol{r}'|/v)}{4\pi\varepsilon\,|\boldsymbol{r} - \boldsymbol{r}'|}\right] dV'. \tag{3.3}$$

Let $\boldsymbol{R} = \boldsymbol{r} - \boldsymbol{r}'$ and then P is specified by the spherical coordinates R, θ, and Φ with reference to the origin at P'. Now the operator ∇^2 is in terms of R, θ, and Φ. But, ∇^2 operates on a scalar function which is dependent on R and not on θ and Φ. Since

$$\nabla^2 = \frac{1}{R^2}\frac{\partial}{\partial R}\left(R^2\frac{\partial}{\partial R}\right) + \frac{1}{R^2\sin\theta}\frac{\partial}{\partial\theta}\left(\sin\theta\frac{\partial}{\partial\theta}\right) + \frac{1}{R^2\sin^2\theta}\frac{\partial^2}{\partial\Phi^2},$$

(3.3) becomes

$$\nabla^2\phi(\boldsymbol{r}, t) = \int_V \frac{1}{R^2}\frac{\partial}{\partial R}\left[R^2\frac{\partial}{\partial R}\left\{\frac{\rho(\boldsymbol{r}', t - R/v)}{4\pi\varepsilon R}\right\}\right] dV'. \tag{3.4}$$

When the differentiation is carried out explicitly, we obtain

$$\frac{1}{R^2}\frac{\partial}{\partial R}\left[R^2\frac{\partial}{\partial R}\left\{\frac{\rho(\boldsymbol{r}', t - R/v)}{4\pi\varepsilon R}\right\}\right] = \frac{1}{4\pi\varepsilon R}\frac{\partial^2}{\partial R^2}\rho\left(\boldsymbol{r}', t - \frac{R}{v}\right) = \frac{1}{4\pi\varepsilon R v^2}\frac{\partial^2}{\partial T^2}\rho(\boldsymbol{r}', T), \tag{3.5}$$

where $T = t - (R/v)$. Hence, (3.4) becomes

$$\nabla^2\phi(\boldsymbol{r}, t) = \frac{1}{4\pi\varepsilon v^2}\int_V \frac{1}{R}\frac{\partial^2}{\partial T^2}\rho(\boldsymbol{r}', T)\, dV'. \tag{3.6}$$

In a similar manner, we may show that

$$\begin{aligned}\frac{1}{v^2}\frac{\partial^2}{\partial t^2}\phi(\boldsymbol{r}, t) &= \frac{1}{4\pi\varepsilon v^2}\int_V \frac{\partial^2}{\partial t^2}\left[\frac{\rho(\boldsymbol{r}', t - R/v)}{R}\right] dV', \\ &= \frac{1}{4\pi\varepsilon v^2}\int_V \frac{1}{R}\frac{\partial^2}{\partial T^2}\rho(\boldsymbol{r}', T)\, dV'. \end{aligned} \tag{3.7}$$

When $\boldsymbol{r}' \neq \boldsymbol{r}$, that is, when $R \neq 0$, if we subtract (3.7) from (3.6) we get

$$\nabla^2\phi(\boldsymbol{r}, t) - \frac{1}{v^2}\frac{\partial^2}{\partial t^2}\phi(\boldsymbol{r}, t) = 0 \qquad (\text{for } \boldsymbol{r}' \neq \boldsymbol{r}). \tag{3.8}$$

When the observation point is inside the region containing the charge distribution, it is possible for $\boldsymbol{r}'$ to become equal to $\boldsymbol{r}$ in the process of integration. Therefore, the integrand of (3.2) has a singularity at $\boldsymbol{r} = \boldsymbol{r}'$. We shall now establish that in the neighborhood of this singularity $\nabla^2\phi(\boldsymbol{r}, t) - (1/v^2)(\partial^2/\partial t^2)\phi(\boldsymbol{r}, t)$ yields a finite result. The corresponding result for $\boldsymbol{r}'$ not in the close neighborhood of $\boldsymbol{r}$ is equal to zero, as we indicated in (3.8). Hence, it is sufficient for us now to restrict attention to a very small spherical volume V_s centered at the observation point specified by $\boldsymbol{r}$. Assuming that $\rho(\boldsymbol{r}', t - (|\boldsymbol{r} - \boldsymbol{r}'|/v))$ is a smooth function, it is legitimate to set

$|\boldsymbol{r} - \boldsymbol{r}'| = 0$ in the arguments of ρ to obtain $\rho(\boldsymbol{r}', t - (|\boldsymbol{r} - \boldsymbol{r}'|/v)) = \rho(\boldsymbol{r}', t)$ everywhere inside the very small spherical volume V_s, since the radius δ of the spherical volume V_s is to be taken to the limiting value zero. Therefore, in the neighborhood of $\boldsymbol{r}' = \boldsymbol{r}$, we may rewrite (3.3) as

$$\nabla^2\phi(\boldsymbol{r}, t) = \frac{1}{4\pi\varepsilon}\int_{V_s}\rho(\boldsymbol{r}', t)\nabla^2\left(\frac{1}{|\boldsymbol{r} - \boldsymbol{r}'|}\right)dV' = \frac{\rho(\boldsymbol{r}, t)}{4\pi\varepsilon}\int_{V_s}\nabla^2\left(\frac{1}{|\boldsymbol{r} - \boldsymbol{r}'|}\right)dV' \qquad (\text{near } \boldsymbol{r}' = \boldsymbol{r}). \tag{3.9}$$

In view of (4.4.70), (3.9) becomes

$$\nabla^2\phi(\boldsymbol{r}, t) = -\rho(\boldsymbol{r}, t)/\varepsilon. \tag{3.10}$$

In a similar manner, in the neighborhood of $\boldsymbol{r}' = \boldsymbol{r}$, (3.7) becomes

$$\begin{aligned}\frac{1}{v^2}\frac{\partial^2}{\partial t^2}\phi(\boldsymbol{r}, t) &= \frac{1}{4\pi\varepsilon v^2}\frac{\partial^2}{\partial t^2}\rho(\boldsymbol{r}, t)\int_{V_s}\frac{1}{R}\,dV', \\ &= \frac{1}{4\pi\varepsilon v^2}\frac{\partial^2}{\partial t^2}\rho(\boldsymbol{r}, t)\int_0^\delta\frac{1}{R}\,4\pi R^2\,dR, \\ &= \frac{\delta^2}{2\varepsilon v^2}\frac{\partial^2}{\partial t^2}\rho(\boldsymbol{r}, t).\end{aligned} \tag{3.11}$$

In the limit of $\delta = 0$, we see from (3.11) that $(1/v^2)(\partial^2/\partial t^2)\phi(\boldsymbol{r}, t) = 0$ and thus gives no additional contribution near $\boldsymbol{r}' = \boldsymbol{r}$. From (3.10) and (3.11), it follows that in the region containing the charge distribution

$$\nabla^2\phi(\boldsymbol{r}, t) - \frac{1}{v^2}\frac{\partial^2}{\partial t^2}\phi(\boldsymbol{r}, t) = -\frac{\rho(\boldsymbol{r}, t)}{\varepsilon}. \tag{3.12}$$

Together with (3.8) and (3.12), we see that $\phi(\boldsymbol{r}, t)$ as given by (3.2) satisfies the differential equation (2.20).

We can obtain the rectangular components of (2.19) as

$$\nabla^2 A_i(\boldsymbol{r}, t) - \frac{1}{v^2}\frac{\partial^2}{\partial t^2}A_i(\boldsymbol{r}, t) = -\mu J_i(\boldsymbol{r}, t) \qquad (i = x, y, z). \tag{3.13}$$

If we compare (3.13) with (2.20) and the solution (3.2) of the differential equation (2.20), we can see that the solution of (3.13) should be of the form

$$A_i(\boldsymbol{r}, t) = \int_V \frac{\mu J_i(\boldsymbol{r}', t - |\boldsymbol{r} - \boldsymbol{r}'|/v)}{4\pi\,|\boldsymbol{r} - \boldsymbol{r}'|}\,dV' \qquad (i = x, y, z). \tag{3.14}$$

In a manner similar to that used for the scalar potential, we can verify that (3.14) is the solution of (3.13). If we add the three rectangular components of $\boldsymbol{A}(\boldsymbol{r}, t)$ as given in (3.14) we obtain

$$\boldsymbol{A}(\boldsymbol{r}, t) = \int_V \frac{\mu\boldsymbol{J}(\boldsymbol{r}', t - |\boldsymbol{r} - \boldsymbol{r}'|/v)}{4\pi\,|\boldsymbol{r} - \boldsymbol{r}'|}\,dV', \tag{3.15}$$

which is the solution of (2.19). When $\boldsymbol{A}(\boldsymbol{r}, t)$ is independent of time, (2.19) reduces to (4.4.33), which is the equation in magnetostatics corresponding to (2.19). The solution (4.4.27) of (4.4.33) bears the same resemblance to (3.15) as the solution (4.2.30) of Poisson's equation (3.1) bears to the time-varying scalar potential given by (3.2).

Since $\mu\varepsilon = 1/v^2$, it follows that the differential equations (2.19) and (2.20) remain unchanged if we change v to $-v$. Therefore, it follows that (3.15) and (3.2) with v changed to $-v$ are also solutions of (2.19) and (2.20) respectively. Hence, the following expressions of $\boldsymbol{A}(\boldsymbol{r}, t)$ and $\phi(\boldsymbol{r}, t)$ are also possible solutions of (2.19) and (2.20) respectively:

$$\boldsymbol{A}(\boldsymbol{r}, t) = \int_V \frac{\mu \boldsymbol{J}(\boldsymbol{r}', t + |\boldsymbol{r} - \boldsymbol{r}'|/v)}{4\pi\, |\boldsymbol{r} - \boldsymbol{r}'|}\, dV', \tag{3.16}$$

and

$$\phi(\boldsymbol{r}, t) = \int_V \frac{\rho(\boldsymbol{r}', t + |\boldsymbol{r} - \boldsymbol{r}'|/v)}{4\pi\varepsilon\, |\boldsymbol{r} - \boldsymbol{r}'|}\, dV'. \tag{3.17}$$

We may verify that (3.16) and (3.17) are respectively the solutions of (2.19) and (2.20) by using a procedure similar to that employed in establishing that (3.2) is a solution of (2.20).

There is a substantial physical difference between the solutions given by (3.15) and (3.2), and those given by (3.16) and (3.17). For example, in (3.2), the scalar potential at a location specified by $\boldsymbol{r}$ and at time t due to the charge distribution at $\boldsymbol{r}'$ depends on the value of the charge density at a *previous* instant of time, the time difference being equal to the propagation time from the source to the observation points, provided v is the velocity of propagation. A corresponding result is also true for the vector potential given by (3.15). The potentials given by (3.2) and (3.15) are called *retarded* potentials since these potentials are dependent on the values of the sources corresponding to earlier times. But, in (3.16) and (3.17), the potentials at a location specified by $\boldsymbol{r}$ and at time t due to the charge or the current distributions at $\boldsymbol{r}'$ depend on their values at a *later* instant of time, the time difference being the same as in the previous cases. Consequently, the potentials given by (3.16) and (3.17) are called *advanced* potentials. The retarded potentials give rise to fields only after the sources are brought into existence, but the advanced potentials give rise to fields only before the current and the charge distributions are brought into existence. In all physical phenomena, the effect should occur after its cause. Consequently, we shall use the *retarded potentials*, which alone conform to the above-mentioned principle, and shall not consider the *advanced potentials* further.

9.4 RETARDED VECTOR AND SCALAR POTENTIALS WITH HARMONIC TIME DEPENDENCE

So far we have assumed no specific time dependence for the vector potential, the scalar potential, and the field vectors. As we have mentioned previously, our main interest is in the fields having the harmonic time dependence. If $f(\boldsymbol{r}, t)$ represents a component of either the potential or the field with harmonic time dependence, then

$$f(\boldsymbol{r}, t) = f(\boldsymbol{r})e^{j\omega t}, \tag{4.1}$$

where the implication is that $f(\mathbf{r}, t)$ equals the real part of the right side of (4.1). The differential equations (2.19) and (2.20), when specialized to the case of time-harmonic potentials, become

$$\nabla^2 \mathbf{A}(\mathbf{r}) + \beta^2 \mathbf{A}(\mathbf{r}) = -\mu \mathbf{J}(\mathbf{r}), \tag{4.2}$$

and

$$\nabla^2 \phi(\mathbf{r}) + \beta^2 \phi(\mathbf{r}) = -\rho(\mathbf{r})/\varepsilon, \tag{4.3}$$

where $\beta = \omega\sqrt{\mu\varepsilon}$. Then, from (2.13) and (2.15), we find the definitions of $\mathbf{B}(\mathbf{r})$ and $\mathbf{E}(\mathbf{r})$ in terms of $\mathbf{A}(\mathbf{r})$ and $\phi(\mathbf{r})$ to be

$$\mathbf{B}(\mathbf{r}) = \nabla \times \mathbf{A}(\mathbf{r}), \tag{4.4}$$

$$\mathbf{E}(\mathbf{r}) = -j\omega\mathbf{A}(\mathbf{r}) - \nabla\phi(\mathbf{r}). \tag{4.5}$$

We can simplify the Lorentz condition (2.18) for the time-harmonic potentials to yield

$$\nabla \cdot \mathbf{A}(\mathbf{r}) + j\omega\mu\varepsilon\phi(\mathbf{r}) = 0. \tag{4.6}$$

With the help of (4.6), we can eliminate $\phi(\mathbf{r})$ in (4.5) to give the following expression of $\mathbf{E}(\mathbf{r})$ in terms of $\mathbf{A}(\mathbf{r})$ alone:

$$\mathbf{E}(\mathbf{r}) = -j\omega\mathbf{A}(\mathbf{r}) + \frac{\nabla[\nabla \cdot \mathbf{A}(\mathbf{r})]}{j\omega\mu\varepsilon}. \tag{4.7}$$

For the time-harmonic charge and current densities, the equation of continuity (2.1) becomes

$$\nabla \cdot \mathbf{J}(\mathbf{r}) + j\omega\rho(\mathbf{r}) = 0. \tag{4.8}$$

As before, we can show that (4.6) is consistent with (4.8). Since for the harmonic time variation

$$\rho\left(\mathbf{r}', t - \frac{|\mathbf{r} - \mathbf{r}'|}{v}\right) = \rho(\mathbf{r}')e^{j\omega(t - |\mathbf{r}-\mathbf{r}'|/v)},$$

from (3.2), the integral expression for the time-harmonic, retarded scalar potential becomes

$$\phi(\mathbf{r}) = \int_V \frac{\rho(\mathbf{r}')e^{-j\beta|\mathbf{r}-\mathbf{r}'|}}{4\pi\varepsilon|\mathbf{r} - \mathbf{r}'|}\, dV'. \tag{4.9}$$

In a similar manner, from (3.15), we can determine the integral expression for the time-harmonic, retarded vector potential as

$$\mathbf{A}(\mathbf{r}) = \int_V \frac{\mu\mathbf{J}(\mathbf{r}')e^{-j\beta|\mathbf{r}-\mathbf{r}'|}}{4\pi|\mathbf{r} - \mathbf{r}'|}\, dV'. \tag{4.10}$$

After obtaining (4.9) and (4.10), we can evaluate $\mathbf{B}(\mathbf{r})$ and $\mathbf{E}(\mathbf{r})$ from (4.4) and (4.5). We should note that the retardation in time associated with the potentials (3.2) and (3.15) having arbitrary time dependence corresponds to the phase lag appearing in the potentials (4.9) and (4.10) having harmonic time dependence.

9.5 RADIATION FROM A CURRENT ELEMENT

We propose to study the radiation characteristics of a current filament of infinitesimal length. The current is assumed (i) to vary harmonically with time and (ii) to have a constant amplitude along the length of the filament. Admittedly, a constant current filament of infinitesimal length cannot be realized in practice. However, such a current filament possesses the general radiation characteristics of a linear antenna on the one hand and serves as the building block for the more complicated, practically realizable antennas on the other. Consequently, it is valuable for us to study the radiation characteristics of a current filament of infinitesimal length known also as the Hertzian dipole.

1. Vector Potential

Let us consider a perfectly conducting cylindrical tube of negligible thickness and of radius a. Let its axis coincide with the z-axis and let it extend from $z = -dl/2$ to $z = dl/2$ as shown in Fig. 9.1. We wish to find the vector potential at P specified by the spherical coordinates r, θ, and ϕ. We assume the current density on the cylindrical tube to be only in the z-direction with the result that the vector potential at P is also in the z-direction. We see the vector potential from (4.10) to be given by

$$A_z(\mathbf{r}) = \int_V \frac{\mu J_z(\mathbf{r}')e^{-j\beta|\mathbf{r}-\mathbf{r}'|}}{4\pi|\mathbf{r} - \mathbf{r}'|} \, dV'. \tag{5.1}$$

The position vector to any point on the cylindrical tube is denoted by $\mathbf{r}'$. If the radius of the cylindrical tube is very small in comparison to a wavelength, that is, $\beta a \ll 1$ and if the length dl of the filament is also infinitesimally small, then we can omit $\mathbf{r}'$ in (5.1). This result is strictly true only in the limiting case in which the radius a and the length dl of the cylindrical tube tend to zero. The so-called current element or the Hertzian dipole corresponds to this limiting case. Hence, for the current element (5.1) becomes

$$A_z(\mathbf{r}) = \frac{\mu e^{-j\beta r}}{4\pi r} \int_V J_z(\mathbf{r}') \, dV'. \tag{5.2}$$

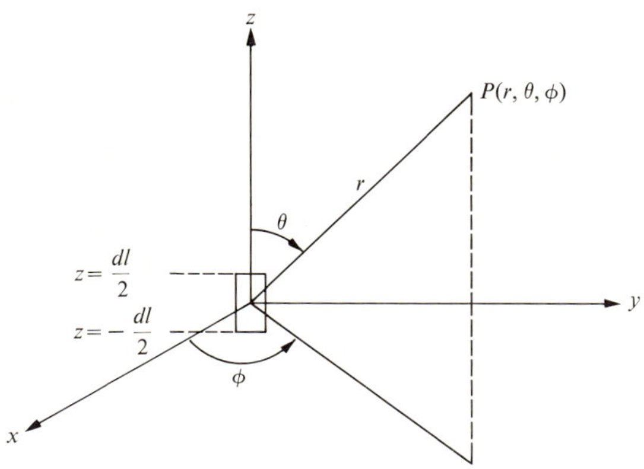

Fig. 9.1 The Hertzian dipole antenna.

The current density $J_z(\mathbf{r}')$ integrated over the cross section of the cylindrical tube yields the total axial current I_0, which we have assumed to be a constant along the length of the filament. Therefore, I_0 integrated over the length of the filament yields $I_0\,dl$. Thus, $\int_V J_z(\mathbf{r}')\,dV' = I_0\,dl$ with the result that (5.2) reduces to

$$A_z(\mathbf{r}) = (I_0\,dl)\,\frac{\mu e^{-j\beta r}}{4\pi r}. \tag{5.3}$$

In (5.3), $I_0\,dl$ is called the current moment.

We can obtain the scalar potential at $P(r, \theta, \phi)$ due to the current element from (4.6). For this purpose, the components of the vector potential in the direction of the spherical coordinates r, θ, and ϕ are needed and these we find from (5.3) and Fig. 9.1 to be given by

$$A_r(\mathbf{r}) = A_z(\mathbf{r})\cos\theta = I_0\,dl\,\frac{\mu e^{-j\beta r}}{4\pi r}\cos\theta, \tag{5.4a}$$

$$A_\theta(\mathbf{r}) = -A_z(\mathbf{r})\sin\theta = -I_0\,dl\,\frac{\mu e^{-j\beta r}}{4\pi r}\sin\theta, \tag{5.4b}$$

$$A_\phi(\mathbf{r}) = 0. \tag{5.4c}$$

2. Scalar Potential

From (4.6), we find that

$$\phi(\mathbf{r}) = -\frac{1}{j\omega\mu\varepsilon}\nabla\cdot\boldsymbol{A}(\mathbf{r}) = -\frac{1}{j\omega\mu\varepsilon}\left[\frac{1}{r^2}\frac{\partial}{\partial r}\{r^2 A_r(\mathbf{r})\} + \frac{1}{r\sin\theta}\frac{\partial}{\partial\theta}\{\sin\theta A_\theta(\mathbf{r})\}\right], \tag{5.5}$$

since $A_\phi(\mathbf{r}) = 0$. Substituting (5.4) in (5.5) and simplifying the resulting expression we obtain

$$\phi(\mathbf{r}) = \left(\frac{I_0}{j\omega}\right) dl\,\frac{e^{-j\beta r}\cos\theta}{4\pi\varepsilon r}\left(j\beta + \frac{1}{r}\right). \tag{5.6}$$

It is instructive for us to consider an alternative method for the determination of the scalar potential. If the charge is concentrated in a very small region near the origin, $\mathbf{r}'$ in (4.9) is negligible throughout the range of integration with the result that (4.9) becomes

$$\phi(\mathbf{r}) = \frac{e^{-j\beta r}}{4\pi\varepsilon r}\int_V \rho(\mathbf{r}')\,dV'. \tag{5.7}$$

The result given by (5.7) is strictly true only when the charge is concentrated at a point. In (5.7), r is the distance from the location of the charge to the location of P where we want the scalar potential. We should note that $\int_V \rho(\mathbf{r}')\,dV' = Q$, the total charge. Hence, we may rewrite (5.7) as

$$\phi(\mathbf{r}) = Qe^{-j\beta r}/4\pi\varepsilon r. \tag{5.8}$$

In the current element, since the current is specified by $I_0 e^{j\omega t}$ throughout the length, we can evaluate the accompanying charge with the help of the equation of continuity. At every point along the length of the current filament other than at the two ends, the divergence of the current is zero and hence there is no charge. Let us assume the charge at the end $z = dl/2$ of the filament to be given by $Q_1 e^{j\omega t}$. Since the current $I_0 e^{j\omega t}$ is directed towards the end $z = dl/2$ of the filament and since the influx of the current at a point is equal to the rate of increase of the charge at that point, it follows that

$$\frac{d}{dt}[Q_1 e^{j\omega t}] = I_0 e^{j\omega t} \qquad \text{or} \qquad Q_1 = \frac{I_0}{j\omega} = Q_0. \tag{5.9}$$

In a similar manner, let $Q_2 e^{j\omega t}$ be the charge at the end $z = -dl/2$ of the filament. Since the current directed towards this end is given by $-I_0 e^{j\omega t}$, it follows that

$$\frac{d}{dt}[Q_2 e^{j\omega t}] = -I_0 e^{j\omega t} \qquad \text{or} \qquad Q_2 = -\frac{I_0}{j\omega} = -Q_0. \tag{5.10}$$

Hence, a current element is accompanied by two equal and opposite charges of magnitude $Q_0 = I_0/j\omega$, separated by an infinitesimal length dl. These charges vary harmonically with time. It is for this reason, that a constant current element is also called an elementary dipole or Hertzian dipole. The magnitude p_0 of the dipole moment of the Hertzian dipole is given by the magnitude of the charge times the separation distance. Hence,

$$p_0 = |Q_0|\, dl = \frac{I_0\, dl}{\omega}. \tag{5.11}$$

We have evaluated the charge distribution along the current element and hence can use (5.8) to obtain the resultant scalar potential at any point P. As shown in Fig. 9.2, let r_1 and r_2 be the distances to P respectively from the ends $z = dl/2$ and $z = -dl/2$ of the filament. Then, from (5.8), we obtain the resultant scalar potential at P as

$$\phi(\mathbf{r}) = \frac{Q_0}{4\pi\varepsilon}\left[\frac{e^{-j\beta r_1}}{r_1} - \frac{e^{-j\beta r_2}}{r_2}\right]. \tag{5.12}$$

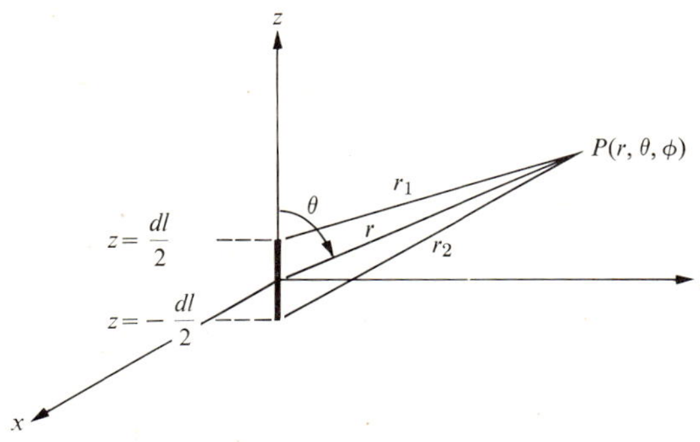

Fig. 9.2 A schematic diagram of a z-directed current element for the evaluation of the scalar potential.

It is seen from Fig. 9.2 that

$$r_1 = \left[r^2 + \left(\frac{dl}{2}\right)^2 - 2r\frac{dl}{2}\cos\theta\right]^{1/2} = r\left[1 - \frac{dl}{r}\cos\theta + \frac{(dl)^2}{4r^2}\right]^{1/2}. \tag{5.13}$$

Since $dl/r \ll 1$, we can expand the square root term in (5.13) into a binomial series and if we retain only the terms up to linear order in dl/r, (5.13) becomes

$$r_1 = r\left(1 - \frac{dl}{2r}\cos\theta\right) = r - \frac{dl}{2}\cos\theta. \tag{5.14}$$

In a similar manner, we can show that

$$r_2 = r\left(1 + \frac{dl}{2r}\cos\theta\right) = r + \frac{dl}{2}\cos\theta. \tag{5.15}$$

Together with (5.14) and (5.15), we can reduce (5.12) to

$$\phi(\mathbf{r}) = \frac{Q_0 e^{-j\beta r}}{4\pi\varepsilon r}\left[\frac{e^{j\beta(dl/2)\cos\theta}}{1 - (dl/2r)\cos\theta} - \frac{e^{-j\beta(dl/2)\cos\theta}}{1 + (dl/2r)\cos\theta}\right]. \tag{5.16}$$

For $dl/r \ll 1$,

$$\left(1 \mp \frac{dl}{2r}\cos\theta\right)^{-1} = 1 \pm \frac{dl}{2r}\cos\theta. \tag{5.17}$$

Hence, (5.16) becomes

$$\phi(\mathbf{r}) = \frac{Q_0 e^{-j\beta r}}{4\pi\varepsilon r}\left[e^{j\beta(dl/2)\cos\theta} - e^{-j\beta(dl/2)\cos\theta} + \frac{dl}{2r}\cos\theta\{e^{j\beta(dl/2)\cos\theta} + e^{-j\beta(dl/2)\cos\theta}\}\right]. \tag{5.18}$$

For dl vanishingly small,

$$e^{\pm j\beta(dl/2)\cos\theta} = 1 \pm j\beta(dl/2)\cos\theta. \tag{5.19}$$

If we substitute (5.19) into (5.18) and retain only the terms which are of linear order in dl, the result is

$$\phi(\mathbf{r}) = \frac{Q_0\, dl e^{-j\beta r}}{4\pi\varepsilon r}\cos\theta\left(j\beta + \frac{1}{r}\right). \tag{5.20}$$

When dl becomes vanishingly small, the terms proportional to the second and the other higher powers of $\beta\, dl$ are negligible with the result that (5.20) is the correct result in the limit of dl tending to zero. Since $Q_0 = I_0/j\omega$, it follows that (5.20) is identical to (5.6), as it should be.

The vector and the scalar potentials due to the current element, as given by (5.4) and (5.6) respectively, are independent of ϕ, that is, $\partial/\partial\phi = 0$. This result is to be expected since the current element has cylindrical symmetry with respect to the direction which coincides with its length.

3. Magnetic and Electric Fields

We can obtain the magnetic field $\boldsymbol{H}(\boldsymbol{r})$ due to the current from (4.4) as

$$\begin{aligned}\boldsymbol{H}(\boldsymbol{r}) &= \frac{1}{\mu}\nabla \times \boldsymbol{A}(\boldsymbol{r}) \\ &= \frac{1}{\mu}\left[\frac{\hat{r}}{r\sin\theta}\left\{\frac{\partial}{\partial\theta}(\sin\theta A_\phi(\boldsymbol{r})) - \frac{\partial}{\partial\phi}A_\theta(\boldsymbol{r})\right\} + \frac{\hat{\theta}}{r}\left\{\frac{1}{\sin\theta}\frac{\partial}{\partial\phi}A_r(\boldsymbol{r})\right.\right. \\ &\quad \left.\left. - \frac{\partial}{\partial r}(rA_\phi(\boldsymbol{r}))\right\} + \frac{\hat{\phi}}{r}\left\{\frac{\partial}{\partial r}(rA_\theta(\boldsymbol{r})) - \frac{\partial}{\partial\theta}A_r(\boldsymbol{r})\right\}\right].\end{aligned} \tag{5.21}$$

Substituting (5.4) into (5.21) we find

$$H_r(\boldsymbol{r}) = 0; \qquad H_\theta(\boldsymbol{r}) = 0; \qquad H_\phi(\boldsymbol{r}) = \frac{I_0\,dl}{4\pi r}e^{-j\beta r}j\beta\sin\theta\left(1 + \frac{1}{j\beta r}\right). \tag{5.22}$$

We can obtain the electric vector either from (4.5) by substituting the expressions for $\boldsymbol{A}(\boldsymbol{r})$ and $\phi(\boldsymbol{r})$ from (5.4) and (5.6) respectively or from (4.7) by substituting the expression for $\boldsymbol{A}(\boldsymbol{r})$ from (5.4). However, since we have determined $\boldsymbol{H}(\boldsymbol{r})$ it is convenient to evaluate $\boldsymbol{E}(\boldsymbol{r})$ directly from Maxwell's equations. Exterior to the current source, (2.3) and (2.4) for the time-harmonic fields yield

$$\begin{aligned}\boldsymbol{E}(\boldsymbol{r}) &= \frac{1}{j\omega\varepsilon}\nabla \times \boldsymbol{H}(\boldsymbol{r}) \\ &= \frac{1}{j\omega\varepsilon}\left[\frac{\hat{r}}{r\sin\theta}\left\{\frac{\partial}{\partial\theta}(\sin\theta H_\phi(\boldsymbol{r})) - \frac{\partial}{\partial\phi}H_\theta(\boldsymbol{r})\right\} + \frac{\hat{\theta}}{r}\left\{\frac{1}{\sin\theta}\frac{\partial}{\partial\phi}(H_r(\boldsymbol{r}))\right.\right. \\ &\quad \left.\left. - \frac{\partial}{\partial r}(rH_\phi(\boldsymbol{r}))\right\} + \frac{\hat{\phi}}{r}\left\{\frac{\partial}{\partial r}(rH_\theta(\boldsymbol{r})) - \frac{\partial}{\partial\theta}H_r(\boldsymbol{r})\right\}\right].\end{aligned} \tag{5.23}$$

We may show the substitution of (5.22) into (5.23) to give the following results:

$$E_r(\boldsymbol{r}) = \frac{I_0\,dl}{2\pi r}e^{-j\beta r}\frac{\eta_{0r}\cos\theta}{r}\left(1 + \frac{1}{j\beta r}\right), \tag{5.24a}$$

$$E_\theta(\boldsymbol{r}) = \frac{I_0\,dl}{4\pi r}e^{-j\beta r}j\beta\eta_{0r}\sin\theta\left(1 + \frac{1}{j\beta r} - \frac{1}{\beta^2 r^2}\right), \tag{5.24b}$$

$$E_\phi(\boldsymbol{r}) = 0, \tag{5.24c}$$

where the intrinsic resistance $\eta_{0r} = \sqrt{\mu/\varepsilon}$.

It is intuitively obvious that a current element should excite waves progressing in the radial direction with the result that the propagation direction coincides with the radial ($\hat{r}$) direction. Since we can see from (5.22) that there is no component of the magnetic vector in the radial direction, it follows that the waves set up by a current element constitute a TM mode. The magnetic vector is in the ϕ-direction only, as we should expect since the elemental current is in the z-direction. We see from (5.24) that the electric vector is perpendicular to the magnetic vector.

4. Far-zone and Near-zone Fields

We can see the magnetic field and the electric field as given by (5.22) and (5.24) respectively to be in the form of a series in inverse powers of βr and this enables us to introduce the concepts of the so-called *far-zone* and *near-zone* fields. The far-zone and the near-zone are defined respectively by the inequalities $\beta r \gg 1$ and $\beta r \ll 1$. The parts of the fields whose amplitude falls off as $1/r$ are called the far-zone fields, which we can obtain from (5.22) and (5.24) as

$$H_\phi(\mathbf{r}) = \frac{I_0\, dl}{4\pi r} e^{-j\beta r} j\beta \sin\theta \qquad \text{[Far-zone]}, \tag{5.25a}$$

$$E_\theta(\mathbf{r}) = \frac{I_0\, dl}{4\pi r} e^{-j\beta r} \eta_{0r} j\beta \sin\theta \qquad \text{[Far-zone]}. \tag{5.25b}$$

All other field components vanish in the far-zone. We see from (5.25) that $E_\theta(\mathbf{r})/H_\phi(\mathbf{r}) = \eta_{0r}$. Since for a lossless medium, η_{0r} is real, it follows that in the far-zone, the electric and the magnetic vectors are in time phase and space quadrature. The electric field, the magnetic field, and the direction of propagation form a triad of a mutually perpendicular right-handed system of vectors; thus, in the far-zone the fields due to a current element constitute a TEM mode. Also, the amplitude of the electric field bears to that of the magnetic field a constant ratio equal to the intrinsic resistance η_{0r} of the medium. In these respects, the far-zone fields of a current element are similar to a plane TEM wave. We should have reasonably expected this result since far from the current element, the spherical phase fronts become approximately equal to the plane phase fronts.

The leading terms in (5.22) and (5.24) in powers of βr constitute the near-zone fields. From (5.22) and (5.24), we obtain the near-zone fields as

$$H_\phi(\mathbf{r}) = \frac{I_0\, dl}{4\pi r^2} \sin\theta; \tag{5.26}$$

$$E_r(\mathbf{r}) = \frac{Q_0\, dl \cos\theta}{2\pi\varepsilon r^3}; \qquad E_\theta(\mathbf{r}) = \frac{Q_0\, dl \sin\theta}{4\pi\varepsilon r^3}. \tag{5.27}$$

In obtaining (5.26) and (5.27), we have expanded $e^{-j\beta r}$ into a power series in βr since $\beta r \ll 1$ and only the leading term has been retained. Also, $I_0/j\omega$ has been replaced by Q_0, as we derived earlier in (5.9). If Biot-Savart's law in magnetostatics as given by (4.4.20) is extended to apply to a time-varying current element, we can deduce the magnetic field due to the current element to be

$$\mathbf{H}(\mathbf{r}) = \frac{I_0\, dl\hat{z} \times \hat{r}}{4\pi r^2} = \hat{\phi}\,\frac{I_0\, dl \sin\theta}{4\pi r^2}, \tag{5.28}$$

which is identical to that obtained in (5.26). Thus, in the near-zone, the magnetic field due to a current element is the same as that obtained by the application of the laws of magnetostatics. If we deduce the scalar potential due to the electric dipole constituted by the constant current element by the application of the law of electrostatics as stated in (4.2.16), the result is given by (4.2.20) which, in the notation used here, becomes

$$\phi(\boldsymbol{r}) = \frac{Q_0\, dl \cos\theta}{4\pi\varepsilon r^2}. \tag{5.29}$$

Then, from (4.2.8), we find the electrostatic field corresponding to (5.29) to be

$$\boldsymbol{E}(\boldsymbol{r}) = -\nabla\phi(\boldsymbol{r}) = -\hat{r}\frac{\partial}{\partial r}\phi(\boldsymbol{r}) - \hat{\theta}\frac{1}{r}\frac{\partial}{\partial\theta}\phi(\boldsymbol{r}) - \frac{\hat{\phi}}{r\sin\theta}\frac{\partial}{\partial\phi}\phi(\boldsymbol{r}). \tag{5.30}$$

We may show the substitution of (5.29) in (5.30) to yield

$$E_r(\boldsymbol{r}) = \frac{Q_0\, dl\cos\theta}{2\pi\varepsilon r^3}; \qquad E_\theta(\boldsymbol{r}) = \frac{Q_0\, dl\sin\theta}{4\pi\varepsilon r^3}; \qquad E_\phi(\boldsymbol{r}) = 0. \tag{5.31}$$

Since (5.31) is identical to (5.27), it follows that in the near-zone the electric field due to a current element is identical to that obtained by the application of the laws of electrostatics. The fields given by (5.26) and (5.27) are called the *quasi-static* fields of a current element since they are identical to those which we obtained by the application of the laws of electrostatics and magnetostatics to the electrodynamical problem of the fields generated by a time-varying constant current element.

If $\beta r \simeq 1$, 1 and $1/\beta r$ are of the same order. Therefore we usually assume $\beta r = 1$ or equivalently $r = \lambda/2\pi$ to define the boundary between the near- and the far-zones.

5. Poynting Vector

We can evaluate the complex Poynting vector $\boldsymbol{S}_c(\boldsymbol{r})$ corresponding to the far-zone fields given by (5.25) as

$$\boldsymbol{S}_c(\boldsymbol{r}) = \tfrac{1}{2}\hat{\theta}E_\theta(\boldsymbol{r}) \times \hat{\phi}H_\phi^*(\boldsymbol{r}) = \frac{\hat{r}\,|I_0|^2\, dl^2}{32\pi^2 r^2}\eta_{0r}\beta^2\sin^2\theta = \boldsymbol{S}(\boldsymbol{r}). \tag{5.32}$$

Since (5.32) is real, it follows that the far-zone fields give rise to transport of time-averaged power only. It is instructive for us to compare (5.32) with the time-averaged Poynting vector $\boldsymbol{S}(\boldsymbol{r})$ corresponding to the actual fields as given by (5.22) and (5.24). Since $\boldsymbol{H}(\boldsymbol{r})$ has only a ϕ-component, $\boldsymbol{S}(\boldsymbol{r})$ has only r and θ components. From (5.22) and (5.24), we obtain that

$$S_\theta(\boldsymbol{r}) = -\tfrac{1}{2}\,\mathrm{Re}\, E_r(\boldsymbol{r})H_\phi^*(\boldsymbol{r}) = \mathrm{Re}\left[\frac{j\,|I_0|^2\,(dl)^2}{16\pi^2 r^3}\eta_{0r}\beta\sin\theta\cos\theta\left(1 + \frac{1}{\beta^2 r^2}\right)\right] = 0. \tag{5.33}$$

In a similar manner, we find that

$$S_r(\boldsymbol{r}) = \tfrac{1}{2}\,\mathrm{Re}\, E_\theta(\boldsymbol{r})H_\phi^*(\boldsymbol{r}) = \mathrm{Re}\left[\frac{|I_0|^2\,(dl)^2}{32\pi^2 r^2}\eta_{0r}\beta^2\sin^2\theta\left(1 + \frac{1}{j\beta^3 r^3}\right)\right]$$

$$= \frac{|I_0|^2\,(dl)^2}{32\pi^2 r^2}\eta_{0r}\beta^2\sin^2\theta. \tag{5.34}$$

If we compare (5.33) and (5.34) with (5.32) we obtain the important result that the time-averaged Poynting vector is contributed to only by the far-zone fields.

The total power flowing through a small spherical surface enclosing the current element is the same as that flowing through a large spherical surface of radius r tending to infinity, since in a lossless medium no power is dissipated in the region between the two spherical surfaces. We obtain the total power flowing through a spherical surface by multiplying the time-averaged Poynting vector by the area of an element of the spherical surface and integrating the result throughout the spherical surface. Since the elemental area is proportional to r^2, it follows that the time-averaged Poynting vector has to be proportional to r^{-2} in order for the time-averaged power flowing through the elemental area and the total time-averaged power flowing through the entire spherical surface to be independent of the radius of the spherical surface and, therefore, to be the same for the small as well as the large spherical surfaces. If the time-averaged Poynting vector is proportional to the third or higher powers of r^{-1}, the total time-averaged power flowing through a spherical surface is proportional to some power of r^{-1} and hence is not the same for the small as for the large spherical surfaces. Consequently, the time-averaged Poynting vector should be proportional to r^{-2}. Hence, the amplitudes of the field components contributing to the time-averaged Poynting vector should fall off as $1/r$ and these, by definition, are the far-zone fields. These arguments confirm the previous result that the time-averaged Poynting vector is contributed to only by the far-zone fields.

6. Radiation Intensity and Radiation Resistance

We shall now introduce the useful concepts of radiation intensity, radiation resistance, and directive gain. Let us consider a spherical surface of radius r and an element of area dS on it lying between the angles θ and $\theta + d\theta$, and ϕ and $\phi + d\phi$. We can deduce with the help of Fig. 9.3 that

$$dS = r^2\, d\Omega = r^2 \sin\theta\, d\theta\, d\phi, \tag{5.35}$$

where $d\Omega$ is the element of the solid angle subtended by dS at the origin. The time-averaged power flowing through the elemental area dS is given by

$$S_r(\mathbf{r})\, dS = S_r(\mathbf{r}) r^2\, d\Omega.$$

Hence, $S_r(\mathbf{r})r^2$ is the time-averaged power flux per unit solid angle. The *radiation intensity* $\Phi(\theta, \phi)$ in a given direction is defined as the time-averaged power flux per unit solid angle in that direction. Hence, from (5.34), we can obtain the radiation intensity of a current element as

$$\Phi(\theta, \phi) = \frac{|I_0|^2\, (dl)^2}{32\pi^2} \eta_{0r} \beta^2 \sin^2\theta. \tag{5.36}$$

Since the characteristics of a current element are independent of ϕ, we shall omit the argument ϕ from now on. We can obtain the total time-averaged power P radiated by the current element by multiplying $\Phi(\theta)$ by the element of the solid angle

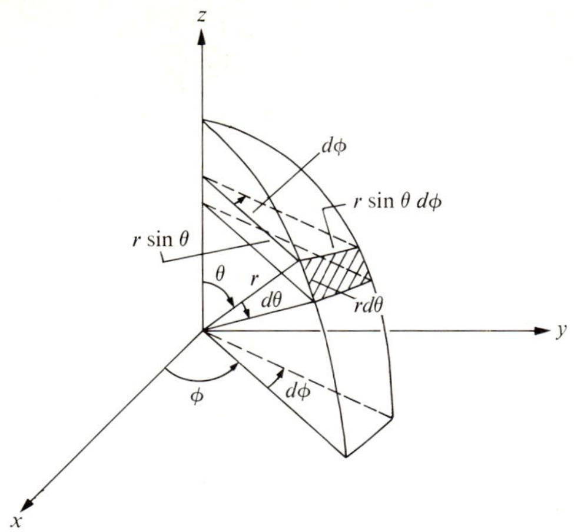

Fig. 9.3 Element of area dS on a spherical surface of radius r lying between the angles θ and $\theta + d\theta$, and ϕ and $\phi + d\phi$.

as given in (5.35) and integrating the result to cover the entire solid angle, that is, from $\phi = 0$ to $\phi = 2\pi$, and from $\theta = 0$ to $\theta = \pi$ (see Fig. 9.3). Thus,

$$P = \int_{\phi=0}^{2\pi}\int_{\theta=0}^{\pi} \Phi(\theta) \sin\theta \, d\theta \, d\phi = \frac{|I_0|^2 (dl)^2}{16\pi} \eta_{0r}\beta^2 \int_{\theta=0}^{\pi} \sin^3\theta \, d\theta. \tag{5.37}$$

Since

$$\int_{\theta=0}^{\pi} \sin^3\theta \, d\theta = \tfrac{1}{4}\int_{\theta=0}^{\pi} [3\sin\theta - \sin 3\theta] \, d\theta = \tfrac{4}{3},$$

(5.37) becomes

$$P = \frac{|I_0|^2 (dl)^2}{12\pi} \eta_{0r}\beta^2. \tag{5.38}$$

The concept of the radiation resistance R_r of the current element is useful since it is a measure of the total time-averaged power radiated. It is the value of the resistance which, when the current in it is I_0, dissipates the same amount of the time-averaged power as that radiated by the antenna. Hence,

$$\tfrac{1}{2}|I_0|^2 R_r = \frac{|I_0|^2 (dl)^2}{12\pi} \eta_{0r}\beta^2 \quad \text{or} \quad R_r = \frac{\eta_{0r}}{6\pi} \beta^2 (dl)^2 = \frac{2\pi\eta_{0r}}{3}\left(\frac{dl}{\lambda}\right)^2, \tag{5.39}$$

since $\beta = 2\pi/\lambda$, where λ is the wavelength. It is usual to approximate the intrinsic resistance η_{0r} by 120π and rewrite (5.39) as

$$R_r = 80\pi^2 (dl/\lambda)^2. \tag{5.40}$$

The value of the radiation resistance depends on the reference current. But, in the case of the current element, there is no ambiguity since the current is a constant along the length of the filament. For an actual antenna, the current varies along its length. It is usual to drive the antenna at its center by a transmission line; the reference current is chosen by convention to be the current at the input to the antenna. Thus, the total time-averaged power radiated is equal to that dissipated in a hypothetical resistance R_r connected to the input to the antenna. For this reason, R_r is also called the input resistance of the antenna. From (5.40), we can see that significant amount of power is radiated only when the dimensions of the radiator become comparable to the wavelength.

7. Directive Gain and Directivity

Let us consider an equivalent hypothetical isotropic antenna radiating equally in all directions the same total time-averaged power as the actual antenna which in this case is the current element. From (5.38), we find the radiation intensity Φ_{iso} of the equivalent isotropic radiator to be

$$\Phi_{iso} = \frac{|I_0|^2 (dl)^2 \eta_{0r} \beta^2}{48\pi^2} = \frac{P}{4\pi}. \tag{5.41}$$

The *directive gain* $g(\theta)$ is defined as the ratio of the radiation intensity to that of the equivalent isotropic radiator. It is a measure of the directional property of the antenna. From (5.36) and (5.41), we obtain

$$g(\theta) = 1.5 \sin^2 \theta, \tag{5.42}$$

which is depicted in Fig. 9.4 as a function of θ. We should note that the directive gain of the current element is independent of ϕ. For $\theta = 60°$, $g(\theta) = 1.125$, which shows that the current element radiates 1.125 times more time-averaged power per unit solid angle in the specified direction than the equivalent isotropic radiator. The maximum value of the directive gain is called the directivity or simply the *gain* and is denoted by g without any arguments. From (5.42), we find the gain of the current element to be 1.5 and this pertains to the broadside direction corresponding to $\theta = \pi/2$.

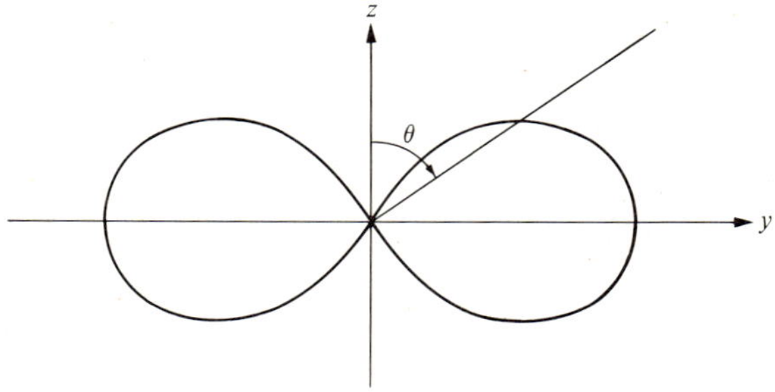

Fig. 9.4 Directive gain $g(\theta)$ of an infinitesimal current element.

9.6 FINITE LENGTH ANTENNA WITH ASSUMED CURRENT DISTRIBUTION

We have pointed out that a significant amount of electromagnetic energy is radiated only when the dimensions of the radiator are comparable to a wavelength. For this reason, it is instructive for us to consider the radiation characteristics of an antenna of length comparable to a wavelength. We assume the antenna to be in the form of a cylinder whose radius is small in comparison to a wavelength with the result that it can be idealized to be in the form of a filament. The filament is along the z-axis of a cylindrical coordinate system ρ, ϕ, and z; it extends from $z = -h$ to $z = h$. The current distribution along the antenna depends on the method of excitation as well as on the geometry of the antenna and is comparatively difficult to obtain. Since we are interested in the examination of the radiation characteristics of the antenna, it is reasonable for us to assume an approximate current distribution. The current should vanish at the two ends of the filamentary antenna. If the antenna is driven at the center, as is usually the case, the current distribution is symmetrical about the center of the antenna. The current distribution which satisfies the above-mentioned requirements and which is a reasonably good approximation for an electrically thin, linear antenna is of the form:

$$\begin{aligned} I(z) = I_m \sin \beta(h - |z|) &= I_m \sin \beta(h - z) \qquad (z > 0), \\ &= I_m \sin \beta(h + z) \qquad (z < 0), \end{aligned} \tag{6.1}$$

where the maximum value of the current I_m is related to the current I_0 at the center as follows:

$$I_m = I_0/\sin \beta h. \tag{6.2}$$

The geometry of the radiator shows that the potentials and the fields are independent of the angle ϕ and hence, it is sufficient for us to determine the fields at any point P in a plane corresponding to one particular value of ϕ, as shown in Fig. 9.5. Let the coordinates of the observation point P be ρ, 0, and z. Let r_{1h}, r, and r_{2h} be the distances from P to the top end, the center and the bottom end of the antenna respectively. Let us consider an element of length dz' of the filament situated at a distance z' from the origin. Let $R(z, z')$ be the distance from the observation point to this element of length at z'. Then, we can obtain with the help of Fig. 9.5:

$$r_{1h} = \sqrt{(z - h)^2 + \rho^2} = R(z, h) = R(-z, -h), \tag{6.3a}$$

$$r_{2h} = \sqrt{(z + h)^2 + \rho^2} = R(z, -h) = R(-z, h), \tag{6.3b}$$

$$r = \sqrt{z^2 + \rho^2} = R(z, 0), \tag{6.3c}$$

$$R(z, z') = \sqrt{(z - z')^2 + \rho^2}. \tag{6.3d}$$

Also, it follows from (6.3a to d) that

$$r_{1h}^2 - (z - h)^2 = r_{2h}^2 - (z + h)^2 = r^2 - z^2 = \rho^2. \tag{6.4}$$

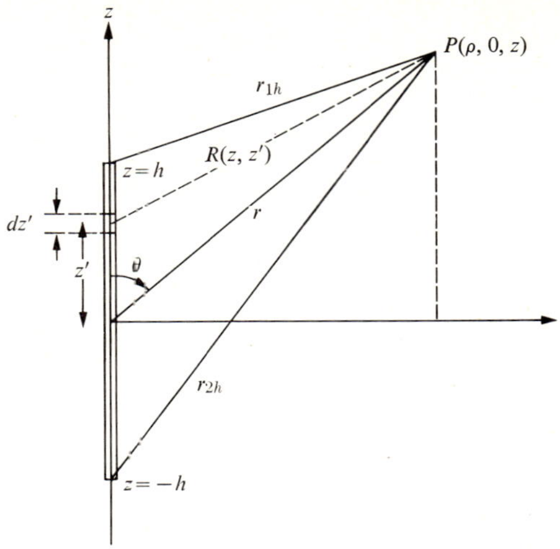

Fig. 9.5 Geometry for the evaluation of the fields of a linear, filamentary antenna with sinusoidal current distribution.

We can obtain the contribution to the vector potential at P due to current element $I(z')\,dz'$ from (5.3) as

$$dA_z(\mathbf{r}) = \mu \frac{I(z')\,dz' e^{-j\beta R(z,z')}}{4\pi R(z, z')}. \tag{6.5}$$

We determine the vector potential at P due to the entire current distribution along the filament from $z = -h$ to $z = h$ by integrating (6.5) from $z' = -h$ to $z' = h$. The result, after substituting the value of $I(z)$ given by (6.1) is

$$A_z(\mathbf{r}) = \frac{\mu I_m}{4\pi}\left[\int_0^h \frac{\sin\beta(h - z')e^{-j\beta R(z,z')}}{R(z, z')}\,dz' + \int_{-h}^0 \frac{\sin\beta(h + z')e^{-j\beta R(z,z')}}{R(z, z')}\,dz'\right]. \tag{6.6}$$

If we change the variable of integration for the second integral in (6.6) from z' to $-z'$, we find on noting $R(z, -z') = R(-z, z')$, that the value of the second integral is obtainable from that of the first by changing z to $-z$. Therefore, we may rewrite (6.6) as

$$A_z(\mathbf{r}) = \frac{\mu I_m}{4\pi}\left[B_1(\rho, z) + B_1(\rho, -z)\right], \tag{6.7}$$

where

$$B_1(\rho, z) = \int_0^h \frac{\sin\beta(h - z')e^{-j\beta R(z,z')}}{R(z, z')}\,dz'. \tag{6.8}$$

It is convenient for us to split the sine function in (6.8) as the sum of two exponentials with the result that (6.8) becomes

$$B_1(\rho, z) = B_{11}(\rho, z) - B_{12}(\rho, z), \tag{6.9}$$

where

$$B_{11}(\rho, z) = \frac{e^{j\beta h}}{2j} \int_0^h \frac{e^{-j\beta[R(z,z')+z']}}{R(z, z')} dz', \tag{6.10}$$

and

$$B_{12}(\rho, z) = \frac{e^{-j\beta h}}{2j} \int_0^h \frac{e^{-j\beta[R(z,z')-z']}}{R(z, z')} dz'. \tag{6.11}$$

Since the vector potential has only a z-component and since $\partial/\partial\phi = 0$, we find from (4.4) that the magnetic field has only a ϕ-component and is given by

$$H_\phi(\mathbf{r}) = -\frac{1}{\mu}\frac{\partial}{\partial\rho} A_z(\mathbf{r}). \tag{6.12}$$

We see from (6.7), (6.9), and (6.12) that $(\partial/\partial\rho)B_{11}(\rho,z)$ and $(\partial/\partial\rho)B_{12}(\rho, z)$ have to be evaluated. From (6.10) and (6.3d), we can deduce that

$$\frac{\partial}{\partial\rho} B_{11}(\rho, z) = \frac{e^{j\beta h}}{2j} \int_0^h \left[\frac{-j\beta\rho e^{-j\beta[R(z,z')+z']}}{R^2(z, z')} - \frac{\rho e^{-j\beta[R(z,z')+z']}}{R^3(z, z')}\right] dz'. \tag{6.13}$$

We can easily verify that the integrand in (6.13) is a perfect differential. Thus,

$$\begin{aligned}\frac{d}{dz'}\left[\frac{\rho e^{-j\beta[R(z,z')+z']}}{R(z, z')[R(z, z') + z' - z]}\right] &= \frac{-j\beta\rho e^{-j\beta[R(z,z')+z']}}{R(z, z')[R(z, z') + z' - z]}\left\{\frac{-(z - z')}{R(z, z')} + 1\right\} \\ &+ \frac{\rho e^{-j\beta[R(z,z')+z']}(z - z')}{R^3(z, z')[R(z, z') + z' - z]} - \frac{\rho e^{-j\beta[R(z,z')+z']}}{R(z, z')[R(z, z') + z' - z]^2}\left\{\frac{-(z - z')}{R(z, z')} + 1\right\}.\end{aligned} \tag{6.14}$$

We can show that the simplification of (6.14) yields

$$\frac{d}{dz'}\left[\frac{\rho e^{-j\beta[R(z,z')+z']}}{R(z, z')[R(z, z') + z' - z]}\right] = \frac{-j\beta\rho e^{-j\beta[R(z,z')+z']}}{R^2(z, z')} - \frac{\rho e^{-j\beta[R(z,z')+z']}}{R^3(z, z')}. \tag{6.15}$$

We see the right-hand side of (6.15) to be precisely the integrand in (6.13). Hence, from (6.15) and (6.3), we can simplify (6.13) to

$$\frac{\partial}{\partial\rho} B_{11}(\rho, z) = \frac{e^{j\beta h}\rho}{2j}\left[\frac{e^{-j\beta[r_{1h}+h]}}{r_{1h}[r_{1h} + h - z]} - \frac{e^{-j\beta r}}{r[r - z]}\right], \tag{6.16a}$$

$$= \frac{e^{j\beta h}}{2j\rho}\left[\left(1 - \frac{h - z}{r_{1h}}\right)e^{-j\beta(r_{1h}+h)} - \left(1 + \frac{z}{r}\right)e^{-j\beta r}\right]. \tag{6.16b}$$

We should note that (6.4) has been used in obtaining (6.16b) from (6.16a). Similarly, from (6.11), we can prove that

$$\frac{\partial}{\partial\rho} B_{12}(\rho, z) = -\frac{e^{-j\beta h}}{2j\rho}\left[\left(1 + \frac{h - z}{r_{1h}}\right)e^{-j\beta(r_{1h}-h)} - \left(1 - \frac{z}{r}\right)e^{-j\beta r}\right]. \tag{6.17}$$

We find that the use of (6.9), (6.16b), and (6.17) yields

$$\frac{\partial}{\partial \rho} B_1(\rho, z) = \frac{1}{j\rho} e^{-j\beta r_{1h}} - \frac{e^{-j\beta r}}{2j\rho}\left[\left(1 + \frac{z}{r}\right)e^{j\beta h} + \left(1 - \frac{z}{r}\right)e^{-j\beta h}\right]. \tag{6.18}$$

From (6.3) and (6.18), it follows that

$$\frac{\partial}{\partial \rho} B_1(\rho, -z) = \frac{1}{j\rho} e^{-j\beta r_{2h}} - \frac{e^{-j\beta r}}{2j\rho}\left[\left(1 - \frac{z}{r}\right)e^{j\beta h} + \left(1 + \frac{z}{r}\right)e^{-j\beta h}\right]. \tag{6.19}$$

Adding (6.18) and (6.19) we get

$$\frac{\partial}{\partial \rho} B_1(\rho, z) + \frac{\partial}{\partial \rho} B_1(\rho, -z) = \frac{1}{j\rho}\left[e^{-j\beta r_{1h}} + e^{-j\beta r_{2h}} - 2\cos(\beta h)e^{-j\beta r}\right]. \tag{6.20}$$

From (6.7), (6.12), and (6.20), it follows that

$$H_\phi(\mathbf{r}) = \frac{jI_m}{4\pi\rho}\left[e^{-j\beta r_{1h}} + e^{-j\beta r_{2h}} - 2\cos(\beta h)e^{-j\beta r}\right]. \tag{6.21}$$

As we mentioned previously, we can obtain the electric field conveniently from (6.21) with the help of the appropriate Maxwell's equation rather than from the scalar potential. Hence, we see that

$$\mathbf{E}(\mathbf{r}) = \frac{\nabla \times \mathbf{H}(\mathbf{r})}{j\omega\varepsilon} = -\frac{\hat{\rho}\,\partial H_\phi(\mathbf{r})}{j\omega\varepsilon\,\partial z} + \frac{\hat{z}}{j\omega\varepsilon}\frac{\partial}{\rho\,\partial\rho}\{\rho H_\phi(\mathbf{r})\}. \tag{6.22}$$

We can deduce with the help of (6.21) and (6.22) that

$$E_\rho(\mathbf{r}) = \frac{j\eta_{0r}I_m}{4\pi\rho}\left[\frac{z-h}{r_{1h}}e^{-j\beta r_{1h}} + \frac{z+h}{r_{2h}}e^{-j\beta r_{2h}} - \frac{2z\cos\beta h}{r}e^{-j\beta r}\right], \tag{6.23}$$

and

$$E_z(\mathbf{r}) = -\frac{j\eta_{0r}I_m}{4\pi}\left[\frac{e^{-j\beta r_{1h}}}{r_{1h}} + \frac{e^{-j\beta r_{2h}}}{r_{2h}} - \frac{2\cos\beta h}{r}e^{-j\beta r}\right]. \tag{6.24}$$

We have deduced the magnetic and the electric fields at any arbitrary distance from a finite-length current filament with a sinusoidal current distribution to be given by (6.21), and (6.23) and (6.24) respectively. We obtain the simple, closed-form results for the fields because the integrand of (6.13) is a perfect differential. This simple situation arises when the current distribution is an exponential function of the distance along the filament or a linear combination of exponential functions such as a sinusoidal function.

The expression for the parallel component $E_z(\mathbf{r})$ of the electric field as given by (6.24) is amenable to the following physical interpretation. The surface of constant phase of that part of $E_z(\mathbf{r})$ corresponding to the first term in (6.24) is given by r_{1h} = constant, which is a spherical surface centered at the end $z = h$ of the filament. The amplitude of this part of $E_z(\mathbf{r})$ falls off as $1/r_{1h}$. Hence, the first term in (6.24) represents a spherical wave emanating from the end $z = h$ of the current filament. In a similar manner, the second term in (6.24) represents a spherical wave emanating from the end $z = -h$ of the current filament. The amplitudes of these isotropic spherical waves are independent of the length of the current filament. The third term in (6.24) also represents a spherical wave that emanates from the center of the current filament. The amplitude of this spherical wave depends on the length of the current filament. For a half-wave length filament $2h = \lambda/2$, $\beta h = (2\pi/\lambda)(\lambda/4) = \pi/2$, and $\cos \beta h = 0$; hence the third spherical wave vanishes.

1. Far-zone Fields

From (6.21), (6.23), and (6.24), we can evaluate the far-zone and the near-zone fields as in the previous section. For example, it is instructive for comparison purposes for us to evaluate the far-zone magnetic field from (6.21). If θ is the angle made by the direction of r with that of z, then it is seen from Fig. 9.5 that

$$\rho = r \sin \theta; \qquad z = r \cos \theta. \tag{6.25}$$

From (6.3a) and (6.25), we find that

$$r_{1h} = (r^2 - 2zh + h^2)^{1/2} = r\left(1 - 2\frac{h}{r}\cos\theta + \frac{h^2}{r^2}\right)^{1/2}. \tag{6.26}$$

If $h/r \ll 1$, we can expand (6.26) into a binomial series in h/r and the result on retaining the first two terms is

$$r_{1h} = r - h \cos \theta. \tag{6.27}$$

In a similar manner, we can show from (6.3b) and (6.25) that

$$r_{2h} = r + h \cos \theta. \tag{6.28}$$

In the far-zone, we can replace r_{1h} and r_{2h} by the expressions given by (6.27) and (6.28) respectively with the result that $H_\phi(\mathbf{r})$ given by (6.21) becomes

$$H_\phi(\mathbf{r}) = \frac{jI_m}{2\pi r \sin\theta} e^{-j\beta r}[\cos\{\beta h \cos\theta\} - \cos\beta h]. \tag{6.29}$$

We should note that (6.29) gives the far-zone magnetic field due to a finite-length current filament with a sinusoidal current distribution. Similarly, from (6.23) and (6.24), we can obtain the far-zone electric field in the cylindrical coordinates and, from these expressions, we can deduce the far-zone electric field components $E_r(\mathbf{r})$ and $E_\theta(\mathbf{r})$ in the spherical coordinates r, θ, and ϕ. For the computation of the radiation intensity and the radiation resistance, it is convenient for us to have the field components in the spherical coordinates. We can also obtain the far-zone electric field in the spherical coordinates from (6.29) and the second of Maxwell's equations in the spherical coordinates.

If we are interested in the far-zone fields alone, it is most convenient for us to introduce the far-zone approximation at the outset in the evaluation of the expression for the vector potential as given by (6.6). Following the procedure we used in obtaining (6.26), we can show from (6.3d) and (6.25) that for $h/r \ll 1$, $R(z, z')$ can be approximated as follows:

$$R(z, z') = r - z' \cos \theta. \tag{6.30}$$

Substituting (6.30) into (6.6) we obtain

$$A_z(\mathbf{r}) = \frac{\mu I_m}{4\pi r} e^{-j\beta r}\left[\int_0^h \frac{\sin \beta(h - z')}{[1 - (z'/r)\cos\theta]} e^{j\beta z'\cos\theta}\, dz' + \int_{-h}^0 \frac{\sin\beta(h + z')}{[1 - (z'/r)\cos\theta]} e^{j\beta z'\cos\theta}\, dz'\right]. \tag{6.31}$$

If we expand $[1 - (z'/r)\cos\theta]^{-1}$ into a binomial series as $1 + (z'/r)\cos\theta$, we find that only the leading term has to be retained for the purpose of the evaluation of the far-zone field whose amplitude is known to fall off as r^{-1}. Then, we can change the variable of integration z' of the second integral in (6.31) to $-z'$ and add the resulting integral to the first integral in (6.31) to yield the following result for the far-zone vector potential:

$$\begin{aligned} A_z(\mathbf{r}) &= \frac{\mu I_m}{4\pi r} e^{-j\beta r}\int_0^h 2\sin\beta(h - z')\cos\{\beta z'\cos\theta\}\, dz', \\ &= \frac{\mu I_m e^{-j\beta r}}{4\pi r}\int_0^h dz'[\sin\{\beta(h - z'(1 - \cos\theta))\} + \sin\{\beta(h - z'(1 + \cos\theta))\}]. \end{aligned} \tag{6.32}$$

From (6.32), we can deduce that

$$\begin{aligned} A_z(\mathbf{r}) &= \frac{\mu I_m e^{-j\beta r}}{4\pi r}\left[\frac{\cos\{\beta(h - z'(1 - \cos\theta))\}}{\beta(1 - \cos\theta)} + \frac{\cos\{\beta(h - z'(1 + \cos\theta))\}}{\beta(1 + \cos\theta)}\right]_0^h, \\ &= \frac{\mu I_m e^{-j\beta r}}{4\pi r\beta\sin^2\theta}[\cos\theta[\cos\{\beta(h - z'(1 - \cos\theta))\} - \cos\{\beta(h - z'(1 + \cos\theta))\}] \\ &\quad + [\cos\{\beta(h - z'(1 - \cos\theta))\} + \cos\{\beta(h - z'(1 + \cos\theta))\}]]_0^h, \\ &= \frac{\mu I_m e^{-j\beta r}}{2\pi r\beta\sin^2\theta}[-\cos\theta\sin\beta(h - z')\sin(\beta z'\cos\theta) \\ &\quad + \cos\{\beta z'\cos\theta\}\cos\beta(h - z')]_0^h, \\ &= \frac{\mu I_m e^{-j\beta r}}{2\pi r\beta\sin^2\theta}[\cos(\beta h\cos\theta) - \cos\beta h]. \end{aligned} \tag{6.33}$$

From (5.21), (5.4b), and (6.33), we can obtain the far-zone magnetic field, that is the part of the magnetic field whose amplitude falls off as $1/r$, as

$$\begin{aligned} H_\phi(\mathbf{r}) &= \frac{1}{\mu r}\frac{\partial}{\partial r}(rA_\theta(\mathbf{r})) = -\frac{\sin\theta}{\mu r}\frac{\partial}{\partial r}\{rA_z(\mathbf{r})\}, \\ &= \frac{j\beta\sin\theta}{\mu} A_z(\mathbf{r}). \end{aligned} \tag{6.34}$$

Also, from (5.23), (6.33), and (6.34), we can obtain the far-zone electric field as

$$E_\theta(\mathbf{r}) = -\frac{1}{j\omega\varepsilon r}\frac{\partial}{\partial r}[rH_\phi(\mathbf{r})] = \frac{j\beta}{j\omega\varepsilon} H_\phi(\mathbf{r}) = \eta_{0r} H_\phi(\mathbf{r}). \tag{6.35}$$

The general expressions (6.34) and (6.35) for the far-zone fields are important and useful. With the help of (6.33) to (6.35), we find that the far-zone magnetic field and the far-zone electric field radiated by a finite-length filament with sinusoidal current distribution are explicitly given by

$$H_\phi(\mathbf{r}) = \frac{jI_m e^{-j\beta r}}{2\pi r \sin\theta}[\cos(\beta h \cos\theta) - \cos\beta h], \tag{6.36}$$

and

$$E_\theta(\mathbf{r}) = \frac{j\eta_{0r} I_m e^{-j\beta r}}{2\pi r \sin\theta}[\cos(\beta h \cos\theta) - \cos\beta h]. \tag{6.37}$$

2. Half-wavelength Antenna

We shall now particularize the analysis to the important case in which the total length of the filament is equal to one half of a wavelength, that is, $2h = \lambda/2$, which yields $\beta h = \pi/2$. Since $\sin\beta h = 1$, it follows from (6.2) that $I_m = I_0$ and the maximum value of the current occurs at the center of a half-wavelength filament. For the half-wavelength filament, we can reduce (6.36) and (6.37) to

$$H_\phi(\mathbf{r}) = \frac{jI_0 e^{-j\beta r}}{2\pi r \sin\theta}\cos\left(\frac{\pi}{2}\cos\theta\right), \tag{6.38}$$

$$E_\theta(\mathbf{r}) = \frac{j\eta_{0r} I_0 e^{-j\beta r}}{2\pi r \sin\theta}\cos\left(\frac{\pi}{2}\cos\theta\right). \tag{6.39}$$

We can evaluate the time-averaged Poynting vector associated with the far-zone fields (6.38) and (6.39) as

$$S_r(\mathbf{r}) = \frac{|I_0|^2 \eta_{0r}}{8\pi^2 r^2}\left[\frac{\cos((\pi/2)\cos\theta)}{\sin\theta}\right]^2. \tag{6.40}$$

We obtain the radiation intensity of a half-wavelength filament from (5.35) and (6.40) as

$$\Phi(\theta) = \frac{|I_0|^2 \eta_{0r}}{8\pi^2}\left[\frac{\cos((\pi/2)\cos\theta)}{\sin\theta}\right]^2. \tag{6.41}$$

The current filament is cylindrically symmetrical with the result that all the field quantities are independent of ϕ, as is explicit in (6.38) to (6.41).

We obtain the total time-averaged power P radiated by the half-wavelength current filament by multiplying $\Phi(\theta)$ by the element of the solid angle as given in (5.35) and integrating the result to cover the entire solid angle. Therefore,

$$\begin{aligned} P &= \int_{\phi=0}^{2\pi}\int_0^{\pi} \frac{|I_0|^2 \eta_{0r}}{8\pi^2} \frac{\cos^2((\pi/2)\cos\theta)}{\sin\theta}\, d\theta\, d\phi, \\ &= \frac{|I_0|^2 \eta_{0r}}{4\pi}\int_0^{\pi} \frac{\cos^2((\pi/2)\cos\theta)}{\sin\theta}\, d\theta. \end{aligned} \tag{6.42}$$

By changing the variable of integration from θ to $\pi - \theta$, we can prove that

$$\int_{\pi/2}^{\pi} \frac{\cos^2((\pi/2)\cos\theta)}{\sin\theta}\, d\theta = \int_0^{\pi/2} \frac{\cos^2((\pi/2)\cos\theta)}{\sin\theta}\, d\theta,$$

with the result that we can rewrite (6.42) as

$$P = \frac{|I_0|^2 \eta_{0r}}{2\pi} \int_0^{\pi/2} \frac{\cos^2((\pi/2)\cos\theta)}{\sin\theta} d\theta. \tag{6.43}$$

We cannot evaluate the integral appearing in (6.43) explicitly but it can be recast into a form whose values may be found in the available tables of mathematical functions. We first rewrite the integral in (6.43) as follows:

$$B_H = \int_0^{\pi/2} \frac{\cos^2((\pi/2)\cos\theta)}{\sin\theta} d\theta = \frac{1}{2} \int_0^{\pi/2} \frac{[1 + \cos(\pi \cos\theta)]}{\sin\theta} d\theta. \tag{6.44}$$

Let $u = \cos\theta$. Therefore, $du = -\sin\theta \, d\theta$ and

$$d\theta/\sin\theta = -du/\sin^2\theta = -du/(1 - u^2).$$

Hence, (6.44) becomes

$$B_H = \frac{1}{2} \int_0^1 \frac{[1 + \cos \pi u]}{(1 - u^2)} du = \frac{1}{4} \int_0^1 (1 + \cos \pi u)\left(\frac{1}{1 + u} + \frac{1}{1 - u}\right) du. \tag{6.45}$$

Since

$$\int_0^1 \frac{(1 + \cos \pi u)}{1 - u} du = \int_{-1}^0 \frac{(1 + \cos \pi u)}{1 + u} du,$$

we see (6.45) to be

$$B_H = \frac{1}{4} \int_{-1}^1 \frac{(1 + \cos \pi u)}{1 + u} du. \tag{6.46}$$

Let $v = \pi + \pi u$ and therefore $dv/v = dv/\pi(1 + u) = du/(1 + u)$. Also, since $\cos \pi u = \cos(v - \pi) = -\cos v$, we can recast (6.46) as

$$B_H = \frac{1}{4} \int_0^{2\pi} \frac{(1 - \cos v)}{v} dv. \tag{6.47}$$

The integral

$$\mathrm{Cin}\,(x) = \int_0^x \frac{(1 - \cos v)}{v} dv, \tag{6.48}$$

is a simple integral to evaluate numerically and its values have been tabulated for a wide range of values of x. We find that $\mathrm{Cin}\,(2\pi) = 2.4377$ and hence from (6.47) $B_H = 0.609$.

It is now simple for us to see from (6.43) that the total time-averaged power radiated by a half-wavelength filament with a sinusoidal current distribution is

$$P = \tfrac{1}{2} |I_0|^2 \frac{\eta_{0r}}{\pi} \times (0.609). \tag{6.49}$$

Since the radiation resistance R_r is defined by the relation $P = \frac{1}{2}|I_0|^2 R_r$, it follows from (6.49) that

$$R_r = \frac{\eta_{0r}}{\pi} \times (0.609) = 73 \text{ ohms}. \tag{6.50}$$

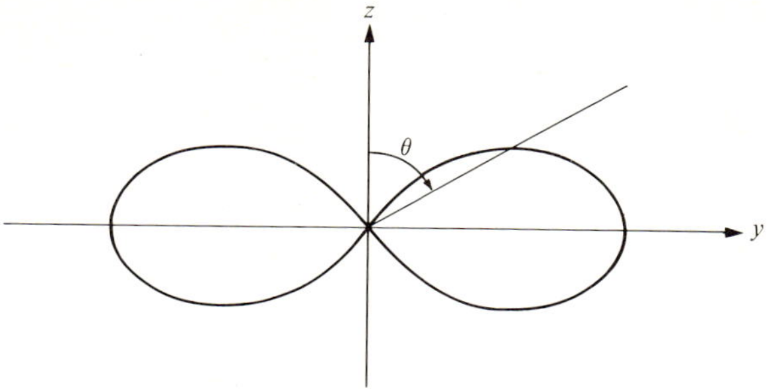

Fig. 9.6 Directive gain $g(\theta)$ of a half-wavelength filamentary antenna with sinusoidal current distribution.

The radiation resistance R_r for the half-wavelength antenna as given in (6.50) is with reference to the current at the center of the antenna or equivalently the maximum value of the current along the antenna.

We find the radiation intensity Φ_{iso} of the equivalent isotropic radiator from (6.49) as

$$\Phi_{\text{iso}} = \frac{P}{4\pi} = \frac{|I_0|^2}{8\pi^2}\eta_{0r} \times (0.609). \tag{6.51}$$

We can determine the directive gain of the half-wavelength filamentary antenna from (6.41) and (6.51) as

$$g(\theta) = \frac{\Phi(\theta)}{\Phi_{\text{iso}}} = 1.64\left[\frac{\cos((\pi/2)\cos\theta)}{\sin\theta}\right]^2, \tag{6.52}$$

which is depicted in Fig. 9.6 as a function of θ. For $\theta = 60°$, $g(\theta) = 1.09$, which shows that the half-wavelength filamentary antenna radiates 1.09 times more time-averaged power per unit solid angle in the specified direction than the equivalent isotropic radiator. We find the maximum value of the directive gain or the directivity g for the half-wavelength filamentary antenna from (6.52) to be 1.64 and this maximum value is attained in the broadside direction. For θ tending to zero,

$$\sin\theta = \theta, \qquad \cos\theta = 1 - \frac{\theta^2}{2},$$

$$\cos\left(\frac{\pi}{2}\cos\theta\right) = \cos\left(\frac{\pi}{2} - \frac{\pi\theta^2}{4}\right) = \sin\frac{\pi}{4}\theta^2 = \frac{\pi\theta^2}{4}.$$

Hence $g(\theta) = 1.64((\pi/4)\theta)^2$ which goes to zero as θ tends to zero. Therefore, the directive gain $g(\theta)$ is zero along the axis of the antenna, as is depicted in Fig. 9.6.

3. Effective Height

We shall now introduce the useful concept of effective height or effective length. Let us consider a linear filamentary antenna extending from $z = -h$ to $z = h$ and let $I(z)$ be the current distribution along the antenna. An examination of (6.6),

(6.30), and the argument that follows (6.31) shows that the far-zone vector potential due to the current distribution $I(z)$ along the filament from $z = -h$ to $z = h$ is given by

$$A_z(\mathbf{r}) = \frac{\mu e^{-j\beta r}}{4\pi r} \int_{-h}^{h} I(z') e^{j\beta z' \cos\theta}\, dz', \tag{6.53}$$

where θ is the angle between the direction of observation and the axis of the antenna as shown in Fig. 9.5. Then, we can deduce the far-zone electric field with the help of (6.34), (6.35), and (6.53) as

$$E_\theta(\mathbf{r}) = j\beta\eta_{0r} \frac{e^{-j\beta r}}{4\pi r} \int_{-h}^{h} \sin\theta I(z') e^{j\beta z' \cos\theta}\, dz', \tag{6.54}$$

which is independent of ϕ. The far-zone electric field in any direction θ, due to a filamentary antenna of actual length $2h$ and with different current distributions, is essentially dependent on the integral in (6.54). This integral in (6.54) is a weighted current moment which has the dimensions of a current multiplied by a length. All filamentary antennas having the same weighted current moment corresponding to a particular direction produce the same far-zone electric field in that direction. Thus, the weighted current moment is a measure of the far-zone electric field in any direction. It is conventional for us to choose the same reference current for all the antennas with different current distributions and this reference current is usually taken to be at the center of the antenna. The weighted current moment when normalized to the reference current yields a quantity which has the dimensions of length and this quantity is called the *effective height*. Thus, the far-zone electric field due to a filamentary antenna with a specified current at its center is proportional to the effective height. Therefore, we define the effective height $2h_e(\theta)$ as

$$2h_e(\theta) = \frac{\sin\theta}{I(0)} \int_{-h}^{h} I(z') e^{j\beta z' \cos\theta}\, dz'. \tag{6.55}$$

It is of interest for us to evaluate the effective height $2h_e(\theta)$ of a filamentary antenna of actual length $2h$ and with a current distribution as given by (6.1) and (6.2). We find from (6.1), (6.2), and (6.55) that

$$2h_e(\theta) = \frac{\sin\theta}{\sin\beta h} \int_{-h}^{h} \sin\beta(h - |z'|) e^{j\beta z' \cos\theta}\, dz'. \tag{6.56}$$

We can evaluate the integral in (6.56) as indicated in (6.32) and (6.33) to yield the following result:

$$2h_e(\theta) = \frac{2}{\beta} \frac{[\cos(\beta h \cos\theta) - \cos\beta h]}{\sin\theta \sin\beta h}. \tag{6.57}$$

The maximum value of the effective height $2h_e(\theta)$ occurs at $\theta = \pi/2$. Thus

$$2h_e = 2h_e\left(\frac{\pi}{2}\right) = \frac{2}{\beta} \frac{[1 - \cos\beta h]}{\sin\beta h}. \tag{6.58}$$

We should note that the maximum value of the effective height is denoted by $2h_e$ without any argument.

For the important special case of a half-wavelength antenna ($\beta h = \pi/2$), (6.57) and (6.58) reduce to the following simple expressions:

$$2h_e(\theta) = \frac{2}{\beta} \frac{\cos((\pi/2)\cos\theta)}{\sin\theta} \qquad (\beta h = \pi/2), \tag{6.59}$$

and

$$2h_e = 2/\beta = \lambda/\pi \qquad (\beta h = \pi/2). \tag{6.60}$$

In terms of the effective height $2h_e(\theta)$, we may rewrite the amplitude of the far-zone electric field given by (6.54) as

$$|E_\theta(r, \theta)| = \frac{\beta\eta_{0r}}{4\pi r} I(0)[2h_e(\theta)]. \tag{6.61}$$

9.7 FINITE LENGTH ANTENNA WITH ASSUMED CURRENT DISTRIBUTION IN THE FORM OF A PROGRESSIVE WAVE

The finite length filamentary antenna which we treated in the preceding section is assumed to have a current distribution in the form of a standing wave. As we discussed in detail in connection with transmission lines, for a standing wave, the amplitude of the current varies along the length of the antenna and the phase does not change linearly with the distance. For a half-wavelength antenna, the phase of the current remains constant and for longer antennas, the phase remains constant except at certain locations along the antenna, where the phase changes abruptly by π radians. These features of the current pertain to the assumed current distribution but in an actual antenna, we expect the phase of the current to vary smoothly along the antenna. In a progressive wave, the amplitude of the current is constant and the phase varies linearly with the distance along the antenna. In the case of transmission lines, if the excitation is at one end and if the terminating impedance at the other end is equal to the characteristic impedance, we obtain a progressive wave along the transmission line. In a similar manner, if a linear antenna is excited at one end and is terminated by a suitable impedance at the other, we obtain a current distribution in the form of a progressive wave along the antenna. The radiation characteristics of a finite length filamentary antenna with an assumed current distribution in the form of a progressive wave are of interest and we shall discuss them in this section.

Let us consider a filamentary antenna which is situated along the z-axis and which extends from $z = -h$ to $z = h$. Let us assume the current distribution along the filament to be of the form

$$I(z) = I_0 e^{-j\beta z}. \tag{7.1}$$

We should note that the phase-shift coefficient of the current along the antenna is the same as that of the plane TEM waves in the medium in which the antenna is situated. Since the current is in the z-direction, the vector potential also is in the

z-direction. From (6.53) and (7.1), we obtain the far-zone vector potential as

$$A_z(\mathbf{r}) = \frac{\mu e^{-j\beta r}}{4\pi r} \int_{-h}^{h} I_0 e^{-j\beta z'(1 - \cos\theta)}\, dz'. \tag{7.2}$$

We can evaluate the integral in (7.2) to yield the following result:

$$A_z(\mathbf{r}) = \frac{\mu e^{-j\beta r}}{4\pi r} I_0 \frac{2}{\beta} \frac{\sin\{\beta h(1 - \cos\theta)\}}{(1 - \cos\theta)}. \tag{7.3}$$

We obtain the magnetic field $H_\phi(\mathbf{r})$ and the electric field $E_\theta(\mathbf{r})$ in the far-zone from (6.34), (6.35), and (7.3) as

$$H_\phi(\mathbf{r}) = \frac{jI_0 e^{-j\beta r}}{2\pi r} \frac{\sin\theta \sin\{\beta h(1 - \cos\theta)\}}{(1 - \cos\theta)}, \tag{7.4}$$

and

$$E_\theta(\mathbf{r}) = \frac{j\eta_{0r} I_0 e^{-j\beta r}}{2\pi r} \frac{\sin\theta \sin\{\beta h(1 - \cos\theta)\}}{(1 - \cos\theta)}. \tag{7.5}$$

We can determine the time-averaged Poynting vector associated with the far-zone fields (7.4) and (7.5) as

$$S_r(\mathbf{r}) = \frac{|I_0|^2 \eta_{0r}}{8\pi^2 r^2} \left[\frac{\sin\theta \sin\{\beta h(1 - \cos\theta)\}}{(1 - \cos\theta)}\right]^2. \tag{7.6}$$

We find the radiation intensity $\Phi(\theta)$ of the finite length antenna with a progressive wave of current along its length from (5.35) and (7.6) to be

$$\Phi(\theta) = \frac{|I_0|^2 \eta_{0r}}{8\pi^2} \left[\frac{\sin\theta \sin\{\beta h(1 - \cos\theta)\}}{(1 - \cos\theta)}\right]^2. \tag{7.7}$$

We evaluate the total time-averaged power P radiated by the current filament by multiplying $\Phi(\theta)$ by the element of the solid angle as given by (5.35) and integrating the result to cover the entire solid angle. Therefore, we find from (7.7) that

$$\begin{aligned} P &= \int_{\phi=0}^{2\pi} \int_{\theta=0}^{\pi} \frac{|I_0|^2 \eta_{0r}}{8\pi^2} \frac{\sin^2\theta \sin^2\{\beta h(1 - \cos\theta)\}}{(1 - \cos\theta)^2} \sin\theta\, d\theta\, d\phi, \\ &= \frac{|I_0|^2 \eta_{0r}}{4\pi} \int_0^{\pi} \frac{\sin^2\theta \sin^2\{\beta h(1 - \cos\theta)\}}{(1 - \cos\theta)^2} \sin\theta\, d\theta. \end{aligned} \tag{7.8}$$

Let $1 - \cos\theta = u$. Therefore, $\sin^2\theta = u(2 - u)$ and $\sin\theta\, d\theta = du$. When we change the variable of integration in (7.8) to u, we can show that

$$\begin{aligned} P &= \frac{|I_0|^2 \eta_{0r}}{4\pi} \int_0^2 \frac{(2 - u)}{u} \sin^2(\beta h u)\, du, \\ &= \frac{|I_0|^2 \eta_{0r}}{4\pi} \int_0^2 \left(\frac{2}{u} - 1\right) \frac{(1 - \cos 2\beta h u)}{2}\, du, \\ &= \frac{|I_0|^2 \eta_{0r}}{4\pi} \left[-\frac{1}{2} \int_0^2 (1 - \cos 2\beta h u)\, du + \int_0^2 \frac{(1 - \cos 2\beta h u)}{u}\, du\right]. \end{aligned} \tag{7.9}$$

We can evaluate the first integral within the square brackets in (7.9) to yield the result $-1 + \sin(4\beta h)/4\beta h$. For the second integral, we change the variable of integration to $v = 2\beta h u$. Consequently, we can simplify (7.9) to

$$P = \frac{|I_0|^2 \eta_{0r}}{4\pi}\left[-1 + \frac{\sin 4\beta h}{4\beta h} + \int_0^{4\beta h} \frac{(1 - \cos v)}{v}\, dv\right],$$

$$= \frac{|I_0|^2 \eta_{0r}}{4\pi}\left[-1 + \frac{\sin 4\beta h}{4\beta h} + \text{Cin}\,(4\beta h)\right], \tag{7.10}$$

where the function Cin (x) is *defined* by (6.48).

We shall now consider the special case for which the total length of the antenna is equal to one half of a wavelength. We should remember, however, that in the practical cases, the length of the antenna with a progressive wave of current along it is usually greater than half of a wavelength. We choose the special case of $\beta h = \pi/2$ here only to enable the characteristics of the antenna with a progressive wave of current to be compared with those of the antenna with a standing wave of current. In the preceding section, we analyzed the latter case in detail for a half-wavelength antenna corresponding to $\beta h = \pi/2$. For $\beta h = \pi/2$, $\sin 4\beta h = 0$, and Cin $(4\beta h) =$ 2.4377. Therefore, (7.10) becomes

$$P = \frac{1}{2}|I_0|^2 \frac{\eta_{0r}}{\pi} \times (0.719). \tag{7.11}$$

Since the radiation resistance $R_r = 2P/I_0^2$, it follows from (7.11) that

$$R_r = \frac{\eta_{0r}}{\pi} \times (0.719) = 86.3 \text{ ohms}. \tag{7.12}$$

If we compare (7.12) with (6.50) we see that a half-wavelength antenna with a progressive wave of current radiates more time-averaged power than the corresponding antenna with a standing wave of current for the same value of current at the centers of the two antennas.

We find the radiation intensity Φ_{iso} of the equivalent isotropic radiator from (7.11) to be

$$\Phi_{\text{iso}} = \frac{P}{4\pi} = \frac{|I_0|^2}{8\pi^2}\eta_{0r} \times (0.719). \tag{7.13}$$

We determine the directive gain from (7.7) and (7.13) as

$$g(\theta) = 1.39\left[\frac{\sin\theta \sin\{(\pi/2)(1 - \cos\theta)\}}{(1 - \cos\theta)}\right]^2 \tag{7.14}$$

which is depicted in Fig. 9.7 as a function of θ. For $\theta = 60°$, $g(\theta) = 2.08$, which shows that the half-wavelength filamentary antenna radiates 2.08 times more time-averaged power per unit solid angle in the direction $\theta = 60°$ than the equivalent isotropic radiator. For $\theta = 120°$, $g(\theta) = 0.23$. This shows that $g(\theta)$ is not symmetrical about $\theta = \pi/2$ and we can verify this fact from (7.14). If we examine Fig. 9.7 we see that the

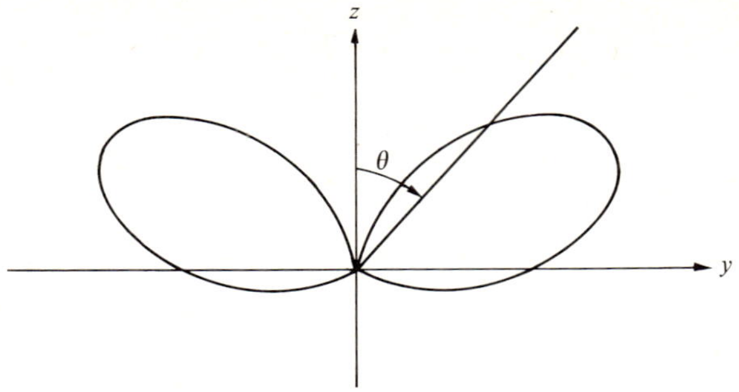

Fig. 9.7 Directive gain $g(\theta)$ of a half-wavelength filamentary antenna with a current distribution in the form of a progressive wave.

radiation intensity is generally stronger for $0 < \theta < \pi/2$, than for $\pi/2 < \theta < \pi$. In other words, the radiation of time-averaged power due to a half-wavelength antenna with a progressive wave of current tends to concentrate in the directions close to the direction of progression of the current along the antenna. We can show the directive gain $g(\theta)$ to be zero for $\theta = 0$ as well as for $\theta = 180°$.

For $\theta = 90°$, we find from (7.14) that $g(\theta) = 1.39$. Since $g(\theta)$ is greater for $\theta = 60°$ than for $\theta = 90°$, it follows that the maximum directive gain does not occur for $\theta = 90°$, as in the case of the half-wavelength antenna with a standing-wave or sinusoidal current distribution. We can find the angle $\theta = \theta_{\text{max}}$ for which the directive gain of a half-wavelength antenna with a current distribution in the form of a progressive wave is a maximum in the following manner. It follows from (7.14) that $g(\theta)$ has a maximum, if

$$f(\theta) = \frac{\sin^2\theta \sin^2\{(\pi/2)(1 - \cos\theta)\}}{(1 - \cos\theta)^2} \tag{7.15}$$

has a maximum. It is convenient for us to set $1 - \cos\theta = u$. Then, we can show that (7.15) reduces to

$$f(u) = \frac{(2 - u)\sin^2((\pi/2)u)}{u}. \tag{7.16}$$

Since θ varies from 0 to π, $\cos\theta$ varies from 1 to -1 and hence the physical range of u is from 0 to 2. The values of u outside the physical range do not yield real values of θ and we need not therefore consider them. From (7.16), we can obtain $(d/du)f(u) = 0$ explicitly with the following result:

$$-\frac{2\sin((\pi/2)u)}{u^2}\left[\sin\left(\frac{\pi}{2}u\right) - \frac{\pi u}{2}(2 - u)\cos\left(\frac{\pi}{2}u\right)\right] = 0. \tag{7.17}$$

The values of u which satisfy (7.17) correspond to the extrema of $f(u)$. For $u = 2$, $\sin((\pi/2)u)$ as well as the term within the square brackets in (7.17) vanish. However, we can show $u = 2$ to correspond to the minimum of $f(u)$; indeed, we see from (7.16) that $f(u) = 0$ for this value of u. We should note that $u = 0$ is not a solution of

(7.17). Consequently, the equation obtained by setting the term within the square brackets in (7.17) equal to zero contains the value of $u = u_{max}$ for which $f(u)$ is a maximum. Therefore, it follows that

$$\tan\left(\frac{\pi}{2} u_{max}\right) = \frac{\pi}{2} u_{max} (2 - u_{max}). \tag{7.18}$$

As we mentioned before, the values of 0 and 2 for u_{max} have to be ruled out since they do not correspond to the maxima of $f(u)$. For $0 < u < 2$, the right side of (7.18) is positive and, therefore, the left side of (7.18) also has to be positive in order that (7.18) be satisfied. Since $\tan \alpha > 0$ for $(n - 1)\pi < \alpha < (n - \frac{1}{2})\pi$ where n is an integer and since it is required that $0 < u_{max} < 2$, it follows that $n = 1$ and hence the solution u_{max} is in the range $0 < u_{max} < 1$. We can find the actual value u_{max} which satisfies (7.18) with the help of the following graphical procedure. The functions $\tan\left((\pi/2)u\right)$ and $(\pi/2)u(2 - u)$ are plotted against u, as shown in Fig. 9.8, for $0 < u < 1$. The value u corresponding to the point of intersection of the graphs of $\tan\left((\pi/2)u\right)$ and $(\pi/2)u(2 - u)$ satisfies (7.18) and hence is the required value of u_{max}. From

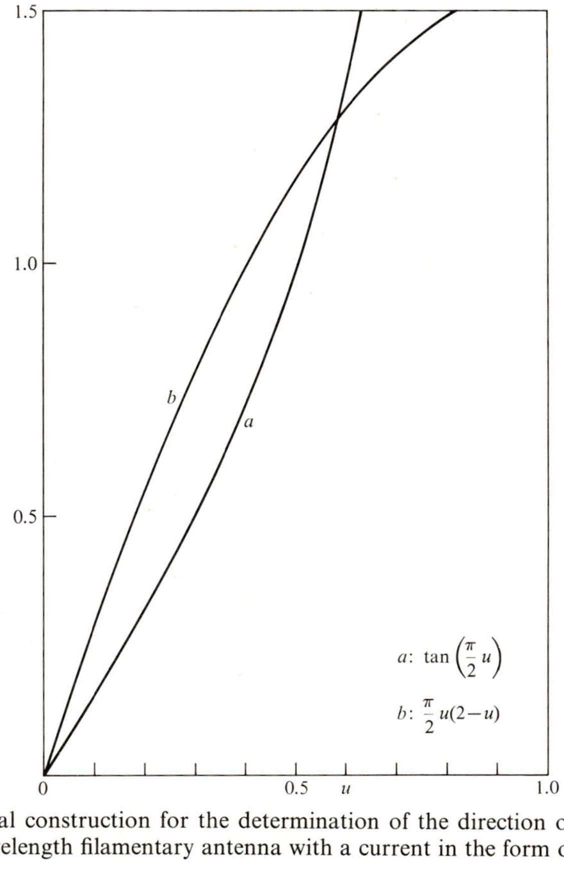

Fig. 9.8 Graphical construction for the determination of the direction of maximum directive gain of a half-wavelength filamentary antenna with a current in the form of a progressive wave.

Fig. 9.8, we find that $u_{max} = 0.582$, and therefore $\theta_{max} = 65.3°$. We can obtain from (7.14) that $g(\theta_{max}) = 2.13$. Therefore, the gain of a half-wavelength antenna with a current distribution in the form of a progressive wave is equal to 2.13 and this maximum directivity corresponds to the direction given by $\theta = \theta_{max} = 65.3°$. If we compare this result with (6.52) we see that the gain of a half-wavelength linear antenna is greater when the current distribution is in the form of a progressive wave than when it is in the form of a standing wave.

9.8 EFFECT OF A PERFECTLY CONDUCTING PLANE ON THE RADIATION FROM ANTENNAS

We discussed the radiation characteristics of some simple antennas in the preceding sections on the assumption that these antennas were isolated to be alone in the medium. However, frequently in practice the antennas operate in the neighborhood of reflecting surfaces such as, for example, the antennas erected over the earth. Since the earth is a good conductor, the study of the effect of a perfectly conducting plane on the radiation characteristics of antennas situated in its neighborhood provides some understanding of the radiation characteristics of antennas situated over the earth. We can show that the effect of a perfectly conducting plane is equivalent to that of an "image" antenna whose radiation augments that due to the original antenna. In this section, we shall study the imaging properties of an infinite perfectly conducting plane.

1. Perpendicular Orientation

For this purpose, it is convenient for us to begin by reexamining the properties of an elemental antenna using the cylindrical coordinates. Let us consider an elemental antenna of current moment $I_1\,dl$ situated on the z-axis at $z = z'$ and oriented parallel to the z-axis. The vector potential due to this elemental antenna is in the z-direction. With the help of (5.3) and (6.5), we obtain the vector potential due to this elemental antenna as

$$A_{1z}(\rho, z; z') = \frac{\mu I_1\,dl e^{-j\beta R(z,z')}}{4\pi R(z, z')}, \tag{8.1}$$

where

$$R(z, z') = \sqrt{(z - z')^2 + \rho^2}. \tag{8.2}$$

We should note that ρ, 0, and z are the coordinates of the point P at which the vector potential is observed. Since the vector potential has only a z-component and since $\partial/\partial\phi = 0$, the magnetic field has only a ϕ-component and is given by

$$H_{1\phi}(\rho, z; z') = -\frac{1}{\mu}\frac{\partial}{\partial\rho} A_{1z}(\rho, z; z'). \tag{8.3}$$

Together with (8.1), (8.3) yields

$$H_{1\phi}(\rho, z; z') = -\frac{I_1\,dl}{4\pi}\frac{\partial}{\partial\rho}\left[\frac{e^{-j\beta R(z,z')}}{R(z, z')}\right]. \tag{8.4}$$

From the second of Maxwell's equations and (8.4) we can see that the electric field has only ρ- and z-components. We obtain the ρ-component of the electric field as

$$E_{1\rho}(\rho, z; z') = -\frac{1}{j\omega\varepsilon}\frac{\partial}{\partial z}H_{1\phi}(\rho, z; z') = \frac{1}{j\omega\varepsilon}\frac{I_1\,dl}{4\pi}\frac{\partial^2}{\partial z\,\partial\rho}\left[\frac{e^{-j\beta R(z,z')}}{R(z, z')}\right]. \tag{8.5}$$

Since z and z' enter in $R(z, z')$ only in the form $z - z'$, we find that in (8.5) $\partial/\partial z = -\partial/\partial z'$ and hence (8.5) becomes

$$E_{1\rho}(\rho, z; z') = -\frac{1}{j\omega\varepsilon}\frac{I_1\,dl}{4\pi}\frac{\partial^2}{\partial\rho\,\partial z'}\left[\frac{e^{-j\beta R(z,z')}}{R(z, z')}\right]. \tag{8.6}$$

In a similar manner, we can evaluate the z-component of the electric field as

$$\begin{aligned} E_{1z}(\rho, z; z') &= \frac{1}{j\omega\varepsilon}\frac{1}{\rho}\frac{\partial}{\partial\rho}[\rho H_{1\phi}(\rho, z; z')], \\ &= -\frac{1}{j\omega\varepsilon}\frac{I_1\,dl}{4\pi}\frac{1}{\rho}\frac{\partial}{\partial\rho}\left[\rho\frac{\partial}{\partial\rho}\left\{\frac{e^{-j\beta R(z,z')}}{R(z, z')}\right\}\right]. \end{aligned} \tag{8.7}$$

Let us consider another elemental antenna of current moment $I_2\,dl$ situated on the z-axis at $z = -z'$ and oriented parallel to the z-axis. The vector potential due to the second current element at the same observation point $P(\rho, 0, z)$ is also in the z-direction and we find it to be given by

$$A_{2z}(\rho, z; z') = \frac{\mu I_2\,dl}{4\pi}\frac{e^{-j\beta R(z,-z')}}{R(z, -z')}. \tag{8.8}$$

The associated magnetic field has only a ϕ-component and we determine this as

$$H_{2\phi}(\rho, z; z') = -\frac{1}{\mu}\frac{\partial}{\partial\rho}A_{2z}(\rho, z; z') = -\frac{I_2\,dl}{4\pi}\frac{\partial}{\partial\rho}\left[\frac{e^{-j\beta R(z,-z')}}{R(z, -z')}\right]. \tag{8.9}$$

As before, the electric field due to the second current element also has only ρ- and z-components. We obtain the ρ-component of the electric field as

$$E_{2\rho}(\rho, z; z') = -\frac{1}{j\omega\varepsilon}\frac{\partial}{\partial z}H_{2\phi}(\rho, z; z') = \frac{1}{j\omega\varepsilon}\frac{I_2\,dl}{4\pi}\frac{\partial^2}{\partial z\,\partial\rho}\left[\frac{e^{-j\beta R(z,-z')}}{R(z, -z')}\right]. \tag{8.10}$$

Since z and z' enter in $R(z, -z')$ only in the form $z + z'$, it follows that in (8.10) $\partial/\partial z = \partial/\partial z'$ and hence we may rewrite (8.10) as

$$E_{2\rho}(\rho, z; z') = \frac{1}{j\omega\varepsilon}\frac{I_2\,dl}{4\pi}\frac{\partial^2}{\partial\rho\,\partial z'}\left[\frac{e^{-j\beta R(z,-z')}}{R(z, -z')}\right]. \tag{8.11}$$

Also, we can determine the z-component of the electric field as

$$\begin{aligned} E_{2z}(\rho, z; z') &= \frac{1}{j\omega\varepsilon}\frac{1}{\rho}\frac{\partial}{\partial\rho}[\rho H_{2\phi}(\rho, z; z')], \\ &= -\frac{1}{j\omega\varepsilon}\frac{I_2\,dl}{4\pi}\frac{1}{\rho}\frac{\partial}{\partial\rho}\left[\rho\frac{\partial}{\partial\rho}\left\{\frac{e^{-j\beta R(z,-z')}}{R(z, -z')}\right\}\right]. \end{aligned} \tag{8.12}$$

Let us suppose that the two current elements are present simultaneously. Since Maxwell's equations are linear, we can apply the principle of superposition. Hence, we obtain the resulting potentials and fields by the superposition of those due to the separate current elements. If the moments of the two current elements are equal, that is, $I_1\,dl = I_2\,dl = I_0\,dl$, then from (8.1), (8.4), (8.6) to (8.9), (8.11), and (8.12), we find the following expressions for the potentials and the fields due to both the current elements:

$$A_z(\rho, z; z') = \frac{\mu I_0\,dl}{4\pi}\left[\frac{e^{-j\beta R(z,z')}}{R(z, z')} + \frac{e^{-j\beta R(z,-z')}}{R(z, -z')}\right], \tag{8.13}$$

$$H_\phi(\rho, z; z') = -\frac{I_0\,dl}{4\pi}\frac{\partial}{\partial\rho}\left[\frac{e^{-j\beta R(z,z')}}{R(z, z')} + \frac{e^{-j\beta R(z,-z')}}{R(z, -z')}\right], \tag{8.14}$$

$$E_\rho(\rho, z; z') = -\frac{1}{j\omega\varepsilon}\frac{I_0\,dl}{4\pi}\frac{\partial^2}{\partial\rho\,\partial z'}\left[\frac{e^{-j\beta R(z,z')}}{R(z, z')} - \frac{e^{-j\beta R(z,-z')}}{R(z, -z')}\right], \tag{8.15}$$

$$E_z(\rho, z; z') = -\frac{1}{j\omega\varepsilon}\frac{I_0\,dl}{4\pi}\frac{1}{\rho}\frac{\partial}{\partial\rho}\left[\rho\frac{\partial}{\partial\rho}\left[\frac{e^{-j\beta R(z,z')}}{R(z, z')} + \frac{e^{-j\beta R(z,-z')}}{R(z, -z')}\right]\right]. \tag{8.16}$$

It is seen from (8.2) that $R(-z, z') = R(z, -z')$ and $R(-z, -z') = R(z, z')$. It follows therefore that the change from z to $-z$ leaves the terms inside the square brackets of (8.13), (8.14), and (8.16) unchanged but changes the sign of the term inside the square brackets of (8.15). Consequently, we obtain the following important results:

$$A_z(\rho, z; z') = A_z(\rho, -z; z'); \qquad H_\phi(\rho, z; z') = H_\phi(\rho, -z; z'); \qquad E_z(\rho, z; z') = E_z(\rho, -z; z'); \tag{8.17}$$

and

$$E_\rho(\rho, z; z') = -E_\rho(\rho, -z; z'). \tag{8.18}$$

From (8.17), we see that $A_z(\rho, z; z')$, $H_\phi(\rho, z; z')$, and $E_z(\rho, z; z')$ are *even* functions of z and from (8.18) we see that $E_\rho(\rho, z; z')$ is an *odd* function of z. Also, we find from (8.15) that $E_\rho(\rho, 0; z') = 0$.

In the entire $z = 0$ plane, the only existing tangential electric field component, $E_\rho(\rho, 0; z')$ is identically zero. Consequently, we can introduce a perfectly conducting plane surface to occupy the plane $z = 0$. The required boundary condition on the perfectly conducting plane is automatically satisfied by the fields given by (8.14) to (8.16). Hence, the introduction of the perfectly conducting surface in the plane $z = 0$ does not alter the fields given by (8.14) to (8.16). The electromagnetic fields cannot penetrate a perfect conductor. Therefore, the electromagnetic fields in the half-spaces $z > 0$ and $z < 0$ are due only to the first and the second current elements respectively. The second current element can now be removed without affecting the fields in the half-space $z > 0$. Thus, in the half-space $z > 0$ terminated at $z = 0$ by an infinite perfectly conducting plane, the fields due to an elementary dipole situated in that half-space are equal to the fields radiated by the given dipole together with those due to an "image" antenna. The "image" antenna is situated in the other half-space $z < 0$ and is symmetrical with respect to the original antenna, as shown in Fig. 9.9. The position of the "image" antenna is at the mirror image of the location of the actual antenna. For the antenna orientation perpendicular to the perfectly conducting plane such as we have considered in this discussion, the "image" antenna has the same perpendicular orientation and the same current moment as the original antenna.

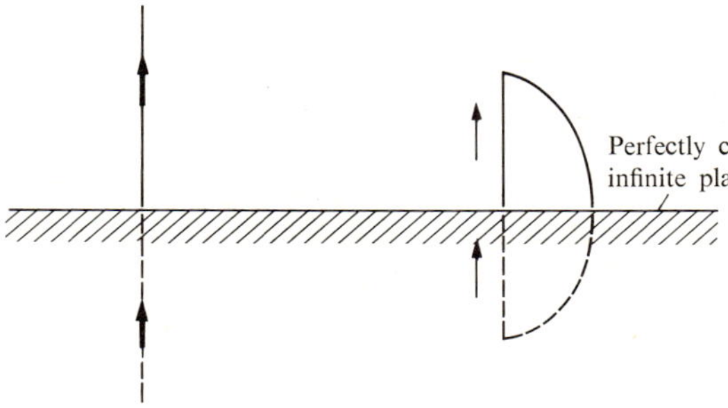

Fig. 9.9 Images of a point dipole and a finite length antenna oriented perpendicular to a perfectly conducting, infinite plane.

Let us consider a quarter-wavelength "monopole" antenna which is situated along the z-axis and which extends from $z = 0$ to $z = h = \lambda/4$, where $z = 0$ is a perfectly conducting surface. Let us assume the current distribution along the antenna, for the sake of definiteness, to be given by $I(z) = I_0 \cos \beta z$ for $0 < z < h$. We should note that the monopole antenna is perpendicular to the perfectly conducting plane. We wish to obtain the electromagnetic fields radiated in the half-space $z > 0$ by the quarter-wavelength monopole antenna situated vertically over but electrically insulated from the perfectly conducting plane. We can decompose the quarter-wavelength monopole antenna into a number of vertical current elements of appropriate current moments and obtain the images of these current elements by using the previously deduced rules for imaging the vertically oriented current elements. When we add all the "image" current elements, another quarter-wavelength monopole antenna situated along the z-axis from $z = -h$ to $z = 0$ results. The current distribution along the "image" monopole antenna is the mirror image of that along the actual monopole antenna, as depicted in Fig. 9.9. Thus, we see that the current distribution along the "image" monopole antenna is given by $I(z) = I_0 \cos \{\beta(-z)\} = I_0 \cos \beta z$ for $-h < z < 0$. We can ascertain that the actual quarter-wavelength monopole antenna together with its image is precisely the half-wavelength "dipole" antenna extending from $z = -h$ to $z = h$ and having the current distribution given by $I(z) = I_0 \cos \beta z$ from $z = -h$ to $z = h$. Since the electromagnetic fields in the half-space $z > 0$ due to the quarter-wavelength monopole antenna situated vertically over the perfectly conducting plane are equal to the fields radiated by the actual monopole antenna as well as those due to the "image" monopole antenna, we find that the electromagnetic fields in the half-space $z > 0$ are the same as those due to the half-wavelength "dipole" antenna extending from $z = -h$ to $z = h$. We should remember, however, that the image monopole antenna gives rise to fields *only* in the half-space $z > 0$. There are no electromagnetic fields in the half-space $z < 0$, since the perfectly conducting plane at $z = 0$ prevents the fields from penetrating into the half-space $z < 0$. We should emphasize that the introduction of the "image" antenna is a convenient operational technique for obtaining the electromagnetic fields due to the actual antenna situated over a perfectly conducting plane but, as is to be expected, the "image" antenna does not give rise to electromagnetic fields in the "image" half-space ($z < 0$).

Even though a quarter-wavelength monopole antenna above a perfectly conducting plane and the corresponding half-wavelength dipole antenna without the perfectly conducting plane give rise to the same electromagnetic fields in the half-space $z > 0$, there are certain differences between these two antennas. The quarter-wavelength monopole antenna above a perfectly conducting plane radiates only in one of the half-spaces whereas the half-wavelength dipole antenna without the perfectly conducting plane radiates in both the half-spaces. In view of the symmetry, the total time-averaged power radiated by the half-wavelength dipole antenna in one of the half-spaces is one half of the total time-averaged power radiated by the antenna. It follows therefore that a quarter-wavelength monopole antenna above a perfectly conducting plane radiates only half as much power as that radiated by the corresponding half-wavelength dipole antenna. We emphasize that the input current is the same for both the antennas. Consequently, the radiation or the input resistance of a quarter-wavelength monopole antenna above a perfectly conducting plane is one half of that of the corresponding half-wavelength dipole antenna without the perfectly conducting plane. For the assumed current distribution of the form $I(z) = I_0 \cos \beta z$, we can see from (6.50) that the radiation resistance of a quarter-wavelength monopole antenna above a perfectly conducting plane is equal to approximately 36.5 ohms.

2. Parallel Orientation

We have deduced the rule for obtaining the image of a current element oriented perpendicular to a perfectly conducting plane. The rule for obtaining the image of a current element whose orientation is parallel to the perfectly conducting plane remains to be derived. For this purpose, it is convenient for us to introduce the cylindrical coordinates ρ, ϕ, and z, as well as the rectangular coordinates x, y, and z with the same z-axis. Let us consider a current element of moment $I_1\, dl$ situated on the z-axis at $z = z'$ and oriented parallel to the y-axis. We can therefore see the vector current element from Fig. 9.10 to be

$$\boldsymbol{I}_1\, dl = \hat{y} I_1\, dl = (\hat{\rho} \sin \phi + \hat{\phi} \cos \phi) I_1\, dl. \tag{8.19}$$

The vector potential due to this current element is in the same direction as the current

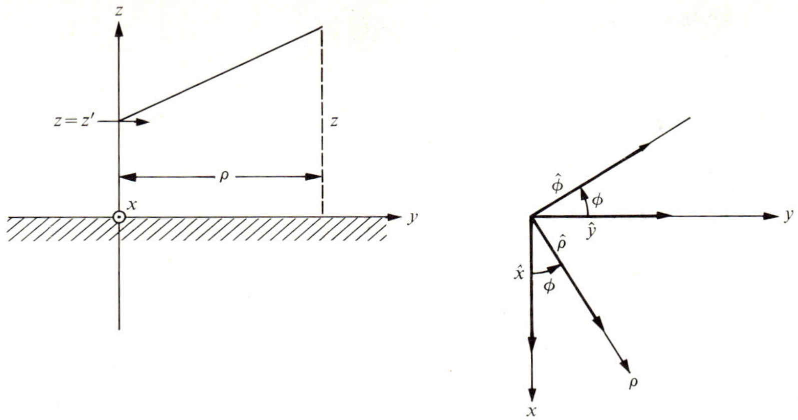

Fig. 9.10 Geometry for the determination of the image of a point dipole oriented parallel to a perfectly conducting, infinite plane.

element and we obtain its nonvanishing components as

$$A_{1\rho}(\rho, \phi, z; z') = \frac{\mu I_1 \, dl \sin\phi}{4\pi} \frac{e^{-j\beta R(z,z')}}{R(z, z')}, \tag{8.20}$$

and

$$A_{1\phi}(\rho, \phi, z; z') = \frac{\mu I_1 \, dl \cos\phi}{4\pi} \frac{e^{-j\beta R(z,z')}}{R(z, z')}, \tag{8.21}$$

where $R(z, z')$ is given by (8.2). The z-component of the vector potential is zero. We observe the vector potential is at $P(\rho, \phi, z)$. We obtain the components of the magnetic field from (4.4), (8.20), and (8.21) as

$$\begin{aligned} H_{1\rho}(\rho, \phi, z; z') &= -\frac{1}{\mu} \frac{\partial}{\partial z} A_{1\phi}(\rho, \phi, z; z'), \\ &= \frac{I_1 \, dl \cos\phi}{4\pi} \frac{\partial}{\partial z'} \left\{ \frac{e^{-j\beta R(z,z')}}{R(z, z')} \right\}, \end{aligned} \tag{8.22}$$

$$\begin{aligned} H_{1\phi}(\rho, \phi, z; z') &= \frac{1}{\mu} \frac{\partial}{\partial z} A_{1\rho}(\rho, \phi, z; z'), \\ &= -\frac{I_1 \, dl \sin\phi}{4\pi} \frac{\partial}{\partial z'} \left\{ \frac{e^{-j\beta R(z,z')}}{R(z, z')} \right\}, \end{aligned} \tag{8.23}$$

$$\begin{aligned} H_{1z}(\rho, \phi, z; z') &= \frac{1}{\mu\rho} \left[\frac{\partial}{\partial \rho} (\rho A_{1\phi}(\rho, \phi, z; z')) - \frac{\partial}{\partial \phi} A_{1\rho}(\rho, \phi, z; z') \right], \\ &= \frac{I_1 \, dl}{4\pi\rho} \cos\phi \left[\frac{\partial}{\partial \rho} \left\{ \rho \frac{e^{-j\beta R(z,z')}}{R(z, z')} \right\} - \frac{e^{-j\beta R(z,z')}}{R(z, z')} \right]. \end{aligned} \tag{8.24}$$

From the second of Maxwell's equations and (8.22) to (8.24) we obtain the following expressions for the components of the electric field:

$$E_{1\rho}(\rho, \phi, z; z') = \frac{1}{j\omega\varepsilon}\left\{\frac{1}{\rho}\frac{\partial}{\partial\phi} H_{1z}(\rho, \phi, z; z') - \frac{\partial}{\partial z} H_{1\phi}(\rho, \phi, z; z')\right\},$$
$$= -\frac{I_1\, dl}{j\omega\varepsilon 4\pi}\sin\phi\left[\frac{1}{\rho^2}\frac{\partial}{\partial\rho}\left(\rho\frac{e^{-j\beta R(z,z')}}{R(z, z')}\right) - \frac{e^{-j\beta R(z,z')}}{\rho^2 R(z, z')} + \frac{\partial^2}{\partial z'^2}\left(\frac{e^{-j\beta R(z,z')}}{R(z, z')}\right)\right], \tag{8.25}$$

$$E_{1\phi}(\rho, \phi, z; z') = \frac{1}{j\omega\varepsilon}\left\{\frac{\partial}{\partial z} H_{1\rho}(\rho, \phi, z; z') - \frac{\partial}{\partial\rho} H_{1z}(\rho, \phi, z; z')\right\},$$
$$= -\frac{I_1\, dl\cos\phi}{j\omega\varepsilon 4\pi}\left[\frac{\partial^2}{\partial z'^2}\left(\frac{e^{-j\beta R(z,z')}}{R(z, z')}\right) + \frac{\partial}{\partial\rho}\left\{\frac{1}{\rho}\frac{\partial}{\partial\rho}\left(\rho\frac{e^{-j\beta R(z,z')}}{R(z, z')}\right) - \frac{e^{-j\beta R(z,z')}}{\rho R(z, z')}\right\}\right], \tag{8.26}$$

$$E_{1z}(\rho, \phi, z; z') = \frac{1}{j\omega\varepsilon\rho}\left\{\frac{\partial}{\partial\rho}\{\rho H_{1\phi}(\rho, \phi, z; z')\} - \frac{\partial}{\partial\phi} H_{1\rho}(\rho, \phi, z; z')\right\},$$
$$= -\frac{I_1\, dl\sin\phi}{j\omega\varepsilon 4\pi\rho}\left[\frac{\partial}{\partial\rho}\left\{\rho\frac{\partial}{\partial z'}\frac{e^{-j\beta R(z,z')}}{R(z, z')}\right\} - \frac{\partial}{\partial z'}\left(\frac{e^{-j\beta R(z,z')}}{R(z, z')}\right)\right]. \tag{8.27}$$

Let us consider another current element of moment $I_2\, dl$ situated on the z-axis at $z = -z'$ and oriented parallel to the y-direction. We obtain the vector potential and the fields due to the second current element from (8.20) to (8.27) by changing z' to $-z'$ and the subscripts from 1 to 2.

Let us suppose that the two current elements are present simultaneously. As before, we can apply the principle of superposition. Let the moments of the two current elements be equal in magnitude and opposite in sign, that is, $I_1\, dl = -I_2\, dl = I_0\, dl$. This is equivalent to the statement that the moments of the two current elements are equal but the first current element is oriented in the $+y$-direction and the second current element is in the $-y$-direction. With the help of (8.20) to (8.27), we can evaluate the vector potential and the fields due to the simultaneous presence of the two current elements to be given by the following expressions:

$$A_\rho(\rho, \phi, z; z') = \frac{\mu I_0\, dl}{4\pi}\sin\phi\; G_{\text{odd}}(z, z'), \tag{8.28}$$

$$A_\phi(\rho, \phi, z; z') = \frac{\mu I_0\, dl}{4\pi}\cos\phi\; G_{\text{odd}}(z, z'), \tag{8.29}$$

$$H_\rho(\rho, \phi, z; z') = \frac{I_0\, dl \cos\phi}{4\pi} \frac{\partial}{\partial z'} G_{\text{even}}(z, z'), \tag{8.30}$$

$$H_\phi(\rho, \phi, z; z') = -\frac{I_0\, dl \sin\phi}{4\pi} \frac{\partial}{\partial z'} G_{\text{even}}(z, z'), \tag{8.31}$$

$$H_z(\rho, \phi, z; z') = \frac{I_0\, dl \cos\phi}{4\pi\rho}\left[\frac{\partial}{\partial\rho}\{\rho G_{\text{odd}}(z, z')\} - G_{\text{odd}}(z, z')\right], \tag{8.32}$$

$$E_\rho(\rho, \phi, z; z') = -\frac{I_0\, dl \sin\phi}{j\omega\varepsilon 4\pi}\left[\frac{1}{\rho^2}\frac{\partial}{\partial\rho}(\rho G_{\text{odd}}(z, z')) - \frac{1}{\rho^2} G_{\text{odd}}(z, z') + \frac{\partial^2}{\partial z'^2} G_{\text{odd}}(z, z')\right], \tag{8.33}$$

$$E_\phi(\rho, \phi, z; z') = -\frac{I_0\, dl \cos\phi}{j\omega\varepsilon 4\pi}\left[\frac{\partial^2}{\partial z'^2} G_{\text{odd}}(z, z') + \frac{\partial}{\partial\rho}\left\{\frac{1}{\rho}\frac{\partial}{\partial\rho}(\rho G_{\text{odd}}(z, z')) - \frac{1}{\rho} G_{\text{odd}}(z, z')\right\}\right], \tag{8.34}$$

$$E_z(\rho, \phi, z; z') = -\frac{I_0\, dl \sin\phi}{j\omega\varepsilon 4\pi\rho}\left[\frac{\partial}{\partial\rho}\left\{\rho \frac{\partial}{\partial z'} G_{\text{even}}(z, z')\right\} - \frac{\partial}{\partial z'} G_{\text{even}}(z, z')\right], \tag{8.35}$$

where

$$G_{\text{odd}}(z, z') = \frac{e^{-j\beta R(z,z')}}{R(z, z')} - \frac{e^{-j\beta R(z,-z')}}{R(z, -z')}, \tag{8.36}$$

and

$$G_{\text{even}}(z, z') = \frac{e^{-j\beta R(z,z')}}{R(z, z')} + \frac{e^{-j\beta R(z,-z')}}{R(z, -z')}. \tag{8.37}$$

With the help of (8.2), we can verify that $G_{\text{odd}}(z, z')$ and $G_{\text{even}}(z, z')$ are respectively odd and even functions of z. Hence we obtain the following important symmetry relations:

Even Functions of z:

$$H_\rho(\rho, \phi, z; z'); \qquad H_\phi(\rho, \phi, z; z'); \qquad E_z(\rho, \phi, z; z'). \tag{8.38}$$

Odd Functions of z:

$$A_\rho(\rho, \phi, z; z'); \qquad A_\phi(\rho, \phi, z; z'); \tag{8.39a}$$

$$E_\rho(\rho, \phi, z; z'); \qquad E_\phi(\rho, \phi, z; z'); \qquad H_z(\rho, \phi, z; z'). \tag{8.39b}$$

Also, since $G_{\text{odd}}(0, z') = 0$, it follows that

$$E_\rho(\rho, \phi, 0; z') = E_\phi(\rho, \phi, 0; z') = H_z(\rho, \phi, 0; z') = 0.$$

In the entire $z = 0$ plane, the tangential electric field components $E_\rho(\rho, \phi, 0; z')$ and $E_\phi(\rho, \phi, 0; z')$ are identically zero. Consequently, we can introduce a perfectly conducting plane surface to occupy the plane $z = 0$ without disturbing the electromagnetic fields as given by (8.30) to (8.35). Moreover, since the electromagnetic fields cannot penetrate a perfect conductor, the fields in the half-spaces $z > 0$ and $z < 0$ are due only to the first and the second current elements respectively. As before, we can now remove the second current element without affecting the fields in the half-space $z > 0$. Therefore, in the half-space $z > 0$ terminated at $z = 0$ by an infinite perfectly conducting plane, the fields due to an elementary dipole situated in that half-space are equal to those radiated by the given dipole together with those due to an "image" antenna. The position of the "image" dipole is at the mirror image of the location of the actual dipole. For the dipole orientation *parallel* to the perfectly conducting plane, as shown in Fig. 9.11, the "image" dipole is also parallel to the perfectly conducting plane, has the same current moment as the original dipole, but is oriented in a direction opposite to that of the actual dipole.

We used the rule for obtaining the image of a point dipole oriented perpendicular to the perfectly conducting plane in determining the image of a finite length antenna situated perpendicular to the perfectly conducting plane. In the same manner, we use the rule for obtaining the image of a point dipole oriented parallel to the perfectly conducting plane in determining the image of a finite length antenna situated parallel to the perfectly conducting plane. We can obtain the image of a half-wavelength antenna whose position corresponds to $x = 0$, $-\lambda/4 < y < \lambda/4$ and $z = d$ and which has the current distribution as specified by $I(y) = I_0 \cos \beta y$ with the help of the rule for the determination of the image of a point dipole with parallel orientation. The image is depicted in Fig. 9.11.

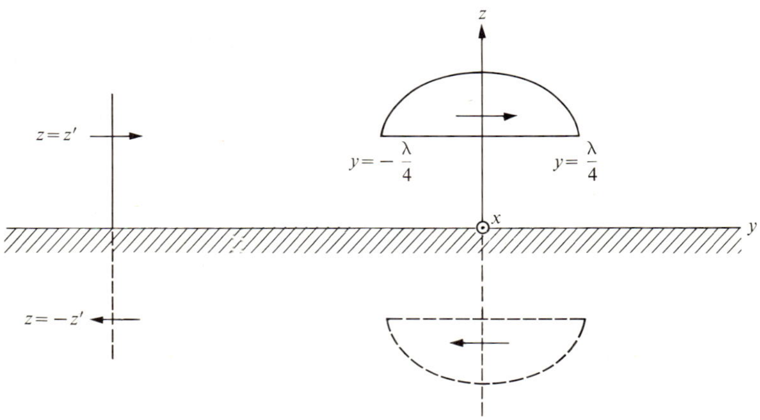

Fig. 9.11 Images of a point dipole and a finite length antenna oriented parallel to a perfectly conducting, infinite plane.

We can also obtain the image of an arbitrarily oriented point dipole in the following manner. We decompose the arbitrarily oriented dipole into two components, one of which is parallel and the other perpendicular to the perfectly conducting plane. We deduce the images of the parallel and the perpendicular components of the actual dipole in accordance with the rules, as discussed previously, for obtaining

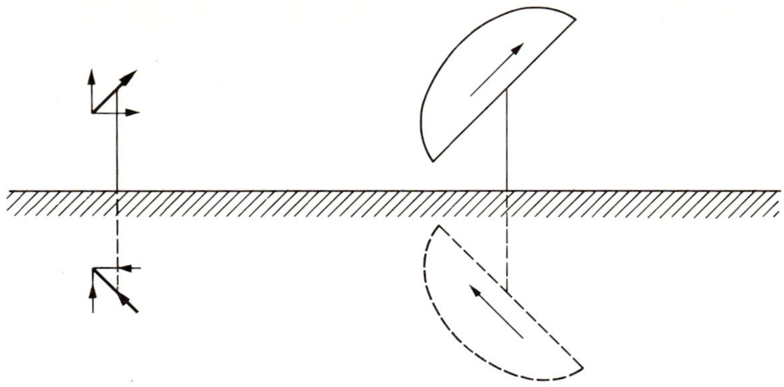

Fig. 9.12 Images of a point dipole and a finite length antenna oriented arbitrarily with respect to a perfectly conducting, infinite plane.

the images for these special orientations of the point dipole. We add the constituent images vectorially, as indicated in Fig. 9.12, to find the image of an arbitrarily oriented dipole. With the help of this procedure for obtaining the image of an arbitrarily oriented dipole, we can also determine the image of an arbitrarily oriented finite length antenna, as is depicted in Fig. 9.12.

The foregoing development of the theory of images is idealized in the sense that it is based on two assumptions which are fulfilled only approximately in practice. The first assumption is that the reflecting plane is made of a *perfect conductor* and the second is that it is *infinite*. Since the earth is a good conductor and not a perfect conductor, we can predict the characteristics of antennas situated over the earth only approximately by the use of the image theory. In the same manner, the reflectors used for the antennas, are finite in extent. If the reflector is a plane and has dimensions much larger than the length of the antenna and the wavelength, we can use the image theory to predict approximately the characteristics of an antenna backed by a plane reflector.

9.9 A SMALL CIRCULAR LOOP ANTENNA

1. Vector Potential

The existence of duality between the radiation characteristics of a small circular loop antenna with a constant current distribution around its circumference and those due to a current element, makes the study of the radiation characteristics of a small circular loop antenna quite interesting. Let us consider a filamentary loop antenna of radius a, with its center coinciding with the origin of the coordinate system. It is convenient for us to use both the rectangular coordinates x, y, and z, and the spherical coordinates r, θ, and ϕ, as shown in Fig. 9.13(a). We assume the axis of the loop antenna to coincide with the z-axis. Let I_0 be the constant current distribution around the circumference of the loop. Since there is cylindrical symmetry, the vector

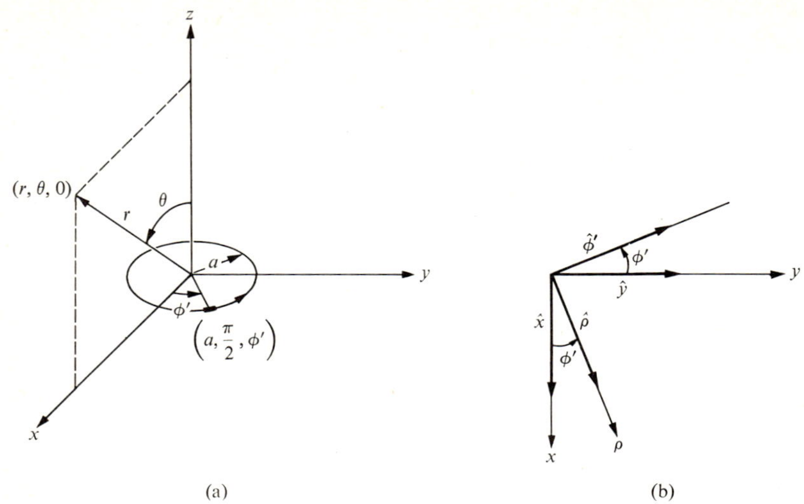

Fig. 9.13 Geometry of a filamentary loop antenna of radius a, indicating the source $(a, \pi/2, \phi')$ and the observation $(r, \theta, 0)$ points.

potential due to the current loop is independent of ϕ. Therefore, without loss of generality, we can evaluate the vector potential at a point P corresponding to $\phi = 0$. The coordinates of the observation point P which we assume to lie in the zx-plane are r, θ, and 0. Let us consider an infinitesimal length of the filamentary loop lying between the angles ϕ' and $\phi' + d\phi'$. The length of the current element is $ad\phi'$. Let $\hat{\phi}'$ be the direction of the current element. Then, from Fig. 9.13(b), we see that

$$\hat{\phi}' = -\hat{x} \sin \phi' + \hat{y} \cos \phi'. \tag{9.1}$$

Since for $\phi = 0$, $\hat{y} = \hat{\phi}$, we may also write (9.1) as

$$\hat{\phi}' = -\hat{x} \sin \phi' + \hat{\phi} \cos \phi'. \tag{9.2}$$

Let the location of the current element be denoted by P'. The coordinates of P' are a, $\pi/2$, and ϕ'. Let $R(\phi')$ be the distance between the source point P' and the observation point P. We obtain the contribution $d\boldsymbol{A}(\boldsymbol{r})$ to the vector potential at P due to the current element at P' as

$$d\boldsymbol{A}(\boldsymbol{r}) = \hat{\phi}' \frac{\mu I_0 a \, d\phi'}{4\pi R(\phi')} e^{-j\beta R(\phi')}. \tag{9.3}$$

We should note that $d\boldsymbol{A}(\boldsymbol{r})$ is in the same direction as the current element at P'. We obtain the vector potential $\boldsymbol{A}(\boldsymbol{r})$ at P due to the entire filamentary current loop by integrating (9.3) with respect to ϕ' from $\phi' = 0$ to $\phi' = 2\pi$ and the result is

$$\boldsymbol{A}(\boldsymbol{r}) = \frac{\mu I_0 a}{4\pi} \int_{\phi'=0}^{2\pi} \hat{\phi}' \frac{e^{-j\beta R(\phi')}}{R(\phi')} \, d\phi'. \tag{9.4}$$

It is important for us to observe from (9.2) that $\hat{\phi}'$ is a function of ϕ' and hence should be retained inside the integral sign in (9.4). With the help of (9.2), it is convenient for us to rewrite (9.4) as follows:

$$\boldsymbol{A}(\boldsymbol{r}) = \hat{x}A_x(\boldsymbol{r}) + \hat{\phi}A_\phi(\boldsymbol{r}), \tag{9.5}$$

where

$$A_x(\boldsymbol{r}) = -\frac{\mu I_0 a}{4\pi}\int_{\phi'=0}^{2\pi} d\phi' \sin\phi' \frac{e^{-j\beta R(\phi')}}{R(\phi')}, \tag{9.6}$$

and

$$A_\phi(\boldsymbol{r}) = \frac{\mu I_0 a}{4\pi}\int_{\phi'=0}^{2\pi} d\phi' \cos\phi' \frac{e^{-j\beta R(\phi')}}{R(\phi')}. \tag{9.7}$$

An explicit expression is now needed for the distance $R(\phi')$. From Fig. 9.13(a), we see the rectangular coordinates of the observation point P to be $x = r\sin\theta$, $y = 0$, and $z = r\cos\theta$. In a similar manner, we find the rectangular coordinates of the source point P' to be $x' = a\cos\phi'$, $y' = a\sin\phi'$, and $z' = 0$. We then deduce the distance between P and P' as

$$\begin{aligned} R(\phi') &= [(x - x')^2 + (y - y')^2 + (z - z')^2]^{1/2}, \\ &= [(r\sin\theta - a\cos\phi')^2 + a^2\sin^2\phi' + r^2\cos^2\theta]^{1/2}, \\ &= [r^2 + a^2 - 2ar\sin\theta\cos\phi']^{1/2}. \end{aligned} \tag{9.8}$$

Since $\cos(2\pi - \phi') = \cos\phi'$, we find that $R(2\pi - \phi') = R(\phi')$. It is now possible for us to show that $A_x(\boldsymbol{r})$ given by (9.6) reduces identically to zero. For this purpose, we rewrite (9.6) as the sum of two parts as

$$\begin{aligned} A_x(\boldsymbol{r}) = &-\frac{\mu I_0 a}{4\pi}\int_0^{\pi} d\phi' \sin\phi' \frac{e^{-j\beta R(\phi')}}{R(\phi')}, \\ &-\frac{\mu I_0 a}{4\pi}\int_{\pi}^{2\pi} d\phi' \sin\phi' \frac{e^{-j\beta R(\phi')}}{R(\phi')}. \end{aligned} \tag{9.9}$$

In the second integral in (9.9), we change the variable of integration from ϕ' to $2\pi - \phi'$. We can verify that $d\phi' \sin\phi'$ is not altered. We pointed out that $R(\phi')$ does not change in the indicated transformation. But the limits of the integral change from $(2\pi, \pi)$ to $(0, \pi)$. Therefore, (9.9) becomes

$$\begin{aligned} A_x(\boldsymbol{r}) = &-\frac{\mu I_0 a}{4\pi}\int_0^{\pi} d\phi' \sin\phi' \frac{e^{-j\beta R(\phi')}}{R(\phi')}, \\ &-\frac{\mu I_0 a}{4\pi}\int_{\pi}^{0} d\phi' \sin\phi' \frac{e^{-j\beta R(\phi')}}{R(\phi')} = 0. \end{aligned} \tag{9.10}$$

Hence, it follows that the vector potential due to a filamentary loop with a constant current distribution is only in the ϕ-direction and its expression is given by (9.7).

2. Far-zone Fields, Radiation Resistance and Gain

First, we wish to obtain for the vector potential as given by (9.7) an approximation which is valid in the far-zone. If we rewrite (9.8) as

$$R(\phi') = r\left[1 - 2\frac{a}{r}\sin\theta\cos\phi' + \frac{a^2}{r^2}\right]^{1/2}, \tag{9.11}$$

and then expand it into a binomial series in a/r for $a/r \ll 1$, we deduce the following result if we retain only the first two terms:

$$R(\phi') = r\left[1 - \frac{a}{r}\sin\theta\cos\phi'\right]. \tag{9.12}$$

Also for $a/r \ll 1$,

$$\frac{1}{R(\phi')} = \frac{1}{r}\left(1 + \frac{a}{r}\sin\theta\cos\phi'\right).$$

Since the far-zone vector potential is defined to be that part of the vector potential whose amplitude falls off as $1/r$, in the amplitude part of the vector potential, we can approximate $R(\phi')$ by r to obtain the far-zone vector potential. Hence, together with (9.12), (9.7) in the far-zone becomes approximately,

$$A_\phi(\mathbf{r}) = \frac{\mu I_0 a e^{-j\beta r}}{4\pi r}\int_0^{2\pi} d\phi' \cos\phi' e^{j\beta a \sin\theta\cos\phi'}. \tag{9.13}$$

Although we can evaluate (9.13) in terms of a well-tabulated special function known as the Bessel function, such an evaluation is not necessary since the main interest is in the case of a small loop antenna for which $\beta a \ll 1$. For this special case we can evaluate the integral in (9.13) approximately in a simple manner. We expand the exponential function into a power series and retain the first two terms with the following result:

$$A_\phi(\mathbf{r}) = \mu I_0 a\frac{e^{-j\beta r}}{4\pi r}\int_0^{2\pi} d\phi' \cos\phi'[1 + j\beta a\sin\theta\cos\phi']. \tag{9.14}$$

The integral of the first term in (9.14) vanishes and hence (9.14) reduces to

$$\begin{aligned} A_\phi(\mathbf{r}) &= \mu I_0 a^2\frac{e^{-j\beta r}}{4\pi r} j\beta\sin\theta\int_0^{2\pi} d\phi' \cos^2\phi', \\ &= \mu(I_0\pi a^2)\frac{e^{-j\beta r}}{4\pi r} j\beta\sin\theta. \end{aligned} \tag{9.15}$$

We obtain the far-zone magnetic field with the help of (5.21) and (9.15) as

$$\begin{aligned} H_\theta(\mathbf{r}) &= -\frac{1}{\mu r}\frac{\partial}{\partial r}\left(rA_\phi(\mathbf{r})\right) = \frac{j\beta}{\mu} A_\phi(\mathbf{r}), \\ &= -(I_0 \pi a^2)\frac{e^{-j\beta r}}{4\pi r}\beta^2 \sin\theta. \end{aligned} \tag{9.16}$$

We find the far-zone electric field with the help of (5.23) and (9.16) to be

$$\begin{aligned} E_\phi(\mathbf{r}) &= \frac{1}{j\omega\varepsilon r}\frac{\partial}{\partial r}\left(rH_\theta(\mathbf{r})\right) = -\eta_{0r} H_\theta(\mathbf{r}), \\ &= (I_0 \pi a^2)\frac{e^{-j\beta r}}{4\pi r}\eta_{0r}\beta^2 \sin\theta. \end{aligned} \tag{9.17}$$

We can evaluate the time-averaged Poynting vector $\boldsymbol{S}(\mathbf{r})$ and the radiation intensity $\Phi(\theta)$ from (9.16) and (9.17) with the following results:

$$S_r(\mathbf{r}) = \frac{(|I_0|\ \pi a^2)^2}{32\pi^2 r^2}\eta_{0r}\beta^4 \sin^2\theta, \tag{9.18}$$

and

$$\Phi(\theta) = \frac{(|I_0|\ \pi a^2)^2}{32\pi^2}\eta_{0r}\beta^4 \sin^2\theta. \tag{9.19}$$

We find the total time-averaged power P radiated by the small current loop to be

$$\begin{aligned} P &= \int_{\phi=0}^{2\pi}\int_{\theta=0}^{\pi}\Phi(\theta)\sin\theta\, d\theta\, d\phi = \frac{(|I_0|\ \pi a^2)^2 \eta_{0r}\beta^4}{16\pi}\int_0^{\pi}\sin^3\theta\, d\theta, \\ &= \frac{(|I_0|\ \pi a^2)^2}{12\pi}\eta_{0r}\beta^4. \end{aligned} \tag{9.20}$$

Hence, the radiation resistance is

$$R_r = \frac{2P}{|I_0|^2} = \frac{\eta_{0r}}{6\pi}\beta^4(\pi a^2)^2 = 320\pi^6\left(\frac{a}{\lambda}\right)^4. \tag{9.21}$$

If we compare (9.21) with (5.40) we see that whereas the radiation resistance of a current element is inversely proportional to the square of the wavelength, that of the small current loop is inversely proportional to the fourth power of the wavelength. As the ratio of the linear dimension of the radiator to the wavelength is reduced, the power radiated by the small circular loop falls off more rapidly than that due to the linear current element.

We find the radiation intensity Φ_{iso} of the equivalent isotropic radiator from (9.20) to be

$$\Phi_{iso} = \frac{P}{4\pi} = \frac{(|I_0|\ \pi a^2)^2}{48\pi^2}\eta_{0r}\beta^4. \tag{9.22}$$

From (9.19) and (9.22), we see the directive gain $g(\theta)$ to be

$$g(\theta) = \frac{\Phi(\theta)}{\Phi_{\text{iso}}} = 1.5 \sin^2 \theta. \tag{9.23}$$

If we compare (9.23) with (5.42) we see that the directive gain and therefore the directivity of a small circular loop with constant current are identical to those of a linear current element situated at the origin and oriented parallel to the axis of the loop.

3. Electromagnetic Fields of a Small Current Loop

There is a similarity between the far-zone fields of a small current loop as given by (9.16) and (9.17) and those of a linear current element as given by (5.25). This suggests that if we are interested only in the *small* current loop, we can evaluate the total fields containing both the near-zone and the far-zone fields of a small current loop in the same manner as for the linear current element. For this purpose, we evaluate the integral in (9.7) for $a/r \ll 1$ and retain all the terms which are up to linear order in a/r. Hence, with the help of (9.12), (9.7) becomes

$$A_\phi(\mathbf{r}) = \frac{\mu I_0 a}{4\pi r} e^{-j\beta r} \int_0^{2\pi} d\phi' \cos \phi' e^{j\beta a \sin\theta \cos\phi'} \left(1 + \frac{a}{r} \sin\theta \cos\phi'\right), \tag{9.24}$$

which is the same as (9.13) except that an additional term which falls off as r^{-2} is retained. Also, we assume that $\beta a \ll 1$, that is, the circumference of the current loop is small in comparison with the wavelength. Therefore, we expand the exponential function in (9.24) into a power series and retain only terms up to linear order in βa. We simplify the integrand by omitting all the terms which are proportional to the product of the small order quantities βa and a/r with the following result:

$$A_\phi(\mathbf{r}) = \frac{\mu I_0 a e^{-j\beta r}}{4\pi r} \int_0^{2\pi} d\phi' \cos \phi' \left[1 + j\beta a \sin\theta \cos\phi' + \frac{a}{r} \sin\theta \cos\phi'\right]. \tag{9.25}$$

The integral of the first term inside the square brackets in (9.25) vanishes. We can evaluate the integrals corresponding to the other terms to obtain the following expression for the vector potential:

$$A_\phi(\mathbf{r}) = \mu I_0 \pi a^2 \frac{e^{-j\beta r}}{4\pi r} j\beta \sin\theta \left(1 + \frac{1}{j\beta r}\right). \tag{9.26}$$

We should note that (9.25) and (9.26) have one additional term as compared to (9.14) and (9.15) respectively. If we substitute (9.26) in (5.21) we obtain the following expressions for the components of the magnetic field:

$$H_r(\mathbf{r}) = I_0 \pi a^2 \frac{e^{-j\beta r}}{2\pi r^2} j\beta \cos\theta \left(1 + \frac{1}{j\beta r}\right), \tag{9.27a}$$

$$H_\theta(\mathbf{r}) = -I_0 \pi a^2 \frac{e^{-j\beta r}}{4\pi r} \beta^2 \sin\theta \left(1 + \frac{1}{j\beta r} - \frac{1}{\beta^2 r^2}\right), \tag{9.27b}$$

$$H_\phi(\mathbf{r}) = 0. \tag{9.27c}$$

Substituting (9.27) into (5.23) we get

$$E_r(\mathbf{r}) = 0, \tag{9.28a}$$

$$E_\theta(\mathbf{r}) = 0, \tag{9.28b}$$

$$E_\phi(\mathbf{r}) = -j\omega\mu I_0 \pi a^2 \frac{e^{-j\beta r}}{4\pi r} j\beta \sin\theta \left(1 + \frac{1}{j\beta r}\right) = -j\omega A_\phi(\mathbf{r}). \tag{9.28c}$$

The small current loop excites waves progressing in the radial direction and hence we identify the radial direction as the propagation direction. Since (9.28) shows that there is no component of the electric vector in the radial direction, the waves set up by a small current loop constitute a TE mode. The electric vector is in the ϕ-direction only and the magnetic vector is perpendicular to the electric vector. If we compare (9.27) and (9.28) with (5.24) and (5.22) respectively we see the dual nature of the fields radiated by the small current loop as compared to those due to the linear current element. We should remember that (9.27) and (9.28) are valid only if $\beta a \ll 1$ and $a/r \ll 1$, that is, the radius of the current loop is small in comparison with the wavelength and the observation distance.

We obtain the far-zone fields of a small current loop from (9.27) and (9.28) as

$$H_\theta(\mathbf{r}) = -I_0 \pi a^2 \frac{e^{-j\beta r}}{4\pi r} \beta^2 \sin\theta, \tag{9.29}$$

and

$$E_\phi(\mathbf{r}) = I_0 \pi a^2 \frac{e^{-j\beta r}}{4\pi r} \eta_{0r} \beta^2 \sin\theta, \tag{9.30}$$

which, as they should be, are identical with (9.16) and (9.17) respectively.

We see the near-zone fields corresponding to $\beta r \ll 1$ from (9.27) and (9.28) to be

$$H_r(\mathbf{r}) = \frac{I_0 \pi a^2}{2\pi r^3} \cos\theta; \qquad H_\theta(\mathbf{r}) = \frac{I_0 \pi a^2}{4\pi r^3} \sin\theta; \tag{9.31}$$

$$E_\phi(\mathbf{r}) = -j\omega\mu \frac{I_0 \pi a^2}{4\pi r^2} \sin\theta. \tag{9.32}$$

If we compare (9.31) with (4.4.62) we see that, in the near-zone, the magnetic field due to a small current loop is the same as that obtained by the application of laws of magnetostatics and, therefore, the near-zone magnetic field due to a small current loop is also called the quasi-magnetostatic field.

The current distribution is a constant around the circumference of the loop and, hence, the use of the equation of continuity (4.8) immediately shows that the accompanying charge distribution is zero. Therefore, the scalar potential vanishes and the quasi-electrostatic field is zero. Consequently, we should not interpret the near-zone field as given by (9.32) as the quasi-electrostatic field. This is due to the fact that the scalar potential is zero and the electric field as indicated in (9.28c) depends directly on the vector potential.

It is convenient for us to summarize the definitions of the far-zone and the near-zone fields. The far-zone fields (9.29) and (9.30) are valid only if $a/r \ll \beta a \ll 1$ or $1 \ll \beta r \ll r/a$ and the near-zone fields (9.31) and (9.32) are valid only if $\beta a \ll a/r \ll 1$ or $\beta r \ll 1 \ll r/a$.

9.10 MAGNETIC DIPOLE

In the equation of continuity (2.1), Maxwell's equations (2.2) and (2.3), and the divergence relations (2.5) and (2.6), the electric charge density $\rho(\boldsymbol{r}, t)$ and the electric current density $\boldsymbol{J}(\boldsymbol{r}, t)$ appear but the corresponding magnetic charge density $\rho_m(\boldsymbol{r}, t)$ and the magnetic current density $\boldsymbol{J}_m(\boldsymbol{r}, t)$ are not present because the magnetic charge and the magnetic current are not physically available. Our study of the radiation characteristics of a small circular current filament has revealed that the electric field and the magnetic field radiated by an electric current element have a close resemblance respectively to the magnetic field and the electric field radiated by a small current loop. Therefore, we may think of the fields radiated by a small current loop as being due to a magnetic analogue of the electric current element, that is, a magnetic current element. This suggests that although magnetic charges and magnetic currents are not physically available, in many physical situations, we may conveniently consider the actual electromagnetic fields as arising from *equivalent* magnetic charges and magnetic currents. In the above-mentioned physical situations, it is often simpler for us to derive the fields from the *equivalent* magnetic current sources than from the *actual* electric current sources. For this purpose, it is helpful for us to develop the theory for determining the electromagnetic fields due to the magnetic currents and the magnetic charges in a manner analogous to that discussed previously for the electric currents and the electric charges.

It is sufficient for us to consider the fields radiated by the equivalent magnetic charges and magnetic currents *alone*. If both the types of sources are present, we can obtain the resulting fields by a linear superposition of the fields associated with each type of sources individually. The conservation of magnetic charge yields the following equation of continuity:

$$\nabla \cdot \boldsymbol{J}_m(\boldsymbol{r}, t) + \frac{\partial}{\partial t} \rho_m(\boldsymbol{r}, t) = 0. \tag{10.1}$$

In the presence of the equivalent magnetic charge alone, the divergence relations (2.5) and (2.6) become

$$\nabla \cdot \boldsymbol{D}(\boldsymbol{r}, t) = 0, \tag{10.2}$$

$$\nabla \cdot \boldsymbol{B}(\boldsymbol{r}, t) = \rho_m(\boldsymbol{r}, t). \tag{10.3}$$

In a similar manner, in the presence of the equivalent magnetic current alone, we should change Maxwell's equations (2.2) and (2.3) to the form:

$$\nabla \times \boldsymbol{E}(\boldsymbol{r}, t) = -\frac{\partial}{\partial t} \boldsymbol{B}(\boldsymbol{r}, t) - \boldsymbol{J}_m(\boldsymbol{r}, t), \tag{10.4}$$

$$\nabla \times \boldsymbol{H}(\boldsymbol{r}, t) = \frac{\partial}{\partial t} \boldsymbol{D}(\boldsymbol{r}, t). \tag{10.5}$$

We can verify the correctness of (10.4) in the following manner. We take the time derivative of both sides of (10.3), and substitute the value of $(\partial/\partial t)\rho_m(\mathbf{r}, t)$ from (10.1). Then, we obtain the following result:

$$\nabla \cdot \left[-\left\{\frac{\partial}{\partial t}\mathbf{B}(\mathbf{r}, t) + \mathbf{J}_m(\mathbf{r}, t)\right\}\right] = 0. \tag{10.6}$$

Since the vector within the square brackets in (10.6) has no divergence, it is expressible as the curl of another vector; a result which is in conformity with the first of Maxwell's equations as given in (10.4). The constitutive relations (2.4) remain unchanged.

As before, it is convenient for us to introduce certain auxiliary functions known as potentials and obtain the field vectors in terms of these potentials. In view of (10.2), it is reasonable for us to set

$$\mathbf{D}(\mathbf{r}, t) = -\nabla \times \mathbf{F}(\mathbf{r}, t), \tag{10.7}$$

where $\mathbf{F}(\mathbf{r}, t)$ is known as the magnetic vector potential. If we substitute (10.7) in (10.5) we find

$$\nabla \times \left[\mathbf{H}(\mathbf{r}, t) + \frac{\partial}{\partial t}\mathbf{F}(\mathbf{r}, t)\right] = 0. \tag{10.8}$$

For any scalar f, $\nabla \times [\nabla f] = 0$. Consequently, we see from (10.8) that

$$\mathbf{H}(\mathbf{r}, t) = -\frac{\partial}{\partial t}\mathbf{F}(\mathbf{r}, t) - \nabla\psi(\mathbf{r}, t), \tag{10.9}$$

where $\psi(\mathbf{r}, t)$ is any scalar function and is known as the magnetic scalar potential. If we substitute (10.7) and (10.9) into (10.4) and (10.3) respectively and use the constitutive relations (2.4) we obtain the following results:

$$-\frac{1}{\varepsilon}[\nabla\{\nabla \cdot \mathbf{F}(\mathbf{r}, t)\} - \nabla^2\mathbf{F}(\mathbf{r}, t)] = \mu\frac{\partial^2}{\partial t^2}\mathbf{F}(\mathbf{r}, t) + \mu\nabla\frac{\partial}{\partial t}\psi(\mathbf{r}, t) - \mathbf{J}_m(\mathbf{r}, t), \tag{10.10}$$

$$-\mu\frac{\partial}{\partial t}\nabla \cdot \mathbf{F}(\mathbf{r}, t) - \mu\nabla^2\psi(\mathbf{r}, t) = \rho_m(\mathbf{r}, t). \tag{10.11}$$

We can remove the arbitrariness of the definition of the vector potential by defining the divergence of $\mathbf{F}(\mathbf{r}, t)$ as follows:

$$\nabla \cdot \mathbf{F}(\mathbf{r}, t) + \mu\varepsilon\frac{\partial}{\partial t}\psi(\mathbf{r}, t) = 0, \tag{10.12}$$

which may be called the Lorentz condition for the magnetic vector and the magnetic scalar potentials. With the help of (10.12), we can reduce (10.10), and (10.11) to two uncoupled differential equations in $\mathbf{F}(\mathbf{r}, t)$ and $\psi(\mathbf{r}, t)$ respectively:

$$\nabla^2\mathbf{F}(\mathbf{r}, t) - \mu\varepsilon\frac{\partial^2}{\partial t^2}\mathbf{F}(\mathbf{r}, t) = -\varepsilon\mathbf{J}_m(\mathbf{r}, t), \tag{10.13}$$

and

$$\nabla^2\psi(\mathbf{r}, t) - \mu\varepsilon\frac{\partial^2}{\partial t^2}\psi(\mathbf{r}, t) = -\frac{\rho_m(\mathbf{r}, t)}{\mu}. \tag{10.14}$$

As in Section 9.2, we can show that (10.12) is consistent with the equation of continuity (10.1).

We should note that (10.13) and (10.14) are very similar to (2.19) and (2.20) respectively. Therefore, we can write down the solutions of (10.13) and (10.14) by an inspection of the solutions for $\boldsymbol{A}(\boldsymbol{r}, t)$ and $\phi(\boldsymbol{r}, t)$ as given in Section 9.3. As before, if we consider only the retarded potentials, the following results are obtained:

$$\boldsymbol{F}(\boldsymbol{r}, t) = \int_V \frac{\varepsilon \boldsymbol{J}_m(\boldsymbol{r}', t - (|\boldsymbol{r} - \boldsymbol{r}'|/v))}{4\pi\, |\boldsymbol{r} - \boldsymbol{r}'|}\, dV', \tag{10.15}$$

and

$$\psi(\boldsymbol{r}, t) = \int_V \frac{\rho_m(\boldsymbol{r}', t - (|\boldsymbol{r} - \boldsymbol{r}'|/v))}{4\pi\mu\, |\boldsymbol{r} - \boldsymbol{r}'|}\, dV'. \tag{10.16}$$

We can particularize the preceding treatment for the case of harmonic time dependence. Since the discussion parallels very closely that given in Section 9.4 for the electric vector potential and the electric scalar potential, we shall not reproduce it here. We shall give only the expressions for the time-harmonic, retarded, magnetic vector potential and magnetic scalar potential. We can deduce these results from (10.15) and (10.16) as

$$\boldsymbol{F}(\boldsymbol{r}) = \int_V \frac{\varepsilon \boldsymbol{J}_m(\boldsymbol{r}') e^{-j\beta|\boldsymbol{r}-\boldsymbol{r}'|}}{4\pi\, |\boldsymbol{r} - \boldsymbol{r}'|}\, dV', \tag{10.17}$$

and

$$\psi(\boldsymbol{r}) = \int_V \frac{\rho_m(\boldsymbol{r}') e^{-j\beta|\boldsymbol{r}-\boldsymbol{r}'|}}{4\pi\mu\, |\boldsymbol{r} - \boldsymbol{r}'|}\, dV'. \tag{10.18}$$

Let us consider a short magnetic current element of magnetic current moment $I_{m0}\, dl$, situated at the origin and oriented parallel to the z-axis. From (10.17) and as in (5.3), we find the magnetic vector potential which has only a z-component to be given by

$$F_z(\boldsymbol{r}) = (I_{m0}\, dl) \frac{\varepsilon e^{-j\beta r}}{4\pi r}, \tag{10.19}$$

where r is the scalar distance from the source point which is the origin, to the point of observation. We should note that the magnetic vector potential $\boldsymbol{F}(\boldsymbol{r})$ is in the same direction as the magnetic current. We can obtain the components of the magnetic vector potential in the direction of the spherical coordinates r, θ, and ϕ with the help of Fig. 9.1 and (10.19) as

$$F_r(\boldsymbol{r}) = F_z(\boldsymbol{r}) \cos\theta = I_{m0}\, dl \frac{\varepsilon e^{-j\beta r}}{4\pi r} \cos\theta, \tag{10.20a}$$

$$F_\theta(\boldsymbol{r}) = -F_z(\boldsymbol{r}) \sin\theta = -I_{m0}\, dl \frac{\varepsilon e^{-j\beta r}}{4\pi r} \sin\theta, \tag{10.20b}$$

$$F_\phi(\boldsymbol{r}) = 0. \tag{10.20c}$$

We can reduce the relation (10.7) for the time-harmonic case to

$$\boldsymbol{E}(\boldsymbol{r}) = -\frac{1}{\varepsilon}\nabla \times \boldsymbol{F}(\boldsymbol{r}). \tag{10.21}$$

From (10.20) and (10.21), we can show the spherical components of the electric field to be

$$E_r(\boldsymbol{r}) = 0; \qquad E_\theta(\boldsymbol{r}) = 0; \qquad E_\phi(\boldsymbol{r}) = -\frac{I_{m0}\,dl}{4\pi r}e^{-j\beta r}j\beta\sin\theta\left(1 + \frac{1}{j\beta r}\right). \tag{10.22}$$

Substituting (10.22) in the first of Maxwell's equations

$$\boldsymbol{H}(\boldsymbol{r}) = -\frac{1}{j\omega\mu}\nabla \times \boldsymbol{E}(\boldsymbol{r}), \tag{10.23}$$

we find the following expressions for the components of the magnetic field:

$$H_r(\boldsymbol{r}) = \frac{I_{m0}\,dl}{2\pi r}e^{-j\beta r}\frac{\cos\theta}{\eta_{0r}r}\left(1 + \frac{1}{j\beta r}\right), \tag{10.24a}$$

$$H_\theta(\boldsymbol{r}) = \frac{I_{m0}\,dl}{4\pi r}e^{-j\beta r}\frac{j\beta\sin\theta}{\eta_{0r}}\left(1 + \frac{1}{j\beta r} - \frac{1}{\beta^2 r^2}\right), \tag{10.24b}$$

$$H_\phi(\boldsymbol{r}) = 0. \tag{10.24c}$$

If we compare (10.22) and (10.24) with (9.28) and (9.27) respectively we see that the fields due to a small electric current loop and a magnetic current element are identical if

$$I_{m0}\,dl = (I_0\pi a^2)j\omega\mu. \tag{10.25}$$

The equivalence between a small electric current loop and a magnetic current element is explicit from (10.25). We should remember, however, that the magnetic current element is normal to the plane of the loop of electric current. The magnetic moment $\boldsymbol{m}$ of an electric current loop is a vector quantity whose magnitude is equal to the product of the current times the area of the loop and its direction, which is perpendicular to the plane of the loop, is given by the right-hand rule. If the fingers of the right hand curl in the direction of the current, the thumb points in the direction of the magnetic moment. Thus, for the electric current loop which we discussed in the previous section, we find the magnetic current moment to be

$$\boldsymbol{m} = \hat{z}(I_0\pi a^2). \tag{10.26}$$

We can also define a vector current moment. Its magnitude is equal to the current times the length for constant current elements and its direction is the same as that of the current. Hence, for the magnetic current element under consideration, the vector current moment $\boldsymbol{p}_i$ is given by

$$\boldsymbol{p}_i = \hat{z}I_{m0}\,dl. \tag{10.27}$$

In view of (10.26) and (10.27), we may recast the equivalence relation (10.25) in the form

$$\boldsymbol{p}_i = \boldsymbol{m}j\omega\mu. \tag{10.28}$$

We have presented a discussion of the method for the determination of the electromagnetic fields due to the equivalent magnetic charges and magnetic currents. We have illustrated this method by the evaluation of the electromagnetic fields due to a magnetic current element. We have shown that the fields due to a small electric current loop are essentially the same as those due to the magnetic current element. Also, we have found that the evaluation of the fields due to the *equivalent* magnetic current element is simpler than that due to the *actual* small electric current loop. This simplicity can be further augmented by the exploitation of the symmetry relations that exist between the field quantities associated with the equivalent magnetic type sources and those associated with the actual electric type sources. A comparison of (10.1), (10.5), (10.4), (10.3), (10.2), (10.7), and (10.9) with (2.1), (2.2), (2.3), (2.5), (2.6), (2.13), and (2.15) respectively reveals the following one-to-one correspondence between the field quantities associated with the actual electric type sources and those associated with the equivalent magnetic type sources:

Electric type	$\boldsymbol{E}$	$\boldsymbol{H}$	$\boldsymbol{J}$	ρ	μ	ε	η_{0r}	$\boldsymbol{A}$	ϕ
Magnetic type	$\boldsymbol{H}$	$-\boldsymbol{E}$	$\boldsymbol{J}_m$	ρ_m	ε	μ	$1/\eta_{0r}$	$\boldsymbol{F}$	ψ

(10.29)

If we apply the symmetry relations (10.29) to the fields due to an electric current element, as given by (5.22) and (5.24), we obtain the corresponding fields due to a magnetic current element, as given by (10.22) and (10.24), in a straightforward manner. We thus clarify the usefulness of the concept of equivalent magnetic charges and magnetic currents.

PROBLEMS

9.1 Show that the magnitude of the far-zone electric field due to a Hertzian dipole is

$$E = 9.49\sqrt{P}\,\frac{\sin\theta}{r} \quad \text{V/m},$$

where P is the time-averaged radiated power in watts, r is the distance in meters from the observation point to the location of the dipole and θ is the angle made by the radius vector $\boldsymbol{r}$ with the axis of the dipole. Assume that the intrinsic resistance of the medium is given by $\eta_{0r} = 120\pi$ ohms.

9.2 Show that the magnitude of the far-zone electric field due to a half-wavelength long filamentary antenna with a sinusoidal current distribution, as treated in Section 9.6 is given by the relation

$$E = 9.93\,\frac{\sqrt{P}}{r}\,\frac{\cos((\pi/2)\cos\theta)}{\sin\theta} \quad \text{V/m},$$

where P, r, and θ have the same meaning as in the previous problem.

9.3 Consider a half-wavelength long filamentary antenna situated in free space of characteristic resistance $\eta_{0r} = 120\pi$ ohms and having a progressive wave current distribution as treated in

Section 9.7. Prove that the magnitude of the far-zone electric field due to this antenna is given by

$$E = 9.14\frac{\sqrt{P}}{r}\frac{\sin\theta\sin\{(\pi/2)(1-\cos\theta)\}}{(1-\cos\theta)} \quad \text{V/m},$$

where P, r, and θ have the same significance as in Problem 9.1.

9.4 Consider a quarter-wavelength long filamentary antenna situated vertically on the surface of the earth. Assume that the earth is a perfect conductor and the medium into which the antenna radiates has a characteristic resistance of $\eta_{0r} = 120\pi$ ohms. If the current along the antenna is sinusoidally distributed, show that the magnitude of the far-zone electric field at the surface of the earth is given by

$$E = 14\frac{\sqrt{P}}{r} \quad \text{V/m},$$

where P is the total time-averaged power radiated in watts and r is the distance of the observation point from the antenna location.

9.5 Let $E(r, \theta)$ be the magnitude of the far-zone electric field due to a linear antenna in free space of characteristic resistance $\eta_{0r} = 120\pi$ ohms, at a distance r in meters and in a direction which makes an angle θ with the axis of the antenna. Deduce that the effective height $2h_e(\theta)$ of the antenna is given by

$$2h_e(\theta) = \frac{\lambda r E(r, \theta)}{60\pi I(0)},$$

where λ is the wavelength and $I(0)$ is the current at the center of the antenna.

9.6 Consider an electrically short ($\beta h \ll 1$) dipole extending along the z-axis from $z = -h$ to $z = h$. The current distribution is assumed to be "triangular" as given by

$$I(z) = I_0\left(1 - \frac{|z|}{h}\right). \tag{a}$$

Find the maximum effective height $2h_e$ occurring in the broadside direction $\theta = \pi/2$, where θ is the angle between the direction of observation and the axis of the dipole.

Show that the effective height $2h_e(\theta)$ for the general direction θ is given by

$$2h_e(\theta) = h\sin\theta\left[\frac{\sin\ ((\beta h/2)\cos\theta\ )}{(\beta h/2)\cos\theta}\right]^2. \tag{b}$$

Under the condition $\beta h \ll 1$, verify that (b) can be approximated by

$$2h_e(\theta) = h\sin\theta. \tag{c}$$

Give a physical explanation of the result expressed by (c).

In view of (c), note that all the physical parameters connected with a short dipole with a "triangular" current distribution are obtained from those of a short dipole with a constant current distribution by replacing dl by the effective height h.

9.7 A short monopole ($h > z > 0$) antenna is situated vertically on the surface ($z = 0$) of the earth, which is assumed to be a perfect conductor and radiates into a medium of characteristic resistance $\eta_{0r} = 120\pi$ ohms. If the current along the antenna has a "triangular" distribution as given by (a) in Problem 9.6, deduce that the magnitude of the far-zone electric field at the surface of the earth is given by

$$E = 13.42 \frac{\sqrt{P}}{r} \quad \text{V/m}, \tag{a}$$

where P and r have the same significance as in Problem 9.4.

9.8 Let the current distribution along the short dipole of Problem 9.6 be "parabolic" as given by

$$I(z) = I_0\left(1 - \frac{z^2}{h^2}\right). \tag{a}$$

Find the maximum effective height $2h_e$ occurring in the broadside direction.

Show that the effective height $2h_e(\theta)$ for the general direction θ is given by

$$2h_e(\theta) = 4h \sin\theta\left[\frac{\sin\alpha}{\alpha^3} - \frac{\cos\alpha}{\alpha^2}\right]; \qquad \alpha = \beta h \cos\theta. \tag{b}$$

Under the condition $\beta h \ll 1$, verify that (b) can be approximated by

$$2h_e(\theta) = \frac{4h}{3}\sin\theta. \tag{c}$$

Note that the physical explanation of the result (c) in Problem 9.6 applies also to the result given by (c) in this problem.

9.9 Examine whether the result given by (a) in Problem 9.7 is changed if the current along the short monopole antenna, instead of having a "triangular" distribution, has a "parabolic" distribution.

9.10 Two short dipole antennas of the same length radiate the same total time-averaged power. The first antenna has a "triangular" current distribution with the current at the center of the antenna given by I_{0t} and the second antenna has a "parabolic" current distribution with the current at the center of the antenna given by I_{0p}. Find the value of the ratio $|I_{0p}|/|I_{0t}|$.

9.11 Consider a linear filamentary antenna situated along the z-axis from $z = -h$ to $z = h$. A "shifted cosine" distribution of current as given by

$$I(z) = I_0 \frac{[\cos\beta h - \cos\beta z]}{[\cos\beta h - 1]}, \tag{a}$$

is assumed to be present along the antenna. Show that the effective height for the general direction θ is given by

$$2h_e(\theta) = h \sin\theta\left[2\cos\beta h\frac{\sin\alpha}{\alpha} - \frac{\sin\alpha_1}{\alpha_1} - \frac{\sin\alpha_2}{\alpha_2}\right], \tag{b}$$

where,

$$\alpha = \beta h\cos\theta; \qquad \alpha_1 = \beta h(1 + \cos\theta); \qquad \alpha_2 = \beta h(1 - \cos\theta). \tag{c}$$

With the help of (b) and (c) prove that for $\beta h \ll 1$

$$2h_e(\theta) = \frac{4h}{3}\sin\theta. \tag{d}$$

Show that the result given by (d) is to be expected from the results of Problem 9.8 by obtaining an approximation to $I(z)$ given by (*a*) for $\beta h \ll 1$.

9.12 Consider a half-wavelength long antenna situated along the z-axis from $z = -h = -\lambda/4$ to $z = h = \lambda/4$. The current distribution along the antenna is of the form

$$I(z) = I_0 \cos^2 \beta z. \tag{a}$$

Note that the current vanishes at the two ends of the antenna. Find the value of the effective height corresponding to the broadside direction $\theta = \pi/2$. Deduce that the effective height for the general direction θ is given by

$$2h_e(\theta) = h \sin\theta \frac{\sin((\pi/2)\cos\theta)}{((\pi/2)\cos\theta)} \frac{1}{(1 - 0.25\cos^2\theta)}. \tag{b}$$

Find the limiting value of (*b*) as θ tends to $\pi/2$ and thus reproduce the value of the effective height for the broadside direction.

9.13 Consider a full-wavelength long filamentary antenna situated in free space of characteristic resistance $\eta_{0r} = 120\pi$ ohms. As discussed in Section 9.6, the current distribution along the antenna is assumed to be sinusoidal. Show that the radiation intensity is given by

$$\Phi(\theta) = \frac{|I_m|^2 \eta_{0r}}{2\pi^2} \frac{\cos^4((\pi/2)\cos\theta)}{\sin^2\theta}, \tag{a}$$

where I_m is the maximum value of the current along the antenna. Prove that the total time-averaged power radiated by the antenna is

$$P = \frac{2|I_m|^2 \eta_{0r}}{\pi} \tilde{B}_H; \qquad \tilde{B}_H = \int_0^{\pi/2} \frac{\cos^4((\pi/2)\cos\theta)}{\sin\theta} d\theta. \tag{b}$$

By employing a procedure similar to that used in the evaluation of B_H in Section 9.6, deduce that

$$\tilde{B}_H = \frac{1}{8} \int_0^{2\pi} \frac{(1 - \cos v)^2}{v} dv. \tag{c}$$

With the help of the relation

$$(1 - \cos v)^2 = 2(1 - \cos v) - \tfrac{1}{2}(1 - \cos 2v),$$

simplify (c) to yield the following result:

$$\tilde{B}_H = \tfrac{1}{2}\,\text{Cin}\,(2\pi) - \tfrac{1}{8}\,\text{Cin}\,(4\pi) = 0.4148. \tag{d}$$

Note that Cin $(2\pi) = 2.4377$; Cin $(4\pi) = 3.1144$.

Establish the following results for the radiation resistance R_r, the directive gain $g(\theta)$, and the directivity g:

$$R_r = 199.1 \text{ ohms},$$

$$g(\theta) = 2.41 \frac{\cos^4((\pi/2)\cos\theta)}{\sin^2\theta},$$

$$g = 2.41.$$

9.14 Show that the magnitude of the far-zone electric field due to a full-wavelength long filamentary antenna discussed in Problem 9.13 is given by the relation

$$E = 12.03 \frac{\sqrt{P}}{r} \frac{\cos^2((\pi/2)\cos\theta)}{\sin\theta} \quad \text{V/m},$$

where P, r, and θ have the same meaning as in Problem 9.1.

9.15 Consider a half-wavelength long antenna situated along the z-axis from $z = -h = -\lambda/4$ to $z = h = \lambda/4$. The current distribution along the antenna is of the form

$$I(z) = I_0 \cos^2 \beta z. \tag{a}$$

Show that the radiation intensity is given by

$$\Phi(\theta) = \frac{|I_0|^2 \eta_{0r}}{32\pi^2} \frac{\sin^2 \theta \sin^2((\pi/2)\cos\theta)}{\cos^2\theta(1 - \frac{1}{4}\cos^2\theta)^2}. \tag{b}$$

Integrate (b) over the entire solid angle, set $u = \cos\theta$ and thus verify that the total time-averaged power radiated by the antenna is given by

$$P = \frac{|I_0|^2 \eta_{0r}}{8\pi} B_H, \qquad B_H = \int_0^1 f(u) \sin^2\left(\frac{\pi}{2} u\right) du, \tag{c}$$

where

$$f(u) = (1 - u^2)/u^2(1 - \tfrac{1}{4}u^2)^2. \tag{d}$$

Expand $f(u)$ into partial fractions as

$$f(u) = \frac{1}{u^2} - \frac{1}{8}\left(\frac{1}{2+u} + \frac{1}{2-u}\right) - \frac{3}{4}\left(\frac{1}{(2+u)^2} + \frac{1}{(2-u)^2}\right). \tag{e}$$

Prove by the method of integration by parts that

$$\int_0^1 \frac{\sin^2((\pi/2)u)}{u^2} du = -1 + \frac{\pi}{2} \operatorname{Si}(\pi), \tag{f}$$

where

$$\int_0^x \frac{\sin v}{v} dv = \operatorname{Si}(x)$$

is the well-known Sine integral which is tabulated extensively.

By setting $2 + u = v$ for the first part and $2 - u = v$ for the second part, show that

$$\int_0^1 \sin^2\left(\frac{\pi}{2}u\right)\left\{\frac{1}{(2+u)^2} + \frac{1}{(2-u)^2}\right\} du = \int_1^3 \frac{\sin^2((\pi/2)v)}{v^2} dv = \frac{2}{3} + \frac{\pi}{2}\{\operatorname{Si}(3\pi) - \operatorname{Si}(\pi)\}. \tag{g}$$

In a similar manner, deduce that

$$\int_0^1 \sin^2\left(\frac{\pi}{2}u\right)\left\{\frac{1}{2+u} + \frac{1}{2-u}\right\} du = \int_1^3 \frac{\sin^2((\pi/2)v)}{v} dv = \tfrac{1}{2}\{\operatorname{Cin}(3\pi) - \operatorname{Cin}(\pi)\}, \tag{h}$$

where Cin (x) is defined in (9.6.48).

With the help of (c) to (h), prove that

$$B_H = -1.5 + \frac{7\pi}{8} \operatorname{Si}(\pi) - \frac{3\pi}{8} \operatorname{Si}(3\pi) - \frac{\operatorname{Cin}(3\pi)}{16} + \frac{\operatorname{Cin}(\pi)}{16} = 1.5464. \tag{i}$$

Note that

$$\operatorname{Si}(\pi) = 1.8519; \qquad \operatorname{Si}(3\pi) = 1.6748;$$
$$\operatorname{Cin}(\pi) = 1.6483; \qquad \operatorname{Cin}(3\pi) = 2.8099.$$

Establish the following results for the radiation resistance R_r, the directive gain $g(\theta)$, and the directivity g:

$$R_r = 46.39 \text{ ohms},$$

$$g(\theta) = 0.6469 \frac{\sin^2 \theta \sin^2((\pi/2)\cos\theta)}{\cos^2\theta(1 - \frac{1}{4}\cos^2\theta)^2},$$

$$g = 1.60.$$

Assume that $\eta_{0r} = 120\pi$ ohms.

9.16 Show that the magnitude of the far-zone electric field due to a half-wavelength long filamentary antenna discussed in Problem 9.15 is

$$E = 6.23 \frac{\sqrt{P}}{r} \frac{\sin\theta \sin((\pi/2)\cos\theta)}{\cos\theta\,(1 - 0.25\cos^2\theta)} \quad \text{V/m},$$

where P, r, and θ have the usual meaning.

9.17 Consider a quarter-wavelength long filamentary antenna ($0 < z < h = \lambda/4$) situated vertically on the surface ($z = 0$) of the earth which is assumed to be a perfect conductor. The antenna radiates into a medium of characteristic resistance $\eta_{0r} = 120\pi$ ohms. The current distribution along the antenna is the same as that given by (*a*) in Problem 9.15. Show that the magnitude of the far-zone electric field at the surface of the earth is

$$E = 13.84 \frac{\sqrt{P}}{r} \quad \text{V/m},$$

where P is the total time-averaged power radiated and r is the distance of the observation point from the antenna location.

9.18 Consider a half-wavelength long filamentary antenna situated along the z-axis from $z = -h = -\lambda/4$ to $z = h = \lambda/4$. The current distribution along the antenna is assumed to be of the form

$$I(z) = I_0 e^{-j\beta\alpha z}. \tag{a}$$

Note that this is a progressive wave of current with a phase-shift coefficient that is different from that of the plane TEM waves in the medium in which the antenna is situated. If a linear antenna is coated with a dielectric, it is possible to increase the phase-shift coefficient of current along the antenna. Show that for this antenna, the radiation intensity is given by

$$\Phi(\theta) = \frac{|I_0|^2 \eta_{0r}}{8\pi^2}\left[\frac{\sin\theta \sin\{(\pi/2)(\alpha - \cos\theta)\}}{(\alpha - \cos\theta)}\right]^2. \tag{b}$$

Integrate (b) over the entire solid angle, set $u = \alpha - \cos\theta$, follow a procedure similar to that used in Problem 9.15 and verify that the total time-averaged power radiated by the antenna is given by

$$P = \frac{|I_0|^2 \eta_{0r}}{4\pi}\left[-2 - \frac{\sin(\pi\alpha)}{\pi} - \cos(\pi\alpha) + \alpha \operatorname{Cin}\{\pi(\alpha + 1)\} - \alpha \operatorname{Cin}\{\pi(\alpha - 1)\} + \frac{\pi}{2}(1 - \alpha^2)\{\operatorname{Si}(\pi(\alpha + 1)) - \operatorname{Si}(\pi(\alpha - 1))\}\right]. \tag{c}$$

If $\eta_{0r} = 120\pi$ ohm and $\alpha = 1.2$, show with the help of (c) that the radiation resistance of the antenna is

$$R_r = 74.24 \text{ ohm}. \tag{d}$$

Note that

$$\text{Cin}(2.2\pi) = 2.4437; \qquad \text{Cin}(0.2\pi) = 0.09709;$$

$$\text{Si}(2.2\pi) = 1.4467; \qquad \text{Si}(0.2\pi) = 0.6147.$$

Obtain the following result for the directive gain:

$$g(\theta) = 1.62\left[\frac{\sin\theta \sin\{(\pi/2)(1.2 - \cos\theta)\}}{(1.2 - \cos\theta)}\right]^2. \tag{e}$$

Set $u = 1.2 - \cos\theta$ and prove that $g(\theta)$ is a maximum at $u = u_{max} = 1.2 - \cos\theta_{max}$ where u_{max} is specified by the following transcendental equation

$$\tan\left(\frac{\pi}{2}u_{max}\right) = \frac{\pi}{2}u_{max}\frac{\{-0.44 + 2.4u_{max} - u_{max}^2\}}{\{-0.44 + 1.2u_{max}\}}. \tag{f}$$

Determine the solution of (f) by a graphical procedure and hence show that

$$u_{max} = 0.71; \qquad \theta_{max} = 60.8°.$$

Establish the following result for the directivity:

$$g = 1.97.$$

9.19 Consider four Hertzian dipoles, A, B, C, and D, situated in the $z = 0$ plane, at the corners of a square. The locations of the dipoles A, B, C, and D are given by $(x = d/2, y = d/2)$, $(x = -d/2, y = d/2)$, $(x = -d/2, y = -d/2)$ and $(x = d/2, y = -d/2)$ respectively. As shown in Fig. P9.1, the dipoles A and D are oriented in the $+y$-direction and the dipoles B and C are oriented in the $-y$-direction. Show that insofar as the electromagnetic fields in the quarter-space $0 < x < \infty$, $0 < y < \infty$, and $-\infty < z < \infty$ are concerned, the dipoles B, C, and D can be replaced by two semi-infinite perfectly conducting planes occupying the regions $(x = 0, 0 < y < \infty, -\infty < z < \infty)$ and $(0 < x < \infty, y = 0, -\infty < z < \infty)$ respectively.

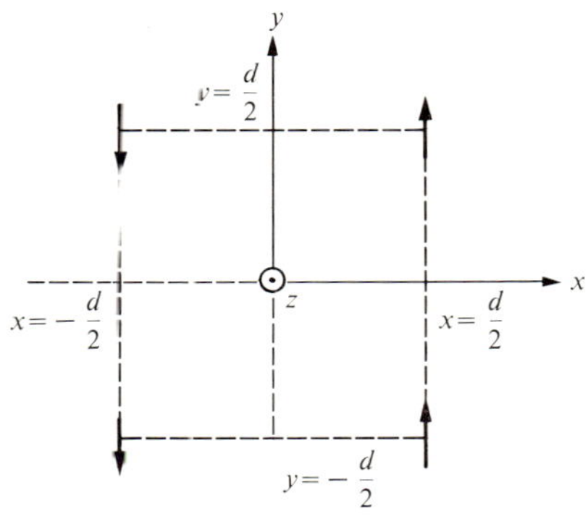

Fig. P9.1 Four Hertzian dipoles situated in the $z = 0$ plane at the corners of a square of side d.

9.20 Consider a square loop antenna $ABCD$ as shown in Fig. P9.2(a) with a constant current distribution around its circumference. It is desired to drive the antenna at the center E of one of its sides AD. In order to prevent the proximity of the driving equipment and the observer from disturbing the environment of the antenna, it is customary to exploit the imaging properties of perfectly conducting planes and use either half of a loop against one imaging plane as shown in Fig. P9.2(b) or quarter of a loop against two image planes located perpendicular to each other as shown in Fig. P9.2(c). The feeding transmission lines extend from the base of the antenna through the imaging plane to the driving equipment. The observer and the driving equipment are located behind the image planes which act as shields and prevent any disturbance being caused to the antenna. Assuming that the loop current is the same, show that the electromagnetic fields in the half-space $(-\infty < x < \infty, 0 < y < \infty, -\infty < z < \infty)$ for the antenna shown in Fig. P9.2(b) and in the quarter-space $(0 < x < \infty, 0 < y < \infty, -\infty < z < \infty)$ for the antenna depicted in Fig. P9.2(c) are the same as those in the corresponding regions for the loop antenna without the imaging planes, as shown in Fig. P9.2(a). Compare the radiation resistances of the three loop antenna configurations.

It is now desired to drive the loop antenna at one of the corners, say, A. In addition, it is desired to use only a quarter of a loop against two imaging, perfectly conducting planes. Sketch the proper arrangement of the antenna together with the image planes that can achieve the desired objective.

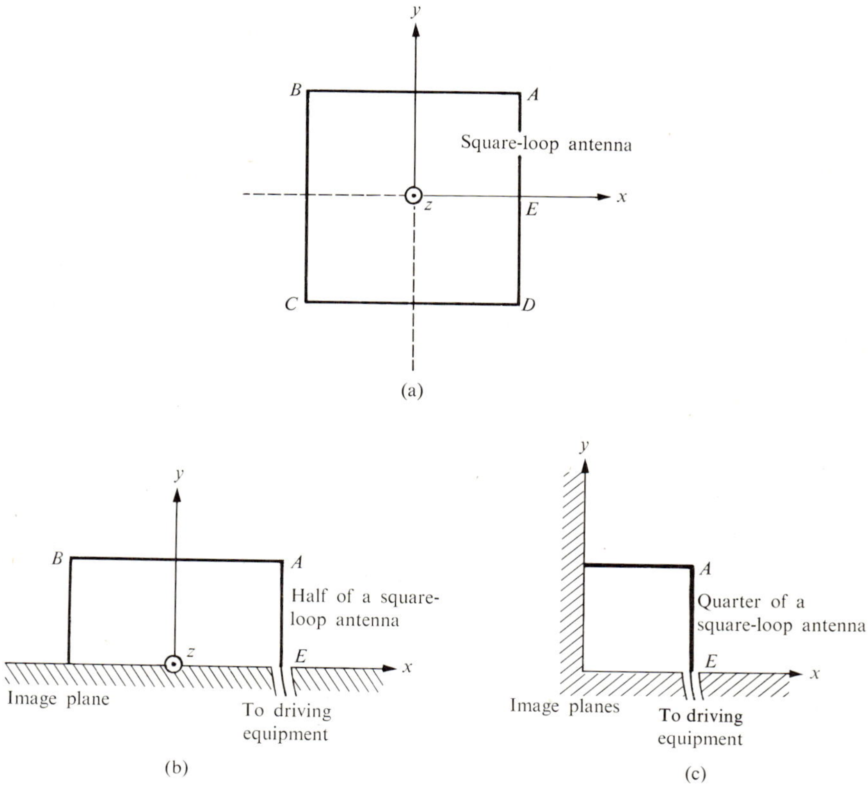

Fig. P9.2 Geometry of the square loop antenna (a) without imaging planes, (b) with one imaging plane, and (c) with two imaging planes.

9.21 Consider a small circular loop antenna with a constant current distribution around its circumference, as shown in Fig. P9.3(a). Also, consider a quarter of a small circular loop antenna placed against two perfectly conducting half-planes located perpendicular to each other as shown in Fig. P9.3(b). The constant current distribution around the circumference of the quarter of a loop shown in Fig. P9.3(b) is assumed to be the same as that for the whole circular loop shown in Fig. P9.3(a). Show that the electromagnetic fields in the quarter space ($0 < x < \infty, 0 < y < \infty, -\infty < z < \infty$) are the same for both the loop antenna configurations. Compare the radiation resistances of the two loop antenna configurations.

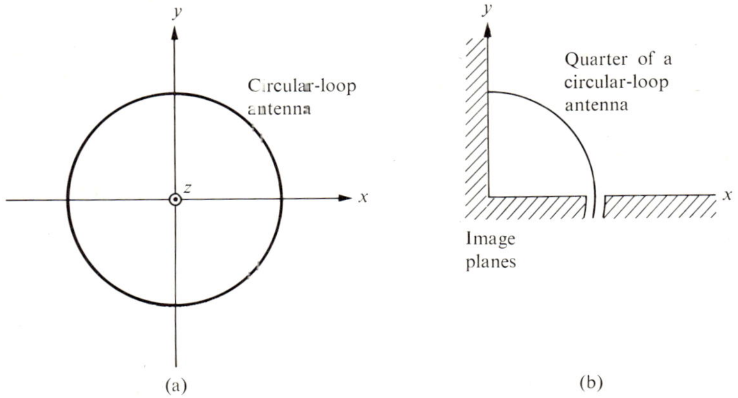

Fig. P9.3 Circular loop antenna (a) without the imaging planes and (b) with two imaging planes.

9.22 Consider a filamentary circular loop of radius $a = 5$ cm radiating in free space of characteristic resistance $\eta_{0r} = 120\pi$ ohms. Let $I_0 = 10$ A be the constant current distribution around the circumference of the loop. The frequency of operation is $f = 10^7$ Hz. Verify that the loop is electrically small. Determine the total time-averaged power radiated by the loop. Evaluate the radiation resistance R_r of the loop. Find the time-averaged power flux density in the plane of the loop at a distance of 200 m from it. Find the directions in which the time-averaged power flux density is 0.75 times the value in the plane of the loop.

9.23 Consider an elementary dipole of length dl having a constant current distribution along its length. The total time-averaged power radiated by the dipole is the same as that of the loop of Problem 9.22. Also, the frequency of operation of the dipole and the current loop are the same. The magnitude of each of the dipole charges is $Q = 0.01$ C. Find the length dl of the dipole.

9.24 Consider two magnetic current elements A and B, situated on the z-axis at $z = z'$ and $z = -z'$ respectively and oriented parallel to the z-axis of a cylindrical coordinate system ρ, ϕ, and z. The current moments of A and B are $I_{m0}\, dl$ and $-I_{m0}\, dl$ respectively. In the entire $z = 0$ plane, show that the tangential components of the electric field are identically zero. Hence, deduce that in the half-space $z > 0$ terminated at $z = 0$ by an infinite perfectly conducting plane, the fields due to the magnetic current element A situated in that half-space are equal to the fields radiated by A in the absence of the perfectly conducting plane, together with those radiated by an "image" antenna also in the absence of the perfectly conducting plane, and that this "image" antenna is the same as B. Note, as illustrated in Fig. P9.4, the differences between the images of electric and magnetic current elements oriented perpendicular to an infinite perfectly conducting plane.

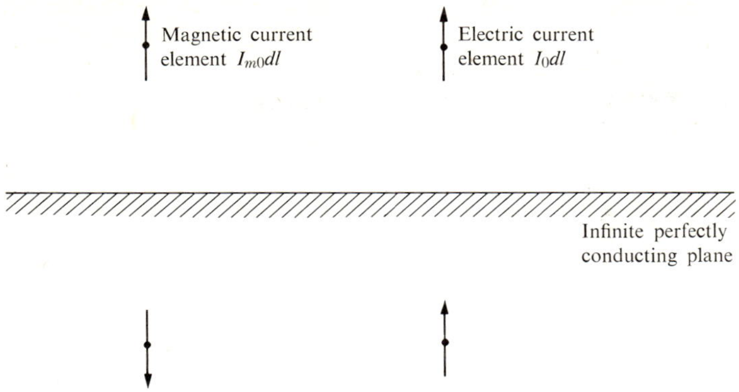

Fig. P9.4 Images of vertically oriented electric and magnetic current elements.

9.25 Consider a small circular loop antenna A of radius a, with its center situated at $z = d$ along the z-axis of a cylindrical coordinate system ρ, ϕ, and z and oriented so that its plane lies in the $z = d$ plane. Let I_0 be the constant current distribution around the circumference of the loop A. Consider another small circular loop antenna B of the same radius a, with its center situated at $z = -d$ along the z-axis and oriented so that its plane lies in the $z = -d$ plane. Let $-I_0$ be the constant current distribution around the circumference of the loop B. In other words, the same constant current exists in loops A and B but in opposite directions: counter-clockwise in loop A and clockwise in loop B. Show that in the entire $z = 0$ plane, the tangential components of the electric field are zero. Note that it is not necessary to actually evaluate the fields; it is sufficient to make use of the symmetry properties of the fields with reference to the $z = 0$ plane. Hence deduce that loop B is the image of loop A for a perfectly conducting infinite surface occupying the $z = 0$ plane.

Replace the small circular loops A and B by their equivalent magnetic current elements A^e and B^e respectively. Verify that the image of the magnetic current element A^e obtained in this manner is consistent with the results of Problem 9.24.

9.26 Consider two magnetic current elements A and B, situated on the z-axis at $z = z'$ and $z = -z'$ respectively and oriented parallel to the y-axis of a Cartesian coordinate system x, y, and z. The current moments of A and B are both equal to $I_{m0}\,dl$. By employing the method used in Section 9.8 for the derivation of the image of an electric current element oriented parallel to

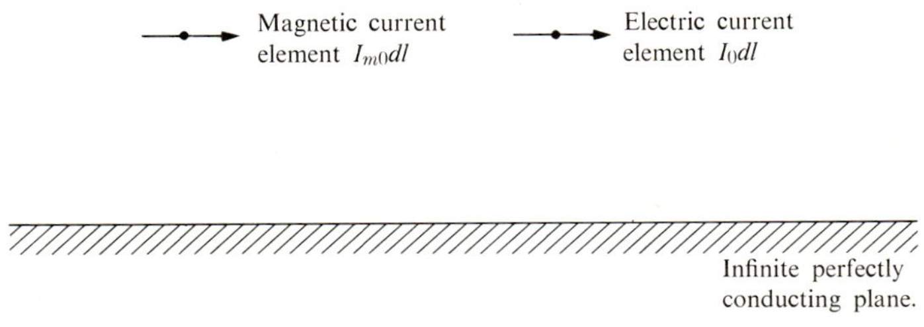

Fig. P9.5 Images of horizontally oriented electric and magnetic current elements.

a perfectly conducting plane, show that in the entire $z = 0$ plane, the tangential components of the electric field are identically zero. Hence, establish that the magnetic current element B is the image of the magnetic current element A for a perfectly conducting infinite surface occupying the $z = 0$ plane.

It is worthwhile to note the differences, as illustrated in Fig. P9.5, between the images of electric and magnetic current elements oriented parallel to an infinite, perfectly conducting plane.

9.27 A linear wire antenna in free space is situated along the z-axis from $z = -h$ to $z = h$. The current distribution along the antenna is assumed to be a constant I_0. Find the radiation intensity $\Phi(\theta)$. Evaluate the total time-averaged power radiated by the antenna.

9.28 A linear antenna in free space is situated along the z-axis from $z = -h$ to $z = h$. Determine the expressions for the charge distributions along the antenna corresponding to the following current distributions:

(a) $I(z) = I_0\left(1 - \dfrac{|z|}{h}\right)$, (b) $I(z) = I_0\left(1 - \dfrac{z^2}{h^2}\right)$,

(c) $I(z) = I_0 \cos \beta z \qquad (\beta h = \pi/2)$,

(d) $I(z) = I_0 \dfrac{(\cos \beta z - \cos \beta h)}{(1 - \cos \beta h)} \qquad (\beta h = \pi/4)$.

Sketch the current and the charge distributions along the antenna from $z/h = -1$ to $z/h = 1$ for all the four different cases.

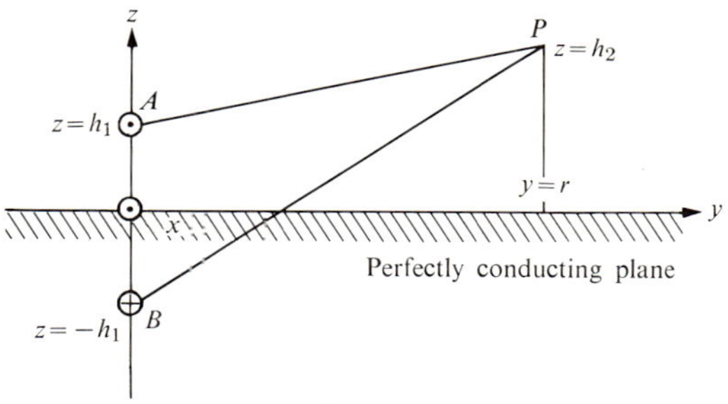

Fig. P9.6 Far-zone field due to a horizontal electric dipole A above a perfectly conducting plane. The source and the observation points are close to the surface of the perfectly conducting plane.

9.29 A Hertzian dipole A is situated on the z-axis at $z = h_1$ above a perfectly conducting plane surface occupying the $z = 0$ plane as shown in Fig. P9.6. The dipole is oriented in the positive x-direction. The observation point P is in the far-zone and is located at $x = 0$, $y = r$, and $z = h_2$. Note that $h_1 \ll r$ and $h_2 \ll r$. The electric field at P is due to the dipole A as well as that due to its image B both in the absence of the perfectly conducting plane. The image dipole B is also situated on the z-axis at $z = -h_1$ and is oriented in the $-x$-direction. Since P is in the far-zone of the dipoles A and B, it is legitimate to assume that the directions from A to P as well as from B to P are the same. Show that the magnitude of the resultant electric field at P is that due to the dipole A in the absence of a perfectly conducting plane multiplied by the factor $2\beta(h_1 h_2/r)$ provided $\beta h_1 h_2/r \ll 1$, where $\beta = 2\pi/\lambda$.

CHAPTER 10

RECEIVING ANTENNAS AND ANTENNA ARRAYS

10.1 INTRODUCTION

In this chapter, we shall treat the characteristics of receiving antennas and simple types of antenna arrays. First, for a receiving antenna, we shall introduce the concepts of effective length, effective area and directive gain. Then, we shall deduce the reciprocity theorem systematically from Maxwell's equations. With the help of the reciprocity theorem and the other well-known network theorems, we shall establish that the directive gain, the input impedance, and the effective height of a receiving antenna are identical to the corresponding quantities of the same antenna used as a transmitter. We shall prove that the ratio of the effective aperture to the directive gain is a universal constant independent of the direction of radiation and the type of the antenna. We shall deduce the Friss formula for transmission from a radiating to a receiving antenna. Then we shall introduce and interpret physically two simple current distributions valid for a linear, filamentary antenna. These current distributions pertain to a transmitting and a reflecting antenna respectively and we shall indicate a method of superposing these two current distributions to obtain the corresponding current distribution along a linear, filamentary antenna with a load impedance at its center.

The simple antennas which we treated in the previous chapter have directional patterns which are rather restricted in type and have in the far-zone a field which has only linear polarization. It is possible to have combination of antennas known as antenna arrays which produce a great flexibility in the available directional patterns and which enable far-zone field with circular and elliptic polarizations to be obtained. We shall investigate a two-element "broadcast" type antenna array and a two-element "point-to-point" type antenna array. We shall then mention the principle of electronic scanning. Finally, we shall discuss in sufficient detail the turnstile array and an array consisting of a small circular current loop with an axial electric dipole situated at its center, which are capable of producing electromagnetic radiation with circular or elliptic polarizations.

10.2 RECIPROCITY THEOREM

Let us consider, for example, a cylindrical antenna extending from $z = h$ to $z = \delta$ and from $z = -\delta$ to $z = -h$, as shown in Fig. 10.1(a). The input ends at $z = \delta$ and $z = -\delta$ are connected by a pair of transmission lines to a voltage generator.

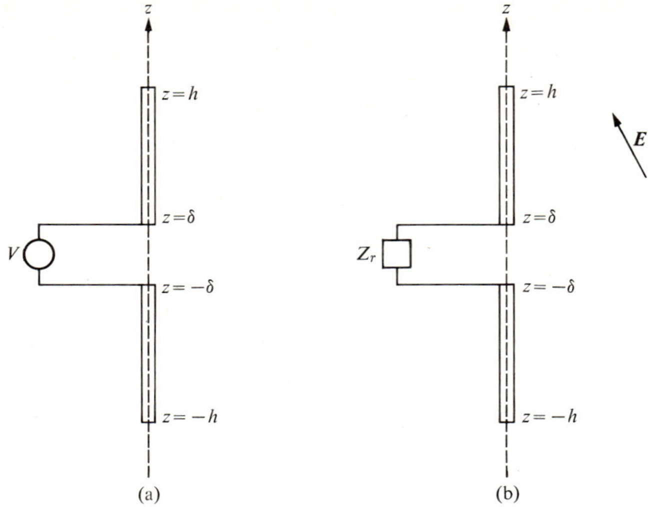

Fig. 10.1 Geometry of (a) the transmitting antenna and (b) the receiving antenna.

This voltage generator sets up time-varying charges and currents on the cylindrical structure, which radiates electromagnetic energy and thus acts like a transmitting antenna. On the other hand, let us consider the situation as illustrated in Fig. 10.1(b) in which the transmission lines are connected to an impedance and the cylindrical structure is placed in the electromagnetic field set up by another antenna. The incident electromagnetic field induces charge and current distributions on the cylindrical structure which in turn produces a voltage across the impedance. Thus, the cylindrical structure acts like a receiving antenna. It is possible for us to deduce the receiving characteristics of an antenna from a knowledge of its transmitting characteristics. The basis for establishing the reciprocity between the transmitting and the receiving properties of an antenna is the Rayleigh-Carson reciprocity theorem which we can deduce in a simple manner from Maxwell's equations.

Let $\boldsymbol{J}'$ be an impressed localized current distribution and let $\boldsymbol{E}'$ and $\boldsymbol{H}'$ be the electric and the magnetic fields due to $\boldsymbol{J}'$. In a similar manner, let $\boldsymbol{J}''$ be another impressed localized current distribution and let $\boldsymbol{E}''$ and $\boldsymbol{H}''$ be the electric and the magnetic fields set up by $\boldsymbol{J}''$. Evidently, $(\boldsymbol{E}', \boldsymbol{H}')$ and $(\boldsymbol{E}'', \boldsymbol{H}'')$ are governed by the following time-harmonic Maxwell's equations:

$$\nabla \times \boldsymbol{E}' = -j\omega\mu\boldsymbol{H}', \tag{2.1}$$

$$\nabla \times \boldsymbol{H}' = j\omega\varepsilon\boldsymbol{E}' + \boldsymbol{J}', \tag{2.2}$$

$$\nabla \times \boldsymbol{E}'' = -j\omega\mu\boldsymbol{H}'', \tag{2.3}$$

$$\nabla \times \boldsymbol{H}'' = j\omega\varepsilon\boldsymbol{E}'' + \boldsymbol{J}''. \tag{2.4}$$

Let us consider the following field quantity: $(\boldsymbol{E}' \times \boldsymbol{H}'' - \boldsymbol{E}'' \times \boldsymbol{H}')$. We can obtain its divergence as

$$\nabla\cdot[\boldsymbol{E}'\times\boldsymbol{H}''-\boldsymbol{E}''\times\boldsymbol{H}']=\boldsymbol{H}''\cdot(\nabla\times\boldsymbol{E}')-\boldsymbol{E}'\cdot(\nabla\times\boldsymbol{H}'')$$
$$-\boldsymbol{H}'\cdot(\nabla\times\boldsymbol{E}'')+\boldsymbol{E}''\cdot(\nabla\times\boldsymbol{H}'). \tag{2.5}$$

If we substitute the curls of the various field vectors from (2.1) to (2.4) into (2.5) we find

$$\begin{aligned}\nabla\cdot[\boldsymbol{E}'\times\boldsymbol{H}''-\boldsymbol{E}''\times\boldsymbol{H}']&=\boldsymbol{H}''\cdot(-j\omega\mu\boldsymbol{H}')-\boldsymbol{E}'\cdot(j\omega\varepsilon\boldsymbol{E}''+\boldsymbol{J}'')\\&\quad-\boldsymbol{H}'\cdot(-j\omega\mu\boldsymbol{H}'')+\boldsymbol{E}''\cdot(j\omega\varepsilon\boldsymbol{E}'+\boldsymbol{J}'),\\&=-\boldsymbol{E}'\cdot\boldsymbol{J}''+\boldsymbol{E}''\cdot\boldsymbol{J}'.\end{aligned} \tag{2.6}$$

If we integrate both sides of (2.6) throughout the volume V enclosed by the surface S and if we apply the divergence theorem to convert the volume integral on the left side of the resulting equation into an equivalent surface integral, we obtain

$$\int_S[\boldsymbol{E}'\times\boldsymbol{H}''-\boldsymbol{E}''\times\boldsymbol{H}']\cdot d\boldsymbol{S}=\int_V[-\boldsymbol{E}'\cdot\boldsymbol{J}''+\boldsymbol{E}''\cdot\boldsymbol{J}']\,dV. \tag{2.7}$$

Let S be the surface area of a sphere of radius r tending to infinity and therefore, in the limit, V is equal to all of space. We may simplify the surface integral I_s on the left side of (2.7) as follows:

$$I_s=\int_S[E_\theta'H_\phi''-E_\phi'H_\theta''-E_\theta''H_\phi'+E_\phi''H_\theta']\,dS. \tag{2.8}$$

In the far-zone, we know that

$$E_\theta'/H_\phi'=-E_\phi'/H_\theta'=E_\theta''/H_\phi''=-E_\phi''/H_\theta''=\eta_0, \tag{2.9}$$

where η_0 is the intrinsic impedance. With the help of (2.9), we may simplify the integrand of (2.8) as follows:

$$\begin{aligned}&E_\theta'H_\phi''-E_\phi'H_\theta''-E_\theta''H_\phi'+E_\phi''H_\theta'\\&\qquad=\eta_0H_\phi'H_\phi''+\eta_0H_\theta'H_\theta''-\eta_0H_\phi''H_\phi'-\eta_0H_\theta''H_\theta'=0.\end{aligned} \tag{2.10}$$

Hence $I_s=0$ and we may recast (2.7) as

$$\int_V\boldsymbol{E}'\cdot\boldsymbol{J}''\,dV=\int_V\boldsymbol{E}''\cdot\boldsymbol{J}'\,dV, \tag{2.11}$$

which is the mathematical statement of the reciprocity theorem.

It is convenient for us to particularize and obtain from (2.11) an equivalent circuit relation which is useful in deducing the receiving properties of an antenna from a knowledge of its transmitting properties. For this purpose, let us consider a perfectly conducting cylindrical antenna 1 of radius a_1, extending from $z_1=h_1$ to $z_1=\delta_1$ and from $z_1=-\delta_1$ to $z_1=-h_1$, as shown in Fig. 10.2. An axial (z_1) electric field which does not vary in a plane perpendicular to z_1 is maintained in the input region, $0<\rho_1<a_1$ and $-\delta_1<z_1<\delta_1$. We assume that δ_1 is vanishingly small and,

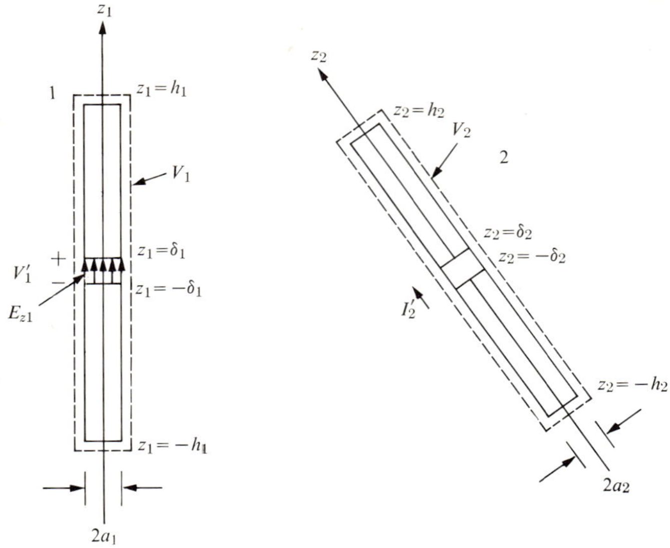

Fig. 10.2 A system of transmitting and receiving antennas for the illustration of the reciprocity theorem.

therefore, the input region is infinitesimally narrow. There is no conduction current in the input region, that is, the input terminals of antenna 1 are *open-circuited.* Let the electric field at the input of antenna 1 be denoted by $E'_1(\mathbf{r}_1)$ where the subscript indicates that the electric field pertains to antenna 1. Let us consider another perfectly conducting cylindrical antenna 2 of radius a_2, extending from $z_2 = h_2$ to $z_2 = \delta_2$ and from $z_2 = -\delta_2$ to $z_2 = -h_2$. As before, we also assume δ_2 to be vanishingly small and, therefore, the output region $0 < \rho_2 < a_2$, $-\delta_2 < z_2 < \delta_2$ is also infinitesimally narrow. The output terminals at $z_2 = \delta_2$ and $z_2 = -\delta_2$ are *short-circuited.* Let $J'_2(\mathbf{r}_2)$ be the axial (z_2) current density and $I'_2(\mathbf{r}_2)$ be the total axial current in the output region. The subscript 2 on $J'(\mathbf{r}_2)$ and $I'(\mathbf{r}_2)$ indicates that these quantities pertain to antenna 2. For convenience, we assume the output terminals to be situated on another antenna but there is no restriction on the output terminals being situated at a different location on the same antenna. Let us note that $J'_2(\mathbf{r}_2)$ is the current density associated with the electric field $E'_1(\mathbf{r}_1)$.

Let us consider another situation in which the terminals at $z_2 = \delta_2$ and $z_2 = -\delta_2$ are *open-circuited.* An axial (z_2) electric field $E''_2(\mathbf{r}_2)$ which does not vary in a plane perpendicular to z_2 is maintained in the output region, $0 < \rho_2 < a_2$ and $-\delta_2 < z_2 < \delta_2$. Now there is no conduction current between the terminals at $z_2 = \delta_2$ and $z_2 = -\delta_2$. We now assume the input terminals at $z_1 = \delta_1$ and $z_1 = -\delta_1$ to be short-circuited. Let $J''_1(\mathbf{r}_1)$ be the axial (z_1) current density and $I''_1(\mathbf{r}_1)$ be the total axial current in the input region. We assume that the radii a_1 and a_2 of the antennas are very small in comparison with a wavelength with the result that the current densities at the input and the output regions are parallel to the axis of the respective antenna, as we have assumed previously. We should note that $J_1''(\boldsymbol{r}_1)$ is the current density associated with the electric field $E_2''(\boldsymbol{r}_2)$.

The reciprocity theorem (2.11), when specialized to the problem of two antennas as detailed in the foregoing can be simplified to:

$$\int_{V_1} E_1'(\boldsymbol{r}_1)J_1''(\boldsymbol{r}_1)\,dV_1 = \int_{V_2} E_2''(\boldsymbol{r}_2)J_2'(\boldsymbol{r}_2)\,dV_2. \tag{2.12}$$

In (2.12) V_1 denotes the region $[\rho_1 \le a_1, |z_1| \le h_1]$ and V_2, the region $[\rho_2 \le a_2, |z_2| \le h_2]$. Clearly, V_1 and V_2 are the volumes occupied by antennas 1 and 2 respectively. Since $J_1''(\boldsymbol{r}_1)$ and $J_2'(\boldsymbol{r}_2)$ vanish outside of V_1 and V_2 respectively, (2.12) follows from (2.11). We may simplify the integral I_L on the left side of (2.12) further to yield

$$I_L = \int_0^{a_1}\int_0^{2\pi}\int_{-\delta_1}^{\delta_1} E_1'(\boldsymbol{r}_1)J_1''(\boldsymbol{r}_1)\rho_1\,d\rho_1\,d\phi_1\,dz_1, \tag{2.13}$$

since $E_1'(\boldsymbol{r}_1) = 0$ for $|z_1| > \delta_1$. Furthermore, δ_1 is vanishingly small and $J_1''(\boldsymbol{r}_1)$ is a smooth function near $z_1 = 0$; therefore, $J_1''(\boldsymbol{r}_1)$ can be taken outside of the sign of integration with respect to z_1. Hence, we may write (2.13) as

$$I_L = \int_0^{a_1}\int_0^{2\pi}\left[J_1''(\boldsymbol{r}_1)\rho_1\,d\rho_1\,d\phi_1 \int_{-\delta_1}^{\delta_1} E_1'(\boldsymbol{r}_1)\,dz_1\right]. \tag{2.14}$$

From (9.4.5) and the fact that $E_1'(\boldsymbol{r}_1)$ is in the z_1-direction, it follows that

$$E_1'(\boldsymbol{r}_1) = -\frac{d}{dz_1}\phi(\boldsymbol{r}_1) - j\omega A_{z1}(\boldsymbol{r}_1), \tag{2.15}$$

where $\phi(\boldsymbol{r}_1)$ is the scalar potential and $A_{z1}(\boldsymbol{r}_1)$ is the z_1-component of the vector potential. In view of (2.15), we obtain

$$\int_{-\delta_1}^{\delta_1} E_1'(\boldsymbol{r}_1)\,dz_1 = [-\{\phi(z_1 = \delta_1) - \phi(z_1 = -\delta_1)\} - j\omega \int_{-\delta_1}^{\delta_1} A_{z1}(\boldsymbol{r}_1)\,dz_1]. \tag{2.16}$$

The z_1-component of the vector potential $A_{z1}(\boldsymbol{r}_1)$ is a smooth function near $z_1 = 0$ and therefore the integral of $A_{z1}(\boldsymbol{r}_1)$ vanishes in the limit of δ_1 tending to zero. Let

$$V_1'(z_1 = 0) = \phi(z_1 = \delta_1) - \phi(z_1 = -\delta_1). \tag{2.17}$$

As we indicated in Fig. 10.2, $V_1'(z_1 = 0)$ is the potential difference between the input terminals in the limiting case of an infinitesimally narrow input region. Since $E_1'(\boldsymbol{r}_1)$ is independent of ρ_1 and ϕ_1, it follows that $V_1'(z_1 = 0)$ is a constant. Together with (2.16) and (2.17), we may show (2.14) to yield

$$I_L = -V_1'(z_1 = 0)\int_0^{a_1}\int_0^{2\pi} J_1''(\boldsymbol{r}_1)\rho_1\,d\rho_1\,d\phi_1 = -V_1'(z_1 = 0)I_1''(z_1 = 0), \tag{2.18}$$

where I_1'' is the total axial (z_1) current at the input region $z_1 = 0$. In a similar manner, we may simplify the integral I_R on the right side of (2.12) to give the following result:

$$I_R = -V_2''(z_2 = 0)I_2'(z_2 = 0), \tag{2.19}$$

where, as indicated in Fig. 10.2, $V_2''(z_2 = 0)$ is the potential difference between the output terminals in the limiting case of an infinitesimally narrow output region and I_2'' is the total axial (z_2) current at the output region. In view of (2.12), (2.18), and (2.19), we obtain the following circuit relation:

$$\frac{V_1'(z_1 = 0)}{I_2'(z_2 = 0)} = \frac{V_2''(z_2 = 0)}{I_1''(z_1 = 0)}. \tag{2.20}$$

The potential differences appearing in (2.20) are *open-circuit voltages* and the currents appearing in (2.20) are *short-circuit currents*. Also, the currents I_2' and I_1'' are due to the voltages V_1' and V_2'' respectively. The reciprocity relation (2.20) states that the ratio of the open-circuit voltage at $z_1 = 0$ to the resulting short-circuit current at $z_2 = 0$ is equal to the ratio of the open-circuit voltage at $z_2 = 0$ to the resulting short-circuit current at $z_1 = 0$.

Before proceeding to apply (2.20) for the determination of the receiving properties of an antenna from a knowledge of its transmitting properties, we shall illustrate the usefulness and the power of the reciprocity theorem as given by (2.20) by solving the following transmission-line problem which we treated previously without employing the reciprocity theorem. Let us consider a lossless transmission line of characteristic resistance R_0, extending from $x = 0$ to $x = l$ (Fig. 10.3(a)). Let $Z_L = R_L + jX_L$ be the load impedance connected to the terminals at $x = l$. A voltage generator V_g in series with a resistor R_0 is connected to the input terminals at $x = 0$. We wish to find the time-averaged power $P(x = 0)$ delivered to the input terminals of the transmission line.

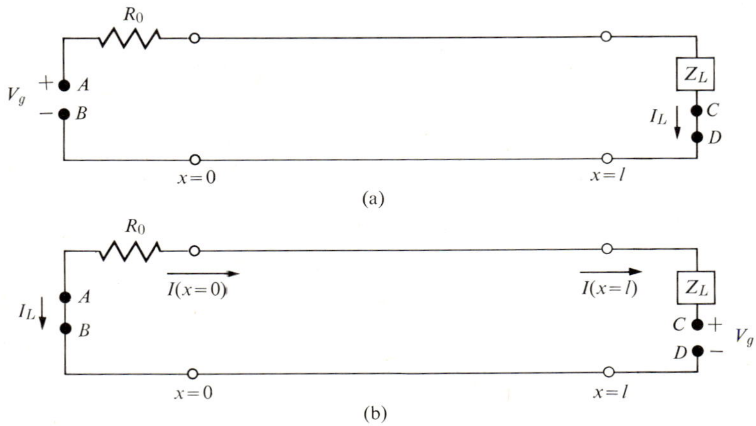

Fig. 10.3 Application of reciprocity theorem to a lossless transmission line driven by a matched generator and terminated in an arbitrary load impedance.

The transmission line is lossless and no power is dissipated along the transmission line and, therefore, the time-averaged power $P(x = 0)$ delivered to the input terminals of the transmission line is the same as that dissipated in the load impedance. Hence

$$P(x = 0) = \tfrac{1}{2}R_L |I_L|^2, \tag{2.21}$$

where I_L is the load current or the current in the short-circuited section CD in series with the load impedance, as shown in Fig. 10.3(a). By the reciprocity theorem (2.20), we find I_L in the following manner. The terminals C and D are open-circuited, the voltage generator which was previously connected to the terminals A and B is removed and is connected to the terminals C and D, as shown in Fig. 10.3(b). Then, the current through the short-circuit connected to the terminals A and B is equal to I_L. The input impedance at $x = l$ of the transmission line extending up to $x = 0$ is equal to R_0 since the transmission line is terminated by its characteristic resistance R_0. We can solve the circuit problem at $x = l$ to yield the following expression for the transmission line current at $x = l$:

$$I(x = l) = \frac{-V_g}{R_0 + Z_L}. \tag{2.22}$$

The current on the transmission line shown in Fig. 10.3(b) is a wave progressing in the $-x$-direction. Therefore, the amplitude of the current remains constant and only the phase changes linearly with the distance along the transmission line. Therefore, we obtain the transmission line current at $x = 0$ from (2.22) as

$$I(x = 0) = \frac{-V_g e^{-j\beta l}}{R_0 + Z_L}, \tag{2.23}$$

where β is the phase-shift coefficient. If we examine the electric circuit at $x = 0$ we see that

$$I_L = -I(x = 0). \tag{2.24}$$

From (2.21), (2.23), and (2.24) we may show that

$$P(x = 0) = \frac{|V_g|^2 R_L}{2[(R_0 + R_L)^2 + X_L^2]}. \tag{2.25}$$

If we use (3.9.25) in (2.25) we obtain

$$P(x = 0) = \frac{|V_g|^2 (1 - |\rho|^2)}{8R_0}, \tag{2.26}$$

where ρ is the terminal reflection coefficient. The result given by (2.26) is the same as that obtained in (3.9.22) which we deduced without the application of the reciprocity theorem. The interchange of the locations of the voltage source and the resulting current yields a transmission-line problem, as illustrated in Fig. 10.3(b) and this new transmission-line problem turns out to be simpler than the original problem in view of the fact that for the new transmission line the terminating impedance is equal to the characteristic resistance of the transmission line. Consequently, if we apply the reciprocity theorem as given by (2.20) we simplify considerably the solution of the given transmission-line problem.

10.3 EFFECTIVE LENGTH AND EFFECTIVE AREA OF A RECEIVING ANTENNA

1. Effective Height

Let us consider a filamentary antenna extending along the z-axis from $z = h$ to $z = \delta$ and from $z = -\delta$ to $z = -h$, as shown in Fig. 10.4(a). We assume the input terminals at $z = \delta$ and $z = -\delta$ to be infinitesimally close to each other. This antenna is in the electromagnetic field of a distant transmitter which is in the direction specified by OP. We assume the receiving antenna to be in the far-zone of the transmitter, as is usually the case. Therefore, the electric field lies in a plane perpendicular to OP. We assume that the electric field is in a plane which contains the axis of the antenna and the direction from the center of the antenna to the distant transmitter. Since the total length $2h$ of the receiving antenna is very small in comparison with the distance from the transmitter to the receiving antenna, it follows that the amplitude of the incident electric field is essentially the same at every point along the length of the antenna. Let E be the amplitude of the incident electric field at the location of the receiving antenna and let V_{OC} be the open-circuit voltage induced between the input terminals at the center of the antenna. The ratio V_{OC}/E has the dimension of a length; it is denoted by $2h_e^r(\theta, \phi)$ and is called the effective height or the effective length of the receiving antenna. For a given amplitude of the incident electric field, the open-circuit voltage induced between the input terminals of the antenna is dependent on the wave normal direction of the incident wave. For example, if the

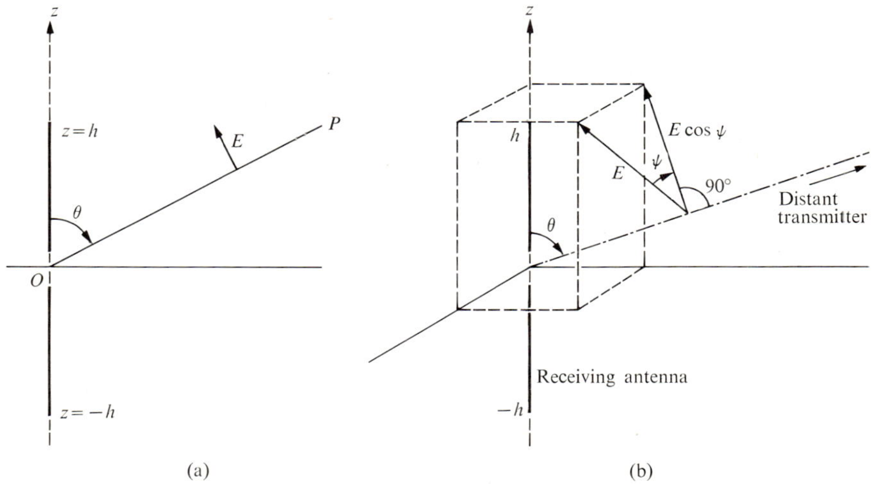

Fig. 10.4 Receiving antenna in an electric field of a distant transmitter. (a) Electric field in the "incident" plane and (b) electric field making an angle ψ to the "incident" plane.

wave arrives in the direction of the axis of the antenna, no voltage is induced since the electric field is perpendicular to the axis of the antenna. If the wave arrives in a direction perpendicular to the length of the antenna, the induced voltage is a maximum since the electric field is parallel to the length of the antenna. Consequently, in general, the effective height $2h_e^r(\theta, \phi)$ of the receiving antenna is a function of θ and ϕ, which specify the direction of arrival of the incident electromagnetic wave. We can restate the defining relation for the effective height $2h_e^r(\theta, \phi)$ of the receiving antenna explicitly as

$$V_{OC} = 2h_e^r(\theta, \phi)E. \tag{3.1}$$

Let us suppose that the electric field is not in a plane which contains the axis of the antenna and the wave normal of the incident wave. For lack of a better term, the plane referred to in the foregoing is called the "incident plane". Let us suppose that the angle between the electric vector and its projection on the "incident plane" is ψ, as shown in Fig. 10.4(b). Then, $E \cos \psi$ is the component of the incident electric field parallel to the "incident plane." Since $2h_e^r(\theta, \phi)$ is *defined* to be the ratio of the open-circuit voltage induced between the input terminals of the antenna to the magnitude of the electric field lying in the so-called "incident plane", it follows that

$$V_{OC} = 2h_e^r(\theta, \phi)E \cos \psi. \tag{3.2}$$

If $\psi = \pi/2$, there is no component of the incident electric field parallel to the "incident plane" and the induced voltage is zero. We should emphasize that the effective length or the effective height $2h_e^r(\theta, \phi)$ of the receiving antenna is the ratio of the induced open-circuit voltage between the input terminals of the antenna to the magnitude of the incident electric field parallel to the plane containing the axis of the antenna and the wave normal direction of the incident wave.

2. Equivalent Circuit of a Receiving Antenna

There is an apparent impedance Z_a^r between the input terminals of the receiving antenna. If the input terminals of the antenna are connected to a transmission line, the measurements on the transmission line can be made to yield the value of the apparent impedance terminating the transmission line and this terminal impedance is the antenna impedance Z_a^r. In general, the antenna impedance has both a resistive part R_a^r and a reactive part X_a^r, and these are dependent on the geometry of the antenna, the frequency of the electromagnetic wave and the characteristics of the environment.

The equivalent circuit of the receiving antenna is shown in Fig. 10.5(a) where Z_L is the equivalent load impedance. In practice, the antenna input terminals are connected to a transmission line whose other end is terminated by the actual load impedance. The "input" impedance of the transmission line at the antenna input terminals is then the equivalent load impedance which is denoted by Z_L. We can obtain the current I_1 in the input terminals of the antenna and the voltage V_1 across the equivalent load impendance Z_L from Fig. 10.5(a) as

$$I_1 = \frac{-V_{OC}}{Z_L + Z_a^r}; \qquad V_1 = \frac{V_{OC}Z_L}{Z_L + Z_a^r}. \tag{3.3}$$

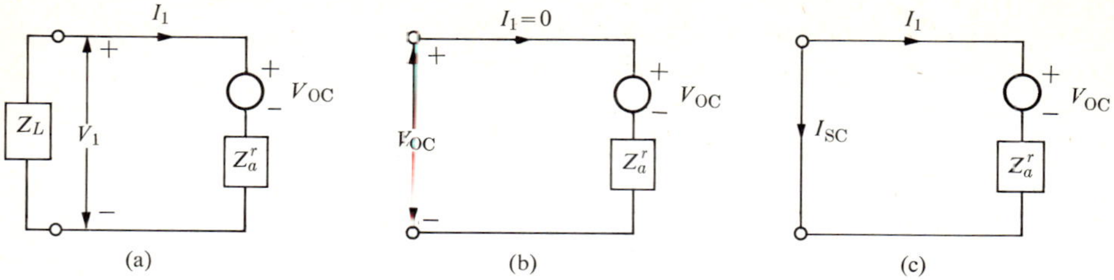

Fig. 10.5 Equivalent circuit of a receiving antenna under (a) loaded, (b) open-circuit, and (c) short-circuit conditions.

For $Z_L = \infty$, (3.3) yields

$$I_1 = 0; \qquad V_1 = V_{OC}. \tag{3.4}$$

It is evident that $Z_L = \infty$ corresponds to the case in which the input terminals of the antenna are open-circuited and the corresponding equivalent circuit is illustrated in Fig. 10.5(b). For $Z_L = 0$, we can see from (3.3) and Fig. 10.5(c) that

$$I_{SC} = -I_1 = V_{OC}/Z_a^r; \qquad V_1 = 0. \tag{3.5}$$

We find from (3.5) that the short-circuit current I_{SC} through the input terminals of the antenna is the ratio of the induced open-circuit voltage to the antenna impedance.

The power delivered to the load impedance depends on the load impedance. We proved in Section 3.10 that the power delivered to the load impedance is a maximum if the load impedance Z_L is the complex conjugate of the generator impedance, which in the present case is equal to the antenna impedance Z_a^r. If the load impedance is adjusted for maximum power transfer, we obtain the current I_1 in the input terminals of the antenna from (3.3) as

$$I_1 = \frac{-V_{OC}}{Z_a^{r*} + Z_a^r} = -\frac{V_{OC}}{2R_a^r}. \tag{3.6}$$

Therefore, we evaluate the maximum time-averaged power P delivered to the load impedance from (3.6) as

$$P = \tfrac{1}{2}R_a^r\,|I_1|^2 = \frac{|V_{OC}|^2}{8R_a^r}. \tag{3.7}$$

3. Effective Area

We see from (3.2) that the induced open-circuit voltage V_{OC} has the largest value for a given direction of arrival of the incident wave if $\psi = 0$, that is, if the electric field lies in a plane containing the axis of the antenna and the wave normal direction of the incident wave. The value of P given by (3.7) has the largest value P_a for $\psi = 0$ and this value of P_a is called the "available" power at the input terminals of the antenna for a given direction of arrival of the incident wave. Therefore, we obtain from (3.2) with $\psi = 0$ and (3.7):

$$P_a = \frac{|2h_e^r(\theta, \phi)|^2}{8R_a^r}\,|E|^2. \tag{3.8}$$

If the receiving antenna is in the far-zone of the transmitter, we may deduce the magnitude of the incident time-averaged Poynting vector S at the location of the antenna to be

$$S = \frac{|E|^2}{2\eta_{0r}}, \tag{3.9}$$

where η_{0r} is the intrinsic resistance of the medium. The ratio P_a/S has the dimensions of an area and is defined as the *effective area* or the *effective aperture* $A_e^r(\theta, \phi)$ of the receiving antenna. Thus

$$P_a = A_e^r(\theta, \phi)S. \tag{3.10}$$

Therefore, the effective area $A_e^r(\theta, \phi)$ of the receiving antenna is a function of the direction of arrival of the incident wave and is the ratio of the maximum time-averaged power transferred to the equivalent load impedance to the magnitude of the time-averaged Poynting vector at the location of the receiving antenna, for the case in which the incident electric vector lies in the plane containing the axis of the antenna and the wave normal direction of the incident wave. From (3.8) to (3.10), we can deduce the effective area $A_e^r(\theta, \phi)$ to be

$$A_e^r(\theta, \phi) = \frac{\eta_{0r}}{4R_a^r}|2h_e^r(\theta, \phi)|^2. \tag{3.11}$$

We see from (3.11) that the effective area $A_e^r(\theta, \phi)$ of the receiving antenna depends on the geometry of the antenna, the frequency, the properties of the environment, and the direction of arrival of the incident electromagnetic wave.

10.4 CHARACTERISTICS OF A RECEIVING ANTENNA

We derived the characteristics of a transmitting antenna from a knowledge of the current distribution along its length. For the idealized case of a filamentary antenna, such as we considered in the previous chapter, we assumed a sinusoidal current distribution. This assumed current distribution is justified for the filamentary antennas on the basis of a rigorous analysis not presented in this treatment. In a similar manner, we can deduce the characteristics of a receiving antenna from a knowledge of the current distribution along its length. The current distribution on a receiving antenna depends on the direction of arrival of the incident electromagnetic wave and hence is considerably more complicated than that on the transmitting antenna. For this reason, it is difficult for us to evaluate the properties of a receiving antenna directly from a knowledge of its current distribution. It is simpler, as has been stated previously, for us to deduce with the help of the reciprocity theorem (2.20) the characteristics of a receiving antenna from a knowledge of its characteristics when used as a transmitting antenna.

1. Directional Pattern

First, we shall show that the directional pattern of a receiving antenna is identical with that of the same antenna used as a transmitter. The directional pattern of a

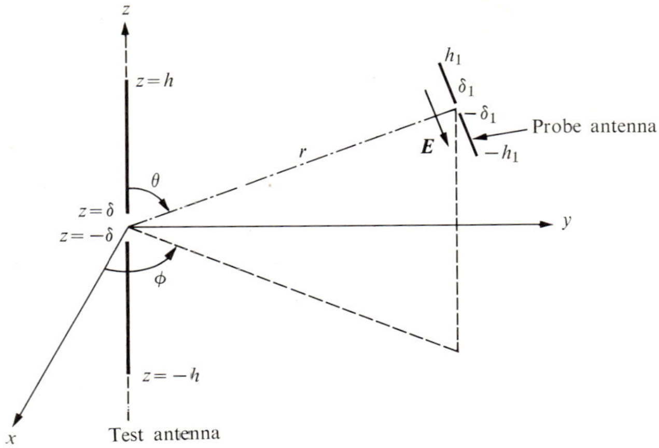

Fig. 10.6 A system of transmitting and receiving antennas for the measurement of the directional pattern.

transmitting antenna indicates the strength of the far-zone electric field at a fixed distance from the center of the transmitting antenna in all different directions in space. The directional pattern of a receiving antenna indicates the response of the antenna, namely the current in or the voltage across the load impedance, to an electromagnetic wave of specified strength and polarization arriving from all different directions in space. Let us consider a linear filamentary antenna with its center at the origin of the coordinate system and extending along the z-axis from $z = h$ to $z = \delta$ and from $z = -\delta$ to $z = -h$. We assume the input terminals at $z = \delta$ and $z = -\delta$ to be infinitesimally close to each other. This antenna is called the test antenna and its directional pattern is desired both as a transmitter and as a receiver. Let us consider another antenna called the probe antenna situated at a large distance r from the test antenna, as shown in Fig. 10.6. The probe antenna is of length $2h_1$ with its input terminals situated symmetrically at its center and separated by an infinitesimally small distance $2\delta_1$. In the far-zone, the electric field radiated by the test antenna is in the plane containing the axis of the test antenna and the line joining the centers of the two antennas. Also, the electric vector is perpendicular to the radius vector r, that is, the electric vector is in the θ-direction, as shown in Fig. 10.6. We adjust the probe antenna to be always parallel to the electric vector. We make two measurements to obtain the data for the directional pattern of the test antenna both as a transmitter and as a receiver. First, we apply a voltage V to the input terminals of the test antenna and measure the short-circuit current I_p at the terminals of the probe antenna. This short-circuit current I_p is a measure of the strength of the electric field at the location of the probe antenna and hence provides data for the directional pattern of the test antenna used as a transmitter. Then, we apply the voltage V to the input terminals of the probe antenna and measure the short-circuit current I_t at the terminals of the test antenna. This short-circuit current I_t is a measure of the response of the test antenna to the electromagnetic field arriving in the direction specified by θ and ϕ, as shown in Fig. 10.6, and hence provides data for the directional pattern of the test antenna used as a receiver. We move the probe antenna to every point on the surface of a large sphere of radius r. Care is taken to ensure that the probe antenna is always in the plane containing the axis of the test antenna and the radius vector to the probe antenna and is always perpendicular to the radius vector. In other words, the probe antenna is oriented so that its length is parallel to the θ-direction. We repeat the two previous measurements for every location of the probe antenna with the same applied voltage. The first set of measurements, that is, the short-circuit current I_p in the probe antenna as a function of θ and ϕ, yield the directional pattern of the test antenna used as a transmitter. The second set of measurements, namely, the short-circuit current I_t in the test antenna as a function of θ and ϕ, which specify the location of the probe antenna and hence the direction of arrival of the electromagnetic wave, give the directional pattern of the test antenna used as a receiver. The reciprocity theorem (2.20) shows that $I_t = I_p$ for every location of the probe antenna. Therefore, the directional patterns of the test antenna are the same irrespective of whether we use it as a transmitter or as a receiver.

If the electric field radiated by an antenna has two components, E_θ and E_ϕ, then the directional patterns for the θ-polarization and the ϕ-polarization are usually shown separately. We obtain the directional pattern for the specified polarization by adjusting the probe antenna to be parallel to the given polarization. The foregoing discussion is restricted to the measurement of the θ-polarization only. For the measurement of the ϕ-polarization, we adjust the probe antenna to be always perpendicular to the plane containing the axis of the test antenna and the radius vector to the probe antenna. By the same arguments as before, for the ϕ-polarization the directional pattern of an antenna is the same irrespective of whether it is used as a transmitter or as a receiver.

Since the directional pattern of a receiving antenna is identical to that of the same antenna used as a transmitter, we can show that the directive gain of a receiving antenna is identical to that of the same antenna used as a transmitter. For the linear filamentary test antenna under consideration, the far-zone electric field is of the form

$$E_\theta(\mathbf{r}) = \frac{e^{-j\beta r}}{4\pi r} f(\theta, \phi), \tag{4.1}$$

where $f(\theta, \phi)$ is defined as the directional pattern. Even though the directional pattern of a filamentary antenna is independent of ϕ, we retain the argument ϕ since the following discussion is also valid for the other types of antennas whose directional patterns depend on ϕ. Since $H_\phi(\mathbf{r})$ is proportional to $E_\theta(\mathbf{r})$, it follows that the radiation intensity $\Phi(\theta, \phi)$ is proportional to $|f(\theta, \phi)|^2$. Hence,

$$\Phi(\theta, \phi) = C\,|f(\theta, \phi)|^2, \tag{4.2}$$

where C is a constant. We can obtain the total time-averaged power radiated by the antenna as

$$P = C \int_{\phi=0}^{\phi=2\pi} \int_{\theta=0}^{\theta=\pi} |f(\theta, \phi)|^2 \sin\theta \, d\theta \, d\phi. \tag{4.3}$$

The radiation intensity Φ_{iso} of the equivalent isotropic radiator is equal to $P/4\pi$ and hence the directive gain of the test antenna used as a transmitter is given by

$$g(\theta, \phi) = \frac{4\pi\,|f(\theta, \phi)|^2}{\left[\int_{\phi=0}^{2\pi}\int_{\theta=0}^{\pi} |f(\theta, \phi)|^2 \sin\theta\, d\theta\, d\phi\right]}. \tag{4.4}$$

Let us suppose that $P(\theta, \phi)$ is the power delivered to the matched load connected to the terminals of the test antenna when used as a receiver of electromagnetic waves arriving in the direction specified by θ and ϕ. If the electromagnetic wave of the same strength and the same polarization arrives from all possible directions, then we can obtain the total power P_r delivered to the matched load as

$$P_r = \int_{\phi=0}^{2\pi}\int_{\theta=0}^{\pi} P(\theta, \phi) \sin\theta\, d\theta\, d\phi. \tag{4.5}$$

For the same total power P_r delivered to the matched load, if the response of the receiving antenna is the same for all directions, that is, if the receiving antenna is isotropic, then we find the power P_{iso} delivered to the matched load for each direction to be

$$P_{iso} = P_r/4\pi. \tag{4.6}$$

P_{iso} may be called the powered delivered to the matched load connected to the terminals of an equivalent isotropic receiver. Then, we define the directive gain $g^r(\theta, \phi)$ of the receiving antenna as

$$g^r(\theta, \phi) = \frac{P(\theta, \phi)}{P_{iso}} = \frac{4\pi P(\theta, \phi)}{\int_{\phi=0}^{2\pi}\int_{\theta=0}^{\pi} P(\theta, \phi) \sin\theta\, d\theta\, d\phi}. \tag{4.7}$$

Since the directional pattern of the receiving antenna is the same as that of the transmitting antenna, we find that the directional pattern of the receiving antenna is proportional to $f(\theta, \phi)$ and hence the power $P(\theta, \phi)$ delivered to the matched load is proportional to $|f(\theta, \phi)|^2$. Therefore,

$$P(\theta, \phi) = C_1\,|f(\theta, \phi)|^2, \tag{4.8}$$

where C_1 is a constant. Substituting (4.8) in (4.7), we find that

$$g^r(\theta, \phi) = g(\theta, \phi), \tag{4.9}$$

that is, the directive gains of the receiving and the transmitting antennas are identical.

2. Antenna Impedance

The apparent impedance between the terminals of an antenna used as a receiver is identical to that of the same antenna used as a transmitter. Let us consider, for the sake of definiteness, a filamentary antenna with its center at the origin of the coordinate system and extending along the z-axis from $z = h$ to $z = \delta$ and from $z = -\delta$ to $z = -h$, as shown in Fig. 10.7(a). Let the electromagnetic field due to any number of transmitting antennas which may be in the near-zone or in the far-zone of the receiving antenna be incident on the receiving antenna. Let V_{OC} be the open-circuit

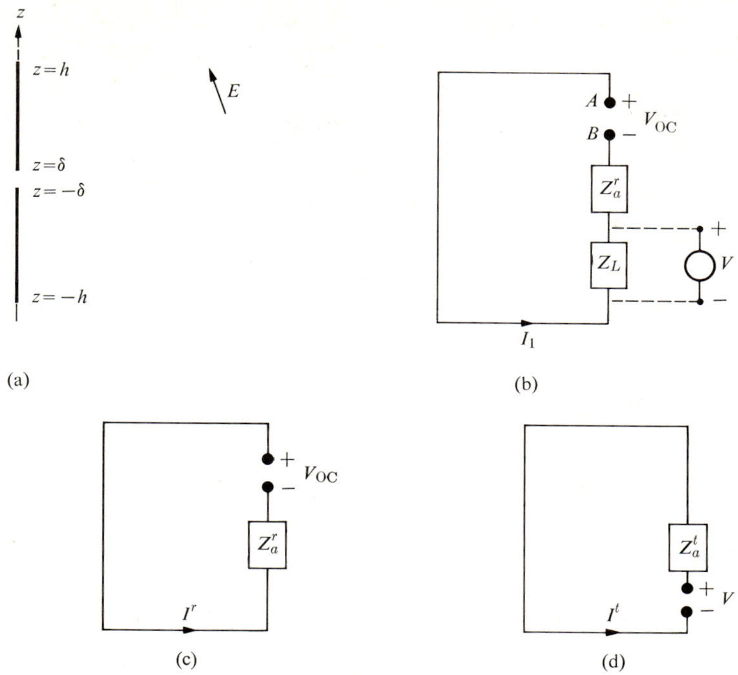

Fig. 10.7 (a) Receiving antenna in an incident electromagnetic field. (b) Equivalent circuit of the receiving antenna indicating the voltage compensating the load impedance. (c) Equivalent circuit of the receiving antenna with only the open-circuit voltage due to the incident field. (d) Equivalent circuit of the receiving antenna with only the compensating voltage for the load impedance.

voltage induced between the input terminals of the receiving antenna. Then, the equivalent electric circuit for the determination of the input current of the antenna is as shown in Fig. 10.7(b) where Z_a^r is the impedance of the antenna used as a receiver and Z_L is the equivalent load impedance as seen at the antenna terminals. Then, we obtain the current I_1 in the antenna terminals from Fig. 10.7(b) as

$$I_1 = \frac{V_{OC}}{Z_L + Z_a^r}. \tag{4.10}$$

The voltage drop V across the equivalent load impedance is depicted in Fig. 10.7(b) and is given by

$$V = -\frac{V_{OC} Z_L}{Z_L + Z_a^r}. \tag{4.11}$$

Now, we can use the compensation theorem to replace the equivalent load impedance by an impedanceless generator whose voltage at every instant is equal to the potential difference across the impedance due to the current in it. Therefore, we can replace the equivalent load impedance Z_L by an impedanceless generator of voltage V as given by (4.11). We can then use the superposition theorem to break up the circuit problem into a superposition of two circuit problems one of which corresponds to the antenna being used as a receiver and the other in which it is used as a transmitter. According to the superposition theorem, the current I_1 in the antenna terminals is equal to the sum of the currents due to the separate presence of the open-circuit voltage V_{OC} and the compensating voltage V. When the open-circuit voltage V_{OC} alone is present, as illustrated in Fig. 10.7(c), the antenna is used as a receiver and therefore the antenna impedance is equal to Z_a^r, the receiving antenna impedance. We obtain the current I_r in the antenna terminals as

$$I_r = \frac{V_{OC}}{Z_a^r}. \tag{4.12}$$

When the compensating voltage V alone is present, as illustrated in Fig. 10.7(d), the antenna is used as a transmitter and therefore the antenna impedance is equal to Z_a^t, the transmitting antenna impedance. We find the current I_t in the terminals in the case of the transmitting antenna with the help of Fig. 10.7(d) and (4.11) to be

$$I_t = -\frac{V_{OC}Z_L}{(Z_L + Z_a^r)Z_a^t}. \tag{4.13}$$

According to the superposition theorem, $I_1 = I_r + I_t$. Hence, it follows from (4.10), (4.12), and (4.13) that

$$\frac{V_{OC}}{Z_L + Z_a^r} = \frac{V_{OC}}{Z_a^r} - \frac{V_{OC}Z_L}{(Z_L + Z_a^r)Z_a^t}. \tag{4.14}$$

Multiplication of both sides of (4.14) by $(Z_L + Z_a^r)/V_{OC}$ yields

$$1 = \frac{Z_L}{Z_a^r} + 1 - \frac{Z_L}{Z_a^t},$$

or

$$Z_a^r = Z_a^t. \tag{4.15}$$

Thus, we have proved that the impedance of an antenna is the same irrespective of whether it is used as a transmitter or as a receiver.

In fact, the result that the impedance of an antenna is the same for transmission as well as for reception is somewhat evident from the very definition of the impedance. The antenna impedance is defined as the ratio of the voltage applied to the antenna terminals to the resulting current and is independent of the voltage or the manner in which it is applied. For the transmitting antenna, in essence, the terminals of a voltage generator are connected to the antenna terminals. For the receiving antenna, a voltage is induced between the antenna terminals by the incident electromagnetic wave. Since the impedance is independent of the manner in which the voltage is applied, it follows that the receiving and the transmitting antenna impedances are identical, a result which we have established previously with the help of the compensation and the superposition theorems.

3. Effective Height

The effective height of an antenna used as a transmitter is equal to that of the same antenna used as a receiver. Let us consider as before, for the sake of definiteness, a filamentary antenna with its center at the origin of the coordinate system and extending along the z-axis from $z = h$ to $z = \delta$ and from $z = -\delta$ to $z = -h$, as shown in Fig. 10.8. We assume the input terminals at $z = \delta$ and $z = -\delta$ to be infinitesimally close to each other. An electromagnetic wave from a transmitter in the far-zone is incident on the receiving antenna from the direction specified by θ and ϕ. We should note that the characteristics of the filamentary antenna are independent of ϕ but we retain ϕ as one of the arguments for the effective height $2h_e(\theta, \phi)$ in order to point out that the equivality of the effective heights, though derived here for the filamentary antennas which have azimuthal symmetry, is a generally valid result. We assume the electric field to lie in the plane containing the axis of the filamentary antenna and the direction of arrival of the incident wave, and to be perpendicular to the wave normal direction, as illustrated in Fig. 10.8. Since the receiving antenna is in the far-zone of the transmitter, the amplitude of the incident electric field is the same throughout the length of the receiving antenna. But, the phase of the incident electric field varies along the length of the antenna. Let us consider an infinitesimal length dz' of the filament situated at a distance z' from the origin. Then, we can see the path difference between the waves arriving at $z = z'$ and $z = 0$ from Fig. 10.8 to be $z' \cos\theta$ and, therefore, the field E at the center of the antenna has to be multiplied by the phase factor $e^{j\beta z' \cos\theta}$ to yield the value of the field at $z = z'$. We should note that the reference for the phase of the incident electric field is taken to be at the center of the antenna or the origin of the coordinate system. Therefore, we obtain the component of the

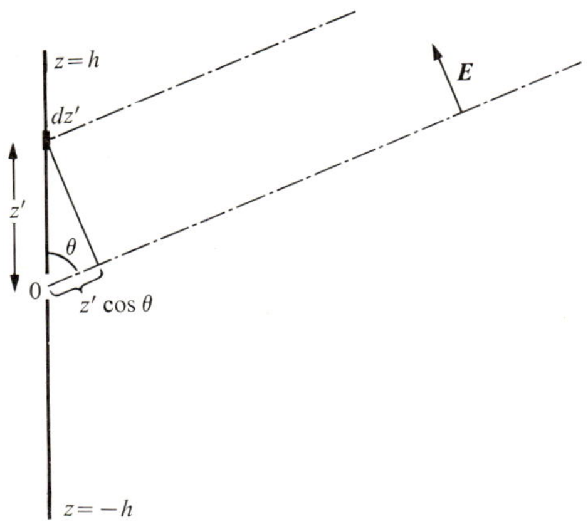

Fig. 10.8 Evaluation of the effective height of a receiving antenna.

electric field parallel to the infinitesimal length dz' of the filament at z' as $E \sin \theta e^{j\beta z' \cos \theta}$. The voltage induced in the infinitesimal length dz' of the filament is $E \sin \theta e^{j\beta z' \cos \theta}\, dz'$. Let dI_{SC} be the corresponding short-circuit current produced at the input terminals of the antenna. Let the antenna be used as a transmitter by the application of voltage V between the input terminals and let $I(z)$ be the resulting current distribution along the antenna. Then, the reciprocity theorem (2.20) yields

$$\frac{dI_{SC}}{E \sin \theta e^{j\beta z' \cos \theta}\, dz'} = \frac{I(z')}{V},$$

or

$$dI_{SC} = \frac{E \sin \theta}{V} I(z') e^{j\beta z' \cos \theta}\, dz'. \tag{4.16}$$

We obtain the total short-circuit current I_{SC} produced at the input terminals of the antenna due to the contributions arising from the voltage induced at every point along the length of the antenna by integrating (4.16) throughout the length of the antenna from $z' = -h$ to $z' = h$. Thus,

$$I_{SC} = \frac{E \sin \theta}{V} \int_{-h}^{h} I(z') e^{j\beta z' \cos \theta}\, dz'. \tag{4.17}$$

Using (9.6.55) in (4.17) we obtain

$$I_{SC} = \frac{EI(0)}{V} \{2h_e(\theta, \phi)\}, \tag{4.18}$$

where $2h_e(\theta, \phi)$ is the effective height of the antenna used as a transmitter. We should note that $V/I(0) = Z_a$, the input impedance of the antenna. From (3.5), it follows that $I_{SC}Z_a = I_{SC}V/I(0)$ is equal to the open-circuit voltage V_{OC} induced between the input terminals of the antenna. Therefore, (4.18) becomes

$$I_{SC} \frac{V}{I(0)} = V_{OC} = E\{2h_e(\theta, \phi)\}. \tag{4.19}$$

If we compare (4.19) with (3.1) we see that

$$2h_e^r(\theta, \phi) = 2h_e(\theta, \phi), \tag{4.20}$$

which brings out the result that the effective height of an antenna used as a receiver is identical to that of the same antenna used as a transmitter.

Since the directive gain $g^r(\theta, \phi)$, the impedance Z_a^r, and the effective height $2h_e^r(\theta, \phi)$ of an antenna used as a receiver are identical with the corresponding quantities of the same antenna used as a transmitter, we omit the superscript r in the remainder of the treatment of the receiving antennas.

10.5 RATIO OF EFFECTIVE APERTURE TO DIRECTIVE GAIN AND FRISS TRANSMISSION FORMULA

There is an important and universally valid relation between the effective aperture and the directive gain of any antenna. Before proceeding to deduce this relation, we shall evaluate the effective areas of few simple antennas.

1. Hertzian Dipole

For the Hertzian dipole, the current is a constant along the dipole, say equal to $I(0)$, the value at the center of the dipole. The total length dl of the dipole is infinitesimally small with the result that it is legitimate to expand the exponential in (9.6.55) into a power series and retain only the leading term of unity. This is equivalent to assuming that the electric field has the same phase everywhere along the dipole and this assumption is valid since the total length of the dipole is infinitesimally small. Therefore, for the Hertzian dipole, we obtain from (9.6.55)

$$2h_e(\theta, \phi) = \frac{\sin\theta}{I(0)} \int_{-dl/2}^{dl/2} I(0)\, dz' = \sin\theta\, dl. \tag{5.1}$$

The resistance of the Hertzian dipole is the same as its radiation resistance R_r as given by (9.5.39). Hence, we can evaluate the effective aperture of the Hertzian dipole from (3.11) as

$$A_e(\theta, \phi) = \frac{\eta_{0r}(dl)^2 \sin^2\theta}{4(2\pi/3)\eta_{0r}(dl/\lambda)^2} = \frac{\lambda^2}{4\pi}\frac{3}{2}\sin^2\theta. \tag{5.2}$$

Since the effective height and the impedance of an antenna used for reception are identical with the corresponding quantities of the same antenna used for transmission, it follows from (3.11) that the result is true also for the effective aperture and hence we omit the superscript r on $A_e(\theta, \phi)$ in (5.2). Substituting (9.5.42) in (5.2) we obtain for the Hertzian dipole

$$A_e(\theta, \phi) = g(\theta, \phi)\frac{\lambda^2}{4\pi}. \tag{5.3}$$

2. Finite Length Antenna; Standing Wave of Current

For a half-wavelength long, linear filamentary antenna with a sinusoidal current distribution, the effective height and the radiation resistance are given by (9.6.59) and (9.6.50) respectively. When these values are substituted in (3.11), we find the effective aperture to be

$$A_e(\theta, \phi) = \frac{\eta_{0r}\pi}{4\eta_{0r}(0.609)}\frac{4}{\beta^2}\left[\frac{\cos((\pi/2)\cos\theta)}{\sin\theta}\right]^2,$$

$$= \frac{\lambda^2}{4\pi}1.64\left[\frac{\cos((\pi/2)\cos\theta)}{\sin\theta}\right]^2. \tag{5.4}$$

Substituting (9.6.52) in (5.4), we get for the half-wave filamentary antenna with a sinusoidal current distribution

$$A_e(\theta, \phi) = g(\theta, \phi)\frac{\lambda^2}{4\pi}. \tag{5.5}$$

3. Finite Length Antenna; Progressive Wave of Current

For a half-wavelength long, linear filamentary antenna with a current distribution in the form of a progressive wave as given by (9.7.1), we can evaluate the effective height with the help of (9.6.55) to be

$$2h_e(\theta, \phi) = \frac{I_0 \sin\theta}{I_0} \int_{-h}^{h} e^{-j\beta z'(1-\cos\theta)}\, dz',$$

$$= \frac{2}{\beta} \frac{\sin\theta \sin\{(\pi/2)(1-\cos\theta)\}}{(1-\cos\theta)}, \tag{5.6}$$

since $\beta h = \pi/2$. We obtained the radiation resistance of this antenna in (9.7.12). From (9.7.12), (5.6), and (3.11), we can determine the effective aperture as

$$A_e(\theta, \phi) = \frac{\eta_{Cr}\pi}{4\eta_{0r}(0.719)} \frac{4}{\beta^2}\left[\frac{\sin\theta \sin\{(\pi/2)(1-\cos\theta)\}}{(1-\cos\theta)}\right]^2,$$

$$= \frac{\lambda^2}{4\pi} 1.39 \left[\frac{\sin\theta \sin\{(\pi/2)(1-\cos\theta)\}}{(1-\cos\theta)}\right]^2. \tag{5.7}$$

We can see that, together with (9.7.14), (5.7) gives for the effective aperture of a half-wave filamentary antenna with a current distribution in the form of a progressive wave:

$$A_e(\theta, \phi) = g(\theta, \phi)\frac{\lambda^2}{4\pi}. \tag{5.8}$$

If we compare (5.3), (5.5), and (5.8) we see that the ratio $A_e(\theta, \phi)/g(\theta, \phi)$ is the same for all the three types of antennas. We can establish the result that the ratio $A_e(\theta, \phi)/g(\theta, \phi)$ is a universal constant equal to $\lambda^2/4\pi$ for any type of antenna with the help of the reciprocity theorem.

4. Ratio of Effective Aperture to Directive Gain

Let us consider an antenna 1 with its center A_1 situated at the origin of the coordinate system 1 and an antenna 2 with its center A_2 situated at the origin of another coordinate system 2. Let the distance A_1A_2 joining the centers of the two antennas be equal to r. Let the direction of A_1A_2 be specified by θ_1 and ϕ_1 with respect to the first coordinate system. Similarly, the direction of A_2A_1 is specified by θ_2 and ϕ_2 with respect to the second coordinate system. The antennas are in the far-zone of each other.

Let antenna 1 be first driven as a transmitter by the application of a voltage between its terminals. Let I_1 be the resulting input current and R_{r1} be the resistance of antenna 1. The total time-averaged power P_t radiated by this antenna is given by

$$P_t = \tfrac{1}{2}|I_1|^2 R_{r1}. \tag{5.9}$$

We then find the radiation intensity Φ_{iso} of the equivalent isotropic radiator to be

$$\Phi_{iso} = P_t/4\pi. \tag{5.10}$$

From the definition of the directive gain $g_1(\theta_1, \phi_1)$ of the first antenna, we can obtain the radiation intensity $\Phi_1(\theta_1, \phi_1)$ of the first antenna in the direction (θ_1, ϕ_1) as

$$\Phi_1(\theta_1, \phi_1) = \frac{P_t}{4\pi} g_1(\theta_1, \phi_1). \tag{5.11}$$

Therefore, the magnitude of the time-averaged Poynting vector $S_1(\theta_1, \phi_1)$ at the location of the second antenna is

$$S_1(\theta_1, \phi_1) = \frac{P_t}{4\pi r^2} g_1(\theta_1, \phi_1). \tag{5.12}$$

If $A_{e2}(\theta_2, \phi_2)$ is the effective aperture of the second antenna in the direction (θ_2, ϕ_2) of the first antenna, we obtain the available power P_r at the terminals of the second antenna as

$$P_r = \frac{P_t}{4\pi r^2} g_1(\theta_1, \phi_1) A_{e2}(\theta_2, \phi_2). \tag{5.13}$$

If V_2 is the open-circuit voltage induced between the terminals of the second antenna, it follows that

$$P_r = \frac{|V_2|^2}{8R_{r2}}, \tag{5.14}$$

where R_{r2} is the resistance of the second antenna. Substitution of (5.9) and (5.14) into (5.13) gives us after some rearrangement

$$\frac{|V_2|^2}{|I_1|^2} = \frac{R_{r1}R_{r2}}{\pi r^2} g_1(\theta_1, \phi_1) A_{e2}(\theta_2, \phi_2). \tag{5.15}$$

Let us suppose now that the second antenna is used as a transmitter and the first antenna is used as a receiver. Let I_2 be the input current of the second antenna when a voltage generator is connected to its terminals and let V_1 be the corresponding open-circuit voltage induced between the terminals of the first antenna. Then, we can determine the relation analogous to (5.15) as

$$\frac{|V_1|^2}{|I_2|^2} = \frac{R_{r2}R_{r1}}{\pi r^2} g_2(\theta_2, \phi_2) A_{e1}(\theta_1, \phi_1), \tag{5.16}$$

where $g_2(\theta_2, \phi_2)$ is the directive gain of the second antenna in the direction (θ_2, ϕ_2) and $A_{e1}(\theta_1, \phi_1)$ is the effective aperture of the first antenna in the direction (θ_1, ϕ_1). By the reciprocity theorem, (5.15) and (5.16) are equal. Therefore, we obtain the following result:

$$\frac{A_{e1}(\theta_1, \phi_1)}{g_1(\theta_1, \phi_1)} = \frac{A_{e2}(\theta_2, \phi_2)}{g_2(\theta_2, \phi_2)}. \tag{5.17}$$

We should note that no assumptions have been made on the relative directions and orientations, nor on the types of the two antennas. Consequently, the ratio $A_e(\theta, \phi)/g(\theta, \phi)$ must be the same for all types of antennas. If we evaluate the ratio $A_e(\theta, \phi)/g(\theta, \phi)$ for any one antenna, the same value obtains for this ratio as for all the other types of antennas as well. For the Hertzian dipole, we proved in (5.3) that $A_e(\theta, \phi)/g(\theta, \phi) = \lambda^2/4\pi$, a constant independent of direction. It follows therefore that for any antenna, $A_e(\theta, \phi)/g(\theta, \phi)$ is a constant independent of direction and is equal to $\lambda^2/4\pi$. We have thus established the universally valid relation:

$$A_e(\theta, \phi) = g(\theta, \phi)\frac{\lambda^2}{4\pi}. \tag{5.18}$$

5. Friss Transmission Formula

We have shown the ratio of the available power P_r at the terminals of the receiving antenna in the direction (θ_2, ϕ_2) to the power P_t radiated by the transmitting antenna in (5.13) to be

$$\frac{P_r}{P_t} = \frac{g_1(\theta_1, \phi_1)A_{e2}(\theta_2, \phi_2)}{4\pi r^2}, \tag{5.19}$$

where the receiving antenna is located in the direction (θ_1, ϕ_1) with respect to the transmitting antenna. In view of (5.18), we can express (5.19) also in terms of the directive gains or the effective apertures of both the antennas as follows:

$$\frac{P_r}{P_t} = g_1(\theta_1, \phi_1)g_2(\theta_2, \phi_2)\left(\frac{\lambda}{4\pi r}\right)^2, \tag{5.20}$$

and

$$\frac{P_r}{P_t} = \frac{A_{e1}(\theta_1, \phi_1)A_{e2}(\theta_2, \phi_2)}{(\lambda r)^2}. \tag{5.21}$$

It is usual to refer to the relations (5.19) to (5.21) as Friss transmission formulas.

10.6 CURRENT DISTRIBUTION

The directive gain, the effective aperture, the impedance, and the effective height of a receiving antenna are identical with the corresponding quantities of the same antenna used as a transmitter. However, as we pointed out previously, the current distribution along a receiving antenna is quite different from that along the same antenna used as a transmitter. The determination of the current distribution along an antenna is a difficult problem and we do not propose to include a treatment of that subject here. However, we shall present here two simple current distributions obtained through a rigorous analysis without any proof. These simple current distributions are reasonably good approximations for a linear, filamentary antenna.

1. Transmitting Antenna

Let us consider a linear, filamentary antenna with its center situated at the origin of the coordinate system and extending along the z-axis from $z = h$ to $z = \delta$ and from $z = -\delta$ to $z = -h$. We assume the input terminals at $z = \delta$ and $z = -\delta$ to be infinitesimally close to each other. When this antenna is driven at its center by connecting a voltage generator to its terminals, we obtain the current distribution along the antenna as

$$I(z) = \frac{I_0 \sin \beta(h - |z|)}{\sin \beta h}. \tag{6.1}$$

We used this current distribution previously in the treatment of the radiation characteristics of a finite length filamentary antenna. We shall now give some useful comments on the method of arriving at (6.1). We can prove that along any portion of a filamentary wire where the external field vanishes, the current distribution is approximately of the form $I(z) = A \cos \beta z + B \sin \beta z$ where A and B are constants and β is the phase-shift coefficient of the electromagnetic waves in the medium surrounding the antenna. Therefore, in both the arms from $z = h$ to $z = \delta$ and from $z = -\delta$ to $z = -h$ where δ tends to zero, the current is sinusoidally distributed; hence

$$I(z) = A_1 \cos \beta z + B_1 \sin \beta z \qquad (z > \delta = +0), \tag{6.2a}$$

$$= A_2 \cos \beta z + B_2 \sin \beta z \qquad (z < -\delta = -0). \tag{6.2b}$$

The current must vanish at the ends $z = \pm h$ of the antenna. This requirement, together with (6.2a,b), yields

$$A_1 \cos \beta h + B_1 \sin \beta h = 0; \qquad A_2 \cos \beta h - B_2 \sin \beta h = 0. \tag{6.3}$$

In view of (6.3), we can express A_1 and B_1 in terms of a new constant I_1; also we can express A_2 and B_2 in terms of another new constant I_2. Thus

$$A_1 = I_1 \sin \beta h; \qquad B_1 = -I_1 \cos \beta h; \tag{6.4a}$$

$$A_2 = I_2 \sin \beta h; \qquad B_2 = I_2 \cos \beta h. \tag{6.4b}$$

Substitution of (6.4a) and (6.4b) respectively in (6.2a) and (6.2b) gives us

$$I(z) = I_1 \sin \beta(h - z) \qquad (z > +0), \tag{6.5a}$$

$$= I_2 \sin \beta(h + z) \qquad (z < -0). \tag{6.5b}$$

The current must be continuous at $z = 0$, that is $I(+0) = I(-0)$. Therefore, we find from (6.5a) and (6.5b) that $I_1 = I_2$, with the result that (6.5a) and (6.5b) become

$$I(z) = I_1 \sin \beta(h - |z|). \tag{6.6}$$

From (6.6), we see that $I_0 = I(0) = I_1 \sin \beta h$ and therefore we may rewrite (6.6) in terms of the current I_0 at the center of the antenna as

$$I(z) = \frac{I_0 \sin \beta(h - |z|)}{\sin \beta h} = I_0 f_V(z), \tag{6.7}$$

which is the approximate current distribution along a center-driven transmitting antenna in the form of a linear filament.

If the length $2h$ of the antenna is very small in comparison with a wavelength, $\beta h \ll 1$. For this case, we can approximate the sine function by its argument with the result that (6.7) becomes

$$I(z) = I_0\left(1 - \frac{|z|}{h}\right) \qquad (\beta h \ll 1). \tag{6.8}$$

The "triangular" current distribution given by (6.8) is a reasonably good approximation for the current distribution along a center-driven linear filamentary antenna of total length which is small in comparison to a wavelength.

2. Reflecting Antenna

Let us consider the case in which the foregoing filamentary antenna is situated in the electromagnetic field of a distant transmitter. Let the terminals of the antenna be short-circuited. The antenna does not abstract any energy from the incident electromagnetic wave and hence this antenna is called a *reflecting* antenna. The current distribution along the reflecting antenna depends on the direction of arrival of the incident field. We shall consider only the simplest special case in which the field arrives normal to the length of the antenna and, also, we shall assume the electric field to be always parallel to the length of the antenna. We may show the current distribution along the reflecting antenna to be given approximately by

$$I(z) = I_0 \frac{[\cos \beta z - \cos \beta h]}{[1 - \cos \beta h]}. \tag{6.9}$$

As before, we shall give here some comments on the derivation of (6.9). We can prove that the current in an infinitely long wire in a uniform electric field parallel to it is a constant. In a finite length filament, the current must vanish at the ends, which must, therefore, act like virtual sources giving rise to additional current waves along the filament. The current distribution due to the virtual sources is, as before, sinusoidal. Thus, the total current along the reflecting antenna is a constant superposed by a sinusoidal distribution. Therefore,

$$I(z) = A + B \cos \beta z + C \sin \beta z, \tag{6.10}$$

where A, B, and C are constants. In view of the special direction of arrival of the incident field, the current distribution is an even function of z, that is, $I(z) = I(-z)$. Substitution of (6.10) in this requirement shows us that $C = 0$. Since the current must vanish at the ends of the filament, $I(h) = I(-h) = 0$ and hence (6.10) yields

$$A + B \cos \beta h = 0 \qquad \text{or} \qquad A = -B \cos \beta h. \tag{6.11}$$

Together with (6.11), (6.10) becomes

$$I(z) = B(\cos \beta z - \cos \beta h). \tag{6.12}$$

From (6.12), we find that $I_0 = I(0) = B(1 - \cos \beta h)$ and therefore we may rewrite (6.12) in terms of the current I_0 at the center of the antenna as

$$I(z) = I_0 \frac{[\cos \beta z - \cos \beta h]}{[1 - \cos \beta h]} = I_0 f_E(z), \tag{6.13}$$

which is the approximate current distribution along a reflecting antenna in the form of a linear filament.

We can obtain a simplified form of (6.13) if $\beta h \ll 1$. For $x \ll 1$, we can use the approximation $\cos x = 1 - (x^2/2)$. With this approximation, we can reduce (6.13) to yield

$$I(z) = I_0 \left(1 - \frac{z^2}{h^2}\right). \tag{6.14}$$

The "parabolic" current distribution given by (6.14) is a good approximation for the current distribution along a reflecting antenna in the form of a linear filament provided the total length of the filament is small in comparison to a wavelength.

3. Receiving Antenna

Let us consider the case of a receiving antenna oriented such that its length is parallel to the electric field E of a distant transmitter, as shown in Fig. 10.9(a). We assume the equivalent load impedance at the terminals of the antenna to be given by Z_L. We wish to find the current distribution $I(z)$ along the receiving antenna. Let I_0 be the current at the center of the antenna. With the help of the compensation theorem, we can reduce the receiving antenna problem to that of an unloaded receiving antenna or a reflecting antenna with an impedanceless generator at its center. According to the compensation theorem, we can replace the equivalent load impedance Z_L by an impedanceless generator of voltage V_L as given by

$$V_L = -I_0 Z_L, \tag{6.15}$$

where V_L is the potential difference across Z_L. The equivalent problem obtained by the substitution of (6.15) is illustrated in Fig. 10.9(b).

At the center of the antenna, we obtain the total current I_0 by the application of the principle of superposition to be given by the sum of the current I_{0V} due to the voltage generator alone and the current I_{0E} due to the incident electric field alone with the antenna terminals short-circuited. Therefore,

$$I_0 = I_{0V} + I_{0E}. \tag{6.16}$$

The current I_{0V} due to the impedanceless generator (6.15) is the same as that in a center-driven transmitting antenna, that is,

$$I_{0V} = \frac{V_L}{Z_a} = -I_0 \frac{Z_L}{Z_a}, \tag{6.17}$$

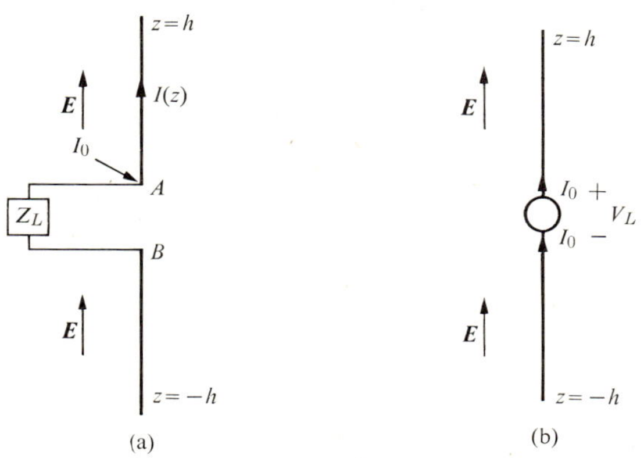

Fig. 10.9 Loaded receiving antenna parallel to the electric field. (a) Actual circuit. (b) Equivalent circuit.

where Z_a is the antenna impedance. From (6.16) and (6.17), we find that

$$I_{0E} = I_0 - I_{0V} = I_0\left(\frac{Z_a + Z_L}{Z_a}\right). \tag{6.18}$$

The currents at the center of the antenna separately due to the impedanceless generator and the incident electric field are given by (6.17) and (6.18) respectively.

We can also find the current at any point along the antenna by the application of the superposition theorem. An examination of (6.7) shows us that in the case of a center-driven antenna, the current at the center of the antenna multiplied by $f_V(z)$ gives the current distribution along the antenna. Therefore, when the impedanceless generator (6.15) alone acts at the center of the antenna, we obtain the corresponding current distribution $I_V(z)$ from (6.17) as

$$I_V(z) = I_{0V} f_V(z) = -I_0 \frac{Z_L}{Z_a} f_V(z). \tag{6.19}$$

In a similar manner, we see from (6.13) that in the case of an unloaded receiving antenna or a reflecting antenna, the current at the center of the antenna multiplied by $f_E(z)$ gives the current distribution along the antenna. Consequently, we obtain the current distribution $I_E(z)$ due to the action of the incident electric field alone from (6.18) as

$$I_E(z) = I_{0E} f_E(z) = I_0\left(\frac{Z_a + Z_L}{Z_a}\right) f_E(z). \tag{6.20}$$

The total current distribution along the loaded receiving antenna is the algebraic sum of the separate currents $I_E(z)$ and $I_V(z)$ due to the electric field and the generator respectively. Hence, we see from (6.19) and (6.20) that

$$I(z) = I_E(z) + I_V(z) = \frac{I_0}{Z_a}\left[(Z_a + Z_L) f_E(z) - Z_L f_V(z)\right]. \tag{6.21}$$

We find the distribution functions $f_E(z)$ and $f_V(z)$ from (6.13) and (6.7) to be

$$f_E(z) = \frac{(\cos\beta z - \cos\beta h)}{(1 - \cos\beta h)}; \qquad f_V(z) = \frac{\sin\beta(h - |z|)}{\sin\beta h}. \tag{6.22}$$

These distribution functions are good approximations for filamentary antennas provided that their lengths are not such as to make the denominators in (6.22) very small.

4. Receiving Antenna with Open-circuited Terminals

When the equivalent load impedance Z_L is very large, the input current I_0 becomes very small. For this case, it is convenient for us to replace I_0 in (6.21) from (6.15) with the following result:

$$I(z) = -\frac{V_L}{Z_a}\left[\left(\frac{Z_a}{Z_L} + 1\right) f_E(z) - f_V(z)\right]. \tag{6.23}$$

In the limiting case in which the terminals of the receiving antenna are open-circuited, $Z_L = \infty$ and $V_L = V_{OC}$, the open-circuit voltage induced between the terminals of the antenna. For this limiting case, we can simplify (6.23) to

$$I(z) = -\frac{V_{OC}}{Z_a}[f_E(z) - f_V(z)]. \tag{6.24}$$

In concluding the discussion of the current distribution along a filamentary antenna, the special case of a half-wave antenna is worthy of mention. For a half-wave antenna $\beta h = \pi/2$ and hence

$$f_E(z) = f_V(z) = \cos \beta z. \tag{6.25}$$

Therefore, the current distributions along the center-driven antenna and the corresponding reflecting antenna are the same. Also, we can reduce (6.21) for any value of Z_L to

$$I(z) = I_0 f_E(z) = I_0 f_V(z) = I_0 \cos \beta z. \tag{6.26}$$

Thus, the current distribution along a half-wavelength long receiving antenna is essentially independent of the load impedance. We should emphasize that in this treatment the receiving antenna is always oriented parallel to the electric field of the distant transmitter.

We mentioned earlier that the determination of the actual current distribution along an antenna is a difficult problem. It is helpful for us to obtain a general understanding of the origin of this difficulty. Let us consider a perfectly conducting cylindrical tubular antenna ($\rho = a$, $|z| \leq h$) of negligible thickness, of radius a and half-length h, where ρ, ϕ, and z form a cylindrical coordinate system. The antenna is in free space and is excited along an infinitesimally narrow circumferential region at $z = 0$ by means of an e.m.f. The excitation is equivalent to maintaining a constant z-component of the electric field around the circumferential gap. Let us suppose that $I_z(z)$ is the current distribution along the antenna. We should note that this current distribution must vanish at both the ends $z = \pm h$ of the antenna. Once we know the current distribution, the vector potential can be computed as an integral of the current. We can then obtain the components of the electric field tangential to the surface of the antenna in terms of the integral of the current distribution. These components must vanish on the surface of the antenna except in the circumferential gap region where the applied electric field should be reproduced. This results in an equation which involves the integral of the current distribution. This integral equation for the current must be solved to obtain the actual current distribution for the specified excitation. The difficulty of obtaining the actual current distribution along an antenna lies in the fact that it is enormously difficult to determine the solution of the integral equation for the current.

10.7 TWO-ELEMENT "BROADCAST" TYPE ARRAY

The transmitting and the receiving characteristics which we have dealt with so far are essentially restricted to those of single antennas. The directional patterns of single antennas can be changed by adjusting the length and the current distribution along the antenna. However, the available directional patterns are limited in type. In order to have greater flexibility in the various possible directional patterns and in order to be able to concentrate the transmission or the reception of electromagnetic energy in the desired directions, it is necessary to resort to an arrangement consisting of combination of antennas or antenna arrays. The relative orientation of, and the spacing between the antennas, the amplitude and the phase of the currents in the individual antennas, as well as the number of antennas are all amenable to adjustment in an antenna array with the result that the variety of available directional patterns is considerably increased. We propose to illustrate the capability of antenna arrays by the treatment of two simple types of antenna arrays consisting of only two elements.

We mentioned that an antenna array can be used to concentrate the electromagnetic radiation in some chosen directions. It is of interest first to examine the physical basis leading to the property of an antenna array which enables the concentration of the radiation in the desired directions. If the sources are of equal intensities, the amplitudes of the fields generated by them tend to become equal in the far-zone. On the other hand, the relative phases of the radiation depend on the separation distances between the sources and on the direction of radiation. In some directions, the waves from the various sources arrive in phase and thus reinforce each other; in others they arrive out of phase and annul each other. Hence, by a proper arrangement of the sources, we can achieve the constructive superposition of waves in the desired directions and the destructive interference of waves in the other directions.

Let us consider two Hertzian dipoles 1 and 2, situated along the z-axis at $z = d/2$ and $z = -d/2$ respectively, and oriented in the same direction, which is parallel to the z-axis (Fig. 10.10). In view of the cylindrical symmetry, the directional pattern

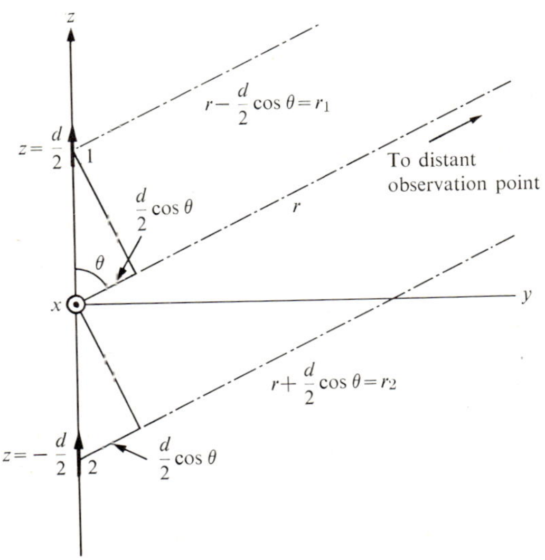

Fig. 10.10 A two-element "broadcast" type antenna array of point dipoles.

of the antenna arrangement shown in Fig. 10.10 does not vary in the azimuthal and hence this antenna array is said to be of the "broadcast" type. Since the fields of this antenna array are independent of ϕ, it is sufficient for us to evaluate the fields for any one value of ϕ, for example $\phi = \pi/2$, as shown in Fig. 10.10. Let the current moments of both the dipoles have the same value equal to $I\,dl$ and let $P(r, \theta, \pi/2)$ be the observation point in the far-zone. Let r_1 and r_2 be the distances from the Hertzian dipoles 1 and 2 to the observation point. Since the observation point is in the far-zone, the directions of r_1, r, and r_2 are all essentially parallel. Therefore, we find with the help of Fig. 10.10 that

$$r_1 = r - \frac{d}{2}\cos\theta; \qquad r_2 = r + \frac{d}{2}\cos\theta. \tag{7.1}$$

For the reasons mentioned previously, we replace r_1 and r_2 by r in the amplitude terms and by those given by (7.1) in the phase terms. We obtain the far-zone electric fields due to the dipoles 1 and 2 from (9.5.25b) as

$$E_{\theta 1}(\mathbf{r}) = \frac{I\,dl}{4\pi r}\eta_{0r} j\beta \sin\theta e^{-j\beta[r-(d/2)\cos\theta]}, \tag{7.2a}$$

$$E_{\theta 2}(\mathbf{r}) = \frac{I\,dl}{4\pi r}\eta_{0r} j\beta \sin\theta e^{-j\beta[r+(d/2)\cos\theta]}. \tag{7.2b}$$

Hence,

$$E_\theta(\mathbf{r}) = E_{\theta 1}(\mathbf{r}) + E_{\theta 2}(\mathbf{r}) = \frac{I\,dl e^{-j\beta r}}{4\pi r}\eta_{0r} j\beta \sin\theta\left[2\cos\left(\frac{\beta d}{2}\cos\theta\right)\right]. \tag{7.3}$$

Let,

$$E_\theta(\mathbf{r}) = \frac{e^{-j\beta r}}{4\pi r} f(\theta, \phi); \tag{7.4}$$

then, as before, $f(\theta, \phi)$ defines the directional pattern of the antenna array. Let us suppose a single Hertzian dipole of current moment $I\,dl$ which is identical to those of the two dipoles in the array is situated at the origin of the coordinate system, then the far-zone electric field is given by

$$E_\theta(\mathbf{r}) = \frac{e^{-j\beta r}}{4\pi r} f_e(\theta, \phi) = \frac{I\,dl}{4\pi r} e^{-j\beta r}\eta_{0r} j\beta \sin\theta, \tag{7.5}$$

where $f_e(\theta, \phi)$ is known as the *element directional pattern*. Let us suppose now that the elements of the original antenna array are replaced by isotropic sources of relative magnitudes which are the same as those of the elements of the antenna array. Therefore, the amplitudes of the two isotropic sources are equal in the present case. We can obtain the resulting field $f(\mathbf{r})$ at the same observation point in a manner similar to that used in finding the scalar potential. If the magnitudes of the sources are taken to be equal to unity, the result is

$$\begin{aligned} f(\mathbf{r}) &= \frac{e^{-j\beta[r-(d/2)\cos\theta]}}{4\pi r} + \frac{e^{-j\beta[r+(d/2)\cos\theta]}}{4\pi r}, \\ &= \frac{e^{-j\beta r}}{4\pi r} 2\cos\left(\frac{\beta d}{2}\cos\theta\right) = \frac{e^{-j\beta r}}{4\pi r} f_a(\theta, \phi), \end{aligned} \tag{7.6}$$

where $f_a(\theta, \phi)$ is the directional pattern of the corresponding array with isotropic elements.

From (7.3) to (7.6), we obtain:

$$f(\theta, \phi) = I\, dl\eta_{0r} j\beta \sin\theta \left[2\cos\left(\frac{\beta d}{2}\cos\theta\right)\right], \tag{7.7a}$$

$$f_e(\theta, \phi) = I\, dl\eta_{0r} j\beta \sin\theta, \tag{7.7b}$$

$$f_a(\theta, \phi) = 2\cos\left(\frac{\beta d}{2}\cos\theta\right). \tag{7.7c}$$

Thus,

$$f(\theta, \phi) = f_e(\theta, \phi) f_a(\theta, \phi), \tag{7.8}$$

which shows that the directional pattern of the antenna array is equal to the product of the directional pattern of the individual elements and that of the corresponding array with isotropic elements. This pattern multiplication theorem is generally valid if the elements of the antenna array are identical and are oriented parallel to each other. This theorem is useful in ascertaining the directional patterns of seemingly complicated arrays of identical elements oriented parallel to each other.

We find the far-zone magnetic field of the two-element "broadcast" type antenna array from (7.3) as

$$H_\phi(\mathbf{r}) = \frac{I\, dl e^{-j\beta r}}{4\pi r} j\beta \sin\theta \left[2\cos\left(\frac{\beta d}{2}\cos\theta\right)\right]. \tag{7.9}$$

From (7.3) and (7.9), we obtain the radiation intensity $\Phi(\theta, \phi)$ as

$$\Phi(\theta, \phi) = \frac{|I|^2 (dl)^2}{8\pi^2} \eta_{0r}\beta^2 \sin^2\theta \cos^2\left(\frac{\beta d}{2}\cos\theta\right). \tag{7.10}$$

We can deduce the total time-averaged power P radiated by this antenna array as

$$P = \int_{\phi=0}^{2\pi}\int_{\theta=0}^{\pi} \frac{|I|^2 (dl)^2}{8\pi^2} \eta_{0r}\beta^2 \sin^2\theta \cos^2\left(\frac{\beta d}{2}\cos\theta\right) \sin\theta\, d\theta\, d\phi,$$

$$= \frac{|I|^2 (dl)^2}{4\pi} \eta_{0r}\beta^2 \int_{\theta=0}^{\pi} [1 - \cos^2\theta] \cos^2\left(\frac{\beta d}{2}\cos\theta\right) \sin\theta\, d\theta. \tag{7.11}$$

Let $(\beta d/2)\cos\theta = u$. Then, (7.11) becomes

$$P = \frac{|I|^2 (dl)^2}{2\pi d} \eta_{0r}\beta \int_{-\beta d/2}^{\beta d/2} \left[1 - \frac{4u^2}{\beta^2 d^2}\right] \cos^2 u\, du,$$

$$= \frac{|I|^2 (dl)^2}{\pi d} \eta_{0r}\beta \int_{0}^{\beta d/2} \left[1 - \frac{4u^2}{\beta^2 d^2}\right] \cos^2 u\, du. \tag{7.12}$$

We should note that

$$\int_0^{\beta d/2} \cos^2 u\, du = \tfrac{1}{4}[\beta d + \sin\beta d]. \tag{7.13}$$

Also,

$$\int_0^{\beta d/2} u^2 \cos^2 u\, du = \int_0^{\beta d/2} \frac{u^2}{2}(1 + \cos 2u)\, du,$$

$$= \frac{\beta^3 d^3}{48} + \tfrac{1}{2}\int_0^{\beta d/2} u^2 \cos 2u\, du. \tag{7.14}$$

We can evaluate the integral in (7.14) by integration by parts as follows:

$$\int_0^{\beta d/2} u^2 \cos 2u\, du = \tfrac{1}{2}\int_0^{\beta d/2} u^2 d(\sin 2u)$$

$$= \frac{1}{2}\left[\frac{\beta^2 d^2}{4}\sin \beta d + \int_0^{\beta d/2} u d(\cos 2u)\right]. \tag{7.15}$$

By another integration by parts, we can simplify (7.15) to give

$$\int_0^{\beta d/2} u^2 \cos 2u\, du = \frac{1}{2}\left[\frac{\beta^2 d^2}{4}\sin \beta d + \frac{\beta d}{2}\cos \beta d - \frac{\sin \beta d}{2}\right]. \tag{7.16}$$

With the help of (7.13), (7.14), and (7.16), we can show (7.12) to give the following result:

$$P = \frac{|I|^2 (dl)^2 \eta_{0r} \beta^2}{6\pi}\left[1 + \frac{3}{\beta^2 d^2}\left(\frac{\sin \beta d}{\beta d} - \cos \beta d\right)\right]. \tag{7.17}$$

From (7.10) and (7.17), we obtain the directive gain of the antenna array as

$$g(\theta, \phi) = \frac{4\pi\Phi(\theta, \phi)}{P} = 3\left[1 + \frac{3}{\beta^2 d^2}\left(\frac{\sin \beta d}{\beta d} - \cos \beta d\right)\right]^{-1} \sin^2 \theta \cos^2\left(\frac{\beta d}{2}\cos \theta\right). \tag{7.18}$$

Let us consider the special case in which the separation distance between the elements is given by $d = \lambda/2$ or $\beta d = \pi$. Then, (7.18) becomes

$$g(\theta, \phi) = \frac{3}{(1 + 3/\pi^2)}\sin^2 \theta \cos^2\left(\frac{\pi}{2}\cos \theta\right); \qquad d = \frac{\lambda}{2}. \tag{7.19}$$

The directive gain (7.19) is independent of ϕ, as it should be. The directive gain as given by (7.19) is depicted in Fig. 10.11, which shows that (7.19) has the maximum value at $\theta = \pi/2$. Therefore, we find the directivity or the gain of a two-element "broadcast" type array of Hertzian dipoles separated by a distance of $\lambda/2$ from (7.19) to be

$$g = \frac{3}{(1 + 3/\pi^2)}. \tag{7.20}$$

The directive gain, as given by (9.5.42) of a single Hertzian dipole is also depicted in Fig. 10.11, which shows the improvement in the concentration of radiation in the broadside ($\theta = \pi/2$) direction obtained by using the two-element array over that of a single Hertzian dipole.

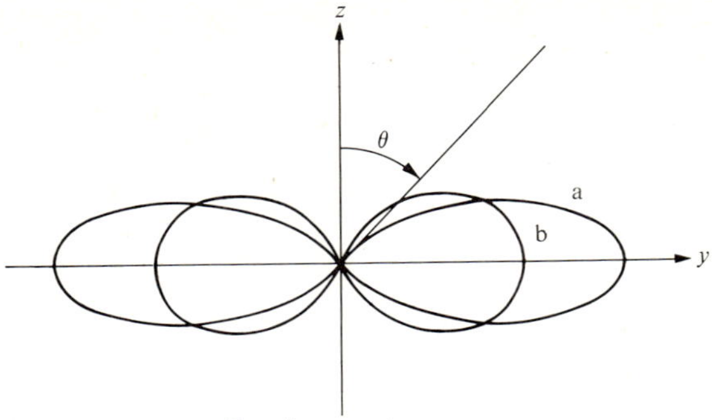

Fig. 10.11 Directive gain of a two-element "broadcast" type antenna array of Hertzian dipoles separated by a half of a wavelength.

The theory of images which we discussed in Section 9.8 has shown that the fields in the half-space $z > 0$, due to two identical Hertzian dipoles oriented along the z-axis and situated at $z = d/2$ and $z = -d/2$ respectively, are equal to that of the single dipole at $z = d/2$, provided an infinite, perfectly conducting surface occupies the plane $z = 0$. Thus, the directional pattern of the two-element "broadcast" type antenna array discussed here is the same as that of a single dipole in front of a perfectly conducting plane provided the dipole is oriented normal to the conducting plane and its distance from the conducting plane is one half of the separation distance between the elements of the array.

We have taken the elements of the array to be Hertzian dipoles. But, we can extend the analysis to the case in which the elements are finite length antennas. Let

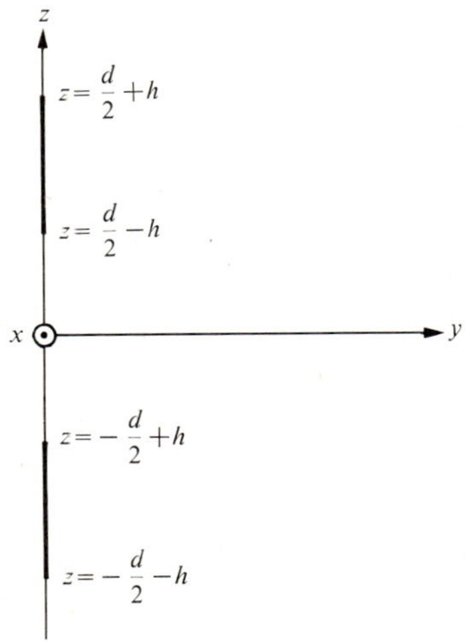

Fig. 10.12 A two-element "broadcast" type antenna array of linear, filamentary antennas of length $2h$.

us consider the "broadcast" type array consisting of two identical filamentary antennas situated along the z-axis, as shown in Fig. 10.12. The antennas are of length $2h$ and have the same sinusoidal current distributions along their lengths. The current distributions along the two antennas are respectively of the form:

$$I_1(z) = I_0 \sin \beta\left\{h - \left|z - \frac{d}{2}\right|\right\}, \tag{7.21a}$$

$$I_2(z) = I_0 \sin \beta\left\{h - \left|z + \frac{d}{2}\right|\right\}. \tag{7.21b}$$

The centers of the two antennas are separated by a distance d which has to be equal to or greater than $2h$. Let us suppose that the individual antennas are half-wavelength long. We can obtain the electric field due to each antenna and, therefore, the resulting electric field from (9.6.39) as

$$E_{\theta 1} = \frac{j\eta_{0r} I_0}{2\pi r} \frac{\cos[(\pi/2)\cos\theta]}{\sin\theta} e^{-j\beta[r-(d/2)\cos\theta]}, \tag{7.22a}$$

$$E_{\theta 2} = \frac{j\eta_{0r} I_0}{2\pi r} \frac{\cos[(\pi/2)\cos\theta]}{\sin\theta} e^{-j\beta[r+(d/2)\cos\theta]}, \tag{7.22b}$$

and

$$E_\theta = E_{\theta 1} + E_{\theta 2} = \frac{j\eta_{0r} I_0 e^{-j\beta r}}{2\pi r} \frac{\cos[(\pi/2)\cos\theta]}{\sin\theta} 2\cos\left(\frac{\beta d}{2}\cos\theta\right). \tag{7.23}$$

It is interesting for us to note that (7.23) could have also been obtained by the use of the pattern multiplication theorem. The remainder of the analysis is the same as for the Hertzian dipoles but is complicated by the fact that the integral for the total time-averaged power corresponding to (7.11) is difficult to evaluate.

We should observe that in the half-space $z > 0$, the directional pattern of the foregoing "broadcast" type array is the same as that of the half-wave filamentary antenna 1 with a sinusoidal current distribution as given by (7.21a) and situated normal to a perfectly conducting plane with the center of the antenna at a distance $d/2$ from the conducting plane.

10.8 TWO-ELEMENT "POINT-TO-POINT" TYPE ARRAY

Let us consider a point dipole of current moment $I\,dl$ situated at the origin and oriented along the z-axis. We obtain the far-zone electric field from (9.5.25b) as

$$E_\theta(\mathbf{r}) = \frac{I\,dl}{4\pi r} e^{-j\beta r} \eta_{0r} j\beta \sin\theta, \tag{8.1}$$

where θ is the angle between the axis of the dipole and the direction of observation. Before proceeding to discuss the two-element "point-to-point" type array, it is necessary for us to evaluate the far-zone electric field due to a point dipole oriented parallel to one of the other two rectangular coordinate axes. Let us consider, for example,

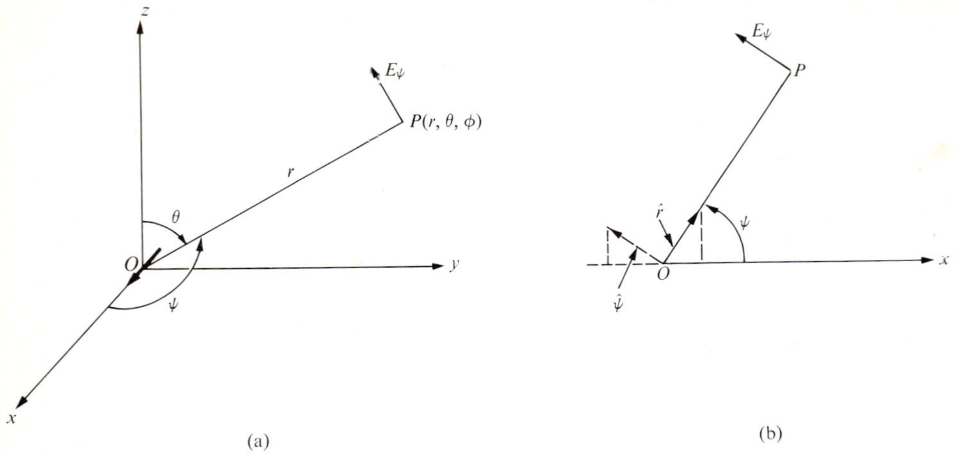

Fig. 10.13 Geometry of a point dipole situated at the origin and oriented parallel to the x-axis.

the dipole oriented along the x-axis, as shown in Fig. 10.13(a). Let $P(r, \theta, \phi)$ be the distant observation point and let ψ be the angle between the axis of the dipole and the direction OP of observation. Comparison with (8.1) shows us that the far-zone electric field due to the point dipole oriented along the x-axis is given by

$$E_\psi(\mathbf{r}) = \frac{I\,dl}{4\pi r} e^{-j\beta r}\eta_{0r} j\beta \sin\psi. \tag{8.2}$$

The direction of the electric vector is in the xOP plane normal to the radius vector OP, as shown in Fig. 10.13(a).

We wish to express $\sin\psi$ and the direction $\hat{\psi}$ of the electric vector in terms of the spherical coordinates θ and ϕ and the directions $\hat{\theta}$ and $\hat{\phi}$ respectively. The unit vector $\hat{r}$ in the radial direction is related to $\hat{x}$, $\hat{y}$, and $\hat{z}$, the unit vectors along the rectangular coordinate axes, as follows:

$$\hat{r} = \hat{x}\sin\theta\cos\phi + \hat{y}\sin\theta\sin\phi + \hat{z}\cos\theta, \tag{8.3}$$

where the coefficients of $\hat{x}$, $\hat{y}$, and $\hat{z}$ in (8.3) are the direction cosines of the unit vector $\hat{r}$. From (8.3), we obtain

$$\hat{x}\cdot\hat{r} = \cos\psi = \sin\theta\cos\phi \qquad \text{or} \qquad \sin\psi = (1 - \sin^2\theta\cos^2\phi)^{1/2}. \tag{8.4}$$

We observe from Fig. 10.13(b) that

$$\hat{x} = -\hat{\psi}\sin\psi + \hat{r}\cos\psi \qquad \text{or} \qquad \hat{\psi} = \frac{1}{\sin\psi}[-\hat{x} + \hat{r}\cos\psi]. \tag{8.5}$$

Also, we can show from Fig. 10.13(a) that

$$\hat{x} = \hat{r}\sin\theta\cos\phi + \hat{\theta}\cos\theta\cos\phi - \hat{\phi}\sin\phi. \tag{8.6}$$

If we substitute $\cos\psi$ and $\sin\psi$ from (8.4), and $\hat{x}$ from (8.6), into (8.5), we find that

$$\hat{\psi} = \frac{[-\hat{\theta}\cos\theta\cos\phi + \hat{\phi}\sin\phi]}{(1 - \sin^2\theta\cos^2\phi)^{1/2}}. \tag{8.7}$$

From (8.2), (8.4), and (8.7), the far-zone electric field due to the point dipole oriented along the x-axis is given by

$$\boldsymbol{E}(\boldsymbol{r}) = I\,dl\frac{e^{-j\beta r}}{4\pi r}\eta_{0r}j\beta[-\hat{\theta}\cos\theta\cos\phi + \hat{\phi}\sin\phi]. \tag{8.8}$$

It is now possible for us to discuss the two-element "point-to-point" type array.

Let us consider two point dipoles 1 and 2 situated along the z-axis at $z = d/2$ and $z = -d/2$ respectively. Let these dipoles be oriented parallel to the x-axis (Fig. 10.14). In a "broadcast" type array, the dipoles are oriented parallel to the line joining the centers of the two elements, but in a "point-to-point" type array the dipoles are oriented perpendicular to the line joining the centers of the two elements. Let r_1 and r_2 be the distances from the distant observation point $P(r, \theta, \phi)$ to the elements 1 and 2 respectively. As before, the approximations given in (7.1) hold good. Also, in the amplitude terms of the fields, we replace r_1 and r_2 by r, but in the phase terms r_1 and r_2 are replaced by the approximations given by (7.1). Let $I_1\,dl$ and $I_2\,dl$, the current moments of the two dipoles, be given by

$$I_1\,dl = I_0\,dle^{-j(\beta d/2)\alpha}, \qquad I_2\,dl = I_0\,dle^{j(\beta d/2)\alpha}, \tag{8.9}$$

where α is a parameter that determines the relative phase of the currents. Thus, the current moments of the two dipoles are equal in magnitude but the phase of the current in the second dipole leads that of the first by $\beta d\alpha$.

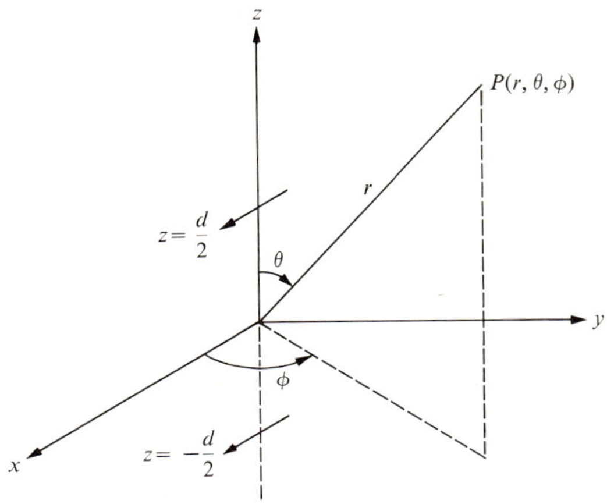

Fig. 10.14 Two-element "point-to-point" type array of Hertzian dipoles.

We can determine the far-zone electric fields due to the dipoles 1 and 2 from (8.8), (7.1), and (8.9) as

$$\boldsymbol{E}_1(\boldsymbol{r}) = \frac{I_0\, dl}{4\pi r} e^{-j\beta r}\eta_{0r} j\beta[-\hat{\theta}\cos\theta\cos\phi + \hat{\phi}\sin\phi]e^{j(\beta d/2)(\cos\theta - \alpha)}, \qquad (8.10a)$$

$$\boldsymbol{E}_2(\boldsymbol{r}) = \frac{I_0\, dl}{4\pi r} e^{-j\beta r}\eta_{0r} j\beta[-\hat{\theta}\cos\theta\cos\phi + \hat{\phi}\sin\phi]e^{-j(\beta d/2)(\cos\theta - \alpha)}. \qquad (8.10b)$$

Hence, we obtain the resultant far-zone electric field of the array as

$$\begin{aligned}\boldsymbol{E}(\boldsymbol{r}) &= \boldsymbol{E}_1(\boldsymbol{r}) + \boldsymbol{E}_2(\boldsymbol{r}),\\ &= \frac{I_0\, dl}{4\pi r} e^{-j\beta r}\eta_{0r} j\beta[-\hat{\theta}\cos\theta\cos\phi + \hat{\phi}\sin\phi] \times 2\cos\left\{\frac{\beta d}{2}(\cos\theta - \alpha)\right\}.\end{aligned} \qquad (8.11)$$

The components of the far-zone magnetic field are related to those of the far-zone electric field by

$$\frac{E_\theta(\boldsymbol{r})}{H_\phi(\boldsymbol{r})} = \eta_{0r} = -\frac{E_\phi(\boldsymbol{r})}{H_\theta(\boldsymbol{r})}. \qquad (8.12)$$

From (8.11) and (8.12), we can evaluate the far-zone magnetic field of the array as

$$\boldsymbol{H}(\boldsymbol{r}) = \frac{I_0\, dl e^{-j\beta r}}{4\pi r} j\beta[-\hat{\theta}\sin\phi - \hat{\phi}\cos\theta\cos\phi] \times 2\cos\left\{\frac{\beta d}{2}(\cos\theta - \alpha)\right\}. \qquad (8.13)$$

From (8.11) and (8.13), we can prove the radiation intensity $\Phi(\theta, \phi)$ of the array to be

$$\begin{aligned}\Phi(\theta, \phi) &= r^2\hat{r}\cdot\tfrac{1}{2}\operatorname{Re}\boldsymbol{E}(\boldsymbol{r}) \times \boldsymbol{H}^*(\boldsymbol{r}),\\ &= \frac{|I_0|^2 (dl)^2\beta^2\eta_{0r}}{8\pi^2}[1 - \cos^2\phi\sin^2\theta]\cos^2\left\{\frac{\beta d}{2}(\cos\theta - \alpha)\right\}.\end{aligned} \qquad (8.14)$$

We can deduce the total time-averaged power P radiated by this antenna array as

$$P = \int_{\phi=0}^{2\pi}\int_{\theta=0}^{\pi} \frac{|I_0|^2 (dl)^2}{8\pi^2}\beta^2\eta_{0r}\cos^2\left\{\frac{\beta d}{2}(\cos\theta - \alpha)\right\}\sin\theta(1 - \cos^2\phi\sin^2\theta)\, d\theta\, d\phi. \qquad (8.15)$$

We can carry out the integration with respect to ϕ in (8.15) with the following result:

$$P = \frac{|I_0|^2 (dl)^2}{8\pi}\beta^2\eta_{0r}\int_{\theta=0}^{\pi}\cos^2\left\{\frac{\beta d}{2}(\cos\theta - \alpha)\right\}\sin\theta(2 - \sin^2\theta)\, d\theta. \qquad (8.16)$$

Let $(\beta d/2)(\cos\theta - \alpha) = u$; then $-(\beta d/2)\sin\theta\, d\theta = du$ and $\cos\theta = \alpha + (2u/\beta d)$. Therefore, we may show (8.16) to become

$$P = \frac{|I_0|^2 (dl)^2}{4\pi d}\beta\eta_{0r}\int_{\beta d(-1-\alpha)/2}^{\beta d(1-\alpha)/2}\left(1 + \alpha^2 + \frac{4\alpha}{\beta d}u + \frac{4}{\beta^2 d^2}u^2\right)\cos^2 u\, du. \qquad (8.17)$$

It is convenient for us to use the relation $\cos^2 u = \frac{1}{2}(1 + \cos 2u)$ and rewrite (8.17) as follows:

$$P = \frac{|I_0|^2 (dl)^2}{8\pi d} \beta \eta_{0r} \left[P_1 + (1 + \alpha^2) P_2 + \frac{4\alpha}{\beta d} P_3 + \frac{4}{\beta^2 d^2} P_4 \right], \tag{8.18}$$

where

$$P_1 = \int_{\beta d(-1-\alpha)/2}^{\beta d(1-\alpha)/2} \left(1 + \alpha^2 + \frac{4\alpha u}{\beta d} + \frac{4}{\beta^2 d^2} u^2 \right) du, \tag{8.19a}$$

$$P_2 = \int_{\beta d(-1-\alpha)/2}^{\beta d(1-\alpha)/2} \cos 2u \, du, \tag{8.19b}$$

$$P_3 = \int_{\beta d(-1-\alpha)/2}^{\beta d(1-\alpha)/2} u \cos 2u \, du, \tag{8.19c}$$

$$P_4 = \int_{\beta d(-1-\alpha)/2}^{\beta d(1-\alpha)/2} u^2 \cos 2u \, du. \tag{8.19d}$$

We can evaluate the integrals in (8.19a) and (8.19b) to yield the following result:

$$P_1 = \tfrac{4}{3}\beta d; \qquad P_2 = \tfrac{1}{2}[\sin\{\beta d(1 - \alpha)\} + \sin\{\beta d(1 + \alpha)\}]. \tag{8.20}$$

We can evaluate the integrals in (8.19c) and (8.19d) by the method of integration by parts, as used in (7.14), and obtain the following results:

$$\begin{aligned} P_3 &= \frac{\beta d}{4}[(1 - \alpha)\sin\{\beta d(1 - \alpha)\} - (1 + \alpha)\sin\{\beta d(1 + \alpha)\}] \\ &\quad + \tfrac{1}{4}[\cos\{\beta d(1 - \alpha)\} - \cos\{\beta d(1 + \alpha)\}], \end{aligned} \tag{8.21}$$

$$\begin{aligned} P_4 &= \frac{\beta^2 d^2}{8}[(1 - \alpha)^2 \sin\{\beta d(1 - \alpha)\} + (1 + \alpha)^2 \sin\{\beta d(1 + \alpha)\}] \\ &\quad + \frac{\beta d}{4}[(1 - \alpha)\cos\{\beta d(1 - \alpha)\} + (1 + \alpha)\cos\{\beta d(1 + \alpha)\}] \\ &\quad - \tfrac{1}{4}[\sin\{\beta d(1 - \alpha)\} + \sin\{\beta d(1 + \alpha)\}]. \end{aligned} \tag{8.22}$$

Substitution of (8.20) to (8.22) into (8.18) gives us after some simplification:

$$P = \frac{|I_0|^2 (dl)^2 \beta^2 \eta_{0r}}{6\pi} P_0(\alpha, \beta d), \tag{8.23}$$

where

$$\begin{aligned} P_0(\alpha, \beta d) &= 1 + \frac{3}{4\beta d}\left[\sin\{\beta d(1 - \alpha)\} + \frac{\cos\{\beta d(1 - \alpha)\}}{\beta d} - \frac{\sin\{\beta d(1 - \alpha)\}}{\beta^2 d^2} \right] \\ &\quad + \frac{3}{4\beta d}\left[\sin\{\beta d(1 + \alpha)\} + \frac{\cos\{\beta d(1 + \alpha)\}}{\beta d} - \frac{\sin\{\beta d(1 + \alpha)\}}{\beta^2 d^2} \right]. \end{aligned} \tag{8.24}$$

From (8.14) and (8.24), we obtain the directive gain $g(\theta, \phi)$ of the antenna array as

$$g(\theta, \phi) = \frac{3}{P_0(\alpha, \beta d)}[1 - \cos^2\phi \sin^2\theta] \cos^2\left\{\frac{\beta d}{2}(\cos\theta - \alpha)\right\}. \tag{8.25}$$

We shall now investigate several special cases.

Case 1. ($d = \lambda/2$; $\alpha = 0$). Let us consider the case in which the currents in the two antennas are not only equal in magnitude but also have the same phase. The separation distance between the antennas is equal to one half of a wavelength, that is, $\beta d = \pi$. We can see from (8.24) that $P_0(0, \pi) = 1 - 3/(2\pi^2)$ and hence we can reduce (8.25) to

$$g(\theta, \phi) = 3\left[1 - \frac{3}{2\pi^2}\right]^{-1}[1 - \cos^2\phi \sin^2\theta] \cos^2\left\{\frac{\pi}{2}\cos\theta\right\}. \tag{8.26}$$

Since the directive gain is a function of both θ and ϕ, its value for all possible directions can be depicted only by a three-dimensional surface. The diagrams obtained from sections of the above-mentioned three-dimensional surface by some typical planes provide sufficient knowledge of the variation of the directive gain with all possible values of θ and ϕ and, therefore, we shall illustrate only some typical planar sections of the three-dimensional, directive gain surface. For the present case, we shall show the patterns of the directive gain in the principal coordinate planes. From (8.26), we obtain

xy-plane: $$g\left(\frac{\pi}{2}, \phi\right) = 3\left[1 - \frac{3}{2\pi^2}\right]^{-1} \sin^2\phi, \tag{8.27}$$

zx-plane: $$g(\theta, 0) = 3\left[1 - \frac{3}{2\pi^2}\right]^{-1} \cos^2\theta \cos^2\left(\frac{\pi}{2}\cos\theta\right), \tag{8.28}$$

yz-plane: $$g\left(\theta, \frac{\pi}{2}\right) = 3\left[1 - \frac{3}{2\pi^2}\right]^{-1} \cos^2\left(\frac{\pi}{2}\cos\theta\right). \tag{8.29}$$

The principal plane patterns given by (8.27) to (8.29) are depicted in Figs. 10.15(a) to 10.15(c) respectively.

Let us consider a single Hertzian dipole with a "vertical" orientation and having a directional pattern with no variation in the azimuth. Let us suppose that in the "horizontal" plane, we wish to concentrate the electromagnetic radiation in the neighborhood of a given direction. We can achieve this objective by forming an array of two "vertical" dipoles, separated by a distance of $\lambda/2$ and with their centers on a line which is perpendicular to the desired direction. If we adjust the currents in the two "vertical" dipoles to be equal in amplitude and phase, then the directional pattern in the "horizontal" plane is the same as that shown in Fig. 10.15(c). In the "horizontal" plane, this two-element array concentrates the radiation in the neighborhood of the direction perpendicular to the line joining the centers of the two dipoles, that is, in the *broadside* ($\theta = \pi/2$) direction. Therefore, this antenna array is called a *broadside array*. The "point-to-point" type antenna array under discussion is

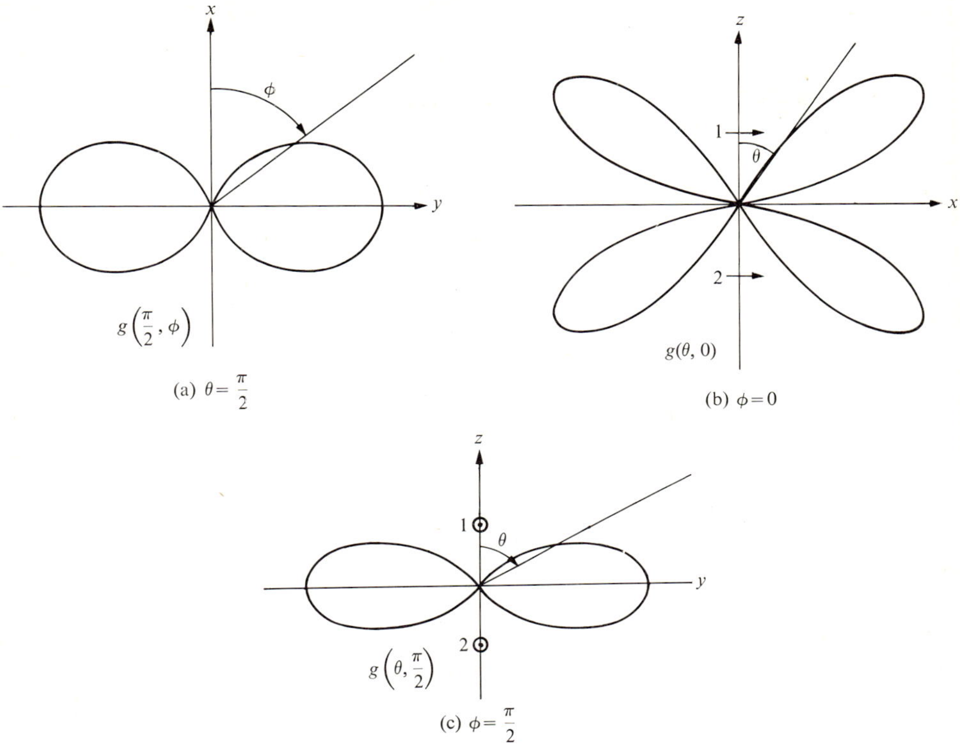

Fig. 10.15 Principal plane patterns of the directive gain of a two-element "point-to-point" type array of Hertzian dipoles separated by a half of a wavelength: (a) $\theta = \pi/2$; (b) $\phi = 0$; (c) $\phi = \pi/2$.

generally used with the "vertical" dipoles in order to be able to concentrate the electromagnetic radiation in the desired directions in the "horizontal" plane. Consequently, the principal plane pattern corresponding to $\phi = \pi/2$, as given by (8.29), is the most important one and, therefore, in the remaining cases to be treated, we shall illustrate only this principal plane pattern.

Case 2. ($d = \lambda/4$; $\alpha = 1$). The antennas are separated by a distance $d = \lambda/4$ and therefore, $\beta d = \pi/2$. From (8.9), it follows that the currents in the two dipoles are equal in magnitude but the phase of the current in the first dipole *lags* that in the second by $\pi/2$ radians. We can see from (8.24) that $P_0(1, \pi/2) = 1$ and hence we can reduce (8.25) to

$$g(\theta, \phi) = 3[1 - \cos^2 \phi \sin^2 \theta] \cos^2 \left\{\frac{\pi}{4} (\cos \theta - 1)\right\}. \tag{8.30}$$

We obtain the pattern of the directive gain in the yz-plane from (8.30) as

$$yz\text{-plane:} \quad g\left(\theta, \frac{\pi}{2}\right) = 3 \cos^2 \left\{\frac{\pi}{4} (\cos \theta - 1)\right\}, \tag{8.31}$$

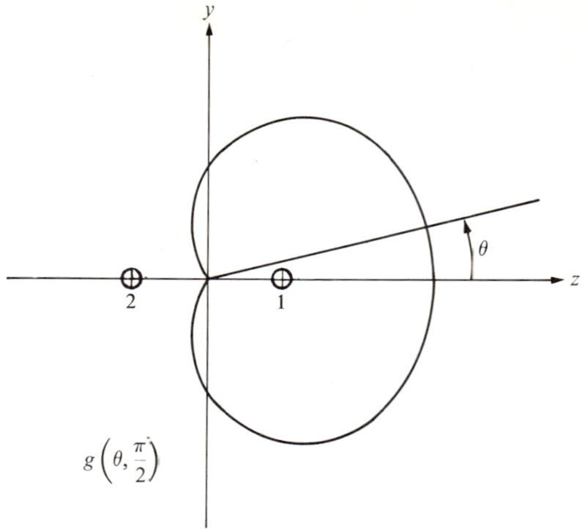

Fig. 10.16 A principal plane ($\phi = \pi/2$) pattern of the directive gain of a two-element "point-to-point" type of an end-fire array of Hertzian dipoles.

which is depicted in Fig. 10.16 as a function of θ. The pattern has a maximum in the direction from antenna 2 to antenna 1 and has a null in the opposite direction from antenna 1 to antenna 2. The direction joining the centers of the two antennas is called the *end-fire* direction. The pattern has the maximum value in the end-fire direction $\theta = 0$ and has the zero value in the end-fire direction $\theta = \pi$. Therefore this antenna array is called a *unidirectional end-fire array.*

It is instructive for us to examine the physical basis which leads to the pattern having the maximum in the direction $\theta = 0$ and the null in the direction $\theta = \pi$. The phase difference between the electromagnetic fields due to two current sources at the respective source locations is the same as the phase difference between the currents. The phase lag of the electromagnetic field in progressing a distance d is equal to βd. Therefore, the electromagnetic field due to antenna 2 at the location of antenna 1 has a phase lag of $\pi/2$ with respect to the current in antenna 2. But, since the current in antenna 1 lags that in antenna 2 by $\pi/2$, it follows that the field due to antenna 2 at the location of antenna 1 as well as the field due to antenna 1 at its location are in the same phase. Hence, it follows that the fields due to both the antennas are in phase in the direction $\theta = 0$; therefore, they interfere constructively to yield a maximum. On the other hand, the electromagnetic field due to antenna 1 at the location of antenna 2 has a phase lag of $\pi/2$ with respect to the current in antenna 1. But, since the current in antenna 2 leads that in antenna 1 by $\pi/2$, it follows that the field due to antenna 1 at the location of antenna 2 lags that due to antenna 2 at its location by π. Hence, the fields due to both the antennas are out of phase in the direction $\theta = \pi$. Since the amplitudes of the fields due to the individual antennas are equal in magnitude in the far-zone, it follows that in the far-zone, in the direction $\theta = \pi$, the fields due to both the antennas interfere destructively to yield a null value.

Case 3. ($d = \lambda/4$; $-1 \leq \alpha \leq 1$). The antennas are separated by a distance $d = \lambda/4$ and therefore $\beta d = \pi/2$. We note that the currents in the two dipoles are equal in magnitude but the relative phase of the currents in the two dipoles is variable. From (8.25), we can determine the pattern of the directive gain in the yz-plane as

$$g\left(\theta, \frac{\pi}{2}\right) = \frac{3}{P_0(\alpha, \pi/2)} \cos^2 \left\{\frac{\pi}{4}(\cos\theta - \alpha)\right\}. \tag{8.32}$$

The directive gain as given by (8.32) has the maximum value, if the $\cos^2\{\ \}$ term is equal to unity or its argument equal to zero. Therefore, it follows that $g(\theta, \pi/2)$ as given by (8.32) has the maximum value at $\theta = \theta_{max}$, which we see to be defined by

$$\cos\theta_{max} = \alpha. \tag{8.33}$$

If $\alpha = 1$, that is, if the current in antenna 1 *lags* that in antenna 2 by $\pi/2$, the directional pattern has one maximum in the end-fire direction $\theta_{max} = 0$. If $\alpha = -1$, that is, if the current in antenna 1 *leads* that in antenna 2 by $\pi/2$, the directional pattern has again only one maximum in the other end-fire direction $\theta_{max} = \pi$. For any other value of α in the range $-1 < \alpha < 1$, since (8.33) is unaltered by changing θ_{max} to $-\theta_{max}$, it follows that the directional pattern has two equal maxima at $\theta_{max} = \pm\cos^{-1}\alpha$. As α varies from 1 through 0 to -1, that is, as the phase of the current in antenna 1 with respect to that in antenna 2 varies from $-\pi/2$ through 0 to $\pi/2$, the position of the maximum of the directional pattern or equivalently the beam of the antenna array varies from the end-fire direction $\theta_{max} = 0$, through the broadside directions $\theta_{max} = \pm\pi/2$ to the other end-fire direction $\theta_{max} = \pm\pi$. This phenomenon of the variation of the angular position of the main beam of the antenna array through all possible directions in the "horizontal" plane is called the scanning of the beam in the "horizontal" plane.

There are two possible methods by which the beam of an array can be scanned. For the sake of definiteness, let it be desired to scan the beam of an array in the horizontal plane from $\theta_{max} = 0$ to $\theta_{max} = \pi$. One possible method is to choose the broadside array of "vertical" dipoles discussed under Case 1, mount the entire array on a frame and rotate the frame mechanically around a vertical axis so that the position of one of the two beams varies from $\theta_{max} = 0$ to $\theta_{max} = \pi$. This method of scanning is called the *mechanical scanning* of the beam. Another possible method is to choose a two-element array of "vertical" dipoles separated by a distance $d = \lambda/4$. If the magnitudes of the currents in the two dipoles are kept constant but the phase of the current in one antenna relative to that of the other is made to vary from $-\pi/2$ to $\pi/2$, again the position of one of the two beams varies in the "horizontal" plane from $\theta_{max} = 0$ to $\theta_{max} = \pi$. This method of scanning is called the *electronic scanning* of the beam since no mechanical rotation of the array is involved. The inertia of the mechanical parts in the mechanical scanning limits its speed and, therefore, the electronic scanning has the advantage of being amenable to higher rates of scanning.

In Fig. 10.17, the directive gain, as given by (8.32) is shown for $\alpha = \frac{1}{2}$ and we see the positions of the maximum of the two beams to occur at $\theta_{max} = \pm\pi/3$.

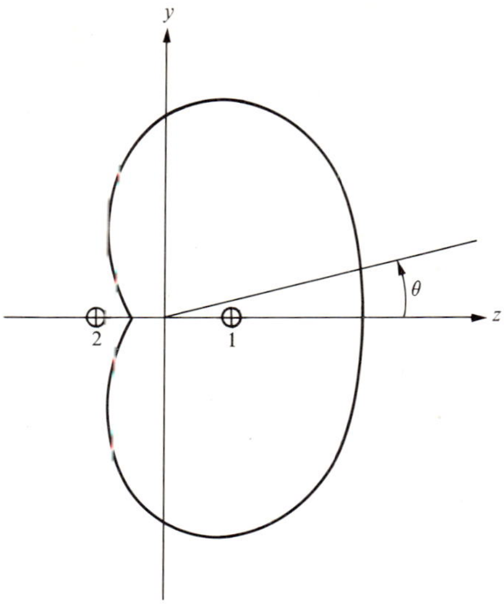

Fig. 10.17 A principal plane ($\phi = \pi/2$) pattern of the directive gain of a two-element "point-to-point" type of an array of Hertzian dipoles; separation distance d = $\lambda/4$; phase difference parameter $\alpha = \frac{1}{2}$.

As in the case of the "broadcast" type array, we can also carry out in principle the analysis of the "point-to-point" type array when the elements of the array are not Hertzian dipoles but are finite length antennas. However, the complication of the resulting integrals makes the treatment too involved for us to present it here.

We have only treated the two-element antenna array with the same amplitude of current in both the elements. By changing the spacing between the elements and by adjusting the relative phase of the current in one element with respect to that of the other, we have found that a large variety of directional patterns is possible even with this restrictive type array. Further flexibility in the available directional patterns can be achieved with arrays of larger number of elements and by permitting the variation of the amplitudes of the currents in the various elements as well. Such antenna arrays find important applications in practice.

For an antenna array of two elements spaced a quarter wavelength apart, the beams of the directional pattern are broad and become broader as they are scanned from the broadside to the end-fire directions. For this reason, the two beams of Fig. 10.17 are not distinct. With antenna arrays of a larger number of elements and with proper adjustments of the amplitudes of the currents in the various elements, we can obtain sharp beams which do not deteriorate significantly as they are scanned from the broadside to the end-fire directions.

10.9 CROSSED DIPOLES

We showed that dipoles which are parallel to each other can be combined to form antenna arrays which enable the concentration of the electromagnetic radiation in the desired directions. In a similar manner, we can use a combination of antennas with suitable orientations and excitations to produce electromagnetic waves with the desired state of polarization besides those obtained with single antennas. The simple antennas which we treated in the previous chapter all produce linearly polarized electric fields in the far-zone. We can use two perpendicularly oriented dipoles with their currents having a phase difference of $\pi/2$ to produce in the far-zone either a circularly or an elliptically polarized electric field in the desired direction. This arrangement of antennas is known as crossed dipoles or a turnstile array.

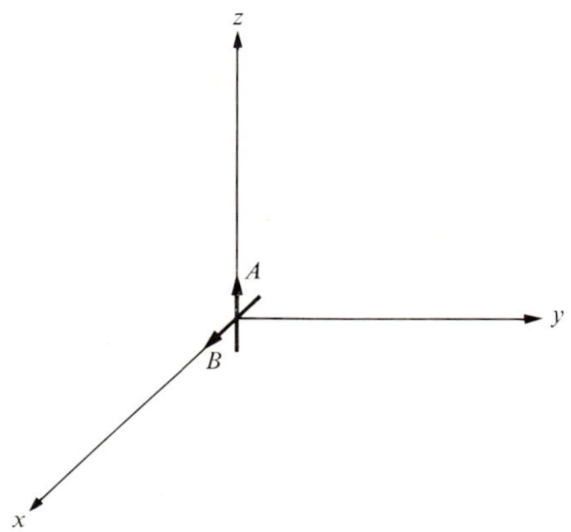

Fig. 10.18 Turnstile array of two Hertzian dipoles.

Let us consider a dipole A of current moment $I_z\,dl$ situated at the origin of the coordinate system and oriented parallel to the z-axis and another dipole B of current moment $I_x\,dl$, also situated at the origin of the coordinate system but oriented parallel to the x-axis, as shown in Fig. 10.18. We obtain the far-zone electric field due to the dipole A from (8.1) to be given by

$$\mathbf{E}_A(\mathbf{r}) = MI_z\,dl[\hat{\theta}\sin\theta], \tag{9.1}$$

where

$$M = \frac{e^{-j(\beta r - \pi/2)}}{4\pi r}\eta_{0r}\beta. \tag{9.2}$$

In a similar manner, we find the far-zone electric field due to the dipole B from (8.8) to be

$$\mathbf{E}_B(\mathbf{r}) = MI_x\,dl[-\hat{\theta}\cos\theta\cos\phi + \hat{\phi}\sin\phi]. \tag{9.3}$$

In Section 5.7, we pointed out that a combination of two linear polarizations which are quadrature in time and space produce circular or elliptic polarization depending on whether or not the two constituent linear polarizations are equal in amplitude. An examination of (9.1) and (9.3) shows us that in order for the resultant electric field in the far-zone due to dipoles A and B to have other than linear polarization, there should be a phase difference between the currents I_z and I_x. Therefore, let us assume that

$$I_z = I_{z0}, \qquad I_x = I_{x0}e^{j\pi/2}, \tag{9.4}$$

where I_{z0} and I_{x0} are real with the result that the current in dipole A lags that of dipole B by $\pi/2$.

The polarization of the resultant far-zone electric field depends on the direction specified by θ and ϕ. We propose to investigate the polarization characteristics only in the three principal coordinate planes.

1. zx-plane; $\phi = 0$.

From (9.1), (9.3), and (9.4), we obtain the resultant far-zone electric field as

$$\boldsymbol{E}(\boldsymbol{r}) = \boldsymbol{E}_A(\boldsymbol{r}) + \boldsymbol{E}_B(\boldsymbol{r}) = M\, dl\hat{\theta}[I_{z0}\sin\theta - I_{x0}e^{j\pi/2}\cos\theta]. \tag{9.5}$$

We see from (9.5) that the polarization in the zx-plane is linear, that is, in the θ-direction.

2. xy-plane; $\theta = \pi/2$.

We assume that the amplitudes of the currents in the two dipoles are equal:

$$I_{z0} = I_{x0} = I_0. \tag{9.6}$$

Then, we evaluate the resultant far-zone electric field with the help of (9.1), (9.3), (9.4), and (9.6) as

$$\boldsymbol{E}(\boldsymbol{r}) = MI_0\, dl[\hat{\theta} + j\hat{\phi}\sin\phi]. \tag{9.7}$$

Comparison of (9.7) with (5.7.13) shows us that the resultant electric field is, in general, left elliptically polarized with respect to the radial ($\hat{r}$) direction for $0 < \phi < \pi$. If the thumb of the left hand points in the radial direction, the fingers curl in the direction of rotation of the instantaneous electric vector. On the other hand, if the current in dipole B instead of leading that in dipole A *lags* that in dipole A, we replace j in (9.7) by $-j$ and then it is evident that the resultant electric field is, in general, right elliptically polarized with respect to the radial ($\hat{r}$) direction for $0 < \phi < \pi$. For the special direction specified by $\phi = 0$ and $\phi = \pi$, we see the electric field from (9.7) to be linearly polarized in the $\hat{\theta}$ $(= -\hat{z})$ direction. Similarly, for the special direction specified by $\phi = \pi/2(3\pi/2)$, we see the electric field from (9.7) to be left (right) circularly polarized. In the xy-plane, as the direction is varied from $\phi = 0$ to $\phi = \pi/2$, the polarization starts out to be linear, then acquires ellipticity which decreases until for $\phi = \pi/2$, the polarization becomes circular. Essentially, this phenomenon repeats itself in the other three quadrants.

3. yz-plane; $\phi = \pi/2$.

As before, we assume the amplitudes of the currents in the two dipoles to be equal. Then, we determine the resultant far-zone electric field with the help of (9.1), (9.3), (9.4), and (9.6) as

$$\boldsymbol{E}(\boldsymbol{r}) = MI_0\, dl[\hat{\theta}\sin\theta + j\hat{\phi}]. \tag{9.8}$$

It is evident from (9.8) that the resultant electric field is, in general, left elliptically polarized with respect to the radial ($\hat{r}$) direction. For the special directions $\theta = 0$ and $\theta = \pi$, we see the electric field from (9.8) to be linearly polarized in the $\hat{\phi}(-\hat{x})$ direction. In a similar manner, for the special direction $\theta = \pi/2$, we observe from (9.8) that the electric field is left circularly polarized. In the yz-plane also, as the direction is varied from $\theta = 0$ to $\theta = \pi/2$, the polarization starts out to be linear, then acquires ellipticity which decreases until for $\theta = \pi/2$ the polarization becomes circular. As the direction is still further varied from $\theta = \pi/2$ to $\theta = \pi$, the polarization becomes elliptical, the ellipticity increases until for $\theta = \pi$ the polarization again becomes linear.

In the normal operation of the turnstile array, the currents in the crossed dipoles are equal in magnitude but differ in phase by $\pi/2$ radians. In the far-zone, in the directions lying in the plane which passes through the axes of the two dipoles, the electric field is linearly polarized. But, in the two directions which are perpendicular to the plane passing through the axes of the two dipoles, the electric field is circularly polarized. In general, in all other directions, the electric field is elliptically polarized.

It is of some interest for us to evaluate the directive gain of the turnstile array. We obtain the resultant electric field from (9.1) to (9.4) and (9.6) as

$$\boldsymbol{E}(\boldsymbol{r}) = \frac{e^{-j(\beta r - \pi/2)}}{4\pi r} \eta_{0r}\beta I_0\, dl[\hat{\theta}\{\sin\theta - j\cos\theta\cos\phi\} + \hat{\phi} j \sin\phi]. \tag{9.9}$$

In view of (8.12), we find the resultant magnetic field in the far-zone to be

$$\boldsymbol{H}(\boldsymbol{r}) = \frac{e^{-j(\beta r - \pi/2)}}{4\pi r} \beta I_0\, dl[-\hat{\theta} j \sin\phi + \hat{\phi}\{\sin\theta - j\cos\theta\cos\phi\}]. \tag{9.10}$$

From (9.9) and (9.10), we deduce the radiation intensity of the turnstile array as

$$\Phi(\theta, \phi) = r^2\hat{r}\cdot\tfrac{1}{2}\,\mathrm{Re}\,\boldsymbol{E}(\boldsymbol{r}) \times \boldsymbol{H}^*(\boldsymbol{r}),$$

$$= \frac{(I_0\, dl)^2\eta_{0r}\beta^2}{32\pi^2}(1 + \sin^2\phi\sin^2\theta). \tag{9.11}$$

We evaluate the total time-averaged power P radiated by this turnstile array as

$$P = \int_{\phi=0}^{2\pi}\int_{\theta=0}^{\pi} \frac{(I_0\, dl)^2\eta_{0r}\beta^2}{32\pi^2}(1 + \sin^2\phi\sin^2\theta)\sin\theta\, d\theta\, d\phi,$$

$$= \frac{(I_0\, dl)^2\eta_{0r}\beta^2}{32\pi}\int_{\theta=0}^{\pi}(2\sin\theta + \sin^3\theta)\, d\theta,$$

$$= \frac{(I_0\, dl)^2\eta_{0r}\beta^2}{32\pi}(2\times 2 + \tfrac{4}{3}) = \frac{(I_0\, dl)^2\eta_{0r}\beta^2}{6\pi}. \tag{9.12}$$

The time-averaged power radiated separately by each dipole is given by (9.5.38) and hence we can determine the sum of the powers radiated by the two dipoles separately from (9.5.38) as

$$P = \frac{(I_0\, dl)^2 \eta_{0r} \beta^2}{6\pi}, \tag{9.13}$$

which is identical to that given in (9.12). Consequently, we may say that the two dipoles of the turnstile array radiate *independently*.

From (9.11) and (9.12), we obtain the directive gain $g(\theta, \phi)$ of the turnstile array as

$$g(\theta, \phi) = \frac{4\pi\Phi(\theta, \phi)}{P} = \tfrac{3}{4}(1 + \sin^2 \phi \sin^2 \theta). \tag{9.14}$$

We see the maximum value of the directive gain from (9.14) to arise for $\phi = \pi/2$, $3\pi/2$, and $\theta = \pi/2$, that is, in the two directions which are normal to the plane containing the axes of the two dipoles. We can determine the directivity or the gain of the turnstile array from (9.14) to be given by $g = 1.5$, which is the same as the directivity of a single Hertzian dipole.

10.10 A SMALL CIRCULAR LOOP ANTENNA WITH AN AXIAL DIPOLE AT ITS CENTER

The far-zone electric field due to a turnstile array is circularly polarized only in the directions which are perpendicular to the plane containing the axes of the two dipoles. In the other directions, the polarization is either linear or elliptical. Sometimes, it is desirable to have an antenna system whose far-zone electric field is either circularly or elliptically polarized but is the same in all directions. We can achieve this objective with the help of an antenna array consisting of a small circular loop antenna having a constant current distribution around its circumference together with an electric dipole situated at the center of the loop and oriented parallel to its axis.

Let us consider a filamentary loop antenna of radius a, with its center coinciding with the origin of the coordinate system. The loop lies in the xy-plane and its axis coincides with the z-axis. We assume the radius of the loop to be very small in comparison to a wavelength. Let I_ϕ be the constant current distribution around the circumference of the loop. Also, let $I_z\, dl$ be the current moment of an Hertzian dipole which is situated at the origin of the coordinate system and oriented parallel to the z-axis (Fig. 10.19).

We obtain the far-zone electric field due to the circular loop from (9.9.17) as

$$\boldsymbol{E}_1(\boldsymbol{r}) = \hat{\phi}(I_\phi \pi a^2) \frac{e^{-j\beta r}}{4\pi r} \eta_{0r} \beta^2 \sin\theta. \tag{10.1}$$

In a similar manner, we find the far-zone electric field due to the electric dipole from (8.1) as

$$\boldsymbol{E}_2(\boldsymbol{r}) = \hat{\theta} I_z\, dl \frac{e^{-j\beta r}}{4\pi r} j\eta_{0r}\beta \sin\theta. \tag{10.2}$$

It is convenient for us to consider three separate cases.

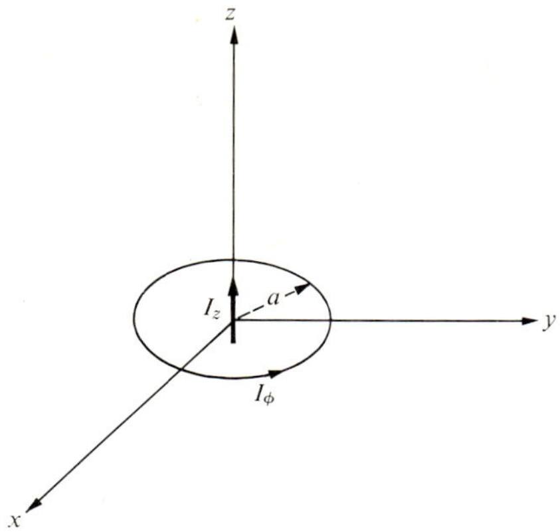

Fig. 10.19 A circular filamentary current loop of radius a with an Hertzian dipole oriented along the z-axis at its center.

Case 1. The current in the dipole differs in phase from that in the loop by $\pi/2$ radians. Hence,

$$I_\phi = I_{\phi 0}, \qquad I_z = I_{z0} e^{j\pi/2}, \tag{10.3}$$

where $I_{\phi 0}$ and I_{z0} are real. Then, we can determine the resultant far-zone electric field from (10.1) to (10.3) as

$$\boldsymbol{E}(\boldsymbol{r}) = \boldsymbol{E}_1(\boldsymbol{r}) + \boldsymbol{E}_2(\boldsymbol{r}) = \frac{e^{-j\beta r}}{4\pi r} \eta_{0r}\beta \sin\theta[-I_{z0}\, dl\hat{\theta} + \beta(I_{\phi 0}\pi a^2)\hat{\phi}]. \tag{10.4}$$

It follows from (10.4) that the polarization of the electric field is linear and is the same in all directions with respect to the radius vector.

Case 2. The currents in the loop and in the dipole are in the same phase. Hence, we can assume that

$$I_\phi = I_{\phi 0}, \qquad I_z = I_{z0}, \tag{10.5}$$

where $I_{\phi 0}$ and I_{z0} are real. For this case, we can obtain the resultant electric field from (10.1), (10.2), and (10.5) as

$$\boldsymbol{E}(\boldsymbol{r}) = \frac{e^{-j\beta r}}{4\pi r} \eta_{0r}\beta \sin\theta[jI_{z0}\, dl\hat{\theta} + \beta(I_{\phi 0}\pi a^2)\hat{\phi}]. \tag{10.6}$$

If we compare (10.6) with (5.7.13) and (5.7.14) we see that the resultant far-zone electric field as given by (10.6) is right elliptically polarized with respect to the radial ($\hat{r}$) direction and the polarization relative to the radius vector is the same in all directions. If the electric dipole instead of being oriented parallel to the $+z$-axis is oriented parallel to the $-z$-axis, in other words, I_z is in opposite phase to I_ϕ, then, j inside the square brackets in (10.6) is changed to $-j$ and the sense of the polarization is reversed.

Case 3. As before, we assume the loop and the dipole currents to be in the same phase and therefore (10.5) continues to hold. Also, we adjust the magnetic moment of the circular loop and the current moment of the Hertzian dipole so that the following relation holds:

$$I_{z0}\, dl = \beta(I_{\phi 0}\pi a^2). \tag{10.7}$$

Then, we obtain the far-zone electric field from (10.6) and (10.7) as

$$\boldsymbol{E}(\boldsymbol{r}) = \frac{e^{-j\beta r}}{4\pi r}\eta_{0r}\beta \sin\theta(I_{z0}\, dl)[j\hat{\theta} + \hat{\phi}]. \tag{10.8}$$

Comparison of (10.8) with (5.7.10) shows us that the resultant far-zone electric field as given by (10.8) is right circularly polarized with respect to the radial ($\hat{r}$) direction and the polarization relative to the radius vector is the same in all directions.

The directive gain of the antenna system consisting of the loop and the dipole is of interest especially for the case in which it produces in the far-zone circularly polarized electric field. For this case, the far-zone electric field is given by (10.8). In view of (8.12), we can evaluate the far-zone magnetic field from (10.8) with the following result:

$$\boldsymbol{H}(\boldsymbol{r}) = \frac{e^{-j\beta r}}{4\pi r}\beta \sin\theta(I_{z0}\, dl)[-\hat{\theta} + j\hat{\phi}]. \tag{10.9}$$

We can determine the radiation intensity from (10.8) and (10.9) as

$$\Phi(\theta, \phi) = r^2\hat{r}\cdot\tfrac{1}{2}\operatorname{Re}\boldsymbol{E}(\boldsymbol{r})\times\boldsymbol{H}^*(\boldsymbol{r}) = \frac{\beta^2\eta_{0r}\sin^2\theta}{16\pi^2}(I_{z0}\, dl)^2. \tag{10.10}$$

We can evaluate the total time-averaged power P radiated by the loop-dipole combination as

$$P = \int_{\phi=0}^{2\pi}\int_{\theta=0}^{\pi}\frac{(I_{z0}\, dl)^2\beta^2\eta_{0r}}{16\pi^2}\sin^3\theta\, d\theta\, d\phi = \frac{(I_{z0}\, dl)^2\beta^2\eta_{0r}}{6\pi}. \tag{10.11}$$

From (9.9.20) and (10.7), we obtain the time-averaged power radiated separately by the loop as

$$P_1 = \frac{(I_{z0}\, dl)^2\eta_{0r}\beta^2}{12\pi}. \tag{10.12}$$

Also, we can see from (9.5.38) that the time-averaged power P_2 radiated separately by the dipole is equal to P_1. Hence, we see that the sum of the powers radiated by the loop and the dipole separately is identical to that given in (10.11). Consequently, the loop and the dipole in the loop-dipole combination radiate independently.

We obtain the directive gain $g(\theta, \phi)$ from (10.10) and (10.11) as

$$g(\theta, \phi) = \frac{4\pi\Phi(\theta, \phi)}{P} = \tfrac{3}{2}\sin^2\theta, \tag{10.13}$$

which is the same as that for the isolated circular loop or the Hertzian dipole alone.

PROBLEMS

10.1 Consider an electrically short ($\beta h = 2\pi(h/\lambda) \ll 1$) dipole antenna in free space extending along the z-axis from $z = -h$ to $z = h$ where $h = 0.05\lambda$. The input terminals of the antenna are situated at its center. When used as a transmitter, the current distribution is assumed to be "triangular" as given by

$$I(z) = I_0\left(1 - \frac{|z|}{h}\right). \tag{a}$$

A plane wave is incident normally on the antenna with the electric field parallel to the length of the antenna. The amplitude of the incident electric field is 1 mV/m. If the equivalent load impedance at the antenna terminals is adjusted to extract maximum power from the incident wave and if the wavelength is 1 m, what is the current at the antenna terminals?

Suppose that the plane wave arrives with its wave normal making an angle $\theta = 30°$ with the antenna axis. If the incident electric field lies in the plane containing the antenna axis and the wave normal direction, find the value of the current at the antenna terminals.

Determine the received power for both the preceding cases.

10.2 Consider a half-wavelength long antenna in free space extending along the z-axis from $z = -h = -\lambda/4$ to $z = h = \lambda/4$, with the input terminals situated at the center of the antenna. When used as a transmitter, the current distribution is assumed to be sinusoidal as given by

$$I(z) = I_0 \cos \beta z. \tag{a}$$

A plane wave is incident on the antenna with the wave normal making an angle θ with the axis of the antenna. The incident electric vector lies in the plane containing the axis of the antenna and the wave normal direction. The amplitude of the incident electric field is 5 mV/m. The antenna impedance is given by

$$Z_a = 73 + j42.5 \text{ ohms}. \tag{b}$$

If the antenna terminals are short-circuited and the wavelength is 1 m, find the current at the input terminals of the antenna for (i) $\theta = 0°$; (ii) $\theta = 60°$; and (iii) $\theta = 90°$.

10.3 Suppose that the half-wave antenna of Problem 10.2, when used as a transmitter, has a current distribution of the form

$$I(z) = I_0 \cos^2 \beta z. \tag{a}$$

The equivalent load impedance at the antenna terminals is adjusted to receive maximum power. Note that the effective height and the antenna resistance for the half-wave antenna with the current distribution as given by (a) are contained in Problems 9.12 and 9.15 respectively. If the plane wave is incident normally on the antenna with the electric vector parallel to the length of the antenna, find the value of the current at the antenna terminals. Assume as in Problem 10.2 that the magnitude of the incident electric vector is 5 mV/m and the wavelength is 1 m.

What is the value of the power delivered to the load impedance?

10.4 A half-wavelength long filamentary antenna in free space is situated along the z-axis from $z = -h = -\lambda/4$ to $z = h = \lambda/4$. When used as a transmitter, the current distribution along the antenna is a progressive wave of the form

$$I(z) = I_0 e^{-j\beta\alpha z}. \tag{a}$$

Assume that $\alpha = 1.0$. Deduce the expression for the effective height of the antenna.

A plane wave is incident on the antenna with the wave normal making an angle of θ with the axis of the antenna. The electric vector lies in the plane containing the antenna axis and the wave normal direction. The value of the time-averaged Poynting vector at the location of the antenna is 1 W/m^2. The wavelength is 2 m.

The equivalent load impedance at the terminals of the antenna is adjusted to receive maximum power. What is the power delivered to the load, if the wave is incident normally on the antenna?

For what angle of incidence is the power delivered to the equivalent load impedance a maximum and what is the value of this maximum power?

Repeat the calculations for $\alpha = 1.2$. Refer to Problem 9.18 for the necessary data.

10.5 A plane electromagnetic wave is incident normally on a dipole receiving antenna of length $2h$. The antenna is situated in free space along the z-axis from $z = -h$ to $z = h$. The incident electric vector is parallel to the length of the antenna whose impedance is assumed to be given by Z_a. The magnitude of the incident electric field is E V/m. Assume that, as a transmitting antenna, the current distribution is sinusoidal; that is,

$$I(z) = I_m \sin \beta(h - |z|).$$

Find the expression for the short-circuit current at the antenna terminals.

10.6 An antenna has an input impedance of $73 + j42.5$ ohms. Its effective area is 0.50 m^2. The antenna operates in a medium of intrinsic impedance of 120π ohms and is in a plane wave with the associated electric field of 600 μV/m. Find the available power at the input terminals of the antenna. What is the value of the equivalent load impedance at the antenna terminals necessary to extract this available power? Instead of the impedance necessary to extract the available power, if the antenna terminals are connected directly to a 50-ohm transmission line which is perfectly matched to a resistance load, what is the power transferred to the load resistance?

10.7 A small linear probe antenna operating in a medium of intrinsic impedance of 120π ohms is in a plane wave with the associated electric field $E = 4\sqrt{15\pi}$ V/m. The antenna length $dl = 0.01$ m and the operating wavelength $\lambda = \pi$ m. Assume that the current distribution along the probe when used as a transmitter is a constant. (a) What is the incident power per unit area? (b) Show that the antenna resistance is 8×10^{-3} ohm.

The input terminals of the probe antenna which is parallel to the electric field are connected to a load which is adjusted to receive maximum power. (c) What is the value of this maximum power? (d) What is the maximum effective area A_e of this probe? (e) Define directivity g and state its relationship to A_e. Find g for this probe.

10.8 An electrically short ($\beta h = 2\pi h/\lambda \ll 1$) transmitting antenna in free space is situated along the z-axis from $z = -h$ to $z = h$. The input terminals of the antenna are situated at its center. The current distribution along the antenna is assumed to be "triangular" as given by

$$I(z) = I_0\left(1 - \frac{|z|}{h}\right).$$

Evaluate the effective aperture $A_e(\theta, \phi)$ and the directive gain $g(\theta, \phi)$. Determine the ratio $A_e(\theta, \phi)/g(\theta, \phi)$.

10.9 An electrically short ($\beta h = 2\pi h/\lambda \ll 1$) reflecting antenna in free space is situated along the z-axis from $z = -h$ to $z = h$. The current distribution is assumed to be "parabolic" as given by

$$I(z) = I_0\left(1 - \frac{z^2}{h^2}\right).$$

Evaluate the effective aperture $A_e(\theta, \phi)$ and the directive gain $g(\theta, \phi)$. Determine the ratio $A_e(\theta, \phi)/g(\theta, \phi)$.

10.10 A half-wavelength long antenna in free space is situated along the z-axis from $z = -h = -\lambda/4$ to $z = h = \lambda/4$. The antenna terminals are situated at its center. The current distribution along the antenna, when used as a transmitter, is sinusoidal as given by $I(z) = I_0 \cos \beta z$. The antenna is illuminated by a plane wave with the wave normal making an angle $\theta = 60°$ with the antenna axis. The electric vector is in the plane containing the antenna axis and the wave normal direction. The operating wavelength is 1 m. The available power at the antenna terminals is 50 μW. What is the magnitude of the incident electric field? Assume that the antenna has an impedance equal to $Z_a = 73 + j42.5$ ohms.

The antenna terminals are connected by means of a quarter-wavelength long transmission line of characteristic impedance $Z_0 = 50$ ohms to a load resistance of 100 ohms. What is the value of the current in the load resistance?

10.11 A half-wavelength long antenna in free space is situated along the z-axis from $z = -h = -\lambda/4$ to $z = h = \lambda/4$. The input terminals of the antenna are situated at the center. The current distribution along the antenna is of the form $I(z) = I_0 \cos^2 \beta z$. Evaluate the effective aperture $A_e(\theta, \phi)$ and the directive gain $g(\theta, \phi)$. Verify that the ratio $A_e(\theta, \phi)/g(\theta, \phi)$ is equal to $\lambda^2/4\pi$.

10.12 A half-wavelength long antenna in free space is situated along the z-axis from $z = -h = -\lambda/4$ to $z = h = \lambda/4$. The wavelength is 10 m. The current distribution along the antenna is a progressive wave of the form $I(z) = I_0 e^{-j\beta\alpha z}$. Assume that $\alpha = 1$. For what direction with respect to the axis of the antenna is the effective aperture a maximum and what is the value of this maximum effective aperture? Repeat the calculations for $\alpha = 1.2$.

10.13 The radiation intensity $\Phi(\theta)$ of an antenna is independent of the angle ϕ and is dependent on θ in the following manner:

$$\Phi(\theta) = \cos^2 \theta \qquad (\text{for } 0 < \theta < \pi/2),$$
$$= 0 \qquad (\text{for } \pi/2 < \theta < \pi).$$

If the wavelength of operation is $\lambda = 0.4\sqrt{\pi}$ m, show that the maximum effective aperture is equal to 0.24 m^2. If the antenna operates in free space where the electromagnetic wave velocity is $v = 3 \times 10^8$ m/sec, what is the operating frequency for which the maximum effective aperture is equal to 1 m^2?

10.14 The radiation intensity $\Phi(\theta)$ of an antenna is independent of the angle ϕ and is dependent on θ in the following manner:

$$\Phi(\theta) = \cos^2 \theta \qquad (\text{for } 0 < \theta < \pi).$$

What is the wavelength of operation for which the maximum effective aperture is equal to 0.24 m^2?

10.15 Calculate the available power at the receiver corresponding to a transmitted power of 500 W under the following conditions: (a) Gain of transmitting antenna = 1.64; effective area of receiver = 0.50 m^2. (b) Effective area of both antennas = 1 m^2; wavelength = 0.04 m. (c) Gain of both antennas = 1.5; wavelength = 0.10 m.

Assume that the distance between the transmitter and the receiver is equal to 10^4 m.

10.16 A wire in free space is situated along the z-axis from $z = -h$ to $z = h$. The length of the wire is small in comparison to a wavelength, that is, $\beta h \ll 1$. A plane wave is incident normally on the wire with the electric vector parallel to the length of the wire. Find the form of the current distribution along the wire antenna. Evaluate the total time-averaged power P reradiated by the antenna.

10.17 Two linear wire antennas are situated along the z-axis from $z = -h$ to $z = 0^-$ and from $z = 0^+$ to $z = h$ respectively. The length of each antenna is small in comparison with the wavelength, that is $\beta h \ll 1$. A plane wave is incident normally on the wires with the electric vector parallel to the length of both the antennas. Find the form of the current distribution along the two antennas. Deduce the total time-averaged power P reradiated by both the antennas. If E_0 is the magnitude of the incident electric field, what is the voltage induced between the terminals ($z = 0^-$ and $z = 0^+$) of the two wires?

10.18 A full-wavelength long reflecting antenna in free space extends along the z-axis from $z = -h$ to $z = h$. Assume that the current distribution along the antenna is of the form $I(z) = I_0 \sin \beta z$. Evaluate the total time-averaged power reradiated by the antenna.

10.19 A half-wavelength long antenna in free space is situated along the z-axis from $z = -h$ to $z = h$. The antenna terminals are situated at its center. A plane wave is incident normally on the antenna with the electric vector parallel to the length of the antenna. The magnitude of the incident electric field is 1 mV/m and the wavelength is 2 m. Determine the expression for and sketch the real and the imaginary parts of the current distribution along the antenna if the antenna terminals are short-circuited. The impedance of the antenna is given by $Z_a = 73 + j42.5$ ohms. Evaluate the total time-averaged power reradiated by the antenna.

Suppose that the antenna terminals are loaded by an equivalent impedance which is adjusted to receive maximum power. What is the power received by the load? Determine the expression for the current distribution along the antenna. Sketch the real and the imaginary parts of the current distribution along the antenna. Determine the total time-averaged power reradiated by the antenna. How does the reradiated power compare with the received power?

10.20 A linear antenna in free space is situated along the z-axis from $z = -h$ to $z = h$. The length of the antenna is such that $\beta h = 1.0$. The impedance of the antenna is given by $Z_a = 25 + j130$ ohms. The antenna terminals are situated at its center.

A plane wave is incident normally on the antenna with the electric vector parallel to the length of the antenna. The magnitude of the incident electric field is 50 μV/m and the wavelength is 10 m. Determine the expression for the current distribution along the antenna, if the antenna terminals are short-circuited. Sketch the real and the imaginary parts of the current distribution along the antenna. Calculate the total time-averaged power reradiated by the antenna.

If the antenna terminals are loaded by an impedance which receives the entire available power, what is the power received by the load? For this case, what is the expression for the current distribution along the antenna and what is the total time-averaged power reradiated by the antenna?

In analogy with the effective aperture of the antenna, it is convenient to introduce the concept of the total scattering cross section σ_t which is the ratio of the total power reradiated by the antenna to the magnitude of the incident Poynting vector at the location of the antenna. Determine the value of the total cross section σ_t both when the antenna terminals are short-circuited and when they are loaded by an impedance to receive the entire available power.

10.21 A linear antenna in free space is situated along the z-axis from $z = -h$ to $z = h$. The antenna is driven asymmetrically from the terminals which are at $z = \xi - 0$ to $z = \xi + 0$. Deduce the expression for the current distribution along the antenna.

10.22 Two Hertzian dipoles 1 and 2 are situated along the z-axis at $z = d/2$ and $z = -d/2$ respectively. These dipoles are oriented parallel to the x-axis. Let $I_1\, dl$ and $I_2\, dl$ be the current moments of the two dipoles. Assume that I_1 and I_2 are equal in magnitude and in phase. Deduce the expression for the field pattern in the horizontal plane (yz-plane). Sketch the field patterns for the following separation distances between the dipoles: (i) $d = \lambda/4$, (ii) $d = \lambda$, and (iii) $d = 2\lambda$.

The field pattern has a number of lobes for $d \geq \lambda/2$. Find the number of lobes for $d = N\lambda$ where N is an integer.

10.23 For the antenna array of problem 10.22, let $I_1 = I_0 e^{+j\alpha/2}$ and $I_2 = I_0 e^{-j\alpha/2}$ where I is real and α is the phase difference between the currents in the two Hertzian dipoles. Deduce the expression for the field pattern in the horizontal plane. Sketch the field pattern for the following cases: (i) $d = \lambda/2$, $\alpha = \pi$; (ii) $d = \lambda/4$, $\alpha = \pi$; and (iii) $d = \lambda$, $\alpha = \pi/2$.

10.24 For the antenna array of Problem 10.22 suppose that the currents in the two Hertzian dipoles are in phase but are unequal in magnitude. When the magnitudes of the currents in the elements of an array are not equal, the array is designated as a nonuniform array. If $I_1 = 2I_2$, find the expression for the field pattern in the horizontal plane of the nonuniform array of two elements. Sketch the field pattern if the separation distance is $d = \lambda/2$.

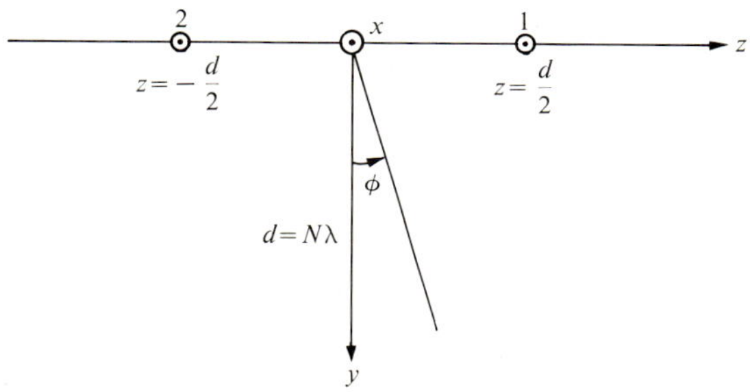

Fig. P10.1 Schematic diagram of the two-antenna radio interferometer.

10.25 A radio interferometer consists of two identical linear antennas 1 and 2 which are situated along the z-axis at $z = d/2$ and $z = -d/2$ respectively. The antennas are oriented in the x-direction. The direction in the horizontal ($x = 0$) plane is measured by the angle ϕ from the broadside (y) direction as shown in Fig. P10.1. The separation distance between the antennas is given by $d = N\lambda$ where N is a large integer. Suppose that the resultant current in the two antennas is observed. If the currents in the two antennas are in phase, the resultant current is a maximum and if they differ in phase by π radians, the resultant current is a minimum. The radio source is at a very great distance with the result that the magnitudes of the currents induced in the two antennas are the same but only the phases of the currents in the two antennas can be different.

Show that the resultant current is a maximum corresponding to the arrival of the wave from the direction $\phi = \phi_m$ as defined by $\sin \phi_m = \pm m/N$ where m is an integer less than or equal to N.

The radio interferometer can be used to measure the angular deviation of the source. Suppose that initially the resultant current is a maximum. As the radio source moves, the resultant current is observed to go through 5 successive minima and then reach a maximum. If $N = 500$, what is the angular deviation of the source? Note that by counting the fringes, that is the minima and the maxima of the resultant current, the angular deviation of the radio source can be obtained.

10.26 Four Hertzian dipoles 1, 2, 3, and 4 are situated in the $x = 0$ plane and are all oriented in the x-direction. The locations of the dipoles 1, 2, 3, and 4 in the $x = 0$ plane are given by ($y = 0$, $z = d/2$), ($y = 0$, $z = -d/2$), ($y = -h$, $z = d/2$) and ($y = -h$, $z = -d/2$) respectively as shown in Fig. P10.2. The currents in all the dipoles are equal in magnitude. Also, the currents in dipoles

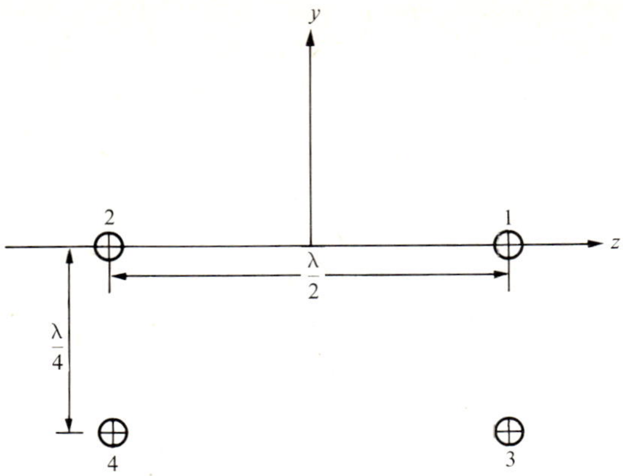

Fig. P10.2 Schematic diagram of a four-element planar antenna array.

1 and 2 are equal in phase and so are the currents in dipoles 3 and 4. But the currents in dipoles 3 and 4 lag those in dipoles 1 and 2 by $\pi/2$ radians. Find the expression for the field pattern in the horizontal (yz-) plane. Sketch the field pattern for $d = \lambda/2$ and $h = \lambda/4$.

10.27 Consider two vertically oriented elementary electric dipoles which are situated in the same horizontal plane at a distance of $d = \lambda$ apart. Let ϕ be the angle measured from the broadside direction and ρ the distance in the horizontal plane to the far-zone measured from midway between the two dipoles. The magnitudes of the current in the two dipoles are the same. (a) What is the direction of the vector potential in (i) the far-zone and (ii) the near-zone? (b) Compute and sketch the field patterns in the horizontal plane when the phases of the currents in the two dipoles are (i) equal and (ii) different by $\pi/2$. (c) What should be the phase difference between the

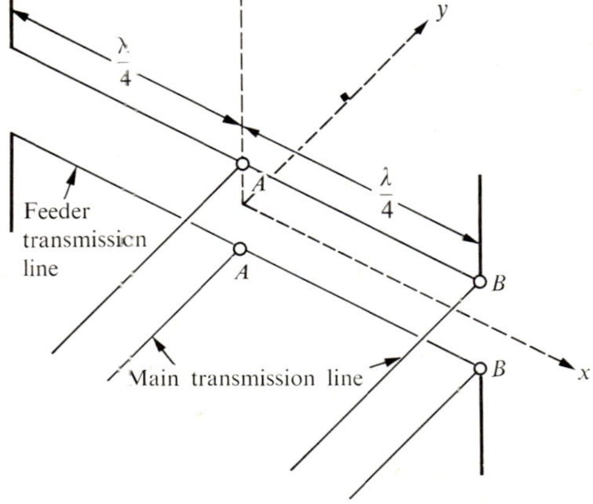

Fig. P10.3 A two-element "point-to-point" type antenna array.

currents in the two antennas in order that a maximum of the field pattern be at an angle of 45° from the broadside direction?

10.28 Two identical antennas are arranged to form a "point-to-point" type array as shown in Fig. P10.3. The antennas have an input impedance of 50 ohms. All the feeding transmission lines have a characteristic resistance of $R_0 = 100$ ohms. (a) When the antenna array is driven by connecting the main transmission line to the feeder transmission line at A, find the voltage standing-wave ratio on the main transmission line and the direction of maximum radiation. (b) When the antenna array is driven by connecting the main transmission line to the feeder transmission line at B, find the relative phases of the currents in the two antennas and the direction of maximum radiation. What is the percentage of the incident power on the main transmission line that is radiated?

10.29 A half-wavelength long filamentary transmitting antenna operates at a frequency $f = 10^9$ Hz. The input terminals of the antenna are situated at its center and are connected by a lossless transmission line of characteristic resistance of 50 ohms and a length of 3.1 m to a voltage generator of 10 V and an internal impedance of 50 ohms. The receiving antenna is also a half-wavelength long filament with its input terminals located at its center. The antennas are situated in free space. The straight line joining the centers of the two antennas is 10^4 m long and is inclined to the axis of the transmitting antenna by 30°. The receiving antenna is oriented to be parallel to the electric field. A load impedance of $30-j40$ ohms is connected by a lossless transmission line of characteristic resistance of 50 ohms and a length of 0.375 m to the input terminals of the receiving antenna.

Find the magnitude of the current in the load impedance.

APPENDIX A

NUMERICAL CONSTANTS

Properties of free space

	Actual values	Approximate values	Unit
Permittivity, ε_0	8.854×10^{-12}	$10^{-9}/36\pi$	farad per meter
Permeability, μ_0	$4\pi \times 10^{-7}$		henry per meter
Velocity of propagation, v	2.998×10^8	3.0×10^8	meter per second
Intrinsic resistance, η_{0r}	376.7	120π	ohm

Physical constants

Mass of electron, m_e,	9.107×10^{-31} kilogram
Mass of proton, m_i,	1.67×10^{-27} kilogram
Charge of electron, e,	-1.601×10^{-19} coulomb
Charge to mass ratio for electron, e/m_e	-1.76×10^{11} coulombs per kilogram

Low-frequency relative permittivity ε_r

Material	ε_r	Material	ε_r
Air	1.00	Plexiglass	2.6 to 3.5
Bakelite	4.9	Polystyrene	2.55
Glass (pyrex)	5.1	Polyethylene	2.25
Lucite	3.2	Porcelain	5.50 to 6.50
Mica	6.0	Quartz (fused)	3.78
Nylon	3.7	Rubber	2.3 to 4.0
Oil	2.3	Styrofoam	1.03
Paper	1.5 to 4.0	Teflon	2.1
Paraffin	2.1	Water (distilled)	80

Relative permeability μ_r

Ferromagnetic materials at small magnetization

Material	μ_r
Nickel	50
Cobalt	60
Cast iron	60
Machine steel	300
Very pure iron	4000

Paramagnetic materials

Material	μ_r
Aluminum	1.00000065
Beryllium	1.00000079

Diamagnetic materials

Material	μ_r
Bismuth	0.99999860
Paraffin	0.99999942
Silver	0.99999981

Low-frequency conductivity σ in mhos per meter

Good conductors

Material	σ	Material	σ
Aluminum	3.54×10^7	Sea Water	3 to 5
Brass	1.57×10^7	Silver	6.17×10^7
Copper	5.80×10^7	Steel	$(0.5 \text{ to } 1.0) \times 10^7$
Gold	4.10×10^7	Tin	9.0×10^6
Nichrome	1.00×10^6	Tungsten	1.81×10^7
Nickel	1.28×10^7	Zinc	1.7×10^7

Good Insulators

Material	σ	Material	σ
Bakelite	10^{-8} to 10^{-10}	Quartz	10^{-17}
Glass	10^{-12}	Rubber (hard)	10^{-14} to 10^{-16}
Mica	10^{-11} to 10^{-15}	Water (distilled)	2×10^{-4}
Porcelain	2×10^{-13}	Water (fresh)	10^{-3}

APPENDIX B:

UNITS AND DIMENSIONS

Fundamental quantities

Quantity	Usual symbol	MKS unit	Dimensions
Length	l	meter	L
Mass	m	kilogram	M
Time	t	second	T
Charge	q	coulomb	Q

Mechanical quantities

Quantity	Usual symbol	MKS unit	Dimensions
Velocity	$\mathbf{v}$	meter per second	LT^{-1}
Acceleration	$\mathbf{a}$	meter per second per second	LT^{-2}
Momentum	$\mathbf{p}$	kilogram-meter per second	MLT^{-1}
Wavelength	λ	meter	L
Frequency	f	hertz	T^{-1}
Force	$\mathbf{F}$	newton	MLT^{-2}
Energy	W	joule	ML^2T^{-2}
Power	P	watt	ML^2T^{-3}

Electrical quantities

Quantity	Usual symbol	MKS unit	Dimensions
Charge density (volume)	ρ	coulomb per cubic meter	$L^{-3}Q$
Charge density (surface)	ρ_s	coulomb per square meter	$L^{-2}Q$
Charge density (linear)	ρ_L	coulomb per meter	$L^{-1}Q$
Voltage	V	volt	$ML^2T^{-2}Q^{-1}$
Electric field	$\mathbf{E}$	volt per meter	$MLT^{-2}Q^{-1}$
Capacitance	C	farad	$M^{-1}L^{-2}T^2Q^2$
Permittivity	ε	farad per meter	$M^{-1}L^{-3}T^2Q^2$
Electric displacement	ψ	coulomb	Q
Electric displacement density	$\mathbf{D}$	coulomb per square meter	$L^{-2}Q$
Electric dipole moment	$\mathbf{p}$	coulomb-meter	LQ
Polarization	$\mathbf{P}$	coulomb per square meter	$L^{-2}Q$

Magnetic quantities

Quantity	Usual symbol	MKS units	Dimensions
Current	$\mathbf{I}$	ampere	$T^{-1}Q$
Current density	$\mathbf{J}$	ampere per square meter	$L^{-2}T^{-1}Q$
Current density (surface)	$\mathbf{J}_s$	ampere per meter	$L^{-1}T^{-1}Q$
Resistance	R	ohm	$ML^2T^{-1}Q^{-2}$
Resistivity	ρ_{res}	ohm-meter	$ML^3T^{-1}Q^{-2}$
Conductivity	σ	mho per meter	$M^{-1}L^{-3}TQ^2$
Magnetic flux	Φ	weber	$ML^2T^{-1}Q^{-1}$
Magnetic flux density	$\mathbf{B}$	weber per square meter	$MT^{-1}Q^{-1}$
Magnetomotive force	m.m.f.	ampere	$T^{-1}Q$
Magnetic field	$\mathbf{H}$	ampere per meter	$L^{-1}T^{-1}Q$
Inductance	L	henry	ML^2Q^{-2}
Permeability	μ	henry per meter	MLQ^{-2}
Magnetic dipole moment	$\mathbf{m}$	amper-square meter	$L^2T^{-1}Q$
Magnetization	$\mathbf{M}$	ampere per meter	$L^{-1}T^{-1}Q$

APPENDIX C:

VECTOR ANALYSIS

We shall present a brief summary of some of the important aspects of vector analysis in this Appendix for the sake of completeness. A scalar quantity is specified completely by its magnitude alone whereas a vector quantity is specified by its magnitude and direction. There are two kinds of operation of multiplication of two vectors. The scalar or the dot product of two vectors $\boldsymbol{A}$ and $\boldsymbol{B}$ is a scalar whose magnitude is equal to the product of the magnitudes of the two vectors and the cosine of the angle θ between them:

$$\boldsymbol{A} \cdot \boldsymbol{B} = |A|\,|B| \cos \theta. \tag{C.1}$$

The vector or the cross product of two vectors $\boldsymbol{A}$ and $\boldsymbol{B}$ is a vector whose magnitude is equal to the product of the magnitudes of the two vectors and the sine of the acute angle θ between them:

$$|\boldsymbol{A} \times \boldsymbol{B}| = |A|\,|B| \sin \theta. \tag{C.2}$$

The direction of $\boldsymbol{A} \times \boldsymbol{B}$ lies normal to the plane containing $\boldsymbol{A}$ and $\boldsymbol{B}$. If the fingers of the right hand curl from the first vector to the second through the acute angle between them, then, by convention, the direction of $\boldsymbol{A} \times \boldsymbol{B}$ is taken to be in the direction of the thumb. If $\hat{x}$, $\hat{y}$, and $\hat{z}$ are the unit vectors in the direction of the axes of a right-handed rectangular coordinates x, y, and z respectively, then it follows that

$$\hat{x} \cdot \hat{x} = \hat{y} \cdot \hat{y} = \hat{z} \cdot \hat{z} = 1; \qquad \hat{x} \cdot \hat{y} = \hat{x} \cdot \hat{z} = \hat{y} \cdot \hat{x} = \hat{y} \cdot \hat{z} = \hat{z} \cdot \hat{x} = \hat{z} \cdot \hat{y} = 0; \tag{C.3}$$

$$\hat{x} \times \hat{x} = \hat{y} \times \hat{y} = \hat{z} \times \hat{z} = 0; \qquad \hat{x} \times \hat{y} = -\hat{y} \times \hat{x} = \hat{z};$$

$$\hat{y} \times \hat{z} = -\hat{z} \times \hat{y} = \hat{x}; \qquad \hat{z} \times \hat{x} = -\hat{x} \times \hat{z} = \hat{y}. \tag{C.4}$$

Any vector can be written in terms of its components along any three mutually perpendicular coordinate axes. Hence,

$$\boldsymbol{A} = \hat{x}A_x + \hat{y}A_y + \hat{z}A_z; \qquad \boldsymbol{B} = \hat{x}B_x + \hat{y}B_y + \hat{z}B_z, \tag{C.5}$$

where $A_x(B_x)$, $A_y(B_y)$, and $A_z(B_z)$ are respectively the x-, y-, and z-components of $\boldsymbol{A}(\boldsymbol{B})$. We can verify from (C.3) and (C.4) that

$$\boldsymbol{A} \cdot \boldsymbol{B} = A_xB_x + A_yB_y + A_zB_z, \tag{C.6}$$

$$\boldsymbol{A} \times \boldsymbol{B} = \hat{x}(A_yB_z - A_zB_y) + \hat{y}(A_zB_x - A_xB_z) + \hat{z}(A_xB_y - A_yB_x). \tag{C.7}$$

In a similar manner, we can define unit vectors along the axes of cylindrical $(\hat{\rho}, \hat{\phi}, \hat{z})$ and spherical $(\hat{r}, \hat{\theta}, \hat{\phi})$ coordinate systems and can express the results of $\boldsymbol{A} \cdot \boldsymbol{B}$ and $\boldsymbol{A} \times \boldsymbol{B}$ in terms of the cylindrical and the spherical components of the two vectors.

The addition, the subtraction, and the scalar multiplication of two vectors obey the commutative law:

$$\boldsymbol{A} + \boldsymbol{B} = \boldsymbol{B} + \boldsymbol{A}; \qquad -\boldsymbol{A} - \boldsymbol{B} = -\boldsymbol{B} - \boldsymbol{A}; \qquad \boldsymbol{A} \cdot \boldsymbol{B} = \boldsymbol{B} \cdot \boldsymbol{A}. \tag{C.8}$$

The cross product of two vectors does not obey the commutative law:

$$\boldsymbol{A} \times \boldsymbol{B} = -\boldsymbol{B} \times \boldsymbol{A}. \tag{C.9}$$

It can be established that the vectors satisfy the associative law of addition:

$$(\boldsymbol{A} + \boldsymbol{B}) + \boldsymbol{C} = \boldsymbol{A} + (\boldsymbol{B} + \boldsymbol{C}). \tag{C.10}$$

The distributive law holds both for the scalar and the vector multiplication:

$$(\boldsymbol{A} + \boldsymbol{B}) \cdot \boldsymbol{C} = \boldsymbol{A} \cdot \boldsymbol{C} + \boldsymbol{B} \cdot \boldsymbol{C}; \qquad (\boldsymbol{A} + \boldsymbol{B}) \times \boldsymbol{C} = \boldsymbol{A} \times \boldsymbol{C} + \boldsymbol{B} \times \boldsymbol{C}. \tag{C.11}$$

We can prove the following two important relations involving the triple product of vectors by expressing the vectors in terms of their rectangular components:

$$\boldsymbol{A} \cdot (\boldsymbol{B} \times \boldsymbol{C}) = \boldsymbol{B} \cdot (\boldsymbol{C} \times \boldsymbol{A}) = \boldsymbol{C} \cdot (\boldsymbol{A} \times \boldsymbol{B}), \tag{C.12}$$

$$\boldsymbol{A} \times (\boldsymbol{B} \times \boldsymbol{C}) = (\boldsymbol{A} \cdot \boldsymbol{C})\boldsymbol{B} - (\boldsymbol{A} \cdot \boldsymbol{B})\boldsymbol{C}. \tag{C.13}$$

There are three important vector differential operations involving functions of position. These are the gradient, the divergence, and the curl operations and are denoted by ∇, $\nabla\cdot$, and $\nabla\times$ respectively. A review of the physical interpretations of these differential operations is useful since they occur frequently in the theory of electromagnetic fields. In rectangular coordinates the vector differential operator ∇, called *del*, is defined as

$$\nabla = \frac{\partial}{\partial x}\hat{x} + \frac{\partial}{\partial y}\hat{y} + \frac{\partial}{\partial z}\hat{z}. \tag{C.14}$$

When ∇ operates on a scalar function ϕ of position, the result is

$$\nabla\phi = \text{grad}\,\phi = \hat{x}\frac{\partial\phi}{\partial x} + \hat{y}\frac{\partial\phi}{\partial y} + \hat{z}\frac{\partial\phi}{\partial z}, \tag{C.15}$$

since $\hat{x}$, $\hat{y}$, and $\hat{z}$ do not vary with x, y, and z. The gradient of a scalar function ϕ is defined by (C.15). In a similar manner, in rectangular coordinates, the divergence and the curl operators are defined by

$$\nabla\cdot = \frac{\partial}{\partial x}\hat{x}\cdot + \frac{\partial}{\partial y}\hat{y}\cdot + \frac{\partial}{\partial z}\hat{z}\cdot, \tag{C.16}$$

$$\nabla\times = \frac{\partial}{\partial x}\hat{x}\times + \frac{\partial}{\partial y}\hat{y}\times + \frac{\partial}{\partial z}\hat{z}\times. \tag{C.17}$$

When (C.16) and (C.17) operate on a vector function $\boldsymbol{A}$ of position, the results are

$$\nabla \cdot \boldsymbol{A} = \text{div}\,\boldsymbol{A} = \frac{\partial}{\partial x}A_x + \frac{\partial}{\partial y}A_y + \frac{\partial}{\partial z}A_z, \tag{C.18}$$

$$\nabla \times \boldsymbol{A} = \text{curl}\,\boldsymbol{A} = \hat{x}\left(\frac{\partial}{\partial y}A_z - \frac{\partial}{\partial z}A_y\right) + \hat{y}\left(\frac{\partial}{\partial z}A_x - \frac{\partial}{\partial x}A_z\right) + \hat{z}\left(\frac{\partial}{\partial x}A_y - \frac{\partial}{\partial y}A_x\right). \tag{C.19}$$

The divergence and the curl of a vector function $\boldsymbol{A}$ are defined by (C.18) and (C.19) respectively. We should note that the gradient of a scalar function is a vector, the divergence of a vector function is a scalar, and the curl of a vector function is a vector. The physical interpretations of the gradient, the divergence, and the curl operations enable us to determine their expressions in other coordinate systems.

Gradient

We can interpret the gradient of a scalar function ϕ as a vector (i) whose direction is the direction of maximum rate of change of the function, and (ii) whose magnitude is the magnitude of the maximum rate of change. Let $\hat{n}$ be an arbitrarily oriented unit vector and let the distances along $\hat{n}$ be denoted by n. Let a_x, a_y, and a_z be the direction cosines of $\hat{n}$ with respect to the coordinate axes x, y, and z respectively:

$$a_x = \hat{n} \cdot \hat{x}; \qquad a_y = \hat{n} \cdot \hat{y}; \qquad a_z = \hat{n} \cdot \hat{z}. \tag{C.20}$$

Therefore,

$$\hat{n} = a_x\hat{x} + a_y\hat{y} + a_z\hat{z} \qquad \text{with} \qquad a_x^2 + a_y^2 + a_z^2 = 1. \tag{C.21}$$

With the help of the method used in Section 1.3, we can show that

$$a_x \frac{\partial}{\partial n}\phi = \frac{\partial}{\partial x}\phi; \qquad a_y \frac{\partial}{\partial n}\phi = \frac{\partial}{\partial y}\phi; \qquad a_z \frac{\partial}{\partial n}\phi = \frac{\partial}{\partial z}\phi. \tag{C.22a,b,c}$$

When (C.22a), (C.22b), and (C.22c) are multiplied by a_x, a_y, and a_z respectively and are added, we find with the help of (C.15), (C.20), and (C.21) that

$$\frac{\partial}{\partial n}\phi = \boldsymbol{G} \cdot \hat{n} = G \cos\theta, \qquad |\boldsymbol{G}| = G, \tag{C.23}$$

where

$$\boldsymbol{G} = \nabla\phi, \tag{C.24}$$

and θ is the angle between the directions of $\hat{n}$ and $\boldsymbol{G}$. We see from (C.23) and (C.24) that the maximum value of $(\partial/\partial n)\phi$ is equal to $|\boldsymbol{G}|$ and that the maximum value is obtained when $\theta = 0$ or equivalently when $\hat{n}$ coincides with the direction of $\boldsymbol{G}$. Thus, according to the physical interpretation stated in the foregoing, $\boldsymbol{G}$ is the gradient of the scalar function ϕ.

It is possible to define the gradient of the scalar function ϕ at a point in space in an alternative method as follows: Let ΔV be an elemental volume centered on the point under consideration and let ΔS be the closed surface enclosing the volume ΔV. Then

$$\nabla\phi = \lim_{\Delta V \to 0} \frac{1}{\Delta V} \int_{\Delta S} \hat{n}\phi \, dS, \tag{C.25}$$

where $\hat{n}$ is an outwardly drawn unit normal to the surface ΔS. If the elemental volume is taken as that bounded by the coordinate planes, x, $x + \Delta x$, y, $y + \Delta y$, z, and $z + \Delta z$, then we can evaluate the right-hand side of (C.25) to yield the result (C.15), thus verifying that the definition as stated in (C.25) yields the same result for $\nabla\phi$ as the previous definition of the gradient operator. Let us consider a finite volume V

bounded by the closed surface S. Then, with the help of (C.25), we can prove the following integral theorem:

$$\int_V \nabla\phi \, dV = \int_S \phi\hat{n} \, dS, \tag{C.26}$$

where, as before, $\hat{n}$ is an outwardly directed unit normal to the surface S.

Divergence

As in (C.25), we can give the divergence of a vector function $\boldsymbol{A}$ the following interpretation:

$$\nabla \cdot \boldsymbol{A} = \lim_{\Delta V \to 0} \frac{1}{\Delta V} \int_{\Delta S} \boldsymbol{A} \cdot \hat{n} \, dS. \tag{C.27}$$

In words, (C.27) states that the divergence of a vector function at a point is the net outward flow per unit volume at the point under consideration. Let an infinitesimal volume of sides of lengths Δx, Δy, and Δz have the sides A, B, C, D, E, and F as shown in Fig. C.1. The outwardly drawn unit vectors normal to the sides A, B, C, D, E, and F are $-\hat{x}$, $\hat{x}$, $-\hat{y}$, $\hat{y}$, $-\hat{z}$, and $\hat{z}$ respectively. The outward flow through any side is the component of vector in the direction of the outward normal integrated throughout the area of the side. Therefore, for the volume of Fig. C.1, we find that

$$\begin{aligned}\frac{1}{\Delta V} \int_{\Delta S} \boldsymbol{A} \cdot \hat{n} \, dS &= [A_x(x + \Delta x, y, z) - A_x(x, y, z)]/\Delta x \\ &\quad + [A_y(x, y + \Delta y, z) - A_y(x, y, z)]/\Delta y \\ &\quad + [A_z(x, y, z + \Delta z) - A_z(x, y, z)]/\Delta z.\end{aligned} \tag{C.28}$$

As Δx, Δy, and Δz tend to zero, $\Delta V = \Delta x \, \Delta y \, \Delta z$ tends to zero. Then, we can see that (C.27) and (C.28) yield

$$\nabla \cdot \boldsymbol{A} = \frac{\partial}{\partial x} A_x + \frac{\partial}{\partial y} A_y + \frac{\partial}{\partial z} A_z, \tag{C.29}$$

which is identical to the previous definition of the divergence of the vector function, as stated in (C.18).

Let us consider a finite volume V bounded by the closed surface S. Let this volume be subdivided into a number of elemental volumes dV bounded by the closed surfaces

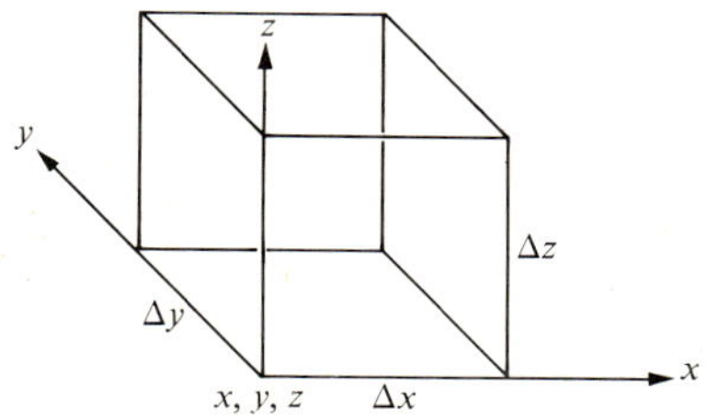

Fig. C.1 Infinitesimal volume of sides of lengths Δx, Δy, and Δz.

dS. For each elemental volume, we see from (C.27) that

$$\nabla \cdot \boldsymbol{A}\, dV = \int_{dS} \boldsymbol{A} \cdot \hat{n}\, dS. \tag{C.30}$$

When (C.30) is summed up for all the elemental volumes, the surface integral for the interior surfaces occur twice corresponding to the adjacent elemental volumes. These two sets of values of the surface integral for the interior surfaces are equal in magnitude but opposite in sign and hence they cancel each other out. Only the surface integral over the bounding surface S remains. Consequently, we obtain the following result from (C.30):

$$\int_V \nabla \cdot \boldsymbol{A}\, dV = \int_S \boldsymbol{A} \cdot \hat{n}\, dS, \tag{C.31}$$

which is the *divergence theorem.*

Curl

The line integral of a vector function taken along a curve is the component of the vector parallel to an elemental length $\boldsymbol{dl}$, integrated throughout the length of the curve. Let us consider an unclosed surface S bounded by the closed curve C. The line integral of a vector taken around the closed curve C is called the circulation of the vector around the surface area S. The curl of a vector function is a vector whose component in a given direction at a point is the circulation of the vector around an elemental surface of area ΔS centered on the point under consideration divided by the area ΔS of the surface. The elemental surface is oriented normal to the given direction. If the thumb of the right hand points in the given direction, the fingers point in the direction in which the line integral around the curve ΔC bounding the elemental area ΔS should be taken. Hence,

$$(\nabla \times \boldsymbol{A}) \cdot \hat{n} = \lim_{\Delta S \to 0} \frac{1}{\Delta S} \int_{\Delta C} \boldsymbol{A} \cdot \boldsymbol{dl}, \tag{C.32}$$

where $\hat{n}$, the normal to the elemental surface ΔS, is related to the direction of integration around the periphery ΔC by the right-hand rule. Let us suppose that we wish to find the x-component of $\nabla \times \boldsymbol{A}$ at a point in space. Then, the elemental area should lie around the point under consideration in the yz-plane, as shown in Fig. C.2. The integration around ΔC should be taken in the counter-clockwise direction. Therefore, for the elemental area ΔS of Fig. C.2, we obtain

$$\begin{aligned} \frac{1}{\Delta S} \int_{\Delta C} \boldsymbol{A} \cdot \boldsymbol{dl} &= [A_z(y + \Delta y, z) - A_z(y, z)]/\Delta y \\ &\quad - [A_y(y, z + \Delta z) - A_y(y, z)]/\Delta z. \end{aligned} \tag{C.33}$$

As Δy and Δz tend to zero, $\Delta S = \Delta y\, \Delta z$ tends to zero. Then, we see that (C.32) and (C.33) give

$$(\nabla \times \boldsymbol{A}) \cdot \hat{x} = \frac{\partial}{\partial y} A_z - \frac{\partial}{\partial z} A_y. \tag{C.34}$$

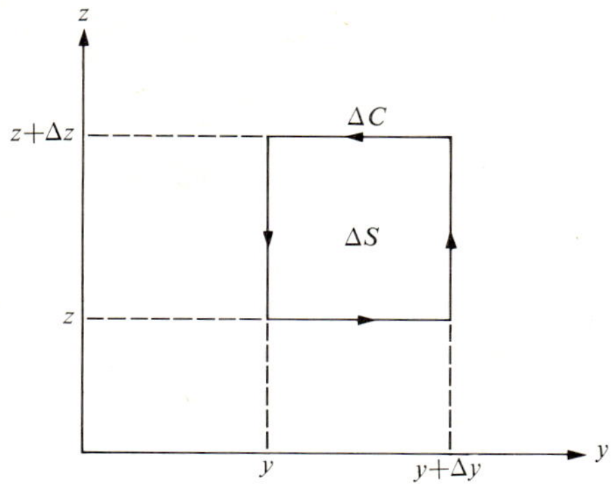

Fig. C.2 Infinitesimal area ΔS bounded by the closed curve ΔC.

In a similar manner, we can evaluate the y- and the z-components of $\nabla \times \boldsymbol{A}$ to yield the following result

$$\nabla \times \boldsymbol{A} = \hat{x}\left(\frac{\partial}{\partial y} A_z - \frac{\partial}{\partial z} A_y\right) + \hat{y}\left(\frac{\partial}{\partial z} A_x - \frac{\partial}{\partial x} A_z\right) + \hat{z}\left(\frac{\partial}{\partial x} A_y - \frac{\partial}{\partial y} A_x\right), \tag{C.35}$$

which is identical to the previous definition of $\nabla \times \boldsymbol{A}$ as contained in (C.19).

Let us consider a finite unclosed area S bounded by the closed curve C. Let this surface be subdivided into a number of elemental surfaces dS bounded by the closed curves ΔC. For each elemental area, we find from (C.32) that

$$(\nabla \times \boldsymbol{A}) \cdot \hat{n}\, dS = \int_{\Delta C} \boldsymbol{A} \cdot d\boldsymbol{l}. \tag{C.36}$$

When (C.36) is summed up for all the elemental areas, the line integral disappears for all internal areas, since a periphery is first traversed in one direction and then in the opposite direction when determining the contribution from an adjacent area. Only the line integral around the bounding curve C is nonvanishing. Therefore, it follows from (C.36) that

$$\int_S \nabla \times \boldsymbol{A} \cdot \hat{n}\, dS = \int_C \boldsymbol{A} \cdot d\boldsymbol{l}, \tag{C.37}$$

which is the statement of *Stokes' theorem.*

We can define the curl of the vector function $\boldsymbol{A}$ at a point in space in the following alternative method. Let ΔV be an elemental volume centered on the point under consideration and let ΔS be the closed surface enclosing the volume ΔV. Then,

$$\nabla \times \boldsymbol{A} = \lim_{\Delta V \to 0} \frac{1}{\Delta V} \int_{\Delta S} \hat{n} \times \boldsymbol{A}\, dS. \tag{C.38}$$

When we evaluate the right-hand side of (C.38) for the elemental volume of Fig. C.1, we find that

$$\nabla \times \boldsymbol{A} = \hat{x} \times \frac{\partial}{\partial x} \boldsymbol{A} + \hat{y} \times \frac{\partial}{\partial y} \boldsymbol{A} + \hat{z} \times \frac{\partial}{\partial z} \boldsymbol{A},$$

$$= \hat{x}\left(\frac{\partial}{\partial y} A_z - \frac{\partial}{\partial z} A_y\right) + \hat{y}\left(\frac{\partial}{\partial z} A_x - \frac{\partial}{\partial x} A_z\right) + \hat{z}\left(\frac{\partial}{\partial x} A_y - \frac{\partial}{\partial y} A_x\right), \quad \text{(C.39)}$$

which is identical to (C.19), as it should be.

With the help of (C.38), it is possible for us to establish the integral theorem

$$\int_V \nabla \times \boldsymbol{A}\, dV = \int_S \hat{n} \times \boldsymbol{A}\, dS, \quad \text{(C.40)}$$

where the surface S bounds the volume V.

We can deduce the expressions for the gradient, the divergence, and the curl in cylindrical and spherical coordinates from the foregoing definitions and it is instructive to carry out the details of these derivations.

Differential Relations

We can prove the following two important vector differential relations by expressing them in component form:

$$\nabla \cdot (\nabla \times \boldsymbol{A}) = 0, \quad \text{(C.41)}$$

and

$$\nabla \times (\nabla \phi) = 0. \quad \text{(C.42)}$$

Another important differential operation is the Laplacian which is defined by

$$\nabla^2 \phi = \nabla \cdot (\nabla \phi). \quad \text{(C.43)}$$

With the help of (C.15) and (C.16), we can show that

$$\nabla^2 \phi = \frac{\partial^2}{\partial x^2} \phi + \frac{\partial^2}{\partial y^2} \phi + \frac{\partial^2}{\partial z^2} \phi. \quad \text{(C.44)}$$

APPENDIX D

VECTOR RELATIONS

1. Relations Involving Sums and Products

$$\boldsymbol{A} + \boldsymbol{B} = \boldsymbol{B} + \boldsymbol{A}$$

$$\boldsymbol{A} \cdot \boldsymbol{B} = \boldsymbol{B} \cdot \boldsymbol{A}$$

$$\boldsymbol{A} \times \boldsymbol{B} = -\boldsymbol{B} \times \boldsymbol{A}$$

$$(\boldsymbol{A} + \boldsymbol{B}) \cdot \boldsymbol{C} = \boldsymbol{A} \cdot \boldsymbol{C} + \boldsymbol{B} \cdot \boldsymbol{C}$$

$$(\boldsymbol{A} + \boldsymbol{B}) \times \boldsymbol{C} = \boldsymbol{A} \times \boldsymbol{C} + \boldsymbol{B} \times \boldsymbol{C}$$

$$\boldsymbol{A} \cdot (\boldsymbol{B} \times \boldsymbol{C}) = \boldsymbol{B} \cdot (\boldsymbol{C} \times \boldsymbol{A}) = \boldsymbol{C} \cdot (\boldsymbol{A} \times \boldsymbol{B})$$

$$\boldsymbol{A} \times (\boldsymbol{B} \times \boldsymbol{C}) = (\boldsymbol{A} \cdot \boldsymbol{C})\boldsymbol{B} - (\boldsymbol{A} \cdot \boldsymbol{B})\boldsymbol{C}$$

2. Relations Involving Differentiation

$$\nabla(\phi + \psi) = \nabla\phi + \nabla\psi$$

$$\nabla(\phi\psi) = \phi\nabla\psi + \psi\nabla\phi$$

$$\nabla \cdot (\boldsymbol{A} + \boldsymbol{B}) = \nabla \cdot \boldsymbol{A} + \nabla \cdot \boldsymbol{B}$$

$$\nabla \times (\boldsymbol{A} + \boldsymbol{B}) = \nabla \times \boldsymbol{A} + \nabla \times \boldsymbol{B}$$

$$\nabla(\boldsymbol{A} \cdot \boldsymbol{B}) = (\boldsymbol{A} \cdot \nabla)\boldsymbol{B} + (\boldsymbol{B} \cdot \nabla)\boldsymbol{A} + \boldsymbol{A} \times (\nabla \times \boldsymbol{B}) + \boldsymbol{B} \times (\nabla \times \boldsymbol{A})$$

$$\nabla \cdot (\phi\boldsymbol{A}) = \boldsymbol{A} \cdot (\nabla\phi) + \phi(\nabla \cdot \boldsymbol{A})$$

$$\nabla \cdot (\boldsymbol{A} \times \boldsymbol{B}) = \boldsymbol{B} \cdot (\nabla \times \boldsymbol{A}) - \boldsymbol{A} \cdot (\nabla \times \boldsymbol{B})$$

$$\nabla \cdot (\nabla \times \boldsymbol{A}) = 0$$

$$\nabla \times (\phi\boldsymbol{A}) = (\nabla\phi) \times \boldsymbol{A} + \phi(\nabla \times \boldsymbol{A})$$

$$\nabla \times (\boldsymbol{A} \times \boldsymbol{B}) = \boldsymbol{A}(\nabla \cdot \boldsymbol{B}) - \boldsymbol{B}(\nabla \cdot \boldsymbol{A}) + (\boldsymbol{B} \cdot \nabla)\boldsymbol{A} - (\boldsymbol{A} \cdot \nabla)\boldsymbol{B}$$

$$\nabla \times (\nabla \times \boldsymbol{A}) = \nabla(\nabla \cdot \boldsymbol{A}) - \nabla^2\boldsymbol{A}$$

$$\nabla \times (\nabla\phi) = 0$$

$$\nabla \cdot (\nabla\phi) = \nabla^2\phi$$

3. Relations Involving Integration

In the following formulas V is a volume bounded by the closed surface S and $\hat{n}$ is a unit normal drawn outwardly to the surface S.

$$\int_V \nabla\phi \, dV = \int_S \phi\hat{n} \, dS$$

$$\int_V \nabla \cdot \boldsymbol{A} \, dV = \int_S \boldsymbol{A} \cdot \hat{n} \, dS \qquad \text{(Divergence theorem)}$$

$$\int_V \nabla \times \boldsymbol{A} \, dV = \int_S \hat{n} \times \boldsymbol{A} \, dS$$

In the following formulas S is an unclosed area bounded by the curve C.

$$\int_S \hat{n} \times \nabla\phi \, dS = \int_C \phi \, d\boldsymbol{l}$$

$$\int_S \nabla \times \boldsymbol{A} \cdot \hat{n} \, dS = \int_C \boldsymbol{A} \cdot d\boldsymbol{l} \qquad \text{(Stokes' theorem)}$$

4. Gradient, Divergence, Curl and Laplacian Operations in Rectangular (x, y, z), Cylindrical (ρ, ϕ, z) and Spherical (r, θ, ϕ) Coordinate Systems

Rectangular coordinates

$$\nabla a = \hat{x}\frac{\partial}{\partial x} a + \hat{y}\frac{\partial}{\partial y} a + \hat{z}\frac{\partial}{\partial z} a$$

$$\nabla \cdot \boldsymbol{A} = \frac{\partial}{\partial x} A_x + \frac{\partial}{\partial y} A_y + \frac{\partial}{\partial z} A_z$$

$$\nabla \times \boldsymbol{A} = \hat{x}\left(\frac{\partial}{\partial y} A_z - \frac{\partial}{\partial z} A_y\right) + \hat{y}\left(\frac{\partial}{\partial z} A_x - \frac{\partial}{\partial x} A_z\right) + \hat{z}\left(\frac{\partial}{\partial x} A_y - \frac{\partial}{\partial y} A_x\right)$$

$$\nabla^2 a = \frac{\partial^2}{\partial x^2} a + \frac{\partial^2}{\partial y^2} a + \frac{\partial^2}{\partial z^2} a$$

$$\nabla^2 \boldsymbol{A} = \hat{x}\,\nabla^2 A_x + \hat{y}\,\nabla^2 A_y + \hat{z}\,\nabla^2 A_z$$

Cylindrical coordinates

$$\nabla a = \hat{\rho}\frac{\partial}{\partial \rho} a + \hat{\phi}\frac{1}{\rho}\frac{\partial}{\partial \phi} a + \hat{z}\frac{\partial}{\partial z} a$$

$$\nabla \cdot \boldsymbol{A} = \frac{1}{\rho}\frac{\partial}{\partial \rho}(\rho A_\rho) + \frac{1}{\rho}\frac{\partial}{\partial \phi} A_\phi + \frac{\partial}{\partial z} A_z$$

$$\nabla \times \boldsymbol{A} = \hat{\rho}\left[\frac{1}{\rho}\frac{\partial}{\partial \phi} A_z - \frac{\partial}{\partial z} A_\phi\right] + \hat{\phi}\left[\frac{\partial}{\partial z} A_\rho - \frac{\partial}{\partial \rho} A_z\right] + \hat{z}\left[\frac{1}{\rho}\frac{\partial}{\partial \rho}(\rho A_\phi) - \frac{1}{\rho}\frac{\partial}{\partial \phi} A_\rho\right]$$

$$\nabla^2 a = \frac{1}{\rho}\frac{\partial}{\partial \rho}\left(\rho\frac{\partial}{\partial \rho} a\right) + \frac{1}{\rho^2}\frac{\partial^2}{\partial \phi^2} a + \frac{\partial^2}{\partial z^2} a$$

Spherical coordinates

$$\nabla a = \hat{r}\frac{\partial}{\partial r}a + \hat{\theta}\frac{1}{r}\frac{\partial}{\partial \theta}a + \hat{\phi}\frac{1}{r\sin\theta}\frac{\partial}{\partial \phi}a$$

$$\nabla \cdot \boldsymbol{A} = \frac{1}{r^2}\frac{\partial}{\partial r}(r^2 A_r) + \frac{1}{r\sin\theta}\frac{\partial}{\partial \theta}(\sin\theta A_\theta) + \frac{1}{r\sin\theta}\frac{\partial}{\partial \phi}A_\phi$$

$$\nabla \times \boldsymbol{A} = \frac{\hat{r}}{r\sin\theta}\left[\frac{\partial}{\partial \theta}(\sin\theta A_\phi) - \frac{\partial}{\partial \phi}A_\theta\right] + \frac{\hat{\theta}}{r}\left[\frac{1}{\sin\theta}\frac{\partial}{\partial \phi}A_r - \frac{\partial}{\partial r}(rA_\phi)\right] + \frac{\hat{\phi}}{r}\left[\frac{\partial}{\partial r}(rA_\theta) - \frac{\partial}{\partial \theta}A_r\right]$$

$$\nabla^2 a = \frac{1}{r^2}\frac{\partial}{\partial r}\left(r^2\frac{\partial}{\partial r}a\right) + \frac{1}{r^2\sin\theta}\frac{\partial}{\partial \theta}\left(\sin\theta\frac{\partial}{\partial \theta}a\right) + \frac{1}{r^2\sin^2\theta}\frac{\partial^2}{\partial \phi^2}a$$

5. Coordinate Transformations

Rectangular ⇄ Cylindrical

$$x = \rho\cos\phi; \qquad y = \rho\sin\phi; \qquad z = z$$

$$A_x = A_\rho\cos\phi - A_\phi\sin\phi; \qquad A_y = A_\rho\sin\phi + A_\phi\cos\phi; \qquad A_z = A_z$$

$$\rho = \sqrt{x^2 + y^2}; \qquad \phi = \tan^{-1}(y/x); \qquad z = z$$

$$A_\rho = A_x\cos\phi + A_y\sin\phi; \qquad A_\phi = -A_x\sin\phi + A_y\cos\phi; \qquad A_z = A_z$$

Rectangular ⇄ Spherical

$$x = r\sin\theta\cos\phi; \qquad y = r\sin\theta\sin\phi; \qquad z = r\cos\theta$$

$$A_x = A_r\sin\theta\cos\phi + A_\theta\cos\theta\cos\phi - A_\phi\sin\phi$$

$$A_y = A_r\sin\theta\sin\phi + A_\theta\cos\theta\sin\phi + A_\phi\cos\phi$$

$$A_z = A_r\cos\theta - A_\theta\sin\theta$$

$$r = \sqrt{x^2 + y^2 + z^2}; \qquad \theta = \tan^{-1}(\sqrt{x^2 + y^2}/z); \qquad \phi = \tan^{-1}(y/x)$$

$$A_r = A_x\sin\theta\cos\phi + A_y\sin\theta\sin\phi + A_z\cos\theta$$

$$A_\theta = A_x\cos\theta\cos\phi + A_y\cos\theta\sin\phi - A_z\sin\theta$$

$$A_\phi = -A_x\sin\phi + A_y\cos\phi$$

Cylindrical ⇄ Spherical

$$\rho = r\sin\theta; \qquad \phi = \phi; \qquad z = r\cos\theta$$

$$A_\rho = A_r\sin\theta + A_\theta\cos\theta; \qquad A_\phi = A_\phi; \qquad A_z = A_r\cos\theta - A_\theta\sin\theta$$

$$r = \sqrt{\rho^2 + z^2}; \qquad \theta = \tan^{-1}(\rho/z); \qquad \phi = \phi$$

$$A_r = A_\rho\sin\theta + A_z\cos\theta; \qquad A_\theta = A_\rho\cos\theta - A_z\sin\theta; \qquad A_\phi = A_\phi$$

BIBLIOGRAPHY

R. F. Harrington, *Time-harmonic Electromagnetic Fields*, New York: McGraw-Hill Book Company, 1961.

W. H. Hayt, *Engineering Electromagnetics*, 2nd Ed., New York: McGraw-Hill Book Company, 1967.

E. C. Jordan and K. G. Balmain, *Electromagnetic Waves and Radiating Systems*, 2nd Ed., Englewood Cliffs, N. J.: Prentice-Hall, Inc., 1968.

R. W. P. King, *Fundamental Electromagnetic Theory*, New York: Dover Publications, Inc., 1963.

R. W. P. King, H. R. Mimno, and A. H. Wing, *Transmission Lines, Antennas and Waveguides*, New York: Dover Publications, Inc., 1964.

J. D. Kraus, *Electromagnetics*, New York: McGraw-Hill Book Company, 1953.

R. Plonsey and R. E. Collin, *Principles and Applications of Electromagnetic Fields*, New York: McGraw-Hill Book Company, 1961.

S. Ramo, J. R. Whinnery, and T. Van Duzer, *Fields and Waves in Communication Electronics*, New York: John Wiley and Sons, Inc., 1965.

S. A. Schelkunoff and H. T. Friis, *Antennas, Theory and Practice*, New York: John Wiley and Sons, Inc., 1952.

S. A. Schelkunoff, *Electromagnetic Waves*, Princeton, N.J.: D. Van Nostrand Company, Inc., 1943.

INDEX

(Page numbers within brackets refer to problems)